DATE DUE

SEP 27			
JAN 11 01			
JE 9 '04			
NO 30 '04			
JY 20 '05			

DEMCO 38-296

Handbook of
OSHA
CONSTRUCTION
SAFETY
and HEALTH

Handbook of
OSHA
CONSTRUCTION
SAFETY
and HEALTH

Charles D. Reese
James V. Eidson

LEWIS PUBLISHERS

Boca Raton London New York Washington, D.C.

Riverside Community College
Library
MAR '00
4800 Magnolia Avenue
Riverside, CA 92506

TH 443 .R434 1999

Reese, Charles D.

Handbook of OSHA
   construction safety and

Library of Congress Cataloging-in-Publication Data

Reese, Charles D.
    Handbook of OSHA construction safety and health / Charles D.
Reese, James V. Eidson.
        p.   cm.
    Includes bibliographical references and index.
    ISBN 1-56670-297-6 (alk. paper)
    1. Building--United States--Safety measures. 2. Construction
industry--Safety regulations--United States.   I. Eidson, James V.
II. Title.
TH443.R434 1999
363.11′969′00973—dc21

98-51604
CIP

This book contains information obtained from authentic and highly regarded sources. Reprinted material is quoted with permission, and sources are indicated. A wide variety of references are listed. Reasonable efforts have been made to publish reliable data and information, but the author and the publisher cannot assume responsibility for the validity of all materials or for the consequences of their use.

Neither this book nor any part may be reproduced or transmitted in any form or by any means, electronic or mechanical, including photocopying, microfilming, and recording, or by any information storage or retrieval system, without prior permission in writing from the publisher.

The consent of CRC Press LLC does not extend to copying for general distribution, for promotion, for creating new works, or for resale. Specific permission must be obtained in writing from CRC Press LLC for such copying.

Direct all inquiries to CRC Press LLC, 2000 N.W. Corporate Blvd., Boca Raton, Florida 33431.

Trademark Notice: Product or corporate names may be trademarks or registered trademarks, and are used only for identification and explanation, without intent to infringe.

© 1999 by CRC Press LLC
Lewis Publishers is an imprint of CRC Press LLC

No claim to original U.S. Government works
International Standard Book Number 1-56670-297-6
Library of Congress Card Number 98-51604
Printed in the United States of America 1 2 3 4 5 6 7 8 9 0
Printed on acid-free paper

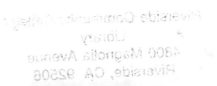
Riverside Community College
Library
4800 Magnolia Avenue
Riverside, CA 92506

PREFACE

The Construction Industry has always been viewed as a unique industry. Although there are many aspects which are the same as in other industries, it certainly has its share of unique hazards. The intent of this book is to provide a tool which can be used to address the occupational safety and health issues faced by those working in the construction industry; this includes contractors, workers, safety and health professionals, project managers, suppliers, and manufactures of equipment and materials.

A vast range of issues are addressed in this book, and some of those issues are as follows: people issues, program development, safety and health program implementation, intervention and prevention of construction incidents, regulatory interpretations, understanding, and compliance, OSHA's expectations, health and safety hazards faced by those working in the construction industry, and sources of information.

From the start of this book it has been my vision to provide a comprehensive approach to construction safety and health. This is manifested by addressing issues which are seldom discussed in the construction arena; some of these issues are such topics as perceptions and motivation. Also included in this book are those issues gleaned from the safety and health disciplines, such as the analyzing of incidents and accident prevention techniques that may be viewed as "stuff" which doesn't apply to construction. A great effort was also undertaken to discuss construction safety and health hazards and the regulations promulgated by the Occupational Safety and Health Administration in order to alleviate these hazards. This comprehensive application of safety and health to the many facets of the construction industry is fostered by a strong belief by the authors that construction safety and health on the jobsite are critical factors in good business practices, productivity, and cost containment.

As a guide and source reference for safety and health in the construction industry, this book becomes the foundation upon which to build stronger safety and health initiatives within the construction industry, while intervening and preventing jobsite deaths, injuries, and illnesses.

Charles D. Reese, Ph.D.

ABOUT THE AUTHORS

CHARLES D. REESE

For over twenty years Dr. Charles D. Reese has been involved with occupational safety and health as an educator, manager, or consultant. In Dr. Reese's early beginnings in occupational safety and health, he held the position of industrial hygienist at the National Mine Health and Safety Academy. He later assumed the responsibility of manager for the nation's occupational trauma research initiative at the National Institute for Occupational Safety and Health's (NIOSH) Division of Safety Research. Dr. Reese has had an integral part in trying to assure that workplace safety and health is provided for all those within the workplace. As the managing director for the Laborers' Health and Safety Fund of North America, his responsibilities were aimed at protecting the 650,000 members of the laborers' union in the United States and Canada.

He has developed many occupational safety and health training programs which run the gamut from radioactive waste remediation to confined space entry. Dr. Reese has written numerous articles, pamphlets, and books on related safety and health issues.

At present Dr. Reese is a member of the graduate and undergraduate faculty at the University of Connecticut, where he teaches courses on OSHA regulations, safety and health management, accident prevention techniques, industrial hygiene, and ergonomics. As Associate Professor of occupational safety and health, he coordinates the bulk of the safety and health efforts at the University and Labor Education Center. He is often called upon to consult with industry on safety and health issues and also asked for expert consultation in legal cases.

JAMES V. EIDSON

Mr. Eidson is currently conducting industrial hygiene and safety investigations in construction and the general industry. He also serves as the director of Professional Health and Safety Consultants. His primary responsibilities are conducting workplace health and safety walk-around inspections, exposure monitoring, and hazard training. Objectives also include developing technical, instructional, and procedural material for training curricula and monitoring a variety of workplace health and safety hazards.

He serves as a master trainer for programs including hazardous waste, hazard communication, lead abatement, radiation safety, blood-borne pathogens, and asbestos abatement. He is an Adjunct Professor for the University of Connecticut specializing in industrial hygiene and OSHA standards classes, and has Connecticut State licenses as an asbestos inspector, management planner, project designer, and project manager.

His previous experience includes three years as Director of Health and Safety with Laborers-AGC, the arm of the International Laborers Union. In this position he wrote training curriculum, served as a master trainer, and monitored the delivery of training programs throughout the University States and Canada.

His experience also includes six years with the U.S. Department of Labor (OSHA) as an Industrial Hygiene and Safety Compliance Officer. In this capacity he was responsible for inspections, worker exposure monitoring, collecting environmental samples, and the development of citations for noncompliance of health and safety regulations. As the senior industrial hygienist, assignments included the more difficult inspections at large companies such as lead smelters and where warrants were needed.

Serving as an industrial hygiene engineer for three years at Kennecott Copper Smelter in Utah, his responsibilities included compliance with OSHA Standards and other industrial hygiene concerns such as ventilation studies, noise abatement, respirator fit testing, quarterly exposure monitoring for lead and arsenic, and follow-up on worker complaints.

His educational achievements include completion of a Bachelor of Science degree specializing in Biology and Electricity/Electronics and a Master of Science degree in Industrial Hygiene. He also is a Certified Safety Professional.

ACKNOWLEDGMENTS

I thank my dedicated wife, Carol, for her patience and perseverance in proofreading and editing my work each step of the way; she makes a great teammate. Also, I want to thank Kay Warren of BarDan Associates, for her efforts in formatting and developing the camera-ready copy; her work has been invaluable.

It certainly took a load off of me to have my co-author, James V. Eidson, a safety and health consultant and previous OSHA Senior Industrial Hygienist write the chapters relevant to occupational health and personal protective equipment.

A special thanks to my longtime friend and colleague, John Forte, who has for many years always been ready to fulfill my requests for help.

It is a pleasure to have Dr. Rodney Allen, who is dedicated to occupational safety and health, support this effort by providing a chapter on workers' compensation which was beyond my expertise.

The previous works of many agencies, organizations, industries and individuals have resulted in the information, illustrations, and materials which make this book possible. Contributions by courtesy or permission have come from:

Building and Construction Trades Department
Bureau of Labor Statistics
The Crosby Group, Inc.
Department of Commerce
Department of Energy
International Union of Operating Engineers, Local No. 487
Laborers-AGC Education and Training Fund
National Institute for Occupational Safety and Health
National Mine Health and Safety Academy
Occupational Safety and Health Administration
O&G Industries, Inc.
Professional Safety and Health Consultants
Scaffold Industry Association, Inc.
Suffolk Construction Company
Walsh Construction Management Company

Certain individual professionals made unique contributions with their experience and knowledge. They were

Donna Civitello, Esq.
John Forte
James Lapping
Bruce Ottman
Nicholas Warren, Sc.D.

DEDICATIONS

This book is dedicated to my deceased parents, Charles R. Reese and Irene M. Reese, in memory of their support through my good and bad times, and to those construction workers who had to give their lives, health, and well-being to generate the need for this book. Last, but by no means least, to that special person, my loving wife, E. Carol Reese, who has shared my journey through life.

TABLE OF CONTENTS

Chapter 1

INTRODUCTION

Charles D. Reese

The construction industry is the builder of our modern world. From the dams to the skyscrapers, all that we have and see was in some way fashioned by construction contractors and workers. But construction can be dangerous work. People in the construction industry not only face the dangers of being the first on a jobsite, but face potential health risks and exposures throughout the building process.

Year after year construction is one of the most dangerous industries, with approximately 1,050 construction workers dying on the job each year. Although construction employment equals just over 5% of the workforce, construction injuries account for in excess of 17% of all occupational deaths.[1] One out of every seven construction workers is injured each year and one out of every fourteen will suffer a disabling injury. These statistics are high for any industrial sector.[2]

The occupational illnesses affecting construction workers have not been accurately measured, but an educated guess is that construction workers suffer both acute (short term) and chronic (long term) illnesses from their exposure to chemicals, dusts, fibers, noise, radiation, vibration, and temperature extremes. For many specific construction trades, specific related occupational illnesses have been documented—such as asbestosis and cancer for asbestos workers—but no complete census as to the prevalence of occupational illnesses among construction workers has been undertaken. Precautions need to be taken to limit exposures which have the potential to cause detrimental health effects to construction workers since accurate exposures often cannot be determined due to the transient nature of the work.

The following introduction to the construction industry sets the stage to delve more deeply into the industry's inherent dangers, to break down its constituents, issues, and problems, and to look more closely at its components. The introduction provides the foundation for the primary function of this book—to assist those who are concerned with the development and implementation of safety and health programs on construction sites in order to protect those working within the construction industry.

CONSTRUCTION INDUSTRY

When the term "construction industry" is relayed, the vision conjured up is usually one of an all-inclusive contractor whose workforce performs all construction functions. But in actuality, construction contractors markedly vary in their areas of specialization, and only the largest companies have the resources and personnel to handle all aspects of construction. The continuum is diverse, ranging from the building of single family dwellings to tunnel construction. This diversity is best seen by studying the Standard Industrial Classification (SIC).

1

The SIC system was developed by the U.S. Department of Commerce. It provides numerical codes to identify the functions of every business, profession, and institution in the United States. The system is organized into 10 major business and professional classifications. It is composed of numeric codes that allow for organizing and retrieving information in a clear logical way. The system classifies industries from the general to the specific based upon how goods and services are provided in the real world.

The SIC code is composed of 1 to 6 digits, and in some categories, an alphabetical character is used. The first 2 digits represent the ten major industry groups. (See Table 1-1.)

Table 1-1

Standard Industrial Classification

2 Digit Classification Code	
01-09	Agriculture, Forestry, & Fishing
10-14	Mining
15-17	Contractors/Construction
20-39	Manufacturing
40-49	Communication, Transportation, & Utilities
50-51	Wholesalers
52-59	Retailers
60-67	Finance, Insurance, & Real Estate
70-89	Services
90-99	Government Offices

Each construction contractor falls into a specific SIC depending upon the type of construction and work most frequently performed. A summary of these SICs can be found in Table 1-2.

Table 1-2

Construction Contractors' SICs[3]

SIC	Type of Construction
15	Building Construction - General Contractors and Operative Builders
152	General Building Contractors - Residential Buildings
1521	General Contractors - Single Family Houses
1522	General Contractors - Residential Buildings, Other Than Single Family
153	Operative Builders
1531	Operative Builders
154	General Building Contractors - Nonresidential Buildings

Table 1-2
Construction Contractors' SICs (*Continued*)

1541	General Contractors - Industrial Building and Warehouses
1542	General Contractors - Nonresidential Buildings, Other than Industrial Buildings and Warehouses
16	Construction Other than Building Construction - General Contractors
161	Highway and Street Construction, Except Elevated Highways
1611	Highway and Street Construction, Except Elevated Highways
162	Heavy Construction, Except Highway and Street Construction
1622	Bridge, Tunnel, and Elevated Highway Construction
1623	Water, Sewer, Pipe Line, Communication and Power Line Construction
1629	Heavy Construction, Not Elsewhere Classified
17	Construction - Special Trade Contractors
171	Plumbing, Heating (Except Electrical), and Air Conditioning
1711	Plumbing, Heating (Except Electrical), and Air Conditioning
172	Painting, Paper Hanging, and Decorating
1721	Painting, Paper Hanging, and Decorating
173	Electrical Work
1731	Electrical Work
174	Masonry, Stonework, Tile Setting, and Plastering
1741	Masonry, Stones Setting, and Plastering
1742	Plastering, Drywall, Acoustical, and Insulation Work
1743	Terrazzo, Tile, Marble, and Mosaic Work
175	Carpentering and Flooring
1751	Carpentering
1752	Floor Laying and Other Floorwork, Not Elsewhere Classified
176	Roofing and Sheet Metal Work
1761	Roofing and Sheet Metal Work
177	Concrete Work
1771	Concrete Work
178	Water Well Drilling
1781	Water Well Drilling
179	Miscellaneous Special Trade Contractors
1792	Structural Steel Erection
1793	Glass and Glazing Work
1794	Excavating and Foundation Work
1795	Wrecking and Demolition Work
1796	Installation or Erection of Building Equipment, Not Elsewhere Classified
1799	Special Trade Contractors, Not Elsewhere Classified

As can be seen, construction contractors widely vary. This variation is noted in many different ways. Some are as follow:

1.   Contractors perform their specialties at unique worksites, such as building sky-scrapers or paving highways.

2.   The equipment varies with the task being performed, from small tools to large earth-moving equipment.

3.   The variety of materials needed to complete the project greatly varies and can include anything from 2x4 studs to large steel I beams.

4.   There is a great variety of procedures used during the work being performed, like tieing re-bar (Figure 1-1) or digging a trench.

Figure 1-1. Ironworker tieing rebar on a lattice

5.   Each construction process has its own safety or health hazards which can dictate different precautions, from the need for fall protection to the need for respirators for asbestos abatement.

6.   The workers or trades (electricians, roofers, etc.) possess special skills and training to perform their specific tasks. This might include operating engineers using cranes or carpenters performing framing.

The above examples are illustrations of why the variations in the industry exist and why a need exists to classify and identify the specific types of work performed by contractors.

As of January 1997, the U.S. Department of Commerce instituted a new industry classification system brought on by the passage of the North American Free Trade Agreement (NAFTA). The North American Industry Classification System (NAICS) was developed to allow for the comparison of industries within Canada, Mexico, and the U.S.A. The new system utilizes a six-digit code maximum instead of the SIC four digit maximum. In order to determine a contractor's NAICS classification and number, use Appendix A – NAICS versus SIC or http://www.census.gov.

CONSTRUCTION AS A BUSINESS

The construction industry is definitely a business. The U.S. Department of Commerce lists the value of the construction industry in excess of $600 billion a year. There have been estimates that at least 850,000 different contractors exist in the United States. As of the 1990 census information, SIC 15 had 186,676 contractors and $220 billion of work per year, SIC 16 employed 37,189 contractors and did $98.6 billion of work per year, and SIC 17 was the largest in number of contractors and dollars with 367,250 and $220 billion dollars, respectively.[4] Contractors employ some 4,500,000 workers. Given such numbers, it seems safe to say construction is an integral part of the U.S. business community, and as such, must adhere to the normal practices of those doing business within the U.S. Many of the larger construction companies also conduct their business in the international arena. With this in mind, the amount of dollars, number of employers, and multitude of workers involved in construction indicates that it should be managed as a business from all aspects, including profit/loss, organizationally, and personnel, as well as the job safety and health component. Thus, the construction industry should be viewed as another responsible U.S. business.

CONTRACTOR LIABILITY

Construction contractors are faced with a myriad of potential problems when a project contract is awarded to the contractor. These problems range from organizational to legal issues. Of course, legal liability is always faced by contractors. This is especially true since construction has the reputation of being among the most dangerous industries.

The majority of contractors on construction sites are subcontractors who have been hired by other entities such as prime contractors, owners, architects, and engineer or construction managers. Subcontractors are often held accountable for the safety of their companies and employees, while the individual who hired them is protected from third party litigation. Thus, there is no shared accountability for safety and health on the jobsite. Recently this has begun to change and prime contractors, owners, and managers have been forced to share accountability and responsibility. All parties need to control safety and health on the construction site. With this shared responsibility comes the following problems:

1. Workers often file liability claims against parties other than just the subcontractor, when the prime contractor, owner, or manager does not expect this action.

2. Courts and juries feel contractors, general contractors (GCs), site managers, and owners who have or should have had control of the construction worksite are accountable and responsible for safety and health.

3. Any parties who fail to exercise control can be legally accountable.

4. OSHA tends to hold all parties responsible for citations and penalties, even though they may not have created the violation(s).

If all parties do not assume control of safety and health on the construction worksite, the owners, construction managers, contractors, and subcontractors may face some very serious problems.

1. Hazardous conditions may go unchecked, which can cause death or serious injuries.
2. Subcontractors who have bad safety records or perform their work in an unsafe manner are very culpable .
3. Courts may take a jaundiced view of prime contractors, owners, or managers who cannot separate their control and responsibility over production and safety.
4. Legal actions can result in large awards.
5. Legal fees, increased insurance costs, and loss of other resources (e.g., lost time, lost production) are outcomes of poor control.

However, there are some actions that can be undertaken to mitigate legal liability.

1. Make sure that all parties are aware of the content of the contract and knowledgeable of everyone's responsibilities regarding safety.
2. Maintain the power to inspect and monitor safety and health and work practices to assure safe completion of the project.
3. Hold each contractor and subcontractor accountable for the daily activities of its workforce and the workers' safety.
4. Employ only reputable contractors who can verify their work quality and safety practices.
5. Have a safety policy in place with job safety and health provisions clearly spelled out.
6. Use Hold Harmless/Indemnity provision to waive worker's compensation while assuring certification of insurance and worker's compensation programs.
7. Comply with OSHA Regulations and require all others to do so.
8. As a general contractor (GC), do not relinquish total control but limit control as much as possible, which creates a "Catch 22" since the GC is ultimately responsible.

As can be seen, a lot of thought should go into organizing safety and health on construction worksites. This includes how much control of production and safety should be exercised by the general contractor and how much responsibility should be entrusted to the subcontractors without relinquishing control of the project. Courts should start assessing liability based upon neglect to control workplace hazards rather than proper assumption of control. Thus, if one of the parties is following proper safety and production procedures or exhibiting control to assure a safe workplace in accordance with the contractual language, then there should be a degree of immunity for that party.

MANAGEMENT/PROFESSIONAL POSITIONS

The construction industry has many employees and a great variety of individuals who have unique skills related to construction. These individuals include estimators, expeditors, safety engineers, civil engineers, architect engineers, field engineers, construction project engineers, draftspersons, and inspectors.

On the management side, many positions are necessary such as project managers, financial managers, office managers, marketing managers, and purchasing agents.

Although every contractor may not employ all of these types of professionals or managers, many employ more than one. The determining factor is often the size of the contractor or magnitude of the project.

As seen, it takes a variety of personnel to staff the construction industry.

SUPERVISORS

Supervisors are the employees of the contractor and may include general forepersons, job superintendents, forepersons, and some lead journeypersons/craftspersons. Contractors need to choose supervisors carefully. Just because individuals have excellent knowledge of the construction processes and are good workers does not assure they will make good supervisors. Surely supervisors need to be knowledgeable about the construction processes, but even more important are their people skills and leadership qualities. Since supervisors are on the front line, they are key people in communicating information in both directions—both from contractors to workers and from workers to contractors.

For example, messages from the contractor are conveyed to workers by the supervisor. Most of what affects construction workers, such as training, job assignments, policy enforcement, decision making, and care of or tending to their needs, comes from the supervisor.

Many times supervisors are given responsibility without much authority, which mitigates their effectiveness. Often their professional development, such as proper training, is overlooked. Supervisors are the work horses on construction projects and should be held accountable for production, safety, and health. But supervisors must be provided with adequate resources to accomplish what is expected of them.

THE COMPETENT PERSON(S)

The construction industry is unique in that the OSHA construction standards (29 CFR 1926) require that contractors have a designated competent person or persons who are to conduct frequent and regular worksite inspections. The term 'designated' means an authorized person(s) who is approved or assigned by the contractor to perform a specific type of duty or duties or to be present at a specific location or locations at the jobsite.

This person should be qualified in the process he or she is overseeing, which means he or she should possess a recognized degree, certification, professional standing, or have extensive knowledge, training, and experience. Also, this individual must have demonstrated the ability to solve or resolve problems related to the subject matter, type of work being performed, or the project.

To be in compliance with the requirements for a competent/qualified person, each project will have a project-competent person capable of identifying existing and predictable hazards with the authority to take prompt corrective action to eliminate them. This individual may designate other competent persons to perform certain job tasks, such as inspecting an excavation or directing scaffold building.

In some cases, the Occupational Safety and Health Administration (OSHA) spells out, within the regulations, the duties and responsibilities of the competent person, but seldom

SAFETY AND HEALTH COMPETENT PERSON ASSIGNMENTS

Project _____

Contractor/Subcontractor _____

Safety Program Coordinator _____

Area Shift _____

Date _____

Senior Supervisor _____

Area/Shift Supervisor _____

Area/Shift Competent Person _____

SUBPARTS		DESIGNATED PERSON	SUBPARTS		DESIGNATED PERSON
1926 .20	Job Site Inspections		1926 .603	Pile Driving – Signalmen	
1926 .50	Medical Services/ First Aid		1926 .650	Excavations – Inspections	
1926 .53	Ionizing Radiation Technician		1926 .651	Excavations – Inspections	
1926 .54	Laser Operators°		1926 .651	Excavations – Design°	
1926 .55	Industrial Hygienist/ Technician		1926 .700	Concrete – Inspection	
1926 .101	Hearing Protection – Fitting		1926 .752	Iron Work – Supervision	
1910 .134	Respiratory Protection		1926 .800	Tunnels –	
1926 .155	Fire Protection		1926 .800	Tunnels – Inspection°	
1910 .184	Sling/Wire Rope Inspections		1926 .800	Tunnels – Equipment	
1926 .302	Powder-Actuated Tools-Trainer		1926 .803	Compressed Air – Senior Designee	
1926 .354	Welding-Industrial Hygienist		1926 .803	Compressed Air – Gauge Tender	
1926 .400	Assured Equip. Grounding Conductor Program		1926 .850	Demolition – Pre-Job Survey	
1926 .451	Scaffolding		1926 .859	Demotion – Site Surveys	
1926 .500	Roofing – MSS		1926 .900	Blaster-in-Charge	
1926 .550	Cranes – Annual Inspection		1926 .900	Blasters°	
1926 .552	Hoists – Inspections & Tests		1926 .900	Blasting – Program	
1926 .556	Aerial lifts – Operations		1926 .950	Power Transmission – Safety Designee	
1926 .601	Motor Vehicles – Daily Inspections		1926 .955	Live-Line Bare-Hand Supervisor°	

° Denotes requirement of Qualified Persons, all others shall be Competent Persons.

Figure 1-2. Safety and health competent person assignment sheet (circa 1982) with Permission of the Building and Construction Trades Department

are the training requirements delineated or discussed. If and when OSHA promulgates a regulation, it should include experience level, responsibilities, and training levels needed by an individual deemed as the competent person for the requirements of that particular regulation. This would be very useful for contractors in determining individuals who would act as competent person(s).

Each project should have a designated competent/qualified person(s) who will be listed on the Safety and Health Competent/Qualified Person Form (see Figure 1-2). Although this form is circa 1982 and additional standards have been added, it is still a viable example. The competent person is often a supervisor, but there are many experienced construction workers who have the knowledge and expertise to act as a competent person on a jobsite.

WOMEN IN CONSTRUCTION

In recent years more and more women have been entering the construction industry. These women are said to be entering nontraditional occupations. The construction industry does not have the luxury of making special allowances for female workers who are performing construction tasks. What is expected, whether a male or female is doing a specific type of work or task, is that both perform as equals (see Figure 1-3).

Figure 1-3. Female carpenter at work on construction site

Of course, male/female hygiene facilities may be needed in many cases. This may include separate showers, changing rooms, toilets, etc. Also, the sizing of personal protective equipment may be an issue at times due to differences in the body structure of males and females.

Whether it is a woman or a man performing the job, each individual must be capable of performing all of the tasks of his or her trade. Work expectations are the same for all workers. Most women or men who enter the construction industry soon discover if this is the type of work they really want to do. If it is, then it makes little difference what the individual's sex is. It is expected more women will find their way into the construction industry for all of the same reasons that men do: the pay is usually higher, benefits exist, high job satisfaction, pride of work accomplished, and the opportunity to become skilled workers.

THE CONSTRUCTION TRADES

Construction workers found in the construction industry are classified according to the unique task they are trained to perform. Many of these trade workers started as apprentices and received training and on-the-job experience before being classified as Journeypersons/ Craftspersons. The trades are classified as follows:

Asbestos Workers/Insulators – Cover pipes and other types of equipment with insulation materials.

Bricklayers – Do all the brick construction such as walls, fireplaces, and chimneys as well as concrete blocks and structural tiles.

Boilermakers – Erect, build and repair boilers, blast furnaces, pressure vessels, and tanks at steel mills, refineries, chemical plants, and power plants.

Carpenters – Cut, fasten, erect, join, mill, assemble, and align with a variety of materials including wood, plastic, metal, fiber, cork, and other structural materials. This is the most versatile trade, including framing walls, building concrete forms, installing dry wall, hanging ceilings, installing cabinets/fixtures, hanging doors, and laying floors.

Cement Masons/Finishers – Produce the finish on freshly poured concrete floors, slabs, and sidewalks.

Construction Laborers – Carry materials to other trades, assist other trades, do excavation and compacting, perform environmental remediation, and perform clean-up tasks (see Figure 1-4).

Figure 1-4. Laborer mixing mud

Electricians – Install the conduit, wiring, and all other items that make up an electrical system.

Elevators Constructors – Install elevators, escalators, and dumbwaiters.

Floor-covering Installers – Place materials such as carpeting, linoleum, and vinyl/rubber tile.

Glaziers – Cut, fit, and install glass in windows, doors, and other types of glass units.

Ironworkers – Assemble or install fabricated structural metal products, usually large beams or columns in the erection of industrial and commercial buildings. Structural ironworkers erect the steel framework for steel-framed buildings. Reinforcing ironworkers place reinforcing steel bars and wire mesh used in reinforced concrete construction (see Figure 1-5).

Figure 1-5. Ironworkers preparing to rig

Lathers – Install the base wall such as wire or perforated gypsum boards to which plaster and stucco is applied.

Mill-cabinet Workers – Build custom cabinets, store fixtures, and other specialized items.

Millwrights – Set up pumps, turbines, generators, conveyors, and other mechanical systems.

Operating Engineers – Operate and maintain heavy construction machinery such as bulldozers, cranes, pumps, pile drivers, derricks, tractors, and ditch-witches.

Painters – Mix paints, prepare surfaces, and apply paints, varnishes, lacquers, shellacs, or similar materials to surfaces using brushes, spray gun, or roller, as well as apply all types of decorative wall coverings, murals, etc.

Pile Drivers – Work on highways, buildings, dams, and bridges including divers who do underwater work.

Pipefitters – Usually work on industrial and commercial facilities installing pipe systems that carry hot water, steam, and various types of liquids and gases.

Plasters – Apply plaster and stucco to buildings.

Plumbers – Place all pipes for water, gas, sewage, and drainage systems. This also includes fixtures such as sinks, bathtubs, and toilets.

Roofers – Apply or install roof coverings on buildings such as wood shingles, asphalt, slate, and tile.

Sheet Metal Workers – Build and install sheet-metal products such as the ducts used for heating and air-conditioning. They also make and install gutters and flashing.

Stonemasons – Work on buildings which have solid stone or stone-veneered walls.

Teamsters – Drive trucks which are entering and leaving the construction site.

Tilesetters – Place ceramic and other types of tiles on floors and walls using materials such as marble and terrazzo.

All of these trades at one time or another are an integral part of the construction industry and the possibility exists that all trades could be represented on one jobsite. Many of the trades are unionized.

UNIONS OF THE BUILDING AND CONSTRUCTION TRADES DEPARTMENT

Although organized labor is responsible, according to 1995 BLS data, for only 17.7% of the construction work occurring in the United States, organized (union) workers are usually more highly represented on large construction projects as well as being affiliated with large national/international construction companies. The 17.7 % figure is higher than for the total organized workforce, which is 14.9%.[5]

The unions' contributions to the construction industry include providing productive, quality, and highly trained workers. All of the Building and Construction Trade Unions are dedicated to selecting the best individuals to enter their apprentice training programs and assuring that those individuals coming out of their programs are highly trained and some of the most skilled and productive workers in the world. A contractor who uses a union crane operator knows that operating a crane is what that worker does for a living on a regular basis and is not a nonunion laborer who only operates a crane at times. Thus, contractors who use union workers have a sense of confidence that they have experienced and trained workers to do specific tasks at the jobsite.

The unions that are part of the Building and Construction Trades Department are as follows:

- International Association of Heat and Frost Insulators and Asbestos Workers International Brotherhood of Boilermakers, Iron Ship Builders, Blacksmiths, Forgers and Helpers
- International Union of Bricklayers and Allied Craftworkers
- United Brotherhood of Carpenters and Joiners of America
- International Brotherhood of Electrical Workers
- International Union of Elevator Constructors
- International Association of Bridge, Structural, and Ornamental Iron Workers
- Laborers' International Union of North America

Figure 1-6. Operating engineer operating heavy equipment

- International Union of Operating Engineers (see Figure 1-6)
- Operative Plasterers' and Cement Masons' International Association of the United States and Canada
- International Brotherhood of Painters and Allied Trades
- United Union of Roofers, Waterproofers, and Allied Workers
- Sheet Metal Workers' International Association
- International Brotherhood of Teamsters
- United Association of Journeymen and Apprentice of the Plumbing and Pipe Fitting Industry of the United States and Canada

WHY THE HAZARDS

It is often asked why construction hazards exist. Construction is a very unique industry and very unlike a stationary or fixed workplace or factory situation, which may help explain the safety and health problems it faces. Some of the reasons construction differs from other industries are

1. Worksites are dynamic (constantly changing) and temporary as work progresses and other construction trade workers enter the process. (See Figure 1-7.)

2. Each worksite may involve several small contractors (subcontractors) performing different types of work in close proximity to each other.

3. Several trades may be present on the worksite at the same time, bringing with them the specific hazards of their trade, such as operating engineers with their

heavy equipment or painters with their exotic paints. Thus, there is the potential for exposure to all workers at the worksite not just those of that specific trade.

4.   At times, on small sites one trade ends up doing all of the tasks usually performed by another trade; therefore, the workers may not be familiar with the hazards involved in performing tasks that are not part of their normal job.

5.   Working surfaces, equipment, machinery, trenching, and scaffolding are regularly being moved, being assembled and disassembled, or modified. Thus, new hazards are constantly emerging.

Figure 1-7. The always changing construction site

6.   Construction workers frequently change worksites and employers over the course of a year. This results in using new procedures and equipment for which they have not been trained.

7.   The work is often seasonal which results in both contractors and workers feeling rushed due to a sense of urgency to quickly complete projects. This increases the chances that an accident might occur or that occupational exposure to harmful agents will go unnoticed.

8.   Construction regulations are often difficult to understand. Sometimes the safety standards or standards governing chemicals do not exist for construction or are designed for factory workplaces, such as the use of asphalt, for which no specific regulation exists.

COST OF ACCIDENTS/INCIDENTS

All accidents/incidents affect the bottom line (profit). Accidents/incidents are those occurrences that result in loss of production, illness or injury, damage to equipment or property, and near misses. Incidents cannot just be measured simply in workers' compensation costs. These incidents damage the continuity of the jobsite, which causes lost time, lost wages, the breakup of a crew or the loss of a key person, property loss in the form of damage to machines and equipment, and culminates in the supervisor's time lost during an incident investigation or an OSHA investigation/ inspection which results in the cost of citations and violations, as well as the cost in the form of legal issues and fees. Also, insurance premiums, as well as worker's compensation experience rates, will be increased. There is also the cost of damaged property, new procedures, new equipment, labor issues, and the contractors' valuable time.

As can be seen, the cost of accidents/incidents is not fully measurable, but the picture is very clear. These costs can definitely impact the efficiency and effectiveness of a construction operation, but often times could be avoided by giving safety and health some attention and support on construction jobsites.

WHY DO ACCIDENTS OCCUR

There are some safety and health factors which are unique to construction and some which are not. Awareness of these factors will help in preventing the occurrence of occupationally related construction incidents.

1. Actual Physical Hazards such as an unstable wall that is being erected.

2. Environmental Hazards such as toxic atmospheres, oxygen deficiency, noise, radiation, and dust.

3. Human Factors such as a supervisor's or worker's failure to follow safe work practices (Figure 1-8).

Figure 1-8. Unsafe act – worker riding in bucket

4.   <u>Lack of or Poorly Designed Safety Standards</u> such as no standard for confined space entry or a construction standard or a chemical that is never used in the construction industry.

5.   <u>Failure to Communicate Within a Single Trade</u> such as one equipment operator not following the standard travelway rules and colliding with another operator's equipment, resulting in potential injuries and damage.

6.   <u>Failure to Communicate Between Two or More Trades</u> such as a crane operator contacting high voltage lines while another trade (construction worker) is guiding the piece with a tagline on the material being hoisted. [6]

CONSTRUCTION DEATHS

The occurrence of traumatic occupational fatalities are unique in that they provide readily accessible information as to their cause and prevention. This is very different than occupational illnesses, which often have a long latency period between exposure and symptoms. There are many reasons why dealing with traumatic deaths has advantages.

- Deaths occur in real time with no latency period, thus, an immediate sequence of events exists.

- The events are usually readily observable, and the reconstruction of minutes or hours rather than days, months, or years is an advantage.

- The roots of basic causes are more clearly identified.

- It is easier to detect cause-and-effect relationships.

- There is no difficulty in diagnosing the outcome (death).

- In most cases, traumatic deaths are highly preventable.

In 1990, OSHA analyzed a group of construction deaths from 1985–1989 and discovered, probably to no one's surprise, that four causes of deaths prevailed in the construction industry. The information indicated that

- 33% occurred from Falls

- 22% occurred from Struck By

- 18% occurred from Caught In/Between

- 17% occurred from Electrocution[7]

When viewing these cases, no relationship was noted between the day of the week, the age group, the size of the company, or whether the workers were union or nonunion when compared with the demographic populations which comprise the construction industry.

There were, however, some differences between the major construction SICs. Special trade contractors (SIC 17) accounted for 53% of the occupational fatalities, while heavy construction, other than building construction (SIC 16), accounted for 34% and building construction (SIC 15) for 13%.

When looking at falls, the greater the fall distance, the greater the probability of death, but there are enough deaths starting at falls from six feet to justify the protection of workers. Most of the falls from elevation are from roof and scaffolds—thus, the emphasis in recent years on fall protection and scaffold safety. These usually occur in the special trades.

Struck by deaths are most frequently caused by heavy construction equipment such as trucks, graders, cranes, or scrapers and also from materials poorly rigged or improperly stored.

The caught in/between fatalities, such as the collapse of excavation or trenches, occurs in trenches between 5 and 15 feet. Caught between a moving piece of equipment and another surface does not occur as often as caught in accidents.

Electrocutions occur at voltages below 480 volts. In most cases this involves contact with live electrical parts. In the study, the contact most frequently was with high voltage lines, usually contacted by a piece of construction equipment such a crane or an aerial lift.

OSHA has begun what they call focused inspections which are keying in on the prevention of these four most frequent types of traumatic construction deaths. When one focuses on these types of accidents, specific known interventions can be taken. As the saying goes, "this is not rocket science."

As examples of this, the following is provided:

1. *Falls* – Use fall protection, practice ladder and scaffold safety, guard openings, practice walking/working surface safety as well as stairway safety, and provide guardrails/hand rails.

2. *Struck by* – Establish traffic patterns, have functional warning devices, train operators, practice proper rigging procedures, stack and store materials properly.

3. *Caught in/between* – Assure that all excavations/trenches are sloped, shored, or properly supported. Train workers to avoid placing themselves between equipment and other objects.

4. *Electrocutions* – Provide electrical safety training to workers, use lockout/tagout, use Personal Protective Equipment (PPE), practice safe operating procedures around voltage sources and especially high voltage lines, and follow safe distance requirements.

The OSHA inspector will use a check sheet to determine if a focused inspection is appropriate. By focusing on these types of construction deaths that could occur on a specific construction project site, preventive approaches can be contracted for and taken prior to beginning the project. Each project has it own hazards which can potentially cause a fatality. Some of the projects may have fall hazards, while others may have electrical hazards, and still others may combine all four hazards on the same project. Thus, it is prudent to address any hazards that have the potential to cause construction deaths, not just those most frequent fatality-causing factors.

CONSTRUCTION INJURIES

Many of the same types of accidents that cause construction deaths also cause serious injuries. Some insights into construction injuries were acquired by OSHA when they analyzed worker's compensation data from 1985-1988. The frequency of occurrence of construction injuries is higher than the frequency rate for the aggregate of all other industries. [8]

There are many similarities between construction deaths and injuries. The major difference is that the most frequent cause of construction injuries is overexertion (24%) followed by stuck by (22%) and then falls from elevation (14%). The remainder is composed of

• Struck against

• Fall (same level)

- Bodily reaction
- Caught in/between
- Rubbed/abraded

As could be expected for the types of injuries occurring within the construction industry, the nature of these injuries in order of decreasing magnitude are

- Sprains/strains
- Cut/laceration/punctures
- Fractures
- Contusions/crushes, bruises
- Scratches/abrasions
- Multiple injuries
- Dislocations

There are differences in the SICs that have accidents which result in injuries on construction sites. The OSHA data indicates that the following percentages of injuries occur in the major SICS:

- Special Trades (SIC 17) – 56%
- Heavy Construction other than Building Construction (SIC 16) – 19%
- Building Construction (SIC 15) – 25%

Within these SICs, the trades (construction workers) that are most frequently injured are the carpenters and laborers. These two trades encompass the greatest numbers on construction worksites. Construction workers who are 20–24 years of age have more injuries than expected and those who have been on the job less than a year have greater numbers of injuries irrespective of their age. The most frequent part of the body injured is the back/trunk followed by lower extremities, fingers, upper extremities, and eyes.

Understanding that each trade has a propensity to be exposed to different hazards and thus suffer injuries unique to their trade, the more that is known regarding the trades, the hazards, and the injuries occurring, the better the opportunity for the development of intervention and prevention strategies which can mitigate and prevent construction injuries.

CONSTRUCTION ILLNESSES

As stated earlier, the numbers of occupational illnesses in construction have not been well documented. From OSHA's workers' compensation analysis, a cursory look at the types of illnesses which are most often reported to workers' compensation indicates that acute illnesses such as dermatitis, toxic poisoning, welder's flash, etc. predominate.

Only 4% of the construction claims reported to workers' compensation are recorded as illnesses. Due to the dynamic and transient nature of construction work, it is very difficult to determine long-term (chronic) affects since the number of, length of, and types of occupational exposures are not easily recordable or documented.

Thus, the concerns regarding the prevalence of respiratory disorders, neurological disorders, damaged vital organs, or cancer from occupational exposures go mostly unanswered by the research community. (See Figure 1-9.)

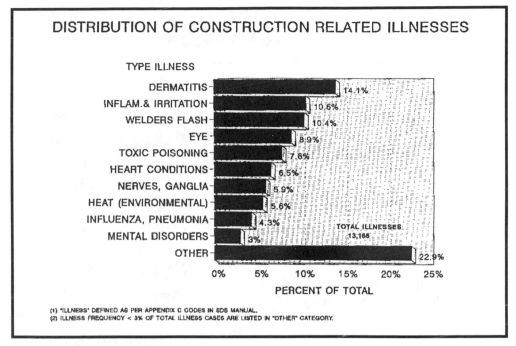

Figure 1-9. Distribution of construction related illnesses. Courtesy of OSHA

Contractors and safety and health professionals should make a concerted effort to assure that construction workers are not exposed to chemicals or other physical factors which result in exposures that might cause long term health effects.

These types of preventive actions can be accomplished by information, training, support for safe work procedures, and utilization of proper Personal Protective Equipment (PPE). More definitive information can be found in the chapter on construction industrial hygiene.

CONSTRUCTION HAZARDS

It is well recognized that the construction industry is one of the more dangerous industries in which to work. The reasons for these dangers are the hazards faced by those in the construction industry. (See Table 1-3.)

Many of these hazards are caused by the equipment used on construction worksites. The equipment used is an integral part of construction. (See Table 1-4.)

Many tools, both hand and power that have the potential to cause injuries to hands, fingers, or eyes are used by workers. These tools vary depending upon the trades on the worksite.

These tools are needed to install and work with the myriad of materials that are used in the construction industry. (See Table 1-5.) There are many construction materials that also make their contribution to the safety and health hazards faced by the construction industry. (See Table 1-6.)

As can be seen, the use of these types of equipment, tools, and materials is fundamental in construction application, procedures, and processes. These highly interrelated activities often manifest themselves in hazardous situations.

Table 1-3

Potential Construction Hazards

• premature explosions	• hand/arm vibration
• rollover	• moving of heavy equipment
• electrocution	• concrete handling
• mounting and dismounting heavy equipment	• working with sharp objects
	• using hand and power tools
• noise	• wet/slippery surfaces
• dust	• mists
• whole body vibration	• ladder and scaffolds
• exhaust emissions and carbon monoxide	• fumes
	• vapors
• whipping air hoses and other hazards of compressed air equipment	• radiation
	• gases
• cave-in	• lifting and carrying heavy materials
• burns	
• working on areas without guards	• working off precarious surfaces

Table 1-4

Construction Equipment

Scrapers	Ditch Witches	Ventilation Fans
Graders	Derricks	Ladders
Bulldozers	Backhoes	Scaffolds
End-Loaders	Fork Lifts	Tampers
Drills	Helicopters	Jack Hammers
Cranes	Pumps	Chain Saws
Trucks	Generators	

Table 1-5

Construction Tools

Radial Saws	Planners	Axes
Table Saws	Riveters	Shovels
Drills	Sanders	Tin Snips
Chippers	Hammers	Pliers
Grinders	Crowbars	Knives
Powder-Actuated Guns	Chisels	Knee Kickers
Table Saws	Screw Drivers	Staplers
Lathes		

Table 1-6

Construction Materials

All Wood Products	Bolts and Nuts	Fiberglass
Steel Beams	Re-bar	Paneling
Stone Products	Sheet Rock	Glass
Bricks	Sheet Metal	Fixtures
Concrete Blocks	Insulation	Concrete
Tiles	Soils	Chemicals
Electrical Wire	Pipes	Compressed Gases
Nails		

CONSTRUCTION'S RELATIONSHIP TO OSHA

The Construction Industry is regulated by the U.S. Department of Labor's Occupational Safety and Health Administration (OSHA) regarding workplace safety and health. As a part of the Occupational Safety and Health Act of 1970, OSHA was mandated to develop a set of regulations specifically for the construction industry that would safeguard those working in the construction industry. These regulations are divided into Subparts from A to Z and are found in Title 29 of the Code of Federal Regulations (CFR) Part 1926. Thus, the regulations are often said to be found in 29 CFR 1926 for short. There are regulations which pertain to construction in other Parts of Title 29 such as Part 1910 (General Industry Standards), while others are found in consensus standards such as American National Standards Institute (ANSI).

These standards and regulations have been developed to guide and direct the construction industry in developing and maintaining safe and healthy workplaces for themselves and their workers. They were developed because of accidents, injuries, illnesses, catastrophes, and deaths within the construction industry.

The standards or regulations are minimum requirements, and the construction industry should strive to go beyond these minimum requirements in order to insure a safe and healthy jobsite. The 29 CFR 1926.20(a) requires a contractor to provide a workplace free of safety and health hazards. This can be accomplished by adhering to and going beyond the intent of the standard.

OSHA's responsibility is to develop, implement, and enforce the standards or regulations. OSHA expects the contractor to follow the intent of these rules. If OSHA comes to your jobsite for an inspection, the OSHA Compliance Officer (Inspector) will be looking for compliance with the construction standards. If the inspector finds violations of the OSHA Standards, you will receive citations which ultimately can be costly both in the form of real dollars and losses to workplace deaths, injuries, and illnesses.

INTENT OF THIS BOOK

This book is written for those individuals interested in making construction worksites safer and healthier. This is accomplished by providing a comprehensive approach to safety and health on the construction sites. All factors that impact safety and health are incorporated within the following chapters.

The first part of this book attempts to structure the foundation for construction safety and health by incorporating the human side, which discusses the perception and attitudes of both employers and employees, as well as how to organize an approach in which everyone at the site knows their responsibility and how they are held accountable for safety and health.

The next section includes the formalization of safety and health on the construction site along with the techniques to analyze the potential hazards prior to starting the project, the site's specific hazards and hazardous jobs. Also incorporated into this section are the tools for assessing incidents, preventing incidents, and tracking incidents at the worksite.

The last section looks specifically at both safety and health hazards, delineates compliance requirements related to OSHA regulations, and denotes some of the mechanisms that can be used to attain safety and health compliance. Applicable OSHA regulations and their requirements are presented as usable tools for construction safety and health. Finally, the sources of information and materials that could be useful as augmentation to the previous material within the chapters are provided.

Thus, the following chapters provide the reader with a multitude of tools to use in developing safe and healthy construction worksites.

REFERENCES

[1] *National Census of Fatal-Occupational Injuries, 1995,* U.S. Bureau of Labor Statistics, Washington, D.C., 1996.

[2] *National Safe Workplace Institute,* Chicago, IL, 1990.

[3] *Standard Industrial Classification Manual,* U.S. Department of Commerce, Washington, D.C., 1972.

[4] *U.S. Census 1990,* U.S. Department of Commerce, U.S. Census Bureau, Washington, D.C., 1992.

[5] *Union Members in 1995,* U.S. Department of Labor, Bureau of Labor Statistics, Washington, D.C. 1996.

[6] *Safety and Health Manual,* Building and Construction Trades Councils, Washington, D.C., pp. 5–8, 1976.

[7] *Analysis of Construction Fatalities – The OSHA Data Base 1985-1989,* U.S. Department of Labor , OSHA, Washington, D.C., 1990.

[8] *Construction Accidents: The Workers' Compensation Data Base 1985-1988,* U.S. Department of Labor, OSHA, Washington, D.C., 1992.

Chapter 2

PERCEPTIONS IN THE CONSTRUCTION INDUSTRY

Charles D. Reese

This chapter discusses the perceptions that both the contractors and construction workers have regarding safety and health within the construction industry. These perceptions are those they have of the industry, itself, as well as the perceptions they have of each other.

When it comes to perceptions, there is always an element of truth in each perception, no matter how absurd it may seem to others, because for the person who holds the perception, it is a reality. Therefore, it is essential that these perceptions be given due consideration when addressing the issues related to construction.

There are three perceptions which exist regarding the construction industry.

- The perceptions by individuals and others, who are outside the construction industry, of the construction industry and its workforce.

- The perception held by construction workers related to their management and the construction industry, as well as the perceived expectations, held by others, for construction workers.

- The perception of management, within the construction industry, of the construction processes and its workers, as well as the general public's view.

Perceptions have the affect of structuring the negative and positive attitudes toward an industry, individuals, or groups which compose an industry. Some of those outside the construction industry may view the industry as different than other businesses, and there may be some who have an inherent distrust of the industry. This may be due to having an encounter with a business which had less than reputable practices and this may even include those individuals who are affiliated with the industry as either owners and/or managers.

In most cases these views are unfounded, but when a contractor underbids a project in order to get the job and then takes shortcuts in the construction process, uses inferior materials, or shortcuts safer work practices, this one bad apple can badly taint and reflect upon the entire industry, especially if faulty construction results in a tragedy.

Much can be done to negate these adverse perceptions. Of course, good/normal business practices do much to belie the poor view held by some. In business, integrity and honesty are of the utmost importance with regard to perception of the character of an industry; it reflects upon the credibility of any industry, including construction. Credibility is important since it is a lot like virginity; once it is gone, it is gone. When someone or a business loses its credibility, it seldom can be regained.

Whenever individuals, groups, or organizations have a perception or view of a company, whether it is a construction company or other business, many aspects besides honesty, integrity, and creditability impact those perceptions. Some are such things as professionalism, caring, cooperation, consideration, security, safety, public image, community-minded, and people-oriented. All of these can be addressed, if a company cares to change the perceptions held by others, by taking a forthright, earnest, and decisive public relations approach.

SURVEY

In an attempt to gain some insight into the perceptions that contractors and construction workers have toward the construction industry safety practices and each other, two similar questionnaires were used which asked virtually the same questions of each group. Although this was not a scientific sample or controlled research study, it had merit because both parties (contractors and workers) had the opportunity to express their perceptions which related to each other, the industry, safety and security, and the dangers involved with construction work.

The survey was actually a pilot testing of the instruments which had been developed, and they were administered to construction workers and contractors. The results of the initial survey of 50 contractors and 300 construction workers is not to be viewed as a valid survey, although the results are seen as providing some insight into the perception held by each group. Examples of the survey instruments and the summary results can be found in Tables 2-1 and 2-2. Thus, the results should be used as tentative perceptions of how construction workers and their contractors appear to view certain issues.

As you can see by the previous perception surveys, some of the outcomes for the construction workers and contractors were at opposite ends of the spectrum, while other responses were the same for both groups. With the construction workers there were a large number of responses which were not viewed as either in agreement or disagreement, but appeared to provide a mixed view of how construction workers felt. The same mixed responses did not occur with the contractors. A possible answer for the large number of mixed responses made by the construction workers may be that they work for many different contractors; some of those contractors may be negatively viewed, while others are positively viewed. This would make it more difficult for the workers to give a definite agree or disagree response to those items.

AGREEMENT

Both groups felt the public did not understand construction or its workers and that the work was dangerous. Each felt that the other needs to improve their professional image. Both the contractor and the construction worker felt that the contractors were concerned with the construction workers' safety and health on the job. Each group noted a sense of time pressures.

When both groups describe the public as not understanding construction and its workers, it may be a signal that some positive actions need to be taken to explain the industry and improve its image. It is a bad omen when a contractor is mentioned and the first question asked is, "have you had problems with them and their integrity?"

Both the contractors and construction workers have a sense of time pressures. Due to the bid process and completion schedules, the contractor knows that to finish behind schedule jeopardizes the profit margin as well as future contracts, if the contractor is viewed as inefficient or inept at completing projects on time. These time pressures are relayed to the construction worker in the form of "get the job done quickly, no matter the cost." This may be perceived by the construction worker as an order to "get the job done quickly," even if it means a disregard of the established safety and health rules and practices in order to meet the deadline.

Table 2-1

Construction Workers' Survey Instrument

SUMMARY OF CONSTRUCTION WORKERS' PERCEPTIONS SURVEY

Only three response categories are used to answer the statements listed below; they are agree, disagree, or mixed. The mixed response indicates that the participant was unable to confidently state whether he/she agreed or disagreed with the statement.

1. I feel contractors show respect to construction workers.	MIXED
2. I feel the contractors are worried about my personal safety.	AGREE
3. I am involved in decisions which affect the jobsite.	DISAGREE
4. I am treated as a skilled tradesman.	AGREE
5. The contractor is concerned about my health on the job.	MIXED
6. Contractors only pay attention to production.	AGREE
7. I feel the I am under time pressures to get the job done.	AGREE
8. Construction workers are different than other workers.	AGREE
9. Contractors respect the construction worker.	MIXED
10. Supervisors are well trained.	MIXED
11. Contractors are open to changing the jobsite when suggestions are made.	MIXED
12. I would like to be involved in making decisions at the jobsite.	AGREE
13. Contractors view construction workers as an expendable commodity.	AGREE
14. I need more information in order to understand my job and what is expected of me.	MIXED
15. Contractors do not trust construction workers.	MIXED
16. I am provided adequate personal protective equipment to do my job.	AGREE
17. The contractor provides enough hazard training to assure my safety and health.	DISAGREE
18. I am loyal to the contractor that I am working for.	AGREE
19. Contractors do listen to construction workers.	MIXED
20. I need to know more about the construction business to appreciate the contractor's position.	MIXED
21. Contractors just don't care about construction workers as real people.	MIXED
22. Construction workers need to conduct themselves in a more professional manner.	AGREE
23. Supervisors are just lackeys for the contractor.	MIXED
24. Construction workers should follow the safety and health rules on the job.	AGREE
25. Construction workers are expected to take care of themselves on the job.	AGREE
26. I know more about construction work than most contractors.	MIXED
27. The contractor will protect you and take care of you on the job.	DISAGREE
28. The Public does not understand construction work and the construction worker.	MIXED
29. Construction is dangerous work.	MIXED
30. Contractors need to improve their professional image.	MIXED

Table 2-2

Contractors' Survey Instrument

SUMMARY OF CONTRACTORS' PERCEPTIONS SURVEY

Only three response categories are used to answer the statements listed below; they are agree, disagree, or mixed. The mixed response indicates that the participant was unable to confidently state whether he/she agreed or disagreed with the statement.

1.	I feel construction workers show respect to contractors.	AGREE
2.	I am worried about the safety and health of construction worries.	AGREE
3.	I involve workers in decisions which affect the jobsite.	AGREE
4.	I treat construction workers as skilled tradesmen.	AGREE
5.	I feel construction workers are concerned about their health safety on the job.	AGREE
6.	My only interest is production.	DISAGREE
7.	I feel that I am under time pressures to get jobs done.	AGREE
8.	Construction workers are different than other workers.	AGREE
9.	I truly respect the construction workers.	AGREE
10.	My supervisors are well trained.	AGREE
11.	I am open to change on the jobsite when workers make suggestions.	AGREE
12.	I allow workers to be involved in making decisions at the jobsite.	AGREE
13.	Construction workers come and go and are an expendable commodity.	DISAGREE
14.	Construction workers do not need to be given information to do their job.	DISAGREE
15.	I do not trust construction workers to do a job right.	DISAGREE
16.	I provide adequate personal protective equipment for the job.	AGREE
17.	I give enough hazard training to workers to assure their safety and health.	AGREE
18.	Construction workers are seldom loyal to their employer.	DISAGREE
19.	I make an effort to listen to construction workers.	AGREE
20.	There is no need for construction workers to know about my construction business.	DISAGREE
21.	It does not make good business sense to invest training and dollars for construction workers.	DISAGREE
22.	Construction workers need to conduct themselves in a more professional manner.	AGREE
23.	I am more likely to hire an unknown worker who has documentation to attest to his/her skills (e.g., certification of testing).	AGREE
24.	Construction workers do follow the safety and health rules on the job.	AGREE
25.	I expect construction workers to take care of themselves on the job.	AGREE
26.	Construction workers need to come to work sober and drug free.	AGREE
27.	Construction workers expect me to play wet nurse maid to them or hold their hand on the job.	AGREE
28.	The Public does not understand construction work and the construction worker.	AGREE
29.	Construction is dangerous work.	AGREE
30.	Contractors need to improve their professional image.	AGREE

According to the survey, construction workers believe contractors treat them as skilled tradespersons, and contractors also felt they treated construction workers with respect. If this is true, it is a very positive factor toward the development of a degree of professionalism between construction workers and their contractors. The hard part will be convincing themselves and others of the professional nature of the construction industry, since the constituencies contractors and construction workers each viewed the other as nonprofessional. This is an area where some effort could be invested.

The reason that contractors and construction workers may have a perception that they are not professionals is because the connotation "professional" is usually used in reference to doctors, lawyers, etc. These perceptions often lead people to believe that they are not considered professionals in their line of work. A description or definition of a professional may help place some criteria on who is considered a professional. Professionals are workers who are involved in an occupation that requires a level of training, competence, and knowledge regarding their occupation. These individuals have knowledge that is unique to their occupation, they have a common vocabulary, and they have a code of ethics or standards by which they operate. They are strongly motivated and have made a lifetime commitment to that career. Professionals also provide a service to people, either directly or indirectly. Professionals have credibility which is defined by their competence, composure, character, and ability to communicate. The professional relationship heavily depends on a kind of mutual trust between the professional and those with or for whom he/she is working.

With this definition, there is the opportunity to strengthen or develop a more professional approach to the contractors' and construction workers' chosen occupation. A lesson can be learned from other industries, such as the automotive industry, where they have changed the title from mechanic to automotive technician. The automotive technician also wears a standard uniform with the technician's name and company logo on it. This gives the automotive technician a more professional appearance. This would be a drastic change in the construction industry (see Figure 2-1).

Figure 2-1. Typical construction workers laying pipe

But, the feeling of being a professional is something that lies deep down in your own personal perceptions of yourself and your profession. If you do not consider yourself worthy of being called a professional contractor or professional construction worker, then you cannot expect others to consider or treat you as a professional either.

DISAGREEMENT

Construction workers felt that they were not involved in the decisions made at the worksite, while contractors thought that they were. Construction workers thought the contractor was interested only in production, while the contractor felt differently about production. Workers believed that they were treated as expendable commodities, but contractors voiced otherwise. Contractors held that they provided adequate safety and health training, while workers stated the contractor did not.

It is interesting that only four areas of major disagreement seem to exist. These areas relate to worker involvement, an emphasis on production, the value of the worker, and the quality of training. One would expect there to be more marked numbers in the areas of disagreements.

From what modern industry knows of motivation, worker involvement is key to a good motivational atmosphere; this is discussed further in Chapter 3. We know that true involvement or having a stake in what is transpiring in the workplace is a powerful motivator. Being able to assist in the decision process or having a degree of control will foster a more positive response by construction workers.

Anyway you cut it, production and the profit it generates pays the bills, but maybe a safe production atmosphere could be developed, and communications on production, quality, safety, and an understanding of profits or a sharing of the profits could be a part of your business approach. All of us have that deep internal need to be of value to our family, friends, and our employers. Contractors can show more concern, caring, and compassion for all their employees. This value of workers needs to be constantly reinforced by contractors. As far as training goes, many employers assume workers are adequately trained, which indeed may not be the case. You may need to do a training-needs assessment. Sometimes just asking if someone needs further training may be helpful in targeting real training needs, and not just perceived training needs.

Where differences exist, the good contractor will see that with a little bit of effort on his/her part and on the part of the supervisors, a large impact may be attained which could result in a more positive work environment.

CONTRACTORS AND MIXED VIEWS

It appears that contractors may have rose colored glasses when it comes to their perceptions of the construction workers. Contractors feel they show construction workers respect, while construction workers hold a mixed view of that respect. Contractors feel that contractors are concerned about the construction worker's job safety and health, while construction workers have mixed feeling regarding the contractor's concern for their safety and health. Contractors universally feel their supervisors are well trained, but construction workers again have a mixed perception of this. The contractors strongly feel that they were open to suggestions from construction workers, but construction workers did not feel either a strong positive or negative response to this issue. Workers were not sure if the contractor did, or did not trust them, or trusted them enough to provide more information about the job or project. Contractors felt that they listened to construction workers, while the workers themselves had a mixed view of this. Construction workers did not seem to have a know-it-all attitude; they did not believe that they knew more about the construction business than the contractor. Contractors seemed to have a very optimistic view of what they were doing, while the construction worker's perception did not support this positive view. A reality check may be necessary, based upon the mixed responses of the construction workers. It would be interesting to see how your workers respond to these same issues.

OTHER AREAS

Construction workers felt the contractor was not going to protect them on the job, while contractors sensed that construction workers wanted to be cuddled or handled like children. These two areas seem to make little sense when other areas of the survey expound respect, concern for safety and health, treatment as a skilled worker, and an expression of loyalty to the contractor.

HOW TO USE THIS SURVEY

If you do not feel that the results of this survey give an accurate portrayal of the perceptions held by contractors or construction workers, you may want to revise the questionnaires given in Tables 2-1 and 2-2.

The following is an example of one way you could revise these surveys to get a more accurate view of the participants' responses. You could use a format similar to the example provided and revise any of the statements to fit your specific or unique needs.

EXAMPLE:

Answer the following questions in an honest and forthright manner.

(You do not need to put your name on this paper; this questionnaire is for survey purposes only.)

Answer each question by circling one of the five responses.

SA = Strongly Agree, A = Agree, N = No Opinion,
D = Disagree, SA = Strongly Disagree

1. I feel contractors show respect to   SA   A   N   D   SD
   construction workers.

2. I feel the contractors are worried about my   SA   A   N   D   SD
   personal safety.

Or, if the previous revisions do not meet all of your own needs and perceptions, you may want to completely revise the instruments.

Once you have administered these instruments to your own supervisors and workers, you can do an analysis and draw your own assumptions regarding the perceptions they hold concerning safety and health in the workplace. Your results should provide insight into where you may want to address certain aspects of safety, health, or communication issues.

USING THE RESULTS

From the results of this survey, it would seem safe to say that some work needs to done to alter the perceptions held by both the contractor and the construction worker. It would be a good idea to build perceptions upon which both parties agree, such as the industry and its workers using a more positive and professional approach.

The next area to address should be the four areas where there are disagreements between the contractors and the workers. The workers indicated that they want more involvement in the decisions made concerning their jobsite; they are concerned about their safety and health while on the job and want the quality and production to be on some sort of an evenly balanced scale; they want to be considered of more value as part of the workplace; and they felt that more training or job instruction was needed. Maybe the workers feel if they are better trained, they will gain the respect of the contractor and, therefore, will be of more value and allowed to be more involved in the decision-making process.

In most cases, what really needs to be done is to open the lines of communications. There needs to be an exchange of information and ideas without prejudgment. Look for areas where input and involvement can fit, and have honest and forthright interchange between both parties. Hopefully, the construction industry, contractors, and workers do not have a defeatist attitude that believes these differences have existed for so long that nothing can or ever will change.

Actually, the goal for the remainder of this book is to get everyone involved in developing a construction safety and health program which will include involvement, training, safe performance, and an atmosphere where everyone supports production in only a safe and healthy manner. Thus, value and concern is extended to protecting everyone on the construction jobsite.

SUMMARY

The use of the previous surveys and their results and analysis is just one example or approach that may be used in determining if there are any areas in the workplace that need to be addressed and improved.

Everyone is reticent to change and, when the normal way or approach of doing a job has been in practice for a long period of time, change is even more difficult. From the contractor's point of view, the jobs that are performed at the jobsite are being done in the most effective manner. But, according to the construction worker, he/she has mixed perceptions of how effective or truly committed the contractors are toward job safety and health.

The changing of perceptions cannot and should not be laid on either the contractor or the construction worker. Everyone involved must be a part of the solution. If changes in perceptions are truly desired, the effort and energy must come from both parties.

Chapter 3

THE PEOPLE ISSUE IN CONSTRUCTION SAFETY AND HEALTH

Charles D. Reese

SETTING THE STAGE

In this chapter the aim is to provide a basis which will allow you to motivate yourself and your employees. When you discuss motivating yourself or others, you are always in search of a blueprint. To most of us the magic formula for motivation is perceived to be composed of plans, tricks, gimmicks, or inducements. This may be the case in some instances or for some individuals, but not the panacea.

The human aspect of safety and health is often brushed aside in the construction industry. After all, construction is different, and the fundamental principles of motivating and dealing with construction workers are, therefore, different also. Granted, people are unique, but all workers have basic needs to be addressed and fulfilled in order for them to work effectively and productively. The principles and examples in this chapter have been used for miners as well as office workers, but are as applicable to construction workers as they arc to any other industry.

The management of people is directly related to understanding the basic principles of motivation. Obtaining good safety and health behavior and work practices can be directly attributable to how effectively you apply good principles of motivation. The intention of this chapter is to provide a practical and insightful view of the workings of motivation in the occupational environment. One must remember that many theories related to motivation exist. Using the principles discussed in this chapter is not a sure-fire guarantee that you will be able to motivate your workers but, if you don't do something, you certainly will not attain the motivation in them you desire. Most of the time managers, contractors, and safety and health professionals want a band-aid solution to their motivational problems in the workplace.

Motivation is a somewhat imprecise science and undertaking. No guarantees for success exist. Each workplace needs to pay close attention to the motivational needs of the individuals who work there or you will have low morale. Since morale is as difficult to define as is motivation, for the context of this chapter, let's use motivation to attain good morale.

There are many legitimate reasons for trying to motivate workers and employees. Some of these reasons may be something as simple as trying to get an employee to work safely, or as complex as fostering safe work teams. It is very challenging to instill motivation where motivation does not presently exist. While you should hesitate to use workers as behavioral guinea pigs, workers are no different than anyone else when it comes to motivational issues. Motivating yourself and others is usually attempted because you care about someone or some group and want to see them accomplish a goal or conquer more than they ever thought was possible. It is rewarding to see a group of workers attain the goal of working 1,000,000 hours without an accident and go on from there.

Although all the principles within this chapter can apply to families, relationships, friendships, coworkers, teams, or employees, most of the successes espoused are those relevant to the workplace. This is, of course, because the majority of our adult life is spent in the workplace.

Defining Motivation

Some people believe motivation has the potential to answer everyone's problems. You may have heard such statements as, "If I were only motivated!," "You should motivate me!," "You should motivate him or her!," "You are not motivated!," "You had better get motivated!" or, " All you have to do is find his/her "hot button!" These statements, however, do not tell us what motivation really is nor do they tell us how we can measure or even understand motivation.

Motivation, in the broadest sense, is self-motivation, complex, and either need or value driven. Someone once stated that they believed "hope" was the secret ingredient to a person being motivated (the hope to accomplish a goal, a dream, or attain a need); there is reason to support this theory. But, possibly a better definition is "motivation presumes valuing, and values are learned behaviors; thus, motivation, at least in part, is learned and can be taught" (Frymier, 1968). This definition provides us with the encouragement we need to go forward and achieve motivation for ourselves and others.

If we want to be successful, we must believe that we can teach someone to be motivated toward specific outcomes (goals) or, at the least, be able to alter some unwanted behavior. On the other hand, we should not want to completely manipulate an individual to the point that they blindly respond to our motivational efforts. If that occurred, we would lose that most important portion of a person, the unique human will.

Thus, motivation is internal. We cannot directly observe or measure it, but a glimpse of the results may be observed when we see a positive change take place in behavior. Such a change might be something as simple as a worker wearing protective eye wear, or something as far reaching as going a full year without an accident or injury. By observing these outward manifested behaviors, you can then be encouraged by even the smallest successes that are related to your motivational techniques.

Summary of the Principles of Motivation

Goals are an integral part of the motivational process and tend to structure the environment in which motivation takes place. The environment in which we find ourselves is many times the springboard to the overall motivational process. You may be fortunate enough to accidentally step into a high energy motivational environment. On the other hand, you may find yourself in an environment that is not at all conducive to motivating others and, thereby, it is very difficult to attain your desired goals. If this is the case, you may need to make a change in the physical environment or possibly even make a change in the work atmosphere (i.e., allowing more independence of individuals in the decision-making process.)

Changing the environment may not affect all individuals in the same way. It has been said that there are three certainties which can be stated regarding people. Those certainties are "People are Unique," "People are Unique," and " People are Unique!" Since people are unique, what motivates one person may be demotivating to another.

If you are that person with the responsibility of trying to motivate another individual or group, you will need to address their motivational needs. Employees fall along a continuum – some need little motivation from you and others need constant attention. It is unrealistic to expect all of them to achieve your level of expectation. The quality of your leadership will be the determining factor of your success with these people.

Why are some leaders more motivational than others? What are the unique talents that these dynamic leaders possess. Some people believe that these individuals were born to be leaders. Most of us do not believe that they are just "born leaders" but individuals who possess a set of talents and have chosen to develop those talents to the maximum. These talents are developed because they have the burning desire (goal) to become leaders and those desires motivate them to learn necessary skills.

To be a successful motivational leader, you must have some sort of plan that will get you from point "A" to point "B." This plan should include your desired goals and objectives, levels of expectations, mechanisms for communication, valuative procedures and techniques for reinforcement, feedback, rewards, and incentives. Any motivational plan is a dynamic tool that must be flexible enough to address changes which may occur over a period of time and take into consideration the universality of people and situations. These plans can use a variety of techniques and gadgetry to facilitate the final desired outcomes or performances which lead to a safer and healthier workplace.

THE MOTIVATIONAL ENVIRONMENT

Everything that surrounds us is part of our motivational environment. Depending upon our environment, we are motivated differently at a given point in time. We actually exist in what we call "micro-motivational" environments. These micro-environments make up the sum total of our motivational environment and are comprised of our work environment, family environment, social environment, team environment, peer environment, or even a nonfunctional environment.

Any one or all of these micro-environments can have an impact on the other. The negative impact of one of a person's micro-environment may cause that person to also react negatively in another one of his or her environments; this can happen even when the environment is, in itself, a positive one. For example, if an individual has problems at home, it may, and many times does, cause that highly motivated employee to become less safety conscious or productive at work.

In illustrating the complexity of this issue, let's think for a moment about problem employees. Many times these employees ask to be moved to a different job because they are either dissatisfied or performing poorly in their current job. Amazingly, once the reassignment is made, their performance vastly improves. It is almost as if this worker becomes a different individual. When they are put into a new and different environment, they get a "new spark" and the new environment becomes their positive motivator; they've been revitalized!

Structuring the Motivational Environment

In order to structure an environment where workers will be motivated to perform their work in a safe and healthy manner, it takes some degree of organization and commitment. The safety and health environment must have a foundation. A written safety and health program is the key component in providing and structuring that foundation. This written program should set the tone for safety and health within the work environment. There are some keys to motivating safety and health (Table 3-1).

Table 3-1

Keys to Motivating Safety and Health Program Functions

1. EXPLAIN – Clarify safety and health performance expectations.
2. INVOLVE people in decision making and problem solving.
3. Describe the CONSEQUENCES of unsafe and poor health performance.
4. Establish attitudinal safety and health GOALS.
5. Provide FEEDBACK on safety and health performance.
6. Provide a SELF-MONITORING system.
7. Recognize and REINFORCE good safety and health performance.
8. Develop a REWARD system.

The organized approach to safety and health should address each of the previous eight keys. Within this chapter you will find each of these keys as an integral part of the occupational safety and health prevention program initiative.

Many tangible and intangible factors comprise the motivational work environment. Something that is tangible could be as simple as moving a piece of equipment in order to create a more desirable environment or something intangible could be your ability to change the way someone feels about you.

You should develop an environment where the majority will be positively motivated to perform. But, this should not prohibit you from making adjustments, when possible, to address individual needs. Furthermore, be sure that you don't allow too much flexibility (i.e., favoritism, etc.) or it could destroy a good situation for the majority.

To assure equal treatment, require all to abide by the rules. For instance, when there is a set group of safety and health rules for your workplace, you should never allow one person to abuse these rules while holding others rigidly to them; this will cause disenchantment with safety and health issues. Top management, supervisors, foremen, and workers should be treated equally and fairly.

As an example, while working with one company as the safety director, I wore moccasins while everyone else was required to wear hard toed shoes. This type of behavior should not be acceptable for one person and not the others. Although you may be the head of an organization or the "boss," you should never consider yourself so lofty that you do not adhere to your own safety and health rules and policies. It is very important that management and supervisors set the tone of the work environment regarding safety and health on the job.

When setting up an environment where you want those involved to be motivated, you should first address the physical needs. For example, in the work environment there may be the need to provide proper tools and personal protective equipment for the workers to safely do their work.

When it comes to developing a safe and productive motivational work atmosphere, the intangible motivational issues are just as important as the tangible ones, but they are also the most time consuming. These challenges run a broad spectrum and, just to list a few, could be some of the following examples: changing the way a person is treated by their peers, colleagues, or supervisors; helping an individual gain a positive perception of the environment around them; or developing a new and positive attitude toward workplace safety and health.

Your ability to structure an environment which provides individual needs and adequate stimulus to motivate each person to his or her "full capacity" is desirable, but not usually possible. In fact, you actually have little chance of setting up the "perfect" environment for every person. There are just too many other environments and factors which compete with what you desire each individual to accomplish. However, do the best you can for each person and then each individual will make a conscious decision as to whether he or she wants to perform safely in the workplace. This is the reason that each worker should know the consequences of any unsafe performance. You should develop mechanisms to assist these individuals perform safely, but also have disciplinary procedures for those who elect not to comply with the safety and health rules.

As part of setting the environment, be assured that each worker understands the expectations regarding working safely. It is also useful to involve them in setting the safety and health rules and goals and to know the expected outcomes. Each worker needs to understand that there will be consequences or penalties for disregarding or violating the safety and health requirements of their work. Therefore, goals are important in setting performance objectives for the company's safety and health program.

Track the progress of the safety and health goals and provide feedback. This allows the workers to monitor their own accomplishments in their work area. Recognize the workers who are progressing towards the safety and health goals and reinforce safe work behaviors.

Reacting to the Motivational Environment

You can provide all of the bells and whistles, but if you do not pay attention to some fundamental characteristics of people, you will not be successful in developing a good motivational environment. Some of the fundamental principles you need to be aware of when working with people are

1. Individuals view themselves as very special. Thus, praise, respect, responsibility, delegated authority, promotion, recognition, bonuses, and raises add to their feelings of high self-esteem and need to be considered when structuring a motivational environment.

2. Instead of criticism, use positive approaches and ask for corrected behavior. Individuals usually react in a positive manner when this approach is used.

3. Verbally attacking (disciplining) individuals tends to illicit a very defensive response (even a mouse, when cornered, will fight back to defend itself). Therefore, it is better to give praise in public and, when necessary, criticize in private.

4. Individuals are unique and, given the proper environment, will astound you with their accomplishments and creativity (even those individuals who you consider noncreative).

Remember, the final outcome lies with the employees; they will decide whether or not to perform safely. But, if the employer has done his or her part, workers will never be able to hold him/her responsible for the decisions they have chosen to make.

There will be people who elect to work unsafely even though the environment may be very motivational to the majority. Thus, when discussing work, pay close attention to the motivational environment, and work at making it the very best! But, when there are those who do fail to perform safely, there should be consequences and discipline administered quickly and fairly. If there is no one enforcing the speed limit, then who will abide by it? Either enforce the rules or lose the effectiveness of your motivational effort.

The key to a successful motivational environment is to pique the interest of people. Let them know you want them to succeed; give them responsibility; and leave them alone to accomplish those goals and succeed. If the above principles are not taken into consideration when setting up a motivational environment, you will be more likely to encounter problems with the success rate. As an illustration, a supervisor noticed that his workers were not giving the performance he expected. He was having difficulty receiving top quality written reports from them and, therefore, had been rewriting each report. When the supervisor was asked if his employees were aware that he was rewriting their reports and, if so, did he think they were putting forth their best effort, he answered, "Oh yes." But after thinking this question through, he decided to go back and ask his workers the same question that he was asked. Their reply to him was, as expected, that they were only giving a half-hearted effort since they knew the report would be rewritten. As you can see from this example, caution is needed so that you don't set yourself up for this type of response.

The way that you structure the motivational environment will allow individuals within the work groups to accomplish safety and health goals and assure they are free from injury and illness. You need to set up an environment where people can be successful. And, in order for that environment and the people within it to succeed, you must demonstrate that you genuinely care about them and the purpose of the mission (goals) they are trying to attain, which in this case is a safe and healthy workplace.

Next, you need to be open to learning from your own experiences, as well as from others. This will facilitate flexibility in your encounters and give you the ability to make the necessary changes. You need to be honestly perceived by everyone as diligently working to prevent workplace incidents and be willing to work at motivating those who are not in tune with your safety and health attempts. It does take an added effort to motivate others.

It is imperative to realize when you have reached a point where you have accomplished as much as you can and have lost the effectiveness of the safety and health environment that you have structured. This may be an indicator that you need to change your approach.

As an example of this type of situation, think for a moment about coaches, especially those who are in the professional ranks or larger collegiate institutions. When they become ineffective, they are forced to move to other coaching positions. But, once they are in their new position, and even though they had become ineffective in their previous one, they often are able to rejuvenate a program which, until their arrival, was unsuccessful. In these cases we realize that the coaches are still the same people, but, because the environment had changed in their previous job and they were unable to adapt to those changes, they became ineffective. Nevertheless, when they were introduced into a new position, they once again became successful.

Motivational Environment Examples

I believe that psychologist Frederick Herzberg was on target with his concept of successful motivators. He said that for individuals to be satisfied with their jobs and remain motivated, they need competent supervision, job security, adequate salary, adequate benefits, and good working conditions; but more importantly, they need the satisfaction of achievement, recognition, responsibility, and challenge. These are the real motivators, the internal ones; the ones that truly satisfy the individual's specific needs.[1]

Thus, when structuring your motivation environment, be sure to load it with as many of these true motivators as possible. They are the most successful incentives and encourage consistent and improved safe performance. Some motivational environment examples are as follows.

A company installed a new air-conditioner for the workers of their appliance assembly in an effort to improve their physical work environment. The new air-conditioner did make the work environment more comfortable, but for some unknown reason, the production of the workers actually decreased, and accidents and rejects increased.

In an attempt to increase production and decrease rejects and workplace injuries, some new incentives were introduced, but to the company's dismay, the production remained low and work-related incidents continued to occur. Finally, the discovery was made that the new air-conditioning system was so loud that the workers could no longer talk or be heard by their fellow workers during the assembly process. Since the workers were doing a repetitive task and were also receiving low wages, their job satisfaction depended upon their social interaction with fellow workers. Once the air-conditioner was shut down and replaced with a quieter unit, productivity improved and rejects and injuries also decreased.

In the past, companies have tried to motivate people by reducing the hours worked, giving longer vacations, increasing wages, increasing benefit packages, providing career counseling services, training supervisors in communications, and organizing interactive groups. However, these incentives have not proven to be highly effective in increasing productivity. Therefore, it is important to determine what affects the satisfaction or dissatisfaction on a job or, for that matter, anywhere else.

In structuring a motivational environment, it is important to help people grow and learn through the task they are asked to perform. Prepare them to stretch their abilities to new and more difficult tasks and help them advance to higher levels of achievement. Help them use and recognize their unique abilities and make sure they can see the results of their efforts. Be sure to recognize when a task is well done; give a promotion or award and provide or reinforce performance with constructive feedback. This is applicable not only at the workplace but is also standard for life situations whether it be school, sports, home, social groups, or peer groups.

In recent years companies such as Ford, Volvo, and General Motors, as well as many others, have found that team approaches to the work environment are very effective. They have found that an increase in quality and overall job satisfaction transpires when a work group is assigned a task and then given control over such decisions as who does what tasks, how the tasks will be accomplished, and has the authority to stop the process if quality is in question.

With this type of system in force, the supervisor is no longer responsible for completion of the task; the group has that responsibility and control. The supervisor's main duty becomes one of advising, providing feedback, and assuring that all materials and tools are available to accomplish the job. This approach has also been very successful with quality circles but may not work in all environments since the end product is not the same for all individuals and in all situations.

When there are barriers to setting up a good motivational environment, put an even greater emphasis on the nontangibles (recognition, achievement, responsibility, and challenge). As an example, let us consider the M.A.S.H. television series. The physical setting was terrible, the wounded were disheartening, and the tools needed to accomplish their mission were often missing. But discipline was not stringent, protocol was lax, individuality and recognition was endeared, and this made the mission not only challenging but also rewarding.

As you can see from these examples, you cannot always predict the way in which individuals will react to a motivational environment, but you can predict with some certainty that if there is no attempt to set up a good motivational environment, an integral part of motivation will be lost. Thus, with this piece of the puzzle missing, the other facets of the motivational plan cannot be effectively applied.

GOAL SETTING

Goals and You

Even though we may not put our goals in writing, it would be safe to say that we do few things in life that are not facilitated by goals and a goal-setting process. Most people set goals in order to accomplish something: to get more money, to get recognition, etc. This is the reason there are so many accomplishments in this world; people are continually setting goals (consciously or unconsciously), attaining those goals, and then moving on and setting new ones.

One company set a goal of decreasing their reportable injuries by one-half; they had had fifty such accidents the previous year. In a six-month period, the number of incidents decreased 75% and the only thing the company had done was set the goal, publicize it, and track and report the progress toward that goal. The company made no program changes.

Setting goals is vital for your safety and health initiative. If you set goals for yourself and your company, you will be amazed; goal setting gets great results.

Rationale Behind Goals

If you and others are to be motivated, then goals must be attainable. A failure to utilize a goal approach will surely result in a failed motivational effort.

Take, for example, the story about the Miami Dolphins and their bid to get to the Super Bowl. As the story goes, a reporter was talking to Nick Buoniconti shortly after the Dolphins' loss in the super bowl. He asked, "What was your team's goal?" and Nick answered, "To get to the Super Bowl." As he spoke, Nick realized that they had achieved their goal but had set it a little short. It should have been to "Get to the Super Bowl and win."

It is not certain that this is a true story but what is vital in this illustration is the fact that it demonstrates the importance of setting goals in order to accomplish what you want. Be aware, though, that at times what you get may not be all that you want. Therefore, be very careful when setting your goals since you can set too high or too low an expectation for yourself.

Goals and Their Many Forms

Some goals may be quickly attained, while others will take a period of time to reach. This is the reason that your safety and health goals should be a mixture of both types of goals.

Groups can be motivated to obtain goals. For instance, a story was told about a factory operation that made cutting machines. The plant manager walked through the shop during the latter part of the day shift and asked the crew how many machines they had finished that day. They told him eight. Without saying a word, he took out a piece of chalk and made a large eight on the concrete floor and walked away. When the next crew started their shift, they asked about the eight on the floor and the day crew explained what had happened. When the day crew came to work the next day, the eight had been erased and a nine had been chalked in its place. This is a good example of goal setting and the effect of peer pressure, competition, recognition, and involvement. You can probably think of other motivational plans as well.

Goal setting is important in all aspects of our life and includes areas we seldom consider as motivational. People who are committed to a cause are motivated. This zeal to a cause, which is in all actuality a goal, has been a strong motivator over the history of civilization. Few people will deny the power of a purpose or cause and the drive to attain it. It has been noted that few people fail if they are driven by a goal. Goals which enhance performance are usually

1.   Specific

2.   Difficult

3.   Challenging

This topic of goals is of great importance. It seems to have been the force behind most of the accomplishments we have seen in this world. I think Tom Lasorda, the Dodger's Baseball Manager, said it best when he said, "Motivation is the ability to communicate to people goals and objectives and show them how to achieve them."

SELF-MOTIVATION

You Are the One

The question that arises is, "who motivates you?" Is it a person, is it peer groups, is it incentives, or is it the environment? It is the contention of this chapter that no matter what, it is you who motivates you. Excuses, blame, and alibiing will not negate this fact. Nobody can motivate you. You must assume the responsibility to motivate yourself within the environment in which you find yourself. Some individuals are motivated by positive happenings within their life, while others succeed through adversity.

People Are Amazing

One person who was motivated by his failures was Dan Jensen, Olympic speed skater. After failing to receive the gold medal in three previous Olympic Games, he went on to become a gold medal winner in the 1994 Winter Olympics Games. He had been expected to win the gold medal in previous Olympic Games, but through disastrous falls or unexpected losses, he was unable to accomplish that goal. But through his trials and failures, he was determined to make them lead him to success and his ultimate goal, a gold medal in the 1994 Winter Olympics' 1000 meter race. Failures can bring success!

On the other hand, what would have happened to Bonnie Blair if she had experienced the same fate as Dan Jensen? She culminated her career with five gold medals and had at least one medal in each of the previous three Olympics. No one can say or even guess the answer since she was motivated by her successes each time, instead of her failures. Mistakes can either have a positive or negative effect on the motivational environment, but tend to be demotivational when punished rather than realizing that people who are doing something are going to make mistakes.

In a recent article, Bill Gates, Chairman and CEO of Microsoft, stated, "Reacting calmly and constructively to a mistake is not the same as taking it lightly. Every employee must understand that management cares about mistakes and is on top of fixing problems. But setbacks are normal, especially among people and companies trying new things." [2]

Regardless of all the efforts made to assure that no accidents, injuries, or work-related illnesses occur, there will still be, at times, mistakes made and negative outcomes that occur. When this happens, this should be an incentive to try even harder; don't trash the safety and health effort over a setback.

Losing the Self in Motivation

The basis for motivation seems to be in our perceptions of ourselves. These perceptions govern our behavior and support the concept of self-motivation. In order for people to motivate themselves, there must be meaning in what they are doing. If they do not perceive that the goal set before them will satisfy their needs, they cannot possibly motivate themselves to accomplish it.

You must realize that no matter how unrealistic a perception may seem to us, it is a reality to the person who holds it. No matter how we try to debunk a perception, there is always some degree of truth and reality within it and, therefore, it is very real to that person.

Individuals will not be motivated to work safely unless they have internalized the goals and expectations of the company. It is not enough for them to know that they will be fired for violation of a safety rule, they need to be motivated to perform their work safely even when there is no one watching them.

People must be inspired to be accountable to themselves. If they put their goals and plans on paper, then they take possession of their own behavior to a greater extent. This motivates them to do something and gives them the time, direction, and a reason to find new or better ways to accomplish their goals and plans. As many experts will advise, you should put your goals or plans in writing. If you can't write it down, then you probably will never achieve it.

DEALING WITH PEOPLE ALONG THE CONTINUUM

The People Issue

Are you unique? All of us like to assume that we are unique individuals; we want our own identity. Then is it any wonder that when someone tries to figure out how to motivate another person, it becomes a difficult task. Think about it for a minute! Why do people behave the way that they do? Of course, these behaviors are what we outwardly see, but behaviors are formed prior to their outward manifestation. Values which are formed early in life affect our attitudes and these attitudes are exhibited in our behavior.

Values are dynamic beliefs that lie at the foundation of your personality. Your values involve the deepest parts of you, so much so that when your values change, you change. Different values mean a different you, which results in different attitudes and behaviors.

Your values and needs are closely related, but they can also be distinguished from each other and the difference between them is important in the motivation arena. Both values and needs affect how you talk and act, but in different ways.

Changing Values

When trying to motivate people, it is important to recognize whether a person is expressing a value or a need. Your values differ from your needs in at least three ways: your values are you; they are something like forces within you; and, to destroy your values is to destroy the "you" you now are.

Could there ever be an opportunity to change these values and, thus, change attitudes which result in new behavior? According to Dr. Morris Massey, a noted national speaker and expert on values, it would take a "significant emotional event" to do this. What is a significant emotional event? It could be something as traumatic as seeing a fellow worker die on the job, or being seriously injured yourself. But, when any one of these events occur in your life, you are changed from that day on and for the rest of your life.

It is true that the most dramatic value changes do occur in people's lives when they experience one of these "significant emotional events." But, there are also less dramatic events that occur in our lives that can also alter our values. These are

1. Values change as we adjust ourselves to life's changes.

2. Old values break down and we gradually replace them with new values.

3. New values will result in new attitudes and behaviors.

Values often gradually or imperceptibly change through the influence of your exposure to the books you read, the things you see, and the experiences you have. Values are a lot like habits. It is impossible to get rid of a habit without replacing it with another habit (e.g., cigarette smoking replaced with an exercise program, etc.); nor do values change unless they are replaced with new ones.

In your safety and health effort, it may be vital for you to change the existing value structure. When you make these changes, expect it to take some period of time before the bad values are replaced with those you desire.

The three most important things to remember about people are: people are different, people are different, and **people are different!** With this in mind, you will need to view each person on a continuum. When trying to figure out what motivates them and how you can begin to get a change, take a look at every aspect concerning that person's life and try to evaluate what is and is not of importance to him or her.

Some individuals are superstars. These individuals are self-motivated and all you have to do is give them support and minimal guidance and then just step back and watch them go. Others, on the other hand, seem to lack any motivation at all. These individuals need to have things structured for them, know exactly what is expected of them, know what happens if they do not perform, and know what the reward or outcome of their performance will be.

You will also find individuals who need to be around other people; they perform best when they are in a social environment and, therefore, are more affected by peer group pressures. Finally, there are people who prefer to work alone. Frequently these individuals are achievement oriented and all they want is your recognition and reinforcement. All it may take to motivate them is to grant their request for something as simple as a tool or piece of equipment which will help them do their job in a safer manner.

Needs Move Mountains and People

Dr. Abraham Maslow of Brandeis University believes that people are motivated not only by their unique personalities and by how they want to fit into their world, but they are also motivated by their own individual needs. The premise which runs through Dr. Maslow's book is "motivation is internal – thus, self-perpetuating."

Dr. Maslow identified five needs. They run the gamut from the basic animal needs to the highly intellectual needs of modern man. They are the physiological, safety, social, ego, and self-fulfillment needs.[3]

In order for you to understand the relationship between these needs and the motivational process, a description of each one follows.

Physiological Needs are the requirements we have for our survival. They encompass the basic needs that are necessary for the body to sustain life or physical well-being. These needs are such things as the food we eat, the clothing we wear, and the shelter we live in. Each of these must be satisfied before other needs can be dealt with. The physiological needs appear in all of the actions each of us take to insure our survival and physical well-being. Individuals who are primarily motivated by these needs will do anything that you ask them to do, no matter how unsafe it might be.

Safety Needs include the requirements for our security. If first the physiological needs are reasonably well satisfied, then people become aware of and start to act to satisfy their safety needs. These needs are such things as having freedom from fear, anxiety, threat, danger, and violence and being able to have stability in their lives. Striving to satisfy these needs might show up in such actions as: (for their safety) avoiding people or situations which are threatening and may be of danger to them, or, (for stability) lobbying for a pension plan or putting money into an agency's credit union. These individuals are concerned about whether you are taking all the precautions to protect them from workplace injury and illness.

Social Needs include the requirements for feeling loved and wanted, and the sense of belonging and being cared for. If the safety needs can be reasonably satisfied, social needs begin to emerge. Some behaviors that take place and indicate a social need of acceptance are: asking the opinion of the group before acting, following group preferences instead of personal preferences, or joining job-related interest groups. These individuals will follow the safety and health pattern set by their workgroup.

Ego Needs include the requirements for self-identity, self-worth, status, and recognition. When social needs are reasonably satisfied, individuals are able to explore the dimensions of who they are and consider how they wish to sell/market themselves. Some examples of ego (esteem) needs are: self-respect, esteem of others, self-confidence, mastery, competence, independence, freedom, reputation, prestige, status, fame, glory, dominance, attention, importance, dignity, and appreciation. These individuals will want to be involved in and a part of the ongoing safety and health effort set by the company.

Self-Fulfillment Needs are composed of the requirements it takes to become all that one is capable of becoming and to fulfill oneself as completely as possible. The self-fulfillment needs are so complex that people never reach a point where they are completely fulfilled. They do what they do, not because they want others to notice or reward them, but because they feel a need to be creative, to grow, to achieve, and to become all that they are capable of becoming. These individuals understand the true importance of safety and health on the job; it is a part of them. They will follow the safety and health rules because they have internalized the true function of the safety and health program; they realize that it is a vital component of the whole operation. These individuals have a sense of needing to help others reach their understanding of the safety and health issue.

Maslow was right when he suggested that needs are motivators for people. As a motivator, you cannot motivate another person by depending upon elements which you deem as important. What you must do is be sensitive to the needs and wants of the people you are trying to motivate.

It is sometimes difficult for many of us to remember where we came from and to relate to someone who has basic needs (physical and security) that are far below our own needs. If you are to be a real motivator, you will need to spend time understanding the real needs of those around you.

There are many ways to motivate people but, for me, I have found that the simplest way has always been accomplished by asking the individual what they want and what presses their button. What makes them go? Involvement is probably the best motivator available to most of us. No matter where a person is along the continuum, they are ego-centered enough to want to be part of the decision-making process that affects their lives.

In summary, this means that each individual you are trying to motivate will need individualized attention. You will need to tailor, as best you can, a motivational plan which will meet their needs and, thus, cause them to be motivated toward the goals which have been developed. A person has their own reasons, based upon their own values, needs, and desires which determines how they apply their own energies. How you accomplish this is not scientific. It may be accomplished by trial and error or, at best, by small successes followed by bigger successes until the goal is reached. It seems safe to say that what works well for one person may fail miserably for another or with modification may also be successful.

MOTIVATIONAL LEADERSHIP

Describing Leadership

Leaders are not born, but some individual personalities are more suited for leadership positions. There may be several potential leaders within an organization, company, or team, but only one of them may meet the criteria or have the leadership skills that are needed for a position at a particular point in time. This does not mean that the other potential leaders are not qualified leaders, but that their unique leadership traits are not appropriate at that time.

There are two types of leaders: those who lead by coercion and those who lead by example. The question that arises is, "which is the right type of leadership?" It seems that there are occasions where leadership by power is appropriate or the only way, but this type of leadership seems more appropriate, for example, in the military. Please keep in mind that power does not, in itself, have to be bad. Many people who have leadership responsibility use power in a responsible manner to help others. Many good leaders use their leadership position and power, not as a divisive tool, but to help people get things done.

Role Models

In order to be a good leader, one must be a good role model and must be willing to sacrifice his or her own wants, desires, needs, and ego. Unfortunately, many leaders are not willing to do this and are, therefore, not good role models. Supervisors and employers should not be privileged to be what they want to be and then expect employees and workers to do as they say, not as they do. The leader must set the example that they want their employees to follow, because most of us follow the models which are set before us.

For example, a safety expert should never go on a worksite without having the proper safety equipment for that particular jobsite. He or she should always ask, prior to going to a jobsite, what the requirements are for that particular jobsite. Although it may be a hassle, a good and responsible safety and health professional or advocate will always carry his or her bulky safety equipment to the jobsite, even if it means carrying it clear across the country on an airplane. How can the safety professional expect others to wear the appropriate safety equipment if he or she does not wear it?

Leadership Characteristics

Leaders should lead by example, but there are additional things leaders can do to facilitate their leadership, motivate others, and achieve their goals.

The motivational leader is capable of building on the strengths of the people he/she deals with by developing (coaching) confidence in others, depending on goodwill, inspiring enthusiasm, and saying things like "WE" and "LET US GO." As a role model for motivation, he/she must build trust, recognize abilities, gain commitment, ensure rewards, and always expect the best of people. Leaders like this are facilitators of the development of human potential and inspire others to adhere to prudent safety and health practices at work.

This is a stark comparison to the old style leader who had all the answers and told people what to do. These leaders were quick to point out weaknesses, inspire fear, use authority, push people, and use words like "I" and "DO THIS" or "DO THAT."

Successful leaders usually possess some fairly distinct characteristics. To paraphrase Dr. James G. Carr, an authority on leadership, these characteristics are having

1.   A clear idea of what is wanted.

2.   A sense of urgency.

3.   A sense of what is right.

4.   A readiness to accept individual responsibility.

5.   A drawing on resources that others lack.[4]

Effective leadership is critical to the structuring of a motivational environment. Managers, supervisors, and others achieve results through the efforts of working with other people. While planning, directing, and controlling receive most of our attention, motivating people is also a critical part of everyone's responsibilities.

Applying Leadership

Motivating people is an ongoing process. It requires a continuing commitment, an objective view of our own style and abilities, and an understanding of the effect our behavior has on others. Recent studies show that a majority of individuals are still motivated by traditional incentives; however, money no longer has the same clout it once did. A significant portion of people today place greater value on positive reinforcers which are related to their accomplishments. They look for more control, responsibility, and meaningful accomplishments that are worthy of their talents and skills.

A good leader is one who is willing to listen. There is nothing more rewarding than to see someone who was considered a lost cause or less than average, blossom just because someone took the time to listen to them. People need someone to listen to their problems and help solve them.

In today's world of diversity, everyone brings a variety of personal experiences to each situation. Failure to show sensitivity to the feelings of others can result in misunderstanding, resentment, anxiety, communication gaps, wasted time, unnecessary work, lower productivity, poor morale, and other negative effects.

Generally, the Golden Rule, "Do unto others as you would have them do unto to you," is a good rule for guidance. This is effective in all situations. There are also some specifics which good leaders need to be cognizant of.

1.   Communications

Always keep people informed of what is going on within their organization. They like to feel they can be trusted with information when it becomes available. Make your expectations clear and follow the old adage, "Say what you mean and mean what you say." Make time available to meet with people, and make that time unhurried and without interruption. Actively listen to what they are saying to you. Get to know the people you are trying to motivate and find out what their goals and aspirations really are.

2.   Involve People

Allow people the flexibility of being involved in the decisions which directly affect them. This will increase their personal commitment and their feelings of having some control over what impacts them. Include individuals in goal setting; this increases their stake in accomplishing the established goals. Let them know what part they play in accomplishing these goals and how they can contribute. It is critical to get everyone involved.

3. Responding to Others

Frequently, provide feedback to others. Whether the comments are positive or negative, do not wait until a specific time or until something is finished. Feedback is most effective when you let people know how they are doing immediately following their performance.

4. Support Others

Help others reach their goals by offering advice and guidance, and recognize and reward good performance. In the work environment, help others get the rewards they deserve and make every effort to get a raise or a promotion for individuals who warrant it.

5. Demonstrate Respect for Others

When meeting with someone, don't disrupt a meeting by answering the telephone; their time is also valuable to them. Avoid canceling or scheduling meetings at the last minute; this is indicative of poor preparation and the lack, again, of consideration for others' schedules. Do not reprimand another person in front of siblings, peers, or fellow workers; this will certainly result in ill will.

6. Be a Role Model

No matter what sacrifice you make, you cannot be an effective motivator unless you demonstrate and live what you expect of others. You need to take the lead by being prompt, conscientious, and consistent if you expect others to mimic your leadership.

Most leaders who have become successful use the following guidelines when working with others. They

1. Develop group goals.

2. Help reach those goals.

3. Coordinate other's activities.

4. Work with others so that they can become part of the group.

5. Have a genuine interest in each individual.

A good leader inspires others to have confidence in him or her; a great leader inspires others to have confidence in themselves.

The Key Person

In maintaining communications, fostering good morale, attaining production goals, and assuring that workers are performing safely, no other person is as important as the first-line supervisor. All that affects workers comes directly from the front line supervisor. This includes all training, job communications, enforcement of safety and health rules, the company line, and feedback on the overall function of the company. Thus, the supervisor sets the tone for the motivational environment and is the role model upon which workers base their own degree of motivation. The supervisor's role is critical to the function of the jobsite and the efficient and effective accomplishment of all work activities. No other person has as much control over the workforce as this individual.

This is the reason that the supervisor's skills as a facilitator of people is more critical than their expertise related to the work being performed. These individuals need more training and support than anyone else on the jobsite. The supervisor is the implementor of all that occurs at the jobsite.

THE EVER-CHANGING MOTIVATIONAL PLAN

The Plan

Can a motivational plan be developed? Yes. But, does it work? Having a plan is no guarantee that it will work, but you have a better opportunity for success if you think through the issues, structure some goals, and develop an action plan.

To obtain the performance or behavior you desire, a systematic procedure can be used with great success. First, identify performance problems that are observable and measurable (such things as attitudes and motivation cannot be measured). Second, measure the present performance. This will provide a baseline (such as the number of homework assignments completed.) Third, analyze the consequences of the performance. (Is good performance reinforced? Is good performance punished? Is performance overlooked? Is nonperformance rewarded?) Fourth, compare the present performance with the established standard or expected performance. Fifth, provide positive feedback on performance (use graphs or a checklist so individuals can self-monitor their progress). And, sixth, give positive reinforcement to strengthen performance (reinforcement must be continuous and ongoing).

An individual's overall performance should be reviewed frequently, or at least quarterly. But, if an individual's performance is poor, it should be reviewed weekly or at least monthly. This will allow for the recognition of those who are performing well, while those who are performing poorly will also be recognized. Performance appraisal is critical to attaining the performance you desire.

Feedback/Reinforcement

A feedback/reinforcement system should always be part of any motivational plan and must be very positive in nature, but feedback/reinforcement will not be effective if it is not properly done.

Kenneth Blanchard's concept of feedback proves to be very interesting. He believes that feedback should be used as a gift and that we should learn to say thank you whether the feedback is positive or negative. His concept of saying thank you, even when negative feedback is given, is especially intriguing. He stresses the fact that we should never make a decision on the feedback that we receive; the decision should be made later after the facts have been gathered and evaluated. Furthermore, he believes that feedback on results is a very important motivator; I concur with this assessment.[5]

Feedback is a way of gaining help in the motivational environment, as well as a mechanism for indicating whether a specific behavior matches the expectations or desired outcomes. When a worker is observed doing a particular task in a safe manner and is told so, they are more likely to repeat that behavior. Some of the guidelines for giving feedback are

1. Make the feedback descriptive and not valuative.

2. Understand that feedback is a powerful communicator. Consider how it will affect your rapport with the people receiving it and also how it will affect them.

3. Direct it toward behavior which individuals can do something about.

4. Try giving feedback to people when you think they are asking or hinting for it, rather than imposing feedback upon them.

5. Make it timely and soon after the desired behavior or outcome has occurred.

6. Ask the individuals involved if your communication has been effective.

7. Make sure that you consult others to insure the accuracy of your feedback.

Some of the things which inhibit feedback are when too much control or status exist as part of the communication system, one person tries to outdo another, feedback is demanded, feedback is overdone or not perceived as legitimate, no recognition or reward is given for providing it, and, finally, the personality of the person receiving it negates the feedback.

Remember, rewards are not positive if they are not useful to those involved (of value to them). Groups and peers must recognize a reward as an actual reward. The number and kind of reinforcements are endless. Some may be, for example, a pat on the back, a verbal announcement, a citation of recognition, or an actual reward of value (e.g., cash, or a gift).

In reinforcement, note the positives and avoid the negatives. Also, remember that goals are crucial in correcting mistakes. Look forward to solutions, look for the underlying problems in emotional situations, and avoid defensiveness. Listen for reinforcement, as well as facts.

Criticism

Providing effective criticism has never been an easy task. People often say that they want to provide constructive criticism, but criticism is criticism whether it is constructive or not. Few people want to be criticized or accept criticism in a positive manner. Criticism can be very demotivational to individuals, as well as groups; thus, great care needs to be taken in giving criticism. Drs. Hendrie Weisinger and Norman Lobsenz in their book, *Nobody's Perfect: How to Give Criticism and Get Results*, provides some insights into giving criticism. Paraphrasing some of their comments may provide you with some guidance.

Criticism often proves ineffective in informing the other person as to what he or she has done "wrong," because we often fail to state clearly what it is we would like him or her to do "right." It is not what you say but how, when, and where you say it that makes the difference between worthwhile criticism and just plain nagging. Think of criticizing as a way of communicating information to others in order to **help** them. Criticism is best expressed as an opinion, not a fact; thus, you won't put the other person on the defensive.

Some other helpful hints made by the authors are

1. Choose the appropriate time and place. Never criticize someone in public or in front of others.

2. Make the criticism as specific as possible and avoid generalizations like "always" and "never."

3. Only criticize behavior that can be corrected. You can't change a left-handed person to a right-handed one.

4. Don't belabor the point; the criticism should not turn into a lecture.

5. Offer incentives or rewards for change and explain how the change will be beneficial to them.

6. Use "I" statements when you inject yourself into the criticism showing it's your opinion, not fact.

7.   Use positives about the person to preface the criticism. Give positive feedback regarding the person's strengths.

8.   Give the person some alternative behaviors which could correct the situation.

9.   The environment and your approach should be as relaxed and calm as possible. Do not lose control of yourself.[6]

Modifying or Changing Behavior

Paul L. Brown and Robert J. Presbie, authors of the book, *Behavior Modification in Business, Industry and Government*, did a good job of simplifying the behavior change or modification approach.

They say that there are existing factors which affect behavior, as well as the consequences for that behavior; thus, in order to change a behavior there are four steps which can normally be followed.

1.   **Pinpoint** the behavior.

2.   **Count** the behavior and **Chart** the behavior.

3.   **Change** the behavior by changing the existing factors, consequences, or both.

4.   **Evaluate** the results.[7]

To pinpoint a behavior is to define the behavior in such a way that anyone listening to the description could see the behavior, count the behavior, or describe the situation in which the behavior occurs. Pinpointing involves being specific about behavior. Sometimes pinpointing can serve as a solution since individuals often do not understand what is expected of them. Pointing out the expected behavior helps the individual target what you expect.

If pinpointing does not seem to change the behavior, you will need to count or chart the behavior which needs improvement. The most effective form of counting or charting is accomplished when it is done by the individual whose behavior needs to change. Any counting done by someone else should be done only to set a baseline prior to attempting to alter the behavior.

At times it may be necessary for you to count the behavior even though it is best if the individual monitors his or her own behavior. Counting can be accomplished by recording the number of undesirable events that have taken place over a certain period of time; or, if it is not possible to record each event, the behavior could be sampled at specific times. These events could be the number of accidents, unsafe acts performed, or unsafe conditions observed. You can also count permanent outcomes which are the results of the pinpointed behavior.

Be aware that it may take a rather long, frustrating period of time to accomplish a behavioral change. If you cannot afford this much time to change an individual's behavior, you may have to reassign an individual to another position (this could be viewed as a form of rewarding nonperformance). Or, if you do not have the time to work with an individual, you may have to dismiss her/him from your organization. But, when you do have the time, behavioral change can be accomplished if you are willing to try.

When pinpointing, or counting and charting behavior proves effective without any additional steps, it is successful because the feedback which was provided to the individual adequately changed the undesired behavior. Feedback or information on performance, along with reinforcement, are two critical factors in improving an individual's performance.

Due to preexisting factors or consequences, there may be times when the social environment needs to be changed in order to affect behavioral change. The existing factors or consequences may be detrimental to your motivational approach. Consequences which have a positive effect are reinforcers which strengthen the probability of the behavior reoccurring, but at times you must examine the consequences to assure that they are not reinforcing the unwanted behavior.

In every work setting there is a "jokester," and even though many within the group do not appreciate the humor in the jokes (no matter what the content of those jokes may be), they laugh. This laughing only reinforces this behavior. If those who do not appreciate this humor would either walk away, or not laugh and quickly change the subject, the jokster would soon quit telling jokes. The negative reinforcement would cause this behavior to become extinct, at least to those who found it objectionable.

Punishment can be used to reduce the possibility of a behavior reoccurring, but punishment can also have some very bad side effects. It may cause enough negatives that, consequently, it may destroy the intent of the behavior modification. Therefore, punishment as a reinforcer should be used only as a last recourse. The old saying of "sometimes discipline, but always love" is usually more effective.

Another approach to behavior modification, and generally more effective, is to change the preexisting factors. This can be accomplished by providing written instructions or signals when new behavior occurs. I like to call this the "prompting" procedure. These "prompting procedures" may be accomplished by using a checklist, writing safe operating procedures, or performing a job safety analysis. This procedure may be used when a task is not being done properly, is done infrequently, or is very dangerous.

As an example of this type of procedure, let me ask you a question. What do you believe is the most important part of the flight for a commercial pilot? Most people will say take-off and landing. Needless to say, the airlines do not leave the behavior for take-off and landing to chance. Pilots use a checklist ("prompting procedure") to get the appropriate behavior.

The same principle applies when you want a particular response. You must not assume that a person understands or knows what is expected of him or her unless he or she is informed about those goals. Providing that information in the form of a checklist is an appropriate way to elicit the desired behavior.

SUPERVISORY MOTIVATIONAL INITIATIVES

Different Approaches

The ability to motivate another individual depends upon our sensitivity to his/her needs and wants. Peoples' needs tend to move them in a certain direction and toward a particular goal. These needs, such as physical comfort, money, curiosity, independence, dominance, individual acceptance, group acceptance, or recognition, may cause a variety of responses and/or patterns to occur in order to fulfill them.

According to Drs. Michael H. and Timothy B. Mescon, a real shock occurs when dealing with the younger generation; they do not respond to the traditional motivators. They are better educated and have less need for security since many have returned to their parent's homes to live. They are motivated more by "outside of work" activities and see work as only a minor part of their lives. These individuals are not satisfied with just having a job; they want to know more about the big picture or the intricacies of the construction business. They want to be involved in their career choices and planning and they want their needs matched with the goals and needs of the workplace.[8]

The precept of this chapter is that of allowing people to become involved in the matters which affect them; this is one of their greatest motivators. The following is an example of this type of motivation. The Texas Instruments Company had contracted to build a radar unit at a fixed price. To the company's dismay, the bid was below the actual cost involved. The company, therefore, tried to rectify the situation by cutting costs, but to no avail. A supervisor who knew of the problem called the workers together. He explained the dilemma to them and asked for their help. The original estimated time to assemble the radar was 138 hours. But by using the workers' suggestions, they were able to cut assembly time to, at first, 100 hours, then 86, 75, 65, 57, and, finally, 32 hours. Because the workers were involved and committed and their self-worth and egos were on the line, it proved to be a real motivator in solving their problem.

On another occasion, the Kaiser Steel Company was having a hard time competing with the Japanese steel producers. The workers asked for an opportunity to comment and be part of the decision process. They recommended buying a new $125,000 saw. Due to the cost involved, management had never considered this option, but they followed the workers' recommendation and bought the saw; productivity increased 32 percent.

If you want to know how to do something better, ask the person doing the job. It is somewhat easy to solicit individuals to become involved, but when groups come into the picture, the world changes.

Joint Labor/ Management Committees [9]

A formal joint labor/management committee is organized to address specific issues such as safety and health or production processes. It is a committee with equal representation which gives both parties an opportunity to talk directly to each other and educate each other concerning the problems faced by either group. In contrast, labor or management only committees have self-serving goals with no consensus on solving problems. They are the only ones who have the authority or power to make changes. Thus, joint committees are aimed at gaining solutions, having equal participation, and having some degree of authority or power.

Joint labor/management (L/M) committees have a different purpose than committees set up by either labor or management alone. When compared to other committees existing at the worksite, L/M committees are different both in their goals and methods of operation. In addition, because of the nature of their goals, they are also much more challenging since they require many different skills from all the participants.

Joint committees provide both parties with the opportunity and structure to discuss a wide range of issues challenging them. Neither partner of the labor/management committee has enough information, commitment, or power to institute the changes that the joint committee eventually identifies as critical to the success of the business.

Thus, a key purpose of joint committees is to gather, review, analyze, and solve problems that are critical to the success of the business and are not appropriate for the collective bargaining process. Another purpose for these committees is the formation of a level playing field which has as its ultimate purpose the success of the business. These committees can help build bridges of cooperation which can lead to increased productivity, quality, efficiency, safety and health, and economic gains shared by all parties.

But the purpose of joint labor/management (L/M) committees goes beyond quality and productivity. They are builders of true and honest relationships which help to realize success through focusing on outcomes, using resources more efficiently, fostering real world flexibility, supporting an information-sharing system, opening communications, and fostering a better working relationship.

In the past, labor/management relationships were built on confrontation, distrust, acrimony, and the perception of loss or gain of control and power. The ultimate goal, to use an over-used phrase, is to attain a "win-win situation."

With all the downsizing, right sizing, and reengineering going on within the workplace, individuals believe that they can help and have an impact on their continued employment and the survival of the company, **if only asked and given the chance**. To successfully do this, it is imperative that they have access to the information needed to solve issues facing them and their employer.

Committee Make-Up

The joint labor/management committee should be composed of at least as many employee members as employer members. Labor must have the sole right to appoint or select its own representatives, just as employers have the right to appoint theirs. Both parties should clearly understand that the members of this committee must not only be risk-takers, but fully capable of making the critical decisions needed to make this process succeed.

In order to cover all facets of the workplace, the labor organization may find it useful to have a broad spectrum of their membership represented. Labor should also allow for turnover in its membership and address this issue by identifying and involving adjunct and alternate members. By doing this you will not compromise the committee's progress by having to introduce new members into the committee who are unfamiliar with the process and are untrained concerning the subject matter.

The chairperson must be elected by the committee and this position should be a rotating position between labor and management. Each committee member should receive training on the joint committee process and also receive other specific instructions which are deemed necessary, such as job-related safety and health training. The labor members should be paid for all committee duties, including attendance at meetings, inspections, training sessions, etc.

Recordkeeping

Each participating party (labor and management) should keep their own notes of all meetings and inspections, as well as copies of agendas. This will insure that agreements and disagreements, time schedules, actions to be taken, etc., are not lost, forgotten, neglected, or misinterpreted. Time has a way of encouraging each one of these things to happen. Good record keeping will also assist in keeping the direction and undertakings of the joint committee in focus.

The agenda could include the following:

- Date and time for the meeting to begin and a projected time for adjournment.
- To's and From's – who the agenda is from and to whom it is sent.
- Location.
- Review of any audits, assessments, unsolved problems, or inspections.
- Old business.
- New business.

A formal set of minutes and reports on inspections should be maintained by the joint committee. Minutes should include

- Employer's name and union for identifiers.
- Date and time of meeting.
- Chairperson(s).
- Members in attendance.
- Old business.
- Actions taken and completion dates since last meeting.

- New business.
- New actions and proposed dates of completion.
- Other business
    1. Outside or OSHA inspections.
    2. Injury and illness incidents.
    3. Educational initiatives.
    4. Administrative activities.
- Joint representative's signatures giving approval of minutes.

Do's and Don'ts of L/M Committees

DO's:

- Always give an agenda to committee members in advance of a meeting; this allows everyone time for preparation.
- Cancel a meeting only for emergencies; hold meeting on schedule.
- Set timelines for solving problems.
- Keep focused on the issues involved.
- Do stay on schedule and stick to the starting and ending times in the agenda.
- Decide on a structured approach to recording and drafting minutes, as well as mechanisms for disseminating them.
- Keep the broader workforce informed of the activities of the committee.
- Keep issues on the agenda until they are resolved to everyone's satisfaction.
- Give worker representatives time to meet as a group and prepare for the meeting.
- Be on time for the meetings.
- Make sure that everyone understands the issues and problems to be discussed.

DON'Ts:

- Tackle the most difficult problems first since some early successes will build a stronger foundation.
- Work on broadly defined issues, but deal with specific problems and concrete corrective actions.
- Allow the meeting to be a gripe session when problem solving is the end result.
- Allow any issue to be viewed as trivial; each issue is important to someone.
- Let individual personalities interfere with the meetings nor the intent of the committee.
- Be a "know-it-all" and assume you know the answer; give everyone an opportunity to participate in solving the problem.
- Neglect to get all the facts before trying to solve an issue or problem.
- Prolong meetings.
- Delay conveying and communicating the solutions to problems and the outcomes or accomplishments which the committee has achieved.

- Expect miraculous successes or results immediately since many of the problems and issues did not occur over night.

Organizing a Joint Committee

When organizing a joint committee, some specifics should be set forth.

- Set up the ground rules or procedural process.

- Have a set place to meet.

- Establish, as a group, the goals, objectives, function, and mission of the committee.

- Select the frequency of the meetings – at a minimum, monthly – as well as setting parameters for the length of meetings.

- Agree to maintain and post minutes of all meetings.

Expectations

Anytime something new is undertaken, such as joint labor/management committees, there are expectations which accompany these new endeavors. Some of the expectations are

- Improved workplaces and work environment.

- Improved working relationships.

- Positive, cooperative approaches.

- A compromise for mutual interests versus self-serving interests.

- A true team approach.

- Sharing of information, thinking, and substantive decision making.

- New/fresh ideas.

- Increased participation and involvement.

Outcomes

A study done by the Work in America Institute lists some of the outcomes you can expect when you have a functional joint labor/management committee. According to this study, both labor and management stands to benefit from joint undertakings. Some of the benefits include

- Economic gains: higher profits, less cost overruns, increased productivity, better quality, greater customer satisfaction, and fewer injuries and illnesses. Working together, workers and supervisors can solve problems, improve product quality, and streamline work processes.

- Improved worker capacities which more effectively contributes to the improvement of the workplace.

- Human resource benefits.

- Innovations at the bargaining table.

- Committee member growth.

- Workplace democracy.

- Employment security.

- Positive perceptions.

Other outcomes which will, in all likelihood, arise from joint L/M committees are

- Shared responsibilities.
- Increased individual involvement.
- Company and labor pro-active with each other.
- Better communication between company and labor.
- Employee ownership of ideas, goals, activities, outcomes, and the company.
- Union leadership and members more challenged.

Joint Labor/Management Occupational Safety and Health Committees

A joint labor/management occupational safety and health committee is a specialized application of the joint labor/management committee and is an excellent format which others can replicate. This type of committee is organized to address specific workplace issues such as

- Monitor the safety and health programs.
- Inspect the workplace to identify hazards.
- Conduct and review accident investigations.
- Recommend interventions and prevention initiatives.
- Review injury and illness data for incident trends.
- Act as a sounding board for workers who are expressing health and safety concerns.
- Become involved in designing and planning for a safe and healthy workplace.
- Make recommendations to the company regarding actions, solutions, and program needs for safety and health.
- Participate and observe workplace exposure monitoring and medical surveillance programs.
- Assure that training and education fully address safety and health issues facing the workplace.

The goals of the joint labor/management occupational safety and health committees are

- To reduce accidents, through a cooperative effort, by eliminating as many workplace hazards as possible.
- To reduce the number of safety and health related complaints filed with regulatory agencies without infringing on the workers' federal and state rights.
- To promote worker participation in all safety and health programs.
- To promote training in the areas of recognition, avoidance, and the prevention of occupational hazards.
- To establish another line of communication whereby the workers can voice their concerns regarding potential hazards and then receive feedback on the status or action being taken.

Summary

What can joint labor/management committees accomplish?

- Increased commitment to achieving the organization's goals or mission.
- Improved productivity, safety and health, customer service, and product quality.
- Joint resolution of problems and issues facing the organization.
- Shared responsibility and accountability for results and outcomes.
- Better and more constructive relationship between labor and management.
- Enhanced employee morale and job satisfaction.
- Heightened communication and information sharing that brings all employees into the decision-making process. This helps them understand the mission, goals, and objectives of the organization and fosters employee support of the organization's undertakings.
- Increased job security and compensation.

In order to make joint labor/management committees work, certain actions must occur and certain procedures must be followed.

- Ensure that upper management supports the joint effort and that this is conveyed both to the union and other company representatives.
- Acknowledge that reservations exist on both sides and try to gradually build trust.
- Keep the committee focused on its goals and mission.
- Strive for a good balance of employee and management representatives who are willing to invest in the process.
- Keep the committee structured; don't allow it to turn into a bull session.
- Remember that the committee is designed to serve all workplace constituencies, not just workers and management.
- Assure that committee leadership is elected or selected, by consensus, to fill various roles.
- Make decisions fairly and use the consensus process.
- Know and work within the guidelines of federal and state regulations.
- Don't raise issues that really must be addressed at the collective bargaining table; it will only undermine the viability and success of this process.

This partnership can succeed only when problems are identified, goals exist, priorities are set forth, and trust is the foundation upon which it is built. Thus, everyone must be willing to work on two-way communications with a forthright exchange of information. Joint solutions and real action will become the visible products of these joint committees.

Peer Pressure

Peer pressure is a very powerful motivator and can be either rewarding or punishing. Peer groups who are doing just enough to get by tend to draw or attract lesser motivated individuals. The lesser motivated individuals tend to identify with the peer group and are governed more by their peer group than are the highly motivated individuals. Normally, highly motivated individuals do not succumb as easily to peer pressure.

Social and family pressure has a role in motivating individuals, but most people sense that peer groups are the prime movers in the workplace motivational arena; thus, in order to motivate an individual who is under the influence of peers, one must spend time trying to change the peer group behavior. This is the only way to achieve the motivation of an individual within that group.

The following is given as an illustration of this type of peer pressure. On one occasion a construction company needed to reduce the number of accidents occurring on a project. Management decided that an incentive program would be installed. In this program the workers were told that if their group went a certain length of time without an accident, the group would receive an award, plus, each individual in the work group would also receive an additional reward. This was visualized as creating a peer pressure situation because everyone began looking out for the each other in order to keep anyone from getting injured.

This incentive program did seem to work and have a positive affect on the workers. This was best illustrated when one of the miners in this group slipped and broke an ankle. As they were loading him into the ambulance, one of his fellow workers yelled to him saying, "tell them you fell and broke it at home." As you can see from this example, there was pressure from the peer group to avoid accidents so they could receive their rewards.

Family Pressure

The following is an example of how family pressure can be used as a positive reinforcement or motivator. In this particular situation the workers in a company were told if they worked safely at their jobsite, they would accumulate points that could be used to purchase items from their company catalog. These items could be purchased for either themselves or their family members.

One worker said his son reminded him daily to work safely. He did this because his father had promised him a bicycle when he accumulated enough points. The father was extra cautious at work because he had made this promise to his son and did not want to let him down. He said this had been a true incentive for him and had motivated him to work safely. So, as you can see by this example, there are times when family pressure can also be a true motivator.

On another occasion a company's safety director sent a letter, written in Braille, to a select group of employees. A few days later another letter followed which said

---

Dear Fellow Worker:

A few days ago you received a letter written in braille. This is what it said:

It has been noticed by your supervisors and coworkers that you often "forget" to wear your safety eye wear. This letter is written in "Braille," the written language of the <u>*blind*</u>*. We suggest that you learn this language very soon so that you will be able to read and write when you can no longer see. Contact your "Federation for the Blind and Visually Handicapped" for instructions.*

In concern for your sight, I remain..........

If this letter did not motivate the worker to wear his/her safety glasses, then the safety director invited the employee and their spouse as his guest for a free weekend, including hotel and meals, to tour the School for the Blind. The safety director also brought along his own blind son. When given the tour of the school, the spouse was told that this would enable him/her to learn how to care for a blind person.

As you can imagine, this particular safety director had a special motivation, a fire, and a zeal for this particular issue; he felt that by involving a family member in this issue, they would be an important motivator and the results would be more successful. Again, family pressure can be a positive motivator.

In most cases this type of approach is not recommended. Pressure and threats are not good motivators; they often cause fear of failure. This often tends to paralyze individuals rather than motivate them, but there are always exceptions to the rule.

The Worker Challenge

Although motivating workers is not a straightforward task, one thing that has been found over the years is that you need to be willing to listen to people and accept and implement their ideas. This is often one of the highest powered motivators. When you are willing to do this, their job becomes more interesting and challenging to them and they are more inclined to perform in a positive manner.

When delegating responsibility, it is important to remember that once that person knows what their responsibility is, let them alone to complete the task. Since they have played such an important part in the decision making, their best motivator will be the desire to see the end product.

Do not arbitrarily assign people to a task but ask them, if possible, what they want to do and what they think they can accomplish. It is well founded that there is more than one way to accomplish a job.

Anyone working on an assignment should feel that he can take a chance on doing it a different way if he feels it will be more effective, as long as it does not endanger him or her safety, or other workers' safety.

It is more important that the person who will be doing the task feel that he or she has had an important part in making the decisions concerning it. Many times it has been found that the suggestions given by others are just as viable, if not more so, than those given by an owner or supervisor. At times there may be mistakes made; nevertheless, it is still more important that those involved in a task be part of the decision-making process—mistakes can normally be corrected without much difficulty!

So, expect mistakes! If people are willing to do something, they will be more exposed to the risk of making a mistake. Remember, only those who are doing nothing will never make a mistake.

Rewards

When positive behavior is accomplished, it is very important to remember to give some form of a reward. This reward does not have to be monetary, but it must be something which is of value to the person receiving it. The reward can be something as simple as granting a person's request for a tool or piece of safety equipment because he or she did the work safely, or approving a shift change for a worker who has been doing a good job and for personal reasons, needs the change. You should always be certain that these things are denoted to these workers as rewards for their positive behavior.

MOTIVATIONAL TECHNIQUES

Are there any tried and proven techniques which will motivate people? The answer is probably. But, if your approach is not well thought out and planned, it will have no long-lasting effect and could merely be thought of as a band-aid approach. You have probably used many of the approaches found in this section, but you may not have thought about them as motivation techniques. Seldom do we organize our motivational approach in such a manner that these techniques are viewed as a part of a complete motivational package.

Training

One of the major techniques in motivating people is training. At the basic roots of training is the structuring of behavior, developing the ability to perform, and instilling a sense of purpose. Training is essential and should be provided in order to achieve a goal or objective. These goals or objectives can be such things as learning to drive a piece of machinery, being able to work safely, teaching a sport, learning to communicate with others, etc.

Most training is behavioral or learning-objective driven. This could be viewed as the major goal in the motivational process. In order to achieve this goal, you need to let workers clearly know what they are expected to do and precisely how you want them to accomplish it. There should be an expected level of attainment, whether it be mastery or just to gain a certain amount of knowledge. And, in order to gain the expected skills, you should allow the workers time to practice the new skills.

Perfect results should not be expected the first time; you should be pleased with small, progressive steps. These steps should be evaluated on a regular basis and constant feedback should be provided concerning their progress. Try dwelling on the positives instead of the failures and work with them until they are able to perform a skill or pass an examination. In the training process, the true reinforcer and reward for an individual or group is being recognized for their ability to perform a skill, pass an examination, or complete a course. Individuals who are trained will perform at a higher level than those who are learning as they go. Training improves efficiency, safety, and the performance of workers.

More About Rewards

Rewards are often short-lived motivators. They are usually not the results of performing an activity, but appear some time after the task is accomplished. Most of us are rethinking our concepts of rewards. It seems that accountability for performance and greater participation in making decisions are more critical than our previous reward structure.

Rewards are often not viewed as rewards by those who receive them. The following is an example of this. A sales manager wanted to keep his sales people working right up until the time of a holiday. The manager, therefore, offered a unique reward. As the first prize, he offered the sales person who had the most sales for that week, one week of free ice fishing. One of his sales staff was heard saying that she thought the second place prize ought to be, "two weeks of free ice fishing." Now who do you think the reward was of value to? Obviously, it was the sales manager, not the sales people.

Reinforcement

Reinforcement is anything that is used to increase the likelihood of a person repeating the desired behavior again. For instance, if you shoot a basketball several times but miss each time, this becomes a negative reinforcer for you. If, on the other hand, you finally make a basket, this becomes a positive reinforcer and you now try a little harder each time to improve your performance.

Be sure when you use reinforcement that it is always used in a positive and not a negative manner. For instance, if a player shoots and makes a basket but the coach takes that player out of the game because he shot the basket at the wrong time, this becomes a negative reinforcer. The coach's intent was not to change the behavior of making baskets but, according to team rules, to reinforce the behavior regarding the proper time to shoot. As you can imagine, this type of negative reinforcement could cause the player to be so intent on shooting a basket at the proper time that his ability to shoot could be considerably diminished.

The threat of something happening to us is neither a positive nor negative reinforcer. It is only after something has actually occurred, that it then and only then, becomes positive or negative. It is the same as trying to train a worker to stop, look, and listen prior to lifting a heavy load. If the worker does not understand the importance of the request, he/she does not realize the danger involved. Only when a back injury has occurred will it become a negative reinforcement.

Then how do we motivate someone to respond to a desired behavior when there is no reinforcing experience to help motivate them? In order to accomplish this, we must state the reason and importance of the request and compliment them when they give the proper response.

Incentives As Rewards

Using an incentive as a reward can be a positive motivator, but, unless you are able to achieve a behavioral change, it may be only a temporary motivator. Therefore, in order to get the behavioral change you desire, you need to be aware that workers hold a more positive attitude toward their work when their supervisor provides them with the reward which they desire and is most meaningful to them. Also, it has been found that employees are very receptive to rewards given to them that were not expected. Rewards of this nature seem to be even more satisfying and are received more enthusiastically than the rewards they knew they were going to receive.

Many of us try to use rewards or incentives to achieve the behavior we would like to see, but, as you have seen from the previous two examples, these have the potential to backfire. Therefore, you can see how very important it is that your incentives and rewards be well thought out and planned so that this does not happen to you.

Incentive Programs

There is no need to try to implement an incentive program if you do not have an implemented safety and health program. All other aspects of safety and health must be in place before utilizing an incentive program. In other words, an incentive program is a component to reinforce what presently exists. If you have not structured an environment in which workers sense the importance of safety, feel involved in the process, have safety conscious leaders and supervisors as role models, and are directed by goals to prevent those incidents that can cause workplace injuries and illnesses, then an incentive program is of no use.

But when an effective incentive program is an integral part of the overall workplace safety and health effort, success after success has been espoused.

SAFETY INCENTIVE PROGRAM FOR XYZ CONTRACTOR

In the hopes of further reducing our lost time accidents, XYZ Contractor has decided to develop an incentive program. This program is designed to show appreciation to our employees who work safely. Providing and maintaining a safe work environment is of the utmost importance to our company, our employees, and their families.

Individuals and work groups who work safely and have no lost-time accidents will be eligible to choose awards from one of four catalogs. Each catalog will offer a wide range of home appliances, shop and yard tools, children toys, sporting goods, and much more. Each catalog will have an upper dollar value on its contents.

Catalog A will have products up to $25.00.

Catalog B will have products up to $50.00.

Catalog C will have products up to $75.00.

Catalog D will have products up to $250.00 (pertains only to groups).

When a worker completes a one-quarter work period with no accidents, he or she will be eligible for an award. At that time, a selection can be made from Catalog A. Two quarters of accident free work will be rewarded with a selection from Catalog B. Three quarters of accident free work will result in a selection from Catalog C. Each additional quarter will result in the worker being eligible to select an additional item from Catalog C.

If a worker should suffer a lost-time accident, the worker will start the process again on the first day of the next quarter. At the end of that quarter if there is no lost time, he or she may make a selection from Catalog A. Also, for a worker to receive an award, they must have no more than one unexcused absence (AWOL) per quarter.

Each member of a distinct workgroup will be eligible for an award when they go a quarter without a lost-time incident. Each member of the workgroup can select an item from Catalog A for that quarter and each ensuing quarter without a lost-time accident. If the work crew completes a full year of accident-free work, each member will receive a selection from Catalog D after the completion of four accident-free quarters.

In addition, a worker who submits an approved suggestion which increases production and/or promotes safety will be eligible for a selection from Catalog C.

It is the company's desire to provide the safest, most productive work environment possible for all of its employee. Production and safety can go hand in hand. Remember "Getting hurt...............HURTS!"

Thank you, and work safely.

The previous incentive program's goals were to address and promote individual safety, peer safety, increase production, and reduce absenteeism. This was done by providing individuals with increasing awards for safe performance, as well as letting them select the types of products which were most valuable to them. Also, absenteeism negated the ability of an individual to receive an award. But, even if a worker had an accident during a quarter, there was still the incentive to start fresh the next quarter rather than having to wait a full year before being eligible to receive another reward; this would have been demotivational. Also, as part of the program, the company hoped that peer pressure would inspire the workers to look out for each other since, as a team, they would be working for a higher reward for four quarters of safe work. Also, in this program the individual worker was encouraged to become involved in both production and safety by being rewarded for giving production and safe work suggestions.

Other types of incentives or rewards that could be given are

- A "title" to those employees who deserve it; remember, titles cost a company nothing.
- Employee of the month plaques.
- Special parking spaces.
- Personal days off (can be used at worker's discretion with approval).
- Compensatory hours.
- Money; it is good for one time and that time only.
- Bonds, gold, or silver; they are usually kept and thus a constant reminder.
- Items with company logo and worker's name.
- Special commemorative items or pins.
- Tangible items (products).
- One-of-a-kind or limited edition items (belt buckles, knives, etc.).

Special Emphasis

In order to address a unique issue or a specific problem, the use of special emphasis programs is appropriate. These programs define a problem or issue, use symbolism (drawings or slogans) to elicit or trigger an appropriate response, indicate a true interest in the special area of emphasis, and use a structured approach to address the special issue. It is used to cause focused attention on a specific area. A good example of this is the organization Mothers Against Drunk Driving (MADD), which has been somewhat successful in decreasing drunk driving.

If you have had a lot of back injuries or failure of individuals to wear personal protective equipment, you can set goals and objectives, develop communication or awareness approaches, design reinforcement/incentives, track progress/evaluate, and revise/revamp when necessary. The target in an emphasis program is focused and very specific; you are looking for observable outcomes such as reducing back injuries on your construction site.

Contests

Some organizations use contests for incentives but, generally, they are found to be unsatisfactory. In many instances, too many negative response factors come into play. Contests are a type of competition and not everyone likes to compete. Many times there is only one winner in the contest and, therefore, many receive no type of reward even though they have worked very hard and to the best of their ability. This can effect an individual's status and, at times, even their morale.

Contests do not increase total productivity, do not increase cooperation, do not motivate everyone, are self-serving, decrease group problem solving, and may cause suspicion and hostility. You can, after reading the previous list of negatives, understand why contests generally are not a recommended practice. As with anything, though, there are always exceptions to the rule.

If you do decide that a contest is good for your particular situation, design the contest so that it will involve all individuals, foster status and pride, involve group competition, and involve the management. Be sure that the same individual does not win all the time; this can be very discouraging to the other participants. If varied skill levels exist, use handicaps. Make sure winning and losing is distributed and give prizes to first, second, and third place winners. Contests can be somewhat effective when properly used, but do not "bet the farm" on them being the answer to your motivational issues.

Gimmicks and Gadgets

Gimmicks and gadgets are novel or unconventional ideas, gifts, or devices which call attention to a desired response and maintain motivation in an unusual manner. Some examples are

- Presenting a crew with monogrammed jackets for their achievement or performance.
- Giving "Spouse Kits" which contain items such as sugar, salt, matches, or lapel pins. These can be used as reminders by putting a message on them and slipping them into a lunch box or pocket.
- Using an olympic weight lifter to demonstrate the right way to lift.
- Utilizing other gadgets such as knives, belt buckles, caps, pens, key chains, patches, rings, tee shirts, or trophies.

All of the above ideas are short-term motivators and should be used only to supplement an existing comprehensive program; they should not be the entire program.

Visuals

Visuals, such as posters and bulletin boards, can be used as motivational tools. They are beneficial in that they serve as a constant reminder of the desired goal you are trying to reach. Bulletin boards posters need to be changed often and kept updated. You can even get your employees who have graphic or artistic talents to develop posters which you can have reproduced and posted. An example of this can be found in Figure 3-1. This way the employees gets some recognition and you do not have the cost of commercially made posters. Videos are also another excellent way of motivating individuals or groups.

When giving a talk, the use of visuals normally increases the effectiveness of that presentation. Using personalized information, written literature, and statistics tends to more readily hold the participants' attention.

Figure 3-1. Example of safety poster

Conferences/Seminars

Conferences and meetings can be used as good motivational tools. In these meetings all members of the group are focused on one topic, problem, or activity and everyone can respond to the same information and materials. In this setting the group can be motivated to act as one entity. Therefore, do not overlook the possibility of using this type of method; it can be very beneficial. When returning from a conference or seminar, individuals or groups are often rejuvenated because of the new ideas they received during the meetings.

Nonfinancial Incentives

Nonfinancial incentives can be such things as the use of praise, knowledge of results (output), competition, experience of progress, experience of achievement, or granting a request. Some of the most powerful motivators are achievement, recognition, a person's work or task, responsibility, and growth potential.

The previously listed motivators can become functional by giving someone more control, but at the same time, holding them accountable. You can make them more accountable by making them responsible for a discrete outcome, or allowing them additional authority. You will also be more successful if you keep the people you are trying to motivate informed by direct communication, instead of through someone else. Challenge them with more difficult tasks and allow them to become specialized in a certain area.

Communication/interpersonal relations, employers/supervisors, promotion/recognition, work conditions, and status are some external factors which motivate people. The personal or internal motivators are those things which give more freedom of choice of activity, freedom from criticism, work environment, choice of peers, fewer status factors, less supervisor or employee conflict, and more opportunity to be oneself.

SUMMARY

It is evident that we spend a large portion of our lives either motivating ourselves or trying to motivate someone else. Thus, hopefully you have gained some insights on ways that you can be more effective at motivating yourself and others. In summary, some of the key traits which I believe are critical to understanding how to motivate people are

1. People are self-motivated.

2. What people do seems logical and rational to them.

3. People are influenced by what is expected of them.

4. To each individual, the most important person is oneself.

5. People support what they create or are involved in.

6. Conflict is natural (normal) and can be used positively.

7. People prefer to keep things the way they are rather than to make a change.

8. People are under-utilized.

With these thoughts in mind, people can be motivated by

1. Allowing them involvement and participation.

2. Delegating to them responsibility with authority.

3. Effectively communicating with them.

4. Demonstrating concern and assisting them with counseling and coaching.

5. Being a good role model to them.

6. Having high expectations of them.

7. Providing rewards and promotions based upon their achievements.

Workers need to know what is expected of them, what happens if they do not perform, and what their rewards, outcomes, or consequences will be. We are motivated by what we think the consequences of our actions will be. Those consequences should be immediate, certain, and positive if we expect them to be motivational.

Practically all our motivational attempts are geared toward peers or employees and this is accomplished through their employers, fellow workers, or supervisors. These motivators are, among other things, a funnel which directs all materials and information to those who need to be motivated. The motivator also directs or carries out the vast majority of learning. Everything that motivates students, children, or employees is applied by them and, obviously, their role is crucial.

People have many abilities and talents that they are unaware of or just don't use. As a motivator, it is your responsibility to bring out those hidden abilities and talents and channel them toward the goals, outcomes, behaviors, and objectives you desire. If this is done as discussed in previous sections, it will give them a new sense of enthusiasm and self-esteem.

Motivation takes a lot of nurturing and caring for both the people involved and the goals to be attained. Many organizations say that people are their most important asset but fail to exhibit that principle by the manner in which they treat their employees.

Motivation is not something which you can schedule for a Thursday at 2:00 P.M. It is a process that requires your continuing commitment and your ability to have an objective view of your own self. You must also have an understanding of your effect on others.

The essence of motivation is to find meaning in what you are doing. Motivation is the predisposition of doing something in order to satisfy a need. In real life, most people rarely have just one need; they have several needs at any one given time and are, consequently, moved to do something about them. Unfortunately, if they have too many needs facing them at one time, they may become indecisive, highly aggressive, negative, or even irrational.

Motivation is internal and can be stimulated by leadership and incentives. But, unless you know something about the needs, desires, and drives of the other person, your leadership and incentives may be completely ineffective. When a person's task or job does not permit him or her to satisfy his own personal needs, he is less likely to work as hard at accomplishing the task you have chosen for him.

It seems safe to say that people do things well if they are excited about their assigned tasks. When their external environment assures that their own needs, wants, and desires will be met, it further enhances their desire to do a good job. You, as a leader/motivator, are also responsible for helping others meet the demands of their world to the level of their capabilities. When each of these aspects are being fulfilled, you will have an excellent motivational situation.

Robert R. Blake and Jane Srygley, management experts, presented some sound principles for behavioral change and motivation in the *Training and Development Journal*. These principles are practical and require creativity by the motivator. They are as follows:

1. Fulfillment through contribution is the motivation that gives character to human interaction and supports productivity.

2. Responsibility for one's own actions is the highest level of maturity and is only possible through widespread delegation of power and authority.

3. Open communication is essential for the exercise of self-responsibility.

4. Shared participation in problem solving and decision making stimulates active involvement in productivity and creative thinking.

5. Conflicts are solved by direct problem-solving confrontation and with understanding and agreement as the basis of a cooperative effort.

6. Management should offer objectives.

7. Merit should be the basis of rewards/promotion /recognition.

8. Learning comes through critique of work experience.

9. Norms and standards that regulate behavior and performance support personal and organizational excellence.[10]

In order to motivate one person or a group of people, you must pay close attention to the fundamentals. But according to an article written by Lee Iacocca, the legendary football coach, Vince Lombardi, said there is another very important ingredient in the motivating process. He believed that the true ingredient for motivating people is having the ability to show a genuine and sincere caring for your team members, your fellow workers, and/or your family members. This caring must be genuine because people can tell the difference![11]

Thus, the perception exists that in order for people to grow they need the freedom to make decisions without the fear of being penalized. When they are given this freedom, they gain confidence in themselves and become more productive workers. Individuals are more apt to be motivated when they sense that you, as their leader, personally care for them and they can approach you without feeling threatened. Be sure when they give you their comments on issues or discuss their needs, that you give them careful consideration; this is vital in having a good working relationship.

It is critical that everyone knows the rules and expectations in each situation. These rules and expectations should apply to everyone; no person should receive preferential treatment. To be a good motivational leader, you must first and foremost set the example you want to be followed. By demonstrating your attitude of being prompt, conscientious, and consistent, others will be more likely to follow your lead.

Communication is imperative in motivational situations! You must keep others informed and be sure to speak in a straightforward manner, using no jargon, slang, or offensive language. Manners are critical and never inappropriate when dealing with people. Make sure that you actively listen to them and get to know those you are trying to motivate: their likes/dislikes, hobbies, families, and special interests.

Involve everyone you are working with. Let them make decisions or, at the very least, be part of the decision process; this includes the goal-setting process as well. Make sure individuals have opportunities to succeed and feel they are contributing.

Interact with the individuals you are working with. This will give you an opportunity to see how they are doing and to gain a good rapport with them. Frequently give them feedback on their progress; don't wait until they have to ask for it or wait for special occasions. Praise their work and provide recognition and rewards when appropriate. Show everyone respect and remember their time is valuable to them and to you. Don't waste it!

Invest as much as you can in those who work with you or for you. You are responsible to help people develop themselves to their fullest potential. Your time is one of the most valuable commodities which you can share with others.

Few motivational techniques can beat the one-on-one contact. People you are trying to motivate do not expect you to act, talk, dress, socialize, or be like them in order for you to be a good motivator. You must be yourself and have a true interest and concern for them. They will respect you for who you are. Emerson said it best,

"What you are or what you do speaks so loudly that I cannot hear what you say,"
Ralph Waldo Emerson

REFERENCES

1. Herzberg, Fredrick. "One More Time: How Do You Motivate Employees?" *Harvard Business Review,* (January-February, 1968), pp. 53-62.

2. Gates, Bill. "The Importance of Making Mistakes," *U.S. Air Magazine*, (July, 1995), pp. 48-49.

3, Maslow, Abraham H. *Motivation and Personality*, Harper and Brothers, New York, NY, 1954.

4. Carr, James. "Leadership...Is It Always Power-motivated?" *Piedmont Airlines*, (November, 1987), pp. 17 and 43.

5. Blanchard, Kenneth. "How To Get Better Feedback," *Success*, (June, 1991), p. 6.

6. Weisinger, Hendric and Lobsewz, Norman. *Nobody's Perfect: How To Give Criticism and Get Results*, Stratford Press, New York, NY, 1981.

7. Brown, Paul L. and Presbie, Robert J. *Behavior Modification in Business, Industry And Government.* Behavior Improvement Associates Inc., Paltz, NY, 1976.

8. Mescon, Michael H. and Timothy S. "On Management: Questions...And Answers," *Sky*, (September, 1990), pp.126-127.

9. Reese, Charles D., *Joint Labor/Management Committees: A Guide for Committee Members*, University of Connecticut Press, Storrs, CT, 1996.

10. Blake, Robert R. and Srygley, Jane. "Principles of Behavior for Sound Management," *Training and Development Journal*, (October, 1979), pp. 26-28.

11. Iacocca, Lee and Novak William, *Iacocca*, Bantam Books, Inc., New York, NY, 1984.

Chapter 4

CONSTRUCTION SAFETY AND HEALTH PROGRAMS

Charles D. Reese

The need for health and safety programs in the construction industry has been an area of controversy for some time. Many in the construction industry feel that written safety and health programs are just more paperwork, a deterrent to productivity, and nothing more than another bureaucratic way of mandating safety and health on the job. But over a period of years, data and information have been mounting in support of the need to develop and implement written safety and health programs in the construction industry.

This perceived need for written programs must be tempered with a view to their practical development and implementation. A very small contractor who employs one to four employees and no supervisors in all likelihood needs only a very basic written plan, along with any other written programs that are required as part of an OSHA regulation. But, as the size of the contractor's business and the number of employees increase, the contractor becomes more removed from the hands-on aspects of what now may be multiple projects.

Now the contractor must find a way to convey support for safety to all those who work with and for him/her. As with all other aspects of business, the contractor must plan, set the policies, apply management principles, and assure adherence to the company's goals in order to facilitate the efficient and effective completion of projects. Job safety and health should be managed the same as any other part of the contractor's business.

The previous paragraph simply states that in order to effectively manage safety and health, a construction company must pay attention to some critical factors. These factors are the essence in managing safety and health on construction worksites. The questions that need to be answered regarding managing safety and health are

1. What is the policy of the contractor regarding safety and health on his/her projects?
2. What are the safety and health goals for the company?
3. Who is responsible for occupational safety and health?
4. How are supervisors and employees held accountable for job safety and health?
5. What are the safety and health rules for this type of construction?
6. What are the consequences of not following the safety rules?
7. Are there set procedures for addressing safety and health on the jobsite?

Specific actions can be taken to address each of the previous questions. The written safety and health program is of primary importance in addressing these items. Have you ever wondered how your company is doing in comparison with a company without a safety and health professional and a viable safety and health program? Well, wonder no more....

In research conducted by the Lincoln Nebraska Safety Council in 1981, the following conclusions were based on a comparison of responses from a survey of 143 national companies. All conclusions have a 95% or more confidence level. Table 4-1 is an abstraction of results from that study.

Table 4-1

Effectiveness of Safety and Health Program Findings

Fact	Statement	Findings
1	Do not have separate budget for safety.	43% <u>more</u> accidents
2	No training for new hires.	52% <u>more</u> accidents
3	No outside sources for safety training.	59% <u>more</u> accidents
4	No specific training for supervisors.	62% <u>more</u> accidents
5	Do not conduct safety inspections.	40% <u>more</u> accidents
6	No <u>written</u> safety program compared with companies that have written programs.	106% <u>more</u> accidents
7	Those using canned programs, not self-generated.	43% <u>more</u> accidents
8	No written safety program.	130% <u>more</u> accidents
9	No employee safety committees.	74% <u>more</u> accidents
10	No membership in professional safety organizations.	64% <u>more</u> accidents
11	No established system to recognize safety accomplishments.	81% <u>more</u> accidents
12	Did not document/review accident reports and reviewers did not have safety as part of their job responsibility.	122% <u>more</u> accidents
13	Did not hold supervisor accountable for safety through merit salary reviews.	39% <u>more</u> accidents
14	Top management did not actively promote safety awareness.	470% <u>more</u> accidents

It seems apparent from the previous research that in order to have an effective safety program, at a minimum, a contractor must

- Have a demonstrated commitment to job safety and health.
- Commit budgetary resources.
- Train new personnel.
- Insure that supervisors are trained.
- Have a written safety and health program.
- Hold supervisors accountable for safety and health.
- Respond to safety complaints and investigate accidents.
- Conduct safety audits.

Other refinements can always be part of the safety and health program, which will help reduce workplace injuries and illnesses. They are more worker involvement (for example, joint labor/management committees); incentive or recognition programs; getting outside help from a consultant or safety association; and setting safety and health goals.

If a decrease in occupational incidents that result in injury, illness, or damage to property is not reason enough to develop and implement a written safety and health program, the other benefits from having a formal safety and health program seem well worth the investment of time and resources. Some of these are

- Reduction of industrial insurance premiums/costs.
- Reduction of indirect costs of accidents.
- Fewer compliance inspections and penalties.
- Avoidance of adverse publicity from deaths or major accidents.
- Less litigation and fewer legal settlements.
- Lower employee payroll deductions for industrial insurance.
- Less pain and suffering by injured workers.
- Fewer long-term or permanent disability cases.
- Increased potential for retrospective rating refunds.
- Increased acceptance of bids – more jobs.
- Improved morale and loyalty from individual workers.
- Increased productivity from work crews.
- Increased pride in company personnel.
- Greater potential of success for incentive programs.

REASONS FOR A COMPREHENSIVE SAFETY PROGRAM[1]

The three major considerations involved in the development of a safety program are

1. **Humanitarian**

    Safe operation of workplaces is a moral obligation imposed by modern society. This obligation includes consideration for loss of life, human pain and suffering, family suffering and hardships, etc.

2. **Legal obligation**

    Federal and state governments have laws charging the employer with the responsibility for safe working conditions and adequate supervision of work practices. Employers are also responsible for paying the costs incurred for injuries suffered by their employees during their work activities.

3. **Economic**

    Prevention costs less than accidents. This fact is consistently proven by the experience of thousands of industrial operations. The direct cost is represented by medical care, compensation, etc. The indirect cost of four to ten times the direct cost must be calculated, as well as the loss of wages to employees and the reflection of these losses on the entire community.

All three of these are good reasons to have a health and safety program. It is also important that these programs be formalized in writing, since a written program sets the foundation and provides a consistent approach to occupational and safety for the construction company. There are other logical reasons for a written safety and health program. Some of them are

*   It provides standard directions, policies, and procedures for all company personnel.
*   It states specifics regarding safety and health and clarifies misconceptions.
*   It delineates the goals and objectives regarding workplace safety and health.
*   It forces the company to actually define its view of safety and health.
*   It sets out in black and white the rules and procedures for safety and health that everyone in the company must follow.
*   It is a plan that shows how all aspects of the company's safety and health initiative work together.
*   It is a primary tool of communication of the standards set by the company regarding safety and health.

BUILDING A SAFETY AND HEALTH PROGRAM

The length of such a written plan is not as important as the content. It should be tailored to the company's needs and the health and safety of its work force. It could be a few pages or a multiple-page document. But, it is suggested you adhere as much as possible with the KISS principle (Keep It Simple Stupid). In order to insure a successful safety program, three conditions must exist. These are management leadership, safe working conditions, and safe work habits by all employees. The employer must

*   Let the employees know that you are interested in safety on the job by consistently enforcing and reinforcing safety regulations.

- Provide a safe working place for all employees; it pays dividends.

- Be familiar with federal and state laws applying to your operation.

- Investigate and report all OSHA recordable accidents and injuries. This information may be useful in determining areas where more work is needed to prevent such accidents in the future.

- Make training and information available to the employees, especially in such areas as first aid, equipment operation, and common safety policies.

- Develop a prescribed set of safety rules to follow, and see that all employees are aware of the rules.

Rules

Employees have the most at stake in terms of losses and must be aware that safety rules are for their own benefit. The employees' responsibilities include

- Abiding by established safety procedures. Rules and safety equipment work only if correctly used.

- Being familiar with federal, state, and company rules.

- Taking advantage of training that would improve his/her knowledge of safety or job skills.

Communications

Many safety problems arise because we assume that everyone knows the proper and safe way to do a job. In actual practice, this is not so. It is imperative that management insures that everyone on the property knows the safety policies of the company and the proper methods to use in performing their job.

A large percentage of injuries occur when people are not aware of policies, methods, or basic skills needed to safely perform a job. The responsibility for communicating these concepts rests with management. A line of communication must be established that is constantly furnishing information to all employees. Some methods of communicating the safety message are

1. Safety meetings.

2. Job training.

3. Safety bulletin boards.

4. Accident investigations.

5. New employee indoctrination.

6. Safe behavior on the part of the supervisors.

7. Job analysis for safety.

8. Tool box safety talks.

Plan your lines of communications and keep them open. A safety program may be good on paper, but unless it is communicated to the workers, it is useless.

The following safety practices should form the basis of a good safety program and delineate safety and health factors that will enhance the overall communication process. Each jobsite or contractor will probably include other standards based on the needs of their own operation.

1. Communications shall be established for emergency purposes.
   A. Telephone.
   B. Two-way radio.

2. Emergency numbers shall be posted.
   A. Doctor.
   B. Ambulance.
   C. Fire.
   D. Federal and state agencies.
   E. Hospital.

3. An emergency plan shall be developed, posted, and practiced. All workers should be trained in emergency procedures.

4. Personal safety protection shall be worn in and around the operations.
   A. Hard hats.
   B. Safety glasses.
   C. Hard-toed shoes/boots.
   D. Special safety equipment needed for specific areas, (e.g., gloves, respirators, ear plugs, safety lines, etc.)

5. Someone shall always be delegated to be in charge in the event the usual supervisor is absent for any reason.

6. No person shall be assigned to work in any area alone, unless he/she can be seen or their cries for help can be heard.

7. No person shall be assigned to perform any job known to be hazardous, unless proper precautions have been taken.

8. Maintenance shall not be performed on machinery until the machinery is stopped and a positive lock-out procedure is followed. A written procedure should be established that is understood by **all** employees.

9. When a guard is removed from machinery for repair and/or maintenance purposes, the guard shall be replaced before the machinery is started and/or energized. Tags stating removal of guard may be used.

10. Vehicles shall be checked for safety prior to starting a shift. If a malfunction occurs during the shift, the operator shall have the vehicle repaired immediately. An equipment checklist good for this purpose for each piece of equipment shall be provided.

11. No person shall work near or under improperly supported (soil) or stored materials (pipe).

12. Areas requiring special protective devices shall be posted, and rules for using equipment shall be strictly enforced.

13. Roadway signs regarding speed and specific driving problems shall be posted to inform drivers of road conditions.

14.   Electrical problems shall not be worked on and adjustments shall not be made by anyone other than a person qualified to do the work.

15.   The storage of explosives in or around an operation must meet the requirements of the Bureau of Alcohol, Tobacco, and Firearms standards.

16.   No person shall be assigned to or allowed to handle explosives for initiating blasts unless they are competent to do so.

Training

Experience shows that a good safety program is based on a well planned, on-going training program. The training program should include both safety and the skills involved in performing the tasks to be accomplished. The training program should include a minimum of classroom-type experiences and continual reinforcement of training concepts in on-the-job training situations. The length of the training sessions is not as important as their quality. A well-planned program will save time and increase the effectiveness of the training. The basic general objectives of the training program should include

1.   An understanding of the company's basic philosophy and genuine concern for safety on the job.

REMEMBER, BEHAVIOR OF THE SUPERVISOR WILL "SAY" MORE THAN ANY TRAINING PROGRAM. SAFE BEHAVIOR OF THE SUPERVISOR IS A MUST FOR ANY SAFETY PROGRAM.

2.   Basic skills training. A skilled worker is a safe worker.

3.   A thorough knowledge of company safety policies and practices.

4.   Indoctrination of new employees.

5.   Annual training required as necessary or any required by law.

A list of suggested training topics for supervisors should include

1.   Organization and operation of a safety program.

2.   Building attitudes favorable to safety.

3.   Knowledge of federal and state laws.

4.   First aid training.

5.   The investigation and methods of reporting accidents to the company and government agencies.

6.   Accident causes and basic remedies.

7.   Job instruction for safety.

8.   Motivating safe work practices.

9   Communicating safe work practices.

10.   Making the workplace safe.

11.   Mechanical safeguarding.

12.   Safe handling of materials.

13.   The number and kinds of accidents.

14.   The supervisor's place in accident prevention.

15.   The cost of accidents and their effect on production.

Subject areas that can sharpen supervision skills also enhance safety effectiveness, especially in the areas of

1.   Giving job instruction.

2.   Supervising employees on the job.

3.   Determining accident causes.

4.   Building safety attitudes.*

*A side benefit of such training is a more effective supervisor for production and cost effectiveness.

Training topics for all employees may include

1.   Technical instruction and job descriptions.

2.   Safety rules and practices.

3.   Method of reporting accidents.

4.   Importance of first aid treatment.

5.   Where to get first aid.

6.   Explanation of policies and responsibilities.

7.   Where to get information and assistance.

8.   Federal and state laws.

9.   First aid training.

10.   Any other subjects of importance.

Training programs may be obtained from a variety of sources.* Much of the training is provided without charge by:

1.   Government agencies, both federal and state.

2.   Professional and industrial associations.

3.   Equipment and product representatives.

4.   Insurance companies.

5.   Films and audio-visual materials (on loan).

6.   Qualified or experienced people in your own company.

7.   Construction safety consultants.

* Use variety to keep your safety training interesting. The use of outside speakers, audio-visual aids, incentives, and awards are all proven techniques.

Attempt to keep safety subjects pertinent. Think of upcoming jobs and anticipate possible safety concerns. <u>An excellent safety meeting may be nothing more than relating some precautions to a crew when giving work instructions</u>. Plan your training program in advance so that the time spent will be worthwhile.

Accident Investigation

The use of real accident outcomes and the company's accident experience are important to good training. The results can be used not only in training sessions but for safety talks and safety meetings.

A complete and thorough investigation should be made to ascertain the cause or causes of every incident. Care should be taken to insure that it is a <u>fact-finding</u> and not a <u>fault-finding</u> operation. Too many investigations are for the express purpose of finding someone to condemn. This is the wrong approach. The only purpose for investigating an accident should be to find exactly what happened so that proper steps can be taken to prevent recurrence.

An investigation can be divided into these five main categories:

WHO? WHAT? WHEN? WHERE? HOW?

The following is a list that should be helpful in answering the five important questions.

1. Who was involved?
2. What was the injured doing at the time of the accident?
3. Was this part of their regular work?
4. How long had the worker been employed in this position?
5. Was the worker trained or experienced in the work being performed?
6. Was/Were the worker(s) qualified to perform the function?
7. Were operating procedures established for the task involved?
8. Were procedures followed, and if not, why not?
9. Was the worker following instructions?
10. Was the work being performed in the customary manner?
11. Were safety rules violated?
12. Did the worker have any known physical defects?
13. Had the worker shown a tendency toward improper safety attitudes?
14. Was adequate safety equipment or proper clothing worn?
15. Why did the accident occur?
16. What caused the situation to occur?
17. What actions contributed to the accident?
18. Was the job or layout properly planned?
19. Was the job being properly supervised?
20. What type of equipment was being used?
21. Was defective equipment a contributing factor?
22. Was proper equipment being used?

23. Was equipment adequately safeguarded?

24. What hazardous arrangement or unsafe process contributed to the accident?

25. Were supervisory personnel aware of the hazards?

26. What were the contributing conditions?

27. Equipment involved was owned by whom?

28. Were weather conditions a contributing factor?

29. Was the time a contributing factor?

30. What should be done?

31. What has been done?

32. Where else might this or a similar situation exist, and how can it be avoided?

Methods for preventing future accidents of a similar nature should be identified. Supervisors and workmen should be aware of hazards at all times. Any good safety program has mechanisms for the identification of hazards. Supervisors should informally inspect the work area at all times; however, a written safety inspection should be performed at least once a week. The following forms the basis of a weekly checklist.

Physical Hazards

1. Machine guarding.

2. Housekeeping.

3. Condition of tools and equipment.

4. Lighting.

5. Soil conditions.

6. Condition of floors, stairways, and walkways.

7. Provisions for safe access to overhead equipment.

8. Electrical hazards.

9. Exit facilities.

10. Personal protective apparel and equipment.

11. Equipment maintenance and check-off sheets.

Not only should supervisors be trained to conduct workplace audits, but workers should also be able to identify not only hazards but unsafe conditions and unsafe acts. Extra pairs of eyes are always helpful and induce everyone to identify and report hazards. This makes the program a dynamic approach to the goal of better job safety and health.

Unsafe Practices (Acts)

1. Improper operation of machines and equipment.

2. Removal or nullification of machine guards or safety devices.

3. Use of defective tools or equipment or makeshift tools. Failure to secure needed tools.

4.   Overloading, overcrowding, improper storage or handling of materials.

5.   Working under suspended loads, near open hatches, riding loads, creating operation traffic hazards.

6.   Repairing or adjusting equipment in motion, under pressure, electrically charged, or containing dangerous materials.

7.   Failure to use, or use of inadequate, personal protective equipment or safety devices.

8.   Horseplay.

Evaluation

The following is a quick review of the overall content of an effective safety and health program. This is a very generalized checklist.

QUICK REVIEW SHEET
FOR
SAFETY AND HEALTH PROGRAMS

COMMUNICATION

_____ A written safety policy exists.

_____ A list of company safety and health rules exist.

_____ Frequent safety meetings are conducted.

SAFE WORK PRACTICES

_____ Trained individuals perform hazardous tasks.

_____ Machine and maintenance checklists are used.

_____ Personal protective equipment is used.

SAFETY INSPECTION

_____ Formal safety inspections are conducted at least weekly.

_____ Daily visual safety inspection takes place.

_____ Follow-up occurs on all safety suggestions.

_____ Job observations are conducted by supervisor.

_____ Health and safety rules are enforced.

TRAINING

_____ Have an approved training plan.

_____ Have outlines for training sessions.

____ Have a systematic approach to task training.

____ Have job safety analysis or safe operating procedures for job classifications.

ACCIDENT INVESTIGATION

____ All accidents are investigated.

____ An accident investigation form is used.

____ Accidents are analyzed.

EXPLAINING THE REQUIREMENTS AND ELEMENTS OF OSHA GUIDELINES FOR A SAFETY AND HEALTH PROGRAM[2]

Although federal regulations do not currently require employers to have a written safety and health program, the best way to satisfy OSHA requirements and reduce accidents is for employers to produce one. In addition, distributing a written safety and health program to employees can increase employee awareness of safety and health hazards while, at the same time, reduce the costs and risks associated with workplace injuries, illnesses, and fatalities.

Federal guidelines for safety and health programs suggest that an effective occupational safety and health program must include evidence of

- Management commitment and leadership.
- Assignment of responsibility.
- Identification and control of hazards.
- Training and education.
- Recordkeeping and hazard analysis.
- Availability of first aid and medical assistance.

If a representative from the Occupational Safety and Health Administration (OSHA) visits a jobsite, he/she will evaluate the safety program using the elements listed above. The compliance officer will review the previous items to assess the effectiveness of the safety and health program.

The Occupational Safety and Health Administration has set some guidelines for effective safety and health programs. Their application and description are presented to assist in the development of a written program.

Management Commitment and Leadership

"Management Commitment and Leadership" includes a policy statement that should be developed and signed by the top person in the company. Safety and health goals and objectives should be included to assist with establishing workplace goals and objectives that demonstrate the company's commitment to safety. An enforcement policy is provided to outline disciplinary procedures for violations of the company's safety and health program. This safety and health, as well as the enforcement policy, should be communicated to everyone on the construction jobsite. Some of the key aspects found under the heading, "Management Commitment and Leadership," are

1. Policy statement: goals established, issued, and communicated to employees.

2. Program should be revised annually.

3. Participation in safety meetings; inspections; safety items addressed in meetings.

4. Commitment of resources is adequate in the form of budgeted dollars.

5. Safety rules and procedures incorporated into jobsite operations.

6. Procedure for enforcement of the safety rules and procedures.

7. Statement that management is bound to adhere to safety rules.

A safety and health policy statement clarifies the policy, standardizes safety within the company, provides support for safety, and supports the enforcement of safety and health within the company. It should set forth the purpose and philosophy of the company, delineate the program's goal, assign responsibility for all company personnel, and be positive in nature. It should be as brief as humanly possible. (See the Model Construction Safety and Health Program (MCS&HP) later in this chapter.)

Goals and objectives are very important and should be directly observable and measurable. They should be reasonable and attainable. The following are some examples of goals and objectives and the issues faced by those using them:

- Zero fatalities or serious injuries. (This is usually a "pie in the sky" or unreachable goal for most contractors. For example, if you had 25 accidents last year, zero is probably not possible.)

- Reduce injuries, lost workday accidents, and workers compensation claims by ____%.

- Prevention of damage or destruction to company property or equipment.

- Increase productivity through reduction of injuries by _____%.

- Reduce workers' compensation costs by decreasing the number of claims to ____ or the cost by _____%.

- Enhance company's image by working safely. Can you measure this in some way?

- Keep safety a paramount part of workers' daily activities. What are indicators of this? Number of near misses, reports of hazards, or number of observable unsafe acts.

- Recognize and reward safe work practices. How is this a goal? What could be the measurable outcome of this objective?

Assignment of Responsibility

"Assignment of Responsibility" identifies the responsibilities of management officials, supervisors, and employees. An emphasis on responsibility for safety and health is more creditable if everyone is held accountable for their safety and health performance as related to established goals.

The company/contractor should designate an individual who knows the site, has knowledge of safety and health, and is accountable for the safety and health function. Also, all supervisors/forepersons must be told their responsibilities for job safety and health. All employees should be informed of the exceptions relevant to safety and health on the jobsite.

Identification and Control of Hazards

"Identification and Control of Hazards" includes those items that can assist you with identifying workplace hazards and determining what corrective action is necessary to control them. These items include jobsite safety inspections, accident investigations, safety and health committees, and project safety meetings. Identification and control of hazards should include periodic site safety inspection programs that involve supervisors and, if you have them, joint labor management committees. Safety inspections should ensure that preventive controls are in place (PPE, guards, maintenance, engineering controls), that action is taken to quickly address hazards, that technical resources such as OSHA, state agencies, professional organizations, and consultants are used, and that safety and health rules are enforced.

The core of an effective safety and health program is hazard identification and control. Periodic inspections and procedures for correction and control provide methods of identifying existing or potential hazards in the workplace and eliminating or controlling them. The hazard control system provides a basis for developing safe work procedures and injury and illness prevention training. Hazards occurring or recurring reflect a breakdown in the hazard control system.

The written safety and health program establishes procedures and responsibilities for the identification and correction of workplace hazards. The following activities are used by this company/contractor to identify and control workplace hazards: jobsite inspections, accident investigation, safe operating procedures, and safety and health committees.

As part of this safety and health program, the responsible site contractor for each company project/jobsite needs to identify "high hazard" areas of operation and determine inspection priorities, establish inspection responsibilities and schedules, and develop a management system to review, analyze, and take corrective action on inspection findings. This is especially true of today's construction environment, where OSHA has gone to great lengths to address safety and health responsibilities on multi-employer jobsites/projects. (See Chapter 13 – OSHA Compliance for Information on Multi-Employer Worksite.)

All accidents should be investigated to determine causal factors and prevent future recurrences of similar accidents. A written report of investigation findings should prepared by the injured employee's immediate supervisor and submitted to the site superintendent for review. In most cases a standard format should be developed to assure that each incident information is consistent and complete. Whenever an accident is reported, the supervisor of the injured worker(s) should respond to the scene of the accident as soon as possible and complete the supervisor's accident report. All witnesses should be interviewed privately as soon as possible after the accident. If possible, the supervisor should interview the worker(s) at the scene of the accident so that events leading up to the accident can be re-enacted.

Photographs should be taken as soon as possible after the accident and should include the time and date taken. Supervisors are required to submit accident investigation reports that answer the questions: Who? What? When? Where? and How? as denoted earlier in this chapter.

Training and Education

"Training and Education" is one of the most important elements of any safety and health program. Each training item should describe methods for introducing and communicating new ideas into the workplace, reinforcing existing ideas and procedures, and implementing your safety and health program into action. The training needs may range from supervisor training, especially work task training, employee updates, and new worker orientation. The content of new worker or new site training should include at least the following topics:

- Company safety and health program and policy.
- Employee and supervisory responsibilities.
- Hazard communication training.
- Emergency and evacuation procedures.

- Location of first aid stations, fire extinguishers, and emergency telephone numbers.
- Site-specific hazard.
- Procedures for reporting injuries.
- Use of personal protective equipment.
- Hazard identification and reporting procedures.
- Review of each safety and health rule applicable to the job.
- Site tour or map where appropriate.

It is a good idea to have follow-up for all training which may include working with a more experienced worker, supervisor coaching, job observations, and reinforced good/safe work practices.

Supervisors/forepersons are responsible for the prevention of accidents for tasks under their direction, as well as for thorough accident prevention and safety training for the employees they supervise. Therefore, all supervisors/forepersons will receive training so that they have a sound theoretical and practical understanding of the site-specific safety program, OSHA construction regulations, and the company's specific safety and health rules. They should also receive training on the OSHA Hazard Communication Standard, site emergency response plans, first aid and CPR, accident and injury reporting and investigation, and procedures for safety communications, such as toolbox safety talks. Beyond these training requirements described previously, additional training might cover the implementation and monitoring of a construction safety program, personnel selection techniques, OSHA recordkeeping requirements, and motivating individuals and groups.

Recordkeeping and Hazard Analysis

"Recordkeeping and Hazard Analysis" identifies the types of records that OSHA requires your company to maintain and who is responsible for maintaining these records. Procedures for conducting hazard analyses are provided to enable you to learn from past experiences and take corrective action to prevent future injuries and illnesses. The records that need to be maintained are employee injuries and illnesses, accident investigations, causes and proposed corrective measures, near-misses, training records, and company-required inspection or maintenance records. Various types of reports are necessary to meet the recordkeeping requirements of OSHA, insurance carriers, and other government regulatory agencies.

Medical records should be maintained for the length of an employee's employment plus 30 years, while exposure records should be kept for 30 years. These records are confidential. Information from an employee's medical record will be released only to the employee or his/her designated representative after written consent from the employee.

Training records should be maintained in each employee's personnel file and should be available for review upon request.

First Aid and Medical Assistance

"First Aid and Medical Assistance" identifies the provisions your company should establish to provide first aid and medical services on your jobsites. Sample emergency procedures are included to respond to various types of emergencies that may occur.

Experience indicates that supervisors/forepersons who receive basic first aid and CPR training are much more safety-conscious and usually have better crew safety performance records. Therefore, all field supervisory personnel should be required to attend basic first aid and CPR training. Each jobsite should maintain a first aid log that includes the following information:

- Injured employee's name.

- Immediate supervisor.

- Date and time of injury.

- Nature and cause of the injury.

- Injured employee's craft.

- Treatment rendered and disposition of employee (returned to work or sent for medical attention).

All employees should be provided with the location(s) of the first aid stations on each project/jobsite. Instructions for using first aid equipment should be located at each station. In the event of an emergency, employees are to contact any supervisor or individual who is trained in first aid.

Emergency and Firefighting Procedures

"Emergency and Firefighting Procedures" should be an integral part of a good safety and health program. There should be guidelines for firefighting, such as all firefighting equipment should be conspicuously located and readily available at all times, all firefighting equipment should be inspected and maintained in operating condition, and all fire protection equipment should be inspected no less than once monthly with documentation maintained for each piece of equipment inspected. Discharged extinguishers or damaged equipment should be immediately removed from service and replaced with operable equipment, and all supervisors and employees should search for potential fire hazards and coordinate their abatement as rapidly as possible.

Individuals assigned safety responsibilities should receive the necessary training to properly recognize fire hazards, inspect and maintain fire extinguishers, and know the proper use of each. A trained and equipped firefighting brigade should be established, as warranted by the project, to assure adequate protection to life.

The emergency procedure should be spelled out as related to response, action, and expectations for workers when such emergencies occur. These may include a standard warning alarm, emergency telephone and communication procedures, and an evacuation plan. These procedures should be practiced by regular drills.

The composition or components of your safety and health program may vary depending upon the complexity of your operations. They should at least include

- Management's commitment and safety and health policy.

- Hazard identification and evaluation.

- Hazard control and prevention.

- Training.

Of course, each of these may have many subparts that address the four elements in some detail. The safety and health program that you develop should be tailored to meet your specific needs. The following is a model safety and health program and should provide a foundation.

MODEL CONSTRUCTION SAFETY AND HEALTH PROGRAM[3]

Management Safety and Health Policy Statements

These are two examples of safety and health policy statements that could be used or tailored to your needs

To all Employees:

<u>Name of the Construction Company</u> has been in business for over ___ years. The company prides itself on the fact that safety is our first priority.

It has always been our policy to provide a safe environment for any contractor or worker at one of our projects. We expect every individual to uphold the standards of the Occupational Safety and Health Act (OSHAct) and the safety measures of the company as presented herein. No priority is to be placed above safety at any time.

Preventive measures and the elimination of any potential hazard are of the utmost importance for the safety of all employees, visitors, and the public in general.

Our safety director and the safety inspectors have the responsibility to immediately report to the project managers, project superintendents, and senior management any potential hazardous conditions. The superintendent and project manager are responsible for immediate actions in order to avoid injury and/or hazard.

It is the responsibility of the corporate safety director to periodically report to senior management the safety status of all projects. It is the responsibility of every employee to support and assist in establishing safety measures. The safety programs shall be implemented and reviewed by the project superintendents.

(Signature)

President

(Typical of a larger company policy)

Or

To All Employees:

It is the policy of this company to ensure that every employee is allowed to work in a safe and productive environment. Due to the nature of our business, it is important to realize the uniqueness of our jobsites and the inherent hazards associated with them. As such, it is important to stress the importance of safety, health preservation, and accident prevention at all times.

To achieve this goal, the company will provide the employee with the knowledge and equipment required to do the job both safely and efficiently. The use of personal protective gear, [hard hats, steel toed shoes, safety glasses, breathing apparatus (respirators as necessary) and hearing protection] is everybody's responsibility. An injury or accident affects us all. Therefore, we all need to watch out for each other.

Specialized safety training will be provided to all employees; specifically HAZWOPER and Confined Space Entry. Generalized safety training will be conducted weekly at the morning Pre-Work Meetings and a quarterly safety seminar will be presented at an informal company meeting by management.

Due to the fact we have few management personnel, safety audits will be performed randomly by the owner or estimator. These audits are not meant to harass any individual employee but rather to encourage a more active participation level in our safety program by all employees.

To reinforce our safety policies, basic knowledge/skills tests will be taken by all new employees and administered by management. As new techniques become available to perform our work, training will be provided to all employees to ensure that we have the skills to match the needs of our profession. A skills inventory will be maintained in each employee's personnel file.

All company equipment, tools, and vehicles will be maintained in good working order with particular attention to safety. If a tool is defective, do not continue to use it, but bring it to the office for repair or replacement. Inspection of all tools will be done daily by employees and monthly by management. An inspection log for each piece of equipment will be maintained by management.

The management of Name of the Construction Company will ensure compliance with all applicable local, state, and federal regulations governing the health and safety of its employees.

*There is no secret to safety, just a common sense approach to our job. If you see someone doing something in an unsafe manner, **STOP THEM, INSTRUCT THEM, and CORRECT A BAD HABIT**. Let's not prove that we have been lucky when it comes to safety, but that we actively pursue safety.*

In concern for your safety and health,

(Signature)

President

(Typical safety policy for a smaller company)

Accountability and Responsibility

Safety and health is a management function that requires management's participation in planning, setting objectives, organizing, directing, and controlling the program. Management's commitment to safety and health is an integral part of every decision the company makes and every action this company takes. Therefore, the management of Name of Company assumes total responsibility for implementing and ensuring the effectiveness of this safety and health program. The best evidence of our company's commitment to safety and health is this written program, which will be fully implemented on each company construction project.

The Name of Person and Title is assigned the overall responsibility and authority for implementing this safety and health program. Company Name fully supports Name of Person and Title and will provide the necessary resources (budget, etc.) and leadership to ensure the effectiveness of this safety and health program.

On each Name of Company project jobsite, the site superintendent will be accountable to management for the successful achievement of targeted Company safety and health goals. Name of Company's project safety and health goals are

(List the Company Objectives and Safety and Health Goals.)

Discipline Policy

The company expects that all workers, including management, will adhere to the company's safety and health rules as well as applicable state and federal regulations. Whenever a violation of safety rules occurs, the following enforcement policy will be implemented:

> *FIRST OFFENSE – Verbal warning and proper instruction pertaining to the specific safety violation. (A notation of the violation may be made and placed in the employee's personnel file.)*
>
> *SECOND OFFENSE – Written warning with a copy placed in the employee's personnel file.*
>
> *THIRD OFFENSE – Dismissal from employment.*
>
> * *The company reserves the right to terminate immediately any employees who flagrantly endanger themselves or others by their unsafe actions while on Name of Company jobsites.*

Supervisory Involvement

Active participation in and support of safety and health programs are essential. Therefore, all management officials of the Name of Company will display their interest in safety and health matters at every opportunity. At least one manager (as designated) will participate in project safety and health meetings, accident investigations, and jobsite inspections. All management personnel are expected to follow the job safety and health rules and enforce them equally. Management personnel safety performance will comprise a significant portion of their annual merit evaluation. The standard safety and health merit evaluation (SSHME) form will be used for evaluating all management personnel's safety and health performance. (See Figure 4-1.)

STANDARD SAFETY AND HEALTH MERIT EVALUATION

Supervisor's Name													
Safety Record													
Setting Good Example													
Compliance with OSHA Standards													
General Safety Attitude													
Housekeeping													
Prompt Correction of Hazards													
Accident Reporting													
Hard Hat Compliance													
Safety Eye Wear Compliance													
Degree of Meeting Safety Goals													
Safety Meeting & Toolbox Talks													
Other Protective Equipment													
Composite Score													
Previous Composite Score													
Change + or − Ranking													

Scoring: 1 = Excellent

2 = Above Average

3 = Average

4 = Fair

5 = Poor (Needs Improvement)

Figure 4-1. Standard safety and health merit evaluation

The first-line supervisor has the key role and primary responsibility in the safety and health of the employees. He/she will be evaluated using the SSHME form to assure that the supervisor provides instruction on safety and health rules, regulations, policies, and procedures by conducting pre-job safety orientations with all workers and reviewing rules as the job or conditions change or when individual workers show a specific need. He/she must require the proper care and use of all necessary personal protective equipment to protect workers from hazards; identify and eliminate job hazards expeditiously through hazard analysis procedure; take initial action on employee suggestions, awards, or disciplinary measures; and conduct foreperson/crew meetings the first five minutes of each work shift to discuss safety matters and work plans for the work day. The supervisor must participate in accident investigations and safety inspections to establish trends and prevent accidents, promote/motivate employee participation in this safety and health program, and exhibit a positive attitude toward workplace safety and health.

Employee Responsibility

Safety is a management responsibility; however, each employee is expected, as a condition of employment, to work in a manner that will not inflict self-injury or cause injury to fellow workers. Each employee must understand that responsibility for his/her own safety is an integral job requirement. Each employee of <u>Name of Company</u> *will*

- *Observe and comply with all safety rules and regulations that apply to his/her trade.*
- *Follow instructions and ask questions of his/her supervisor when in doubt about any phase of his/her operation.*
- *Report all unsafe conditions or situations that are potentially hazardous.*
- *Report all on-the-job accidents and injuries to his/her supervisor immediately.*
- *Report all equipment damage to his/her supervisor immediately.*
- *Help to maintain a safe and clean work area.*

(Note: this list can be easily expanded by referring to Chapter 12 or Appendix G)

Jobsite Inspections

Safety audits/inspections of the jobsite will be conducted, usually on a monthly basis, or when conditions change, or when a new process or procedure is implemented. The inspections are to identify and correct potential safety and health hazards. A standard site evaluation worksheet (see example in Figure 4-2) will be used to conduct these jobsite safety inspections. Safe Operating Procedures will be used to determine the effectiveness of safety and health precautions. These audits/inspections are to be used to improve jobsite safety and health.

SAMPLE JOBSITE INSPECTION FORM

Check if no unsafe act/conditions exist. Otherwise denote the extent of the problem.

Jobsite: _____ Date: _____

_____ Housekeeping. Explain:_____

_____ No Protruding Nails Exist. Explain: _____

_____ Adequate Illumination. Explain: _____

_____ Floor Openings are Covered or Guarded. Explain: _____

_____ All Stairways are in Good Condition. Explain:_____

_____ Ventilation is Adequate. Explain: _____

_____ Fire Extinguishers Present and Accessible. Explain :_____

_____ All Equipment Guards are in Place. Explain: _____

_____ All Ladders are in Good Condition. Explain: _____

_____ All Gas Cylinders are Secured. Explain: _____

_____ No Open Access to Energized Electrical Circuits. Explain: _____

_____ GFCIs are Being Used. Explain: _____

_____ Guardrail Systems are in Place. Explain :_____

_____ Hard Hats Are Being Worn. Explain: _____

_____ All Chemical Containers are Labeled. Explain: _____

_____ Trenches and Excavations are Inspected by Competent Person. Explain: _____

_____ Workers are Following Safe Lifting Practices. Explain: _____

_____ Compressed Air is Below 30 psi. Explain: _____

_____ First Aid Supplies Exist and are Stocked. Explain: _____

Figure 4-2. Sample jobsite inspection form*

*   *This is only a short, not comprehensive, example of a jobsite inspection or audit instrument.
    A more detailed description of job inspection instruments, as well as recommendations for
    designing your own construction site evaluation instrument for your company, can be found
    in Chapter 6.*

Name of Company jobsite will establish a safety and health committee to assist with the implementation of this program and the control of identified hazards. The Safety and Health Committee will be comprised of employees and management representatives. The committee should meet regularly but not less than once a month. Written minutes from safety and health committee meetings will be available and posted on the project bulletin board for all employees to see.

The safety and health committee will participate in periodic inspections to review the effectiveness of the safety program and make recommendations for improvement of unsafe and unhealthy conditions. This committee will be responsible for monitoring the effectiveness of this program. The committee will review safety inspection and accident investigation reports, and where appropriate, submit suggestions for further action.

All employees, from supervisors to workers, will receive safety training on all phases of work performed by Name of Company. The following safety education and training practices will be implemented and enforced at all company projects/jobsites.

Accident Investigations

Supervisors/forepersons will conduct an investigation of any accident/incident that results in death, injury, illness, or equipment damage. The supervisor will use the company's standard investigation form (see, for example, Figure 4-3). The completed accident investigation report will be submitted to the individual assigned responsibility for occupational safety and health.

Recordkeeping

The Occupational Safety and Health Administration (OSHA) requires Name of Company to record and maintain injury and illness records. These records are used by management to evaluate the effectiveness of this safety and health program. A summary of all recordable injuries and illnesses will be posted during the month of February on an OSHA 200 Log Form for all employees to see.

Training

New employees and current employees who are transferred from another project must attend a project-specific, new-hire safety orientation. This program provides each employee the basic information about the project/jobsite-specific safety and health plan, federal and state OSHA standards, and other applicable safety rules and regulations. Attendance is mandatory prior to working on the construction project. The site superintendent documents attendance using the Company Training Form (see, for example, Figure 4-4) and all training records will be maintained by the company and placed into each worker's personnel file.

ACCIDENT INVESTIGATION FORM

VICTIM INFORMATION:

Name:_____ Sex: _____ Male _____ Female

Age: _____ Job Classification: _____

Experience at the Job Classification: _____

Total Construction Experience :_____

What Activity was Being Performed at Time of Accident? _____

Victim's Experience at This Activity: _____

Was the Victim Trained in This Task? _____

Is There a Record of the Victim's Training? _____ (Attach if Answer is "Yes")

SUPERVISOR INFORMATION:

Supervisor's Name: _____

Experience of Supervisor: _____

Total Construction Experience: _____

When was He/She Last Present at the Accident Location? _____

What did the Supervisor do?_____

Were Instructions Issued Relative to the Accident? _____

When was the Last Time the Supervisor Contacted the Victim?_____

Did the Supervisor see any Unsafe Acts or Conditions? _____

ACCIDENT INFORMATION:

Date of Accident:_____ Time of Accident: _____

Injury:_____ Equipment Damaged: _____

Location of Accident: _____

Description of Accident:_____

_____

Cause of Accident: _____

_____

Recommendations for Prevention: _____

_____

Disciplinary Action Taken:_____

Estimate Disability: _____Days   Equipment Damage: $_____

Witnesses:

_____   _____

Figure 4-3. Accident investigation form

COMPANY TRAINING FORM

WORKER'S NAME:_____

SOC. SEC.# _____

DATE(S) OF TRAINING: _____

LENGTH OF TRAINING: _____HOURS

SUBJECTS COVERED *(Check all that apply):*

_____ New Hire Orientation   _____ Fall Protection

_____ Company Rules and Policies   _____ Electrical Safety

_____ Hazard Communications   _____ Use and Care of Hand Tool

_____ Fire Safety and Fire Prevention   _____ Ladder Safety

_____ Scaffolding Safety   _____ Vehicle Safety

_____ Machine and Equipment Guarding   _____ Trench/Excavation Safety

_____ Steel Erection   _____ Rigging

_____ Material Handling   _____ Explosive/Blasting

_____ Personal Protective Equipment   _____ Confined Space Entry

_____ Respirator Use   _____ Asbestos Abatement

_____ Lead Abatement   _____ Hazardous Waste Remediation

Other:

_____

_____

_____

WORKER'S SIGNATURE:_____DATE:_____

Figure 4-4. Company training form

To maintain awareness, updated training, and convey important safety and health information, supervisors/forepersons will conduct at least weekly toolbox safety meetings, usually prior to the start of work. These toolbox meetings may be held more frequently, depending upon the circumstances (i.e., fatality, injury, new operations, etc.). Each supervisor will complete the Toolbox Meeting Form (see, for example, Figure 4-5), which includes the topic and attendees.

TOOLBOX MEETING FORM

SUBJECT: _____

PRESENTER: _____

DATE: _____

LENGTH OF TIME:_____

WORKER'S NAME	**SIGNATURE**	**SOC. SEC #**
_____	_____	_____
_____	_____	_____
_____	_____	_____

NOTE: Staple any handouts or materials used during the toolbox meeting.

Figure 4-5. Toolbox meeting form

First Aid and Medical Availability

All employees will be informed by posted notice of the existence, location, and availability of medical or exposure records at the time of initial employment and at least annually thereafter. Name/Title of Individual is responsible for maintaining and providing access to these records.

Each Name of Company project/jobsite will have adequate first aid supplies and certified, trained personnel or readily available medical assistance in case of injury. It is also imperative that all treatments be documented in the construction first aid log (see, for example, Figure 4-6). Each Name of Company project/jobsite will have medical services available either on the jobsite or at a location nearby. Emergency phone numbers will be posted on the jobsite for employees to call in the event of an injury or accident on the jobsite. Nurses will be available from _____ a.m. until _____ p.m. to respond to medical emergencies. First aid will be available from the Name Fire Department at all other times.

FIRST AID LOG
FOR

(COMPANY NAME)

INJURED WORKER'S NAME:_____

TRADE OF INJURED WORKER:_____

IMMEDIATE SUPERVISOR:_____

DATE:_____ AND TIME _____ OF INJURY

CAUSE OF INJURY:_____

BODY PART INJURED:_____

NATURE OF INJURY:_____

TREATMENT RENDERED:_____

_____RETURNED TO WORK _____SENT HOME _____SENT TO HOSPITAL

INJURED WORKER'S NAME:_____

TRADE OF INJURED WORKER:_____

IMMEDIATE SUPERVISOR:_____

DATE:_____ AND TIME _____ OF INJURY

CAUSE OF INJURY:_____

BODY PART INJURED:_____

NATURE OF INJURY:_____

TREATMENT RENDERED:_____

_____ RETURNED TO WORK _____ SENT HOME _____ SENT TO HOSPITAL

INJURED WORKER'S NAME:_____

TRADE OF INJURED WORKER:_____

IMMEDIATE SUPERVISOR:_____

DATE:_____ AND TIME _____ OF INJURY

CAUSE OF INJURY:_____

BODY PART INJURED:_____

NATURE OF INJURY:_____

TREATMENT RENDERED:_____

_____ RETURNED TO WORK _____ SENT HOME _____ SENT TO HOSPITAL

INJURED WORKER'S NAME:_____

TRADE OF INJURED WORKER:_____

IMMEDIATE SUPERVISOR:_____

DATE:_____ AND TIME _____ OF INJURY

CAUSE OF INJURY:_____

BODY PART INJURED:_____

NATURE OF INJURY:_____

TREATMENT RENDERED:_____

_____RETURNED TO WORK _____ SENT HOME _____ SENT TO HOSPITAL

Figure 4-6. First aid log form

Emergency Procedures and Response

Fire is one of the most hazardous situations encountered on a project/jobsite because of the potential for large losses. Prompt reaction to and rapid control of any fire is essential. Name of Company is responsible to provide fire protection procedures for each project/jobsite and assure that they are followed. It is the supervisor's/foreperson's responsibility to review all aspects of the firefighting and fire prevention program with his/her workers. The program should provide for effective firefighting equipment to be available without delay and be designed to effectively meet all fire hazards as they occur.

Some emergencies may require company personnel to evacuate the project/ jobsite. In the event of an emergency that requires evacuation from the workplace, the signal will be a Describe the Actual Sound that will be Used, all employees are required to go to the area adjacent to the project that has been designated as the "safe area." The safe area for this project is located Description of Location.

This is a sample program to provide a foundation on which to build your own safety and health program. It is now up to you to develop and implement an effective safety and health program. You can build a more comprehensive program or pare down the model to meet your specific needs.

SAFETY AND HEALTH PROGRAM EVALUATION

It is always appropriate to evaluate your present safety and health program. It is also appropriate to assess your safety and health programs progress as it evolves. It is useful to have an instrument or checklist which can be applied in a consistent manner to assist in this evaluation process. The following instrument found on pages 97 and 98 can be used or adapted to serve this purpose.

EVALUATION OF YOUR RESPONSES

If you have answered "YES" to between:

> 58 to 64 items — You have an excellent program that needs minimal effort to improve it.
>
> 51 to 57 items — You have a very good program, but you should strengthen it by addressing those areas answered "NO."
>
> 45 to 50 items — Your program is average, and definitely needs review and improvement.
>
> 38 to 44 items — Your program is very weak and needs immediate attention.
>
> Less than 37 — Your program is unacceptable and the lack of safety and health at your jobsites will cost you money.

This is a rough evaluation of your safety and health program. You may have addressed some items on the checklist in other ways and thus have a better program than is indicated by the checklist. You may have had a high score on this checklist, which may indicate that you have a very good program on paper, but you may fail to effectively implement your program.

SAFETY AND HEALTH PROGRAM EVALUATION

In order to gain some insight into the comprehensiveness of your company's safety and health program, answer the following questions "Yes" or "No." It is imperative that you answer as accurately as possible since the only person you would be fooling is yourself.

Health and Safety Program Management:

_____ 1. Is there a safety policy signed and dated by top management?
_____ 2. Is someone assigned responsibility for health and safety?
_____ 3. Does a health and safety manual or handbook exist?
_____ 4. Is a set time devoted to health and safety during management meetings?
_____ 5. Are health and safety rules and regulations established for all employees and/or specific jobs?
_____ 6. Are supervisors held accountable for health and safety during merit pay evaluations?
_____ 7. Are there a set of specific goals for safety and health (revised yearly)?

Inspections/Audits:

_____ 8. Are safety and health inspections conducted?
_____ 9. Are H&S inspections conducted on at least a monthly basis?
_____ 10. Are unsafe conditions or hazards found and corrected immediately?
_____ 11. Is a preventive equipment maintenance program in place?
_____ 12. Do operators of equipment perform daily inspections?
_____ 13. Is good housekeeping prevalent on all jobsites?
_____ 14. Does monitoring for health hazards occur?
_____ 15. Are written inspection reports completed?
_____ 16. Are inspection reports disseminated and open to everyone?

Job Observations:

_____ 17. Are job observations done?
_____ 18. Do job observations result in new work practices, workplace design, training, retraining, task analysis, or JSAs or SOPs.
_____ 19. Are job observations done to improve work practices?

Illness and Injury Investigations:

_____ 20. Are all incidents involving injury or illness investigated?
_____ 21. Are all incidents of equipment damage investigated?
_____ 22. Are all near misses investigated?
_____ 23. Are written reports generated for all incidents?
_____ 24. Are preventive recommendations made?
_____ 25. Are preventive recommendations implemented?
_____ 26. Do employees review incident reports?
_____ 27. Is incident data analyzed to determine illness and injury trends?

Task Analysis:

_____ 28. Do inspections, job observations, and incident investigations result in a task analysis?
_____ 29. Do task analyses result in changes in work practices or workplace design?

SAFETY AND HEALTH PROGRAM EVALUATION (continued)

_____ 30. Does task analysis facilitate the development of JSAs or SOPs?

_____ 31. Does task analysis result in new training or retraining?

Training:

_____ 32. Do <u>all</u> employees receive health and safety training?

_____ 33. Do employees receive site-specific training?

_____ 34. Are employees given job-specific or task-specific training?

_____ 35. Do both classroom and on-the-job (OJT) training occur?

_____ 36. Do management and supervisors receive health and safety training?

_____ 37. Are training records maintained?

Personal Protective Equipment (PPE):

_____ 38. Does the work require the use of PPE?

_____ 39. Is proper PPE available?

_____ 40. Have employees been trained in the use of PPE?

_____ 41. Do you have a respirator program (29CFR 1910.134), if needed?

_____ 42. Are the rules and use of PPE enforced?

Communication/Promotion of Health and Safety:

Communication

_____ 43. Are health and safety measures visible?

_____ 44. Are company/contractor health and safety goals communicated?

_____ 45. Are health and safety meetings held (i.e., toolbox)?

_____ 46. Do health and safety talks convey relevant information?

_____ 47. Are personal health and safety contacts made?

_____ 48. Are bulletin boards used to communicate health and safety issues?

_____ 49. Do those responsible for health and safety request feedback?

_____ 50. Are health and safety suggestions given consideration and/or used?

_____ 51. Are supervisors interested in health and safety?

Promotion

_____ 52. Is there an award/incentive program tied to safety and health?

_____ 53. Are health and safety exhibits or posters used?

_____ 54. Are paycheck stuffers on safety and health used?

_____ 55. Have safety and health handouts been used?

_____ 56. Are employees recognized for contributions toward the health safety program?

Personal Perception:

_____ 57. Do you, the company/contractor, extend considerable effort to assure an effective health and safety program?

_____ 58. Do supervisors support and enforce all aspects of the health and safety program?

_____ 59. Do employees insist on doing all tasks in a safe and healthy manner?

Off-the-Job Health and Safety:

_____ 60. Is off-the-job health and safety promoted as part of the total health and safety program?

_____ 61. Does the company/contractor provide a wellness or fitness program?

_____ 62. Does the company/contractor have an employee assistance program (EAP)?

_____ 63. Does the company/contractor foster and encourage healthier lifestyles?

OTHER REQUIRED WRITTEN PROGRAMS

Many of the OSHA regulations have requirements for written programs that coincide with the regulations. This may become a bothersome requirement to many within the construction industry. But, failure to have these programs in place and written is a violation of the regulations and will result in a citation for the company. At times it is difficult to determine which regulations require a written program but, in most cases, the requirements are well known. For example, everyone should know that a written "Hazard Communication Program" is required, since this has been one of the most frequent violations cited by OSHA.*

The Hazard Communication Standard's requirement for a written program must include the following:

- A listing of hazardous chemicals on the jobsite.

- The method the employer will use to inform employees of the hazards associated with nonroutine tasks involving hazardous chemicals.

- How the employer plans to provide employees or other companies on the jobsite with Material Safety Data Sheets (MSDSs), such as making them available at a central location.

- The method the employer will use to inform employees of other companies on the jobsite about their labeling system.

- How the employer will inform his/her workers of the company's labeling system.

- How the employer intends to train workers on hazardous chemicals.

A sample written Hazard Communication Program can be found in Appendix C.

Some of the other OSHA regulations that require written programs are

1. Process Safety Management of Highly Hazardous Chemicals (29 CFR 1926.64).

2. Bloodborne Pathogens/Exposure Control Plan (29 CFR 1910.1030).

3. Emergency Action Plan (29 CFR 1926.35).

4. Respirator Program (29 CFR 1910.134).

5. Lockout/Tagout/Energy Control Program (29 CFR 1910.147).*

6. Hazardous Waste and Emergency Response/Site Specific S&H Program, Training Program, and Personal Protective Equipment Program 29 CFR 1926.65).

7. Hazard Communication Program (29 CFR 1926.59).

8. Fall Protection Plan (29 CFR 1926.500).

9. Confined Space "Permit Entry" Plan (29 CFR 1910.146).*

* Not a construction industry regulation.

The specific requirements for the content of written programs vary with the regulation. The respirator standard requires that the following exist:

- Written Standard Operating Procedures.

- Program Evaluation Procedures.

- Respirator Selection Procedures.

- Training Program.

- Fit Testing Requirements and Procedures.

- Inspection, Cleaning, Maintenance, and Storage Procedures.

- Provision for Medical Examinations.

- Process for Work Area Surveillance.

- The Acceptable Air Quality Standards.

- The Use of Approved Respirators.

The Bloodborne Pathogen Standard requires an Exposure Control Plan in which the employer is to identify workers performing routine tasks and procedures in the workplace who may be exposed to blood or other potentially infectious materials. The employer is to develop a schedule of how and when the provisions of the standard will be implemented and develop a plan of action to be taken when an exposure incident occurs. These guidelines give employers more flexibility in developing their plan with fewer required elements than the Respirator Program.

Although a Fall Protection Plan is not called for by the regulations, you may still want to develop one if your workers are exposed to falls. The following is a sample Fall Protection Plan.

FALL PROTECTION PLAN

OSHA - 29 CFR 1926.500, 501, 502, & 503

I.   INTRODUCTION

In accordance with OSHA Regulations (29 CFR 1926.500, 501, 502, & 503), the following plan has been developed for employees of the Name of the Construction Company.

This plan has been established to provide compliance guidelines for the employer's effort to meet the requirements of OSHA Fall Protection Standard 1926.500, 501, 502, & 503. The program applies to all employees who are required to climb and work at heights or elevated work areas.

II.   POLICY

The Name of Construction Company is concerned with the safety and health of all employees. The hazards and risks involved with climbing and working at heights is of specific concern in regard to this program.

The Name of Construction Company expects that all employees shall follow the procedures and policies set forth in this Fall Protection Plan.

The company is committed to ensuring that all employees are trained, are familiar with the applicable regulations, can recognize the common hazards, are provided the appropriate personal protective equipment, and follow fall protection procedures.

III. RESPONSIBILITY

Responsible Person's Name, Title, and Telephone Number is responsible for the implementation, administration and enforcement of this Fall Protection Plan. Supervisors are responsible for enforcing all aspects of the Fall Protection Plan.

IV. RULES

1. *Personal Fall Protection Equipment shall be used when a free-fall hazard exceeds six feet.*

2. *A full body harness and shock-absorbing lanyards shall be worn if you can free fall more than six feet with your equipment and where immediate post-fall self-recovery is not possible.*

3. *All employees, or the group site safety officer, shall complete the <u>Working at Heights Checklist</u> (see Figure 4-7) to determine the safety of the climb or elevated work.*

4. *Fall protection devices that have been subjected to shock loading that was imposed during fall arresting shall be removed from service. Notify the group supervisor and the Health and Safety Coordinator immediately.*

V. PROCEDURES

1. *Upon arriving at a jobsite, survey the area for potential hazards.*

2. *Complete the <u>Working at Heights Checklist</u>.*

   (Note: Each employee or supervisor should maintain a three-ring binder of all completed checklists.)

3. *Assure that you have all the equipment needed to accomplish the assigned task.*

4. *Make an informed decision on the safety of completing this task or not attempting the task.*

5. *Contact your supervisor immediately if you cannot attempt the task or have questions regarding safety or health issues.*

VI. TRAINING

<u>The Name of the Company's</u> workers who have work responsibilities that require climbing and working at heights shall receive training on the following topics. This training will be for all employees, new, transferred, or reassigned employees, and any time that the work procedures are changed, a new hazard has been identified, or new equipment or tasks are implemented. In addition, periodic refresher training will be provided.

- *Hazard Recognition.*
- *Policies and Procedures.*
- *Regulatory Requirements.*
- *Safe Climbing Procedures.*
- *Working at Heights Procedures.*
- *Fall Protection Systems.*
- *Use of Personal Protective Equipment.*
- *Donning and Doffing to Fall Protection Equipment.*

- *Inspection and Care of Fall Protective Equipment.*
- *Emergency Procedure and Rescue Procedures.*

VII.   HAZARD REPORTING

Employees who witness hazards or have health and safety concerns are asked to complete a Hazard Report Form (see Figure 4-8) which will include the type and the location of the hazard, and suggestions for intervention or prevention. (See Hazard Report Form.)

VIII.   INCIDENT/NEAR-MISS INVESTIGATIONS

All fall incidents will be investigated by your supervisor, Health and Safety Coordinator, and OSHA Review Officer. This includes actual fall incidents, as well as near misses. The investigation will help identify hazards, assist in revising any faulty procedures, and implementing any interventions. You shall report all fall or near-miss incidents to your supervisor. Use the Report of Incident/Near-Miss Investigation Form (see Figure 4-9).

IX.   EMERGENCIES

Follow emergency procedures that are set forth by the jobsite where you are working. If no emergency procedures exist, you must identify the closest rescue service and provide the telephone number to an authorized standby person who can summon assistance if needed.

The previous written fall protection plan has most, if not all, of the components of a good safety and health program or plan. It states the policy of the company and fixes accountability/responsibility, denotes the rules and safe operating procedures to follow, delineates the training requirements, provides for hazard recognition/reporting and incident investigation, and describes how to handle emergencies. The fall protection plan could be shortened or lengthened to meet your specific needs.

Any written program or plan should be tailored for your operation. It should be long enough to serve your purpose but concise enough to be implemented, maintained, and enforceable. Do not develop a written program or plan if there does not truly exist a need within your company. If you must develop a mandated written program or plan, then put it to its intended use and make it an asset to your company.

WORKING AT HEIGHTS CHECKLIST

Name of Climber(s): _____ Date: _____

_____

_____

_____

Jobsite: _____

Work Area: _____

Worked to be Performed: _____

_____

Answer "Yes" or "No" to the following:

____ Has area been barricaded or fenced? (Caution is acceptable)
____ Are there potential or existing hazards? (Explain any checks made below)
____ Electrical
____ Mobile Equipment
____ Housekeeping
____ Poorly Constructed Scaffolds
____ Unsafe Fixed Ladder
____ Unsafe Ladder or Improperly Used Ladder
____ Lack of or Inadequate Guardrails
____ Overhead Hazard From Falling Objects
____ Unsecured Footing
____ Uncovered Opening
____ Gases, Vapors, Mists, or Fumes
____ Obstructed Walkways
____ Uneven Surfaces
____ Confined Spaces

Explanation: _____

_____

_____

____ Do you need chemically protective PPE?
____ Chemical Resistant Suit
____ Gloves
____ Rubber Boots
____ Protective Goggles
____ Is a respirator needed or required?
____ Are you wearing all required personal protective equipment?
____ Have you inspected all your fall protection equipment?
____ Is fall arrester protection available?
____ Is there adequate anchorage for tying off?
____ Have you looked up/down visually for hazards?
____ Have you checked with the facility management regarding unique hazards?
____ Is fall rescue provided or at least readily available in case of a fall (radio, emergency numbers, etc.)?
____ Are adverse weather conditions a safety factor?
____ Have you evaluated the jobsite as safe to perform your assigned work?
____ Did you decide the site was too hazardous to accomplish your work?*   Why?

_____

_____

_____

* CONTACT YOUR SUPERVISOR IMMEDIATELY

I certify that I have reviewed all potential hazards and have used this checklist to evaluate the safety and health of this worksite and task.

Signature: _____

Title: _____

Figure 4-7. Working at heights checklist

HAZARD REPORT FORM

*EMPLOYEE NAME:*_____*DATE:*_____

*JOB TITLE:*_____ *JOBSITE:*_____

TELEPHONE NUMBER: _____

1. Describe the Hazard that Exists:

_____

2. Where is the Hazard Located?

_____

3. Does the Hazard Violate Known State or Federal Regulations?
    ___ *Yes*   ___*No*
    (If Yes, Explain)

_____

4. Has the Hazard Already had an Adverse Effect on Any Workers?
    ___*Yes*   ___*No*
    (If Yes, Explain)

_____

*5. What are the Duties Performed that Expose Workers to the
Hazard?*

_____

6. How Long has the Hazard Been Present?

_____

7. What Action Should be Taken to Correct the Hazard?

_____

SUBMIT COMPLETED FORM TO YOUR SUPERVISOR.

(If you need additional space, use the back of this form.)

Figure 4-8. Hazard report form

REPORT OF INCIDENT/NEAR-MISS INVESTIGATION FORM

Employee Name(s): _____

Jobsite: _____ *Area :* _____

Title(s): _____

Date of Incident: _____ *Time:* _____ A.M. or P.M.

Day of Week: _____

When Was Supervisor Notified _____

_____

1. Description of Incident (Include Place):

_____

2. Description of Injury(ies) to Employee(s) or Other(s)

_____

3. Description and Estimate of Damage to Property, if Any:

_____

4. Probable Cause of Incident:

_____

5. Corrective Action Needed to Prevent Recurrence of Incident:

_____

6. Witnesses, if Any:

_____

7. Include a Sketch if Appropriate:

_____

* SUBMIT COMPLETED FORM TO YOUR SUPERVISOR

(If you need additional space, use the back of this form)

Figure 4-9. Report of incident/near-miss investigation form

REFERENCES

[1] Reese, Charles D., *Mine Safety and Health for Small Surface Sand/Gravel/Stone Operations,* UConn/LEC, Storrs, CT, 1997.

[2] *OSHA Instruction STD 3-1.1.*, Office of Construction and Maritime Compliance Assistance, OSHA, Washington, D.C., June 22, 1987.

[3] *Model Construction Safety and Health Program,* Laborers' Health and Safety Fund of North America, Washington, D.C., 1994.

Chapter 5

ANALYZING CONSTRUCTION HAZARDS AND ACCIDENTS/INCIDENTS

Charles D. Reese

Analysis often implies mathematics, but calculating math equations is not the major emphasis when attempting to address hazards or accidents/incidents which occur within the construction industry.

Analysis in the context of this chapter means taking time to systematically examine the construction worksite's existing or potential hazards. This can be accomplished in a variety of ways.

Prior to the beginning of a project, the owner, construction manager, general contractor, and subcontractors may meet to assess hazards. At this time the safety and health aspects of the entire project can be worked out regarding the requirements, responsibilities, safety and health goals, and the initiatives required to provide a safe workplace. However, this type of analysis should not replace an actual worksite analysis.

The worksite analysis may utilize the site superintendent, the forepersons, and the jobsite safety and health officer. A worksite analysis may include the entire site, or be broken down into specific operations or jobs. As work progresses on a jobsite, analysis may need to be an ongoing process. At times the foreperson may feel that unique jobs or work practices need to be examined in more detail.

At times there may be a need for a detailed analysis of a specific job or task. This is called a job safety analysis (JSA). With this type of analysis, each step of a particular job or task is looked at in detail and the potential hazards which do or could exist are identified. Even when these site and job analyses are done, accidents and incidents still may occur. When they do occur, they need to be examined and analyzed to determine how to prevent them.

If a number of accidents or incidents transpire, you will need to analyze these events to determine if similarities or trends are occurring. This is why it is important to investigate all accidents, incidents, or even near misses. This is the most mathematical of all the analysis techniques since you will be using the number of occurrences, comparing number of events, and evaluating against national statistics.

This chapter provides information on each of these types of analyses related to construction. It presents each analysis as a tool to be used when needs arise regarding these varied facets of the construction process.

PRECONSTRUCTION CONFERENCE

The construction manager, who may be the general contractor, prime contractor, owner, engineer, or architect, needs to qualify the subcontractors and conduct a preconstruction conference to deal with the most serious hazards which may be encountered on the project. Planning must be done to eliminate or develop preventions/interventions regarding these hazards. The conference also deals with who will supply the materials and equipment and when they are to be in place at the project site.

Regarding safety, it is imperative to understand who is responsible for performing specific safety activities (e.g., confined space entry.) This could affect other subcontractors and cause work to get out of sequence. In these instances, all contractors and subcontractors are required to cooperate in order to assure that a safe operation occurs and the schedule is brought back into the normal project cycle.

All subcontractors should be required to submit an approved safety plan which would include identifying possible serious hazards, as well as the persons responsible for the safety of the subcontractors' workers. The plan details how worksite hazards will be eliminated, made safe, or safeguarded. Also, the subcontractors need to provide training on the hazards which the workers will encounter on the project and explain how hazard information will be transmitted on this project (e.g., tool box talks, bulletin boards, or training).

The construction manager must take control of the worksite safety and assure subcontractors that he intends to evaluate and enforce compliance with the agreements made in the preconstruction safety meeting. The construction manager should agree to create and maintain a safety status board which informs subcontractors and workers of the changing safety status and any issues concerning the project site. The safety officer, who should be identified to the workers by the construction manager, needs to assure that all work crews receive meaningful and specific safety information.

The safety officer may have a number of specific functions such as qualifying each subcontractor according to previously developed criteria, maintaining a daily project safety bulletin board, convening preplanned safety meetings for the prime contractor and subcontractors, scheduling work practice observations and safety inspections to determine compliance with preplanning safety practices, assisting with the planning, recognition, and control of safety and health hazards, and preparing a site-specific safety plan which includes the safety of equipment and work practices. This should require all contractors to achieve maximum injury/illness prevention, assure compliance with OSHA and state occupational safety and health regulations, act as a resource for management regarding safety and health, and develop and maintain a hazard communication program.

During the preconstruction conference it is a good idea to include labor and each trade representative who will be involved in the project. For everyone's benefit, the topics that should be covered are those which delineate the steps and procedures to follow in order to insure jobsite safety. Everyone should know who is responsible for safety and health, as well as the duties of supervisors and forepersons. This is the time for safety and health suggestions and also the time to discuss the procedures for reporting hazards and how remedies will occur. It is far better to identify and correct hazards than to have complaints filed with OSHA.

The accident investigation procedures should be outlined for everyone at this time and, if joint labor/management safety committees are part of the project, all those involved in the project need to know their extent of authority, responsibility, and purpose. The scheduling/intent of safety meetings and how often they will be held should be discussed. If there is a need for safety training in order to address specific hazards on the project or, if unique special training or precautions are needed for a certain phase, this should be accomplished during this conference.

Conference participants should discuss the requirements for safety equipment and when and who must use it, as well as specifics on its proper use. In addition, discussion should include the care of ill or injured workers, the workers' compensation carrier, and the return-to-work policy following an injury. What constitutes "light duty" following an injury should also be established.

The preconstruction conference is the time to address all safety and health concerns, spell out the expectations and responsibilities, and delineate the safety and health policies and procedures to be followed during the project period, as well as the enforcement, punishment, and disciplinary procedures for noncompliance with project site safety and health. It is a good idea to have a preconstruction conference checklist to assure that all aspects of project safety and health have been covered (Figure 5-1).

WORKSITE HAZARD ANALYSIS

Worksite analysis is the process of identifying hazards related to a project's construction activities and the construction site. Identify the workplace hazards before determining how to protect employees. In performing worksite analyses, consider not only hazards that currently exist in the workplace, but also those hazards that could occur because of changes in operations or procedures or because of other factors, such as concurrent work activities. First, perform hazard analyses of all construction projects prior to the start of work, determine the hazards involved with each phase of the project, and perform regular safety and health site inspections.

Second, require supervisors and employees to inspect their workplace prior to the start of each workshift or new activity, investigate accidents and near misses, and analyze trends in accident and injury data.

When performing a worksite analysis, all hazards should be identified. This means conducting comprehensive baseline worksite surveys for safety and health and periodic comprehensive update surveys. You must analyze planned and new facilities, processes, materials, and equipment, as well as perform routine job hazard analyses. This also means that regular site safety and health inspections need to be conducted so that new or previously missed hazards and failures in hazard controls are identified.

A hazard reporting/response program should be developed to utilize the employees' insight and experience in safety and health protection. The employees' concerns should be addressed, and a reliable system should be provided whereby employees, without fear of reprisal, may notify management personnel about conditions that appear hazardous. These notifications should receive timely and appropriate responses and the employees should be encouraged to use this system.

Another way to maintain a worksite hazard analysis is to investigate accidents and "near miss" incidents so that their causes and means for their prevention are identified. By analyzing injury and illness trends over a period of time, patterns with common causes can be identified and prevented.

Construction projects may require different types of hazard analyses, depending on the company's role in the project, the project's size and complexity, and the nature of associated hazards. You may choose to use a project hazard analysis, a phase hazard analysis, and/or job safety assessment.

A project hazard analysis (preliminary hazard analysis) should be performed for each project prior to the start of work and should provide the basis for the project-specific safety and health plan. The project hazard analysis should identify the following:

- The anticipated phases of the project.

- The types of hazards likely to be associated with each anticipated phase.

- The control measures necessary to protect site workers from the identified hazards.

- Those phases and specific operations of the project for which activities or related protective measures must be designed, supervised, approved, or inspected by a registered professional engineer or competent person.

- Those phases and specific operations of the project that will require further analyses are those that have a complexity of hazards or unusual activities involved,

DATE OF CONFERENCE:_____

LOCATION OF CONFERENCE:_____

CONSTRUCTION MANAGER'S:

    Name:_____

    Address: _____

    _____

    Telephone: _____

    Jobsite Telephone: _____

    Superintendent: _____

JOBSITE:

    Location: _____

    Date of Start: _____

    Duration of the Project:_____

PERSONNEL INVOLVEMENT:

    _____Subcontractor.

    _____Forepersons.

    _____Workers.

EXCAVATION:

    _____Equipment.

    _____Type of Soil.

    _____Shoring System.

    _____Trench Box.

    _____Blasting.

FOUNDATION:

    _____Footing for Forms.

    _____Type of Forms.

    _____Pouring Procedures.

    _____Curing Times.

    _____Schedule for Form Removal.

    _____Backfill and Soil Stability and Solidification.

ERECTION:

    _____Types of Cranes and Hoists.

    _____Safety Harnesses, Nets and Decking Requirements, and Procedures.

    _____Worksite Access (Personnel Hoists, Stairways, Walkways).

    _____Scaffolding Procedures and Responsibilities.

    _____Welding Procedures.

SANITATION:

    _____Sanitary Facilities.

    _____Potable Water.

    _____Shower and Change Areas.

ACCIDENT PREVENTION:

    _____New Employee Orientation to Hazards and Safe Work Practices.

    _____Accident Prevention Training Programs.

    _____Safety Bulletin Boards.

    _____Location of Emergency Telephone Numbers.

    _____Monthly Safety Meetings.

    _____First Aid Supplies Locations.

    _____Electrical Grounding Procedures.

    _____Equipment Maintenance Program.

    _____Machine and Equipment Guarding.

    _____Emergency Procedures and Warning Signals.

    _____Posting of Employer Responsible for OSHA Compliance.

    _____Hazardous Materials Response and Cleanup.

    _____Housekeeping.

PROTECTIVE EQUIPMENT:

    _____Safety Toed Shoes.

    _____Eye Protection.

    _____Hearing Protection.

    _____Respirators.

    _____Toxic Dust, Fumes Vapors, Mists, or Aerosols (Ventilation or Respirators).

FIRE PREVENTION:

    _____Fire Prevention Plan.

    _____Fire Extinguishers.

    _____Emergency Evacuation Plan.

Figure 5-1. Preconstruction conference checklist

there is uncertainty concerning the site conditions that will be present at the time of construction, or there is concern for construction methods that will be used to complete the phase or operation.

A phase hazard analysis may be performed for those phases of the project for which the project hazard analysis has identified the need for further analysis, and for those phases of the project for which construction methods or site conditions have changed since the project hazard analysis was completed. The phase hazard analysis is performed prior to the start of work on that phase of the project and is expanded, based on the results of the project hazard analysis, by providing a more thorough evaluation of related work activities and site conditions. As appropriate, the phase hazard analysis should include

- Identification of the specific work operations or procedures.

- An evaluation of the hazards associated with the specific chemicals, equipment, materials, and procedures used or present during the performance of that phase of work.

- An evaluation of the safety and health impacts of any changes in the schedule, work procedures, or site conditions that have occurred since the performance of the project hazard analysis.

- Identification of specific control measures necessary to protect workers from the identified hazards.

- Identification of specific operations for which protective measures or procedures must be designed, supervised, approved, or inspected by a registered professional engineer or competent person.

A job safety assessment or analysis should be performed at the start of any task or operation. The designated competent person should evaluate the task or operation to identify potential hazards and determine the necessary controls. This assessment should focus on actual worksite conditions or procedures that differ from or were not anticipated in the related project or phase hazard analysis. In addition, the competent person shall ensure that each employee involved in the task or operation is aware of the hazards related to the task or operation and of the measures or procedures that they must use to protect themselves. Note: the job safety assessment is not intended to be a formal, documented analysis, but instead is more of a quick check of actual site conditions and a review of planned procedures and precautions. A more detailed explanation of job safety analysis is provided in the next section.

ACCIDENT/INCIDENT ANALYSIS

Accident/incident analysis is a technique which has been used by the National Mine Safety and Health Academy in Beckley, West Virginia to examine mining related accidents and incidents. There are many similarities between the mining and construction industries; therefore, it makes this analysis technique also applicable to the construction industry.

In the construction industry, thousands of accidents and incidents occur each year throughout the United States. Most are caused by the failure of people, equipment, supplies, or surroundings to interact as expected. Investigations are made to determine how and why these failures occurred. By using the information gained through an investigation, a similar, perhaps more disastrous, accident may be prevented. Thus, accident and incident investigations or analyses are conducted with prevention in mind.

Many factors contribute to the causes of construction accidents. There are contributing factors, causal factors, mitigating factors, and multiple-cause outcome factors. For the purpose of this book, accidents and incidents are used interchangeably to avoid philosophical debates.

The following are working definitions of accident and incident:

Accident is any undesired event resulting in personal injury and/or property damage and/or equipment failure.

Incident includes all of the above, as well as adverse production effects.

Accidents and incidents are usually complex. They may result from ten or more individual or casual events. Elimination of one or more of those events may result in no accident at all (or a less serious one). Analysis of an accident or incident requires knowledge of many factors. For example, accident investigators may want to know

1. The location
2. Time of day
3. Accident type
4. Victim
5. Nature of injury
6. Released energy.
7. Equipment being used

8. Hazardous materials
9. Unsafe acts
10. Unsafe conditions
11. Policies and decisions
12. Personal factors
13. Environmental factors
14. Impact of others on the incident

All of these may have contributed to the accident. The intent of this section is to systematically guide an accident investigator through a three-level approach to accident cause, identification, and analysis. Once accident causes have been identified, recommendations, and specific preventive measures for each casual factor may be developed.

Accidents: Why They Happen

In order for work to progress in the workplace, certain components must interact. These components are people (WORKERS), equipment (MACHINES), and supplies (MATERIALS). These three interacting components use established procedures (METHODS) to accomplish the job (TASK). When these components interact according to planned methods, the result will most likely be efficient or safe production. However, at some point, something unplanned may happen. This usually results from a change in the workers, machines, and material interaction. This change causes deviation from normal procedure and the result may be an ACCIDENT/INCIDENT. To prevent this incident, it is necessary to identify the exact point(s) where change from normal procedures occurred.

Construction worksites deal with all these components which may be factors in the large numbers of accidents/incidents occurring on construction projects.

Tasks are not performed in isolation. The PHYSICAL ENVIRONMENT constantly influences the interaction of workers, machines, and materials. These influences may be either helpful or harmful. Other factors also affect task performance. These factors are part of the SOCIAL ENVIRONMENT (e.g., government agencies, unions, families and friends, company management). Social environment factors may, likewise, be helpful or harmful. In this accident/incident analysis concept, the effects of all the interactions may result in either a successful job completion or an accident. The following are examples of the consequences of these types of influences.

After Amy Rice had worked in her job for a number of years, Amy's supervisor decided to have her change several of her regular procedures. Because she did not understand the instructions, she made a mistake in the operation of her machine and was injured. In this example, the key interaction factor was communication between the supervisor and employee. Failure to communicate contributed to Amy's accident.

Tom Moore, an equipment operator, was upset and distracted because of a family problem. He lost his concentration on the job, altered his routine, and damaged an expensive piece of equipment. Interaction between the worker and his family was a key factor in this case.

In both examples, changes in procedure preceded the accident. All changes in procedures/methods that preceded an accident should be identified. Identification of procedural changes and what caused them may help initiate specific preventive measures.

Analyzing Accidents/Incidents

To analyze an accident or incident it is necessary to know, at a minimum, the type of accident, kind/nature of any personal injury, property damage, or equipment failure. Some examples of accident types include

1. Struck by.

2. Struck against.

3. Caught in, under, or between.

4. Rubbed or abraded.

5. Bodily reaction.

6. Overexertion.

7. Contact with electricity.

8. Contact with temperature extremes.

9. Contact with radiation, caustics, toxic, or noxious substances.

10. Fall from elevation.

11. Fall to same level.

12. Public transportation.

13. Motor vehicle.

Personal injuries include broken bones, lacerations, etc. Property is damaged by fire, water, collisions, etc. Equipment failures include hydraulic leaks, metal fatigue, etc. All these have occurred at construction worksites.

Direct Causes

When making a detailed analysis of an accident or incident, consider the release of energy and/or hazardous material as a direct cause. Energy or hazardous material is considered to be the force which results in injury or other damage at the time of contact. It is important to identify the direct cause(s). In order to prevent injury, it is often possible to redesign equipment or facilities and provide personal protection against energy release or release/contact with hazardous materials. Some examples of direct causes in the form of energy or hazardous materials sources are found in Table 5-1.

Table 5-1

Sources of Direct Causal Agents

ENERGY SOURCES	HAZARDOUS MATERIALS

1. Mechanical:
     Machinery
     Tools
     Noise
     Explosives
     Moving objects
     Strain (Self)
2. Electrical:
     Uninsulated conductors
     High voltage sources
3. Thermal:
     Flames
     Hot surfaces
     Molten metals
4. Chemical:
     Acids
     Bases
     Fuels
     Explosives
5. Radiation:
     Lasers
     X-rays
     Microwave
     Radiation Sources
     Welding
	1. Compressed or liquefied gas:
	     Flammable
	     Non-flammable
	2. Corrosive material
	3. Flammable material:
	     Solid
	     Liquid
	     Gas
	4. Poison
	5. Oxidizing material
	6. Dust

Figure 5-2. Electrical energy sources at construction worksites are powerlines

Indirect Causes

Unsafe acts and/or unsafe conditions comprise indirect causes of accidents and/or incidents. These indirect causes can inflict injury, property damage, or equipment failure. They allow the energy and/or hazardous material to be released. Unsafe acts can lead to unsafe conditions and vice-versa. Examples of unsafe acts and unsafe conditions are found in Table 5-2.

Table 5-2

Unsafe Acts and Conditions

UNSAFE ACTS	UNSAFE CONDITIONS
1. Failure to wear personal protective equipment	1. Congested work areas
2. Failure to warn coworkers or to secure equipment	2. Defective machinery/tools
3. Ignoring equipment/tool defects	3. Improperly stored explosive or hazardous materials
4. Improper lifting	4. Poor illumination
5. Improper working position	5. Poor ventilation
6. Improper use of equipment: At excessive speeds Using defective equipment Servicing moving equipment	6. Inadequate supports/guards
7. Operating equipment without authority	7. Poor housekeeping
8. Horseplay	8. Radiation exposure
9. Making safety devices inoperable	9. Excessive noise
10. Drug misuse	10. Hazardous atmospheric conditions
11. Alcohol use	11. Dangerous soil conditions
12. Violation of safety and health rules	12. No firefighting equipment
13. Failure to wear assigned PPE	13. Unstable work areas/platforms

Basic Causes

Some accident investigations result only in the identification and correction of indirect causes, but indirect causes of accident are symptoms that some underlying causes exist which are often termed basic causes. By going one step further, accidents can best be prevented by identifying and correcting the basic causes. Basic causes are grouped in Table 5-3.

Figure 5-3. An unsafe/congested work area

Table 5-3

Basic Causes

POLICIES AND DECISIONS

    1.   Safety policy is not
- in writing
- signed by top management
- distributed to each employee
- reviewed periodically

    2.   Safety procedures do not provide for
- written manuals
- safety meetings
- job safety analysis
- housekeeping
- safety audits/inspections

    3.   Safety is not considered in the procurement of
- supplies
- equipment
- services

    4.   Safety is not considered in the personnel practices of:
- selection
- hiring
- training
- placement
- medical surveillance
- authority
- responsibility
- accountability

Table 5-3

Basic Causes (*Continued*)

PERSONAL FACTORS
    1.   Physical
        • inadequate size
        • inadequate strength
        • inadequate stamina
    2.   Experiential
        • insufficient knowledge
        • insufficient skills
        • accident records
        • unsafe work practices
    3.   Motivational
        • needs
        • capabilities
    4.   Attitudinal
        • toward others —
          people
          company
          job
        • towards self —
          alcoholism
          drug use
          emotional upset
    5.   Behavioral
        • risk taking
        • lack of hazard awareness

ENVIRONMENTAL FACTORS

    1.   Unsafe facility design
        • poor mechanical layout
        • inadequate electrical system
        • inadequate hydraulic system
        • crowded limited access ways
        • insufficient illumination
        • insufficient ventilation
        • lack of noise control
    2.   Unsafe operating procedures
        • normal
        • emergency
    3.   Weather
    4.   Geographical area

When basic causes are eliminated, unsafe acts/unsafe conditions may not occur. (For example: Charlie South used a broken hammer because no replacement hammer existed in the tool bin.) In Charlie's case, the basic cause, lack of inventory control procedures, set up his subsequent unsafe act.

Accidents, thus, have many causes. Basic causes lead to unsafe acts and unsafe conditions (indirect causes). Indirect causes may result in a release of energy and/or hazardous material (direct causes). The direct cause may allow for contact, resulting in personal injury and/or property damage and/or equipment failure (accident).

An analysis of an accident does not stop with the identification of the direct, indirect, and basic causes of the accident or incident. In order to make positive gains from the event, changes should be made in the interaction of man, machines, materials, methods, and physical and social environments. These changes should result from the recommendations which are derived from the causes identified during the investigation. The goal of these changes is the prevention of future accidents and/or incidents similar to the one investigated.

When an accident investigation is received, the report should contain itemized information about the accident. As a minimum, the accident report should include the

- Location.
- Date.
- Time.
- Name(s) of injured.
- Accident/incident type.
- Description of accident/incident.
- List of property damage.
- List of equipment involved.
- Direct causes.
- Indirect causes.
- Basic causes.
- Recommendations for prevention of direct causes.
- Recommendations for prevention of indirect causes.
- Recommendations for prevention of basic causes.

The following Accident Report Form in Figure 5-4 may be used to guide an accident/incident analysis through the three-level approach in identifying causes and making recommendations for the prevention of future accident/incidents. This Accident Report Form will be useful only when you or your safety committee completes it for each accident/incident that occurs. Granted, this is a rather simplistic approach, but the causes of accidents and their prevention is not "rocket science." It does, however, require more than words of commitment and support.

Accident Report Form

DEPARTMENT: _____
DATE OF ACCIDENT: _____
TIME OF ACCIDENT: _____

EMPLOYEE NAME: _____
EMPLOYEE AGE: _____
EMPLOYEE OCCUPATION: _____

LOCATION OF ACCIDENT: _____
ACCIDENT TYPE: _____
ACCIDENT CLASSIFICATION: _____

DESCRIPTION OF ACCIDENT/INCIDENT: _____

Itemize Personal Injury Involved: _____
Itemize Property Damage Involved: _____
Itemize Tools/Equipment Involved: _____

CAUSES

DIRECT

Energy Sources   Hazardous Materials

INDIRECT

Unsafe Acts   Unsafe Conditions

BASIC

Inadequate Policy and/or Decisions   Environmental and/or Personal Factors

RECOMMENDATIONS

DIRECT LEVEL

INDIRECT LEVEL

BASIC LEVEL

Figure 5-4. Accident report form – Courtesy of the National Mine Health and Safety Academy

The use of an accident/incident analysis approach affords you the opportunity to dissect your accident reports into the actual causal factors and then make recommendations for intervention and prevention. This type of analysis may trigger the need to more closely analyze a job or task which has been identified as having a high risk of producing hazards or injuries.

JOB SAFETY ANALYSIS

Job Safety Analysis (JSA) provides a mechanism by which a contractor, safety professional, or supervisor may take a detailed look at how a task or job is performed and its inherent hazards. Job Safety Analysis is a basic approach to develop improved accident prevention procedures. It employs the firsthand experience and cooperation of the workers and supervisor in the recognition, evaluation, and control of hazards. Make it an integral and important component of your safety and health program.

Job Safety Analysis (JSA) is a good technique to use for reviewing a job. Its purpose is to uncover inherent or potential hazards which may be encountered in the work environment. When properly used, the JSA will be an effective tool for training and orienting the new employee into the work environment. A JSA may also be used to retrain the older employee.

During the development of a JSA, those who supervise will learn more about the jobs they supervise. Workers who are encouraged to participate will develop a better attitude and knowledge of safety. The program will develop safer job procedures and create a better working environment.

A Job Safety Analysis includes five steps

1. Select a job.
2. Break the job down into steps.
3. Identify the hazards or determine the necessary controls for them.
4. Apply the controls to the hazards.
5. Evaluate the controls.

To increase the effectiveness of a JSA, a complete program should be developed in which management is involved with the workers, supervisors, and safety and health professionals. The following five major elements are necessary to establish and maintain the JSA program:

1. Management controls.
2. Identification of a method for job selecting.
3. Analysis of the job (perform JSA).
4. Incorporation of the JSA into the operational systems.
5. Monitoring the JSA program.

Management Controls

Management controls are needed to establish an effective JSA program. The entire program must be directed and supported through the management controls. Some aspects of management controls are

1.   Policies.

2.   Directives.

3.   Responsibilities (line and staff).

4.   Vigor and example.

5.   Accountability.

6.   Budget.

Establishing a Method for Selecting Jobs

Potential jobs for analysis should be tasks that have sequential steps and the work goal will be attained when these steps have been performed. Thus, the jobs for analysis should be those that are single-task oriented and not ones which require a multitude of tasks and machines/equipment.

Some jobs can be defined in general terms (e.g., constructing a building, building a bridge, digging a tunnel). Such broadly defined jobs are not suitable for a JSA. Similarly, a job may be defined in terms of a single action (e.g., pulling a switch, tightening a screw, pushing a button). These narrowly defined jobs are also sometimes not suitable for a JSA but occasionally, pulling a switch or tightening a screw may be very critical; in such instances, the job should be broken down into single action steps.

To effectively use the JSAs, a method must be established to select and prioritize the jobs to be analyzed. One method for selecting tasks to be analyzed is for the supervisor and employees to list the jobs performed. The supervisor and workers then select the jobs which represent the greatest injury potential and these are analyzed. Other approaches can be undertaken; this, of course, depends upon the organizational structure and objectives.

Selection of Tasks for Analysis

The structure and objectives of a company will determine the individual who selects the tasks. If the employee turnover rate is high, the management might choose the supervisor or foreperson to determine potential tasks for analysis. An alternative to this approach is for the safety and health professional or safety director to develop the list, although in some companies the safety professional is responsible for more than one project and may not be sufficiently familiar with each area and, thus, would not be the one best suited to suggest jobs for analysis. In this case, the workers, supervisor, and safety professional might combine their knowledge and effectively develop a list. A group discussion method will usually benefit the overall safety program. The most knowledgeable individuals should be used to determine potential tasks for analysis, regardless of the method used.

Jobs suitable for JSA are the ones assigned by a line supervisor. Operating a machine, installing duct work, and stacking lumber can be good subjects for job safety analysis. They are neither too broad nor too narrow. Assignments, which require the performance of a number of sequential steps in order to accomplish the task, are good candidates for job safety analysis. The following are examples of these types of jobs:

1.   Material handling (heavy drums).

2.   Work on electrical systems.

3.   Acid/caustic cleaning.

4.   Crane repair.

5.   Crane operation.

6.   Trenching and excavating.

7.   Erection and use of scaffolding.

Prioritization of Tasks

The next step is the establishment of priorities for the jobs selected. The tasks must be ranked in the order of greatest accident potential (injury and illness or property damage). The tasks with the highest risks should be analyzed first. To achieve this ranking of tasks, the following criteria should be used. The use of more than one selection criterion will optimize decisions. Caution must be exercised when selecting jobs for analysis if the employees are inexperienced. Analyses should begin with a simple task and proceed to more complex jobs. Prioritization can be based upon some or any of the following types of information:

1.   Accident frequency for jobs that have repeatedly produced accidents. The greater the number of accidents associated with the job, the greater its priority should be.

2.   The severity of the accidents that the job has produced, such as disabling injuries and number of lost workdays. The severity of the injuries prove that the preventive action taken prior to their occurrence was not successful. There may also be some jobs which do not have a history of accidents but have a potential for a severe injury.

3.   The experience and knowledge of the workplace may cause a judgment to be made regarding the initiation of a JSA. Many jobs qualify for immediate job safety analysis because of the potential hazard involved. Such types of hazards might be identified as "It hasn't happened yet, but when it does—watch out." Good examples of these types of hazards are jobs that involve explosion possibilities or ones involving lifting exceptionally heavy equipment. When you have a job which has fatality possibilities, even if a fatality has never occurred, it is sound judgment to develop a JSA for this particular type of job.

4.   The need to look at routine or repetitive tasks which have inherent hazards where the employee is repeatedly exposed to these hazards. For example, the constant lifting of concrete blocks or the use of a power-actuated tool.

5.   Changes in processes, equipment, or materials may introduce new hazards which may not be apparent. Changes can also increase the workload of an employee and overtax his ability. It is not necessary to wait until there is an accident on such jobs before making a job safety analysis study.

These five criteria help establish the sequence in which jobs are to be analyzed. They also help to identify some of the hazards associated with the job.

Conducting a Job Safety Analysis

After a job has been selected and the JSA has been initiated, use a worksheet similar to the Job Safety Analysis Worksheet (Figure 5-5) to list the basic steps, their corresponding hazards, and the safe operation procedures for each step.

Job Safety Analysis Worksheet

Title of Job/Operation _____

Position/Title(s) of Person(s) Who Does Job _____

Department _____

Section _____

Date _____

Name of Employee Observed _____

Analysis Made By _____

Analysis Approved By _____

No. _____

Sequence of Basic Job Steps	Potential Accidents or Hazards	Recommended Safe Job Procedures

1. Struck By (SB)
2. Struck Against (SA)
3. Contacted By (CB)
4. Contact With (CW)

5. Caught On (CO)
6. Caught In (CI)
7. Caught Between (CBT)
8. Fall - Same Level (FS)

9. Fall To Below (FB)
10. Overexertion (OE)
11. Exposure (E)

Figure 5-5. Job safety analysis worksheet – Courtesy of the National Mine Health and Safety Academy

The form requires that you briefly describe what is done for each step. Each step should be in the sequence in which it is accomplished. Be careful not to make the breakdown of steps too detailed; this results in a large number of unnecessary steps. At the same time, do not be so general as to leave out or omit some of the basic steps. If there are a large number of steps, generate more than one JSA. Now is the time to actually observe the job being performed. It is a good idea to involve an employee who performs the job or is very familiar with the job. Also, if possible, observe more than one worker performing the same job. To get started, ask the employee to list the steps of the job. Work back and forth with each other until agreeing upon a list of job steps.

For example, for a front-end loader the first step of a JSA might be a walk-around inspection, followed by a check of fluids, then visually checking the controls and area around the loader, starting the loader, etc. There should be between ten and fifteen basic steps. It is important that these steps accurately describe the work to be done. Once all the basic steps are identified, the hazards for each step needs to be identified. Developing a list of hazards for each step will result in finding some hazards which are more likely to occur than others and some which are more likely to produce more serious injuries. To make hazard identification easier and generally more understandable, a system of eleven accident types has been developed. Using this system, a hazard is identified by the potential accident(s) it can cause. The basic accident types are

1. **STRUCK-BY (SB)** – A person is forcefully struck by an object; the force of contact is provided by the object.

2. **STRUCK-AGAINST (SA)** – A person forcefully strikes an object; the person provides the force.

3. **CONTACT-BY (CB)** – Contact by a substance or material that by its very nature is harmful and causes injury.

4. **CONTACT-WITH (CW)** – A person comes in contact with a harmful material; the person initiates the contact.

5. **CAUGHT-ON (CO)** – A person or part of his clothing or equipment is caught on an object that is either moving or stationary. This causes the person to lose his or her balance and fall, be pulled into a machine, or suffer some other harm.

6. **CAUGHT-IN (IN)** – A person or a part of that person is trapped, stuck, or otherwise caught in an opening or enclosure.

7. **CAUGHT-BETWEEN (CBT)** – A person is crushed, pinched, or otherwise caught between either a moving object and stationary object or between two moving objects.

8. **FALL-SAME LEVEL (FS)** – A person slips or trips and falls to the surface he or she is standing or walking on.

9. **FALL-TO-BELOW (FB)** – A person slips or trips and falls to a level below the one on which he or she was walking or standing.

10. **OVER-EXERTION (OE)** – A person over extends or strains himself or herself while doing a job.

11. **EXPOSURE (E)** – Over a period of time, someone is exposed to harmful conditions.

For example, the walk-around for a front-end loader could result in a fall to the same level; checking fluids could result in exposure or contact with hazardous materials; and starting a loader could result in a fall to a lower level. As you examine each step, look for one hazard at a time.

Considering Human Problems in the JSA Process

While working with potential accidents or hazards, consider the following points related to human problems:

- What effects could there be if equipment is used incorrectly?

- Can the worker take shortcuts to avoid difficult, lengthy, or uncomfortable procedures?

Other personnel factors to consider include the following:

1. Personnel tend to take shortcuts to avoid arduous, lengthy, uncomfortable, or unintelligible procedures.

2. Equipment that is difficult to maintain will suffer from lack of maintenance.

3. Requirements for special employee training should be kept to a minimum.

4. Written procedures which are difficult to understand without further explanation may result in employee confusion.

5. Stress in the work environment and in private life may have a negative impact on workplace safety.

Normally the job steps for a JSA are listed in a logical sequence for one reason or another. For example, one worker may choose to check fluids prior to doing the walk-around inspection. This type of flexibility is good for worker morale and productivity. But on the other hand, there are times when the sequence of job steps or deviations from the job steps may be critical to safe performance on the job. An example of this would be the walk-around for the loader. The inspection must be made and safety deficiencies corrected before the loader is put into service for the day or an accident/incident may occur. It would not be safe or proper to do the walk-around inspection after the loader has been put in service. If the sequence of job steps or the deviation from established job steps are critical to the safe performance, this should be noted in the JSA.

Hazards Elimination or Control

The final step in evaluating a job is to develop a safe job procedure to eliminate or reduce the potential accidents or hazards. The following criteria should be considered:

1. Find a less hazardous way by using an engineering revision to find an entirely new and safe way to do a job. Determine the work goal and analyze the various ways of reaching this goal to establish which way is safest. Consider work-saving tools and equipment. An example of this would be to install gauges on the loader which would visually indicate the fluid level and prevent contact with the fluids.

2. Change the physical conditions that created the hazard. If a new, less hazardous way of doing the job cannot be found, try to change the physical conditions which are creating the hazards. Changes made in the use of tools, materials, equipment, or the environment could eliminate or reduce the identifying

hazards. But when changes are found, study them carefully to determine the potential benefits. Consider if the changes possess latent, inherent hazards which may be equally as hazardous as the original condition. In this case, assess both conditions to determine which will be less hazardous and refer the decision to the proper level of management for approval and acceptance. An example of this would be the development of a new loader operator cab with better visibility for safer operations.

3.   To eliminate those hazards which cannot be engineered out of the job, change the job procedure. Changes in job procedures, which are developed to help eliminate the hazards, must be carefully studied. If the job changes are too arduous, lengthy, or uncomfortable, the employee will take risky shortcuts to circumvent these procedures. Caution must be exercised when changing job procedures to avoid creating additional hazards. To help determine the effectiveness of procedural changes, two questions which might be asked are (1) In order to eliminate this particular hazard or prevent this potential accident, what should be the action of the employee? (2) How should the employee accomplish this? Answers must be specific and concrete if new procedures are to be useful and effective. Answers should precisely state what to do and how to do it. This might mean changing the loader walk-around inspection to include fluid inspection at the same time, thus saving time and possibly decreasing the time of exposure.

4.   Try to reduce the necessity of doing the job, or at least reduce the frequency that it must be performed. Often maintenance jobs require frequent service or repair of the equipment which is hazardous. To reduce the necessity of such a repetitive job, ask "What can be done to eliminate the cause of the condition that makes excessive repairs or service necessary?" If the cause cannot be eliminated, then try to improve the condition.

5.   Finally, use personal protective equipment (PPE) for potentially dangerous conditions. Use gloves, aprons, and goggles to avoid acid splashing. Wear ear plugs for protection from high noise levels and wear respirators to protect against toxic chemicals. The use of PPE should be the last consideration in eliminating or reducing the hazards the employee is subjected to because PPE can be heavy, awkward, uncomfortable, and expensive to maintain. Therefore, try to engineer the identified hazards out of the job.

A lengthy format of the JSA chart is less effective than a short and concise format. A format, listed on a single page with the job steps, hazards, and how to eliminate hazards, is more beneficial than a JSA with one page listing job steps, another page listing hazards, and a third page illustrating two common formats. Employees will not read lengthy explanations; brevity is desirable. A completed sample JSA appears in Figure 5-6 for your review and use.

Methods of Performing a JSA

The two methods most used in performing the JSA are group discussion and direct observation. In the group discussion, the supervisors meet with their employees to perform analysis of a job. The job is broken into a step-by-step process with the identification of all associated hazards. Changes in the work procedures or in the physical environment will eliminate or reduce the identified hazards. The following basic advantages are obtained from the group discussion method:

1.   The sharing of experiences by a group of employees will generally produce a more thorough analysis of potential hazards. One employee may identify a potential hazard which has been overlooked by others. The combined experience is also valuable in analyzing possible changes for the reduction or elimination of the identified hazards.

2.   Group discussions will serve as an effective safety training program. The employees will learn "potential hazards" from each other. They share ideas and usually every person knows a little more safety as a result of the discussions.

3.   The employee assumes an active role in the safety program. The safety program will benefit because the employee will be more inclined to accept the procedures in the JSA.

Of course these advantages are not automatic; they depend upon how skillfully the discussions are conducted. If they are poorly handled and without preparation or skill, the results will be meager.

In the direct observation method, either the supervisor or safety director observes the employee performing the task from start to finish. The job must be broken down step by step and the hazards associated with the job must be identified. Once the hazards have been identified, they should be eliminated or reduced. There is no reason why the supervisor or safety engineer cannot consult with the employee doing the job. The advantage of the direct observation method is that the supervisor or safety director does not have to try to recall or visualize how the job is done because the performance is seen.

The organizational structure and personnel determine the method most beneficial for a company. A combination of both methods may also be used.

Review the Analysis and Potential Solutions

After the JSA is completed, qualified individuals who did not participate in the analysis must review it. For example, if the direct observation method was used, the JSA performed by the supervisor can be reviewed by the safety professional. But, in the case where the safety professionals participate in the job safety analysis, then he/she must not review the JSA. This independent review will help ensure a higher quality analysis.

Implementing the JSA in an Operational System

In the past, JSAs have been developed and then filed away and never used. After some period of time, an operation accumulates a great number of JSAs and unless an effective filing system is implemented, they will be lost in the shuffle. The completed, reviewed, and approved JSA must be incorporated into the operation—not filed away and forgotten.

When a JSA is distributed, the supervisor's first responsibility is to explain its contents to the workers and, if necessary, to provide individual training. The entire JSA must be reviewed with the involved workers so they will know how the job is to be done without accidents. The original JSA may have been revised to a step-by-step procedure which includes Detailed Operating Procedures (DOP) and Standard Operating Procedures (SOP) (see Chapter 6). The step-by-step procedure may be developed from the JSA if the task is to be repeated a number of times, or JSA information may be integrated into an existing operational procedure.

Two benefits of the JSAs are that they provide the workers with information about the hazards in the jobs they perform and allow training of the basic job steps to both new and old employees. The employees must be taught to recognize the hazards associated with each job

Job Safety Analysis Worksheet

Title of Job/Operation CHANGE TIRE ON PICKUP TRUCK   Date _____   No. _____

Position/Title(s) of Person(s) Who Does Job DRIVER   Name of Employee Observed _____

Analysis Made By _____

Department _____

Section _____   Analysis Approved By _____

Sequence of Basic Job Steps	Potential Accidents or Hazards	Recommended Safe Job Procedures
1. Prepare materials to change tire.	1a. Struck by vehicle	1a. Engine must be shut off.
	1b. Slip and fall to same level	1b. Observe area. Remove tripping or stumbling hazards or move vehicle to better location.
	1c. Exposure to cold, frostbite	1c. Wear gloves and other appropriate clothing.
	1d. Overexertion removing spare tire from carrier	1d. Follow instructions in owner's manual or posted near jack.
	1e. Overexertion loosening lug nuts	1e. Use tire tool provided or large 4-way wrench. Use leg muscles to break lug nuts.
2. Raise vehicle.	2a. Struck by vehicle	2a. Put vehicle in gear or park. Set parking brake. Chock wheel on opposite end of vehicle (both sides).

1. Struck By (SB)
2. Struck Against (SA)
3. Contacted By (CB)
4. Contact With (CW)

5. Caught On (CO)
6. Caught In (CI)
7. Caught Between (CBT)
8. Fall - Same Level (FS)

9. Fall To Below (FB)
10. Overexertion (OE)
11. Exposure (E)

Sequence of Basic Job Steps	Potential Accidents or Hazards	Recommended Safe Job Procedures
2. Raise vehicle. (continued)	2b. Struck by jack	2b. Follow jacking instructions in owner's manual or posted near jack.
	2c. Contact with hot exhaust	2c. Exhaust system may be very hot - do not touch.
	2d. Overexertion using jack	2d. Use jack as described in instructions/owner's manual. Examine jack and handle for defects.
3. Change tire.	3a. Caught between tire and ground or caught between vehicle body and ground	3a. Check jack stability before any work is performed on the raised vehicle. Grasp tire on sides to remove from hub.
	3b. Overexertion lifting tire off of or on to hub	3b. Use proper lifting techniques. Get help if needed.
	3c. Contact with hot hub wheel or lug nuts	3c. Wear gloves when removing tire. Wheel and lug nuts may be hot.
	3d. Exposure to cold (frostbite)	3d. Wear gloves and other appropriate clothing.
4. Lower vehicle.	4a. Same as Step 2	4a. Same as Step 2.
5. Replace materials.	5a. Overexertion putting flat tire in truck bed	5a. Same as Step 3.
	Same as Step 1	

1. Struck By (SB)
2. Struck Against (SA)
3. Contacted By (CB)
4. Contact With (CW)

5. Caught On (CO)
6. Caught In (CI)
7. Caught Between (CBT)
8. Fall - Same Level (FS)

9. Fall To Below (FB)
10. Overexertion (OE)
11. Exposure (E)

Figure 5-6. Sample of a completed JSA – Courtesy of National Mine Health and Safety Academy

step and learn the necessary precautions to prevent injuries. A well-prepared JSA can be vital in this training.

Monitoring a JSA Program

It is not enough to just develop and utilize a JSA for a potentially hazardous job; a monitoring program must be established and the following questions should be asked:

1. How effective is the JSA in eliminating or reducing the hazards associated with the task?

2. Does the JSA need to be revised to update changes in the physical environment which may have introduced new and unforeseen hazards?

3. Is the JSA followed by the employees performing the task? If not, then why not? What hazards are introduced?

The following techniques may be used to monitor a JSA:

1. Observation of task by management, safety professional, or others concerned.

2. Review of accident records.

3. Feedback.

The observation of a task in a JSA provides management and safety with an opportunity to review the task in actual practice. It also shows the employees that management and safety care about well-being. Reviewing accident records may reveal possible deficiencies in a job which has been analyzed, or indicate jobs which should be analyzed.

The feedback technique allows the worker, supervisor, and safety professional to discuss possible deficiencies in the JSA. It also gives the employee an active role in safety practices.

HAZARD ANALYSIS

Hazard analysis is a technique used to examine the workplace for hazards with the potential to cause accidents. Hazard analysis, as envisioned in this section, is a worker-oriented analysis. The workers are asked to provide the information for the analysis. Management is not as close to the actual work being performed as are those performing the work. The workers have hazard concerns and have devised ways to mitigate the hazards, thus preventing injuries and accidents. This type of information is invaluable when analyzing workplace hazards.

This type of hazard analysis does not require that it be conducted by someone with special training and could also be accomplished by the use of a short fill-in-the-blank questionnaire. This technique works well where management is open and genuinely concerned about the safety and health of its workforce. The most time-consuming portion of this process is analyzing the results of the interviews or questionnaires.

Only four responses are required.

1. What are three to five hazards involved with your specific task?

2. Of the hazards you have identified, how would you rank them? Rank them from most hazardous to least hazardous.

3. While doing your task, what do you do to prevent injuries from these hazards?

4. What could management do to reduce each hazard and make your job safer?

The information obtained by the hazard analysis provides the foundation for making decisions on which jobs should be altered in order for the worker to perform safer and expeditiously. Also, this process allows workers to become more involved in their own destiny. For some time, involvement has been recognized as a key motivator of people. This is also a positive mechanism in fostering labor/management cooperation.

It is important to remember that a worker may perceive something as a hazard, when in fact it may not be a true hazard; the risk may not match the ranking that the worker placed on it. Also, even if hazards exist, you need to prioritize them according to the ones which can be handled quickly; which may take time; or which will cost money above your budget. If the correction will cause a large capital expense and the risk is real but does not exhibit an extreme danger to life and health, you might need to wait until next year's budget cycle. An example of this would be when workers complained of a smell and dust created by a chemical process. The dust was not above accepted exposure limits and the smell was not overwhelming. Therefore, the company elected to install a new ventilation system, but not until the next year because of budgetary constraints.

Hazard analysis is a process which is controlled by management. Management must assess the outcome of the hazard analysis process and determine if immediate action is necessary or if, in fact, there is an actual hazard involved. When you do not view a reported hazard as an actual hazard, it is a good idea to inform the worker of your opinion and explain why.

The expected benefits are a decrease in the incidents of injuries, a decrease in lost workdays and absenteeism, a decrease in workers compensation cost, better productivity, and an increase in cooperation and communication. The baseline for determining the benefit of the hazard analysis can be formulated from existing company data on occupational injuries/illnesses, workers' compensation, attendance, profit, and production.

ANALYZING ACCIDENT DATA

Many companies conduct accident investigations and keep accident records and other data on the company's safety and health initiatives. If a company has a sufficient number of accidents/incidents and enough detail in their occupational injury/illness investigation data, the company can begin to examine trends or emerging issues relevant to their safety and health intervention/prevention effort. The analysis of this data can be used to evaluate the effectiveness of safety and health on various projects and jobsites or groups of workers. The safety and health data can be used to compare your company to companies that perform similar work, employee a comparable workforce, or bid on the same size of projects on a state, regional, or national basis.

By analyzing your accidents/incidents, you are in a better position to compare apples to apples rather than apples to oranges. You will be able to identify not only the types of injuries, accidents, and causes, but you will also be able to intervene and provide recommendations for preventing these accidents/incidents in the future. You will be able to say with confidence that "I do" or "I do not" have a safety and health problem. If you find that you have a problem, your analysis and data will be essential if you try to elicit advice on how to address your health and safety needs.

Gathering and analyzing accident/incidents data are not your entire safety and health program, but a single element. It provides feedback and evaluative information as you proceed toward accomplishing your safety and health goals; thus, it is an important element.

The two most frequent statistical pieces of information which are designed to allow you to compare your company's safety and health performance with others is the Incident Rate and the Severity Rate. These two rates, respectively, answer the questions of "How often or frequently are accidents occurring? and "How bad are the injuries/illnesses which are occurring?" The number of times that occupational injuries/illnesses happen is the determinant for the Incident Rate while the number of days (lost-work days) is the prime indicator of the Severity Rate. Both of these rates provide unique information regarding your safety and health effort.

To find the Incident Rate, count the number of distinct events which resulted in injuries/illness. To compare your Incident Rate to other construction companies, you must normalize your data. This is accomplished by using a constant of 200,000 work hours, which was established by the Bureau of Labor Statistics. The 200,000 work hours is the number of hours which 100 full-time workers would work during fifty weeks at forty hours per week. Thus, you can calculate your Incident Rate in the following manner:

$$\text{Incident Rate} = \frac{\text{Number of Your OSHA Recordable Injuries/Illnesses} \times 200,000 \text{ (work hour constant)}}{\text{Total Number of Hours which Your Employees Worked During the Year}}$$

The Incident Rate could be a rate calculated for recordable (combined) injuries and illnesses, recordable injuries, recordable illnesses, all injuries with lost work days, all illnesses with lost work days, injuries requiring only medical treatment, or first aid injuries. These rates would not normally be calculated on a national basis, but could be used to compare your progress on a yearly basis, or between jobsites or projects.

The Severity Rate, which is often called the "lost-time workday rate," is used to determine how serious the injuries and illnesses are. A company may have a low Incident Rate or few injuries and illnesses but, if the injuries and illnesses which are occurring result in many days away from work, the lost-workday cases can be as costly or more costly than having a large number of no lost-workday injuries which have only medical costs associated with them. Lost workday cases can definitely have a greater impact upon your worker compensation costs and premiums.

Calculation of the Severity Rate is similar to the Incident Rate except that the total number of lost-time workdays is used in place of the number of OSHA recordable injuries/illness. Thus, you can calculate your Severity Rate in the following manner:

$$\text{Severity Rate} = \frac{\text{Number of Your Lost-Time Workdays} \times 200,000 \text{ (work hour constant)}}{\text{Total Number of Hours which Your Employees Worked During the Year}}$$

The Incident and Severity Rates are both expressed as a rate per 100 full-time workers. This provides a standard comparison value for a company whether it has 20 or 1000 workers. Thus, both the 20 employee company and 1000 employee company can compare their safety and health performance to each other.

Sometimes the temptation exists to focus on only the lost-time workday cases, but how can you identify which injury or illness is going to result in a lost-time workday? At times, the difference between a medical treatment injury and a lost-time workday injury may be only a matter of inches or chance. Thus, it is more logical to address your total injury problem.

SUMMARY

The analysis of construction-related hazards and the accident/incidents which they cause is an important step in the overall process of reducing construction-related injuries, illnesses, and deaths. Only after a systematic look at the hazards and accidents can you hope to integrate the accident prevention techniques and tools which can have an impact upon your company's safety and health initiative.

REFERENCES

[1] *Accident Prevention Accident/Incident Analysis,* National Mine Health and Safety Academy, U.S. DOL, Beckley, WV, 1980.

[2] *The Job Safety Analysis Process: A Practical Approach,* National Mine Health and Safety Academy, U.S. DOL, Beckley, WV, 1990.

Chapter 6

CONSTRUCTION ACCIDENT PREVENTION TECHNIQUES

Charles D. Reese

Accident prevention techniques (APT) are tools which can be used as an integral part of any safety and health program. Each technique serves a unique function. It will be your job to decide which techniques to incorporate into your safety and health effort.

The first types of APT you may want to use are safety and health management, safety communication, tool box talks, and training. These techniques set a foundation for accident prevention. Once established, identify the hazards and plan procedures for avoiding those hazards. Recognize the need to call on a specialist, engineer, expert, or to have some form of a special program to address your company's specific needs (e.g., maintenance program, eye protection program, etc.). Finally, evaluate your efforts using safety and health audits, job safety observations, and accident investigations.

This chapter explains each technique and denotes its use, including positive and negative aspects. For specific problems, remember that using these techniques as stand-alone entities may have only a band-aid effect for specific problems. The implementation of a complete approach is a huge step in assuring a functional safety and health initiative. This completeness requires a written program that addresses workers' needs and concerns, hazard recognition and analysis, and compliance with OSHA standards and requirements. Only after you have an organized safety and health program can you expect to effectively apply the APTs listed in this chapter; they are only a part of a complete safety and health program. Using this type of approach should afford a workplace free of hazards, worker injuries, and illnesses.

Within this chapter, thirteen accident prevention tools are addressed in detail. The application and implementation of these tools, as applied to the construction industry, will help assure a good safety and health program for your company and employees.

When a construction contract/project is in the implementation stage, the prime contractor/owner needs to address safety and health. This assures that all general contractors and subcontractors comply with the project's job safety and health requirements. Use of the tools presented in this chapter will help meet those requirements.

SAFETY AND HEALTH MANAGEMENT

Where does safety and health on a construction site/project begin? The answer is, of course, from day one. When does it stop? The answer is "never." Safety and health management should be considered from the design inception, through the bid process, the contract award, and the initial start. Then the management of safety and health continues through the project period and becomes the foundation for the next and all other ensuing projects that your company undertakes.

The phrase "safety and health management" needs a definition which is both practical and relevant to the construction industry. Thus, in the broadest sense, safety and health management means taking control of the existing hazards. Develop a systematic or programmed approach which allows you to effectively, efficiently, and productively direct, mesh, control, and intervene in a complex set of variables and contributing factors (e.g., risks, hazards, job tasks, other contractors, employees, machines/equipment, the environment). These factors may impact your attempts to complete a project in a safe, healthy, and responsible manner.

COMMUNICATIONS

Communication is the key to occupational safety and health. The message of accident prevention must be constantly reinforced. Use constant reminders in the form of message boards, fliers, newsletters, paycheck inserts, posters, the spoken word, and face-to-face encounters. Make the message consistent with the policies and practices of the company so it is believable and credible. It is important to communicate safety and health goals and provide feedback progress toward accomplishing those goals. Most workers want their information in short, easily digested units.

Face-to-face interactions personalize the communications, provides information immediately (no delays), and allows for two-way communications which improves the accuracy of the message. Finally, these interactions provide for performance and real-time feedback and reinforcement regarding safety and health.

TOOLBOX TALKS

Toolbox safety talks are especially important to the supervisors on your jobsites and projects because they afford the supervisor the opportunity to convey, in a timely manner, important information to the workers. Toolbox talks may not be as effective as the one on one, but still surpass a memorandum or written message. In the five to ten minutes prior to the workday, a shift, at a break, or as needed, this technique helps communicate time-sensitive information to a department, crew, or work team.

In these short succinct meetings, convey changes in work practices, short training modules, facts related to an accident or injury, specific job instructions, policies and procedures, rules and regulation changes, or other forms of information which the supervisor feels are important to every worker under his supervision.

Although toolbox talks are short, these types of talks should not become just a routine part of the workday. Thus, in order to be effective, they must cover current concerns or information, be relevant to the job, and have value to the workers. Carefully plan toolbox talks to effectively transmit a specific message and a real accident prevention technique. Select topics applicable to the existing work environment; plan the presentation and focus on one issue at a time. Use materials to reinforce the presentation and clarify the expected outcomes.

Some guidelines are

1. Plan a toolbox training schedule in advance and post a notice.

2. Prepare supporting materials in advance.

3. Follow a procedure in the presentation: explain goals; try to answer questions; restate goals; and ask for action.

4. Make attendance mandatory.

5. Make each employee sign a log for each session.

6.   Ask for feedback from employees on the topic or other proposed topics.

7.   Involve employees by reacting to suggestions or letting them make presentations when appropriate.

8.   Reinforce the message throughout the work week.

No matter how effectively you communicate with your workforce, you still need to assure that your workforce has the competence to perform the basic skills for the tasks they have been assigned.

TRAINING

You may offer all types of programs and use many of the accident prevention techniques, but, without workers who are trained in their trade and have safe work practices, your efforts to reduce and prevent accidents and injuries will result in marginal success. If a worker has not been trained to do his/her job in a productive and safe manner, a very real problem exists.

Do not assume that a worker knows how to do his or her job and will do it safely unless he or she has been trained to do so. Even with training, some may resist safety procedures and then you have a deportment or behavioral problem and not a training issue.

It is always a good practice to train newly hired workers or experienced workers who have been transferred to a new job. It is also important that any time a new procedure, new equipment, or extensive changes in job activities occur, workers receive training. Well-trained workers should be more productive, more efficient, and safer.

Training for the sake of documentation is a waste of time and money. Training should be purposeful and goal or objective driven. An organized approach to on-the-job safety and health will yield the proper ammunition to determine your real training needs. These needs should be based on accidents/incidents, identified hazards, hazard/accident prevention initiatives, and input from your workforce. You may then tailor training to meet the company needs and that of workers.

Look for results from your training. Evaluate those results by looking at the number of reduced accidents/incidents, improved production, and good safety practices performed by your workforce. Evaluate the results by using job safety observation and safety and health audits, as well as statistical information on the numbers of accidents and incidents.

Many OSHA regulations have specific requirements on training for fall protection, hazard communication, hazardous waste, asbestos and lead abatement, scaffolding, etc. It seems relatively safe to say that OSHA expects construction workers to have training on general safety and health provisions, hazard recognition, as well as task-specific training. Training workers regarding safety and health is one of the most effective accident prevention techniques.

HAZARD IDENTIFICATION

In order to prevent accidents, identify existing and potential hazards which can prevail upon the worksite. When looking at specific jobs, break the job down into a step-by-step sequence and identify potential hazards associated with each step. Consider the following questions:

1.   Is there a danger of striking against, being struck by, or otherwise making injurious contact with an object? (For example, being struck by a suspended drill casing or piping as it is moved into place.)

2. Can the employee be caught in, on, or between objects? (For example, an unguarded v-belt, gears, or reciprocating machinery.)

3. Can the employee slip, trip, or fall on same level, or to another level? (For example, slipping in an oil-changing area of a garage, tripping on material left on stairways, or falling from a scaffold.)

4. Can the employee strain him/herself by pushing, pulling, or lifting? (For example, pushing a load into place or pulling it away from a wall in a confined area.) Back injuries are common in every type of industrial operation; therefore, do not overlook the lifting of heavy or awkward objects.

5. Does the environment have hazardous toxic gas, vapors, mist, fumes, dust, heat, or ionizing or nonionizing radiation? (For example, arc welding on galvanized sheet metal produces toxic fumes and nonionizing radiation.)

Review each step as many times as necessary to identify all hazards or potential hazards. With the identification of the hazards, take steps to prevent the accidents or incidents from occurring. If you know the hazard, it is easier to develop interventions which mitigate the risk potential. These interventions may be in the form of safe operating procedures.

At times a hazard hunt is appropriate. This constitutes providing a form (see Figure 6-1) to your workers and asking them to list anything which they perceive to be a hazardous condition on the jobsite. You will be able to review these hazard hunt forms to determine if hazards truly exist or are just perceived. If you use the hazard hunt as a hazard identification process, it is important to provide feedback to the workers on whether a hazard does or does not exist. It is never wise to disregard this important communication. Hazard identification has always been an effective accident prevention technique.

SAFE OPERATING PROCEDURES (SOPs)

Workers may not automatically understand a task just because they have experience or training. Thus, many jobs, tasks, and operations are best supported by a Safe or Standard Operating Procedure (SOP). The SOP walks the worker through the steps of how to do a task or procedure in a safe manner and calls attention to the potential hazards at each step.

You might ask why an SOP is needed if the worker has already been trained to do the job or task. As you may remember from the previous chapter, a job safety analysis usually keys on those particular jobs that pose the greatest risk of injury or death. They are the high risk type of work activities that definitely merit the development and use of a SOP.

There are times when an SOP or step-by-step checklist ARE useful. Some of these times are when

1. A new worker is performing a job or task for the first time.

2. An experienced worker is performing a job or task for the first time.

3. An experienced worker is performing a job which he/she has not done recently.

4. A mistake could cause damage to equipment or property.

5. A job is done on an intermittent or infrequent basis.

6. A new piece of equipment or different model of equipment is obtained.

HAZARD HUNT FORM

Worker's Name (Optional): _____ Date:_____

Jobsite:_____ Job Titles:_____

---

1. Describe the hazard that exists.

---

2. Has the hazard been discussed with your supervisor?   When? (Date)

    _____Yes   _____No

---

3. What preventive action needs to be taken?

---

4. Do you know what state or federal safety regulations is being violated?
    _____Yes _____No
    (Please list it if your answer is yes.)

---

5. Have there already been injuries or illnesses caused by the hazard? _____Yes _____No
    (List the names, dates, events, or symptoms)

---

6. How long has the hazard been present?

Manager's or Supervisor's Response to Hazard Concern Identified.

NOTE: Use a Separate Form for Each Hazard Identified.

Figure 6-1. Hazard hunt form

When airline pilots fly, the most critical parts of the job are take-offs and landings. Since these are two very critical aspects of flying, a checklist for proceeding in a safe manner is used to mitigate the potential for mistakes. It is critical to provide help when a chance for error can result in grave consequences.

Few people or workers want to admit that they do <u>not</u> know how to perform a job or task. They will <u>not</u> ask questions, let alone ask for help in doing an assigned task. This is the time when a plasticized SOP or checklist could be placed at the worksite or attached to a piece of equipment. This can prove to be a very effective accident prevention technique. It can safely walk a worker through the correct sequence of necessary steps and thus avoid the exposure to hazards which can put the worker at risk of injury, illness, or death.

These safe operating procedures could be used when helicopters are used for lifting, industrial fork lifts are used, materials are moved manually, etc. These types of SOPs should list the sequential steps required in order to safely perform the job or task, the potential hazards involved, and the personal protective equipment needed. Each step in the SOP should provide all the information needed to accomplish the task in a safe manner.

If you do not have annual training, the use of SOPs may instill in the workers the confidence that they have a way of refreshing their memory on how to do their tasks. When changing procedures, update the SOP to assure that the changed procedure is safely performed. The checklist is one form of an SOP. The checklist is very effective when you are trying make sure that every step is being accomplished.

SOPs are useful only when they are up to date and readily accessible at the actual job or task site. Since we now have the ability to store SOPs in the computer, revision is made simple and, therefore, the SOP can be easily changed if you receive suggestions from supervisors or workers. Figure 6-2 shows a picture of an excavator and Figure 6-3 shows an example of an SOP for an excavator. Using the format from this example, develop your own SOPs for procedures, jobs, tasks, or equipment.

Figure 6-2. An excavator

**SAFE OPERATING PROCEDURES
FOR AN
EXCAVATOR**

WHAT TO DO (Steps in sequence)	HOW TO DO IT (Instructions)	KEY POINTS (Items to be emphasized)
1) Climb onto engine compartment.	1) Place hand on rail of operator's compartment and foot on step of track frame. Gently pull yourself up onto track. Continue around operator's compartment following the steps and rails.	1) Be sure to use rails and steps or a fall may occur causing serious injury. Make sure shoes are free of mud for proper traction.
2) Unlock engine compartment and check all fluids. (Engine oil, hydraulic oil, coolant, and batteries.)	2) Place key in lock of engine compartment handle. Turn key and remove. Turn handle and lift to open. Remove dip sticks to check fluids. Fluid should register in the shaded area of the stick. If low, fill immediately. Wear safety glasses while checking fluids.	2) Engine compartment is spring loaded. Be sure to stand back while opening. Fluid levels are very important. Low fluid levels means machine will not operate properly and could cause injury or loss of productivity.
3) Unlock and open operator's door.	3) Place key in lock and turn. Open door and securely latch it so that it doesn't slam closed during operation.	3) Machine can be operated with door closed, but will not be able to hear coworkers outside which is very important due to the number of blind spots while operating.
4) Climb into operator's seat.	4) Grab handle and place foot on step of track frame and pull yourself up onto the track. Step inside operator's compartment and sit in seat.	4) Seat should be adjusted to fit operator's posture.

Figure 6-3. A safe operating procedure for an excavation

WHAT TO DO (Steps in sequence)	HOW TO DO IT (Instructions)	KEY POINTS (Items to be emphasized)
5) Check warning lights on instrument panel.	5) Place key in ignition and push test button to illuminate all warning lights and sound all warning buzzers.	5) Warning signals are very important and should always be carefully checked. They will alert the operator to problems or failures within the machine itself. They could prevent serious injury from occurring.
6) Start Engine.	6) Turn ignition key to the left for thirty seconds for engine preheat. Then turn key to the right until engine starts. Be sure engine throttle is at low before starting.	6) Diesel motors work on compression and not spark. The engine preheat heats up the cylinders so the fuel can be compressed and start the motor. Failure to do so could cause battery failure.
7) Check to make sure that there is good oil and hydraulic pressure.	7) Gauges on console should all be in the green level.	7) If levels are not in the green, shut machine off by pushing throttle to off position and check the operator's manual.
8) Let machine warm up for a minimum of five minutes, fifteen if the weather is below thirty-five degrees Fahrenheit.	8) Pull throttle level to 1/4 and leave it until warm up is complete.	8) Warm up allows the oil in the engine to get loose before operating at full throttle.
9) While machine is warming up, climb out and check ground for possible fluid leaks from broken hoses or loose fittings.	9) Do not climb under machine while it is operating. If there is a leak, it will be seen from a distance due to the amount of liquid that is utilized.	9) If a leak is spotted, climb into operator's compartment, push throttle to off position, and contact mechanic for repairs. Do not attempt repair immediately due to the temperature of the fluids being released. Be sure to wear safety goggles while checking for leaks to prevent eye damage.

Figure 6-3. A safe operating procedure for an excavation (continued)

WHAT TO DO (Steps in sequence)	HOW TO DO IT (Instructions)	KEY POINTS (Items to be emphasized)
10) Check exterior for possible exposures.	10) Look on all sides of machine for possible obstacles that could interfere with safe operations. Also look up to check for power lines and tree limbs. Check the ground for utility marks to show possible underground utilities.	10) Any obstacle that could interfere with safe operation could result in serious injury to the operator and/or coworkers. Never operate near power lines. No excavation should proceed without calling CBYD (Call Before You Dig).
11) Warm up hydraulic oil, if possible.	11) Climb into operator's seat and release lever locks by pulling them back. Pull throttle back to full throttle. Rest arms on arm rest and use just hands to operate machine. Directions for lever operations are on console. Lift boom and spin machine in circles. If the temperature is above 35 degrees, this is not necessary.	11) Be aware of all obstacles while spinning machine in circles. The spinning allows the hydraulic oil to loosen up and reduce excess pressure. This is caused by cold temperatures that make the oil thick.
12) Operate machine.	(12) Operate machine while staying aware of all personnel on the site.	12) Different hazards are always interfering with operations. A good operator is always aware of such hazards and is ready and able to deal with the unexpected ones.
13) Engine cool down.	13) After operations are complete, push throttle to 1/4 and allow engine to cool down for a minimum of 5 minutes.	13) This allows the turbochargers to slow down while there is still oil being used. Failure to cool down will cause the turbo to spin without lubrication and could cause failure of the turbo and inability to use the machine.

Figure 6-3. A safe operating procedure for an excavation (continued)

WHAT TO DO (Steps in sequence)	HOW TO DO IT (Instructions)	KEY POINTS (Items to be emphasized)
14) Put boom and dipper perpendicular to the ground.	14) Turn bucket flat and dipper (part attached to the bucket) straight up and down.	14) This allows the machine to rest with the least amount of stress on the the steel.
15) Shut down engine.	15) After cool down, push throttle all the way forward to the off position and turn key to the left.	15) Be sure to take key out of ignition and replace locks on operation levers.
16) Lock machine and secure.	16) Close door and lock it by turning key. Check to make sure the lock is secure.	16) A locked and secure machine will prevent children from gaining entry and possibly hurting themselves or others.
17) Clean tracks and rollers.	17) Use hand shovel to clean dirt and mud from underneath tracks. Be sure to get around the rollers on each side.	17) Cleaning away the mud and dirt allows the track to move freely. Without this, the track will be under strain and will cause unneeded strain on the drive sprocket. The rollers keep the track spinning in a straight line. If the rollers are not free of debris, the track could jump off the drive sprocket and cause injury to anyone near it and will also cause unneeded down time.
18) Fuel machine.	18) Fuel machine with treated diesel fuel. There is a tube on outside of tank to show level.	18) Keeping the fuel level full will keep moisture out of the tank. Be sure to wear safety glasses and gloves while refueling.

Figure 6-3. A safe operating procedure for an excavator. (continued)

A SOP is only one accident prevention technique or component of any safety and health initiative. There are specific jobs or tasks which lend themselves well to this approach. Make sure that you use SOPs when they benefit your type of work the most and not as a cure-all for all your accidents and injuries. Use it as one of the many tools for accident prevention.

WHEN AN ENGINEER IS NEEDED[1]

The Occupational Safety and Health Administration's existing and proposed standards for the construction industry contain stated and implied requirements for engineers. Contractors often assign professional engineers to perform or inspect jobsite tasks or activities for the assurance of the safety of employees. The role of construction engineers is changing as a result of the federal government's emphasis on "performance" rather than the traditional "specification" standards and new enforcement strategies.

Construction contractors want the flexibility to choose the means and methods to perform their operations. Their position is that each project process and design is unique, and compliance with a rigid set of rules is not feasible. This position has moved the regulatory agencies to adopt policies and regulations that rely on the professional expertise and experience of qualified individuals. In addition, investigations into major catastrophes in construction have shown that a lack of planning and engineering oversight has contributed to the accidents.

Focused Inspections

The Occupational Safety and Health Administration (OSHA), a federal regulatory agency, is beginning to change its rules and enforcement strategies. One of the strategies adopted by OSHA is the Focused Inspection Initiative. In a focused inspection, the OSHA Inspector conducts an abbreviated inspection of the project. Instead of the usual one to three days inspecting a project, the time spent could be as little as one to three hours. The Focused Inspection Initiative recognizes the efforts of responsible contractors to implement effective safety and health programs and this requires less OSHA time.

Focused inspections concentrate on the project safety and health program/plan and the four leading hazards that account for most fatalities and serious injuries in the construction industry: falls; electrical hazards; caught in/between hazards (such as trenching); and "struck-by" hazards (such as materials handling equipment and construction vehicles). OSHA inspectors will review the project plan which includes the designation of qualified personnel responsible for the implementation of the plan.

Identify the Need for an Engineer

Focused inspections must determine if the contractor has identified work/activities requiring planning, design, inspection, or supervision by an engineer, competent person, or other professional. If a contractor does not identify work/activities that require an engineer, the contractor will be subject to a comprehensive jobsite inspection and will not qualify for the 25% reduction in penalties available to contractors with effective programs.

Engineer Requirements

In addition to OSHA's enforcement strategy, existing construction standards include specific requirements for contractors to assign engineers to perform work/activities. Engineers are required in the OSHA standards for scaffolds, fall arrest systems, cranes, material hoists, excavations, cast-in-place concrete, lift slabs, asbestos, etc.

Stated Engineering Requirements

The following are locations within the construction regulations (29 CFR 1926) where there exist requirements for engineers to be involved.

Subpart L – Scaffolds

1926.451(d) (3) (i)	Mason's multipoint suspension scaffold connections shall be designed by an engineer experienced in such scaffold design.
(d)(5)	Scaffolds shall not be moved horizontally while employees are on them unless they have been specifically designed by a registered professional engineer for such movement.
1926.452:(a) (10)	Pole scaffolds over sixty feet in height shall be designed by a registered professional engineer and shall be constructed and loaded in accordance with such design.
(b)(10)	Tube and coupler scaffolds over 125 feet in height shall be designed by a registered professional engineer and shall be constructed and loaded in accordance with such design.
(c) (5)	Brackets used to support cantilevered loads shall
	(iii) Be used only to support personnel, unless the scaffold has been designed for other loads by a qualified engineer and built to withstand the tipping forces caused by those other loads.
(i)	Outrigger scaffolds.
	(8) Scaffolds and scaffold components shall be designed by a registered professional engineer and shall be constructed in accordance with such design.

Subpart M – Fall Protection

Appendix C – Personal fall arrest systems
ll. Tie-off considerations.
(h)(1)(i) Properly planned anchorages should be used if they are available. In some cases, anchorages must be installed immediately prior to use. In such cases, a registered professional engineer with experience in designing fall protection systems, or another qualified person with appropriate education and experience, should design and anchor the point to be installed.

Subpart N – Cranes, Derricks, Hoists, Elevators, and Conveyors

1926.550	Cranes and derricks.
(g)(4)(i)	The personnel platform and the suspension system shall be designed by a qualified engineer or a qualified person competent in structural design.
1926.552	Material hoists, personnel hoists, and elevators.

(a)(1)	...Where the manufacturer's specifications are not available, the limitations assigned to the equipment shall be based on the determination of a professional engineer competent in this field.
(b)(7)	All material hoist towers shall be designed by a licensed professional engineer.
	(17)(i) Personnel hoists used in bridge tower construction shall be approved by a registered professional engineer and erected under the supervision of a qualified engineer competent in this field.

Subpart P – Excavations

1926.651(i)(2)(iii)	A registered professional engineer has approved the determination that the structure is sufficiently removed from the excavation so as to be unaffected by the excavation activity; or (iv) A registered professional engineer has approved the determination that such excavation work will not pose a hazard to employees.
1926.652(i)	Sloping and benching systems not utilizing Option (1) shall be approved by a registered professional engineer. (d) (3) When material or equipment that is used for protective systems is damaged, a competent person shall examine the material or equipment and evaluate its suitability for continued use. If the competent person cannot assure the material or equipment is able to support the intended loads or is otherwise unsuitable for safe use, then such equipment shall be removed from service, and shall be evaluated by a registered professional engineer before being returned to service.
1926.650-652	Appendix B – Sloping and shoring. Sloping or benching for excavations greater than 20 feet deep shall be designed by a registered professional engineer.

Subpart 0 – Concrete and Masonry Construction

1926.703(a)(S)(i)	Design of shoring shall be prepared by a qualified engineer and the erect shoring shall be inspected by an engineer qualified in structural design.
1926.705(a)	Lift-slab operations shall be designed and planned by a registered professional engineer who has experience in lift-slab construction.
	(k)(1) ...The phrase "reinforced sufficiently to insure integrity" used in this paragraph means that a registered professional engineer, independent of the engineer who designed and planned the lifting operation, has determined that if there is a loss of support at any jack location, that loss will be confined to that location and the structure as a whole will remain stable.

Implied Engineering Requirements

Implied engineering requirements are those OSHA standards that describe a task/ activity, design, or decision that must be based upon the knowledge and expertise of an engineer. In most states these tasks/activities, designs, or decisions are granted to the engineers registered and licensed to practice in that state. Generally, when an OSHA standard refers to design, use, operation, etc. in accordance with the manufacturer's specifications/recommendations, this implies that these specifications/recommendations were prepared by an engineer. Examples of the standards where requirements for engineers are implied include the following.

Subpart E – Personal Protective and Life Saving Equipment

1926.104(b)   Lifelines shall be secured above the point of operation to an anchorage or structural member capable of supporting a minimum dead weight of 5,400 pounds.

Subpart L – Scaffolds:

1926.451(a)(6)   Scaffolds shall be designed by a qualified person and shall be constructed and loaded in accordance with that design.

Subpart M – Fall Protection

1926.502(d)(8)   Horizontal lifelines shall be designed, installed, and used under the supervision of a qualified person, as part of a complete personal fall arrest system, which maintains a safety factor of at least two.

Subpart 0 – Concrete and Masonry Construction

Construction loads. (a) No construction loads shall be placed on a concrete structure or portion of a concrete structure unless the employer determines, based on information received from a person who is qualified in structural design, that the structure or portion of the structure is capable of supporting the locals.

Subpart P – Excavations

(c)(1)(i) Structural ramps that are used solely by employees as a means of access or egress from excavations shall be designed by a competent person. Structural ramps used for access or egress of equipment shall be designed by a competent person qualified in structural design, and shall be constructed in accordance with the design.

Proposed OSHA Standards

The OSHA Steel Erection Negotiated Rule-making Advisory Committee (SENRAC) is nearing completion on a proposed rule that will be published in the Federal Register and will modify the current Subpart R – Steel Erection. SENRAC is a special OSHA committee that represents all segments of the steel erection industry owners, contractors, subcontractors, manufacturers, labor, etc. The latest public draft document from this committee specifies the duties of contractors to assign engineers to ensure the stability and safety during the erection of steel structures.

SENRAC proposed that a Project Structural Engineer of Record (SER) be established The SER is defined as "the registered, licensed professional responsible for the design of structural steel framing and whose seal appears on the structural contract documents." The committee has assigned the following responsibilities to the SER:

"Anchor bolts shall not be repaired, replaced, or field modified without the approval of the Project Structural Engineer of Record" and "Modifications shall not be made to steel joists that affect the strength of the joist without the approval of the Project Structural Engineer of Record." The SENRAC has included in appendices to their proposal options that may be adopted by the controlling contractor that if implemented would be considered to be in compliance with the standard. The provisions in the appendices are:

1. In multi-story structures, the Project Structural Engineer of Record may facilitate the ease of erecting perimeter safety cable, where structural design allows, by placing column splices sufficiently high so as to accommodate safety cables located at 42-45 inches above the finished floor."

2. Pre-construction conference(s) and site inspection(s) should be held between the erector and the controlling contractor, and others such as the project engineer and the fabricator, prior to the start of steel erection to both develop and review the site-specific erection plan which will meet the requirements of this section."

Consensus Standards

In addition to OSHA requirements, numerous consensus standards provide that engineers will perform certain duties. These standards are either referenced by OSHA standards or can be cited in support of citations for violations of the Occupational Safety and Health Act General Duty Clause.

The ANSI A10.9 "Concrete and Masonry Work Safety Requirements," revised June 1997, standard includes provisions that address engineers. Some examples are:

1.8 Design and Drawings. Structural concrete, not on grade, and vertical walls, except for single-story residential basement walls, shall require formwork drawings prepared by an engineer, a copy of which shall be available at the jobsite. Design, specifications, erection, and reshoring drawings shall be prepared or approved by an engineer.

1.9.2 The plan shall list all instances where an engineer or competent person is required to design, plan, supervise, test, or perform inspections of materials, including structural members. The plan shall describe the specific precautions to be taken to ensure stability of structures during the construction process.

1.9.5 The plan shall describe procedures to identify individuals responsible to address conditions that could create hazards to employees. Special inspections shall be conducted by an engineer or a person experienced with safety in hazardous conditions. This person shall make notes in the project log describing the methods to protect employees under such conditions.

2. Definitions

Engineer. A licensed or registered professional engineer.

6. Vertical Shoring

6.4.1 Shoring or reshoring equipment shall not be removed until the specified required tests indicate that the specified concrete strength has been reached as established by an engineer.

10.2 Designing and Planning. Lift-slab operations shall be designed, planned, and supervised by an engineer. Such plans and designs shall include detailed instructions and sketches indicating the prescribed method of erection.

The ANSI A10.33-1992 "Safety and Health Program Requirements for Multi-Employer Projects" standard includes provisions that address engineers.

3.9   Critical Structures and Complex Processes. The Project Constructor shall determine whether the project or parts of the project are critical structures or complex processes that require planning, design, inspection, and/or supervision by a licensed professional engineer.

Inconsistent Reference to Engineers

The above examples describe where and when a contractor is required to engage an engineer. These examples illustrate that inconsistent terms are used to describe engineers. The terms include licensed, registered, qualified, and competent. An example is found in the OSHA regulation 29 CFR 1926 Subpart N – Cranes, Derricks, Hoists, Elevators, and Conveyors standard.

1926.552 (c) (17)(i) Personnel hoists used in bridge tower construction shall be approved by a registered professional engineer and erected under the supervision of a qualified engineer competent in this field.

The checklist in Figure 6-4 will help to determine when to use an engineer in construction work.

SPECIAL PROGRAMS

Special emphasis programs have been previously mentioned but should be reinforced as an effective accident prevention technique. Any time you institute a special program which targets a unique safety and health issue, you have developed an organized approach in prevention. The benefits of instituting a special program are the potential hazard is kept on everybody's mind, you receive feedback, and the workers receive reinforcement for the desired performance. You can develop a program in any area where you feel the need. Some areas of focus could be ladder safety, back injuries, vehicle or equipment safety, power tool incidents, etc. For success, the program may contain goals to attain, rewards to receive, or even consequences to be enforced, if the rules of the program are not followed. By setting up a program you are at least taking action to target accidents and prevent their occurrence.

PREVENTIVE MAINTENANCE PROGRAMS

A Preventive Maintenance Program (PMP) depends heavily on an inspection form or checklist to assure that a vehicle or equipment inspection procedure has been fully accomplished and its completion documented. It has long been noted that a PMP has benefits which extend beyond caring for equipment. Of course, equipment is expensive and if cared for properly and regularly, will last a lot longer, cost less to operate, operate more efficiently, and have fewer catastrophic failures.

Remember, properly maintained equipment is also safer and there is a decreased potential risk of accidents occurring. The degree of pride for having safe operating equipment will transfer to the workers in the form of better morale and respect for equipment. Well-maintained equipment sends a strong message regarding safe operating equipment.

Engineering Checklist

The following checklist is provided as a reference to the construction standards that may require the services of a professional engineer. Use this checklist as a guide to help plan for and implement the engineer requirements for various construction activities.

<u>Yes</u> <u>No</u> <u>NA</u>

Process Safety Management

❒ ❒ ❒ 1. Process safety information on equipment design – 1926.64(d)(2)(iii)
❒ ❒ ❒ 2. Quality assurance for mechanical integrity tests – 1926.64(j)(6)(ii)

Fire Protection and Prevention

❒ ❒ ❒ 1. Design of automatic extinguishing systems for indoor flammable/combustible liquid storage areas – 1926.152(b)(4)(ii)
❒ ❒ ❒ 2. Tank design for flammable/combustible liquids – 1926.152(i)(1)(i)(C)-(D) and (F), (i)(5)(ii)-(vii), (8)(7)
❒ ❒ ❒ 3. Design of piping, valves, and fittings for flammable/combustible liquids – 1926.152(j)(1)(i)
❒ ❒ ❒ 4. Design of LPG storage container systems using non-DOT containers – 1926.153(m)(3)(vi)

Scaffolds

❒ ❒ ❒ 1. Scaffold capacity – 1926.451(a)(6)
❒ ❒ ❒ 2. Scaffold components from different manufacturers not intermixed – 1926.451(b)(10)
❒ ❒ ❒ 3. Supporting surfaces on suspension scaffolds able to support loads – 1926.451(d)(3)(i)
❒ ❒ ❒ 4. Use of scaffolds in accordance with design – 1926.451(f)(3), (5)
❒ ❒ ❒ 5. Design of pole scaffolds – 1926.452(a)(10)
❒ ❒ ❒ 6. Design of tube and coupler scaffolds – 1926.452(b)(10)
❒ ❒ ❒ 7. Design of fabricated frame scaffolds – 1926.452(c)(5)(iii), (6)
❒ ❒ ❒ 8. Design of outrigger scaffolds – 1926.452(i)(8)
❒ ❒ ❒ 9. Design of two point adjustable scaffolds – 1926.452(p)(1)
❒ ❒ ❒ 10. Design of interior hung scaffolds – 1926.452(t)(2)

Fall Protection

❒ ❒ ❒ 1. Design of safety net systems – 1926.502(c)(4)(ii)
❒ ❒ ❒ 2. Design of personal fall arrest systems – 1926.502(d)(8), (19), Appendix C (II), Appendix E (IV)(A)(4)

Cranes, Derricks, Hoists, Elevators, and Conveyers

❒ ❒ ❒ 1. Cranes and Derricks, compliance with manufacturers specifications and limitations – 1926.550(a)(1)
❒ ❒ ❒ 2. Design, inspection, and use of crane or derrick suspended personnel platforms – (126.550(g)(2), (4)(i), (5)(iv), (5)(vi), (6)

Figure 6-4. Engineering checklist

Yes	No	NA		
☐	☐	☐	3.	Material hoists, personnel hoists, and elevators, compliance with manufacturers specifications and limitations – 1926.552(a)(1)
☐	☐	☐	4.	Design of material hoists – 1926.552(b)(7)
☐	☐	☐	5.	Design of personnel hoists – 1926.552(c)(16), (17)(i)

Motor Vehicles, Mechanized Equipment, and Marine Operations

Yes	No	NA		
☐	☐	☐	1.	Stability of pile driving equipment – General – 1926.603(a)(12)

Excavations

Yes	No	NA		
☐	☐	☐	1.	Design of access and egress ramps – 1926.651(c)(1)(i)
☐	☐	☐	2.	Protection from water accumulation hazards – 1926.651(h)(1)
☐	☐	☐	3.	Stability of adjacent structures – 1926l651(i)(1) – (3)
☐	☐	☐	4.	Design of sloping and benching systems – 1926l.652(b)(3)(i), (iii), (4)(i), (ii), and Appendix B (c)(3)(iii)
☐	☐	☐	5.	Design of support systems – 1926.652(c)(3)(i), (iii), (4)(i), (iii)(B)
☐	☐	☐	6.	Inspection of damaged materials and equipment – 1926.652(d)(3)
☐	☐	☐	7.	Installation and removal of support systems – 1926.652(e)(1)(iii)
☐	☐	☐	8.	Design of aluminum hydraulic shoring for trenches – 1926.652 Appendix D, Tables D1.3 and D1.4, and (d)(3)(ii)(A)
☐	☐	☐	9.	Design of protective systems – 1926.652 Appendix F

Concrete and Masonry

Yes	No	NA		
☐	☐	☐	1.	Inspection of concrete structures prior to loads being placed on them – 1926.701(a)
☐	☐	☐	2.	Concrete pumping systems provided with pump supports – 1926.702(e)(1)
☐	☐	☐	3.	Design of formwork for cast-in-place concrete – 1926.703(a)(1)
☐	☐	☐	4.	Damaged or weakened shoring and reshoring equipment – 1926.703(b)(4)
☐	☐	☐	5.	Design of vertical slip forms – 1926.703(c)(1)-(8)
☐	☐	☐	6.	Support of precast concrete walls – 1926.704(a)
☐	☐	☐	7.	Design of lift-slab construction operations – 1926.705(a), (k)(1), and Appendix

Steel Erection

Yes	No	NA		
☐	☐	☐	1.	Design of permanent flooring, skeleton steel construction – 1926.750(a)(1)
☐	☐	☐	2.	Removal of plumbing up guys – 1926.752(d)(4)
☐	☐	☐	3.	Lifelines secured to member capable of supporting 5,400 pounds – 1926.104(b)

Underground Construction, Caissons, Cofferdams, and Compressed Air

Yes	No	NA		
☐	☐	☐	1.	Check-in/check-out procedures in place – 1926.800(c)

Figure 6-4. Engineering checklist (continued)

Yes	No	NA		
❒	❒	❒	2.	Ground stability in subsidence and underground areas – 1926.800(o)(2), (3)(i)(A), (iv)(A)-(B)
❒	❒	❒	3.	Medical locks designed for working pressure of 75 psig – 1926.803(b)(10)(vii)

Demolition

Yes	No	NA		
❒	❒	❒	1.	Survey conducted prior to demolition activities – 1926.850(a)
❒	❒	❒	2.	Design of chutes – 1926.852(g)
❒	❒	❒	3.	Removal of floor arches not more than 25 feet above grade and not endangering stability of structure – 1926.857(d)
❒	❒	❒	4.	Dismembered steel not over stressed – 1926.858(d)
❒	❒	❒	5.	Inspection during mechanical demolition – 1926.859(g)

Rollover Protective Structures: Overhead Protection

Yes	No	NA		
❒	❒	❒	1.	Minimum performance criteria for rollover protective structures – 1926.1001(a)
❒	❒	❒	2.	Design of overhead protection for operators of agricultural and industrial tractors – 1926.1002(a)(1)

Stairways and Ladders

Yes	No	NA		
❒	❒	❒	1.	Design of Ladders – 1926.1053(a)(22), (23)

Toxic and Hazardous Substances

Yes	No	NA		
❒	❒	❒	2.	Evaluation of asbestos work areas – 1926.1011(g)(6)(ii), Appendix F

Figure 6-4. Engineering checklist (continued)

The first aspect of a PMP is to have a schedule for regular maintenance of all your equipment; second, have supervisors assure that all operators are conducting daily inspections; third, assure all defects are reported immediately; and fourth, make sure repairs are made prior to operating vehicles or equipment. If this is impossible, the equipment should be tagged and removed from service.

If you allow operators to use equipment, machinery, or vehicles that are unsafe or in poor operating condition, you send a negative message which says, "I don't value my equipment, machinery, or vehicles and I don't value my workforce." A structured PMP will definitely foster a much more positive approach regarding property and the workforce.

The reasons for a PMP are

1. Improved operating efficiency of equipment, machinery, or vehicles.

2. Improved attitudes toward safety by maintaining good/safe operating equipment.

3. Fosters involvement of not only maintenance personnel but also supervisors and operators which forces everyone to have a degree of ownership.

4. Decreased risk for incidents or mishaps.

What you will need is

1. A maintenance department carrying out a regular and preventive maintenance schedule.

2. Accountability and responsibility of both the supervisor and operator.

3. A preshift checklist for each type of equipment, machinery, or vehicle. An example of a vehicle checklist can be found in Figure 6-5.

4. An effective response system when defects or hazards are discovered.

5. A commitment by you that this is important and will be done.

Preventive Maintenance Programs are another example of an accident prevention technique. Of course, this technique would apply only if your company has equipment, machinery, or vehicles. A preventive maintenance program is also an integral part of a fleet safety program.

FLEET SAFETY PROGRAM

Fleet safety may be viewed as vehicle safety or mechanical safety, but this type of safety depends upon both maintenance and operators. An operator's job would include pre-operation inspections and, upon completing his/her use of the vehicle, the reporting of any defects. This should be normal operating procedure for any preventive maintenance program. Nevertheless, these types of expectations are dependent upon the quality of the operators. Your operators will be the true king pins of the fleet safety program

Thus, in a fleet safety program, the importance of operator selection is vital to the prevention of accidents, incidents, vehicle damage, and injuries; operator selection is paramount to an effective fleet safety program. The selection process should involve access to operators' past employment history, driving record (including accidents), accommodations, or awards, as well as previous operators' experience, if any, on your types of equipment.

As a condition of employment and based upon the criteria in a written job description, all potential operators should be able to pass a physical and mental examination and an alcohol/drug test. You will also need a well-written job description and you may want your legal counsel and others to review it prior to its use.

PRESHIFT EQUIPMENT CHECKLIST

Check any of the following defects prior to operating equipment or vehicles and report those defects to your supervisor or maintenance department.

Type of Equipment:_____ Identification Number:_____

Date:_____

1. Walk Around:
_____ Broken Lights
_____ Oil Leaks
_____ Hydraulic Leaks
_____ Tires
_____ Tracks
_____ Damaged Hose
_____ Bad Connections/Fittings
_____ Cracks in Windshields or Other Glass
_____ Damaged Support Structures
_____ Damage to Body Structures
_____ Fluid Levels
_____ Oil
_____ Hydraulic
_____ Brake
_____ Mirrors

3. While Underway:
_____ Engine Knocks, Misses, Overheats
_____ Brakes do not Operate Properly
_____ Steering Loose, Shimmy, Hard, etc.
_____ Transmission Noisy, Hard Shifting,
Jumps Out of Gear, etc.
_____ Speedometer
_____ Speed Control

4. Emergency Equipment:
_____ First Aid Kit
_____ Fire Extinguisher
_____ Flags, Flares, Warning Device
_____ Reflectors
_____ Tire Chains if Needed

2. Operation:
_____ Engine Starts
_____ Oil Pressure
_____ Air Pressure or Vacuum Gauge
_____ Brakes
_____ Parking Brakes
_____ Horn
_____ Front Lights
_____ Back Lights
_____ Directional Lights
_____ Warning Lights
_____ Back-up Alarm
_____ Noises or Malfunctioning
_____ Engine
_____ Clutch
_____ Transmission
_____ Axles
_____ Fuel Level
_____ Instrument Panel
_____ Windshield Wipers or Washers
_____ Heater or Defroster
_____ Mirrors

_____ Seat Belts
_____ Steering Wheel Play
or Alignment

5. Cargo Related Equipment:
_____ Tie Downs
_____ Cargo Nets
_____ Tarps

6. Other Items:
_____ Hand Tools
_____ Spare Parts
_____
_____
_____

No Defects Noted: Operator Signature:_____ Date:_____ Time:_____
Describe Any Defects Notes:_____
_____
_____
_____

Defects Corrected: Defect Correction Unnecessary: Defects Corrected By: Date:
(Initials)   (Initials)   (Signature)

Defects Corrected: Operators Signature   Date:
_____Yes _____No
_____

Figure 6-5. Preshift vehicle checklist

All operators should undergo training related to company and government policies and procedures. This training should include recordkeeping, accident/incident reporting, driving requirements, and defensive driving. After classroom training, each operator should be required to take a supervised driving test or hands-on supervised operational drive, to determine his/her competence. This should be done before the operator is released for work-related driving assignments. Even after the operator is released from training status, he/she may have a supervisor accompany him/her on their work assignment.

Operators should be observed and evaluated on a periodic basis and retrained, if necessary, or more closely supervised. If your vehicles are in proper and safe operating condition, your operators become the key to your fleet safety program. Good, conscientious operators can prevent accidents from occurring; they are the focus of your fleet safety program.

ACCIDENT INVESTIGATION

It is important to have some mechanism in place to analyze accidents/incidents to determine the basis of cause-and-effect relationships. You may determine these types of relationships only when you actively investigate all accidents and incidents which result in injuries, illnesses, or property/ equipment/machinery damage. You must have a system in place for these investigations.

Once the types of accidents/incidents which are transpiring have been determined you can undertake prevention and intervention activities to assure that there will be no recurrences. Even if you are not experiencing large numbers of accidents/incidents, you still need to implement activities which actively search for, identify, and correct the risk from hazards on jobsites. Reasons to investigate accidents/incidents include

1. To know and understand what happened.

2. To gather information and data for present and future use.

3. To deter cause and effect.

4. To provide answers for the effectiveness of intervention and prevention approaches.

5. To document the circumstances for legal and workers' compensation issues.

6. To become a vital component of your safety and health program.

If there are only a few accidents/incidents, you might want to move down one step to examine near misses and first aid related cases. It is only a matter of luck or timing that separates the near miss or first aid event from being a serious, recordable, or reportable event. The truth is you probably have been lucky by seconds or inches. (A second later and it would have hit someone or an inch more and it would have cut off a finger.) Truly, it pays dividends to take time to investigate these accidents and incidents occurring in the workplace.

Purpose of Accident Investigations

An important element of a safety program is accident investigation. Although it may seem to be too little, too late, accident investigations serve to correct the problems which contribute to an accident and it also reveals accident causes which might otherwise remain uncorrected. The kinds of accidents which should be investigated and reported are

1. Disabling injury accidents.

2. Nondisabling injury accidents that require medical treatment.

3. Circumstances that have contributed to acute or chronic occupational illness.

4. Noninjury, property damage accidents that exceed a job project specified cost.

5. Near accidents with a potential for serious injury or property damage.

Potential Risks

Potential is a key word in accident investigation. Potential means looking at how bad it could be. Of course, anything can be stretched to the extreme (the slightest cut could become infected and the person could die). The clue is to look at what the results would normally be and what the normal risks are—and act accordingly. Let's look at two examples regarding accident potential.

A construction worker is walking between some earth-moving equipment. A sudden gust of wind blows a small speck of sand into his eyes and it's irritating. He goes to the nurse and gets it removed and in the process he loses thirty minutes from the job.

Sometime later another worker is sharpening a tool. A small piece of metal is thrown off by the grinding wheel. It lodges in his eye; it's also irritating. He, too, goes to the nurse, gets it removed, and loses thirty minutes from the job.

Both accidents are the same in terms of actual loss. In terms of potential, there is a big difference. The first has little potential unless winds are unusually strong and a wind "tunnel" is created by the positions of the equipment, or there is frequent travel between the pieces of equipment. The second accident has a normally high potential for serious injury because pieces that fly off a grinding wheel usually hit an object with a great amount of energy. For that reason, the second case should be investigated even though there was no serious injury at that particular time.

When accidents are not reported, their causes usually go uncorrected. This allows the chance for the same accident to happen again. Every accident, if properly investigated, serves as a learning experience to the people involved. However, the investigation should avoid becoming a mechanical routing. It should strive to establish what happened, why it happened, and what must be done to prevent a recurrence. An accident investigation must be conducted to learn the facts, not to place blame.

In any accident investigation, consider the aspect of multiple causation. The contributing factors surrounding an accident, as well as the unsafe acts and unsafe conditions, should be considered. If only the unsafe acts and conditions are considered when investigating an accident, little will be accomplished toward accident prevention effort because the "root causes" still remain. This allows for the possibility of an accident to recur. The "root causes," such as management, policies, and decisions, are the personal and environmental factors which could effect permanent results when corrected.

In 1969 an analysis was made of 1,753,498 accidents reported by 297 cooperating companies. These companies represented twenty-one different industrial groups with 1,750,000 employees who worked over 3 billion man-hours during the analyzed exposure period. The study revealed the following ratios of accidents reported: 1 major injury, 10 minor injuries, 30 property damage accidents, 600 incidents with no apparent injury or property damage. For every reported injury (resulting in death, disability, lost time, or medical treatment), there were 9.8 reported minor injuries (requiring only first aid). For the ninety-five companies which further analyzed their major injuries, the ratio was one lost-time injury for each fifteen medical treatment injuries.

Forty-seven percent of the companies indicated that they investigated all property damage accidents, while eighty-four percent stated that they investigated only major property damage accidents. The final analysis indicated that 30.2 property damage accidents were being reported for each major injury.

Part of the study also involved 4,000 hours of worker interviews. These interviews were done by trained supervisors and the subject was on the occurrence of incidents that, under slightly different circumstances, could have resulted in injury or property damage.

In referring to the 1-10-30-600 ratio, it should be remembered that this represents reported accidents and incidents and not the total number of accidents or incidents that actually occurred. As we consider the ratio, we observe that thirty property damage accidents were reported for each serious or disabling injury. Property damage accidents cost billions of dollars annually and yet they are frequently misnamed and referred to as "near-miss" accidents. Ironically, this line of thinking recognized the fact that each property damage situation could possibly have resulted in a personal injury. This term is a hold-over from earlier training and misconceptions that lead supervisors to relate the term accident to injuries only.

The 1-10-30-600 relationship in the ratio indicates quite clearly how foolish it is to direct our total effort at the relatively few events terminating in serious or disabling injury. The fact that 630 property damage or no-loss incidents also occurred provides a much larger basis for more effective control of the total accident losses.

Accident investigations become more effective when all levels of management, particularly top management, takes a personal interest in controlling accidents. Management adds a contribution when it actively supports accident investigations. It is normally the responsibility of line supervisors to investigate all accidents, and in cases where serious injury or equipment damage results, other personnel such as department managers, as well as an investigation team, might become involved.

Let's take a look at the reasons why the line supervisor should make the investigation. An accident happens because there is a problem somewhere and the problem exists because

1. A deficiency is not known.

2. The risk involved in the deficiency was thought to be less than it really is.

3. In spite of the deficiency, someone without authority decided to go ahead.

4. Someone with authority decided that the cost was more than the money they had available, or the cost outweighed the benefit of correcting the deficiency.

When taking enough time, anyone can find out what the exact problem is. For every problem there is an answer—a single best way to correct the problem and for several reasons the supervisor is the best person to put this all together.

1. The supervisor has a personal interest to protect. The supervisor is responsible for the area and everything that goes on in it. Supervisory performance is affected by mistakes, defects, breakdowns, absences, and the time it takes to recover from injuries.

2. The supervisor knows the most about the workers and conditions. The supervisor knows better than anyone else the skill and personality of the people who work for him or her. He or she knows the equipment in use, past problems with the equipment which relate to the accident, and why the decision was made to use that equipment instead of some other model. The supervisor knows why the work materials were selected and the nature of the materials and how they could affect the work. He or she knows the work area, the lighting, temperature control, layout, and countless things that affect the way the job is done.

3. The supervisor knows best how to get information. The supervisor knows and understands the people who work for him or her. He or she knows how they

think and how to talk to them. Anyone else would have to take a lot of their time away from the job to establish rapport and communicate with them.

Supervisors also know where the technical information on the equipment is; where the training records are; where purchasing specifications for materials are; who comes in the area for maintenance; who delivers material; and who does the outside jobs which are related to the accident. Supervisors can check all the information related to the accident in less time than it would take you to tell another investigator how and where to find it.

4. The supervisor will take the action anyway. When the investigation is all over, who do you think is going to have to take the corrective action? That's right: the supervisor. Anything that's going to be done to correct the problem is going to involve the supervisor.

The supervisor can take a lot of time trying to make someone else's ideas work, but then have to write a report on why the recommended action won't work and what he or she suggests should be done. Or, he or she can save all that time and find out what the real problem is and correct it in a way he or she knows will work.

Investigations Benefit the Supervisor

There are some other benefits a line supervisor can gain from making the accident investigation.

Those benefits are

1. An increase in production time.

2. An evidence of the supervisor's concern for the workers.

3. A reduction in operating costs.

4. A demonstration of the supervisor's control.

The supervisor who gets directly into an investigation and takes immediate action to prevent other accidents shows the workers that he or she really cares for their physical welfare and their effectiveness in doing the job. The supervisor who lets others do his or her work makes a poor impression on his or her people.

One of the common effects of an accident is that it interrupts work. Effective investigations prevent repeated accidents and control work interruptions and inefficiencies. The cost of delays, injuries, property damage, and insurance all add to the cost of the job. Control of costs is one way a supervisor is measured in job performance.

Persistent efforts to solve problems reflect the image of a capable supervisor who is in control of things. Actions speak louder than words. People like to work for a supervisor who takes control of any situation; who shows he or she is capable of dealing with any problem.

Reporting Accidents

The first step in effective investigation is the prompt reporting of accidents. You can't respond to accidents, evaluate their potential, and investigate them, if they are not reported when they happen. Prompt reporting is the key to effective accident investigations. Hiding small accidents doesn't help prevent the serious accidents that kill people, put the company out of business, and take away jobs. If workers don't report accidents to the supervisor, they are stealing part of the supervisor's authority to manage his or her job.

The results-oriented supervisor recognizes that the real value of investigation can be achieved only when his or her people report every problem, incident, or accident that they know of. In order to promote conscientious reporting, it may be helpful to know some of the reasons why workers fail to, or avoid, reporting accidents.

Ten Reasons for Not Reporting Accidents

1. Fear of Discipline: Punishment for contributory actions or negligence when property is damaged, people are injured, or problems are encountered, is an age old management abuse. Good discipline is always in order, but discipline should not mean punishment or making others the scapegoat for inadequacies. The punishment approach only drives the problem underground.

2. Concern about the Record: Whether it is a job delay with a production's record missed, or an injury reported and an injury-free record broken, most workers don't want to spoil the contractors' and crews' record. Workers get caught up in the spirit of competition or personal satisfaction in being part of the best team, and thereby avoid reporting accidents.

3. Concern for Reputation: Few people want to be labeled an unsafe worker; called "accident prone," or marked as an accident repeater.

4. Fear of Medical Treatment: It is surprising how many people dread the thought of medical care, even simple first aid. Many are absolutely certain they will never have a medical problem if they stay away from doctors. Consequently, they suffer severe complications of minor injuries.

5. Dislike of Medical Personnel: Sometimes medical personnel develop abrupt or even gruff manners. As a result, people don't want to be treated by them and injuries never get reported.

6. Desire to Keep Personal Record Clear: Many companies give bonuses or other recognition to people with accident-free records. Others keep accident data in personnel records for drug-control, security evaluations, or other purposes. Misunderstanding of the use of this information leads to defensive concealment.

7. Avoidance of Red Tape: Extensive forms, lengthy interviews, preparation of statements, and other excessive administrative measures lead workers to avoid getting involved in reporting.

8. Desire to Prevent Work Interruptions: Many, if not most, workers are sincerely interested in getting the job done. Taking the time to go for treatment or to have seemingly small damages evaluated just doesn't seem practical to them. Others recognize that an entire work crew could be held up by their absence to report an accident.

9. Concern about Attitudes of Others: Workers want their fellow workers and their supervisors to have a favorable attitude toward them. If acceptance is thought to require not reporting accidents that reflect on their peers and supervisors, they will avoid reporting accidents.

10. Poor Understanding of Importance: Most workers, studies have shown, neglect to report accidents because they either are not aware of the benefit that can come from an investigation, or they have not seen any benefit come from previous reports.

How can we combat these reporting problems?

1. React in a more positive way.

2. Give more attention to prevention and control.

3. Recognize individual performance.

4. Develop the value of reporting.

5. Demonstrate belief by action.

6. Do not make mountains out of molehills.

Let the worker know that you appreciate his or her reporting promptly. Inquire about his or her knowledge of the accident. Don't interrogate or "grill" the person. Stress the value of knowing about problems while they are still small. Focus on accident prevention and loss control. Emphasize compliance with practices, rules, protective equipment, and promptly commend good performance. Pay attention to the positive things workers do and give sincere, meaningful recognition where and when it is deserved. Use compliments as often as you use warnings. Use group and personal meetings to point out and pass on knowledge gained from past accidents. Make an accident example an important part of every job instruction. Show that you believe what you say by taking corrective action promptly. You can always do something right at the moment, even if permanent correction requires time to develop new methods, buy new equipment, or modify the building.

Good Accident Investigations

A complete investigation includes an objective evaluation of all the facts, opinions, and statements, and related information, as well as an action plan or steps to prevent or control a similar recurrence.

The quality and usefulness of the information is directly related to the degree of thoroughness and conscientiousness of the investigation. During the entire course of the investigation, it is very important that you keep a positive attitude. The whole purpose of the investigation is to prevent or control a future recurrence of a similar event. Blame fixing, fault-finding, and witch-hunting exercises destroy this purpose. One of the key objectives in investigating is to find the cause of the accident.

Remember these three basic accident investigation principles:

1. All accidents/incidents should be investigated, including those that result only in first aid type injuries or property damage.

2. The reporting system should be easy to use. One form should be used to collect all the information needed (similar to the example in Chapter 4).

3. The key person in any accident prevention program is the supervisor.

A complete accident investigation requires extensive documentation, as well as problem-solving skills. It is similar to the investigations conducted by police officers and those of detectives trying to solve crimes. The documentation process requires the use of a variety of tools which need to be available to an investigator. These tools should include, but not be limited to

1. photographic cameras
2. accident investigation forms
3. rulers and measuring tapes
4. tape recorder
5. calipers, micrometers, feeler gauges
6. sampling containers
7. sealable plastic bags
8. caution/barrier tape
9. video camera
10. interview forms

11. flashlights
12. stopwatch
13. special tools needed
14. sampling devices/instruments

15. electrical or other testing equipment
16. special personal protective equipment
17. latex gloves

In order to prevent a recurrence of an accident, all aspects of the accident scene must be completely documented and studied to determine what transpired prior and during the accident sequence.

The specific circumstances of an accident investigation cannot always be anticipated. By planning to ensure that proper procedures will be followed, many problems faced by the investigator may be eliminated. This will assure the accuracy of the investigation and the usefulness of the investigation as an accident prevention technique.

Determining Accident Types

A decision must initially be made as to what types of accidents are most probable at your jobsite or projects. A construction operation has many different accident-producing hazards. An industrial operation also has various associated hazards which are different from the construction operation. These hazards, inherent to the particular operation, must be determined.

Designing a Report/Investigation Form

An accident report form must be designed to cover all accidents/incidents likely to occur. This form will serve as a vehicle for future use in decision making. A report form should be grouped into specific categories, such as

1. Background information.

2. Account of the accident.

3. Discussion (analysis of the accident).

4. Recommendations (to prevent a recurrence).

At times, due to the type of accidents, number of incidents, or number of injuries which are occurring, you may need to develop a form which addresses the specifics of that accident, incident, or injury. It is not unusual for a company to have a supplementary form which is used in conjunction with their standard accident investigation form. This supplementary form may be specifically designed for the collection of information on back injuries, if large numbers are occurring or you cannot determine the cause of them. An example of a form which is designed for a specific problem area is found in Figure 6-6.

Organizing/Assigning Responsibilities

Organizing a division of responsibility toward accident investigations should include a broad range of experience, responsibility, and authority. Although more serious accidents or equipment damage should be investigated primarily by a person with more authority, upper-level management personnel, department supervisors, and first-line supervisors should still play an active role in accident investigations.

Specific accident investigation responsibilities should be assigned to various people. Serious injury or major equipment damage accidents should be investigated by a committee. The committee should involve all levels of management, and should involve a range of technical competencies. Most accidents that are potentially serious or have the potential for major

---

ELECTRICAL RELATED ACCIDENTS

1. What is the worker's job title?_____

2. What was the worker's job at the time of the accident?_____

3. Did the worker have electrical safety training?_____

4. Was the worker qualified to do electrical work?_____

5. How long had the worker been with the company?   _____yrs. _____mos.

6. How much electrical experience does the worker have?   _____yrs. _____mos.

7. What did the worker contact? _____

8. Was there equipment involved? _____

9. What was the operating voltage? _____

10. What was the current (amperage)?_____

11. Was the equipment damaged by the accident? _____

       Extent of Damage:_____

12. Was there a fault with the equipment?_____

       Describe: _____

13. Was the worker wearing personal protective equipment? _____

14. Was the worker wearing electrical protective equipment?_____

       Describe: _____

15. Describe the Accident: _____

    _____

16. Describe the worker's injuries: _____

    _____

17. What was the cause(s) of the accident? _____

    _____

18. How could it have been prevented? _____

    _____

Figure 6-6. Specific accident type investigation form

equipment damage should be investigated by the department's supervisor in conjunction with his/her own supervisors. Minor injuries and minor equipment damage accidents should normally be investigated by the responsible first-line supervisor.

Explaining Accident Investigations

The purpose of communicating the program is to help shape the workers' attitudes and acquaint them with management's intent toward maintaining a safe working environment through properly conducted accident investigations. Explain to the workers through meetings, posters, newsletters, etc., how accident investigation procedures will operate. Ways to encourage workers to report minor injury accidents/incidents are

1. Indoctrinate workers on the importance of reporting all accidents.

2. Make sure everyone knows what kinds of accidents should be reported.

3. Show disapproval of injuries neglected and not reported.

Controlling and Follow-Up

A follow-up should be imposed on all investigations. One way of doing this is to have the department supervisor report the implementation of all recommended corrective actions. Other management personnel can also perform follow-ups of the proposed corrective actions; this can be accomplished by conducting a simple inspection to insure that the corrective actions have taken place.

Monthly safety meetings are also a good place to determine whether corrections have been implemented.

Sources of Information

In order to effectively analyze and take good and effective action when an accident occurs, you need to know where to get the appropriate information. There are four sources of information and they are

1. People.

2. Positions.

3. Parts.

4. Papers.

People are the source of eye or ear witness testimony. Evidence is recorded through their powers of observation and memory, but it is most fragile because it is subject to forgetfulness, rationalization, influence, personal conflicts, and manner of the interview. All of these can induce subtle distortions of the evidence.

Sources of evidence are also fragile because sometimes things have been moved by medical teams, fire fighters, or other people who were involved in the accident.

Parts evidence has many facets: internal defects, proper installation, manufacturing standards and tolerances, use of correct parts and materials, and willful or inadvertent abuse. They are also subject to pilferage, corrosion, maring, and misplacement.

Paper evidence are such things as job standards, training records, operating instructions, maintenance records, etc. These are far more reliable than other evidence and more difficult to change. Nevertheless, they are often overlooked or are not impounded and therefore, may be subject to misplacement or alteration.

The Interview Process

When interviewing workers, don't immediately come out with the full story. There are a few tricks to a good interview.

1. Visual orientation. Survey the accident scene to get "the big picture." This helps you talk with the witness from a point of common knowledge. While making this survey, you should outline in your mind what questions need to be answered. Make notes to help remember the things that look different or out of place so you can ask questions about the incident. Try to avoid moving or disturbing anything at the scene; this might alter the evidence.

2. Select a suitable place for the interview; this gets things off to a good start. It should be a place where the witness will feel comfortable. If the scene of the accident is not too dangerous or unsuitable because of noise, weather, or work activity, an interview at the scene can help stimulate the witness's memory. It can help communication through visually relating to things at the scene. If the accident site is not suitable, a neutral location will help the witness feel at ease.

   It is not a good idea to meet with the witness at your office, especially if you are sitting behind your desk. It is psychologically uncomfortable for him or her since this is where he or she is used to getting orders or maybe even criticism. Because of this, the witness would be more reluctant to give information which could imply criticism of anyone. If you must use your office, get out from behind the desk and sit next to the witness.

   The interview should always be private, or at least out of the hearing of others. This will allow the witness to respond from his or her own observations and not be influenced by the presence of others. When witnesses are interviewed in the presence of others, they react to the strongest personality in the group and don't speak up when they disagree.

3. Control the impulse to get right to the facts of the accident. A few moments to show consideration of the witness and to inquire about his or her injuries or general welfare will save time in the long run. Greeting the witness by name and with a handshake can help overcome nervousness and get the interview moving smoothly.

4. Next, explain that the purpose of the interview is to find out what happened and why, so the problem can be corrected before another accident occurs.

5. Actual testimony should start by asking the witness to tell, in his or her own words, what was observed, seen, heard, or known about the accident. Don't ask for a chronological order. What the witness remembers best is what is most vivid and meaningful to him; this in itself is meaningful. If you attempt to force thoughts into a sequence, you will cause the witness to forget things or to make up things in order to create a smooth story. When the witness is asked to remember something he or she didn't actually see, this makes him or her uncomfortable. When this happens, the witness may add to his or her story just to fill the gaps, while at the same time honestly believing that he or she actually remembers those "facts."

   Put things in order after you get the facts but remember that your memory is not better than the witness's, so major points of the testimony must be recorded in brief notes. Don't try to write everything word for word; this forces either a slow pace or excessive repetition. To insure understanding and provide the witness with time to reflect and consider, give periodic feedback. Give short summaries of the main points, as you understand them; this corrects misunderstandings.

6. While getting the initial facts, let the witness have silent periods to collect his or her thoughts. Don't prompt or question the witness until he or she appears to have exhausted his or her memory. What he or she remembers without prompting is the most meaningful. After the witness has exhausted his or her memory, you can ask questions to

    A. Expand detail or earlier testimony.

    B. Answer predetermined questions you formed while getting the big picture, or that were raised by previous witnesses' comments.

7. The questions should be neutral in form so they require the witness to form answers in his or her own words. Questions that can be answered yes or no or that offer multiple choices should be avoided as they lead the witness into distortions. Questions should also be objective rather than asking for the witness's opinion.

8. When closing the interview, the investigator can employ control questions to help evaluate the witness's ability to observe and remember. His or her answers to questions on well-established facts can show you his or her capability to observe and be objective.

9. Control questions can include

    A. Time and location of the accident.

    B. Weather, lighting, temperature, noise.

    C. Identification of other witnesses.

    D. Actions of people and emergency teams after the accident.

Another important question to ask the witness is, "what attracted your attention to the accident?" This answer will define the point where he or she actually started observing the accident. During the interview, keep in mind that people often offer good solutions to problems; be sure to get their ideas. They are more likely to support changes when they have had a part in those changes.

Finally, end the interview so that the witness will come back if he or she thinks of anything else about the accident. Thank witnesses for their time and effort and mention how their information will help correct the problem.

Reenacting Accidents

Frequently, a supervisor gets a repeat of an accident when the person who was involved in the accident tries to reenact what took place. Many times the supervisor did not ask for a demonstration but simply asked, "What do you mean?" or "How did that happen?" Reenactment is a critical situation and needs to be tightly controlled.

Reenactment of accidents should only take place when

1. Information cannot be gained in another way.

2. It is vital to know the precise act or sequence in order to develop remedial action.

3. It is necessary to verify conflicting facts.

When circumstances justify the reenactment, the following preliminary precautions should be taken:

1. Make sure the worker understands that he or she is to first explain how the accident happened, step by step, without any motions. Try to find any problems that might come up in a demonstration.

2. Make sure the worker understands that he or she is not to repeat the step that took place just before the accident occurred. Have him or her act out, but not actually operate, each step in slow motion, as he or she explains it a second time. Have him or her explain and then perform the step. Again, look for problems.

3. Before you start, make sure the worker is emotionally fit and has no objection to the reenactment.

In summary, the following points should be covered in an interview:

1. Remind the person of the investigation's purpose.

2. Ask him or her to give their complete version.

3. Ask him or her questions to fill in the gaps.

4. Check your understanding of the accident.

5. Discuss how to prevent an accident recurrence.

The Accident Report

An accident report is the basic component of an accident investigation. There are various types of report formats. A good form will accommodate all types of accidents/incidents that an operation might encounter. Whatever form is used, it must obtain concise report data to help management make various determinations (e.g., insurance or compensation benefits; statistical data for accident analysis; and recommendations for corrective action).

Information from accident investigation reports help to

1. Provide all the facts of an accident in a condensed form.

2. Give management a way to assess where failure occurred and what preventative actions to take to prevent a recurrence.

3. Provide a way to make periodic appraisals of the safety program.

In order to structure an accident investigation report, it is important that all information is gathered and analyzed. Some of the other components needed for an effective accident investigation program are found in Chapter 4 – Accident Investigation, and Chapter 5 – Accident/Incident Analysis.

Investigations of all accidents will provide a barometer of the effectiveness of any safety and health initiative. Gathering facts alone on accidents is not an effective accident prevention technique. The best outcomes are accomplished by utilizing the recommendations and prevention strategies gathered from these types of investigations. By reviewing your accident investigations, you will be able to evaluate the effectiveness of interventions. When the same types of accidents aren't repeated, you will realize how important these types of investigations are. In this light, accident investigations are, in fact, a type of accident prevention technique, as well as an evaluation tool.

JOB SAFETY OBSERVATIONS

There are many categories for accident causes and also many terms used to describe these causes. To precisely determine the causes for one category, the terms "person causes" and "environmental causes" are often used. The "actual" and "potential" causes of accidents are generally accepted as the key factors in a successful loss prevention effort. Actual causes—direct and indirect—can be considered only after an accident has occurred. They can be found by asking the question, "What caused the accident?" Potential causes may be avoided before an accident actually occurs by asking the question, "What unsafe conditions (environmental causes) and/or unsafe procedures (person causes) <u>could</u> cause an accident?" Working with actual causes is similar to firefighting, after-the-fact analysis, or hindsight. Understanding, determining, and correcting potential causes is comparable to fire prevention or foresight.

All categories of accident causes must be considered and used in any complete loss prevention program. Safety observations and safety inspections are necessary phases in the overall safety effort. Making a safety <u>observation</u>—watching a person perform a specific job will detect unsafe behavior (person causes); making a safety <u>inspection</u> by visually examining the work area and work equipment – will detect unsafe conditions (environmental causes). Detecting and eliminating the <u>potential</u> causes of accidents may best be accomplished when supervisors understand safety observations and safety inspections as separate phases of the loss prevention work. This section primarily deals with making safety observations.

The safety observation phase is initiated when a written program of procedures is prepared by management and safety personnel. The program should include the use of prepared job safety analyses or safe job procedures and must include the training of all supervisors in observation procedures. Objectives must be established for each step of the program. The establishment of definite goals at all levels of management will give direction to the safety effort. Management should outline

1. What is to be accomplished by safety observations.

2. How the observations are to be made.

3. Who is responsible for each step of the program.

Each supervisor must set priorities as to workers and jobs to be observed. When instructions have been given, safety observations must be planned to cover, on a systematic basis, all workers through job observations. First line supervisors should be required to conduct a sensible number of observations per week or per month. The required number must be set according to the number of employees and jobs involved, the need for the observations, and the need to adhere to the systematic pattern of the program. Safety personnel should be responsible for the training of supervisors, coordinating the recordkeeping requirements, and assisting all levels of supervisions, as well as making some safety observations.

All employees must be well informed about the observation program. They must realize that it is a phase of the overall construction safety program and they must clearly understand the reasons for observations. When the procedure is established, each employee must be told of the plans for the future. New employees must receive this information as a part of the orientation program. Only by doing this can management expect this phase to be successful and obtain the cooperation, acceptance, and proper relationship needed from the employees.

Benefits of Job Safety Observations

The benefits of the Job Safety Observation effort are

1. <u>A means of checking the effectiveness of training and detecting violations:</u> Written JSAs, safe job procedures, company safety rules, and specific safety regulations should exist and be used in the safety orientation and training phase of accident prevention. When this procedure is followed and employees are properly instructed in safe job procedures, any deviation from these practices is a violation. Violations can be detected and corrected through safety observations.

    The first-line supervisor must observe the job to determine whether the worker has absorbed and retained the instructions. Workers do not always understand all that they have been told and taught; workers sometimes forget. This means that the worker may not always have the know-how that the supervisor thinks he or she should have concerning safe job procedures. The supervisor can ask for a verbal account of how to go about doing a job, but this is an unreliable check. Even when a worker can recite how a job should be done, it does not mean that the job will be done that way. The most convincing way to check the person and to find out if he or she knows how to do a job safely is to watch it being done. The person either knows how to do it and performs the job safely, or does not know how to do it and cannot demonstrate the correct procedure.

2. <u>A way to detect unsafe practices and lack of personal protective equipment:</u> The JSAs, safe job procedures, company safety rules, etc. may not cover all conceivable actions that could take place on the part of a worker. This could result in an unexpected injury. Furthermore, all employees may not have received specific instructions in certain safety rules and procedures. If an employee has not received training for a certain hazardous behavior and is observed doing a job in a hazardous way, those actions should not be considered a violation. Nonetheless, this unsafe practice must be corrected. This is also true if the hazardous behavior has not been covered in existing rules, regulations, and procedures. These unsafe practices, even though not previously covered, must be detected and addressed in order to eliminate potential accidents.

All violations are indeed unsafe practices, but all unsafe practices should not be considered violations. This distinction is necessary because a violation requires enforcement action, while an unsafe practice, where a violation did not occur, shows only the need for training, not disciplinary action. A safety observation will not only expose and highlight violations of safety rules and procedures but will also enable the supervisor to detect other unsafe practices.

Through safety observations, the supervisor can verify that all required personal protective equipment is being used, as well as determine that the hazards are understood and each construction worker is using all safety know-how at all times.

3. <u>A way for the supervisor to gain a good understanding of the workers and their work practices:</u> The supervisor who systematically observes employees at work will learn a great deal about them. One type of worker may always be in too much of a hurry, another may show impatience and anger if all does not go as planned, another one may become upset and confused in an emergency, and still others may be clowns or practical jokers (those who do not seem to realize that by their actions they endanger themselves and others). All supervisors must know how necessary it is to understand their employees and their work practices. The importance of this cannot be overemphasized.

When a first-line supervisor takes time to deliberately watch employees work, the workers will learn to expect such observations and also learn that the supervisor will not tolerate unsafe practices or violations of his or her instructions. Workers soon learn that the supervisor is serious in his or her efforts to prevent accidents. They interpret the supervisor's safety observations as intentions to enforce compliance with all safety instructions and eliminate unsafe behavior.

4. <u>To show the need for giving recognition and retraining:</u> The safety observation provides a natural opportunity to let workers know that their efforts and safe behavior are appreciated. If a first-line supervisor is in the habit of letting his or

her workers know when they are wrong, why not also let them know when they are right? When observing employees working safely, the supervisor should take advantage of the situation and compliment them. Everyone appreciates a sincere, deserved compliment, and most employees respond with a greater willingness to cooperate when they are given credit for safe, effective performance. Many supervisors do not compliment their employees enough. This may be true because they have not learned to compliment them without feeling uncomfortable.

Safety observations often point out which employees present safety problems, as well as revealing which ones need further training. Instruction of any kind is generally more effective when it is done at the jobsite. Therefore, retraining, or making corrections as a result of safety observations, has its advantages. It provides an opportunity to re-instruct or correct the worker on the spot. This is more effective because the instruction is personal and immediately deals with what the worker did or failed to do. If a worker is seen unsafely performing some phase of work, the act should be reviewed and the hazards pointed out. If the unsafe action is not in violation of the safety regulations or specific instructions for the job, it still should be pointed out and the job safety analysis should be modified in light of this observation.

5. <u>It leads to improved job procedures and safer job behavior:</u> Job observation is an excellent means of stimulating ideas. The supervisor may see something in the job itself that encourages shortcuts. It may be some point that was not dealt with when the job safety analysis was developed. It may be an unsafe condition—or more likely an action—inherent in the job procedure which up to this point had not been recognized as being unsafe. Any supervisor who makes it a point to observe a person on a specific job often gains an important by-product. In addition to finding a safer way of doing the job, the supervisor probably will find ways to improve procedures that will eliminate waste, improve quality, and save time. This often leads to a needed revision of the job safety analysis.

A first-line supervisor must frequently think and plan on the go. Responsibilities as a supervisor take him or her through work areas quite often. Detecting and correcting any unsafe behavior is an important responsibility of any supervisor and he or she must give this item as much attention as possible.

While on the go, the supervisor should occasionally pause here and there to talk and watch the employees work. And by a casual glance, the supervisor should also be looking for safe and unsafe behavior; such actions are incidental safety observations. Experience has taught most supervisors that it pays to be watchful, so incidental safety observations are second nature to them.

Planned Safety Observations

In a planned observation, the supervisor selects the person and the job that will be observed. The supervisor also decides the most appropriate time. On some construction sites, the superintendent will want to make assignments for the planned safety observations, but in most cases the decision is the duty of the first-line supervisor. The basic tool for making a planned observation is the written JSA or SOP for the job selected. Otherwise, the supervisor must be thoroughly familiar with the job steps, job hazards, and safe job procedure. When possible, the supervisor will observe the worker doing the job from start to finish. He or she will do this with full attention given to watching for safe or unsafe procedures and conditions. All safe practices noted should result in a sincere compliment to the worker involved, while any unsafe actions will call for appropriate corrective measures. In either instance, the supervisor should make a record of the observations. A program for making planned safety observations is a valuable tool for loss prevention.

Frequency and Extent of Observations

Most work that is done in and around construction worksites requires workers to do a number of distinctly different jobs. Some jobs warrant more attention than others because of the hazards involved. When the decision is made to observe a particular worker, the supervisor should first observe the worker doing the jobs that are potentially more dangerous, or have a higher accident history, and then observe the employee doing the jobs that are relatively hazard free. It may even be wise to observe, for a second time, an employee doing an exceptionally hazardous job <u>before</u> the worker is observed doing a job with little potential for accidents.

The first-line supervisor should have a reason for deciding to observe a particular worker. Some workers are more of a safety problem than others and it would be wrong to devote an equal number of planned safety observations to all workers. Some reasons for observing a person are

1. The employee is relatively new on the job.
2. The worker is repeatedly involved in accidents.
3. He or she is doing exceptionally hazardous jobs.
4. As a result of incidental observations, it is discovered that there is a need for a planned observation.

Well-kept and frequently analyzed accident report records will show just who the repeat offenders are. The supervisor should be assigned, or should select, this type of person for regular planned safety observations. Such observations will show several reasons why some workers have more accidents than others. Detecting this behavior on the job will necessitate corrective action on the part of the supervisor. Some clues as to why these workers are accident repeaters are

1. They may show emotional or impatient behavior when everything does not go right on the job.
2. They may try to work too fast.
3. They may seem to be preoccupied, even in a "mental fog."
4. They may appear clumsy or very inept in the procedures they follow in doing the job.
5. They may just seem to have a special affinity for accidents.

A first-line supervisor who spends time with his or her employees will soon know a great deal about them individually and as a work crew. Individually, some may be less cooperative than others. They may have received thorough safety training but still take unnecessary chances, violate safety rules, and develop their own improper methods of working. It is necessary to not only observe these workers, but also to use the necessary corrective actions that have been established by the company's policies. Once corrective action has been taken, you should record the observation and the corrective measures taken. When training is given on how to do the work safely and an explanation is given as to why it must be done according to safe job procedures, most workers will not fall into this category. The few who are chronic unsafe workers will become less enthusiastic in displaying unsafe behavior once they realize they have been selected for closer supervision. This close supervision is given most effectively by using the planned safety observation.

A supervisor has an added responsibility when there are workers who are suspected of being physically or mentally incapable of safe work on some jobs. By performing a series of planned safety observations and conscientiously studying each person in this class, it should reveal whether any of these workers do or do not fall into this category. The observation may establish the fact that an employee's condition in no way interferes with his or her ability to work safely, or it may confirm the belief that a worker is a hazard to others, as well as to himself or herself, while doing a specific job.

It is important for a supervisor to acknowledge that the term "inexperienced" has a broader application than just to newly employed young persons. Any worker should be considered inexperienced on a job until the supervisor's knowledge and personal observations prove otherwise. Certainly a worker who has never worked a specific position, or who has only worked the position on and off, may be regarded as inexperienced. What about an employee who has been in a certain position a long time but has never handled the present job before? This worker may also be regarded as inexperienced.

In the case of an inexperienced worker, it is necessary to quickly detect unsafe behavior. Quickly spotting unsafe practices will allow the supervisor to take corrective measures at once. When this is done, accidents are prevented and unsafe work habits are not developed. It is much easier and more effective to correct a worker immediately when unsafe actions are observed rather than just trying to convince him or her that this way of working is unsafe.

Inexperienced workers require frequent planned safety observations in order to ascertain exactly what they do or do not know about performing their work safely. Further instruction and guidance is usually the required procedure after the observations.

Far too many accidents still occur by employees who are experienced in all the jobs of their position. There are many reasons for this and these reasons may be found through planned safety observations. Some workers who have been doing a job for many years will often develop shortcuts and effort-saving practices that are hazardous. Because no accidents have occurred yet from following their methods, they insist that it is not unsafe and is the best procedure. In most cases, these workers learn the hard way when the law of averages finally catches up with them.

Inexperienced workers will tend to follow the examples set by experienced workers. When the old timer on the job works in an unsafe manner, it has an influence on the new worker on the job. The unsafe practices of the experienced workers must be found and corrected for many reasons. Planned safety observations are the logical first steps in helping the experienced workers fulfill their responsibilities in the field of accident prevention. All workers must be included in the safety observation program, but employees who fall into the five listed categories need more frequent planned safety observations than do others. A problem that may result from these observations is the risk of some workers feeling that they are being singled out and "picked on." To lessen this possibility avoid using, for comparative purposes, a safe worker as an example.

The Incidental Safety Observation

A first-line supervisor would never be labeled "efficient" or "effective" if he or she had a "one track mind" on the job. In fulfilling his or her many duties, the supervisor must be alert as he or she travels through the work area. While going from place to place, the supervisor should not only think and plan, but should also do a lot of casual listening and looking. This casual listening and looking for unsafe practices or violations, and the actions that are taken as a result of what is seen or heard, are the incidental safety observations. Many accidents are prevented by supervisors who refuse to let themselves get too wrapped up in their thoughts to hear about or observe unsafe behavior.

When making the incidental observation, ask yourself some of the following questions concerning the workers:

1. Do they show concern for doing a good job?
2. Do they cooperate with supervisors and other workers?
3. How do they handle unexpected problems?
4. How do they handle tools and equipment?
5. Do they appear to be preoccupied with thoughts other than job duties?

There are other questions that should be added to the above list, but the point is that the supervisor must be alert and active in his or her approach.

Indirect Types of Unsafe Procedures

Any specific unsafe behavior on the job which is observed can usually be categorized under one of these indirect types of unsafe procedures when conducting a JSO.

1. Operating or using equipment or materials without authority.
2. Failure to secure equipment or materials against unexpected movement.
3. Operating or working at unsafe speed.
4. Using unsafe tools and equipment.
5. Using safe tools and equipment unsafely.
6. Failure to warn or signal as required.
7. Assuming an unsafe position or unsafe posture.
8. Removing or making safety devices inoperable.
9. Repairing, servicing, or riding equipment in a hazardous manner.
10. Failure to wear required personal protective equipment.
11. Wearing unsafe personal attire.
12. Violation of known safety rules and safe job procedures.
13. Engaging in other unsafe practices (not violations).
14. Indulging in horseplay, practical jokes, fighting, sleeping, creating distractions, and so on.

All supervisors must be continually alert to notice and correct violations and other unsafe practices. This habitual alertness will convince workers that all levels of management are serious and sincere about employee safety. As part of the observation, the supervisor should compliment, correct, warn, or reprimand according to his or her findings, and fair judgment should be based on the established company policy.

Employees who are observed working unsafely should be corrected immediately but with tact and understanding. The supervisor who sees a worker acting unsafely and fails to correct him or her is doing the employee a disservice. The supervisor is letting the person think he or she can act unsafely and get away with it without any consequences (such as an accident or supervisory correction). When a worker repeatedly engages in unsafe practices and gets away with it, the worker will believe he or she is immune to accidents, or that his or her behavior is not really unsafe.

Any supervisor who does not follow through on observations will weaken his or her status as a supervisor. If he or she "turns his or her back" instead of making the correction, displays reluctance in giving a warning, or, apologetically and half-heartedly reprimands an offender, the supervisor's effectiveness will be diminished and so will the respect the workers have for him or her.

Another effective time for making incidental safety observations is at the beginning of each shift. An alert supervisor can deal with some possible causes of accidents or injuries by making a quick check of the overall physical appearance of his workers. To derive the benefit from observation, make the observation before the workers start on any job. The most obvious things to detect will be hazardous clothing and inadequate personal protective equipment. In some instances, what would be considered hazardous clothing may depend upon the type of work to be done, the work area, weather, and/or company regulations. Workers doing work of any type must be kept aware that there is a right and wrong way to dress for the job. This type of incidental observation will give the supervisor a means of detecting such things as

1. Ill-fitting, loose, or ragged clothing that could be caught in moving machinery.

2. Work clothing that has become excessively greasy or oil soaked and could catch fire or cause skin infections.

3. Long baggy trousers or coveralls with cuffs that could be a tripping hazard.

4. Rings, bracelets, necklaces, or wristwatches that could be hazardous when climbing, using certain tools, working around heat or chemicals, or doing electrical work.

5. Extremely long hair or beards that could catch in or on moving machinery or that could become oil soaked and be harmful from the standpoint of flying sparks.

6. Wearing dark safety glasses when not needed.

With a casual glance at the workers, the first-line supervisor can quickly check for all the required items of personal protective equipment. The condition of the equipment must be also noticed. Some of the general items to check for are

1. Proper head protection, eye protection, and foot protection.

2. Proper gloves, respirators, and suitable ear protection, when needed.

3. Respirators and other items required for specific areas.

In addition to checking for hazardous clothing and inadequate personal protective equipment, there are many other things to consider when making the quick preshift check of the overall physical appearance of the crew. This incidental safety observation will enable the supervisor to look and listen for the following possibilities:

1. Persons suffering from temporary illness.

2. Persons who have suffered a personal injury off the job.

3. Persons who are intoxicated.

4. Persons who are suffering from the after effects of intoxication.

5. Persons who have alcoholic beverages in their possession.

6. Persons who are under the influence of narcotics.

A supervisor knows his or her employees and is aware of the normal look and activity of each individual. A specific action may be normal for one worker and abnormal for another.

By knowing each employee and using practical judgment, the supervisor should be able to determine the worker's general condition without depending on the knowledge of medical symptoms for the above mentioned possibilities.

The Planned Safety Observation

A planned safety observation program gives the first-line supervisor a positive means of determining the effectiveness of the safety instructions that have been given. It also serves as a learning tool for the supervisor. The supervisor learns more about each job, more about each worker, and more about supervising. The supervisor will learn to be more perceptive in all areas of responsibility; this will result in the best distribution of supervisory time and will allow the supervisor to devote time and effort where it is needed most.

The prime requisite to accomplish the above is by starting with a written program of procedures for the planned safety observation. The program must be tailored to the needs of the type of construction work and should be written and include, but not be limited to, the following:

1. The reasons for the plan to be followed. (Must include orientation of all employees as to purpose of program.)

2. The objectives of the program

3. Training requirements for all levels of supervision.

4. Safety department's responsibilities.

5. Preparation procedures for making an observation.

6. How to make a planned safety observation.

7. Recording the observation

8. Holding a post-observation conference.

9. Follow-up procedures, including enforcement policy.

Since supervising workers means observing people at work, the first-line supervisor is the member of management who is most qualified to make safety observations. The supervisor knows his or her workers and the training they have received. He or she knows the jobs under his or her supervision and how these jobs should be done. With proper preparation, the supervisor will conduct a more thorough observation. He or she will be able to make frequent planned safety observations and more efficiently cover all employees. The supervisor's responsibility is not limited to higher management, but he or she has a responsibility to each employee. Therefore, other members of management may periodically participate in the planned safety observations, but the supervisor must be responsible for observing his or her workers.

In construction operations, some jobs involve only a few simple routine tasks, but most occupations in this industry involve doing many jobs; some more often than others. To help decide which job a worker should be observed doing, a "Job Safety Observation Record" form should be maintained for each employee. The form shown in Figure 6-7 keeps more than observation records. When it is partially prepared in advance, it allows the supervisor to pick the job, for a reason, and without unnecessary duplication.

This form will tell each supervisor just what all supervisors have done with respect to a particular worker. The form contains the employee's name, work location, and position and should also list the jobs performed under that position. To choose the job the supervisor should observe, the supervisor should consider such things as: (1) does it involve some new procedures because of a recent JSA revision; (2) is it an exceptionally hazardous job; or (3) is it a job done infrequently, but is complex?"

JOB SAFETY OBSERVATION FORM

1. Job being observed:_____

2. Worker observed (name):_____

3. Experience of worker at job or task _____yrs. _____ mos.

4. Is worker dressed appropriately for the job? _____yes _____no

    Comments: _____

5. Is worker wearing all required personal protective equipment? _____yes _____no

    Comments:_____

Steps in performing the job or task should be marked (S) Satisfactory, (R) Re-observe, or (U) Unsatisfactory.

Steps (Describe)   Hazard Involved   Worker Performance

1.

2.

3.

4.

5.

6. Did the worker perform the job according to the safety operating procedure?
    _____yes _____no

    Comments:_____

7. Did the worker follow the safety and health rules? _____yes _____no

    Comments:_____

8. Did the worker perform the job or task safely? _____yes _____no

    Comments:_____

9. Did the worker have a good safety attitude? _____yes _____no

    Comments:_____

10. Does the worker need training?_____

11. Should disciplinary action be taken? _____

12. Should the worker be removed from the job or task?_____

13. Was the worker told about any deficiencies?_____

Figure 6-7. Job safety observation form

Supervisory Preparation for Job Safety Observations

Preparation for the observation starts with the selection of the employee and the job. Before the supervisor observes the person selected, he or she must be thoroughly familiar with the job. If there is a JSA, it should be carefully studied. If no written safe job procedure exists, the supervisor should review the job. It would help to make notes as to the steps, hazards, safety procedures, and personal protective equipment needed. In this preparation time, the supervisor must determine when the worker will be doing the chosen job. Some supervisors approach safety observations on a "catch-as-catch-can" basis. In other words, when the supervisor decides to make an observation, any job the worker happens to be doing at that time is the job observed and recorded. This may be easier, but it is not what is meant by a planned observation. Such a method defeats the plan of systematically observing all jobs that a person does and, most importantly, giving additional emphasis and observations to jobs that are known to be hazardous. When the supervisor studies the job selected and observes that specific job, he or she will know exactly what to look for. In due time, such studying will strengthen familiarity with the job to the point that further study may not be necessary.

It is important to keep in mind that the worker who will be observed should be aware that he or she will be observed from time to time. This awareness must exist from the time of employment, or from the time the plan was first put into practice. All employees must be told, as part of their orientation, that planned safety observations are a phase of the safety program of the operation. Most important, all employees must understand the reason for incidental and planned observations. They need to realize that these safety observations help reduce injuries by eliminating the potential causes of accidents. And remember, tact needs to be used when explaining this to all employees.

With this in mind, there is another point to consider in preparing for the planned safety observation. Should the worker be informed in advance of each observation? Under certain conditions the worker should be told in advance, but this depends upon the kind of worker the supervisor intends to observe and the purpose of the observation. The accepted rule is: "If you want to find out what a person knows and does not know about working the job safely, tell him or her in advance that he or she is going to be observed doing that job." No normal worker will intentionally show unsafe behavior when they have been told that they will be observed for their ability to do a job safely. If workers know that they are being observed to see how safely they work, they will naturally work as safely as they know how. If they reveal any unsafe practices under this condition, the supervisor can conclude that they do not know better. So, if the supervisor tells the person in advance, he or she does their best and the supervisor will learn what the worker knows and does not know about working the job safely.

If the worker is new, one who has relatively little experience with a certain job, he or she should be told in advance. This is also true for a person who may have considerable experience but who has never before been checked out with a planned safety observation.

The other accepted principle is "If the objective of the observation is to learn how the job is normally done, do not tell the worker of the planned observation." When the supervisor knows, as a result of past observation, that a person knows how to work a job safely, the supervisor must then find out how the person works when no one is standing over him or her. To determine this, the worker should not be told in advance of the safety observation. Most employees work safely under the eyes of the supervisor, especially when they have been told that they are being observed. If the person reveals no unsafe practices when he or she has been told of the observation, the supervisor can assume that the employee usually works safely. If the same person is later observed doing some part of the job unsafely, the supervisor must believe that the employee also works unsafely at other times. In this instance, the problem to be dealt with is this: The supervisor is aware that the worker knows how to work the job safely –

past observation has shown this – but now the supervisor has learned the person is not putting that knowledge into practice.

The supervisor should never attempt to observe a worker from a concealed position. It would be childish to try to hide behind equipment. Something may be learned, but what is lost in terms of respect and human relations is not worth it. The person should be observed from a distance. The procedure would be to observe other workers, equipment, or the environment momentarily then switch back to the person who is the subject of the planned safety observation. The supervisor should conduct himself or herself in such a way that the worker is unaware that he or she is being singled out.

When the worker is to be told in advance that he or she will be observed, the employee must be approached in a friendly manner. If he or she appears ill at ease, a brief informal discussion of a favorite sport or hobby may help to relax the employee. A smile and friendly manner go a long way toward putting a worker at ease. Remind the employee that the observation is a regular part of the safety program. Explain again the reasons for the importance of making periodic observations.

Making the Observation

Even when the worker has been told that he or she will be observed, the supervisor must stand clear of the worker. The worker should be given plenty of room to efficiently do his or her job. The worker's line of vision for doing the job should not be blocked by the supervisor. This could create a hazard for the worker. Also, by standing away, the supervisor will minimize the possibility of distracting the worker with his or her presence.

When available, the written JSA should be used as a guideline for the observation. With the JSA in his or her possession, the supervisor should watch each step of the job and be on the alert for violations of the safe job procedures that have been taught. The supervisor should also watch for other unsafe procedures and conditions that may need to be added to the JSA.

When there is no job safety analysis program, the supervisor should make notes during the preparation for the planned observation. The notes should include a list of unsafe procedures that could be detected and, during the observation, the supervisor needs to watch for any other potential unsafe procedures that are not on the list. Most checklists do not cover every conceivable unsafe action. A general checklist, similar to that presented in this chapter, could be used on all jobs when the supervisor has learned to detect specific circumstances under the indirect unsafe procedures.

The supervisor should avoid interruptions unless absolutely necessary. Few people can stand repeated interruptions of their work without becoming upset. It is better to save all minor corrections until the job has been completed, but if the worker is seen doing something in an unsafe manner, he or she should be stopped immediately. The supervisor should explain what has been done that is incorrect, and explain why it is unsafe. And when a worker has been trained under a JSA Program, the worker's attention should be called to that written JSA.

The supervisor should check to make sure the worker fully understands what has been explained. This can often be done by having the person tell and show the safe way to do the job. Remember that the way the worker is corrected will determine the good that comes out of it. The supervisor should be firm, but by all means calm and friendly; the person should not be embarrassed. If the worker is embarrassed, he or she will resent it and this could affect how the rest of the job is done. When a supervisor intentionally or unintentionally embarrasses a worker, it will certainly affect the supervisor-worker relationship and this relationship is vital in having an effective safety program.

Recording the Observation

When a supervisor knows the number of observations required each month, he or she must plan them so that they cover, on a systematic basis, all the jobs for which each worker has received instruction. This planning is accomplished by using the form described in Figure 6-7, which contains the employee's name, occupation or position, and jobs performed, and gives a permanent record of what has been done and what needs to be done in the field of observation.

When a worker is observed on one of those jobs, the supervisor enters the date of the observation and initials the block after that particular job. If the worker has completed the job safely, a check is placed under the "S" which indicates "satisfactory." A minor error would still allow for a satisfactory rating. When the employee has worked unsafely and in a manner that calls for reinstruction or correction, the proper block under "R," "re-observe," should be checked. All spaces marked "R" will show a need to re-observe, within one week, the same employee on the same job.

The comment space on the form is used to record any specific unsafe practices that have been observed, or it may be used to record some particular impression the supervisor had while observing the worker. When several supervisors have the same worker under their supervision, the remarks in the comment space will keep them all informed. It is best for all supervisors to work from a single record; this can be done by having each supervisor date, initial, rate, and comment after each job observation.

The records of the results of the observations are used to make sure that safety communications between supervisors and employees—basic training and individual contacts—have been adequate. When these records (Job Safety Observation Forms) are maintained in an up-to-date fashion, management can verify that all the required protective equipment is being used, safe job procedures are being followed, hazards are understood, and each worker is using safety know-how at all times.

Post Observation Conference

A brief post observation conference should always be held with the worker who has been observed. One of two things should be done at this time. When the job was done as required, the worker should be complimented. The employee should be told that his or her efforts are appreciated and he or she should be encouraged to continue his or her important part in the company safety program. Personal recognition, when it is deserved and sincerely given, will improve safety behavior; this is positive motivation. Undeserved compliments and personal recognition may appear to improve the supervisor-worker relationship for a brief period of time, but, in reality, much harm can be done by using this approach.

Any unsafe behavior should have been promptly handled during the observation, but it should be briefly summarized during the conference to emphasize the point. If there were minor things noted that need correcting, plans for reinstruction must be followed. If the worker performed some phase of the work unsafely, the fact must be established as to just what was observed. The supervisor should get the worker's reasons for doing the job as he or she did. Regardless of the reasons, the supervisor should not argue with the worker. The concern is not with what has been done, but with changing the worker's behavior so the same thing is not done in the future. The supervisor must take whatever corrective action is required, but should give the reasons behind all precautionary measures enforced. All corrections and reinstruction should be accomplished in a firm but friendly manner. Even though it is deserved, most people are not eager for correction. Some workers are afraid of it because they think it reflects on their abilities. Remember, the person probably did his or her best while being observed. The post-observation conference should always end with the feeling that things are right between the supervisor and the worker, in spite of any mistakes made or corrections given.

This type of post-observation conference should be held even when the worker was not notified in advance of the observation. An important point to remember is that all workers will eventually know they have been observed. However, deliberately watching people at work over a long period of time could create a problem with some workers. They may resent being a part of a "cat and mouse" game. And, some may resent being secretly observed, especially when the observation shows that they are doing something wrong. This is all the more reason for all employees to understand that safety observations are a necessary part of the complete safety program. And, above all, they must be told—and they must believe—that it is not a method of spying but a means of preventing injuries and other losses from preventable accidents. This does not mean that safety observations should not be made. It does mean that the supervisor must handle himself or herself skillfully when talking to a worker after an observation. This is the kind of situation in which the supervisor should know his or her workers, particularly those who are sensitive to observation.

Follow-Up Procedures

As explained earlier, if the worker is seen doing something in an unsafe manner, he or she should be stopped immediately and proper steps taken to make the correction. In some instances, the post-observation conference will be the time to make corrections and give brief steps of reinstruction. Performance noted and practices seen in some observations will call for a more detailed plan for retraining. It may even be necessary to give the complete basic job instruction – using the JSA – over again.

When the necessary corrective action or retraining is complete and the worker has again been on the job for several days, re-observation is required. It should be done within a week to determine whether the worker has responded to the reinstruction. The person should not be told in advance that he or she is going to be re-observed. If the re-observation indicates that the worker is working safely, he or she should be complimented; this gives the worker added incentive to continue to work safely. If the re-observation shows that he or she is still working unsafely, steps must be taken to find out why.

Dealing with Unsafe Performance

Dealing with unsafe performance may be helpful in determining why the unsafe practices are continuing. You should institute a step-by-step procedure to follow to correct the performance. The supervisor should record the specific unsafe performance that has been observed and determine the type of accident that could result from this action. Since the supervisor has already discussed this action with the worker, both during and after the first observation, and the possible results, the supervisor must now decide what caused this action to continue. Three basic causes of unsafe behavior are (1) lack of skill or knowledge, (2) lack of motivation, or (3) a mental or physical handicap.

If the worker does not show proper skill or knowledge after a period of training and retraining, it may be because he or she will not try to do what is necessary to perform the job safely. Motivation could be wanting. All workers want safety, especially their own. The motivation problem, as it relates to safety, is usually not the lack of desire for safety; the problem is that workers are not always willing to do what is necessary to assure their own safety. There are many conflicting motivations that sometimes win over the motivation to work safely. These conflicting motivations – to save time, to avoid extra effort, to be physically comfortable, to attract attention, to gain group approval, and so on—do exist and are troublesome when they conflict with the motivation to work safely.

One of the greatest challenges of any first-line supervisor is to develop and apply ways of motivating employees to work safely <u>all the time</u>. The best approach involves three principles.

1. The supervisor can develop positive safety attitudes by personal example.

2. The supervisor must develop a foundation for gaining the cooperation of workers by building good personal relations.

3. The supervisor must deal properly with the employee who is not cooperative in safety. This can be accomplished through education and/or enforcement.

Since most employees will reflect the safety behavior of their supervisor, it follows that the supervisor's behavior is all important. The supervisor who accepts safety responsibilities grudgingly, who tolerates violations of safety rules, who covers up accident facts, and who neglects good housekeeping, is quickly cataloged by his or her workers. They soon learn he or she is not serious or sincere about accident prevention, so he or she will have little success in developing proper safe behavior in his or her employees. The opposite is true of the supervisor who, day after day, demonstrates by his or her actions that safety is a matter of utmost concern and importance to all.

Since people – workers and supervisors – differ in behavior, in knowledge, in personality, and so on, a worker may be motivated by one supervisor and not by another. Assigning an employee to work under a different supervisor, who defines the objectives and removes obstacles, may result in the same training being more effective. One supervisor may be able to emphasize the positive features associated with working safely and, consequently, make the safe way of doing a job more attractive to a particular worker.

A worker may have been properly motivated to follow all safe job procedures but have some slight physical or mental handicap that prevents this from happening. There may also be other reasons why a person cannot safely do a specific job. The solution to this problem, in some situations, would be to place the worker on a different job and start the training process over again for the new job. The supervisor should maintain as close a contact as possible with this worker and continue to make incidental safety observations. To keep the records current, the follow-up procedure should be recorded on the "Job Safety Observation Form."

Since first-line supervisors are basically responsible for making sure the safety rules are followed, the safety observation program is a worthwhile method for accomplishing this. Safety observations naturally supplement the job safety analysis and the safety orientation and training programs. This develops three phases of the loss prevention work.

The primary purpose of the incidental and planned safety observations is to instill in each worker a willingness to comply with safe job procedures. Where violations and unsafe practices are found, the object is not merely to stop the workers from continuing this behavior; the real objective is to get all employees to willingly comply with the procedure in question, and at the same time, remain productive and cooperative workers.

Job safety observations are used to evaluate individual workers as they perform their tasks. They determine the effectiveness of the workers' training in safe operating procedures as they are performing their assigned tasks. They detect unsafe acts and unsafe performance of job-related tasks prior to an accident or injury. When used in this manner, they definitely fit the category of an accident prevention technique.

SAFETY AND HEALTH AUDITS

The safety and health audits (inspections), which are often conducted on worksites and projects, serve a number of evaluative purposes. Audits or inspections can be performed to

1. Check compliance with company rules and regulations.
2. Check compliance with OSHA rules.
3. Evaluate supervisors' safety and health performance.
4. Evaluate workers' safety and health performance.
5. Evaluate progress regarding safety and health issues and problems.
6. Determine the effectiveness of new processes or procedural changes.

First, determine what needs to be audited. You might want to audit a specific trade (iron workers), task (welder), topic (electrical), unit/crew (masonry workers), worker (crane operator), part of the jobsite (excavation activities), compliance with an OSHA regulation (Hazard Communication Standard), or the complete jobsite. You may want to perform an audit if any of the previous lists or activities have unique identifiable hazards, new tasks involved, increased risk potential, changes in job procedures, areas with unique operations, or areas where comparison can be made regarding safety and health factors.

In the process of performing audits, you may discover hazards that are in a new process, hazards occurring once the process has been instituted, a need to modify or change processes or procedures, or situational hazards that may not exist at all times. These audits may also verify that job procedures are being followed, identify work practices that are both positive or negative, detect exposure factors both chemical and physical, and determine monitoring and maintenance methods and needs.

At times audits are driven by injury frequency, a potential for injury, the severity of injuries, new or altered equipment, processes, and operations, and excessive waste or damaged equipment. These audits may be continuous, ongoing, planned, periodic, intermittent, or depending upon your specific needs. Audits may also determine the comprehension of procedures and rules, the effectiveness of training, assess the work climate or perceptions held by workers and others, or evaluate the effectiveness of a supervisor regarding his or her commitment to safety and health.

On many active construction sites or projects, daily site inspections are performed by the competent person or supervisor in order to detect hazardous conditions, equipment, materials, or unsafe work practices. At other times, periodic site inspections are conducted by the site safety and health officer. The frequency of the inspections is established in the project-specific safety and health plan.

The project supervisor, in conjunction with the site safety and health officer, determines the required frequency of these inspections. This is based on the level and complexity of the anticipated construction activities and on the hazards associated with these activities. When addressing site hazards and protecting site workers, in addition to a review of worksite conditions and activities, these inspections should include an evaluation of the effectiveness of the project-specific safety and health plan. The site safety and health officer should revise the project-specific safety and health plan, as necessary, to ensure the plan's continued effectiveness.

Prior to the start of each workshift or new activity, a workplace and equipment inspection should take place. This should be done by the work crews, supervisor, and other qualified employees. At a minimum, they should check the equipment and materials that they will be using during the operation or workshift, for damage or defects that could present a safety hazard. In addition, they should check the work area for new or changing site conditions or activities that could also present a safety hazard.

All employees should immediately report any identified hazards to their supervisors. All identified hazardous conditions should be immediately eliminated or controlled. When this is not possible

1. Interim control measures should be immediately implemented to protect workers.

2. Warning signs should be posted at the location of the hazard.

3. All affected employees should be informed of the location of the hazard and the required interim controls.

4. Permanent control measures should be implemented as soon as possible.

When a supervisor is not sure how to correct an identified hazard, or is not sure if a specific condition presents a hazard, he or she should seek technical assistance from the designated competent person, site safety and health officer, or project supervisor.

Safety and health audits should be an integral part of safety and health efforts. Anyone conducting a safety and health audit must know the jobsite or project, procedures or processes being audited, previous accident history, and the company's policies and operations. This person should also be trained in hazard recognition and interventions regarding construction safety and health.

The complexity of the worksite and the myriad of areas, equipment, tasks, materials, and requirements can make the content of most audits overwhelming. As you can see in Table 6-1, the audit topics which could be targeted on a construction worksite are expansive.

Table 6-1

Construction's Audit Topics

Acids	Fork Lifts	Signs
Alarms	Fumes	Scaffolds
Barriers	Gases	Slings
Barricades	Generators	Solvents
Blasting	Hand Tools	Stairways
Buildings	Hazardous Chemical Processes	Steel Erection
Caustics	Heavy Equipment	Storage Facilities
Chains	Hoists	Transportation Equipment
Chemicals	Hoses	Trucks
Compressed Gas Cylinders	Housekeeping	Unsafe Conditions
Dusts	Materials	Unsafe Acts
Concrete Forms	Ladders	Ventilation
Conveyors	Lifting	Walkways and Roadways
Cranes	Lighting	Walls and Floor Openings
Confined Spaces	Loads	Warning Devices
Demolition	Lockout/Tagout	Welding and Cutting
Derricks	Machines	Wire Ropes
Electrical Equipment	Mists	Working Surfaces
Elevators and Manlifts	Nets	
Emergency Procedures	Noise	
Environmental Factors	Platforms	
Excavations	Personal Protective Equipment	
Explosives	Personal Services and First Aid	
Fall Protection	Power Sources	
Fibers	Power Tools	
Fire Extinguishers	Radiation	
Fire Protection	Rigging	
Flammables	Respirators	

The topics of audits and the form used will depend upon the type of construction work. There are three primary audit formats: checklists, evaluation, or narrative. A short example of a checklist is shown below. This checklist requires a simple yes or no response.

    ____ Yes   ____ No   Hardhats are being worn by workers.
    ____ Yes   ____ No   Hardhats are in good condition.
    ____ Yes   ____ No   Hardhats meet the ANSI standard's requirements.

The evaluation format requires the person completing the audit to complete the form as follows: 1 - excellent; 2 - very good; 3 - good, 4 - fair; and 5 - very poor. An example of this type of format is:

    1 2 3 4 5 Hardhats are being worn by workers. (All workers or some)
    1 2 3 4 5 Hardhats are in good condition.
    1 2 3 4 5 Hardhats meet the ANSI standard's requirements.

A narrative format requires a written response. An example of this type of format is

Hardhats are in use on the project site. Explain:_____

_____

Hardhats are in good condition. Explain: _____

_____

Hardhats meet the ANSI standard's requirements. Explain:_____

_____

      A comprehensive construction safety and health audit form can be found in Appendix D. This comprehensive form may be expanded or to meet your needs, only portions of it used. It does not cover every facet of the regulatory compliance requirements for each Subpart of 29 CFR 1926, but it does cover a majority of the most common audit considerations on a construction jobsite.

      In order to prevent accidents on the jobsite, you definitely should develop audits which meet your company's individual needs and include the components for your particular type of construction work. Packaged audits developed by others in all likelihood will not be very usable to you since each construction company is unique and has its own hazards and procedures

REFERENCES

[1] Lapping, James E., "OSHA Standards that Require Engineers," Occupational Safety and Health Administration, Special Assistant for Construction and Engineering, 200 Constitution Ave. NW, Room N3306, Washington, D.C. 20210.

[2] *Accident Prevention Techniques, Job Observation,* National Mine Health and Safety Academy, U.S. DOL, Beckley, WV, 1980.

[3] *Accident Prevention Techniques, Accident Investigation,* National Mine Health and Safety Academy, U.S. DOL, Beckley, WV, 1980.

[4] "Ten Tips for Toolbox Talks," *Outlook,* pp. 38-41, November, 1992.

Chapter 7

CONSTRUCTION SAFETY: A THROUGH H

Charles D. Reese

Safety and the potential risk of serious injuries in the construction industry are often interrelated or even synergistic with other hazards associated with construction work. As efficiently as one might try to address each safety hazard on a construction worksite or within the construction industry, some hazards or some facet of a construction hazard may be overlooked. Often the unique work experiences which a contractor, health and safety professional, supervisor, or worker has had make them aware of a hazard or risk which could not be foreseen without the special knowledge of the jobsite or work experience, which those individuals within the construction industry possess. It is often impossible to duplicate that practical experience on the written page.

The next two chapters cover the most common types of construction risks or hazards which may impact the safety and health of construction workers. These chapters address each topic in alphabetical order but do not delve into subtle nuances which can occur on a construction worksite or project. Each chapter does try to address the OSHA requirements for protecting workers on construction jobsite and projects.

One of the complaints often voiced by many is that practical real life construction is often lacking in discussions on construction safety and its hazards. These chapters mesh the regulations, common sense, and practical construction work aspects as much as is logically possible within these few pages.

The safety hazards are presented in alphabetical order. This is to facilitate quick reference to the specific topic of interest to the reader. Each narrative relevant to a specific construction safety risk or hazard should not be the only source that one consults regarding that hazard. Use the appropriate regulations, OSHA-generated materials, industry-generated materials, and manufacturers' materials to try to assure a complete understanding of the hazards, the risks, and potential intervention strategies to mitigate accidents and injuries.

ABRASIVE GRINDING (1926.303)

All grinding machines are to be supplied with sufficient power to maintain the spindle speed at safe levels under all conditions of normal operation. Grinding machines are equipped with safety guards in conformance with the requirements of the American National Standards Institute, B7.1-1970, Safety Code for the Use, Care, and Protection of Abrasive Wheels. The safety guard covers the spindle end, nut, and flange projections. The safety guard is mounted so as to maintain proper alignment with the wheel, and the strength of the fastenings exceeds the strength of the guard.

Floor-stand and bench-mounted abrasive wheels used for external grinding are provided with safety guards (protection hoods). The maximum angular exposure of the grinding wheel periphery and sides are not more than 90 degrees except when work requires contact with the wheel below the horizontal plane of the spindle, then the angular exposure shall not exceed 125 degrees. In either case, the exposure begins not more than 65 degrees above the horizontal plane of the spindle. Safety guards are to be strong enough to withstand the effect of a bursting wheel. Floor and bench-mounted grinders are provided with work rests which are rigidly supported and readily adjustable. Such work rests are kept at a distance not to exceed one-eighth inch from the surface of the wheel. (See Figure 7-1 for an example of abrasive grinder.) All abrasive wheels should be closely inspected and ring-tested before mounting to ensure that they are free from cracks or defects. All employees using abrasive wheels are protected by eye protection equipment in accordance with the requirements except when adequate eye protection is afforded by eye shields which are permanently attached to the bench or floor stand.

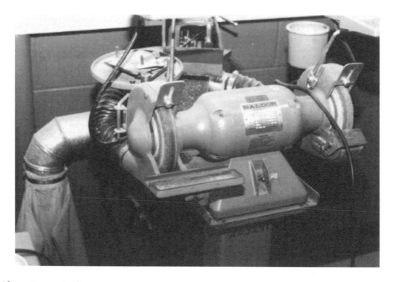

Figure 7-1. Abrasive grinder

AERIAL LIFTS (1926.556)

Aerial lifts are required to meet design and construction guidelines of Vehicle Mounted Elevated and Rotating Platforms (ANSI A 92.2-1969). Aerial lifts include the following:

1. Extensible boom platforms.

2. Aerial ladders.

3. Articulating boom platforms.

4. Vertical tower.

5. A combination of any of the previous.

They may be constructed of a wide variety of materials and may be powered or manually operated. These aerial lifts can be field modified if approved by the manufacturer, testing laboratory, or certified engineer. (See Figure 7-2 for an example of an aerial lift.)

Figure 7-2. Aerial lift with worker wearing a safety belt

Prior to movement of an aerial lift, the boom or ladder is to be secured and locked into place to assure that the outriggers are properly stowed. Prior to use, lift controls are to be tested each day. Aerial lifts used to transport workers must have both lower and upper controls which are of easy access to operators. The lower controls should be designed to override the upper controls. All controls are to be labeled according to their function. The lower controls should never be used unless permission is given by the operator in the basket, or for an emergency. Only authorized workers shall operate a lift.

The manufacturer's load limits for the boom or basket should be followed. When outriggers are set on pads or solid surfaces, the brakes are to be set, as well as wheel chocks in place. Aerial lift trucks are not to be moved when workers are in the basket unless equipment is so designed for such movement.

Never alter the insulating capacity of the boom or an aerial lift. All electrical and hydraulic bursting factors must comply with ANSI A92.2-1969. All components must have a bursting factor of 2 to 1.

Workers in a basket should have a body harness with a lanyard attached to the boom or basket but never attached to an adjacent pole or other equipment. While working in a basket, the worker is to remain completely within the basket. Climbers should never be worn in an aerial lift.

AIR RECEIVERS (1926.306)

Compressed air receivers and other equipment are used for providing and utilizing compressed air when performing operations such as cleaning, drilling, hoisting, and chipping. All new air receivers, which are installed after the effective date of the air receiver regulations, are to be constructed in accordance with the 1968 edition of the A.S.M.E. Boiler and Pressure Vessel Code Section VIII, and all safety valves must also be constructed, installed, and maintained in accordance with Section VIII.

Air receivers should be installed such that all drains, handholes, and manholes therein are easily accessible. Under no circumstance is an air receiver buried underground or located in an inaccessible place. A drain pipe and valve are to be installed at the lowest point of every air receiver in order to provide for the removal of accumulated oil and water. Adequate automatic traps may be installed in addition to drain valves. The drain valve on the air receiver is to be opened and the receiver completely drained, frequently, and at such intervals as to prevent the accumulation of excessive amounts of liquid in the receiver.

Every air receiver must be equipped with an indicating pressure gage (so located as to be readily visible) and with one or more spring-loaded safety valves. The total relieving capacity of such safety valves is such as to prevent pressure in the receiver from exceeding the maximum allowable working pressure by more than 10 percent. No valve of any type is placed between the air receiver and its safety valve or valves. Safety appliances such as safety valves, indicating devices, and controlling devices are to be constructed, located, and installed so that they cannot be readily rendered inoperative by any means, including the elements. All safety valves are tested frequently and at regular intervals to determine whether they are in good operating condition.

ALARMS (1926.159 AND .602)

Although an alarm is not normally considered a hazard, the absences of a unique alarm system can be a hazard. An alarm system needs to be present on the jobsite. There needs to be a unique alarm signal which alerts workers to specific hazards or actions which need to be taken when an emergency exists, such as evacuation. These alarms should be recognizable and different from commonly used signals such as for breaks, lunch, or quitting time.

All emergency employee alarms installed to meet a particular OSHA standard are to be maintained, tested, and inspected. This applies to all local fire alarm signaling systems used for alerting employees, regardless of the other functions of the system. The employee alarm system must provide a warning for necessary emergency action, as called for in the emergency action plan, or for reaction time for safe escape of employees from the workplace or the immediate work area, or both. The employee alarm is to be capable of being heard above ambient noise or light levels by all employees in the affected portions of the workplace. Tactile devices may be used to alert those employees who would not otherwise be able to recognize the audible or visual alarm. The employee alarm must be distinctive and recognizable as a signal to evacuate the work area or to perform actions designated under the emergency action plan.

The employer is to explain to each employee the preferred means of reporting emergencies, such as manual pull box alarms, public address systems, radios, or telephones. The employer shall post emergency telephone numbers near telephones, employee notice boards, and other conspicuous locations when telephones serve as a means of reporting emergencies. Where a communication system also serves as the employee alarm system, all emergency messages have priority over all nonemergency messages.

The employer is to establish procedures for sounding emergency alarms in the workplace. For those employers with ten or fewer employees in a particular workplace, direct voice communication is an acceptable procedure for sounding the alarm, provided all employees can hear the alarm. Such workplaces need not have a back-up system. The employer must assure that all devices, components, combinations of devices, or systems constructed and installed, are approved. Steam whistles, air horns, strobe lights or similar lighting devices, or tactile devices that meet the requirements of this section, are considered to meet this requirement for approval.

The employer must restore all employee alarm systems to normal operating conditions as promptly as possible after each test or alarm. Spare alarm devices, and components that are subject to wear or destruction, are to be available in sufficient quantities at each location for prompt restoration of the systems. The employer must make sure that all employee alarm systems are maintained in operating condition except when undergoing repairs or maintenance.

A test of the reliability and adequacy of nonsupervised employee alarm systems is made every two months. A different actuation device is used in each test of a multi-actuation device system so that no individual device is used for two consecutive tests. The employer is to maintain or replace power supplies as often as is necessary to assure a fully operational condition. Back-up means of alarm, such as employee runners or telephones, are provided when systems are out of service.

The employer assures that all supervised employee alarm systems are tested at least annually for reliability and adequacy. The employer assures that the servicing, maintenance, and testing of employee alarms are done by persons trained in the designed operation and functions necessary for reliable and safe operation of the system. The employer makes sure that the manually operated actuation devices, used in conjunction with employee alarms, are unobstructed, conspicuous, and readily accessible.

ARC WELDING AND CUTTING (1626.351)

Burns and electricity are the hazards confronted by the arc welder. Thus, only manual electrode holders which are specifically designed for arc welding and cutting, and are of a capacity capable of safely handling the maximum rated current required by the electrodes, are to be used. Any current-carrying parts passing through the portion of the holder which the arc welder or cutter grips in his hand, and the outer surfaces of the jaws of the holder, must be fully insulated against the maximum voltage encountered to ground. All arc welding and cutting cables are to be completely insulated, flexible, capable of handling the maximum current requirements of the work in progress, and take into account the duty cycle under which the arc welder or cutter is working (see Figure 7-3).

Figure 7-3. An arc welder

Cable that is free from repair or splices must be used for a minimum distance of ten feet from the cable end to which the electrode holder is connected, except cables having standard insulated connectors, or splices with insulating quality equal to that of the cable, are permitted. When it becomes necessary to connect or splice lengths of cable one to another, substantial insulated connectors of a capacity at least equivalent to that of the cable shall be used. If connections are effected by means of cable lugs, they must be securely fastened together to give good electrical contact, and the exposed metal parts of the lugs must be completely insulated.

Cables in need of repair are not used. When a cable becomes worn to the extent of exposing bare conductors, the exposed portion must be protected by means of rubber and friction tape or other equivalent insulation. A ground return cable is to be of safe current carrying capacity equal to or exceeding the specified maximum output capacity of the arc welding or cutting unit which it services. When a single ground return cable services more than one unit, its safe current-carrying capacity is to equal or exceed the total specified maximum output capacities of all the units which it services.

Pipelines containing gases or flammable liquids, or conduits containing electrical circuits, are not to be used as a ground return. For welding on natural gas pipelines, the technical portions of regulations issued by the Department of Transportation, Office of Pipeline Safety, 49 CFR Part 192, Minimum Federal Safety Standards for Gas Pipelines, apply.

When a structure or pipeline is employed as a ground return circuit, it must be determined that the required electrical contact exists at all joints. The generation of an arc, sparks, or heat at any point causes rejection of the structures as a ground circuit. When a structure or pipeline is continuously employed as a ground return circuit, all joints are to be bonded, and periodic inspections must be conducted to ensure that no condition of electrolysis or fire hazard exists by virtue of such use.

The frames of all arc welding and cutting machines need to be grounded either through a third wire in the cable containing the circuit conductor, or through a separate wire which is grounded at the source of the current (see Figure 7-4). Grounding circuits, other than by means of the structure, are checked to ensure that the circuit between the ground and the grounded power conductor has resistance low enough to permit sufficient current to flow to cause the fuse or circuit breaker to interrupt the current. All ground connections are to be inspected to ensure that they are mechanically strong and electrically adequate for the required current.

Figure 7-4. Grounding for an arc welder

Workers are to be instructed in the safe means of arc welding and cutting as follows:

1. When electrode holders are to be left unattended, the electrodes must be removed and the holders are so placed or protected that they cannot make electrical contact with employees or conducting objects.

2. Hot electrode holders are not dipped into water; to do so may expose the arc welder or cutter to electric shock.

3. When the arc welder or cutter has occasion to leave his work or stop work for any appreciable length of time, or when the arc welding or cutting machine is to be moved, the power supply switch to the equipment is opened.

4. Any faulty or defective equipment is reported to the supervisor.

Whenever practical, all arc welding and cutting operations are shielded by noncombustible or flameproof screens which will protect welders and other persons working in the vicinity from the direct rays of the arc.

BARRICADES (1926.202)

These structures provide a substantial deterrent to the passage of individuals or vehicles. Barricades for the protection of workers must conform to the National Standards Institute D6.1-1971, *Manual on Uniform Control Devices for Streets and Highways*, Barricades Section. The most familiar type of barricade is the infamous "jersey barrier" which is made of reinforced concrete, but barricades can be made of other materials, if it provides an adequate protection factor. See Figure 7-5 for an example of a barricade.

Figure 7-5. The use of jersey barriers as barricades

BARRIERS

Barriers are used to warn workers and others of existing hazards and are usually not as substantial as barricades. They may be made of rope, wire ropes, warning tapes, or plastic fencing. A warning sign should accompany such barriers, either detailing the hazard or cautioning not to enter the area cordoned off by the barrier. See Figure 7-6 for an example of a barrier.

Figure 7-6. A barrier marked with caution tape

BATTERIES (1926.441)

Batteries of the unsealed type should be located in enclosures with outside vents or in well-ventilated rooms and arranged so as to prevent the escape of fumes, gases, or electrolyte spray into other areas. Ventilation is to provide diffusion of the gases from the battery and to prevent the accumulation of an explosive mixture, especially hydrogen gas. When racks and trays are used to store batteries, they should be substantial and treated to make them resistant to the electrolyte. An example of a poorly designed battery charging area can be seen in Figure 7-7. Floors should be of acid resistant construction unless protected from acid accumulations. Face shields, aprons, and rubber gloves should be provided for workers handling acids or batteries. Also, facilities must be provided for the quick drenching of eyes and the body and be within 25 feet (7.62 m) of battery-handling areas.

Facilities must be provided for flushing and neutralizing spilled electrolyte and for fire protection. Designated areas need to be provided for the purpose of battery charging installations and they must be located in areas designated for that purpose and charging apparatus must be protected from damage by trucks. When batteries are being charged, the vent caps must be kept in place to avoid electrolyte spray. Vent caps are to be maintained in functioning condition.

CHANGE ROOMS (1926.51)

When workers are required to wear PPE, there is the possibility of contamination with toxic materials. When this is the case, there must be a change room with storage for street clothes, so the contaminated PPE can be removed or disposed.

COMPRESSED AIR, USE OF   (1926.302)

Compressed air is not to be used for cleaning purposes, except when reduced to less than 30 psi, and then only with effective chip guarding and personal protective equipment which meets the requirements of Subpart I. The 30 psi requirement does not apply for concrete forms, mill scales, and similar cleaning purposes.

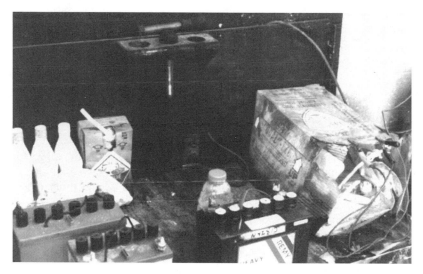

Figure 7-7. Poorly designed battery charging area

COMPRESSED AIR, WORKING UNDER

When working under compressed air, no employee is permitted to enter a compressed air environment until a physician has examined and reported him/her to be physically qualified to engage in such work. At least one physician is to be available at all times, while work is in progress, to provide medical supervision of employees engaged in compressed air work. If an employee is absent from work for 10 days or more, they must be reexamined by the physician prior to returning to work in a compressed air environment.

Employees continuously employed in compressed air will be reexamined by the physician within one year to determine if they are still physically qualified to engage in compressed air work. Examination records will be maintained by the physician.

A fully equipped first aid station will be provided at each tunnel project, regardless of the number of persons employed. A medical lock must be established and maintained in working order whenever air pressure in the chamber is increased above the normal atmosphere.

An identification badge will be provided to any employee working in a compressed air environment. The badge must give the employee's name, address of the medical lock, telephone number of the licensed physician for the compressed air project, and contain instructions for rushing wearer to the medical lock. The badge must be worn at <u>all</u> times – off the job, as well as on the job.

Effective and reliable means of communication (e.g., bells, whistles, or telephones) will be continuously maintained at the following locations:

- The working chamber face.

- The working chamber side of the man lock near the door.

- The interior of the man lock.

- Lock attendant's station.

- The compressor plant.

- The first aid station.

- The emergency lock (if required).

- The special decompression chamber (if required.)

A record must be kept of employees who work under compressed air. This record must be kept for each eight-hour shift and is to be located outside the lock and near the entrance. Every employee going under air pressure for the first time must receive instruction on how to avoid excessive discomfort. Except in an emergency, no employee working in compressed air will be permitted to pass from the working chamber to atmospheric pressure until after decompression.

Lock attendants who are in charge of a man lock will be under the direct supervision of the physician. They must be stationed at the lock controls on the free air side during the period of compression and decompression, and remain at the lock control whenever employees are in the working chamber or main lock. Lighting in compressed air chambers must be by electricity exclusively and use two independent sources of supply. The emergency source must automatically operate in the event of failure of the regular source. Lighting must not be less than 10 foot-candles on any walkway, ladder, stairway, or working level.

Firefighting equipment must be available and in good working condition at all times.

COMPRESSED GAS CYLINDERS (1926.350)

Compressed gas cylinders have the potential to become a guided missile if broken or damaged and will release a tremendous amount of energy. Thus, compressed gas cylinders should be treated with respect. Most compressed gas cylinders are approximately 1/4 inch in thickness, weigh 150 pounds or more, and are under some 2,200 pounds per square inch (psi) of pressure.

When transporting, moving, or storing compressed gas cylinders, valve protection caps should be in place and secured. When cylinders are hoisted, they must be secured on a cradle, slingboard, or pallet. (See Figure 7-8 for an example of a compressed gas lifting cradle.) They should never be hoisted or transported by means of magnets or choker slings. The valve protection caps are never used for lifting cylinders from one vertical position to another. Bars should not be used under valves or valve protection caps to pry cylinders loose when frozen. Warm, not boiling, water is used to thaw cylinders loose.

Compressed gas cylinders can be moved by tilting and rolling them on their bottom edges, but care must be taken to not drop or strike them together. This is especially true while transporting them by powered vehicles since they must be secured in a vertical position. In most cases, regulators are removed, and valve protection caps put in place, before cylinders are moved. (See Figure 7-9 which shows cylinders that have been secured against falling.)

Oxygen cylinders that are put in storage are to be separated from fuel-gas cylinders or combustible materials (especially oil or grease). They must be separated by a minimum distance of 20 feet (6.1 m), or by a noncombustible barrier of at least 5 feet (1.5 m) high; this barrier must have a fire-resistance rating of at least one-half hour. Cylinders should be stored

Figure 7-8. Cradle for lifting compressed gas cylinders

in assigned places that are away from elevators, stairs, or gangways; in areas where they will not be knocked over or damaged by passing or falling objects; or in areas where they are subject to tampering by unauthorized persons.

Cylinders are to be kept far enough away from the actual welding or cutting operation so that sparks, hot slag, or flame will not reach them. When this is impractical, fire resistant shields must be provided.

Also, cylinders should be placed where they cannot become part of an electrical circuit. Cylinders containing oxygen, acetylene, or other fuel gas must not be taken into confined spaces. Cylinders, whether full or empty, are not to be used as rollers or supports. No damaged or defective cylinders are to be used.

All employees should be trained in the safe use of fuel gas. The training should include how to crack a cylinder valve to clear dust or dirt prior to connecting a regulator. The worker cracking the valve should stand to the side and make sure the fuel gas is not close to an ignition source. The cylinder valve is always opened slowly to prevent damage to the regulator. For quick closing, valves on fuel gas cylinders are not opened more than 1 1/2 turns. When a special wrench is required, it is left in position on the stem of the valve. It is kept in this position while the cylinder is in use, so that the fuel gas flow can be quickly shut off in case of an emergency. In the case of manifolded or coupled cylinders, at least one such wrench is always available for immediate use. Nothing is placed on top of a fuel gas cylinder when it is in use; this may damage the safety device or interfere with the quick closing of the valve.

Fuel gas is not used from cylinders unless a suitable regulator is attached to the cylinder valve or manifold. Before a regulator is removed from a cylinder valve, the cylinder valve is always closed and the gas released from the regulator. If, when the valve on a fuel gas cylinder is opened, there is found to be a leak around the valve stem, the valve is closed and the gland nut tightened. If this action does not stop the leak, the use of the cylinder is discontinued, and it shall be properly tagged and removed from the work area. In the event that fuel gas should leak from the cylinder valve rather than from the valve stem, and the gas cannot be shut off, the cylinder is properly tagged and removed from the work area. If a regulator attached to a cylinder valve will effectively stop a leak through the valve seat, the cylinder need not be removed from the work area. But, if a leak should develop at a fuse plug or other safety device, the cylinder is removed from the work area.

COMPRESSED GAS WELDING

When using fuel gas and oxygen for compressed gas welding, certain precautions are necessary.

Fuel gas and oxygen manifolds should have the name of the substance they contain written in letters at least 1 inch high. This should be painted on the manifold or on a sign permanently attached to it. Fuel gas and oxygen manifolds are to be placed in safe, well-ventilated, and accessible locations. They are not to be located within enclosed spaces. The manifold hose connections, including both ends of the supply hose that leads to the manifold, are to be designed so that the hoses cannot be interchanged between fuel gas and oxygen manifolds and supply header connections. Adapters are not to be used to permit the interchange of hoses. Hose connections must be kept free of grease and oil. When not in use, manifold and header hose connections are to be capped. Nothing is to be placed on top of a manifold when in use, which will damage the manifold or interfere with the quick closing of the valves (see Figure 7-9).

Figure 7-9. Compressed gas welding set up with chain-secured gas cylinders

Fuel gas and oxygen hoses are to be easily distinguishable from each other. The contrast may be made by different colors or by surface characteristics readily distinguishable by the sense of touch. Oxygen and fuel gas hoses are not to be interchangeable. A single hose having more than one gas passage is not to be used. When parallel sections of oxygen and fuel gas hoses are taped together, not more than 4 inches out of 12 inches may be covered by tape.

All hoses in use, carrying acetylene, oxygen, natural, or manufactured fuel gas, or any gas or substance which may ignite or enter into combustion, or be in any way harmful to employees, should be subject to inspection at the beginning of each working shift. Defective hoses must be removed from service. Hoses which have been subject to flashback, or which show evidence of severe wear or damage, must be tested at twice the normal pressure to which it is subject, but in no case less than 300 psi Defective hoses, or hoses in doubtful condition, should not be used.

Any hose couplings must be of the type that cannot be unlocked or disconnected by means of a straight pull without rotary motion. Any boxes used for the storage of gas hoses must be ventilated. Hoses, cables, and other equipment are to be kept clear of passageways, ladders, and stairs.

Clogged torch tip openings should be cleaned with suitable cleaning wires, drills, or other devices designed for such purpose. Torches in use are to be inspected at the beginning of each working shift for leaking shutoff valves, hose couplings, and tip connections. Defective torches are to be removed from service. All torches are to be lighted by friction lighters or other approved devices, and not by matches or from hot work.

Oxygen and fuel gas pressure regulators, including their related gauges, must be in proper working order while in use. Oxygen cylinders and fittings are kept away from oil or grease. Cylinders, cylinder caps and valves, couplings, regulators, hoses, and apparatus are kept free from oil or greasy substances and are not handled with oily hands or gloves. Oxygen is not directed at oily surfaces, greasy clothes, or within a fuel oil or other storage tank or vessel. Oxygen must never be used for ventilation. Do not use oxygen to blow off clothes or clean welds; serious burns or death can occur

Additional rules – For additional details not covered in this subpart, applicable technical portions of American National Standards Institute, Z49.1-1967, Safety in Welding and Cutting, apply.

CONCRETE CONSTRUCTION (1926.701)

Compliance with concrete construction regulations will help prevent injuries and accidents that occur too frequently during concrete and masonry construction. No construction loads are placed on a concrete structure or portion of a concrete structure unless the employer determines, based on information received from a person who is qualified in structural design, that the structure or portion of the structure is capable of supporting the loads.

Employees are required to wear a safety belt or equivalent fall protection, when placing or tieing reinforcing steel more than six (6) feet above any working surface. All protruding reinforcing steel, onto which workers could fall, must be guarded to eliminate the hazard of impalement. (See Figure 7-10.)

During concrete construction, workers should not be permitted behind the jack during tensioning operations, and signs and barriers must be erected during tensioning operations to limit access to that area.

Concrete buckets, equipped with hydraulic or pneumatic gates, must have positive safety latches or similar safety devices installed to prevent premature or accidental dumping. Concrete buckets are to be designed to prevent concrete from hanging up on the top and sides. Workers are prohibited from riding in concrete buckets and from working under concrete buckets while the buckets are being elevated or lowered into position. If at all possible, elevated concrete buckets are to be routed so that no employee, or the fewest number of employees, are exposed to the hazards associated with falling concrete buckets. (See Figure 7-11.)

Figure 7- 10. Protected rebar to guard against impalement

Figure 7-11. Concrete bucket in use on construction site

There are specific requirements for equipment used in concrete construction work. Bulk storage bins, containers, and silos must be equipped with conical or tapered bottoms and a mechanical or pneumatic means of starting the flow of material. No employee should be permitted to enter storage facilities unless the ejection system has been shut down, locked out, and tagged to indicate that the ejection system is not to be operated. Lifelines/harness are be used when workers must entry bins, etc. that need to be unclogged. Workers are not permitted to perform maintenance or repair work on equipment (such as compressors, mixers, screens, or pumps used for concrete and masonry construction activities) where inadvertent operation of the equipment could occur and cause injury, unless all potentially hazardous energy sources have been locked out and tagged. Tags must read "Do Not Start," or similar language, to indicate that the equipment is not to be operated.

Concrete mixers, with one cubic yard or larger loading skips, must be equipped with a mechanical device to clear the skip of materials; guardrails must also be installed on each side of the skip.

Powered and rotating type concrete troweling machines, that are manually guided, must be equipped with a control switch that will automatically shut off the power whenever the hands of the operator are removed from the equipment handles. (See Figure 7-12.) Concrete buggy handles may not extend beyond the wheels on either side of the buggy.

Figure 7-12. Worker using a powered trowel

Concrete pumping systems which use discharge pipes must have pipe supports designed for 100 percent overload. Compressed air hoses, used on concrete pumping systems, must be provided with positive fail-safe joint connectors to prevent separation of sections when pressurized. Sections of tremies and similar concrete conveyances are secured with wire rope (or equivalent materials), in addition to the regular couplings or connections. No worker is permitted to apply a cement, sand, or water mixture through a pneumatic hose unless the employee is wearing protective head and face equipment.

Bull float handles, used where they might contact energized electrical conductors, are constructed of nonconductive material or insulated with a nonconductive sheath, whose electrical and mechanical characteristics provide the equivalent protection of a handle constructed of nonconductive material (see Figure 7-13). Masonry saws are guarded with a semicircular enclosure over the blade and a method for retaining blade fragments is incorporated in the design of the saw's semicircular enclosure.

Figure 7-13. Construction worker using a bull float

CONCRETE CAST-IN-PLACE (1926.703)

Formwork is designed, fabricated, erected, supported, braced, and maintained so that it will be capable of supporting, without failure, all vertical and lateral loads that are anticipated to be applied to the formwork (see Figure 7-14). Formwork must be designed, fabricated, erected, supported, braced, and maintained in conformance with strength specifications. These specifications must include the appropriate drawings or plans, including all revisions for the jack layout, formwork (including shoring equipment), working decks, and scaffolds that need to be available at the jobsite.

Figure 7-14. Formwork for cast-in-place concrete

All shoring equipment (including equipment used in reshoring operations) is to be inspected prior to erection to determine that the equipment meets the requirements specified in the formwork drawings

Shoring equipment, which is found to be damaged and its strength reduced to less than the required standards, must be immediately reinforced. All erected shoring equipment should be inspected immediately prior to, during, and immediately after concrete placement; if found to be damaged or weakened, after erection, such that its strength is reduced to less than that required, it must immediately be reinforced.

The sills for shoring are sound, rigid, and capable of carrying the maximum intended load. All base plates, shore heads, extension devices, and adjustment screws are in firm contact, and secured when necessary, with the foundation and the form. Eccentric loads on shore heads and similar members are prohibited unless these members have been designed for such loading.

Whenever single post shores are used one on top of another (tiered), the employer must comply with the following specific requirements, in addition to the general requirements for formwork:

1.   The design of the shoring is prepared by a qualified designer and the erected shoring is inspected by an engineer qualified in structural design.

2.   The single post shores are vertically aligned.

3.   The single post shores are spliced to prevent misalignment.

4.   The single post shores are adequately braced in two mutually perpendicular directions at the splice level. Each tier is also diagonally braced in the same two directions.

Adjustment of single post shores to raise formwork is not made after the placement of concrete. Reshoring is erected, as the original forms and shores are removed, whenever the concrete is required to support loads in excess of its capacity. The steel rods or pipes on which jacks climb, or by which the forms are lifted, are to be specifically designed for that purpose and adequately braced when not encased in concrete.

All forms must be designed to prevent excessive distortion of the structure during the jacking operation. All vertical slip forms are provided with scaffolds or work platforms where employees are required to work or pass.

Jacks and vertical supports should be positioned in such a manner that the loads do not exceed the rated capacity of the jacks. Whenever failure of the power supply or lifting mechanism could occur, the jacks or other lifting devices are to be provided with mechanical dogs or other automatic holding devices to support the slip forms. The form structure is to be maintained within all design tolerances specified for plumbness during the jacking operation, and the predetermined safe rate of lift is not to be exceeded.

Any reinforcing steel for walls, piers, columns, and similar vertical structures must be adequately supported to prevent overturning and collapse. Also, employers are to take measures to prevent unrolled wire mesh from recoiling. Such measures may include, but are not limited to, securing each end of the roll, or turning over the roll.

Forms and shores (except those used for slabs on grade and slip forms) are never removed until the employer determines that the concrete has gained sufficient strength to support its weight and superimposed loads. Such determination shall be based on compliance with one of the following:

1.　The plans and specifications stipulate conditions for removal of forms and shores, and such conditions have been followed.

2.　The concrete has been properly tested with an appropriate ASTM standard test method designed to indicate the concrete compressive strength, and the test results indicate that the concrete has gained sufficient strength to support its weight and superimposed loads.

Reshoring is not removed until the concrete being supported has attained adequate strength to support its weight and all loads placed upon it.

Finally, formwork which has been designed, fabricated, erected, braced, supported, and maintained in accordance with Sections 6 and 7 of the *American National Standard for Construction and Demolition Operations Concrete and Masonry Work*, ANSI A10.9-1983, shall be deemed to be in compliance with the provision of 1926.703(a)(1).

CONFINED SPACES (1910.146 AND 1926.21)

Improper entry into confined spaces has resulted in approximately 200 lives lost each year. Confined spaces are not of adequate size and shape to allow a person to enter easily, have limited openings for workers to enter and exit, and are not designed for continuous human occupancy. Examples of confined spaces are storage tanks, silos, pipelines, manholes, and underground utility vaults. (See Figure 7-15 for an example of a confined space.) In evaluating confined space accidents in the past, certain scenarios predominated the event. These included the failure to recognize an area as a confined space; test, evaluate, and monitor for hazardous atmospheres; train workers regarding safe entry; and establish rescue procedures.

With the promulgation of the "Permit-Required Confined Space Entry" standard (29 CFR 1910.146) (Note: This standard is not directly applicable to construction.), all of these failures are addressed. First, a written and signed permit is required prior to entry, if the space contains or has the potential to contain a hazardous atmosphere (oxygen deficient, flammable, toxic); if it contains materials that could engulf the entrant; if the space's configuration could cause the entrant to become entrapped or asphyxiated by converging walls; or if it contains any other recognized serious safety (electrical) or health (pathogen bacteria) hazard. This regulation requires that all permit-required spaces be identified, evaluated, and controlled.

The flow chart in Figure 7-16 can be used to determine whether a space is a permit-required confined space.

Figure 7-15. A typical confined space with a barrier and warning sign

If a permit entry confined space exists then a written permit is needed which details and assures that procedures for entry exist; appropriate equipment and training for the authorized entrant(s) is provided; entry supervisors and attendants are trained and present; a written/signed entry permit (see Figure 7-17 for an example of a entry permit) exists prior to entry; trained and available rescue personnel exists (over one-half of the deaths in confined spaces are rescuers); the space has been posted with warning signs; barriers have been erected; and personal protective and rescue equipment are provided.

Work in confined spaces can be safely accomplished by paying attention to the key areas prior to entry. These are to identify, evaluate/test and monitor, train, and plan for rescue. Do not enter confines spaces without prior approval or a permit. Follow confined space entry procedures. Never enter a space unless you are trained. Wear required confined space entry PPE. Make sure, prior to entry, that arrangements have been made in case rescue is needed.

One of the least recognized, and most dangerous hazards on a construction jobsite, is working in confined spaces. Entry into confined spaces without the proper precautions could result in injury and/or impairment, or death due to

- An atmosphere that is flammable or explosive.

- Lack of oxygen to support life.

- Toxic materials, that upon contact or inhalation, could cause injury, illness, or death.

- General safety hazards such as steam, high pressure systems, or other work area hazards.

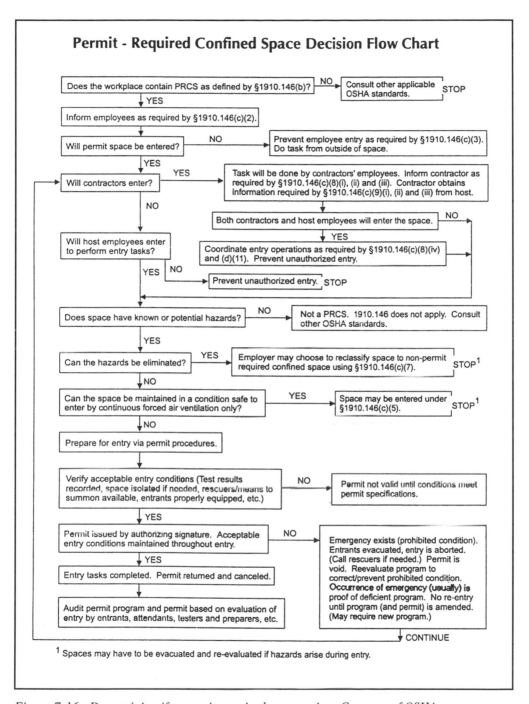

Figure 7-16. Determining if a permit-required space exists. Courtesy of OSHA

ENTRY PERMIT

PERMIT VALID FOR 8 HOURS ONLY. ALL COPIES OF PERMIT WILL REMAIN AT JOB SITE UNTIL JOB IS COMPLETED

DATE:_____ SITE LOCATION and DESCRIPTION _____

PURPOSE OF ENTRY _____

SUPERVISOR(S) in charge of crews Type of Crew Phone #

_____

_____

COMMUNICATION PROCEDURES _____

RESCUE PROCEDURES (PHONE NUMBERS AT BOTTOM) _____

_____

* BOLD DENOTES MINIMUM REQUIREMENTS TO BE COMPLETED AND REVIEWED PRIOR TO ENTRY*

REQUIREMENTS COMPLETED	DATE	TIME
Lock Out/De-energize/Try-out	____	____
Line(s) Broken-Capped-Blanked	____	____
Purge-Flush and Vent	____	____
Ventilation	____	____
Secure Area (Post and Flag)	____	____
Breathing Apparatus	____	____
Resuscitator - Inhalator	____	____
Standby Safety Personnel	____	____
Full Body Harness w/"D" ring	____	____
Emergency Escape Retrieval Equip	____	____
Lifelines	____	____
Fire Extinguishers	____	____
Lighting (Explosive Proof)	____	____
Protective Clothing	____	____
Respirator(s) (Air Purifying)	____	____
Burning and Welding Permit	____	____

Note: Items that do not apply enter N/A in the blank.

**RECORD CONTINUOUS MONITORING RESULTS EVERY 2 HOURS

CONTINUOUS MONITORING**	Permissible	
TEST(S) TO BE TAKEN	Entry Level	
PERCENT OF OXYGEN	19.5% to 23.5%	__ __ __ __ __ __ __
LOWER FLAMMABLE LIMIT	Under 10%	__ __ __ __ __ __ __
CARBON MONOXIDE	+35 PPM	__ __ __ __ __ __ __
Aromatic Hydrocarbon	+ 1 PPM * 5PPM	__ __ __ __ __ __ __
Hydrogen Cyanide	(Skin) * 4PPM	__ __ __ __ __ __ __
Hydrogen Sulfide	+10 PPM *15PPM	__ __ __ __ __ __ __
Sulfur Dioxide	+ 2 PPM * 5PPM	__ __ __ __ __ __ __
Ammonia	*35PPM	__ __ __ __ __ __ __

* Short-term exposure limit:Employee can work in the area up to 15 minutes.

+ 8 hr. Time Weighted Avg.:Employee can work in area 8 hrs (longer with appropriate respiratory protection).

REMARKS:_____

GAS TESTER NAME & CHECK #	INSTRUMENT(S) USED	MODEL &/OR TYPE	SERIAL &/OR UNIT #
_____	_____	_____	_____
_____	_____	_____	_____

SAFETY STANDBY PERSON IS REQUIRED FOR ALL CONFINED SPACE WORK

SAFETY STANDBY PERSON(S)	CHECK #	CONFINED SPACE ENTRANT(S)	CHECK #	CONFINED SPACE ENTRANT(S)	CHECK #
_____	_____	_____	_____	_____	_____
_____	_____	_____	_____	_____	_____

SUPERVISOR AUTHORIZING – ALL CONDITIONS   SATISFIED_____

DEPARTMENT/PHONE _____

AMBULANCE _____ FIRE _____ Safety _____ Gas Coordinator _____

Figure 7-17. Sample written entry permit. Courtesy of OSHA

In an effort to prevent injury or death when working in confined spaces, the contractor should implement and enforce the following safe work procedures:

1.   A confined space entry permit must be completed and signed by an authorized person prior to entry into a confined space.

2.   A hazard evaluation must be conducted before any work is started in a confined space.

3.   Technically competent personnel (i.e., industrial hygienist, safety specialist, etc.) must test the atmosphere within the confined space with an appropriate gas detector and approved oxygen testing equipment before employees enter.

4.   If combustible gases are detected, employees are prohibited from entering the confined space until the source has been isolated and the space flushed or purged to less than 10% of the lower explosive limit.

5.   If an oxygen-deficient atmosphere (less than 19.5% by volume) is present, positive ventilation techniques, including fans and blowers, may be used to increase the oxygen content. If further testing indicates the atmosphere is still oxygen deficient, Self-Contained Breathing Apparatus (SCBAs), or other air supplied respiratory protection will be provided.

6.   When toxic or chemical materials are detected or suspected, the following actions should be taken:

     • Any piping that carries or may carry hazardous materials to the confined space, will be isolated.

     • Empty the hazardous substance from the space until safe limits are reached.

     • Provide adequate ventilation and personal protective equipment for the eyes, face, and arms, if welding, burning, cutting, or heating operations, which may generate toxic fumes and gases, are performed.

     • Employees must wear eye and other appropriate protective equipment to prevent possible contact with corrosive materials.

7.   An emergency plan of action that provides alternate life support systems and a means of escape from confined spaces must be developed and communicated to all employees engaged in work in confined spaces.

8.   For evacuation purposes, each employee entering a confined space should wear a safety belt equipped with a lifeline, in case of an emergency. (See Figure 7-18 for example of a confined space retrieval system.)

9.   Emergency equipment (e.g., lifelines, safety harnesses, fire extinguishers, breathing equipment, etc.) appropriate for the situation should be ready and immediately available.

10. All persons engaged in the confined space activity must receive training in the use of the life support system, rescue system, and emergency equipment.

11. An attendant, trained in first aid and respiration, must remain outside the entrance to the confined space. The attendant must be ready to provide assistance, if needed, by utilizing a planned, and immediately available, communications means (radio, hand signals, whistle, etc.). The attendant should never enter the confined space in an attempt to rescue workers until additional rescue team personnel have arrived.

Figure 7-18. Confined space worker retrieval system with attendant

CONSTRUCTION MASONRY (1926.706)

A limited access zone is to be established whenever a masonry wall is being constructed. The limited access zone shall conform to the following:

1. The limited access zone is established prior to the start of construction of the wall.

2. The limited access zone is equal to the height of the wall to be reconstructed plus four feet, and must run the entire length of the wall.

3. The limited access zone is established on the side of the wall which will be unscaffolded.

4. The limited access zone is restricted to entry by employees actively engaged in constructing the wall; no other employees are permitted to enter the zone.

5. The limited access zone remains in place until the wall is adequately supported to prevent overturning or collapse, unless the height of wall is over eight feet, in which case the limited access zone shall remain in place until the requirements of paragraph (b) of this section have been met.

All masonry walls over eight feet in height are to be adequately braced to prevent overturning or collapse unless the wall is adequately supported so that it will not overturn or collapse. The bracing shall remain in place until permanent supporting elements of the structure are in place.

CONVEYORS (1926.555)

Since conveyors present a moving hazard to workers, great care is to be taken to assure that operators have access to the on/off switch and an audible warning exists prior to the conveyor starting.

Workers are prohibited from riding moving chains, conveyors, or similar equipment. Many workers have been caught between the belt and rollers which has resulted in loss of fingers, hands, arms, or other severe injuries and even death.

Thus, conveyors and other such equipment need to be equipped with emergency cut-off switches, and workers should be protected from the conveyors' moving parts by guards. All nip points and noise or tail pieces should be protected so workers cannot come into contact with moving pulleys, gears, belts, or rollers (see Figure 7-19). The conveyor should be shut down and locked out/tagout during maintenance and repair work.

Figure 7-19. Example of conveyor hazards with some guarding

If material can fall from the conveyor, overhead protection needs to be in place, or the walkway under the conveyor should have barriers or barricades to prevent individuals from passing through. For specific questions, consult ANSI B20.1-1957.

CRANES AND DERRICKS (1926.550)

Rated Loads

The employer shall comply with the manufacturer's specifications and limitations applicable to the operation of any and all cranes and derricks. Where manufacturer's specifications are not available, the limitations assigned to the equipment are based on the determinations of a qualified engineer competent in this field, and such determinations will be appropriately documented and recorded. Attachments used with cranes must not exceed the capacity, rating, or scope recommended by the manufacturer.

Rated load capacities, recommended operating speeds, special hazard warnings, or instructions are to be conspicuously posted on all equipment. Instructions or warnings must be visible to the operator while he/she is at his/her control station. All employees are kept clear of suspended loads or loads about to be lifted.

Cranes must deploy outriggers in order to widen the base to be able to lift their intended loads. Warning barriers are erected to keep workers out of the area or operation and the swing radius of the crane itself (see Figure 7-20).

Figure 7-20. Crane with outriggers deployed and warning barriers

Hand Signals

Hand signals to crane and derrick operators are those prescribed by the applicable ANSI standard for the type of crane in use. An illustration of the signals must be posted at the jobsite. (See Figure 7-21.)

Crane Inspections

The employer must designate a competent person who inspects all machinery and equipment prior to each use, and during use, to make sure it is in safe operating condition. (See Figure 7-22 for an example of a crane inspection form.) Any deficiencies are repaired, or defective parts replaced, before continued use. A thorough annual inspection of the hoisting machinery is made by a competent person, or by a government or private agency recognized by the U.S. Department of Labor. The employer maintains a record of the dates and results of inspections for each hoisting machine and piece of equipment.

Wire Rope

Wire rope safety factors are in accordance with American National Standards Institute B 30.5-1968 or SAE J959-1966. Wire rope is taken out of service when any of the following conditions exist:

1. In running ropes, six randomly distributed broken wires in one lay, or three broken wires in one strand in one lay.

2. Wear of one-third of the original diameter of outside individual wires. Kinking, crushing, bird caging, or any other damage resulting in distortion of the rope structure.

3. Evidence of any heat damage from any cause.

4. Reductions from nominal diameter of more than one-sixty-fourth inch for diameters up to, and including five-sixteenths inch; one-thirty-second inch for diameters three-eighths inch to, and including one-half inch; three-sixty-fourths inch for diameters nine-sixteenths inch to, and including three-fourths inch; one-sixteenth inch for diameters seven-eighths inch to one and one-eighth inch

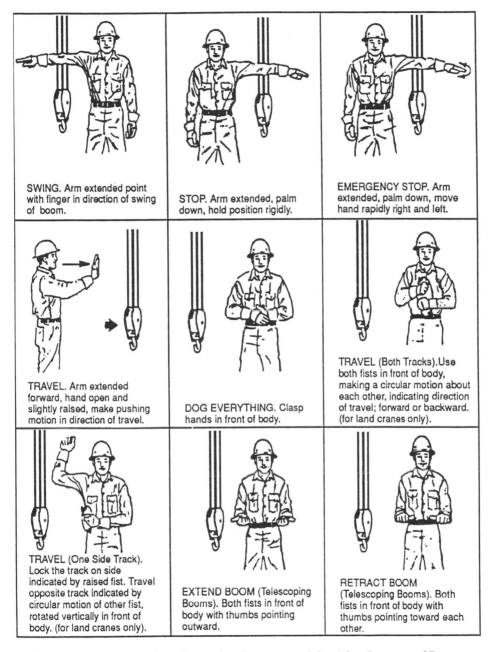

Figure 7-21. Hand signals to be used with cranes and derricks. Courtesy of Department of Energy

inclusive; three-thirty-seconds inch for diameters one and one-fourth to one and one-half inch inclusive.

5. In standing ropes, more than two broken wires in one lay in sections beyond end connections, or more than one broken wire at an end connection.

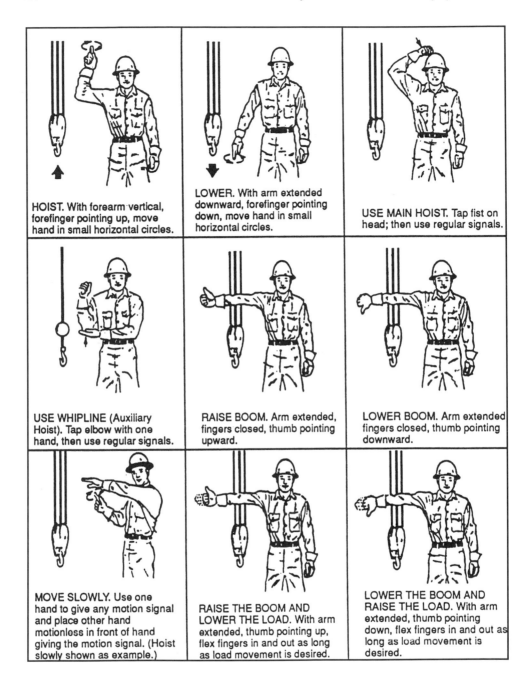

Figure 7-21.   Hand signals to be used with cranes and derricks. Courtesy of Department of Energy (continued)

MONTHLY CRANE INSPECTION FORM

Make: _____Model:_____ Serial #_____

Date of Inspection: _____Crane Location:_____

Job # _____Inspected by:_____

AREA	OK	R/N	N/A	R/C
General:				
Appearance				
Paint				
Glass				
Cab				
Fire Extinguisher (5BC)				
Manuals (Parts & Operators)				
Load Charts				
Hand Signals				
Grease/Oil				
Boom				
Angle Indicator				
Warning Signs				
Steps				
Guards in Place				
Hands and Grab Rails				
Access to Roof				
Anti-Lock Block				
Back-up Alarm				
Engine:				
Operating Condition				
Oil Level and Condition				
Hour Meter				
Engine Instruments				
Cooling System				
Anti-Freeze				
Battery Condition				
Hose Condition				
Clamps				
Air System				
Pressure				
Belt Condition				
Filter				
Compressor				
Converters				
Engine Clutch				
Day Tank (Converters)				
Hydraulic Reservoir				
Oil Pressure				
Operating Temperature				
Transmission Case				
Electrical & Control System:				
Electrical Components & Assemblies				
Emergency Stop Switch				
Overload Switch				
Master Switches or Drum Controller				
Switches, Contacts & Relays				
Main Control Station				
Remote Control Station & Bucket/Ground Controls				

AREA	OK	R/N	N/A	R/C
Tracks:				
Chains				
Sprockets				
Idlers				
Pins				
Track Adjustments				
Roller Path				
Travel Brake				
Carrier:				
Tire Condition				
Brakes				
Steering				
Outriggers & Pads				
Glass				
Controls				
Fire Extinguisher (5BC)				
Horn				
Turn Signals				
Lights				
Transmission				
Frame				
Battery				
License No. (Current)				
Boom:				
No. or Type				
Length				
Tagline				
Swing System				
Clutch				
Brake				
House Rollers				
Hook Rollers				
Swing Gears				
Drum Shaft				
Clutches				
Main Line				
Aux. Line				
Brakes				
Main Line				
Aux. Line				
Boom:				
Boom Hoist				
Worn Gear				
Brass Gear				
Brake				
Clutches				
Hydraulic System				
Pawl				
Lubrication				
Guard in Place				

Figure 7-22. Crane inspection form

MONTHLY CRANE INSPECTION FORM *(CONTINUED)*

AREA	OK	R/N	N/A	R/C	AREA	OK	R/N	N/A	R/C
Indicators: Levels: Boom Angle and Length					**Cables & Wire Rope:**				
Drum Rotation: Load					Boom Hoist				
					Load Line				
Boom & Attachments:					Aux./Whip Line				
Number/Type					Ringer Boom Line				
Boom Inventory					Ringer Aux./Whip Line				
Point					Load Line Wedge Socket				
Heel					Aux./Whip Line Wedge Socket				
10 Feet					Cable Clamps				
20 Feet					Kinks & Broken Strands (Any Abuse)				
30 Feet					Jib Pendants				
40 Feet									
50 Feet									
Load Block (Capacity)					**Records:**				
Hook					Current Certification Posting				
Safety Latch					Operator Instructions				
Boom Pins/Cotter Pins					Preventive Maintenance				
Gantry					Properly Marked Operator Controls, Levels & Diagrams				
Equalizer					Info. & Warning Decals				
Cable Rollers									
Cable Guides									
Spreader Bar									
Jib (Type & Length)									
Lubrication									
Jib Inventory									
Point									
10 Feet									
15 Feet									
20 Feet									
Headache Ball (Capacity)									
Cable Rollers									
Pendant Lines									
Off-Set Links									
Boom Stops									
Auto Boom Stop									
Counterweights									

OK - Satisfactory   R/N - Repairs Needed   N/A - Not Applicable   R/C - Repairs Completed

Figure 7-22   Crane inspection form (continued)

Guarding

Belts, gears, shafts, pulleys, sprockets, spindles, drums, fly wheels, chains, or other reciprocating, rotating, or moving parts or equipment are guarded if such parts are exposed to contact by employees, or otherwise create a hazard. Guarding must meet the requirements of the American National Standards Institute B 15.1-1958 Rev., Safety Code for Mechanical Power Transmission Apparatus.

Accessible areas within the swing radius of the rear of the rotating superstructure of the crane, either permanently or temporarily mounted, are barricaded in such a manner as to prevent an employee from being struck or crushed by the crane.

All exhaust pipes are guarded or insulated in areas where contact by employees is possible in the performance of normal duties. Whenever internal combustion engine-powered equipment exhausts in enclosed spaces, tests are made and recorded to see that employees are not exposed to unsafe concentrations of toxic gases or oxygen-deficient atmospheres.

All windows in cabs are of safety glass, or equivalent, that introduces no visible distortion that will interfere with the safe operation of the machine.

Where necessary for rigging or service requirements, a ladder or steps are provided to give access to a cab roof. Guardrails, handholds, and steps are provided on cranes for easy access to the car and cab, conforming to American National Standards Institute B30.5. Platforms and walkways are to have anti-skid surfaces.

Fueling

Fuel tank filler pipe is located in such a position, or protected in such manner, as to not allow spill or overflow to run onto the engine, exhaust, or electrical equipment of any machine being fueled. An accessible fire extinguisher of 5BC rating, or higher, is available at all operator stations or cabs of equipment. All fuels are transported, stored, and handled to meet the rules of the construction fire safety regulation. When fuel is transported by vehicles on public highways, Department of Transportation rules contained in 49 CFR Parts 177 and 393 concerning such vehicular transportation are considered applicable.

Electrical Concerns

Except where electrical distribution and transmission lines have been deenergized and visibly grounded at the point of work, or where insulating barriers, not a part of, or an attachment to the equipment or machinery, have been erected to prevent physical contact with the lines, equipment, or machines, are to be operated proximate to power lines only in accordance with the following:

1. For lines rated 50 kV. or below, minimum clearance between the lines and any part of the crane or load shall be 10 feet.

2. For lines rated over 50 kV., minimum clearance between the lines and any part of the crane or load shall be 10 feet plus 0.4 inch for each 1 kV. over 50 kV., or twice the length of the line insulator, but never less than 10 feet.

3. In transit with no load and boom lowered, the equipment clearance is a minimum of 4 feet for voltages less than 50 kV., and 10 feet for voltages over 50 kV., up to and including 345 kV., and 16 feet for voltages up to and including 750 kV.

4. A person is designated to observe clearance of the equipment and give timely warning for all operations where it is difficult for the operator to maintain the desired clearance by visual means.

5. Cage-type boom guards, insulating links, or proximity warning devices may be used on cranes, but the use of such devices does not alter the requirements of any other regulation of this part, even if such device is required by law or regulation.

6. Any overhead wire is considered to be an energized line, unless and until the person owning such line or the electrical utility authorities indicate that it is not an energized line and it has been visibly grounded.

7. Prior to work near transmitter towers, where an electrical charge can be induced in the equipment or materials being handled, the transmitter is deenergized or tests are made to determine if electrical charge is induced on the crane. The

following precautions are taken, when necessary, to dissipate induced voltage: the equipment is provided with an electrical ground directly to the upper rotating structure supporting the boom, and ground jumper cables are attached to materials being handled by the boom equipment when an electrical charge is induced while working near energized transmitters. Crews are provided with nonconductive poles having large alligator clips, or other similar protection to attach the ground cable to the load. Combustible and flammable materials are removed from the immediate area prior to operations.

Modifications

No modifications or additions, which affect the capacity or safe operation of the equipment, are made by the employer without the manufacturer's written approval. If such modifications or changes are made, the capacity, operation, and maintenance instruction plates, tags, or decals are changed accordingly. In no case is the original safety factor of the equipment to be reduced. The employer shall comply with Power Crane and Shovel Association Mobile Hydraulic Crane Standard No. 2. Sideboom cranes, mounted on wheel or crawler tractors, must meet the requirements of SAE J743a-1964. Illustrations of the different types of cranes and derricks can be found in Appendix M.

Crawler, Locomotive, and Truck Cranes

Crawler, locomotive, and truck cranes must have positive stops on all jibs to prevent their movement of more than 5 degrees above the straight line of the jib and boom on conventional type crane booms. The use of cable type belly slings does not constitute compliance with this rule. All crawler, truck, or locomotive cranes in use must meet the applicable requirements for design, inspection, construction, testing, maintenance, and operation as prescribed in the ANSI B30.5-1968, Safety Code for Crawler, Locomotive and Truck Cranes. However, the written, dated, and signed inspection reports and records of the monthly inspection of critical items prescribed in Section 5-2.1.5 of the ANSI B30.5-1968 standard are not required. Instead, the employer shall prepare a certification record which includes the date the crane items were inspected, the signature of the person who inspected the crane items, and a serial number, or other identifier, for the crane inspected. The most recent certification record shall be maintained on file until a new one is prepared (see Appendix M).

Hammerhead Tower Cranes

For hammerhead tower cranes, adequate clearance is to be maintained between moving and rotating structures of the crane and fixed objects, in order to allow for safe passage of employees. Each employee required to perform duties on the horizontal boom of the hammerhead tower cranes is protected against falling by guardrails, or by a personal fall arrest system. Buffers are provided at both ends of travel of the trolley. Cranes mounted on rail tracks are equipped with limit switches limiting the travel of the crane on the track, and stops or buffers at each end of the tracks. All hammerhead tower cranes in use are required to meet the applicable requirements for design, construction, installation, testing, maintenance, inspection, and operation as prescribed by the manufacturer (see Appendix M).

Overhead and Gantry Cranes

For overhead and gantry cranes, the rated load of the crane is plainly marked on each side of the crane, and if the crane has more than one hoisting unit, each hoist has its rated load marked on it or its load block; this marking is clearly legible from the ground or floor. Bridge trucks are equipped with sweeps which extend below the top of the rail and project in front of the truck wheels. Except for floor-operated cranes, a gong or other effective audible warning signal is provided for each crane equipped with a power traveling mechanism. All overhead and gantry cranes in use meet the applicable requirements for design, construction, installation, testing, maintenance, inspection, and operation as prescribed in the ANSI B30.2.0-1967, Safety Code for Overhead and Gantry Cranes (see Appendix M).

Derricks

All derricks in use meet the applicable requirements for design, construction, installation, inspection, testing, maintenance, and operation as prescribed in American National Standards Institute B30.6-1969, Safety Code for Derricks. For floating cranes and derricks, when a mobile crane is mounted on a barge, the rated load of the crane does not exceed the original capacity specified by the manufacturer.

A load rating chart, with clearly legible letters and figures, is provided with each crane, and securely fixed at a location easily visible to the operator. When load ratings are reduced to stay within the limits listed for a barge with a crane mounted on it, a new load rating chart is to be provided (see Appendix M).

Floating Cranes and Derricks

Mobile cranes on barges must be positively secured. For permanently mounted floating cranes and derricks, when cranes and derricks are permanently installed on a barge, the capacity and limitations of use are based on competent design criteria. A load rating chart, with clearly legible letters and figures, is provided and securely fixed at a location easily visible to the operator. Floating cranes and floating derricks in use must meet the applicable requirements for design, construction, installation, testing, maintenance, and operation as prescribed by the manufacturer. The employer shall comply with the applicable requirements for protection of employees working onboard marine vessels.

Crane- and Derrick-Suspended Personnel Platforms

The use of a crane or derrick to hoist employees on a personnel platform is prohibited, except when the erection, use, and dismantling of conventional means of reaching the worksite, such as a personnel hoist, ladder, stairway, aerial lift, elevating work platform, or scaffold would be more hazardous, or is not possible because of structural design or worksite conditions. Hoisting of the personnel platform is performed in a slow, controlled, cautious manner with no sudden movements of the crane, derrick, or the platform.

Platform Operations

Load lines must be capable of supporting, without failure, at least seven times the maximum intended load, except where rotation resistant rope is used; when they are used, the lines must be capable of supporting, without failure, at least ten times the maximum intended load. The required design factor is achieved by taking the current safety factor of 3.5 and applying the 50 percent derating of the crane capacity. The total weight of the loaded personnel platform and related rigging does not exceed 50 percent of the rated capacity for the radius and configuration of the crane or derrick. Load and boom hoist drum brakes, swing brakes, and locking devices, such as pawls or dogs, are to be engaged when the occupied personnel platform is in a stationary position.

The use of machines having live booms (booms in which lowering is controlled by a brake without aid from other devices which slow the lowering speeds) is prohibited. Cranes and derricks with variable angle booms are equipped with a boom angle indicator readily visible to the operator. Cranes with telescoping booms are equipped with a device to indicate clearly to the operator, at all times, the boom's extended length or an accurate determination of the load radius, which will be used during the lift, is made prior to hoisting personnel.

The crane must be uniformly level within one percent of level grade and located on firm footing. Cranes equipped with outriggers must be fully deployed, following the manufacturer's specifications, insofar as applicable, when hoisting employees.

A positive acting device is used which prevents contact between the load block or overhaul ball and the boom tip (anti-two-blocking device), or a system is used which deactivates the hoisting action before damage occurs in the event of a two-blocking situation (two-block damage prevention feature).

The load line hoist drum has a system or device on the power train, other than the load hoist brake, which regulates the lowering rate of speed of the hoist mechanism (controlled load lowering); free fall is prohibited.

Platform Specifications

The personnel platform and suspension system shall be designed by a qualified engineer or a qualified person competent in structural design. The suspension system is designed to minimize tipping of the platform due to movement of employees occupying the platform. The personnel platform, except for the guardrail and personnel fall arrest system anchorages, must be capable of supporting, without failure, its own weight and at least five times the maximum intended load. The criteria for guardrail systems and personal fall arrest system anchorages are described under fall protection. Each personnel platform is equipped with a guardrail system and is enclosed at least from the toeboard to the mid-rail, with either solid construction or expanded metal, having openings no greater than 1/2 inch. A grab rail is to be installed inside the entire perimeter of the personnel platform. Access gates, if installed, are not to swing outward during hoisting. All access gates, including sliding or folding gates, are equipped with a restraining device to prevent accidental opening.

Headroom is to be provided which allows employees to stand upright on the platform. In addition to the use of hard hats, employees are protected by overhead protection on the personnel platform when employees are exposed to falling objects.

All rough edges exposed to contact by employees are to be surfaced or smoothed in order to prevent puncture or laceration injuries to employees. All welding of the personnel platform, and its components, must be performed by a qualified welder who is familiar with the weld grades, types, and material specified in the platform design.

The personnel platform is to be conspicuously posted with a plate, or other permanent markings, which indicate the weight of the platform and its rated load capacity or maximum intended load. The personnel platform must not be loaded in excess of its rated load capacity. When a personnel platform does not have a rated load capacity, then the personnel platform is not loaded in excess of its maximum intended load.

The number of employees occupying the personnel platform cannot exceed the number required for the work being performed. Personnel platforms are used only for employees, their tools, and the materials necessary to do their work; when not hoisting personnel, they are not to be used to hoist only materials or tools.

Materials and tools used during a personnel lift must be secured, to prevent displacement, and evenly distributed within the confines of the platform while the platform is suspended.

Rigging

When a wire rope bridle is used to connect the personnel platform to the load line, each bridle leg is connected to a master link or shackle in such a manner to ensure that the load is evenly divided among the bridle legs. Hooks on overhaul ball assemblies, lower load blocks, or other attachment assemblies, shall be of a type that can be closed and locked, eliminating the hook throat opening. Alternatively, an alloy anchor type shackle with a bolt, nut, and retaining pin may be used.

Wire rope, shackles, rings, master links, and other rigging hardware must be capable of supporting, without failure, at least five times the maximum intended load applied or transmitted to that component. Where rotation resistant rope is used, the slings are capable of supporting, without failure, at least ten times the maximum intended load. All eyes in wire rope slings are fabricated with thimbles.

Bridles and associated rigging for attaching the personnel platform to the hoist line are only used for the platform and the necessary employees, their tools, and the materials necessary to do their work; they are not used for any other purpose when not hoisting personnel.

Inspection and Proof Test

A trial lift, with the unoccupied personnel platform loaded at least to the anticipated lift weight, is made from ground level, or any other location where employees will enter the platform to each location at which the personnel platform is to be hoisted and positioned. This trial lift is performed immediately prior to placing personnel on the platform. The operator determines that all systems, controls, and safety devices are activated and functioning properly, that no interferences exist, and that all configurations necessary to reach those work locations, will allow the operator to remain under the 50 percent limit of the hoist's rated capacity. Materials and tools to be used during the actual lift can be loaded in the platform for the trial lift. A single trial lift may be performed at one time for all locations that are to be reached from a single setup position.

The trial lift is repeated prior to hoisting employees whenever the crane or derrick is moved and set up in a new location, or returned to a previously used location. Additionally, the trial lift is repeated when the lift route is changed, unless the operator determines that the route change is not significant (i.e., the route change would not affect the safety of hoisted employees). After the trial lift and just prior to hoisting personnel, the platform is hoisted a few inches and inspected to ensure that it is secure and properly balanced. Employees are not hoisted unless the following conditions are determined to exist:

1. Hoist ropes arc free of kinks.

2. Multiple part lines are not twisted around each other.

3. The primary attachment is centered over the platform.

4. The hoisting system is inspected, if the load rope is slack, to ensure all ropes are properly stated on drums and in sheaves.

A visual inspection of the crane or derrick, rigging, personnel platform, and the crane or derrick base support, or ground, must be conducted by a competent person immediately after the trial lift. This must be done to determine whether the testing has exposed any defect or produced any adverse effect upon any component or structure. Any defects found during inspections, which create a safety hazard, are to be corrected before hoisting personnel.

At each jobsite, prior to hoisting employees on the personnel platform and after any repair or modification, the platform and rigging are proof tested to 125 percent of the platform's rated capacity by holding it in a suspended position for five minutes with the test load evenly distributed on the platform. (This may be done concurrently with the trial lift.) After proof testing, a competent person must inspect the platform and rigging. Any deficiencies found are to be corrected and another proof test conducted. Personnel hoisting is not conducted until the proof testing requirements are satisfied.

Work Practices

Workers are to keep all parts of the body inside the platform during raising, lowering, and positioning. This provision does not apply to an occupant of the platform performing the duties of a signal person. Before employees exit or enter a hoisted personnel platform that has not landed, the platform is to be secured to the structure where the work is to be performed, unless securing to the structure creates an unsafe situation. Tag lines must be used unless their use creates an unsafe condition. The crane or derrick operator is to remain at the controls at all times when the crane engine is running and the platform is occupied. Hoisting of employees is promptly discontinued upon indication of any dangerous weather conditions or other impending danger.

Employees being hoisted must remain in continuous sight of, and in direct communication with, the operator or signal person. In those situations where direct visual contact with the operator is not possible, and the use of a signal person would create a greater hazard for the person, direct communication, such as by radio, may be used.

Except over water, employees occupying the personnel platform must use a body harness system with lanyard appropriately attached to the lower load block or overhaul ball, or attached to a structural member within the personnel platform which is capable of supporting an employee who might fall while using the anchorage. No lifts are made on another of the crane's or derrick's loadlines while personnel are suspended on a platform.

Traveling

Hoisting of employees while the crane is traveling is prohibited, except for portal, tower, and locomotive cranes, or where the employer demonstrates that there is no less hazardous way to perform the work. Under any circumstances where a crane would travel while hoisting personnel, the employer must implement the following procedures to safeguard employees:

1. Crane travel is restricted to a fixed track or runway.

2. Travel is limited to the load radius of the boom used during the lift.

3. The boom must be parallel to the direction of travel.

A complete trial run is performed to test the route of travel before employees are allowed to occupy the platform. This trial run can be performed at the same time as the trial lift which tests the route of the lift. If travel is done with a rubber tired-carrier, the condition and air pressure of the tires must be checked. The chart capacity for lifts on rubber is to be used for application of the 50 percent reduction of rated capacity. Outriggers may be partially retracted, as necessary, for travel.

Prelift Meeting

A meeting, attended by the crane or derrick operator, signal person(s) (if necessary for the lift), employee(s) to be lifted, and the person responsible for the task to be performed, is held to review the appropriate requirements and procedures to be followed. This meeting is held prior to the trial lift at each new work location, and is repeated for any employees newly assigned to the operation.

Accidents involving cranes are often caused by human actions or failure to act. Therefore, the company must employ competent operators who are physically and mentally fit and thoroughly trained in the safe operation of crane and rigging equipment, and the safe handling of loads. Upon employment, the crane operator must be assigned to work with the crane and rigging foreman, and the new operator should work only on selective jobs which will be closely monitored for a period of not less than one week.

DEMOLITION (1926.850)

When undertaking demolition work, an engineering survey must be completed prior to the start of the demolition project to determine condition of the framing, floors, and walls. Once work is ready to start, all utilities or energy sources will be turned off, all floors or walls to be demolished will be shored or braced, and all wall or floor openings must be protected. If hazardous chemicals, gases, explosives, flammable materials, or similarly dangerous substances have been used in pipes, tanks, or other equipment on the property, testing and purging must be performed to eliminate the hazard prior to demolition. During demolition involving combustible materials, charged hose lines, supplied by hydrants, water tank trucks with pumps, or equivalent, are made available. During demolition or alterations, existing automatic sprinkler installations are retained in service as long as reasonable. The operation of sprinkler control valves are permitted only by properly authorized persons. Modification of sprinkler systems, to permit alterations or additional demolition, should be expedited so that the automatic protection may be returned to service as quickly as possible. Sprinkler control valves are checked daily at close of work to ascertain that the protection is in service.

Only use stairways, passageways, and ladders which are designated as a means of access to the structure of a building. Stairs, passageways, ladders, and incidental equipment must be periodically inspected and maintained in a clean and safe condition. Stairwells must be properly illuminated and completely and substantially covered over at a point not less than two floors below the floor on which work is being performed.

When balling or clamming is being performed, never enter any area which may be adversely affected by demolition operations unless you are needed to perform these operations. During demolition, a competent person must make continued inspections as the work progresses to detect any hazards resulting from weakened or deteriorated floors, walls, or loosened material.

Chutes (1926.852)

No material is dropped outside the exterior wall of the structure unless the area is effectively protected. All chutes or sections thereof which are at an angle of more then 45 degrees from horizontal, are entirely enclosed, except for openings equipped with closures or about floor level for insertion of materials. Openings must not exceed 48 inches in height (measured along the wall of the chute). All stories below the top floor opening are to be kept closed except when in use. A substantial gate is to be installed in each chute near the discharge end. A competent person must be assigned to control the gate and the backing and loading of trucks. Any chute opening into which debris is dumped shall be protected by a guardrail 42 inches above the floor or surface on which workers stand to dump materials. Any space between the chute and edge of opening around the chute shall be covered. When material is dumped from mechanical equipment or a wheelbarrow, a toeboard or bumper, not less than four inches thick and six inches high is to be provided around the chute opening. Chutes must be strong enough to withstand the impact of materials dumped into it.

Removal of Materials through Floor Openings (1926.853)

Any openings in the floor are to be less than 25% of the aggregate of the total floor unless lateral supports of the floor remain in place. Any weakened floor is to be shored to support demolition loads.

Removal of Walls, Masonry Sections, and Chimneys (1926.854)

Masonry walls, or other masonry sections, are not permitted to fall upon the floors of buildings in such amount as to exceed the load limit of the floor. No wall section which is more than one story in height is allowed to stand without lateral bracing, unless originally designed to do so, and still has that integrity. No unstable wall shall exist at the end of a shift. Workers are not to work on top of walls when weather constitutes a hazard. No structural or lead-supporting member are to be cut or removed until all stories above have been demolished or removed, except for disposal openings or equipment installation. Floor openings within 10 feet of any wall being demolished must be planked solid or workers must be kept out of the area below.

In skeletal-steel constructed buildings, the steel frame may be left in place during the demolition of masonry, but all beams, girders, or other structural supports are to be kept clear of debris and loose materials. Walkways and ladders must be provided so workers can safely reach or leave any scaffold or wall. Walls which retain or support earth or adjoining structures are not removed until the earth or adjoining structure is supported.

Walls against which debris is piled must be capable of supporting that load.

Manual Removal of Floors (1926.855)

Openings cut into a floor must extend the full span of the arch between supports. Before floor arches are removed, debris and other material are to be removed. Two by ten planks must be full size, undressed, and have open spaces no more than 16 inches in order for workers to safely stand on them during the break down of floor arches between beams. Walkways not less than 18 inches wide, formed of wooden planks no less than 2 inches thick, or equivalent strength, if metal, are to be provided to workers to prevent them from walking on exposed beams to reach work points.

Stringers of ample strength must be installed to support flooring planks. These stringers are supported by floor beams or girders and not floor arches. Planks are to be laid together over solid bearings with ends overlapped at least one foot. Employees are not allowed in the area directly below where floor arches are being removed. This area is barricaded to prevent access. Floor arches are removed only after the surrounding area for a distance of 20 feet has been cleared of debris and other unnecessary materials.

Removal of Walls, Floors, and Material with Equipment (1926.856)

All floors and work surfaces are to be of sufficient strength to support any mechanical equipment being used. Curbs or stop-logs are put in place to prevent equipment from running over the edges.

Storage (1926.857)

The allowable floor loads are not to be exceeded by the waste or debris being stored there. Floor boards may be removed in buildings having wooden floor construction not more than one floor above grade. This will provide storage space for debris and assure that no falling material will endanger the stability of the structure. Wood floor beams, which brace interior walls or support free-standing exterior walls, must be left in place until equivalent support can be installed. Storage space, into which materials are dumped, is blocked off, except for the openings necessary for material removal. These openings are to be kept closed when not in use.

Removal of Steel Construction (1926.858)

Planking shall be used when floor arches have been removed and workers are engaged in razing the steel frame. Cranes, derricks, and other hoisting equipment must meet the construction specifications.

Steel construction is to be dismantled column length by column length, and tier by tier (columns may be in two-story lengths). Structural members being dismembered are not to be overstressed.

Mechanical Demolition (1926.859)

Workers are not permitted in areas where balling and clamming is being performed. Only essential workers are permitted in those areas at any other time. The weight of the demolition ball shall not exceed 50% of the crane's rated load, based on the length of the boom and the maximum angle of operation, and shall not exceed 25% on the nominal break strength of the line by which it is suspended, whichever is the lesser value. The crane load line is to be kept as short as possible. The ball must be attached to the load line with a swivel-type connection to prevent twisting, and it must be attached in such a manner that it cannot accidentally become disconnected. All steel members of walls must be cut free prior to pulling over a wall or portion thereof. Roof cornices or other ornamental stonework have to be removed prior to pulling over walls. To detect hazards resulting from weakened or loosened materials, a competent person must conduct inspections as demolition progresses. Workers are not permitted to work where such hazards exist until corrective action is taken.

Selective Demolition by Explosives (1926.860)

Demolition, using explosives, is conducted following the specifications of Subpart U.

DISPOSAL CHUTES (1926.252)

Disposal chutes are installed to prevent the hazards which exist from materials, etc., dropped from heights greater than 20 feet (see Figure 7-23). When materials are dropped to the exterior of a building, the chutes must be enclosed. The disposal chute should mitigate the danger from overhead hazard from falling materials.

At times materials or debris are dropped through holes in the floor. The landing area for this type of waste must have a 42 inch high barricade which is six feet back from the edge on all sides of the floor opening's edges. Also, warnings are to be posted at each level denoting the fall materials hazard.

DIVING (1926.1071)

Most construction companies do not conduct diving operations, but, of course, there are those who specialize in this type of operation and, thus, must comply with 29 CFR 1926

Figure 7-23. Enclosed disposal chute on exterior of building

Subpart Y. This is the same standard as used for commercial diving in general industry, and is found in 29 CFR 1910 Subpart T. This standard applies to diving and related support operations conducted in connection with all types of work and employments, including general industry, construction, ship repairing, shipbuilding, shipbreaking, construction, and longshoring.

Each dive team member must have the experience or training necessary to perform the assigned tasks in a safe and healthful manner. Each dive team member shall have experience or training in the following:

1. The use of tools, equipment, and systems relevant to assigned tasks.

2. Techniques of the assigned diving mode.

3. Diving operations and emergency procedures.

All dive team members are to be trained in cardiopulmonary resuscitation and first aid (American Red Cross standard course or equivalent). Dive team members who are exposed to, or control the exposure of others to hyperbaric conditions, are to be trained in diving-related physics and physiology. Each dive team member is assigned tasks in accordance with the employee's experience or training, except that limited additional tasks may be assigned to an employee undergoing training provided that these tasks are performed under the direct supervision of an experienced dive team member.

The employer does not require a dive team member be exposed to hyperbaric conditions against the employee's will, except when necessary to complete decompression or treatment procedures. The employer does not permit a dive team member to dive or be otherwise exposed to hyperbaric conditions for the duration of any temporary physical impairment or condition which is known to the employer and is likely to adversely affect the safety or health of a dive team member. The employer, or an employee designated by the employer, is to be at the dive location and in charge of all aspects of the diving operation affecting the safety and health of dive team members. The designated person-in-charge must have experience and training in the conduct of the assigned diving operation.

The employer develops and maintains a safe practices manual which is made available to each dive member at the dive location. The safe practices manual contains a copy of this standard and the employer's policies for implementing the requirements of the diving standard. The safe practices for each type of diving is included in the manual as follows:

1. Safety procedures and checklists for diving operations.

2. Assignments and responsibilities of the dive team members.

3. Equipment procedures and checklists.

4. Emergency procedures for fire, equipment failure, adverse environmental conditions, medical illness, and injury.

Prior to each diving operation, a list of telephone or call numbers shall be made and kept at the following dive locations:

1. An operational decompression chamber (if not at the dive location).

2. Accessible hospitals.

3. Available physicians.

4. Available means of transportation.

5. The nearest U.S. Coast Guard Rescue Coordination Center.

6. A first aid kit, appropriate for the diving operation and approved by a physician, must be available at the dive location. When used in a decompression chamber or bell, the first aid kit must be suitable for use under hyperbaric conditions. In addition to any other first aid supplies, an American Red Cross standard first aid handbook or equivalent, and a bag-type manual resuscitator with transparent mask and tubing, is to be available.

When planning a diving operation, a safety and health assessment must be made and include the following:

1. Diving mode.

2. Surface and underwater conditions and hazards.

3. Breathing gas supply (including reserves).

4. Thermal protection.

5. Diving equipment and systems.

6. Dive team assignments and physical fitness of dive team members (including any impairment known to the employer).

7. Repetitive dive designation or residual inert gas status of dive team members.

8. Decompression and treatment procedures (including altitude corrections).

9. Emergency procedures.

To minimize hazards to the dive team, diving operations are coordinated with other activities in the vicinity which are likely to interfere with the diving operation. Dive team members are briefed on the tasks to be undertaken, safety procedures for the diving mode, any unusual hazards or environmental conditions likely to affect the safety of the diving operation, and any modifications to operating procedures necessitated by the specific diving operation.

Prior to making individual dive team member assignments, the employer must inquire into the dive team member's current state of physical fitness, and indicate to the dive team member the procedure for reporting physical problems or adverse physiological effects during and after the dive. The breathing gas supply system, including reserve breathing gas supplies, masks, helmets, thermal protection, and bell handling mechanism (when appropriate), are to be inspected prior to each dive. When diving from surfaces other than vessels, and diving into areas capable of supporting marine traffic, a rigid replica of the international code flag "A," which is at least one meter in height, must be displayed at the dive location. It is to be displayed in such a manner that it allows all-round visibility, and must be illuminated during night diving operations.

During a dive the following requirements are applicable to each diving operation, unless otherwise specified. There must be: a means provided which is capable of supporting the diver when entering and exiting the water, and that means must extend below the water surface to enable the diver an easy exit of the water; a means to assist an injured diver from the water or into a bell; and an operational two-way voice communication system between each surface-supplied air or mixed-gas diver and a dive team member at the dive location or bell (when provided or required at the bell and dive location).

At each dive location it is expected that

1. An operational, two-way communication system is available at the dive location to obtain emergency assistance.

2. Decompression, repetitive, and no-decompression tables (as appropriate) are at the dive location.

3. Depth-time profile, including, when appropriate, any breathing gas changes, shall be maintained for each diver during the dive, including decompression.

4. Hand-held electrical tools and equipment are deenergized before being placed into, or retrieved from the water.

5. Hand-held power tools are not supplied with power from the dive location until requested by the diver.

6. A current supply switch, which interrupts the current flow to the welding or burning electrode, is to be tended by a dive team member in voice communication with the diver performing the welding or burning, and it must be kept in the open position except when the diver is welding or burning. Also, the welding machine frame is to be grounded. Welding and burning cables, electrode holders, and connections shall be capable of carrying the maximum current required by the work and must be properly insulated, and insulated gloves must be provided to divers performing welding and burning operations. Prior to welding or burning on closed compartments, structures, or pipes which contain a flammable vapor, or where a flammable vapor may be generated by the work, they shall be vented, flooded, or purged with a mixture of gases which will not support combustion.

7. Employers who transport, store, and use explosives must comply with the provisions of and 1926.912 of Title 29 of the Code of Federal Regulations. Electrical continuity of explosive circuits are not to be tested until the diver is out of the water. Explosives must not be detonated while the diver is in the water. A blaster shall conduct all blasting operations, and no shot is fired without his approval. Loading tubes and casings of dissimilar metals are not to be used because of the possibility of electric transient currents occurring from the galvanic action of the metals and water. Only water-resistant blasting caps and detonating cords are to be used for all marine blasting. Loading shall be done through a nonsparking metal loading tube when a tube is necessary. No blast is to be fired while any vessel under way is closer than 1,500 feet of the blasting area. Those on board vessels or craft moored or anchored within 1,500 feet must be notified before a blast is fired. No blast shall be fired while any swimming or diving operations are in progress in the vicinity of the blasting area. If such operations are in progress, signals and arrangements are agreed upon to assure that no blast is fired while any person is in the water. Blasting flags must

be displayed. The storage and handling of explosives aboard vessels, which are used in underwater blasting operations, must be handled according to provisions outlined therein. When more than one charge is placed under water, a float device is to be attached to an element of each charge in such a manner that it will be released by the firing. Misfires are handled in accordance with the requirements of 1926.911.

The working interval of a dive shall be terminated when: a diver requests termination; a diver fails to respond correctly to communications or signals from a dive team member; communications are lost and cannot be quickly reestablished with the diver; communications between a dive team member at the dive location, and the designated person-in-charge or the person controlling the vessel in liveboating operations; or a diver begins to use diver-carried reserve breathing gas or the dive-location reserve breathing gas.

After the completion of any dive, the employer shall check the physical condition of the diver, instruct the diver to report any physical problems or adverse physiological effects, including symptoms of decompression sickness, advise the diver of the location of a decompression chamber which is ready for use, and alert the diver to the potential hazards of flying after diving.

For any dive which is outside the no-decompression limits and is deeper than 100 fsw, or uses mixed gas as a breathing mixture, the employer must instruct the diver to remain awake and in the vicinity of the decompression chamber, which is at the dive location, for at least one hour after the dive (including decompression or treatment, as appropriate). A decompression chamber that is capable of recompressing the diver, at the surface, to a minimum of 165 fsw (6 ATA), must be available at the dive location if: surface-supplied air diving reaches depths deeper than 100 fsw but is less than 220 fsw; mixed gas diving is less than 300 fsw; or, diving is outside the no-decompression limits and is less than 300 fsw.

The following information is recorded and maintained for each diving operation: names of the dive team members, including designated person-in-charge, date, time, location, diving modes used, general nature of work performed, approximate underwater and surface conditions (visibility, water temperature, and current), and maximum depth and bottom time for each diver.

For each dive outside the no-decompression limits, which is deeper than 100 fsw or uses mixed gas, the following additional information must be recorded and maintained: depth-time and breathing gas profiles, decompression table designation (including modification), and elapsed time since last pressure exposure, if less than 24 hours or repetitive dive designation for each diver.

Any time decompression sickness is suspected or symptoms are evident, the following additional information shall be recorded and maintained: description of decompression sickness symptoms (including depth and time of onset), and description and results of treatment.

Each incident of decompression sickness must be investigated and evaluated based on the recorded information, consideration of the past performance of decompression table used, and individual susceptibility. Corrective action should take place to reduce the probability of a recurrence, and an evaluation of the decompression procedure assessment, including any corrective action taken, should be written within 45 days of the incident.

Also, records are to be kept on the occurrence of any diving-related injury or illness which requires any dive team member to be hospitalized for 24 hours or more; specify the circumstances of the incident and the extent of any injuries or illnesses. Records and documents are required to be retained by the employer for the following period of time:

1. Dive team member medical records (physician's reports) (1910.411) – 5 years.

2. Safe practices manual (1910.420) – current document only.

3. Depth-time profile (1910.422) – until completion of the recording of dive, or until completion of decompression procedure assessment where there has been an incident of decompression sickness.

4. Recording of dive (1910.423) – 1 year, except 5 years where there has been an incident of decompression sickness.

5. Decompression procedure assessment evaluations (1910.423) – 5 years.

6. Equipment inspections and testing records (1910.430) – current entry or tag, or until equipment is withdrawn from service.

7. Records of hospitalizations (1910.440) – 5 years.

For specifics on diving modes such as scuba diving, surface supplied air diving, mixed-gas diving, lifeboating, and diving equipment, consult the standard 29 CFR 1926 Subpart Y.

DRINKING WATER

See potable water

EATING AND DRINKING AREAS (1926.51)

No food or drink is allowed to be consumed in the vicinity of toilets, or in areas exposed to toxic materials.

EGRESS (1926.34)

In every building or structure, exits are to be so arranged and maintained as to provide free and unobstructed egress from all areas, at all times, when occupied. No lock or fastening device shall be installed which prevents free escape from the inside of any building, except in mental, penal, or corrective institutions. In these institutions, supervisory personnel must be on duty at all times, and effective provisions must be made to remove occupants in case of fire or other emergencies. All exits must be marked by a readily visible sign, and where access to reach the exists are not immediately visible, they must, in all cases, be marked with readily visible directional signs. The means of egress is to be continually maintained and free of all obstructions or impediments, which would inhibit an easy and immediate exit of the building or structure, in case of a fire or other emergency.

ELECTRICAL (1926.400)

The electrical safety requirements are those necessary for the practical safeguarding of employees involved in construction work. These installation safety requirements are divided into four major areas: the electric equipment and installations used to provide electric power and light on jobsites; safety-related work practices which cover hazards that may arise from the use of electricity at jobsites, and hazards that may arise when employees accidentally contact energized lines, direct or indirect, that are above or below ground, or are passing through or near the jobsites; safety-related maintenance and environmental considerations; and safety requirements for special equipment.

These requirements cover the installation safety requirements for electrical equipment and the installations used to provide electric power and light at the jobsite. This includes installations, both temporary and permanent, used on the jobsite, but does not apply to existing permanent installations that were in place before the construction activity commenced. If the electrical installation is made in accordance with the National Electrical Code ANSI/NFPA 70-1984, it is deemed to be in compliance with the construction electrical safety requirements. The generation, transmission, and distribution of electric energy, including related communication, metering, control, and transformation installations, are not part of the following electrical safety requirements.

General Requirements (1926.403)

The general requirements state that all electrical conductors and equipment must be approved types. The employer must ensure that electrical equipment is free from recognized hazards that are likely to cause death or serious physical harm to employees. The safety of equipment is to be determined on the basis of the following considerations:

1. Suitability for installation and use in conformity with the electrical safety provisions. Suitability of equipment for an identified purpose may be evidenced by listing, labeling, or certification for that identified purpose.

2. Mechanical strength and durability, including parts designed to enclose and protect other equipment, and adequacy of the protection thus provided.

3. Electrical insulation, heating effects under conditions of use, and arcing effects.

4. Classification by type, size, voltage, current capacity, and specific use.

5. Other factors which contribute to the practical safeguarding of employees using, or likely to come in contact with the equipment.

Listed, labeled, or certified equipment is to be installed and used in accordance with instructions included in the listing, labeling, or certification. Equipment intended to break current must have an interrupting rating at system voltage which is sufficient for the current that must be interrupted. Electrical equipment shall be firmly secured to the surface on which it is mounted. Wooden plugs, driven into holes in masonry, concrete, plaster, or similar materials, are not to be used.

Electrical equipment, which depends upon the natural circulation of air and convection principles for cooling of exposed surfaces, is to be installed so that room for air flow over such surfaces is not prevented by walls or by adjacent installed equipment. For equipment designed for floor mounting, clearance between top surfaces and adjacent surfaces shall be provided to dissipate rising warm air. Electrical equipment provided with ventilating openings is to be installed so that walls or other obstructions do not prevent the free circulation of air through the equipment.

If conductors are spliced or joined, it is to be done by splicing devices designed for that use, or by brazing, welding, or soldering with a fusible metal or alloy. Soldered splices are first spliced or joined so as to be mechanically and electrically secure without solder, and then soldered. All splices and joints and free ends of conductors are to be covered with an insulation equivalent to that of the conductors, or with an insulating device designed for the purpose.

Parts of electrical equipment, which in ordinary operation produce arcs, sparks, flames, or molten metal, are to be enclosed or separated and isolated from all combustible material.

No electrical equipment is to be used unless the manufacturer's name, trademark, or other descriptive marking is in place. The organization responsible for the product must be identified on the equipment and unless other markings are provided giving voltage, current, wattage, or other ratings as necessary. The markings must be of sufficient durability to withstand the environment involved.

Each disconnecting means for motors and appliances are to be legibly marked to indicate its purpose, unless located and arranged so the purpose is evident. Also, each service, feeder, and branch circuit, at its disconnecting means or overcurrent device, is to be legibly marked to indicate its purpose, unless located and arranged so the purpose is evident. These markings shall be of sufficient durability to withstand the environment involved.

Working Distances

When a working space has electrical equipment with 600 volts, nominal, or less, there must be sufficient access. The working space must be well maintained and provide adequate space around all electrical equipment in order to permit ready and safe operation and maintenance of such equipment. When there are live circuits that require examination, adjustment, servicing, or maintenance, the safe working distances found in Table 7-1 shall be followed. In addition to the dimensions shown, the working space is not to be less than 30 inches wide in front of the electrical equipment. Distances are measured from the live parts, if they are exposed, or from the enclosure front or opening, if the live parts are enclosed. Walls constructed of concrete, brick, or tile are considered to be grounded. Working space is not required in the back of assemblies of dead-front switchboards or motor control centers, where there are no renewable or adjustable parts, such as fuses or switches on the back, and where all connections are accessible from locations other than the back.

Table 7-1

Working Clearances
Courtesy of OSHA

	Minimum clear distance for nominal voltage to ground conditions[1]		
	(a) Feet	(b) Feet	(c) Feet
0-150	3	3	3
151-600	3	3 1/2	4

[1] Conditions (a), (b), and (c) are as follows: (a) Exposed live parts on one side and no live or grounded parts on the other side of the working space, or exposed live parts on both sides effectively guarded by insulating material. Insulated wire or insulated busbars operating at not over 300 volts are not considered live parts. (b) Exposed live parts on one side and grounded parts on the other side. (c) Exposed live parts on both sides of the workplace [not guarded as provided in Condition (a) with the operator between].

Working space is not used for storage. When normally enclosed live parts are exposed for inspection or servicing, the working space, if in a passageway or general open space, is guarded.

At least one entrance is to be provided to give access to the working space around electric equipment.

Where there are live parts normally exposed on the front of switchboards or motor control centers, the working space in front of such equipment is not to be less than three feet. The minimum headroom of working spaces about service equipment, switchboards, panelboards, or motor control centers is 6 feet 3 inches (see Figure 7-24).

Figure 7-24. Electrician with adequate working clearance

Figure 7-25. Electrical enclosure accessible to qualified persons

Guarding Electrical Equipment

Live parts of electric equipment, operating at 50 volts or more, are to be guarded against accidental contact by using cabinets or other forms of enclosures, or by any of the following means:

1. By location in a room, vault, or similar enclosure that are accessible only to qualified persons (see Figure 7-25).

2. By partitions or screens so arranged that only qualified persons will have access to the space within reach of the live parts. Any openings in such partitions or screens must be so sized and located that persons are not likely to come into accidental contact with the live parts, or bring conducting objects into contact with them.

3. By locating them on a balcony, gallery, or platform, and elevated and arranged so that it excludes unqualified persons.

4. By elevation of 8 feet or more above the floor or other working surface, and so installed as to exclude unqualified persons.

In locations where electric equipment would be exposed to physical damage, enclosures or guards are to be so arranged, and of such strength, as to prevent such damage. Entrances to rooms and other guarded locations containing exposed live parts, are to be marked with conspicuous warning signs which forbid unqualified persons to enter.

Conductors Exceeding 600 Volts

Conductors and equipment used on circuits exceeding 600 volts, nominal, must comply with all applicable previous requirements, as well as with provisions which supplement or modify those requirements, but does not apply to the supply side of those conductors.

Electrical installations in a vault, room, closet or in an area surrounded by a wall, screen, or fence, access to which is controlled by lock and key or other equivalent means, are considered to be accessible to qualified persons only. A wall, screen, or fence less than 8 feet in height is not considered adequate to prevent access, unless it has other features that provide a degree of isolation equivalent to an 8-foot fence. The entrances to all buildings, rooms, or enclosures containing exposed live parts or exposed conductors operating at over 600 volts, nominal, are kept locked, or are kept under the observation of a qualified person at all times. Installations are to be accessible to qualified persons only. Electrical installations having exposed live parts are to be accessible to qualified persons only, and they must comply with workspace requirements.

Sufficient workspace is to be provided and maintained around electrical equipment to permit ready and safe operation and maintenance of such equipment. Where energized parts are exposed, the minimum clear workspace is not to be less than 6 feet 6 inches high (measured vertically from the floor or platform), or less than 3 feet wide (measured parallel to the equipment). The depth is as required in Table 7-2. The workspace must be adequate to permit at least a 90-degree opening of doors or hinged panels.

The minimum clear working space in front of electric equipment such as switchboards, control panels, switches, circuit breakers, motor controllers, relays, and similar equipment, is not to be less than specified in Table 7-2 unless otherwise specified. Distances shall be measured from the live parts, if they are exposed, or from the enclosure front or opening, if the live parts are enclosed. However, working space is not required in back of such equipment as deadfront switchboards or control assemblies, where there are no renewable or adjustable parts (such as fuses or switches) on the back, and where all connections are accessible from locations other than the back. Where rear access is required to work on deenergized parts on the back of enclosed equipment, a minimum working space of 30 inches, horizontally, is provided.

Table 7-2

**Minimum Depth of Clear Working Space in Front of Electric Equipment
Courtesy of OSHA**

Nominal voltage to ground	Conditions[1]		
	(a)	(b)	(c)
	Feet	Feet	Feet
601 to 2,500	3	4	5
2,501 to 9,000	4	5	6
9,001 to 25,000	5	6	9
25,001 to 75 kV	6	8	10
Above 75kV	8	10	12

[1] Conditions (a), (b), and (c) are as follows: (a) Exposed live parts on one side and no live or grounded parts on the other side of the working space, or exposed live parts on both sides effectively guarded by insulating materials. Insulated wire or insulated busbars, operating at not over 300 volts, are not considered live parts. (b) Exposed live parts on one side and grounded parts on the other side. Walls constructed of concrete, brick, or tile are considered to be grounded surfaces. (c) Exposed live parts on both sides of the workspace [not guarded as provided in Condition (a)] with the operator between them.

Installations Accessible to Unqualified Persons

Electrical installations that are open to unqualified persons are made with metal-enclosed equipment, are enclosed in a vault, or are in an area where access is controlled by a lock. Metal-enclosed switchgear, unit substations, transformers, pull boxes, connection boxes, and other similar associated equipment shall be marked with appropriate caution signs. If equipment is exposed to physical damage from vehicular traffic, guards shall be provided to prevent such damage. Ventilating or similar openings in metal-enclosed equipment are to be designed so that foreign objects inserted through these openings will be deflected from energized parts.

Lighting Outlets

The lighting outlets are to be so arranged that persons changing lamps or making repairs on the lighting system will not be endangered by live parts or other equipment. The points of control shall be so located that persons are not likely to come in contact with any live or moving part of the equipment while turning on the lights. Unguarded live parts, above working space, are to be maintained at elevations not less than specified in Table 7-3.

Table 7-3
Elevation of Unguarded Energized Parts Above Working Space
Courtesy of OSHA

Nominal voltage between phases	Minimum Elevation
601-7,500	8 feet 6 inches.
7,501-35,000	9 feet.
Over 35kV	9 feet + 0.37 inches per kV above 35kV.

At least one entrance, not less than 24 inches wide and 6 feet 6 inches high, must be provided to give access to the working space at the location of electric equipment. On switchboard and control panels exceeding 48 inches in width, there shall be one entrance at each end of such board, where practicable. Where bare energized parts at any voltage or insulated energized parts above 600 volts are located adjacent to such entrance, they are to be guarded.

Wiring Design and Protection (1926.404)

A conductor used as a grounded conductor is to be identifiable and distinguishable from all other conductors. A conductor used as an equipment grounding conductor must be identifiable and distinguishable from all other conductors. No grounded conductor shall be attached to any terminal or lead so as to reverse designated polarity. A grounding terminal or grounding-type device on a receptacle, cord connector, or attachment plug is not to be used for purposes other than grounding.

The employer must use either ground fault circuit interrupters, or an assured equipment grounding conductor program to protect employees on construction sites. These requirements are in addition to any other requirements for equipment grounding conductors.

Ground-Fault Circuit Interrupters

All 120-volt, single-phase 15- and 20-ampere receptacle outlets on construction sites, which are not a part of the permanent wiring of the building or structure and are in use by employees, must have approved ground-fault circuit interrupters for personnel protection. Receptacles on a two-wire, single-phase portable or vehicle-mounted generator rated not more than 5kV, where the circuit conductors of the generator are insulated from the generator frame and all other grounded surfaces, need not be protected with ground-fault circuit interrupters (see Figure 7-26).

Figure 7-26. Example of a ground-fault circuit interrupter connected to a drill

Assured Grounding Program

The employer must establish and implement an assured equipment grounding conductor program on construction sites. This program must cover all cord sets, receptacles which are not a part of the building or structure, and equipment connected by a cord and plug and are available for use, or used by employees (see Table 7-4). This program must comply with the following minimum requirements:

1. A written description of the program, including the specific procedures adopted by the employer, is to be available at the jobsite for inspection and copying by the Assistant Secretary and any affected employees. (See Figure 7-27.)

2. The employer shall designate one or more competent persons to implement the program.

3. Each cord set, attachment cap, plug and receptacle of cord sets, and any equipment connected by a cord and plug, except cord sets and receptacles which are fixed and not exposed to damage, is to be visually inspected, before each day's use, for external defects, such as deformed or missing pins or insulation damage, and for indications of possible internal damage. Equipment found damaged or defective is not to be used until repaired.

4. The following tests must be performed on all cord sets, receptacles which are not a part of the permanent wiring of the building or structure, and cord- and plug-connected equipment required to be grounded (see Table 7-4).

All equipment grounding conductors are to be tested for continuity and be electrically continuous. Each receptacle and attachment cap or plug must be tested for correct attachment of the equipment grounding conductor. The equipment grounding conductor is to be connected to its proper terminal. All required tests are performed before first use, before equipment is returned to service following any repairs, before equipment is used after any incident which can be reasonably suspected to have caused damage (for example, when a cord set is run over), and at intervals not to exceed 3 months, except cord sets and receptacles which are fixed and not exposed to damage, they are to be tested at intervals not exceeding 6 months. The employer is not to make available, or permit any employee to use any equipment which has not met these requirements. Tests performed are required to be recorded. These test records shall identify

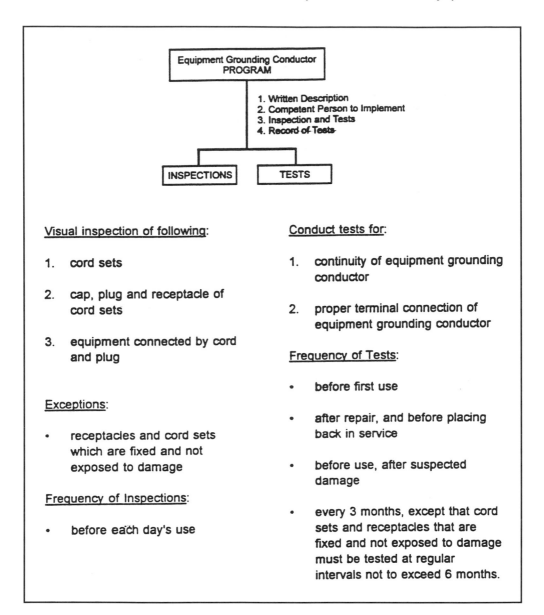

Figure 7-27. Assured grounding program. Courtesy of OSHA

each receptacle, cord set, and cord- and plug-connected equipment that passed the test, and indicate the last date it was tested, or the interval for which it was tested. This record is to be kept by means of logs, color coding, or other effective means and is to be maintained until replaced by a more current record. The record must be made available on the jobsite for inspection by the Assistant Secretary and any affected employees.

Table 7-4

Verifying Inspection

To verify inspection and testing, a piece of color coded tape will be affixed each time equipment is inspected. Four colors of tape will be used, one for each quarter of the year. The color coding system is as follows:

Color	Quarter	Expiration Date
White	First	March 31
Green	Second	June 30
Red	Third	September 30
Orange	Fourth	December 31

(Brown will be used to verify that repair is needed.)

Inspection tape will not be used for any other purpose. Storage of tape will be strictly controlled by the site superintendent. Only persons designated by the site superintendent are authorized to remove inspection tape. Unauthorized removal or defacing of inspection tape will be cause for disciplinary action.

Requirements for Outlets

Outlet devices must have an ampere rating not less than the load to be served and comply with the following:

1. A single receptacle, installed on an individual branch circuit, must have an ampere rating of not less than that of the branch circuit.

2. When connected to a branch circuit supplying two or more receptacles or outlets, receptacle ratings are to conform to set values (see Table 7-5).

3. The rating of an attachment plug or receptacle used for cord- and plug-connection of a motor to a branch circuit, shall not exceed 15 amperes at 125 volts, or 10 amperes at 250 volts, if individual overload protection is omitted.

Outdoor Conductors

Any branch circuit, feeder, and service conductors rated 600 volts, nominal, or less, and run outdoors as open conductors supported on poles, must provide a horizontal climbing space not less than 30 inches for power conductors below communication conductors. For power conductors alone or above communication conductors, the following must be provided: 300 volts or less–24 inches; more than 300 volts–30 inches. For communication conductors that are below power conductors and the power conductors have 300 volts or less, 24 inches must be provided, and when they have more than 300 volts, 30 inches are to be provided.

Open conductors must be at least 10 feet above finished grade, sidewalks, or from any platform or projection from which they might be reached; 12 feet over areas subject to vehicular traffic other than truck traffic; 15 feet over areas, other than those specified, that are subject to truck traffic; and 18 feet over public streets, alleys, roads, and driveways (see Figure 7-28).

Table 7-5

Receptacle Ratings for Various Size Circuits
Courtesy of OSHA

Circuit rating amperes	Receptacle rating amperes
15	Not over 15
20	15 or 20
30	30
40	40 or 50
50	50

Figure 7-28. Conductors over a typical construction site

Conductors must have a clearance of at least 3 feet from windows, doors, fire escapes, or similar locations. Conductors that run above the top level of a window are considered to be out of reach from that window and, therefore, do not have to be 3 feet away. Conductors above roof space, which are accessible to employees on foot, must have not less than 8 feet of vertical clearance from the highest point of the roof surface for insulated conductors, not less than 10 feet vertical or diagonal clearance for covered conductors, and not less than 15 feet for bare conductors, except where the roof space is also accessible to vehicular traffic, then the vertical clearance is not to be less than 18 feet. Where the roof space is not normally accessible to employees on foot, fully insulated conductors must have a vertical or diagonal clearance of not less than 3 feet.

Where the voltage between conductors is 300 volts or less and the roof has a slope of not less than 4 inches in 12 inches, the clearance from roofs is to be at least 3 feet. Where the voltage between conductors is 300 volts or less and the conductors do not pass over more than 4 feet of the overhang portion of the roof and they are terminated at a through-the-roof raceway or support, the clearance from roofs shall be at least 18 inches.

Lamps for outdoor lighting are to be located below all live conductors, transformers, or other electric equipment, unless such equipment is controlled by a disconnecting means that can be locked in the open position, or unless adequate clearances or other safeguards are provided for relamping operations.

Disconnects

Means are to be provided to disconnect all conductors, in a building or other structures, from the service-entrance conductors. The disconnecting means must plainly indicate whether it is in the open or closed position and must be installed at a readily accessible location nearest the point of entrance of the service-entrance conductors. Each service disconnecting means must simultaneously disconnect all ungrounded conductors.

The following requirements apply to services over 600 volts, nominal, that are service-entrance conductors and installed as open wires. They must be guarded so that no one except qualified persons has access to them. There must also be signs warning of high voltage and they are to be posted where unauthorized employees might come in contact with live parts.

Overcurrent Protection

When overcurrent protection of circuits rated 600 volts, nominal, or less is used, the conductors and equipment are to be protected from overcurrent in accordance with their ability to safely conduct current. Conductors must have sufficient ampacity to carry the load. Except for motor-running overload protection, overcurrent devices shall not interrupt the continuity of the grounded conductor, unless all conductors of the circuit are opened simultaneously. Except for devices provided for current-limiting on the supply side of the service disconnecting means, all cartridge fuses which are accessible to other than qualified persons, and all fuses and thermal cutouts on circuits over 150 volts to ground, are to be provided with a disconnecting means. These disconnecting means are to be installed so that the fuse or thermal cutout can be disconnected from its supply without disrupting service to equipment and circuits unrelated to those protected by the overcurrent device. Overcurrent devices are to be readily accessible. Overcurrent devices are not to be located where they could create an employee safety hazard by being exposed to physical damage, or located in the vicinity of easily ignitable material.

Fuses and Circuit Breakers

Fuses and circuit breakers must be so located or shielded that employees will not be burned or otherwise injured by their operation. Circuit breakers shall clearly indicate whether they are in the open (off) or closed (on) position. Where circuit breaker handles on switchboards are operated vertically rather than horizontally or rotationally, the up position of the handle is the closed (on) position. If used as switches in 120-volt, fluorescent lighting circuits and circuit breakers are to be marked "SWD." Feeders and branch circuits over 600 volts, nominal, must have short-circuit protection.

Grounding

The following systems which supply premises wiring are to be grounded such that all 3-wire DC systems have their neutral conductor grounded and two-wire DC systems, operating at over 50 volts through 300 volts between conductors, are to be grounded unless they are rectifier-derived from an AC system. AC circuits of less than 50 volts must be grounded if they are installed as overhead conductors outside of buildings, or if they are supplied by transformers and the transformer's primary supply system is ungrounded or exceeds 150 volts to ground. AC systems of 50 volts to 1000 volts are to be grounded under any of the following conditions, unless exempted:

1. If the system can be so grounded that the maximum voltage to ground on the ungrounded conductors does not exceed 150 volts.

2. If the system is nominally rated 480Y/277 volt, 3-phase, 4-wire in which the neutral is used as a circuit conductor.

3. If the system is nominally rated 240/120 volt, 3-phase, 4-wire in which the midpoint of one phase is used as a circuit conductor.

4. If a service conductor is uninsulated.

At certain times AC systems of 50 volts to 1000 volts are not required to be grounded, if the system is separately derived and is supplied by a transformer that has a primary voltage rating less than 1000 volts and provided all of the following conditions are met: the system is used exclusively to control circuits; the conditions of maintenance and supervision assure that only qualified persons will service the installation; the continuity of control power is required; and ground detectors are installed on the control system.

Separately Derived Systems

The requirements found in Ground Connections apply when the need exists for the grounding of wiring systems where power is derived from a generator, transformer, or converter windings, and there is no direct electrical connection, including a solidly connected grounded circuit conductor, to supply conductors originating in another system.

Portable- and Vehicle-Mounted Generators

Under the certain conditions, the frame of a portable generator need not be grounded and may serve as the grounding electrode for a system supplied by the generator (see Figure 7-29). Under these conditions the generator is to supply only equipment mounted on the generator and/or cord- and plug-connected equipment through receptacles mounted on the generator, and the noncurrent-carrying metal parts of equipment, and the equipment grounding conductor terminals of the receptacles, are to be bonded to the generator frame.

Figure 7-29. Example of portable electrical generator

For vehicle mounted generators, the frame of a vehicle may serve as the grounding electrode for a system supplied by a generator located on the vehicle, if certain conditions are met.

1. The frame of the generator is bonded to the vehicle frame.

2. The generator supplies only equipment located on the vehicle and/or cord- and plug-connected equipment through receptacles mounted on the vehicle or generator.

3. The noncurrent-carrying metal parts of equipment, and the equipment grounding conductor terminals of the receptacles, are bonded to the generator frame.

4. The system complies with all other provisions.

Neutral Conductor Bonding

A neutral conductor is to be bonded to the generator frame, if the generator is a component of a separately derived system. No other conductor need be bonded to the generator frame. For AC premises wiring systems the identified conductor must be grounded.

Ground Connections

For a grounded system, a grounding electrode conductor is used to connect both the equipment grounding conductor and the grounded circuit conductor to the grounding electrode. Both the equipment grounding conductor and the grounding electrode conductor is connected to the grounded circuit conductor on the supply side of the service disconnecting means, or on the supply side of the system disconnecting means or overcurrent devices, if the system is separately derived.

For an ungrounded service-supplied system, the equipment grounding conductor is connected to the grounding electrode conductor at the service equipment. For an ungrounded, separately derived system, the equipment grounding conductor is connected to the grounding electrode conductor at, or ahead of, the system disconnecting means or overcurrent devices. The path to ground from circuits, equipment, and enclosures is to be permanent and continuous.

Supports and Enclosures for Conductors

Metal cable trays, metal raceways, and metal enclosures for conductors are grounded, except that metal enclosures, such as sleeves, that are used to protect cable assemblies from physical damage need not be grounded. Metal enclosures for conductors, added to existing installations of open wire, knob-and-tube wiring, and nonmetallic-sheathed cable, need not be grounded, if all of the following conditions are met:

1. Runs are less than 25 feet.

2. Enclosures are free from probable contact with ground, grounded metal, metal laths, or other conductive materials.

3. Enclosures are guarded against employee contact.

Metal enclosures for service equipment shall be grounded. Exposed noncurrent-carrying metal parts of fixed equipment, which may become energized, are to be grounded under any of the following conditions: if within 8 feet vertically or 5 feet horizontally of ground or grounded metal objects and subject to employee contact; if located in a wet or damp location and subject to employee contact; if in electrical contact with metal; if in a hazardous (classified) location; if supplied by a metal-clad, metal-sheathed, or grounded metal raceway wiring method; and if equipment operates with any terminal over 150 volts to ground. However, there is no need for grounding: if enclosures for switches or circuit breakers, are used for other than service equipment and are accessible to qualified persons only; if metal frames of electrically heated appliances are permanently and effectively insulated from ground; or if the cases of distribution apparatus, such as transformers and capacitors, are mounted on wooden poles at a height exceeding 8 feet aboveground or grade level.

Any exposed noncurrent-carrying metal parts of cord- and plug-connected equipment, which may become energized, are to be grounded: if located in a hazardous (classified) location; if operated at over 150 volts to ground, except for guarded motors and metal frames of electrically heated appliances, if the appliance frames are permanently and effectively insulated from ground; if the equipment is a hand-held motor-operated tool; if cord- and plug-connected equipment is used in damp or wet locations or is used by employees standing on the ground or metal floors or working inside metal tanks or boilers; if portable and mobile X-ray and associated equipment are used; if tools are likely to be used in wet and/or conductive locations; if portable hand lamps are used; or if tools are likely to be used in wet and/or conductive locations. They need not be grounded if supplied through an isolating transformer with an ungrounded secondary of not over 50 volts. Listed or labeled portable tools and appliances protected by a system of double insulation, or its equivalent, need not be grounded. If such a system is employed, the equipment shall be distinctively marked to indicate that the tool or appliance utilizes a system of double insulation.

Nonelectrical Equipment

The metal parts of the nonelectrical equipment are to be grounded. These include such things as frames and tracks of electrically operated cranes, frames of nonelectrically driven elevator cars to which electric conductors are attached, hand-operated metal shifting ropes or cables of electric elevators, and metal partitions, grill work, and similar metal enclosures around equipment of over 1kV between conductors.

Noncurrent-carrying metal parts of fixed equipment, if required to be grounded, are grounded by an equipment grounding conductor which is contained within the same raceway, cable, or cord, or runs with or encloses the circuit conductors. For DC circuits only, the equipment grounding conductor may be run separately from the circuit conductors. A conductor used for grounding fixed or movable equipment must have the capacity to conduct safely any fault current which may be imposed on it.

Effective Grounding

Electric equipment is considered to be effectively grounded if it is secured to, and in electrical contact with, a metal rack or structure that is provided for its support. The metal rack or structure is to be grounded by the method specified for the noncurrent carrying metal parts of fixed equipment, or metal car frames are to be supported by metal hoisting cables attached to, or running over metal sheaves or drums of grounded elevator machines.

Bonded Conductors

Bonding conductors are used to assure electrical continuity. They have the capacity to conduct any fault current which may be imposed.

Made Electrodes

If made electrodes are used, they are to be free from nonconductive coatings, such as paint or enamel, and, if practicable, they are to be embedded below the permanent moisture level. A single electrode, consisting of a rod, pipe, or plate, which has a resistance to ground greater than 25 ohms, must be augmented by one additional electrode and installed no closer than 6 feet to the first electrode.

Grounded High Voltage

The grounding of systems and circuits of 1000 volts and over (high voltage) are usually in compliance, if they meet the previous grounding requirements. When grounding systems that supply portable or mobile equipment, the systems which are supplying portable or mobile high voltage equipment, other than substations installed on a temporary basis, must assure that the portable and mobile high voltage equipment is supplied from a system having its neutral grounded through an impedance. If a delta-connected high voltage system is used to supply the equipment, a system neutral shall be derived. Also, exposed noncurrent-carrying metal parts of portable and mobile equipment are to be connected by an equipment grounding conductor to the point at which the system's neutral impedance is grounded.

The use of ground-fault detection and relaying shall be provided and it shall automatically deenergize any high voltage system component which has developed a ground fault. The continuity of the equipment grounding conductor must be continuously monitored so as to deenergize, automatically, the high voltage feeder to the portable equipment upon loss of continuity of the equipment grounding conductor. The grounding electrode, to which the portable or mobile equipment system neutral impedance is connected, is to be isolated from, and separated in the ground by at least 20 feet from any other system or equipment grounding electrode; there must be no direct connection between the grounding electrodes, such as buried pipe, fence, or like objects. All noncurrent-carrying metal parts of portable equipment and fixed equipment, including their associated fences, housings, enclosures, and supporting structures, are to be grounded.

However, equipment which is guarded by location, and isolated from ground, need not be grounded. Additionally, pole-mounted distribution apparatus, at a height exceeding 8 feet aboveground or grade level, need not be grounded.

Wiring Methods, Components, and Equipment for General Use (1926.405)

The requirements in this section do not apply to conductors which form an integral part of equipment such as motors, controllers, motor control centers, and like equipment.

General Requirements

Metal raceways, cable armor, and other metal enclosures for conductors are metallically joined together into a continuous electric conductor and are so connected to all boxes, fittings, and cabinets as to provide effective electrical continuity. No wiring systems of any type are to be installed in ducts used to transport dust, loose stock, or flammable vapors, and no wiring systems of any type are to be installed in any duct used for vapor removal, or in any shaft containing only such ducts.

Temporary Wiring

Temporary electrical power and lighting wiring methods which may be of a class less than would be required for a permanent installation. Except as noted, all other requirements for permanent wiring apply to temporary wiring installations. Temporary wiring is removed immediately upon completion of construction, or removed once the purpose for which the wiring was installed is completed.

General Requirements for Temporary Wiring

Feeders are to originate in a distribution center. The conductors must run as multiconductor cord or cable assemblies, or within raceways. Where they are not subject to physical damage, they may be run as open conductors on insulators that are not more than 10 feet apart.

Branch circuits are to originate in a power outlet or panelboard. Again, conductors run as multiconductor cord or cable assemblies, open conductors, or run in raceways. All conductors are to be protected by overcurrent devices at their ampacity. Runs of open conductors shall be located where the conductors will not be subject to physical damage, and the conductors are to be fastened at intervals not exceeding 10 feet. No branch-circuit conductors shall be laid on the floor. Each branch circuit that supplies receptacles or fixed equipment must contain a separate equipment grounding conductor, if the branch circuit is run as open conductors.

Receptacles Must Be of the Grounding Type

Unless installed in a complete metallic raceway, each branch circuit must contain a separate equipment grounding conductor, and all receptacles are to be electrically connected to the grounding conductor. Receptacles, used for other than temporary lighting, must not be installed on branch circuits which supply temporary lighting. Receptacles shall not be connected to the same ungrounded conductor of multiwire circuits which supply temporary lighting.

Disconnecting switches or plug connectors shall be installed to permit the disconnection of all ungrounded conductors of each temporary circuit.

Temporary Lights

All lamps used for general illumination are to be protected from accidental contact or breakage. Metal-case sockets must be grounded. Temporary lights shall not be suspended by their electric cords unless cords and lights are designed for this means of suspension (see Figure 7-30). Portable electric lighting used in wet and/or other conductive locations, as for example, drums, tanks, and vessels, are to be operated at 12 volts or less. However, 120-volt lights may be used if protected by a ground-fault circuit interrupter.

Figure 7-30. Protected temporary lighting

Boxes

A box is used wherever a change is made to a raceway system or cable system which is metal clad or metal sheathed.

Flexible Cords and Cables

Flexible cords and cables are to be protected from damage. Sharp corners and projections need to be avoided. Flexible cords and cables may pass through doorways or other pinch points, if protection is provided to avoid damage. Extension cord sets, used with portable electric tools and appliances, must be of three-wire type and designed for hard or extra-hard usage. Flexible cords, used with temporary and portable lights, shall be designed for hard or extra-hard usage. The types of flexible cords designed for hard or extra-hard usage are: hard service types S, ST, SO, STO, and junior hard service cord types SJ, SJO, SJT, SJTO; these are acceptable for use on construction sites.

Flexible cords and cables are to be suitable for the conditions of use and location. Flexible cords and cables are only used for pendants, wiring of fixtures, connection of portable lamps or appliances, elevator cables, wiring of cranes and hoists, connection of stationary equipment to facilitate their frequent interchange, prevention of the transmission of noise or vibration, and appliances where the fastening means and mechanical connections are designed to permit removal for maintenance and repair. The attachment plugs for flexible cords, if used, are to be equipped with an attachment plug and energized from a receptacle outlet.

Flexible cords and cables are not to be used as a substitute for the fixed wiring of a structure, for runs through holes in walls, ceilings, or floors, for a run through doorways, windows, or similar openings, for attachment to building surfaces, or for concealment behind building walls, ceilings, or floors. A conductor of a flexible cord or cable, that is used as a grounded conductor or an equipment grounding conductor, is to be distinguishable from other conductors.

Flexible cords are used only in continuous lengths without splice or tap. Hard service flexible cords, No. 12 or larger, may be repaired if spliced so that the splice retains the insulation, outer sheath properties, and usage characteristics of the cord being spliced. Flexible cords are to be connected to devices and fittings so that strain relief is provided which prevents the pull from being directly transmitted to joints or terminal screws. Flexible cords and cables must be protected by bushings or fittings when they are passed through holes in covers, outlet boxes, or similar enclosures.

Guarding

For temporary wiring over 600 volts, nominal, there must be fencing, barriers, or other effective means provided to prevent access of other than authorized and qualified personnel (see Figure 7-31).

Cabinets, Boxes, and Fittings

Conductors entering boxes, cabinets, or fittings are to be protected from abrasion, and openings through which conductors enter must be effectively closed. Unused openings in cabinets, boxes, and fittings must also be effectively closed.

All pull boxes, junction boxes, and fittings shall be provided with covers. If metal covers are used, they are grounded. In energized installations, each outlet box shall have a cover, faceplate, or fixture canopy. Covers of outlet boxes which have holes through which

Figure 7-31. Typical dog house as a barrier from electricity for unqualified persons

flexible cord pendants passed, must provide bushings designed for the purpose, or have smooth, well-rounded surfaces on which the cords may bear.

In addition to other requirements for pull and junction boxes, the following applies for systems over 600 volts, nominal:

1. Boxes are to be provided a complete enclosure for the contained conductors or cables.

2. Boxes shall be closed by covers securely fastened in place. Underground box covers that weigh over 100 pounds must meet this requirement. Covers for boxes must be permanently marked "HIGH VOLTAGE." The marking is to be on the outside of the box cover and be readily visible and legible.

Knife Switches

Single-throw knife switches are to be connected so that the blades are dead when the switch is in the open position. Single-throw knife switches are to be placed so that gravity will not tend to close them. Single-throw knife switches, approved for use in the inverted position, are to provide a locking device that will ensure that the blades remain in the open position, when so set. Double-throw knife switches may be mounted so that the throw will be either vertical or horizontal. However, if the throw is vertical, a locking device must be provided to ensure that the blades remain in the open position, when so set. Exposed blades of knife switches must be dead when open.

Switchboards and Panelboards

Switchboards that have any exposed live parts are to be located in permanently dry locations and accessible only to qualified persons. Panelboards shall be mounted in cabinets, cutout boxes, or enclosures designed for the purpose and must be dead front. However, panelboards, other than the dead front externally-operable type, are permitted where accessible only to qualified persons.

Wet or Damp Locations

Cabinets, cutout boxes, fittings, boxes, and panelboard enclosures, which are in damp or wet locations, must be installed so as to prevent moisture or water from entering and accumulating within the enclosures. In wet locations, the enclosures are to be weatherproof. Switches, circuit breakers, and switchboards, installed in wet locations, must be enclosed in weatherproof enclosures.

Conductors

All conductors used for general wiring are to be insulated with few exceptions. The conductor insulation is to be of a type that is suitable for the voltage, operating temperature, and location of use. Insulated conductors must be distinguished as being grounded conductors, ungrounded conductors, or equipment grounding conductors by using appropriate colors, or other means. Multiconductor portable cable, used to supply power to portable or mobile equipment at over 600 volts, nominal, must consist of No. 8, or larger, conductors employing flexible stranding. Cables operated at over 2000 volts are to be shielded for the purpose of confining the voltage stresses to the insulation, and grounding conductors must be provided. Connectors for these cables are to be of a locking type, with provisions to prevent their opening or closing while energized. Strain relief must be provided at connections and terminations. Portable cables are not operated with splices unless the splices are of the permanent molded, vulcanized, or other equivalent type. Termination enclosures must be marked with a high voltage hazard warning, and terminations shall be accessible only to authorized and qualified personnel.

Fixtures

Fixture wires must be suitable for the voltage, temperature, and location of use. A fixture wire which is used as a grounded conductor is to be identified. Fixture wires may be used for installation in lighting, fixtures, and similar equipment when enclosed or protected, when not subject to bending or twisting while in use, and when used for connecting lighting fixtures to the branch-circuit conductors supplying the fixtures. Fixture wires are not designed as branch-circuit conductors except as permitted for Class 1 power-limited circuits.

Fixtures, lampholders, lamps, rosettes, and receptacles are, normally, to have no live parts exposed to employee contact. However, rosettes and cleat-type lampholders, and receptacles located at least 8 feet above the floor, may have exposed parts. Fixtures, lampholders, rosettes, and receptacles are to be securely supported. A fixture that weighs more than 6 pounds, or exceeds 16 inches in any dimension, shall not be supported by the screw shell of a lampholder.

Portable lamps are to be wired with flexible cord and have an attachment plug of the polarized or grounding type. If the portable lamp uses an Edison-based lampholder, the grounded conductor shall be identified and attached to the screw shell and the identified blade of the attachment plug. In addition, portable handlamps must have a metal shell; paperlined lampholders are not to be used; handlamps must be equipped with a handle of molded composition or other insulating material; handlamps are to be equipped with a substantial guard attached to the lampholder or handle; and metallic guards shall be grounded by the means of an equipment grounding conductor which runs within the power supply cord.

Lampholders of the screw-shell type are to be installed for use as lampholders only. Lampholders installed in wet or damp locations must be of the weatherproof type. Fixtures installed in wet or damp locations must be identified for the purpose and installed so that water cannot enter or accumulate in wireways, lampholders, or other electrical parts.

Receptacles

Receptacles, cord connectors, and attachment plugs are to be constructed so that no receptacle or cord connector will accept an attachment plug with a different voltage or current rating, other than that for which the device is intended. However, a 20-ampere T-slot receptacle or cord connector may accept a 15-ampere attachment plug of the same voltage rating. Receptacles connected to circuits, having different voltages, frequencies, or types of current (ac or dc) on the same premises, must be of such design that the attachment plugs used on these circuits are not interchangeable. A receptacle installed in a wet or damp location must be designed for the location.

Appliances

Appliances, other than those in which the current-carrying parts at high temperatures are necessarily exposed, shall, normally, have no live parts exposed to employee contact. A means is to be provided to disconnect each appliance. Each appliance is to be marked with its rating in volts and amperes, or volts and watts.

Motors

Motors, motor circuits, and controllers may require that one piece of equipment shall be "in sight from" another piece of equipment: one is visible and not more than 50 feet from the other. A disconnecting means is to be located, in sight, from the controller location. The controller disconnecting means for motor branch circuits over 600 volts, nominal, may be out of sight of the controller, if the controller is marked with a warning label that gives the location and identification of the disconnecting means; this means is to be locked in the open position.

The disconnecting means must disconnect the motor and the controller from all ungrounded supply conductors, and be designed so that no pole can be operated independently. If a motor and the driven machinery are not in sight from the controller location, the installation must comply with one of the following conditions: the controller disconnecting means must be capable of being locked in the open position; a manually operable switch must be placed in sight of the motor location that will disconnect the motor from its source of supply; the disconnecting means must plainly indicate whether it is in the open (off) or closed (on) position; the disconnecting means must be readily accessible; if more than one disconnect is provided for the same equipment, only one need be readily accessible; or an individual disconnecting means must be provided for each motor.

For a specific group of motors, a single disconnecting means may be used under any one of the following conditions:

1. If a number of motors drive special parts of a single machine or piece of apparatus, such as a metal or woodworking machine, crane, or hoist.

2. If a group of motors is under the protection of one set of branch-circuit protective devices.

3. If a group of motors is in a single room in sight from the location of the disconnecting means.

Motors, motor-control apparatus, and motor branch-circuit conductors are to be protected against overheating due to motor overloads, failure to start, short circuits, or ground faults. These provisions do not require overload protection that will stop a motor, when a shutdown is likely to introduce additional or increased hazards, as in the case of fire pumps, or where continued operation of a motor is necessary for a safe shutdown of equipment, or where process and motor overload sensing devices are connected to a supervised alarm.

Stationary motors, which have commutators, collectors, and brush rigging located inside of motor end brackets and are not conductively connected to supply circuits operating at more than 150 volts to ground, need not have such parts guarded. Exposed live parts of motors and controllers, operating at 50 volts or more between terminals, are to be guarded against accidental contact by any of the following:

1. By installation in a room or enclosure that is accessible only to qualified persons.

2. By installation on a balcony, gallery, or platform, so elevated and arranged as to exclude unqualified persons.

3. By elevation 8 feet or more above the floor.

Where there are live parts of motors, or controllers are operating at over 150 volts to ground, they are to be guarded against accidental contact only by location. And, where adjustments or other attendance may be necessary during the operation of the apparatus, insulating mats or platforms are to be provided so that the attendant cannot readily touch live parts, unless standing on the mats or platforms.

Transformers

The only transformers that are not addressed in this section are the current transformers, dry-type transformers which are installed as a component part of other apparatus, transformers which are an integral part of an X-ray, high frequency, or electrostatic-coating apparatus, or transformers used with Class 2 and Class 3 circuits, signs, outline lighting, electric discharge lighting, and power-limited fire-protective signaling.

The operating voltage of exposed live parts of transformer installations are to be indicated by warning signs or visible markings on the equipment or structure. Dry-type, high fire point liquid-insulated, and askarel-insulated transformers installed indoors and rated over 35 kV must be placed in a vault. If oil-insulated transformers present a fire hazard to employees, they are to be installed indoors in a vault.

Fire Protection

Combustible material, combustible buildings and parts of buildings, fire escapes, and door and window openings are to be safeguarded from fires which may originate in oil-insulated transformers attached to, or adjacent to a building, or combustible material. Transformer vaults are to be constructed so as to contain fire and combustible liquids within the vault and to prevent unauthorized access. Locks and latches must be so arranged that a vault door can be readily opened from the inside.

Transformer Guidelines

Any pipe or duct system which is foreign to the vault installation shall not enter or pass through a transformer vault. Materials must not be stored in transformer vaults.

Capacitors

All capacitors, except surge capacitors or capacitors included as a component part of other apparatus, are to be provided with an automatic means of draining the stored charge and maintaining the discharged state after the capacitor is disconnected from its source of supply. Capacitors rated over 600 volts, nominal, have other requirements and, therefore, need isolating or disconnecting switches (with no interrupting rating) which are interlocked with a load interrupting device or provided with prominently displayed caution signs that prevent switching load current. For series capacitors, proper switching must be assured by using at least mechanically sequenced isolating and bypass switches, interlocks, or a switching procedure which is prominently displayed at the switching location.

Specific Purpose Equipment and Installation (1926.406)

The installation of electric equipment and wiring used in connection with cranes, monorail hoists, hoists, and all runways is addressed in this section. A readily accessible disconnecting means is to be provided between the runway contact conductors and the power supply. The disconnecting means must be capable of being locked in the open position, and is provided in the leads from the runway contact conductors or other power supply on any crane or monorail hoist. If this additional disconnecting means is not readily accessible from the crane or monorail hoist operating station, a means must be provided at the operating station to open the power circuit to all motors of the crane or monorail hoist. The additional disconnect may be omitted if a monorail hoist or hand-propelled crane bridge installation meets all of the following:

1.   The unit is floor controlled.

2.   The unit is within view of the power supply disconnecting means.

3.   No fixed work platform has been provided for servicing the unit.

A limit switch or other device shall be provided to prevent the load block from passing the safe upper limit of travel of any hoisting mechanism. The dimension of the working space in the direction of access to live parts which may require examination, adjustment, servicing, or maintenance while alive, is to be a minimum of 2 feet 6 inches. Where controls are enclosed in cabinets, the door(s) are to open at least 90 degrees, be removable, or the installation must provide equivalent access.

All exposed metal parts of cranes, monorail hoists, hoists, and accessories, including pendant controls, are to be metallically joined together into a continuous electrical conductor so that the entire crane or hoist will be grounded. Moving parts, other than removable accessories or attachments, having metal-to-metal bearing surfaces are considered to be electrically connected to each other through the bearing surfaces for grounding purposes. The trolley frame and bridge frame are considered as electrically grounded through the bridge and trolley wheels and its respective tracks unless conditions, such as paint or other insulating materials, prevent reliable metal-to-metal contact. In this case, a separate bonding conductor shall be provided.

Elevators, Escalators, and Moving Walks

Elevators, escalators, and moving walks must have a single means for disconnecting all ungrounded main power supply conductors for each unit. If control panels are not located in the same space as the drive machine, they are to be located in cabinets with doors or panels capable of being locked closed.

Electric Welder Disconnects

Motor-generators, AC transformers, and DC rectifier arc welders must have a disconnecting means, in the supply circuit, for each motor-generator arc welder, and for each AC transformer and DC rectifier arc welder, which is not equipped with a disconnect that is mounted as an integral part of the welder. A switch or circuit breaker is to be provided by which each resistance welder and its control equipment can be isolated from the supply circuit. The ampere rating of this disconnecting means must not be less than the supply conductor ampacity.

X-ray Equipment

A disconnecting means for X-ray equipment is to be provided in the supply circuit. The disconnecting means is to be operable from a location readily accessible from the X-ray control. For equipment connected to a 120-volt branch circuit of 30 amperes or less, a grounding-type attachment plug cap and receptacle of proper rating may serve as a disconnecting means. If more than one piece of equipment is operated from the same high-voltage circuit, each piece or each group of equipment as a unit must be provided with a high-voltage switch or equivalent disconnecting means. This disconnecting means shall be constructed, enclosed, or located so as to avoid contact by employees with its live parts. All radiographic and fluoroscopic-type equipment are to be effectively enclosed or have interlocks that de-energize the equipment automatically to prevent ready access to live current-carrying parts.

Hazardous (Classified) Locations (1926.407)

The requirements for electric equipment and wiring depends on the location of the properties and the potential for the presence of a flammable or combustible concentration of vapors, liquids, gases, combustible dusts, or fibers. Each room, section, or area shall be considered individually in determining its classification. These hazardous (classified) locations are assigned six designations (see Figure 7-32).

Class I, Division 1
Class I, Division 2
Class II, Division 1
Class II, Division 2
Class III, Division 1
Class III, Division 2

(See Appendix H for detailed description on hazard locations.)

Equipment, wiring methods, and installations of equipment in hazardous (classified) locations are to be approved as intrinsically safe, or approved for the hazardous (classified) location. Requirements for each of these options are that the equipment and associated wiring, which is approved as intrinsically safe, is permitted in any hazardous (classified) location included in its listing or labeling, and is approved for the hazardous (classified) location.

Equipment must be approved not only for the class of location, but also for the ignitable or combustible properties of the specific gas, vapor, dust, or fiber that will be present. The NFPA 70, the National Electrical Code, lists or defines hazardous gases, vapors, and dusts by "Groups" characterized by their ignitable or combustible properties.

Equipment must not be used unless it is marked to show the class, group, and operating temperature or temperature range, based on operation in a 40-degree C ambient, for which it is approved. The temperature marking must not exceed the ignition temperature of the specific gas, vapor, or dust to be encountered. However, the following provisions modify this marking requirement for specific equipment:

1. Equipment of the non-heat-producing type (such as junction boxes, conduit, and fittings) and equipment of the heat-producing type having a maximum temperature of not more than 100 degrees C (212 degrees F) need not have a marked operating temperature or temperature range.

2. Fixed lighting fixtures marked for use only in Class I, Division 2 locations need not be marked to indicate the group.

Summary of Class I, II, III Hazardous Locations			
CLASSES	**GROUPS**	**DIVISIONS**	
		1	**2**
I Gases, vapors, and liquids (Art. 501)	A: Acetylene B: Hydrogen, etc. C: Ether, etc. D: Hydrocarbons, fuels, solvents, etc.	Normally explosive and hazardous	Not normally present in an explosive concentration (but may accidentally exist)
II Dusts (Art. 502)	E: Metal dusts (conductive,* and explosive) F: Carbon dusts (some are conductive, and all are explosive) G: Flour, starch, grain, combustible plastic or chemical dust (explosive)	Ignitable quantities of dust normally are or may be in suspension, or conductive dust may be present	Dust not normally suspended in an ignitable concentration (but may accidentally exist). Dust layers are present.
III Fibers and flyings (Art. 503)	Textiles, wood-working, etc. (easily ignitable, but not likely to be explosive)	Handled or used in manufacturing	Stored or handled in storage (exclusive of manufacturing)

Figure 7-32. Hazardous locations

3.   Fixed general-purpose equipment in Class I locations, other than lighting fixtures, which is acceptable for use in Class I, Division 2 locations need not be marked with the class, group, division, or operating temperature.

4.   Fixed dust-tight equipment, other than lighting fixtures, which is acceptable for use in Class II, Division 2 and Class III locations need not be marked with the class, group, division, or operating temperature.

Equipment which is safe for the location is to be of a type and design which the employer demonstrates will provide protection from the hazards arising from the combustibility and flammability of vapors, liquids, gases, dusts, or fibers. The National Electrical Code, NFPA 70, contains a guidelines form determining the type and design of equipment and installations which will meet this requirement.

All conduits are to be threaded and made wrench-tight. Where it is impractical to make a threaded joint tight, a bonding jumper shall be utilized.

Special Systems (1926.408)

Systems over 600 volts, nominal, must meet the general requirements for all circuits and equipment operated at over 600 volts. Wiring methods for fixed installations containing above-ground conductors are to be installed in rigid metal conduit, in intermediate metal conduit, in cable trays, in cable bus, in other suitable raceways, or as open runs of metal-clad cable designed for this use and purpose. However, open runs of non-metallic-sheathed cable, or of bare conductors or busbars, may be installed in locations which are accessible only to qualified persons. Metallic shielding components, such as tapes, wires, or braids for conductors, are to be grounded. Open runs of insulated wires and cables having a bare lead sheath or a braided outer covering must be supported in a manner designed to prevent physical damage to the braid or sheath.

Installations Emerging from the Ground

Conductors emerging from the ground need to be enclosed in raceways. Raceways installed on poles are to be of rigid metal conduit, intermediate metal conduit, PVC schedule 80, or the equivalent and extend from the ground line up to a point 8 feet above finished grade. Conductors entering a building shall be protected by an enclosure from the ground line to the point of entrance. Metallic enclosures are to be grounded.

Interrupting and Isolating Devices

Circuit breakers located indoors must consist of metal-enclosed or fire-resistant, cell-mounted units. In locations accessible only to qualified personnel, open mounting of circuit breakers is permitted. A means of indicating the open and closed position of circuit breakers is to be provided.

Fused cutouts, installed in buildings or transformer vaults, are to be of a type identi-fied for the purpose. They shall be readily accessible for fuse replacement. A means must be provided to completely isolate equipment for inspection and repairs. Isolating means which are not designed to interrupt the load current of the circuit shall be either interlocked with a circuit interrupter or provided with a sign warning against opening them under load.

Mobile and Portable Equipment

A metallic enclosure is to be provided on the mobile machine for enclosing the termi-nals of the power cable. The enclosure must include provisions for a solid connection for the ground wire(s) terminal to ground, effectively, the machine frame. The method used for cable termination must prevent any strain or pull on the cable from stressing the electrical connec-tions. The enclosure is to have provisions for locking so that only authorized qualified persons may open it, and it must be marked with a sign warning of the presence of energized parts.

Guarding Live Parts

All energized switching and control parts are to be enclosed in effectively grounded metal cabinets or enclosures. Circuit breakers and protective equipment shall have the operat-ing means projecting through the metal cabinet or enclosure so these units can be reset without the locked doors being opened. Enclosures and metal cabinets are to be locked so that only authorized qualified persons have access, and must be marked with a sign warning of the presence of energized parts. Collector ring assemblies on revolving-type machines (shovels, draglines, etc.) are to be guarded.

Tunnel Installations

The installation and use of high-voltage power distribution and utilization equipment which are associated with tunnels and are portable and/or mobile, such as substations, trailers, cars, mobile shovels, draglines, hoists, drills, dredges, compressors, pumps, conveyors, and underground excavators, are to have conductors installed in one or more of the following:

1. Metal conduit or other metal raceway.

2. Type MC cable.

3. Other suitable multiconductor cable. Conductors are also to be so located or guarded as to protect them from physical damage. Multiconductor portable cables may supply mobile equipment. An equipment grounding conductor must run with circuit conductors inside the metal raceway or inside the multiconductor cable jacket. The equipment grounding conductor may be insulated or bare.

Bare terminals of transformers, switches, motor controllers, and other equipment are to be enclosed to prevent accidental contact with energized parts. Enclosures for use in tunnels are to be drip-proof, weatherproof, or submersible as required by the environmental conditions.

A disconnecting means that simultaneously opens all ungrounded conductors is to be installed at each transformer or motor location. All nonenergized metal parts of electric equipment and metal raceways and cable sheaths shall be grounded and bonded to all metal pipes and rails at the portal and at intervals not exceeding 1000 feet throughout the tunnel.

Classification. Class 1, Class 2, or Class 3 Remote Control, Signaling, or Power-Limited Circuits

Class 1, Class 2, or Class 3 remote control, signaling, or power-limited circuits are characterized by their usage and electrical power limitation which differentiates them from light and power circuits. These circuits are classified in accordance with their respective voltage and power limitations.

A Class 1 power-limited circuit is to be supplied from a source having a rated output of not more than 30 volts and 1000 volt-amperes. A Class 1 remote control circuit or a Class 1 signaling circuit shall have a voltage which does not exceed 600 volts; however, the power output of the source need not be limited.

Power for Class 2 and Class 3 circuits shall be limited either inherently (in which no overcurrent protection is required), or by a combination of a power source and overcurrent protection. The maximum circuit voltage is to be 150 volts AC or DC for a Class 2 inherently limited power source, and 100 volts AC or DC for a Class 3 inherently limited power source. The maximum circuit voltage is to be 30 volts AC and 60 volts DC for a Class 2 power source limited by overcurrent protection, and 150 volts AC or DC for a Class 3 power source limited by overcurrent protection. The maximum circuit voltages apply to sinusoidal AC or continuous DC power sources, and where wet contact occurrence is not likely. A Class 2 or Class 3 power supply unit is not to be used unless it is durably marked and plainly visible to indicate the class of supply and its electrical rating.

Communications Systems

The provisions for communication systems apply to such systems as central-station-connected and non-central-station-connected telephone circuits, radio receiving and transmitting equipment, outside wiring for fire and burglar alarm, and similar central station systems. Communication circuits so located as to be exposed to accidental contact with light or power conductors operating at over 300 volts, must have each circuit, so exposed, provided with an approved protector.

Each conductor of a lead-in from an outdoor antenna must be provided with an antenna discharge unit or other means that will drain static charges from the antenna system.

Receiving distribution lead-in or aerial-drop cables attached to buildings, and lead-in conductors to radio transmitters, are to be so installed as to avoid the possibility of accidental contact with electric light or power conductors. The clearance between lead-in conductors and any lightning protection conductors is to be not less than 6 feet. Where practicable, communication conductors on poles shall be located below the light or power conductors. Communications conductors must not be attached to a crossarm that carries light or power conductors.

Indoor antennas, lead-ins, and other communication conductors attached as open conductors to the inside of buildings are to be located at least 2 inches from conductors of any light, power, or Class 1 circuits unless a special and equally protective method of conductor separation is employed.

Outdoor metal structures which support antennas, as well as self-supporting antennas such as vertical rods or dipole structures, are to be located as far away from overhead conductors of electric light and power circuits of over 150 volts to ground, as necessary, to avoid the possibility of the antenna or structure falling into, or making accidental contact with such circuits.

If lead-in conductors are exposed to contact with electric light or power conductors, the metal sheath of aerial cables entering buildings must be grounded or interrupted close to the entrance to the building by an insulating joint or equivalent device. Where protective devices are used, they are to be grounded. Masts and metal structures supporting antennas must be permanently and effectively grounded without splice or connection in the grounding conductor.

Transmitters are to be enclosed in a metal frame or grill, or separated from the operating space by a barrier, and all metallic parts shall be effectively connected to ground. All external metal handles and controls which are accessible to the operating personnel shall be effectively grounded. Unpowered equipment and enclosures are considered grounded when connected to an attached coaxial cable with an effectively grounded metallic shield.

Electrical Work Practices (1926.416)

No employer shall permit an employee to work in proximity of any part of an electric power circuit where the employee could contact the electric power circuit in the course of work, unless the employee is protected against electric shock by deenergizing the circuit and grounding it, or by guarding it effectively by insulation or other means.

In work areas where the exact location of underground electric powerlines is unknown, employees using jack-hammers, bars, or other hand tools which may contact a line, are to be provided with insulated protective gloves.

Before work is begun, the employer must ascertain, by inquiry, direct observation, or instruments, whether any part of an energized electric power circuit, exposed or concealed, is so located that the performance of the work may bring any person, tool, or machine into physical or electrical contact with the electric power circuit. The employer must post and maintain proper warning signs where such a circuit exists. The employer shall advise employees of the location of such lines, the hazards involved, and the protective measures to be taken.

Barriers, or other means of guarding are to be provided to ensure that the workspace for electrical equipment will not be used as a passageway during periods when energized parts of electrical equipment are exposed. Working spaces, walkways, and similar locations are to be kept clear of cords so as not to create a hazard to employees.

In existing installations, no changes in circuit protection shall be made to increase the load in excess of the load rating of the circuit wiring. When fuses are installed or removed with one or both terminals energized, special tools insulated for the voltage are to be used.

Worn or frayed electric cords or cables must not be used. Extension cords shall not be fastened with staples, hung from nails, or suspended by wire.

Lockout/Tagging of Circuits (1926.417)

Controls that are to be deactivated during the course of work on energized or deenergized equipment or circuits are to be tagged. Equipment or circuits that are deenergized are to be rendered inoperative and have tags attached at all points where such equipment or circuits can be energized. Tags must be placed to identify plainly the equipment or circuits being worked on (see Figure 7-33).

Figure 7-33. Lockout/tagout example

Safety-Related Maintenance and Environmental Considerations

Maintenance of Equipment (1926.431)

The employer shall ensure that all wiring components and utilization equipment in hazardous locations are maintained in a dust-tight, dust-ignition-proof, or explosion-proof condition, as appropriate. There shall be no loose or missing screws, gaskets, threaded connections, seals, or other impairments to a tight condition.

Environmental Deterioration of Equipment (1926.432)

Unless identified for use in the operating environment, no conductors or equipment shall be located

1.   Where locations are damp or wet.

2.   Where exposed to gases, fumes, vapors, liquids, or other agents having a deteriorating effect on the conductors or equipment.

3.   Where exposed to excessive temperatures.

Control equipment, utilization equipment, and busways, approved for use in dry locations only, are protected against damage from the weather during building construction.

Metal raceways, cable armor, boxes, cable sheathing, cabinets, elbows, couplings, fittings, supports, and support hardware are of materials appropriate for the environment in which they are to be installed.

EMPLOYEE EMERGENCY ACTION PLANS (1926.35)

The emergency action plan is to be in writing and cover those designated actions employers and employees must take to ensure employee safety from fire and other emergencies. The emergency action plan must contain these elements, at a minimum:

1. Emergency escape procedures and emergency escape route assignments.

2. Procedures to be followed by employees who remain to operate critical plant operations before they evacuate.

3. Procedures to account for all employees after emergency evacuation has been completed.

4. Rescue and medical duties for those employees who are to perform them.

5. The preferred means of reporting fires and other emergencies

6. Names or regular job titles of persons or departments who can be contacted for further information or explanation of duties under the plan.

The employer must establish an employee alarm system which complies with 1926.159. If the employee alarm system is used for alerting fire brigade members, or for other purposes, a distinctive signal for each purpose is to be used.

Before implementing the emergency action plan, the employer must designate and train a sufficient number of persons to assist in the safe and orderly emergency evacuation of employees. The employer must review the plan with each employee covered by the plan at the following times:

1. Initially when the plan is developed.

2. Whenever the employee's responsibilities or designated actions under the plan change.

3. Whenever the plan is changed.

The employer must review with each employee, upon initial assignment, those parts of the plan which the employee must know to protect the employee in the event of an emergency. The written plan is to be kept at the workplace and made available for employee review. For those employers with 10 or fewer employees, the plan may be communicated orally to employees and the employer need not maintain a written plan.

EXCAVATIONS /TRENCHES (1926.650)

Excavation operations are among the first undertaken at a construction site. Accidental cave-ins of the earth that has been excavated account for a large number of construction site related deaths. All openings made in the earth's surface are considered excavations. Excavations are defined to include trenches.

In almost all cases of excavation/trench accidents, the resulting accident occurred because the known regulations and safe work practices were violated. This is not a statement which can be made about most construction accidents.Trenching and excavation work presents a serious risk to all employees. The greatest risk, and one of primary concern, is a cave-in. A cubic yard of soil, or other material weighs more than 2000 pounds and usually more than a cubic yard of material is involved. When a cave-in occurs, this type of weight is beyond the physical capabilities of workers to protect themselves. Cave-in accidents are much more likely to result in worker fatalities than any other excavation-related accident (see Figure 7-34).

Figure 7-34. Workers in an unprotected trench

Specific Excavation Requirements (1926.651)

All surface encumbrances, that are located so as to create a hazard to employees, are to be removed or supported, as necessary, to safeguard employees.

<u>Utilities</u>

The estimated location of utility installations, such as the sewer, telephone, fuel, electric, or water lines, or any other underground installation that may be expected to be encountered during excavation work, is to be determined prior to opening an excavation. Prior to the start of the actual excavation, utility companies or owners must be contacted within the established or customary local response times, advised of the proposed work, and asked to establish the location of the utility underground installations. When utility companies or owners cannot respond to a request to locate underground utility installations within 24 hours (unless a longer period is required by state or local law), or cannot establish the exact location of these installations, the employer may proceed, provided the employer does so with caution, and provided detection equipment, or other acceptable means to locate utility installations are used. When excavation operations approach the estimated location of underground installations, the exact location of the installations shall be determined by safe and acceptable means. While the excavation is open, underground installations are to be protected, supported, or removed, as necessary, to safeguard employees (see Figure 7-35).

Figure 7-35. Unsupported trench and underground installations

Egress Ramps and Runways

Structural ramps, that are used solely by employees as a means of access or egress from excavations, are to be designed by a competent person. Structural ramps, used for access or egress of equipment, are to be designed by a competent person qualified in structural design, and are to be constructed in accordance with the design. Ramps and runways constructed of two or more structural members must have the structural members connected together to prevent displacement. Structural members, used for ramps and runways, must be of uniform thickness. Cleats, or other appropriate means used to connect runway structural members, are to be attached to the bottom of the runway, or attached in a manner to prevent tripping. Structural ramps, used in lieu of steps, must provide cleats or other surface treatments on the top surface to prevent slipping. A stairway, ladder, ramp, or other safe means of egress must be located in trench excavations that are 4 feet or more in depth so as to require no more than 25 feet of lateral travel for employees (see Figure 7-36).

Figure 7-36. An open excavation with a ladder for egress

Equipment and Loads

No worker is permitted underneath loads handled by lifting or digging equipment. To avoid being struck by any spillage or falling materials, employees are required to stand away from any vehicle being loaded or unloaded. Operators may remain in the cabs of vehicles being loaded or unloaded when the vehicles are equipped to provide adequate protection for the operator during loading and unloading operations.

When mobile equipment is operated adjacent to an excavation, or when such equipment is required to approach the edge of an excavation, and the operator does not have a clear and direct view of the edge of the excavation, a warning system is to be utilized; these warning systems can be such things as barricades, hand or mechanical signals, or stop logs. If possible, the grade should be away from the excavation.

Hazardous Atmospheres

To prevent exposure to harmful levels of atmospheric contaminants, and to assure acceptable atmospheric conditions, the following requirements apply:

1. Where oxygen deficiency (atmospheres containing less than 19.5 percent oxygen), or ahazardous atmosphere exists, or could reasonably be expected to exist, such as in excavations in landfill areas, or excavations in areas where hazardous substances are stored nearby, the atmospheres in the excavation is to be tested before employees enter excavations greater than 4 feet in depth.

2. Adequate precautions must be taken to prevent employee exposure to atmospheres containing less than 19.5 percent oxygen, and other hazardous atmospheres.

3. Adequate precaution is to be taken, such as providing ventilation to prevent employee exposure to the atmosphere containing a concentration of a flammable gas in excess of 20 percent of the lower flammable limit of the gas.

4. When controls are used that are intended to reduce the level of atmospheric contaminants to acceptable levels, testing shall be conducted as often as necessary to ensure that the atmosphere remains safe.

Emergency rescue equipment, such as breathing apparatus, a safety harness and line, or a basket stretcher, must be readily available where hazardous atmospheric conditions exist, or may reasonably be expected to develop during work in an excavation. This equipment must be attended when in use.

Employees entering bell-bottom pier holes, or other similar deep and confined footing excavations, must wear a harness with a lifeline securely attached to it. The lifeline is to be separate from any line used to handle materials, and is to be individually attended at all times while the employee, wearing the lifeline, is in the excavation.

Water Accumulation

Employees are not to work in excavations in which there is accumulated water, or in excavations in which water is accumulating, unless adequate precautions have been taken to protect employees against the hazards posed by water accumulation. The precautions necessary to adequately protect employees vary with each situation, but could include special support or shield systems to protect from cave-ins, water removal to control the level of accumulating water, or the use of a safety harness and lifeline.

If water is controlled, or prevented from accumulating by using water removal equipment, the water removal equipment and operations are to be monitored by a competent person to ensure proper operation.

If excavation work interrupts the natural drainage of surface water (such as streams), diversion ditches, dikes, or other suitable means are to be used to prevent surface water from entering the excavation. Adequate drainage of the area adjacent to the excavation must also be provided. Excavations subject to runoff from heavy rains are required to be inspected by a competent person .

Below Level Excavations

Where the stability of adjoining buildings, walls, or other structures are endangered by excavation operations, support systems such as shoring, bracing, or underpinning must be provided for the protection of the employees and to ensure the stability of such structures. Excavation below the level of the base, or footing of any foundation, or retaining walls that could be reasonably expected to pose a hazard to employees, must not be permitted except when

1. A support system, such as underpinning, is provided to ensure the safety of employees and the stability of the structure.

2. The excavation is in stable rock.

3. A registered professional engineer has approved and determined that the structure is sufficiently removed from the excavation so as to be unaffected by the excavation activity.

4. A registered professional engineer has approved and determined that such excavation work will not pose a hazard to employees.

Sidewalks, pavements, and appurtenant structures are not to be undermined unless a support system or other method of protection is provided to protect employees from the possible collapse of such structures.

Loose Materials

Adequate protection must be provided to protect employees from loose rock or soil that could pose a hazard by falling or rolling from an excavation face. Such protection consists of scaling to remove loose material; installation of protective barricades at intervals, as necessary on the face, to stop and contain falling material; or other means that provide equivalent protection. Employees are to be protected from excavated or other materials or equipment that could pose a hazard by falling or rolling into excavations. Protection shall be provided by placing and keeping such materials or equipment at least 2 feet from the edge of excavations, by using retaining devices that are sufficient to prevent materials or equipment from falling or rolling into excavations, or by a combination of both, if necessary.

Inspections

Daily inspections of excavations, the adjacent areas, and protective systems must be made by a competent person for any evidence of a situation that could result in possible cave-ins, failure of protective systems, hazardous atmospheres, or other hazardous conditions. An inspection is to be conducted by the competent person prior to the start of work, and as needed throughout the shift. Inspections shall also be made after every rainstorm or other hazard which may increase the occurrence of a hazardous condition. These inspections are only required when employee exposure can be reasonably anticipated. When the competent person finds evidence of a situation that could result in a possible cave-in, indication of possible failure of protective systems, hazardous atmospheres, or other hazardous conditions, exposed employees are to be removed from the hazardous area until the necessary precautions have been taken to ensure their safety.

<u>Walkways and Barriers</u>

Walkways are to be provided where employees or equipment are required, or permitted to cross over excavations. Guardrails need to be provided where walkways are 6 feet or more above lower levels. Adequate physical barrier protection must be provided at all remotely located excavations. All wells, pits, shafts, etc. must be barricaded or covered. Upon completion of exploration and other similar operations, temporary wells, pits, shafts, etc. are to be backfilled.

Requirements for Protective Systems 1926.652)

Each employee in an excavation must be protected from cave-ins by an adequate protective system except when

1.   Excavations are made entirely in stable rock.

2.   Excavations are less than 5 feet in depth and examination of the ground, by a competent person, provides no indication of a potential cave-in.

Protective systems must have the capacity to resist, without failure, all loads that are intended, or could reasonably be expected to be applied or transmitted to the system.

<u>Slopes</u>

The slopes and configurations of sloping and benching systems are to be selected and constructed by the employer or his designee, and must be in accordance with the requirements as follows:

1.   Excavations must be sloped at an angle not steeper than one and one-half horizontal to one vertical (34 degrees measured from the horizontal), unless the employer uses one of the other options listed below (see Figure 7-37).

Figure 7-37. Example of sloping and excavation

2. Slopes specified are to be excavated to form configurations that are in accordance with the slopes shown for Type C soil in Appendix B of 29 CFR 1926, Subpart P.

3. Maximum allowable slopes, and allowable configurations for sloping and benching systems, are to be determined in accordance with the conditions and requirements set forth in Appendices A and B of this Subpart P.

4. Designs of sloping or benching systems shall be selected from, and in accordance with tabulated data, such as tables and charts. The tabulated data is to be in written form and shall include the identification of the parameters that affect the selection of a sloping or benching system drawn from such data. The identification of the limits of use of the data must include the magnitude and configuration of the slopes determined to be safe, any explanatory information, as may be necessary, to aid the user in making a correct selection of a protective system from the data, and at least one copy of the tabulated data which identifies the registered professional engineer who approved the data. This information must be maintained at the jobsite during construction of the protective system. After that time, the data may be stored off the jobsite.

5. Sloping and benching systems must be designed by a registered professional engineer. The designs shall be in written form and shall include at least the magnitude of the slopes that were determined to be safe for the particular project, the configurations that were determined to be safe for the particular project, and the identity of the registered professional engineer approving the design.

Support, Shield, and Other Protective Systems

Designs of support systems, shield systems, and other protective systems are selected and constructed by the employer or his designee as follows:

1. Designs using Appendices A, C, and D. Designs for timber shoring in trenches are determined in accordance with the conditions and requirements set forth in Appendices A and C of Subpart P (see Figure 7-38). Designs for aluminum hydraulic shoring are to be in accordance with Appendix D of Subpart P.

Figure 7-38. A timber shored wall of an excavation

2. Design of support systems, shield systems, or other protective systems that are drawn from manufacturer's tabulated data, are to be in accordance with all specifications, recommendations, and limitations issued or made by the manufacturer. Deviation from the specifications, recommendations, and limitations issued, or made by the manufacturer, is only allowed after the manufacturer issues specific written approval.

3. Manufacturer's specifications, recommendations, and limitations, and manufacturer's approval to deviate from the specifications, recommendations, and limitations must be in written form at the jobsite during construction of the protective system. After that time this data may be stored off the jobsite.

4. Designs of support systems, shield systems, or other protective systems are selected from and are to be in accordance with tabulated data, such as tables and charts. The tabulated data must be in written form and include identification of the parameters that affect the selection of a protective system drawn from such data, identification of the limits of use of the data, and explanatory information, as may be necessary, to aid the user in making a correct selection of a protective system from the data. At least one copy of the tabulated data, which identifies the registered professional engineer who approved the data, must be maintained at the jobsite during construction of the protective system.

5. The systems must be designed by a registered professional engineer.

Materials and equipment used for protective systems are to be free from damage or defects that might impair their proper function. Manufactured materials and equipment used for protective systems are to be used and maintained in a manner that is consistent with the recommendations of the manufacturer, and in a manner that will prevent employee exposure to hazards.

When material or equipment that is used for protective systems is damaged, a competent person must examine the material or equipment and evaluate its suitability for continued use. If the competent person cannot assure the material or equipment is able to support the intended loads, or is otherwise suitable for safe use, then such material or equipment shall be removed from service, and shall be evaluated and approved by a registered professional engineer before being returned to service.

Members of support systems are to be securely connected together to prevent sliding, falling, kickouts, or other predictable failure. Support systems must be installed and removed in a manner that protects employees from cave-ins, structural collapses, or from being struck by members of the support system. Individual members of support systems are not to be subjected to loads exceeding those which those members were designed to withstand.

Before temporary removal of individual members begins, additional precautions must be taken to ensure the safety of employees, such as installing other structural members to carry the loads imposed on the support system. Removal begins at, and progresses from the bottom of the excavation. Members are released slowly so as to note any indication of possible failure of the remaining members of the structure, or possible cave-in of the sides of the excavation. Backfilling is to progress together with the removal of support systems from excavations.

Excavation of material, to a level no greater than 2 feet below the bottom of the members of a support system, is permitted, but only if the system is designed to resist the forces calculated for the full depth of the trench, and if there are no indications, while the trench is open, of a possible loss of soil from behind or below the bottom of the support system.

Installation of any support system is to be closely coordinated with the excavation of trenches. Employees are never permitted to work on the faces of sloped or benched excavations at levels above other employees, except when employees at the lower levels are adequately protected from the hazard of falling, rolling, or sliding material or equipment.

Shield Systems

Shield systems must not be subjected to loads exceeding those which the system was designed to withstand. Shields are to be installed in a manner which restricts lateral, or other hazardous movement of the shield, in the event of the application of sudden lateral loads. Employees are to be protected from the hazard of cave-ins when entering or exiting the areas protected by shields. Employees are not allowed in shields when shields are being installed, removed, or moved vertically. Some additional requirements for the use of shields in trenches or excavations are when earth material reaches a level not greater than 2 feet below the bottom of a shield are permitted, only if the shield is designed to resist the forces calculated for the full depth of the trench, and where there are no indications, while the trench is open, of a possible loss of soil from behind or below the bottom of the shield (see Figure 7-39).

Figure 7-39. Worker using a trench box as a shield with a ladder for egress

EXPLOSIVES AND BLASTING (1926.900)

Although there are not as many tragedies as in the past from working with explosives, the dangers still exist, and from time to time explosive/blasting related accidents are reported. Much of the blasting which now occurs is carried out by contract blasters who specialize in the use and handling of explosives. All employers, workers, and contract blasters are required to follow the regulatory requirements for use of explosives.

Blaster Qualifications (1926.901)

To begin with, only authorized and qualified persons are permitted to handle and use explosives. These individuals must be able to understand and give written and oral orders. Blasters are to be in good physical condition and not be addicted to narcotics, intoxicants, or similar types of drugs. They must be qualified, by reason of training, knowledge, or experience, in the field of transporting, storing, handling, and using explosives, and have a working knowledge of state and local laws and regulations which pertain to explosives. Blasters are required to furnish satisfactory evidence of competency in the handling explosives, and performing in a safe manner the type of blasting that will be required. They are also required to be knowledgeable and competent in the use of each type of blasting method used.

General Provisions

Smoking, firearms, matches, open flame lamps, other fires, flame or heat producing devices and sparks, are prohibited in or near explosive magazines, or while explosives are being handled, transported, or used. No person is allowed to handle or use explosives while under the influence of intoxicating liquors, narcotics, or other dangerous drugs.

All explosives must be accounted for at all times. Explosives not being used are to be kept in a locked magazine, unavailable to persons not authorized to handle them. The employer shall maintain an inventory and have records of all explosives. Appropriate authorities must be notified of any loss, theft, or unauthorized entry into a magazine. In no case are explosives or blasting agents abandoned.

Employees authorized to prepare explosive charges or conduct blasting operations, must use every reasonable precaution including, but not limited to, visual and audible warning signals, flags, or barricades, to ensure employee safety. When blasting is done in congested areas, or in proximity to a structure, railway, highway, or any other installation that may be damaged, the blaster shall take special precautions in the loading, delaying, initiation, and confinement of each blast by using mats (see Figure 7-40), or other methods so as to control the throw of fragments; this will prevent bodily injury to employees.

Figure 7-40. Example of blasting mats

Insofar as possible, blasting operations above ground are to be conducted between sunup and sundown. Precautions are to be taken to prevent accidental discharge of electric blasting caps from current induced by radar, radio transmitters, lightning, adjacent powerlines, dust storms, or other sources of extraneous electricity. These precautions shall include

1. Detonators are to be short-circuited in holes which have been primed and shunted until wired into the blasting circuit.

2. The suspension of all blasting operations, and removal of persons from the blasting area during the approach and progress of an electric storm.

3. The prominent display of adequate signs, warning against the use of mobile radio transmitters on all roads within 1,000 feet of blasting operations. When-

ever adherence to the 1,000-foot distance would create an operational handicap, a competent person must be consulted to evaluate the particular situation, and alternative provisions may be made which are adequately designed to prevent any premature firing of electric blasting caps. A description of any such alternatives must be in writing and certified by the competent person consulted, as meeting the purposes of this subdivision. The description shall be maintained at the construction site during the duration of the work, and shall be available for inspection by representatives of the Secretary of Labor.

Blasting operations, in the proximity of overhead power lines, communication lines, utility services, or other services and structures, are not to be carried on until the operators and/or owners have been notified and measures have been taken for safe control. Cables in the proximity of the blast area must be deenergized and locked out by the blaster.

Great care needs to be taken to ensure that mobile radio transmitters, which are less than 100 feet away from electric blasting caps and are in other than original containers, are deenergized and effectively locked. There must be compliance with the recommendations of The Institute of the Makers of Explosives with regard to blasting in the vicinity of radio transmitters and as stipulated in Radio Frequency Energy-A, Potential Hazard in the Use of Electric Blasting Caps, IME Publication No. 20, March 1971.

Transporting Explosives (1926.902)

When moving explosives, only the original containers or Class II magazines are to be used for taking detonators and other explosives from storage magazines to the blasting area. Transportation of explosives must meet the provisions of the Department of Transportation regulations, contained in 46 CFR Parts 146-149, Water Carriers; 49 CFR Parts 171-179, Highways and Railways; 49 CFR Part 195, Pipelines; and 49 CFR Parts 390-397, Motor Carriers. Motor vehicles or conveyances transporting explosives shall only be driven by, and be in the charge of, a licensed driver who is physically fit. Drivers are to be familiar with the local, State, and Federal regulations governing the transportation of explosives. No person can smoke, carry matches, or any other flame-producing device, nor shall firearms or loaded cartridges be carried while in or near a motor vehicle or conveyance transporting explosives. Explosives, blasting agents, and blasting supplies are not to be transported with other materials or cargoes. Blasting caps (including electric) shall not be transported in the same vehicle with other explosives. Vehicles used for transporting explosives must be strong enough to carry the load without difficulty, and must be in good mechanical condition. When explosives are transported by a vehicle with an open body, a Class II magazine or original manufacturer's container is to be securely mounted on the bed to contain the cargo. All vehicles used for the transportation of explosives shall have tight floors, and any exposed spark-producing metal on the inside of the body shall be covered with wood, or other nonsparking material, to prevent contact with containers of explosives.

Every motor vehicle or conveyance used for transporting explosives shall be marked or placarded on both sides, the front and the rear, with the word "Explosives" in red letters on a white background, and not less than 4 inches in height (see Figure 7-41). In addition to such marking or placarding, the motor vehicle or conveyance may display, in such a manner that it will be readily visible from all directions, a red flag 18 inches by 30 inches, with the word "Explosives" painted, stamped, or sewed thereon, in white letters, and at least six inches in height. Each vehicle used for transportation of explosives is to be equipped with a fully charged fire extinguisher, in good condition. An Underwriters Laboratory-approved extinguisher of not less than 10-ABC rating will meet the minimum requirement. The driver must be trained in the use of the extinguisher on his or her vehicle. No fire is to be fought where the fire is in imminent danger of contact with explosives. All employees are to be removed to a safe area and the fire area guarded against intruders.

Figure 7- 41. Typical vehicle used by those transporting explosives

Motor vehicles or conveyances carrying explosives, blasting agents, or blasting supplies, are not to be taken inside a garage or shop for repairs or servicing. No motor vehicle transporting explosives shall be left unattended.

Use of Explosives (1926.904 and 905)

Explosives and related materials are to be stored in approved facilities required under the applicable provisions of the Bureau of Alcohol, Tobacco, and Firearms regulations contained in 27 CFR part 55. Blasting caps, electric blasting caps, detonating primers, and primed cartridges are not stored in the same magazine with other explosives or blasting agents. Smoking and open flames are not permitted within 50 feet of explosives and detonator storage magazine.

The loading of explosives and blasting agents must follow procedures that permit safe and efficient loading before loading is started. All drill holes must be sufficiently large enough to admit freely the insertion of the cartridges of explosives. Tamping is done only with wood rods or plastic tamping poles without exposed metal parts, but non-sparking metal connectors may be used for jointed poles. Violent tamping must be avoided. The <u>primer</u> is never tamped. Holes are not loaded except those to be fired in the next round of blasting. After loading, all remaining explosives and detonators are immediately returned to an authorized magazine.

New drilling is not started until all remaining butts of old holes are examined for unexploded charges, and if any are found, they are refired before work proceeds. No person is allowed to deepen drill holes which have contained explosives or blasting agents. Machines and all tools not used for loading explosives into bore holes are removed from the immediate location of holes before explosives are delivered. No equipment is to be operated within 50 feet of loaded holes. No activity of any nature, other than that which is required for loading holes with explosives, is permitted in a blast area (see Figure 7-42).

Holes are checked prior to loading to determine depth and conditions. Where a hole has been loaded with explosives but the explosives have failed to detonate, there shall be no drilling within 50 feet of the hole. When loading a long line of holes with more than one loading crew, the crews must be separated by a practical distance consistent with efficient operation and supervision of crews. All blast holes in open work are to be stemmed to the collar or to a point which will confine the charge. Warning signs, indicating a blast area, are to be maintained at all approaches to the blast area. The warning sign lettering shall not be less than 4 inches in height and be on a contrasting background. A bore hole is never sprung when it is adjacent to or near a hole that is loaded. Flashlight batteries are not used for springing holes. Drill holes which have been sprung or chambered, and which are not water-filled, must be allowed to cool before explosives are loaded. No loaded holes shall be left unattended or unprotected. When loading blasting agents pneumatically over electric blasting caps, a semiconductive delivery hose is to be used, and the equipment is to be bonded and grounded.

Figure 7-42. Drilled blasting area ready for loading

Electrical Blasting (1926.906)

When initiating a blast, electric blasting caps are not to be used where sources of extraneous electricity make the use of electric blasting caps dangerous. Blasting cap leg wires shall be kept short-circuited (shunted) until they are connected into the circuit for firing. Before adopting any system of electrical firing, the blaster is to conduct a thorough survey for extraneous currents, and all dangerous currents shall be eliminated before any holes are loaded. In any single blast using electric blasting caps, all caps are to be of the same style or function, and the same manufacturer. Electric blasting must be carried out by using blasting circuits or power circuits in accordance with the electric blasting cap manufacturer's recommendations, or by an approved contractor or his designated representative. When firing a circuit of electric blasting caps, care must be exercised to ensure that an adequate quantity of delivered current is available, in accordance with the manufacturer's recommendations. Connecting wires and lead wires are to be insulated single solid wires of sufficient current-carrying capacity. Also, bus wires must be solid single wires of sufficient current-carrying capacity. When firing electrically, the insulation on all firing lines is to be adequate and in good condition. The power circuit used for firing electric blasting caps must not be grounded. When firing from a power circuit, the firing switch must be locked in the open or "Off" position at all times, except when firing. It must be so designed that the firing lines to the cap circuit are automatically short-circuited when the switch is in the "Off" position. Keys to this switch are to be entrusted only to the blaster.

Blasting machines in use must be in good condition and the efficiency of the machine tested periodically to make certain that it can deliver power at its rated capacity. When firing with blasting machines, the connections are to be made as recommended by the manufacturer of the electric blasting caps used. The number of electric blasting caps connected to a blasting machine are not to be in excess of its rated capacity. Furthermore, in primary blasting, a series circuit shall contain no more caps than the limits recommended by the manufacturer of the electric blasting caps in use. The blaster is in charge of the blasting machines, and no other person shall connect the leading wires to the machine. Whenever the possibility exists that a leading line or blasting wire might be thrown over a live powerline by the force of an explosion, care must be taken to see that the total length of wires are kept too short to hit the lines, or that the wires are securely anchored to the ground. If neither of these requirements can be satisfied, a non-electric system is to be used. In electrical firing, only the person making leading wire connections shall fire the shot. All connections are to be made from the bore hole back to the source of firing current, and the leading wires must remain shorted and not connected to the blasting machine or other source of current until the charge is to be fired. After firing an electric blast from a blasting machine, the leading wires must immediately be disconnected from the machine and short-circuited.

Safety Fuse (1926.907)

At times when extraneous electricity makes the use of electric blasting caps dangerous, safety fuses are to be used. The use of a fuse that has been hammered or injured in any way is forbidden. The hanging of a fuse on nails or other projections which will cause a sharp bend to be formed in the fuse is prohibited. Before capping a safety fuse, a short length is to be cut from the end of the supply reel so as to assure a fresh cut end in each blasting cap. Only a cap crimper of approved design is to be used for attaching blasting caps to a safety fuse. No unused cap or short capped fuse shall be placed in any hole to be blasted; such unused detonators are removed from the working place and destroyed. No fuse is capped, or primers made up, in any magazine or near any possible source of ignition, and no one is permitted to carry detonators or primers of any kind on his person. The minimum length of a safety fuse to be used in blasting shall not be less than 30 inches. At least two persons must be present when multiple cap and fuse blasting is done by the hand-lighting methods. Not more than 12 fuses are to be lighted by each blaster when hand-lighting devices are used. However, when two or more safety fuses in a group are lighted as one, by means of igniter cord or other similar fuse-lighting devices, they may be considered as one fuse. The so-called "drop fuse" method of dropping or pushing a primer or any explosive with a lighted fuse attached is forbidden. Caps and fuses are not used for firing mudcap charges unless charges are separated sufficiently to prevent one charge from dislodging other shots in the blast. When blasting with safety fuses, consideration must be given to the length and burning rate of the fuse. Sufficient time, with a margin of safety, must always be provided for the blaster to reach a place of safety.

Using Detonating Cord (1926.908)

When a detonating cord is used care must be taken to select a detonating cord consistent with the type and physical condition of the bore hole and stemming, and the type of explosives used. A detonating cord is to be handled and used with the same respect and care given other explosives. The line of a detonating cord, extending out of a bore hole or from a charge, is to be cut from the supply spool before loading the remainder of the bore hole, or placing additional charges. Detonating cord connections are to be competent and positive in accordance with the approved and recommended methods. Knot-type, or other cord-to-cord connections, are to be made only with a detonating cord in which the explosive core is dry. All detonating cord trunklines and branchlines are to be free of loops, sharp kinks, or angles that direct the cord back toward the oncoming line of detonation. All detonating cord connections must be inspected before firing the blast. When detonating cord millisecond-delay connectors or short-interval-delay electric blasting caps are used with a detonating cord, the practice shall conform strictly to the manufacturer's recommendations. When connecting a blasting cap or an electric blasting cap to a detonating cord, the cap is to be taped, or otherwise attached securely along the side or end of the detonating cord, with the end of the cap containing the explosive charge pointed in the direction in which the detonation is to proceed. Detonators for firing the trunkline must not brought to the loading area, nor attached to the detonating cord, until everything else is in readiness for the blast. A code of blasting signals equivalent to Table 7-6, are to be posted in one or more conspicuous places at the operation, and all employees are required to familiarize themselves with the code and conform to it. Danger signs must be placed at suitable locations.

TABLE 7-6

Blasting Signals

Courtesy of OSHA

WARNING SIGNAL	– A 1-minute series of long blasts 5 minutes prior to blast signal.
BLAST SIGNAL	– A series of short blasts 1 minute prior to the shot.
ALL CLEAR SIGNAL	– A prolonged blast following the inspection of the blast area.

Firing a Blast (1926.909)

Before a blast is fired, a loud warning signal is given by the blaster in charge, who has made certain that all surplus explosives are in a safe place and all employees, vehicles, and equipment are at a safe distance, or under sufficient cover. Flagmen are to be safely stationed on highways which pass through the danger zone, so as to stop traffic during blasting operations. It is the duty of the blaster to fix the time of blasting. Before firing an underground blast, warning must be given, and all possible entries into the blasting area and any entrances to any working place where a drift, raise, or other opening is about to hole through, must be carefully guarded. The blaster shall make sure that all employees are out of the blast area before firing a blast. Immediately after the blast has been fired, the firing line is to be disconnected from the blasting machine, or where power switches are used, they are to be locked open or in the off position.

Sufficient time must be allowed, not less than 15 minutes in tunnels, for the smoke and fumes to leave the blasted area before returning to the shot. Before employees are allowed to return to the operation, an inspection of the area and the surrounding rubble must be made by the blaster to determine if all charges have been exploded, and an inspection must be made in the tunnels, after the muck pile has been wetted down.

Handling Misfires (1926.911)

If a misfire is found, the blaster must provide proper safeguards for excluding all employees from the danger zone. No other work shall be done except that necessary to remove the hazard of the misfire, and only those employees necessary to do the work shall remain in the danger zone. No attempts are to be made to extract explosives from any charged or misfired hole; a new primer shall be put in and the hole reblasted. If refiring of the misfired hole presents a hazard, the explosives may be removed by washing out with water or, where the misfire is under water, blown out with air. If there are any misfires while using a cap and fuse, all employees must remain away from the charge for at least 1 hour. Misfires shall be handled under the direction of the person in charge of the blasting. All wires are to be carefully traced and a search made for unexploded charges. No drilling, digging, or picking is permitted until all missed holes have been detonated or the authorized representative has approved that work can proceed.

General Guidelines

Empty boxes, paper, and fiber packing materials which have previously contained high explosives, are not to be used again for any purpose; they are to be destroyed by burning at an approved location. Explosives, blasting agents, and blasting supplies that are obviously deteriorated or damaged are not to be used. The use of black powder is prohibited. All loading and firing shall be directed and supervised by competent persons who are thoroughly experienced in this field. Buildings used for the mixing of blasting agents must conform to the requirements of this section. Buildings are to be constructed of noncombustible construction, or sheet metal on wood studs. Floors in a mixing plant must be concrete or of other nonabsorbent materials. All fuel oil storage facilities must be separated from the mixing plant and located in such a manner, that in case of tank rupture, the oil will drain away from the mixing plant building. The building is to be well ventilated. Heating units which do not depend on combustion processes, when properly designed and located, may be used in the building. All direct sources of heat must be provided exclusively from units located outside the mixing building. All internal-combustion engines used for electric power generation are to be located outside the mixing plant building, or must be properly ventilated and isolated by a firewall. The exhaust systems on all such engines are to be located so that any spark emission will not be a hazard to any materials in, or adjacent to the plant.

In summary, only authorized and qualified individuals are permitted to handle and use explosives. Smoking, firearms, matches, open flame lamps, other fires, flames, heat producing devices, and sparks are prohibited in or near explosive magazines. Explosives not being used must be kept in a locked magazine. The blasters must maintain an up-to-date record of explosives, blasting agents, and blasting supplies used and stored on the jobsite. Appropriate authorities must be notified of any loss, theft, or unauthorized entry into a magazine.

EYE AND FACE PROTECTION (1926.102)

If a potential risk of eye or face injury exists from machines or operations, appropriate eye and face protection must be provided. All eye and face protection must meet the requirements of ANSI Z87.1 – 1968, Practices for Occupational and Industrial Eye and Face Protection. If a worker has to wear prescription lenses, the worker needs to use safety approved prescription glasses or goggles which cover, or are incorporated within them, the prescriptive lenses.

For selection of appropriate protective eye and face wear consult Table E-1 in 29 CFR 1926.102. Eye and face protection should provide the worker with

    1.   Adequate protection against the hazard.

    2.   Proper fit and comfort.

    3.   Durability.

    4.   The capability of being disinfected and cleaned.

Welding workers will need to be furnished with the proper shade of filter lenses for a specific welding process; consult Table E-2 in 29 CFR 1926.102 for the filter lens selection. When using laser safety goggles for eye protection from laser beams, consult Table E-3 in 29 CFF 1926,102 for selection of laser protective eyewear (see Figure 7-43).

Figure 7-43. Worker using a chop saw with eye and hearing protection

FALL PROTECTION (1926.500 - 503)

Scope, Application, and Definitions Applicable to this Subpart (1926.500)

Falls on construction sites are the leading cause of death to construction workers; thus the need for fall protection in construction workplaces. Fall protection is not required when workers are making an inspection, investigation, or assessment of workplace conditions prior to the actual start of construction work, or after all construction work has been completed. Requirements relating to fall protection, as described in this section, does not apply to scaffolds, cranes and derricks, steel erection, ladders and stairways, and tunneling operations, or to electrical power transmission and distribution construction, each of which has its own requirements. The major components of fall protection, described herein, are for installation, construction, and the proper use of body belts, lanyards, and lifelines, and the requirements for the training of fall protection. Some general do's and don'ts for climbing and working at heights are found in Table 7-7 and Table 7-8.

Duty to have Fall Protection (1926.501)

It is the employers responsibility to determine if the walking/working surfaces on which its employees are to work have the strength and structural integrity to support employees safely. Employees are allowed to work on those surfaces only when the surfaces have the requisite strength and structural integrity. Any time a worker is on a walking/working surface (horizontal and vertical surface), or constructing a leading edge with an unprotected side or edge which is 6 feet or more above a lower level, the worker must be protected from falling by using guardrail systems, safety net systems, or personal fall arrest systems. If the employer can demonstrate that it is not feasible or creates a greater hazard to use these systems, the employer shall develop and implement a fall protection plan (consult Appendix E in Subpart M of 29 CFR 1926).

Table 7- 6

Do's for Working at Heights

- Do Close and Latch All Hatches, Security Gates, and Hinged Walkways, to Seal Openings Where There Is Fall Potential.
- Do Barricade or Fence Around Chimneys or Stacks When Working on These Structures.
- Do Tie Off to a Secure Anchor. Ask Yourself if You Would Hang Your New Pickup Truck From This Anchor Point.
- Do Inspect All Fall Protection Equipment Prior to Use.
- Do Wear All Other Types of Personal Protective Equipment Such As Hardhats, Gloves, Hearing Protection, Rain Suits, Respirators, or Protective Eyewear.
- Do Maintain Three Points of Contact at All Times While Climbing (Two Hand and One Foot, or Two Feet and One Hand).
- Do Work Cautiously and Slowly.
- Do Follow All Safety and Health Procedures for Working at Heights.
- Do Wear a Pair of Leather Gloves While Climbing.

Table 7-7

Don'ts for Working at Heights

- Don't Climb Without Fall Protection.
- Don't Overlook Potential Hazards.
- Don't Climb or Work at Heights During Adverse Weather.
- Don't Assume Someone Else Has Assured Your Safety.
- Don't Wear Jewelry Which Could Catch On Other Objects.
- Don't Climb or Work at Heights if You Evaluate the Conditions as Unsafe.
- Don't Use Any Fall Protection Which Is Worn or Has Not Been Inspected.
- Don't Walk on Roofs Unless You Can Verify They Are Strong Enough to Support You.
- Don't Work at Heights Where There Is a Chance of Falling Without Tying Off.
- Don't Carry Tools or Other Objects, Which Could Slip or Fall, While Tucked Into Your Safety Belt or Harness.
- Don't Hoist Heavy Loads to Your Work Area Without Being Tied Off.
- Don't Climb or Work on Faulty Built Scaffolds.
- Don't Climb or Work on Ladders Which Are Not on Stable Ground, Chocked, or Secure at Base and Secured at the Top.
- Don't Climb a Stack, Chimney, Etc. Unless There Is a Securely Anchored Ladder or Rungs.
- Don't Use Safety Belts, Use Only Full Body Harnesses.
- Don't' Carry Any Loads or Objects in Your Hands as You Climb.

Each worker on a walking/working surface 6 feet or more above a lower level where leading edges are under construction, but where the worker is not engaged in the leading edge work, must be protected from falling by using a guardrail system, safety net system, or personal fall arrest system. If a guardrail system is chosen to provide the fall protection, and a controlled access zone has already been established for leading edge work, the control line may be used in lieu of a guardrail along the edge that parallels the leading edge.

Workers in a hoist area are to be protected from falling 6 feet or more to lower levels by using guardrail systems or personal fall arrest systems. If guardrail systems, (chains, gates, or guardrails), or portions thereof, are removed to facilitate the hoisting operation (e.g., during landing of materials), and the worker must lean through the access opening, or out over the edge of the access opening (to receive or guide equipment and materials, for example), that employee must be protected from fall hazards by using a personal fall arrest system (see Figure 7-44).

Figure 7-44. Worker with personal fall protection in a hoist area

Each worker on a walking/working surface is to be protected from tripping in, or stepping into or through holes (including skylights), and from objects falling through holes (including skylights), by using covers.

Each employee on the face of formwork or reinforcing steel shall be protected from falling 6 feet or more to lower levels by using personal fall arrest systems, safety net systems, or positioning device systems. Also, workers on ramps, runways, and other walkways are to be protected from falling 6 feet or more to lower levels by using guardrail systems; workers at the edge of an excavation 6 feet or more in depth are to be protected from falling by using guardrail systems, fences, or barricades; and, workers at the edge of a well, pit, shaft, and similar excavation which is 6 feet (1.8 m) or more in depth must be protected from falling by using guardrail systems, fences, barricades, or covers.

Workers less than 6 feet above dangerous equipment are to be protected from falling into or onto the dangerous equipment by the use of guardrail systems or equipment guards. When workers are 6 feet or more above dangerous equipment, they must be protected from fall hazards by using guardrail systems, personal fall arrest systems, or safety net systems.

When workers are performing overhand bricklaying and related work 6 feet or more above lower levels, they must be protected from falling by using guardrail systems, safety net systems, personal fall arrest systems, or they must work in a controlled access zone. If these workers must reach more than 10 inches below the level of the walking/working surface on which they are working, they are to be protected from falling by using a guardrail system, safety net system, or personal fall arrest system.

Workers engaged in roofing activities on low-slope roofs, with unprotected sides and edges 6 feet or more above lower levels, are to be protected from falling by using guardrail systems, safety net systems, or personal fall arrest systems; or, they are to be protected by using a combination of a warning line system and guardrail system, a warning line system and safety net system, a warning line system and personal fall arrest system, or a warning line system and safety monitoring system. When on a steep roof with unprotected sides and edges 6 feet (1.8 m) or more above lower levels, workers must be protected from falling by using guardrail systems with toeboards, safety net systems, or personal fall arrest systems.

During the erection of precast concrete members (including, but not limited to the erection of wall panels, columns, beams, and floor and roof "tees"), and related operations such as grouting of precast concrete members, those working 6 feet or more above lower levels are to be protected from falling by using guardrail systems, safety net systems, or personal fall arrest systems, unless the employer can demonstrate that it is not feasible, or creates a greater hazard to use these systems. The employer must then develop and implement a fall protection plan. *Note:* There is a presumption that it is feasible and will not create a greater hazard to implement at least one of the above-listed fall protection systems. Accordingly, the employer has the burden of establishing an appropriate fall protection plan.

When workers are engaged in residential construction activities 6 feet or more above lower levels, they must be protected by guardrail systems, safety net systems, or personal fall arrest systems unless the employer can demonstrate that it is not feasible, or creates a greater hazard to use these systems.

Each employee working on, at, above, or near wall openings (including those with chutes attached), where the outside bottom edge of the wall opening is 6 feet or more above lower levels, and the inside bottom edge of the wall opening is less than 39 inches above the walking/working surface, must be protected from falling by using a guardrail system, safety net system, or personal fall arrest system.

When workers are exposed to falling objects, the employer must have each employee wear a hard hat and implement one of the following measures:

1.   Erect toeboards, screens, or guardrail systems to prevent objects from falling from higher levels.

2.   Erect a canopy structure and keep potential fall objects far enough from the edge of the higher level so that those objects would not go over the edge if they were accidentally displaced (see Figure 7-45).

Figure 7-45. Canopy for overhead protection

3. Barricade the area to which objects could fall; prohibit employees from entering the barricaded area; and keep objects that may fall far enough away from the edge of a higher level so that those objects would not go over the edge if they were accidentally displaced.

Fall Protection Systems Criteria and Practices (1926.502)

Employers are to provide and install all fall protection systems before an employee begins the construction work that necessitates the fall protection. The fall protection system, selected by the employer, is to be the one which the employer deems is most appropriate for protecting the workforce.

Guardrail Systems

Guardrail systems are to be composed of the top rail, midrail, and toeboard. The top edge height of top rails, or equivalent guardrail system members, are to be 42 inches, plus or minus 3 inches, above the walking/working level. When conditions warrant, the height of the top edge may exceed the 45-inch height, provided the guardrail system meets all other criteria. *Note:* When employees are using stilts, the top edge height of the top rail, or equivalent member, shall be increased an amount equal to the height of the stilts. Guardrail systems are to be capable of withstanding, without failure, a force of at least 200 pounds applied within 2 inches of the top edge, in any outward or downward direction, at any point along the top edge. When the 200-pound test load is applied in a downward direction, the top edge of the guardrail must not deflect to a height less than 39 inches above the walking/working level. Guardrail system components are to be selected and constructed in accordance with the Appendix B to Subpart M.

Midrails, screens, mesh, intermediate vertical members, or equivalent intermediate structural members are to be installed between the top edge of the guardrail system and the walking/working surface, when there is no wall or parapet wall at least 21 inches high. Midrails, when used, are to be installed at a height midway between the top edge of the guardrail system and the walking/working level. Screens and mesh, when used, must extend from the top rail to the walking/working level, and along the entire opening between the top rail supports. Intermediate members (such as balusters), when used between posts, must not be more than 19 inches apart. Other structural members (such as additional midrails and architectural panels) shall be installed such that there are no openings in the guardrail system that are more than 19 inches (.5 m) wide. Midrails, screens, mesh, intermediate vertical members, solid panels, and equivalent structural members shall be capable of withstanding, without failure, a force of at least 150 pounds applied in any downward or outward direction at any point along the midrail or other member (see Figure 7-46).

Guardrail systems must be so surfaced as to prevent injury to an employee from punctures or lacerations, and to prevent snagging of clothing. The ends of all top rails and midrails are not to overhang the terminal posts, except where such overhang does not constitute a projection hazard. Steel banding and plastic banding are not to be used as top rails or midrails. Top rails and midrails shall be at least one-quarter inch nominal diameter or thickness to prevent cuts and lacerations. If wire rope is used for top rails, it must be flagged at not more than 6-foot intervals, with high-visibility material. When manila, plastic, or synthetic ropes are being used for top rails or midrails, they are to be inspected as frequently as necessary to ensure that they continue to meet the strength requirements.

When guardrail systems are used at hoisting areas, a chain, gate or removable guardrail section must be placed across the access opening between guardrail sections, when hoisting operations are not taking place. When guardrail systems are used at holes, they are to be

Figure 7-46. Guardrail system on formwork

erected on all unprotected sides or edges of the hole. When guardrail systems are used around holes, and the guardrails are also used for the passage of materials, the guardrails must not have more than two sides with removable guardrail sections which will be used for the passage of the materials. When the hole is not in use, it must be closed over with a cover, or a guardrail system must be provided along all unprotected sides or edges. When guardrail systems are used around holes which are used as points of access (such as ladderways), they must be provided with a gate, or be so offset that a person cannot walk directly into the hole. Guardrail systems used on ramps and runways are to be erected along each unprotected side or edge.

Safety Nets

Safety nets are to be installed as close as practicable under the walking/working surface on which employees are working, but in no case more than 30 feet below such level. When nets are used on bridges, the potential fall area from the walking/working surface to the net is to be unobstructed. Safety nets shall extend outward from the outermost projection of the work surface as follows:

Vertical distance from working level to horizontal plane of net	Minimum required horizontal distance of outer edge of net from the edge of the working surface
Up to 5 feet	8 feet
More than 5 feet up to 10 feet	10 feet
More than 10 feet	13 feet

Safety nets are to be installed with sufficient clearance under them to prevent contact with the surface or structures below, when subjected to an impact force equal to the drop test. The safety nets and safety net installations are to be drop-tested at the jobsite after the initial installation. When safety nets are relocated, they must be retested (drop-tested) before they are used again as a fall protection system. They must also be drop-tested after a major repair, and at 6-month intervals, if left in one place. The drop-test consists of a 400-pound bag of sand 30 + or – 2 inches in diameter, dropped into the net from the highest walking/working surface at which employees are exposed to fall hazards, but not from less than 42 inches above that level. When the employer can demonstrate that it is unreasonable to perform the drop-test required, then the employer (or a designated competent person) must certify that the net and net installation is in compliance the strength requirements, by preparing a certification record prior to the net being used as a fall protection system. The certification record must include an identification of the net and net installation for which the certification record is being prepared; the date that it was determined that the identified net and net installation were in compliance; and the signature of the person making the determination and certification. The most recent certification record for each net and net installation is to be available for inspection at the jobsite (see Figure 7-47).

Figure 7-47. Example of a perimeter safety net

Defective nets are not to be used. Safety nets must be inspected at least once a week for wear, damage, and other deterioration. Defective components shall be removed from service and safety nets shall also be inspected after any occurrence which could affect the integrity of the safety net system. Materials, scrap pieces, equipment, and tools which have fallen into the safety net are to be removed as soon as possible from the net, and at least before the next work shift.

The maximum size of each safety net mesh opening shall not exceed 36 square inches (230 cm), nor be longer than 6 inches on any side, and the opening, measured center-to-center of mesh ropes or webbing, shall not be longer than 6 inches. All mesh crossings are to be secured to prevent enlargement of the mesh opening. Each safety net (or section of it) must have a border rope, for webbing, with a minimum breaking strength of 5,000 pounds. Connections between safety net panels are to be as strong as integral net components, and must be spaced not more than 6 inches apart.

Personal Fall Arresting System

Effective January 1, 1998, body belts are not acceptable as part of a personal fall arrest system. Note: The use of a body belt in a positioning device system is acceptable. Connectors are to be drop forged, pressed or formed steel, or made of equivalent materials. Connectors must have a corrosion-resistant finish, and all surfaces and edges shall be smooth to prevent damage to interfacing parts of the system. Ropes and straps (webbing) used in lanyards, lifelines, and strength components of body belts and body harnesses are to be made from synthetic fibers (Figure 7-48).

Figure 7-48. Personal fall protection system

Dee-rings and snaphooks must have a minimum tensile strength of 5,000 pounds. Dee-rings and snaphooks are to be proof-tested to a minimum tensile load of 3,600 pounds without cracking, breaking, or taking permanent deformation. Snaphooks shall be sized to be compatible with the member to which they are connected in order to prevent unintentional disengagement of the snaphook by depression of the snaphook keeper by the connected member; or, they are to be a locking type snaphook designed and used to prevent disengagement of the snaphook by the contact of the snaphook keeper by the connected member (see Figure 7-49). Effective January 1, 1998, only locking type snaphooks are to be used. Unless the snaphook is a locking type and designed for the following connections, snaphooks are not to be engaged:

1. Directly to webbing, rope, or wire rope.

2. To each other.

3. To a dee-ring to which another snaphook or other connector is attached.

4. To a horizontal lifeline.

5. To any object which is incompatibly shaped or dimensioned in relation to the snaphook such that unintentional disengagement could occur by the connected object being able to depress the snaphook keeper and release itself.

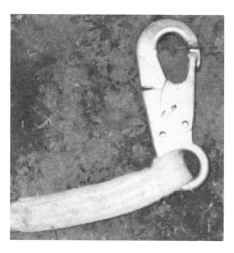

Figure 7-49. Double locking snaphook

On suspended scaffolds, or similar work platforms with horizontal lifelines which may become vertical lifelines, the devices used to connect to a horizontal lifeline are to be capable of locking in both directions on the lifeline. Horizontal lifelines must be designed, installed, and used, under the supervision of a qualified person, as part of a complete personal fall arrest system, which maintains a safety factor of at least two. Lanyards and vertical lifelines must have a minimum breaking strength of 5,000 pounds. Except when vertical lifelines are used, each employee is to be attached to a separate lifeline. During the construction of elevator shafts, two employees may be attached to the same lifeline in the hoistway, provided both employees are working atop a false car that is equipped with guardrails; the strength of the lifeline must be 10,000 pounds [5,000 pounds per employee attached]. Lifelines are to be protected against being cut or abraded.

Self-retracting lifelines and lanyards, which automatically limit free fall distance to 2 feet or less, are to be capable of sustaining a minimum tensile load of 3,000 pounds applied to the device with the lifeline or lanyard in the fully extended position. Self-retracting lifelines, ripstitch lanyards, and tearing and deforming lanyards, where the free fall distance is greater that 2 feet, must be capable of sustaining a minimum tensile load of 5,000 pounds applied to the device with the lifeline or lanyard in the fully extended position.

Anchorages used for attachment of personal fall arrest equipment are to be independent of any anchorage being used to support or suspend platforms, and must be capable of supporting at least 5,000 pounds per employee attached, or designed, installed, and used as follows:

1. As part of a complete personal fall arrest system which maintains a safety factor of at least two.

2. Under the supervision of a qualified person.

Any personal fall arrest system that is used to stop a fall must limit the maximum arresting force on an employee to 900 pounds when it is used with a body belt. When it is used with a body harness, limit the maximum arresting force on an employee to 1,800 pounds. The body harness must be rigged such that an employee can neither free fall more than 6 feet (1.8 m), nor contact any lower level, and it must bring an employee to a complete stop and limit the maximum deceleration distance an employee travels to 3.5 feet. The system must also have sufficient strength to withstand twice the potential impact energy of an employee free falling a distance of 6 feet, or the free fall distance permitted by the system, whichever is less. *Note:* If the personal fall arrest system meets the criteria and protocols contained in Appendix C to Subpart M, and if the system is being used by an employee having a combined person and tool weight of less than 310 pounds, the system will be considered to be in compliance. If the system is used by an employee having a combined tool and body weight of 310 pounds or more, then the employer must appropriately modify the criteria and protocols of Appendix C to provide proper protection for such heavier weights or the system will not be deemed to be in compliance

Personal fall arrest systems are to be inspected prior to each use, for wear, damage, and other defects. The attachment point of the body harness is to be located in the center of the wearer's back and near shoulder level, or above the wearer's head. Full body harnesses and components are to be used only for employee protection (as part of a personal fall arrest system or positioning device system), and not to hoist materials. Personal fall arrest systems and components subjected to impact loading must be removed immediately from service and must not be used again for employee protection until inspected and determined, by a competent person, to be undamaged and suitable for reuse. Deteriorated and defective components are to be removed from service.

Personal fall arrest systems must not be attached to guardrail systems, nor shall they be attached to hoists. When a personal fall arrest system is used at hoist areas, it is to be rigged to allow the movement of the employee only as far as the edge of the walking/working surface. In the event of a fall, the employer is to provide for prompt rescue of employees or assure that employees are able to rescue themselves.

Positioning device systems, and their use, must conform to the following: they are to be rigged such that an employee cannot free fall more than 2 feet and they are to be secured to an anchorage capable of supporting at least twice the potential impact load of an employee's fall or 3,000 pounds, whichever is greater. The connectors, dee-rings, snaphooks, inspection, and synthetic webbing have the same requirements as did the full body harness.

Warning Lines

Warning lines are to be erected around all sides of the roof work area. When mechanical equipment is not being used, the warning line must be erected not less than 6 feet from the roof edge, but when mechanical equipment is being used, the warning line is to be erected not less than 6 feet from the roof edge which is parallel to the direction of mechanical equipment operation, and not less than 10 feet from the roof edge which is perpendicular to the direction of mechanical equipment operation.

Points of access, materials handling areas, storage areas, and hoisting areas shall be connected to the work area by an access path formed by two warning lines. When the path to a point of access is not in use, a rope, wire, chain, or other barricade, equivalent in strength and height to the warning line, must be placed across the path at the point where the path intersects the warning line erected around the work area, or the path must be offset such that a person cannot walk directly into the work area.

Warning lines consisting of ropes, wires, chains, and supporting stanchions are to be flagged at not more than 6-foot intervals with high-visibility material and are to be rigged and supported in such a way that its lowest point (including sag) is no less than 34 inches from the walking/working surface; its highest point is to be no more than 39 inches from the walking/working surface.

Once erected with the rope, wire, or chain attached, stanchions must be capable of resisting, without tipping over, a force of at least 16 pounds applied horizontally against the stanchion. It must be 30 inches above the walking/working surface, perpendicular to the warning line, and in the direction of the floor, roof, or platform edge. The rope, wire, or chain must be a minimum tensile strength of 500 pounds, and after being attached to the stanchions, must be capable of supporting, without breaking, the previously described loads. The line shall be attached at each stanchion in such a way that pulling on one section of the line between stanchions will not result in slack being taken up in adjacent sections before the stanchion tips over.

No worker is allowed in the area between a roof edge and a warning line unless the employee is performing roofing work in that area. Mechanical equipment on roofs is to be used or stored only in areas where employees are protected by a warning line system, guardrail system, or personal fall arrest system.

These controlled access zones are to be areas where leading edge and other operations are taking place and the controlled access zone must be defined by a control line, or by any other means that restricts access. When control lines are used, they are to be erected not less than 6 feet, nor more than 25 feet from the unprotected or leading edge, except when erecting precast concrete members. When erecting precast concrete members, the control line is to be erected not less than 6 feet, more than 60 feet, or half the length of the member being erected, whichever is less, from the leading edge. The control line must extend along the entire length of the unprotected or leading edge, and must be approximately parallel to the unprotected or leading edge. The control line is to be connected, on each side, to a guardrail system or wall. When used to control access to areas where overhand bricklaying and related work is taking place,

1. The controlled access zone is to be defined by a control line erected not less than 10 feet, nor more than 15 feet from the working edge.

2. The control line must extend for a distance sufficient for the controlled access zone to enclose all employees performing overhand bricklaying and related work at the working edge, and is to be approximately parallel to the working edge.

3. Additional control lines are to be erected at each end to enclose the controlled access zone.

4. Only employees engaged in overhand bricklaying or related work are permitted in the controlled access zone.

Control lines consist of ropes, wires, tapes, or equivalent materials, and supporting stanchions for overhand bricklaying and are to be flagged, or otherwise clearly marked, at not more than 6-foot intervals with high-visibility material, and are to be rigged and supported in such a way that its lowest point (including sag) is not less than 39 inches from the walking/ working surface. Its highest point shall not be more than 45 inches (50 inches when overhand bricklaying operations are being performed) from the walking/working surface. Each line must have a minimum breaking strength of 200 pounds (see Figure 7-50). On floors and roofs where guardrail systems are not in place prior to the beginning of overhand bricklaying operations, controlled access zones are to be enlarged, as necessary, to enclose all points of access, material handling areas, and storage areas. On floors and roofs where guardrail systems are in place, but need to be removed to allow overhand bricklaying work or leading edge work to take place, only that portion of the guardrail necessary to accomplish that day's work is to be removed.

Safety Monitoring System

When an employer designates a competent person to monitor the safety of other employees, that person is responsible for monitoring and recognizing fall hazards, and warning the workers when it appears that they are unaware of a fall hazard, or are acting in an unsafe

Figure 7-50. Control lines on the perimeter of an open floor

manner. The person monitoring the other workers is to be on the same walking/working surface as the employees being monitored; is to be within visual sighting distance of the employees being monitored; is to be close enough to communicate orally with the employees; and must not have other responsibilities which could take the monitor's attention from the monitoring function.

Mechanical equipment is not to be used or stored in areas where safety monitoring systems are being used to monitor employees engaged in roofing operations on low-slope roofs. No employee, other than an employee engaged in roofing work (on low-sloped roofs), or an employee covered by a fall protection plan, is allowed in an area where an employee is being protected by a safety monitoring system.

Each employee working in a controlled access zone must comply immediately with any fall hazard warnings from safety monitors. See Table 7-9 which provides guidance regarding which fall protection is appropriate for different types of construction work.

<u>Covers</u>

Covers are to be used for holes in floors, roofs, and other walking/working surfaces. Covers located in roadways and vehicular aisles shall be capable of supporting, without failure, at least twice the maximum axle load of the largest vehicle expected to cross over the cover. All other covers are to be capable of supporting, without failure, at least twice the weight of the employees, equipment, and materials that may be imposed on the cover at any one time. All covers, when installed, must be secured so as to prevent accidental displacement by the wind, equipment, or employees (see Figure 7-51). Covers must be color coded or marked with the word "HOLE" or "COVER" to provide warning of the hazard. *Note:* This does not apply to cast iron manhole covers, or steel grates used on streets or roadways.

<u>Overhead Protection</u>

Protection from falling objects can mitigate when toeboards are used Toeboards are to be erected along the edge of an overhead walking/working surface for a distance sufficient to protect employees below. Toeboards shall be capable of withstanding, without failure, a force of at least 50 pounds applied in any downward or outward direction at any point along

Table 7-9
Fall Protection Safeguards

Types	Guardrails	Safety Nets	Personal Fall Arrest Systems	Other
Open Sided Floors	X	X	X	
Leading Edge	X	X	X	Fall Protection Plan (FPP)
Hoist Areas	X	X	X	
Holes (Roofs, Floors)	X	X	X	Secure Covers Marked Danger
Formwork/Rebar		X	X	> 2ft. Positioning Device
Ramps, Runways, Walkways	X			
Wells, Pits, Shafts	X			Fences, Covers, Barricades
Above Dangerous Equipment	X	X	X	Equipment Guards
Overhand Bricklaying	X	X	X	Controlled Access Zone
Roofing				
Low Pitch	X	X	X	Warning Lines 6ft from edge, guardrails, safety nets, PFAS, Safety Monitors (most common)
Steep Pitch	X	X	X	
Precast	X	X	X	FPP
Residential (24 feet)	X	X	X	Slope < 8 in. 12 in. Safety Monitor, FPP
Wall Openings <39 in. High, > 18 in. Wide	X	X	X	

the toeboard. The minimum vertical height for a toeboard is 3 1/2 inches from the top edge to the level of the walking/working surface, and there must not be more than 1/4 inch clearance above the walking/working surface. Toeboards are to be solid, or have openings not over 1 inch at the greatest dimension. Where tools, equipment, or materials are piled higher than the top edge of a toeboard, paneling or screening is to be erected from the walking/working surface or toeboard and to the top of a guardrail system's top rail or midrail, for a distance sufficient to protect workers below.

Guardrail systems, when used for falling object protection, must have all openings small enough to prevent passage of potential falling objects. During the performance of overhand bricklaying and related work, no materials or equipment, except masonry and mortar, are to be stored within 4 feet of the working edge. In order to keep work areas clear, excess mortar, broken or scattered masonry units, and all other materials and debris are to be removed, at regular intervals, from the work areas.

During the performance of roofing work, materials and equipment are not to be stored within 6 feet of a roof edge unless guardrails are erected at the edge, and materials which are piled, grouped, or stacked near a roof edge are stable and self-supporting.

Canopies, when used as falling object protection, are to be strong enough to prevent collapse and prevent penetration by any objects which may fall onto the canopy.

Figure 7-51. Covered and guarded floor opening (Note: although it cannot be seen, "Hole" was painted on this cover.)

Fall Protection Plan

Employers engaged in leading edge work, precast concrete erection work, or residential construction work, who can demonstrate that it is not feasible, or would create a greater hazard to use conventional fall protection equipment, can elect to develop and implement an alternate fall protection plan, but, the fall protection plan must conform to the following:

1. The fall protection plan is to be prepared by a qualified person and developed specifically for the site where the leading edge work, precast concrete work, or residential construction work is being performed; and, the plan must be maintained up to date.

2. Any changes to the fall protection plan is to be approved by a qualified person.

3. A copy of the fall protection plan, with all approved changes, are to be maintained at the job site.

4. The implementation of the fall protection plan is to be under the supervision of a competent person.

5. The fall protection plan must document the reasons why the use of conventional fall protection systems (guardrail systems, personal fall arrest systems, or safety nets systems) are not feasible, or why their use would create a greater hazard.

6. The fall protection plan must include a written discussion of other measures that will be taken to reduce or eliminate the fall hazard for workers who cannot be provided with protection from the conventional fall protection systems. For example, the employer must discuss the extent to which scaffolds, ladders, or vehicle-mounted work platforms can be used to provide a safer working surface and thereby reduce the hazard of falling.

7. The fall protection plan must identify each location where conventional fall protection methods cannot be used. These locations shall then be classified as controlled access zones.

8. Where no other alternative measure has been implemented, the employer must implement a safety monitoring system.

9   The fall protection plan must include a statement which provides the name or other method of identification for each employee who is designated to work in controlled access zones. No other employees may enter the controlled access zones.

10. In the event an employee falls, or some other related, serious incident occurs, (e.g., a near miss), the employer must investigate the circumstances of the fall, or other incident, to determine if the fall protection plan needs to be changed (e.g., new practices, procedures, or training), and shall implement those changes to prevent similar types of falls or incidents.

Training Requirements (1926.503)

Employers are required to provide a training program for each employee who might be exposed to fall hazards. The program must enable each worker to recognize the hazards of falling, and be trained in the procedures to be followed in order to minimize these hazards. Each employee must be trained, as necessary, by a competent person qualified in the following areas:

1. The nature of fall hazards in the work area.

2. The correct procedures for erecting, maintaining, disassembling, and inspecting the fall protection systems to be used.

3. The use and operation of guardrail systems, personal fall arrest systems, safety net systems, warning line systems, safety monitoring systems, controlled access zones, and other protection to be used.

4. The role of each employee in the safety monitoring system when this system is used.

5. The limitations on the use of mechanical equipment during the performance of roofing work on low-sloped roofs.

6. The correct procedures for the handling and storage of equipment and materials, and the erection of overhead protection.

7.   The role of employees in fall protection plans.

8.   The standards contained in this subpart.

The employer must prepare a written certification record. The written certification record is to contain the name or other identity of the employee trained, the date(s) of the training, and the signature of the person who conducted the training, or the signature of the employer. If the employer relies on training conducted by another employer, or training completed prior to the effective date of this section, the certification record must indicate the date the employer determined the prior training was adequate, rather than the date of actual training. The latest training certification is to be maintained.

When an employer has reason to believe that an employee, who has already been trained, does not have the understanding or required skills, the employer shall retrain that employee. Circumstances where retraining is required include, but are not limited to, situations where

1.   Changes in the workplace render previous training obsolete.

2.   Changes in the types of fall protection systems, or equipment to be used, render previous training obsolete.

3.   Inadequacies in an employee's knowledge, or use of fall protection systems or equipment indicate that the employee has not retained the requisite understanding or skill.

FIRE PROTECTION AND PREVENTION (1926.150)

The possibility of fire on a construction site needs to be given careful consideration since the potential exists due to hot work, flammable and combustible materials, and the presence of ignition and fuel sources which are omnipresent. Efforts must be undertaken to prevent the occurrence of fires. See Table 7-10 for a summary of fire protection and prevention.

The employer must provide a fire protection program, as well as firefighting equipment, and it must be followed throughout all phases of the construction and demolition work. Access to all available firefighting equipment is to be maintained at all times. All firefighting equipment, provided by the employer, is to be conspicuously located. All firefighting equipment must be periodically inspected, and maintained in operating condition. Defective equipment is to be immediately replaced. Depending upon the project, the employer shall have a trained and equipped firefighting organization (Fire Brigade) to assure adequate protection to life.

A temporary or permanent water supply, of sufficient volume, duration, and pressure, which will properly operate the firefighting equipment, must be made available. If underground water mains are to be used for firefighting, they are to be installed, completed, and made available for use as soon as practicable.

A fire extinguisher, rated not less than 2A, is to be provided for each 3,000 square feet of the protected building area, or major fraction thereof. Travel distance from any point of the protected area to the nearest fire extinguisher, shall not exceed 100 feet (see Figure 7-52). Carbon tetrachloride and other toxic vaporizing liquid fire extinguishers are prohibited. Portable fire extinguishers are to be inspected periodically and maintained in accordance with Maintenance and Use of Portable Fire Extinguishers, NFPA No. 10A-1970.

Table 7-10

General Rules for Fire Protection and Prevention

Fire Protection

1. Access to all available firefighting equipment will be maintained at all times.

2. Firefighting equipment will be inspected periodically and maintained in operating condition. Defective or exhausted equipment must be replaced immediately.

3. All firefighting equipment will be conspicuously located at each jobsite.

4. Fire extinguishers, rated not less than 2A, will be provided for each 3,000 square feet of the protected work area. Travel distance from any point of the protected area to the nearest fire extinguisher must not exceed 100 feet. One 55-gallon open drum of water, with two fire pails, may be substituted for a fire extinguisher having a 2A rating.

5. Extinguishers and water drums exposed to freezing conditions must be protected from freezing.

6. Do not remove or tamper with fire extinguishers installed on equipment or vehicles, or in other locations, unless authorized to do so or in case of fire. If you use a fire extinguisher, be sure it is recharged or replaced with another fully charged extinguisher.

TYPES OF FIRES

- Class A (wood, paper, trash) - use water or foam extinguisher.
- Class B (flammable liquids, gas, oil, paints, grease) - use foam, CO_2, or dry chemical extinguisher.
- Class C (electrical) - use CO_2 or dry chemical extinguisher.
- Class D (combustible metals) - use dry powder extinguisher only.

Fire Prevention

1. Internal combustion engine powered equipment must be located so that exhausts are away from combustible materials.

2. Smoking is prohibited at, or in the vicinity of operations which constitute a fire hazard. Such operations must be conspicuously posted: "No Smoking or Open Flame."

3. Portable battery powered lighting equipment must be approved for the type of hazardous locations encountered.

4. Combustible materials must be piled no higher than 20 feet. Depending on the stability of the material being piled, this height may be reduced.

5. Keep driveways between and around combustible storage piles at least 15 feet wide and free from accumulation of rubbish, equipment, or other materials.

6. Portable fire extinguishing equipment, suitable for anticipated fire hazards on the jobsite, must be provided at convenient, conspicuously accessible locations.

7. Fire fighting equipment must be kept free from obstacles, equipment, materials and debris that could delay emergency use of such equipment. Familiarize yourself with the location and use of the project's fire fighting equipment.

8. Discard and/or store all oily rags, waste, and similar combustible materials in metal containers on a daily basis.

9. Storage of flammable substances on equipment or vehicles is prohibited unless such unit has adequate storage area designed for such use.

Table 7-10

General Rules for Fire Protection and Prevention *(continued)*

Flammable and Combustible Liquids

1. Explosive liquids, such as gasoline, shall not be used as cleaning agents. Use only approved cleaning agents.

2. Store gasoline and similar combustible liquids in approved and labeled containers in well-ventilated areas free from heat sources.

3. Handling of all flammable liquids by hand containers must be in approved type safety containers with spring closing covers and flame arrestors.

4. Approved wooden or metal storage cabinets must be labeled in conspicuous lettering: "Flammable-Keep Fire Away."

5. Never store more than 60 gallons of flammable, or 120 gallons of combustible liquids in any one approved storage cabinet.

6. Storage of containers shall not exceed 1,100 gallons in any one pile or area. Separate piles or groups of containers by a 5-foot clearance. Never place a pile or group within 20 feet of a building. A 12-foot wide access way must be provided within 200 feet of each container pile to permit approach of fire control apparatus.

Figure 7-52. Typical fire extinguisher on a construction site

One 55-gallon open drum of water with two fire pails may be substituted for a fire extinguisher. Also, a 1/2-inch diameter garden-type hose line, not to exceed 100 feet in length, equipped with a nozzle with a capacity of 5 gallons per minute, may be substituted for a fire extinguisher, if the hose stream can be applied to all points in the area.

Each floor must have one or more fire extinguishers. On multi-story buildings, at least one fire extinguisher must be located adjacent to stairway. Extinguishers and water drums, subject to freezing, are to be protected from freezing.

A fire extinguisher, rated not less than 10B, is to be provided within 50 feet of wherever more than 5 gallons of flammable or combustible liquids, or 5 pounds of flammable gas are being used on the jobsite. This requirement does not apply to the integral fuel tanks of motor vehicles.

Fire hose and connections, with one hundred feet or less of 1 1/2-inch hose, and with a nozzle capable of discharging water at 25 gallons or more per minute, may be substituted, in a designated area, for a fire extinguisher rated not more than 2A. This may be done if the hose line can reach all points in the area. When fire hose connections are not compatible with local firefighting equipment, the contractor must provide adapters, or the equivalent, to permit connections.

If the facility being constructed includes the installation of an automatic sprinkler protection, and the installation closely follows the construction, and is placed in service as soon as applicable, laws permit following completion of each story, and this is considered fixed firefighting equipment.

In all structures where standpipes are required, or where standpipes exist in structures being altered, they shall be brought up as soon as applicable laws permit, and are maintained as construction progresses in such a manner that they are always ready for fire protection use. The standpipes are to be provided with Siamese fire department connections on the outside of the structure, at street level, and be conspicuously marked. There must be at least one standard hose outlet at each floor.

An alarm system (e.g., telephone system, siren, etc.) is to be established by the employer whereby employees on the site, and the local fire department, can be alerted of an emergency. The alarm code and reporting instructions must be conspicuously posted at the phones and employee entrances.

Fire walls and exit stairways, required for the completed buildings, are to be given construction priority. Fire doors, with automatic closing devices, are to be hung on openings, as soon as practicable. Fire cutoffs are to be retained in buildings undergoing alterations or demolition until operations necessitate their removal.

Fire Prevention – Storage (CFR 1926.151)

Internal combustion engine powered equipment shall be so located that the exhausts are well away from combustible materials. When the exhausts are piped to the outside of a building under construction, a clearance of at least 6 inches is to be maintained between such piping and combustible material.

Smoking is prohibited at, or in the vicinity of operations which constitute a fire hazard, and are to be conspicuously posted: "No Smoking or Open Flame."

Portable battery powered lighting equipment, used in connection with the storage, handling, or use of flammable gases or liquids, are to be of the type approved for the hazardous locations.

The nozzle of air, inert gas, and steam lines or hoses, when used in the cleaning or ventilation of tanks and vessels that contain hazardous concentrations of flammable gases or vapors, are to be bonded to the tank or vessel shell. Bonding devices are not to be attached or detached in hazardous concentrations of flammable gases or vapors.

No temporary building shall be erected where it will adversely affect any means of exit. Temporary buildings, when located within another building or structure, must be of either noncombustible construction, or combustible construction having a fire resistance of not less than one hour.

Temporary buildings, located other than inside another building and not used for storage, handling, or use of flammable or combustible liquids, flammable gases, explosives, blasting agents, or similar hazardous occupancies, are to be located at a distance of not less than 10 feet from another building or structure. Groups of temporary buildings, not exceeding 2,000 square feet in aggregate, are, for the purposes of this part, to be considered a single temporary building.

Combustible materials are to be piled with due regard to the stability of piles, and in no case higher than 20 feet. Driveways between and around combustible storage piles are to be at least 15 feet wide and maintained free from the accumulation of rubbish, equipment, or other articles or materials. Driveways must be so spaced that a maximum grid system unit of 50 feet by 150 feet is produced. Entire storage sites are to be kept free from accumulation of unnecessary combustible materials. Weeds and grass are to be kept down, and a regular procedure must be provided for the periodic cleanup of the entire area. When there is a danger of an underground fire, that land is not to be used for combustible or flammable storage. The method used for piling is, wherever possible, to be solid and in orderly and regular piles. No combustible material is to be stored outdoors within 10 feet of a building or structure.

Indoor storage must not obstruct, or adversely affect the means of exit, and all materials are to be stored, handled, and piled with due regard to their fire characteristics. Noncompatible materials, which may create a fire hazard, are to be segregated by a barrier which has a fire resistance of at least one hour. Material shall be piled to minimize the spread of fire internally, and to permit convenient access for firefighting. Stable piling must be maintained at all times. Aisle space is to be maintained to safely accommodate the widest vehicle that may be used within the building for firefighting purposes. A clearance of at least 36 inches is to be maintained between the top level of the stored material and the sprinkler deflectors. Clearance shall be maintained around lights and heating units to prevent ignition of combustible materials. A clearance of 24 inches is to be maintained around the path of travel of fire doors, unless a barricade is provided, in which case no clearance is needed. Material must not be stored within 36 inches of a fire door opening.

Flammable and Combustible Liquids (1926.152)

When storing or handling flammable and combustible liquids, only approved containers and portable tanks are to be used (see Figure 7-53). Approved metal safety cans are to be used for the handling and use of flammable liquids in quantities greater than one gallon, except that this does not apply to those flammable liquid materials which are highly viscid (extremely hard to pour); they may be used and handled in original shipping containers. For quantities of one gallon or less, only the original container or approved metal safety cans can be used for storage, use, and handling of flammable liquids.

Flammable or combustible liquids are not to be stored in areas used for exits, stairways, or normally used for the safe passage of people. No more than 25 gallons of flammable or combustible liquids are to be stored in a room outside of an approved storage cabinet. Quantities of flammable and combustible liquid in excess of 25 gallons are to be stored in an acceptable or approved cabinet, and meet the following requirements.

Specially designed and constructed wooden storage cabinets, which meet unique criteria, can be used for flammable and combustible liquids. Approved metal storage cabinets are also acceptable. Cabinets are to be labeled in conspicuous lettering, "Flammable–Keep Fire Away." Not more than 60 gallons of flammable, or 120 gallons of combustible liquids are to be stored in any one storage cabinet. Not more than three such cabinets may be located in a single

Figure 7-53. Containers for flammable or combustible liquids

storage area. Quantities in excess of this must be stored in an inside storage room. Inside storage rooms are to be constructed to meet the required fire-resistive rating for their use. Such construction must comply with the test specifications set forth in Standard Methods of Fire Test of Building Construction and Material, NFPA 251-1969.

Where an automatic extinguishing system is provided, the system must be designed and installed in an approved manner. Openings to other rooms or buildings must be provided with noncombustible liquid-tight raised sills or ramps at least 4 inches in height, or the floor in the storage area must be at least 4 inches below the surrounding floor. Openings are to be provided with approved self-closing fire doors. The room is to be liquid-tight where the walls join the floor. A permissible alternate to the sill or ramp is an open-grated trench, inside of the room, which drains to a safe location. Where other portions of the building, or other buildings are exposed, windows are to be protected as set forth in the Standard for Fire Doors and Windows, NFPA No. 80-1970, for Class E or F openings. Wood of at least one inch nominal thickness may be used for shelving, racks, dunnage, scuffboards, floor overlay, and similar installations. Materials which will react with water and create a fire hazard are not to be stored in the same room with flammable or combustible liquids.

Electrical wiring and equipment, located in inside storage rooms, are to be approved for Class I, Division 1, Hazardous Locations. Every inside storage room is to be provided with either a gravity or a mechanical exhausting system. Such system shall commence not more than 12 inches above the floor, and be designed to provide for a complete change of air within the room at least 6 times per hour. If a mechanical exhausting system is used, it must be controlled by a switch located outside of the door. The ventilating equipment and any lighting fixtures are to be operated by the same switch. An electric pilot light shall be installed adjacent to the switch, if flammable liquids are dispensed within the room. Where gravity ventilation is provided, the fresh air intake, as well as the exhausting outlet from the room, must be on the exterior of the building in which the room is located.

In every inside storage room there shall be maintained one clear aisle at least 3 feet wide. Containers over 30 gallons capacity are not to be stacked one upon the other. Flammable and combustible liquids in excess of that permitted in inside storage rooms are to be stored outside of buildings.

The quantity of flammable or combustible liquids kept in the vicinity of spraying operations is to be the minimum required for operations, and should ordinarily not exceed a supply for one day, or one shift. Bulk storage of portable containers of flammable or combustible liquids are to be in a separate, constructed building detached from other important buildings, or cut off in a standard manner.

Storage of containers (not more than 60 gallons each) shall not exceed 1,100 gallons in any one pile or area. Piles or groups of containers are to be separated by a 5-foot clearance. Piles or groups of containers are not to be closer than 20 feet to a building. Within 200 feet of each pile of containers, there should be a 12-foot-wide access way to permit the approach of fire control apparatus. The storage area is to be graded in a manner as to divert possible spills away from buildings or other exposures, or is to be surrounded by a curb or earth dike at least 12 inches high. When curbs or dikes are used, provisions must be made for draining off accumulations of ground or rain water, or spills of flammable or combustible liquids. Drains shall terminate at a safe location, and must be accessible to the operation under fire conditions.

Portable tanks are not to be closer than 20 feet from any building. Two or more portable tanks, grouped together, having a combined capacity in excess of 2,200 gallons, are to be separated by a 5-foot-clear area. Individual portable tanks exceeding 1,100 gallons shall be separated by a 5-foot-clear area. Within 200 feet of each portable tank, there must be a 12-foot-wide access way to permit the approach of fire control apparatus.

Portable tanks, not exceeding 660 gallons, are to be provided with emergency venting and other devices, as required by Chapters III and IV of NFPA 30-1969, The Flammable and Combustible Liquids Code.

Portable tanks, in excess of 660 gallons, must have emergency venting and other devices, as required by Chapters II and III of The Flammable and Combustible Liquids Code, NFPA 30-1969. At least one portable fire extinguisher, having a rating of not less than 20-B units, is to be located outside of, but not more than 10 feet from, the door opening into any room used for storage of more than 60 gallons of flammable or combustible liquids, and not less than 25 feet, nor more than 75 feet, from any flammable liquid storage area located outside.

At least one portable fire extinguisher, having a rating of not less than 20-B:C units, is to be provided on all tank trucks or other vehicles used for transporting and/or dispensing flammable or combustible liquids.

Areas in which flammable or combustible liquids are transferred at one time, in quantities greater than 5 gallons from one tank or container to another tank or container, are to be separated from other operations by 25-feet distance, or by construction having a fire resistance of at least one hour. Drainage or other means are to be provided to control spills. Adequate natural or mechanical ventilation must be provided to maintain the concentration of flammable vapor at, or below 10 percent of the lower flammable limit.

Transfer of flammable liquids from one container to another is done only when containers are electrically interconnected (bonded). When flammable or combustible liquids are drawn from, or transferred into vessels, containers, or tanks within a building or outside. Only through a closed piping system, from safety cans, by means of a device drawing through the top, or from a container, or portable tanks, by gravity or pump, through an approved self-closing valve. Transferring by means of air pressure on the container or portable tanks is prohibited. Any dispensing units are to be protected against collision damage. Dispensing devices and nozzles for flammable liquids are to be of an approved type. The dispensing hose must be an approved type, as well as the dispensing nozzle must be an approved automatic-closing type without a latch-open device. Clearly identified, and easily accessible switch(es) are to be provided at a location remote from dispensing devices, in order to shut off the power to all dispensing devices, in the event of an emergency.

Heating equipment, of an approved type, may be installed in the lubrication or service area where there is no dispensing or transferring of flammable liquids, provided the bottom of the heating unit is at least 18 inches above the floor and is protected from physical damage. Heating equipment installed in lubrication or service areas, where flammable liquids are dispensed, is to be of an approved type for garages, and is to be installed at least 8 feet above the floor. No smoking or open flames are permitted in the areas used for fueling, servicing fuel systems for internal combustion engines, or receiving or dispensing flammable or combustible liquids, and conspicuous and legible signs, prohibiting smoking, are to be posted.

The motors of all equipment being fueled are to be shut off during the fueling operation. Each service or fueling area is to be provided with at least one fire extinguisher having a rating of not less than 20-B:C, and located so that an extinguisher will be within 75 feet of each pump, dispenser, underground fill pipe opening, and lubrication or service area. More information regarding specifics on flammable and combustible liquid storage tanks can be found in 29 CFR 1926.152. Also, further information on Fixed Extinguishing Systems (1926.156), Fixed Extinguishing Systems, Gaseous Agent (1926.157), and Fire Detection System (1926.158) in 29 CFR 1926.

FLAGPERSON (1926.201)

The flagperson has a high exposure potential for injury due to moving motor vehicles and equipment The flagperson must maintain constant vigil since traffic flow is always changing and the flagperson must never leave their work post till properly relieved. The flagperson is to use flags, at least 18 inches square, or sign paddles when hand signaling and must be outfitted with a red or orange reflectorized warning vest while flagging. The flagperson should be trained in directing traffic flow and how to place themselves in a safe position as well as have adequate communications with the work crew and other individuals conducting flagging task (see Figure 7-54).

Figure 7-54. A flagperson directing traffic flow

All required signs and symbols must be visible at all times when work is being done, and removed promptly when the hazard no longer exists. Signaling directions must conform to ANSI D6.1-1971, *Manual on Uniform Traffic Control Devices for Streets and Highways.*

FLOOR AND WALL OPENINGS (1926.501)

See 29 CFR 1926.501.

FOOD HANDLING (1926.51)

Food handling and food service must meet applicable laws, ordinances, or regulations, as well as sound hygiene principles.

FOOT PROTECTION (1926.96)

Although construction companies do not require safety-toed shoes to be worn by all workers, it would seem that the habit of wearing safety-toed shoes which meet the ANSI Standard Z41.1-1967 should be a mandatory part of the construction personal protective equipment program. The variety and comfort of these shoes today offset the previous complaints, and definitely foster good foot safety on construction sites. The use of safety-toed shoes should be in the same vane as mandatory hardhats and eye protection, when considering the risk reduction of potential foot injuries.

HAND PROTECTION

In regards to hand protection, a good pair of leather gloves is a must for construction. Leather gloves should be worn when climbing, handling materials, being around sharp or jagged materials, when needed as antivibration protection, and when near low voltage electrical circuits (dry leather gloves provide a degree of insulation from electrical shock). If construction workers do not remove their rings prior to work, the wearing of gloves helps prevent the rings from catching or snagging on objects which can cause cuts or contusion of the fingers. Construction workers' hands and fingers are in constant use and thus, a high level of exposure and risk of injury exist. Hand protection makes a lot of sense.

At times, special hand protection may be needed when exposure to chemical or high voltage electricity exists. This is when the proper selection of hand protection is so vital. The use of commercially generated glove selection guides, provided by manufacturers, need to be consulted to assure that workers are using the right glove for the job.

With the high number of hand and finger injuries, it is well worth requiring the use of hand protection in the form of gloves.

HAND AND POWER TOOLS/GUARDING (1926.300)

All hand and power tools, and similar equipment, whether furnished by the employer or the employee, are to be maintained in a safe condition. When power-operated tools are designed to accommodate guards, they must be equipped with such guards when in use. Any belts, gears, shafts, pulleys, sprockets, spindles, drums, fly wheels, chains, or other reciprocating, rotating, or moving parts of equipment are to be guarded, if such parts are exposed to contact by workers, or otherwise create a hazard. Guarding must meet the requirements as set forth in American National Standards Institute, B15.1-1953 (R1958), *Safety Code for Mechanical Power-Transmission Apparatus.*

One or more methods of machine guarding must be provided to protect the operator and other employees in the machine area from hazards such as those created by point of operation, in going nip points, rotating parts, flying chips, and sparks. Examples of guarding methods are: barrier guards, two-hand tripping devices, electronic safety devices, etc. Point of operation is the area on a machine where work is actually performed upon the material being processed. The point of operation of machines where operation exposes a worker to injury, is to be guarded. The guarding is to be in conformity with any appropriate standards therefore, or in the absence of applicable specific standards, is so designed and constructed as to prevent the operator from having any part of his body in the danger zone during the operating cycle.

Special handtools for placing and removing material are to be such as to permit easy handling of material without the operator placing a hand in the danger zone. Such tools are not to be used in lieu of other guarding required by this section, but can only be used to supplement protection provided. The following are some of the machines which usually require point of operation guarding: guillotine cutters, shears, alligator shears, power presses, milling machines, power saws, jointers, portable power tools, forming rolls, and calenders. Machines designed for a fixed location shall be securely anchored to prevent walking or moving.

When the periphery of the blades of a fan is less than 7 feet (2.128 m) above the floor or working level, the blades must be guarded. The guard must have openings no larger than 1/2 inch (1.27 cm).

Employees using hand and power tools, and exposed to the hazard of falling, flying, abrasive, and splashing objects, or exposed to harmful dusts, fumes, mists, vapors, or gases are to be provided with the particular personal protective equipment necessary to protect them from the hazard. All personal protective equipment shall meet the requirements and be maintained according to Subparts D and E of 29 CFR 1926.

All hand-held powered platen sanders, grinders with wheels 2-inch diameter or less, routers, planers, laminate trimmers, nibblers, shears, scroll saws, and jigsaws with blade shanks one-fourth of an inch wide or less, are to be equipped with only a positive "on-off" control. All hand-held powered drills, tappers, fastener drivers, horizontal, vertical, and angle grinders with wheels greater than 2 inches in diameter, disc sanders, belt sanders, reciprocating saws, saber saws, and other similar operating powered tools, are to be equipped with a momentary contact "on-off" control, and may have a lock-on control, provided that turnoff can be accomplished by a single motion of the same finger, or fingers that turn it on. All other hand-held powered tools, such as circular saws, chain saws, and percussion tools without positive accessory holding means, are to be equipped with a constant pressure switch that will shut off the power when the pressure is released. This paragraph does not apply to concrete vibrators, concrete breakers, powered tampers, jack hammers, rock drills, and similar hand-operated power tools.

HAND TOOLS (1926.301)

Tools are such a common part of construction work that it is difficult to remember that they may pose hazards. In the process of removing or avoiding the hazards, workers must learn to recognize the hazards associated with the different types of tools, and the safety precautions necessary to prevent injury from those hazards. Therefore, in an effort to minimize accidents resulting from the use of hand tools, certain precautions need to be taken, such as:

1. Do not use broken, defective, burned, or mushroomed tools. Report defective tools to your supervisor and turn tools in for replacement.

2. Always use the proper tool and equipment for any task you may be assigned to do. For example: do not use a wrench as a hammer, or a screwdriver as a chisel.

3. Do not leave tools on scaffolds, ladders, or any overhead working surfaces. Racks, bins, hooks, or other suitable storage space must be provided and arranged to permit convenient arrangement of tools.

4. Do not strike two hardened steel surfaces together (i.e., two hammers, or a hammer and hardened steel shafts, bearings, etc.).

5. Do not throw tools from one location to another, from one worker to another, or drop them to lower levels; this is prohibited. When necessary to pass tools or material under the above conditions, suitable containers and/or ropes must be used.

6. Wooden tool handles must be sound, smooth, and in good condition, and securely fastened to the tool (see Figure 7-55).

7. Sharp-edged or pointed tools should never be carried in employee's pockets.

8. Only non-sparking tools shall be used in locations where sources of ignition may cause a fire or explosion.

9. Tools requiring heat treating should be tempered, formed, dressed, and sharpened by workmen experienced in these operations. Wrenches, including adjustable, pipe, end, and socket wrenches are not to be used when jaws are sprung to the point that slippage occurs.

10. Any defective tool should be removed from service and tagged indicating it is not to be used.

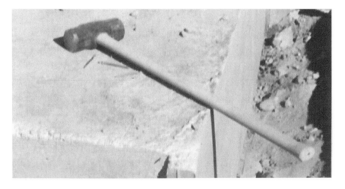

Figure 7-55. Wooden handled sledge hammer in good repair

HAZARD COMMUNICATIONS (1926.59)

OHSA has established regulations for the general industry, and the construction industry, called the Hazard Communication Standard (29 CFR 1926.59 & 29 CFR 1910.1200). These standards require that manufacturers of hazardous chemicals inform employers about the hazards of those chemicals. Also, it requires employers to inform employees of the identities, properties, characteristics, and hazards of chemicals they use, and the protective measures they can take to prevent adverse effects. The standard covers both physical hazards (e.g., flammability), and health hazards (e.g., lung damage, cancer). Knowledge acquired under the Hazard Communication Standard will help employers provide safer work places for workers, establish proper work practices, and help prevent chemical-related illnesses and injuries. Employers are required to do the following:

- The employer must develop a written hazard communication program.

- The employer must provide specific information and training to workers.

- All employers on a multiple employer site must provide information to each other so that all employees can be protected.

- The owner must provide information to contractors about hazardous materials on the job site.

The specific requirements for each of the four main provisions are summarized as follows.

Written Hazard Communication Program

1. List of hazardous chemicals on the jobsite.
2. The method the employer will use to inform employees of the hazards associated with non-routine tasks involving hazardous chemicals.
3. How the employer plans to provide employees of other companies on the jobsite with the material safety data sheets (MSDSs), such as making them available at a central location.
4. The method the employer will use to inform employees of other companies on the jobsite about their labeling system.
5. How the employer will inform workers about their labeling system.
6. How the employer plans to provide workers with MSDSs. More information on MSDs can be found in Chapter 9.
7. How the employer intends to train workers on hazardous chemicals.

Information Provided by the Employer

1. Hazardous chemicals used on the job.
2. How to recognize these hazardous chemicals.
3. How those chemicals might affect worker safety and health.
4. How workers can protect themselves from those chemicals (see Figure 7-56).

Training Provided by the Employer

1. Requirements of the OSHA Hazard Communication Standard.
2. Operations at the worksite where hazardous chemicals are present.
3. The location and availability of the
    *   Written Hazard Communication Program.
    *   List of all hazardous chemicals.
    *   MSDSs for all hazardous chemicals used on the jobsite.

Figure 7-56. Many different chemicals are found on construction sites

4. Methods and observations workers can use to detect the presence or release of hazardous chemicals in your work area {e.g., labels, color, form (solid, liquid, gas), and order}.

5. The physical and health hazards workers may be exposed to from the hazardous chemicals on the job.

6. Methods of protecting oneself, such as work practices, personal protective equipment, and emergency procedure.

7. Details of the hazardous communication program used by the employer.

8. Explanation of how workers can obtain and use hazard information.

Multiple Employer Sites

1. All employers on a multiple employer site must supply information to each other, so all employees will be protected.

2. The hazard communication program must specify how an employer will provide other employers with a copy of the MSDSs, or make it available at a central location in the work place, for each hazardous chemical the other employer(s)' employees may be exposed to while working.

3. The employers must provide the procedures for informing other employer(s) of any precautionary measures that need to be taken to protect employees during the worksite's normal working operating conditions, and of any foreseeable emergencies.

4. An employer must provide the mechanism to inform other employer(s) of his or her labeling system.

Owner's/Contractor's Responsibilities

In summary, employers are responsible to develop a hazard communication program and provide information to employees and other employer's employees and provide training to employees. All workers, as well as other employees on multiple employer worksites, must be provided with information regarding any hazardous chemicals to which workers might be exposed to at the employers' place of work.

Site owners, such as owners of chemical companies or paper pulp mills, must inform the contractor about the hazardous chemicals used at their plant which could result in exposure to the construction workers who will be working at their worksite. All employers in construction, general industry, etc. must comply with the hazardous communication regulations.

HAZARDOUS WASTE OPERATIONS (1926.65)

Hazard waste operations include the following operations, unless the employer can demonstrate that the operation does not involve employee exposure, or the reasonable possibility for employee exposure to safety or health hazards.

1. Clean-up operations required by a governmental body, whether Federal, state, local, or others involving hazardous substances, which are conducted at uncontrolled hazardous waste sites (including, but not limited to, the EPA's National Priority Site List (NPL), state priority site lists, sites recommended for the EPA NPL, and initial investigations of government identified sites which are conducted before the presence or absence of hazardous substances has been ascertained).

2. Corrective actions involving clean-up operations at sites covered by the Resource Conservation and Recovery Act of 1976 (RCRA) as amended in (42 U.S.C. 6901 *et seq.*).

3. Voluntary clean-up operations at sites recognized by Federal, state, local, or other governmental bodies as uncontrolled hazardous waste sites.

4. Operations involving hazardous wastes that are conducted at treatment, storage, and disposal (TSD) facilities regulated by 40 CFR parts 264 and 265 pursuant to RCRA; or by agencies under agreement with U.S.E.P.A. to implement RCRA regulations.

5. Emergency response operations for releases of, or substantial threats of releases of, hazardous substances without regard to the location of the hazard.

Employers must comply with paragraph (p) of 1926.65 unless conducting emergency response operations, then employers must comply with paragraph (q) of 1926.65. Hazardous waste operations pose not only chemical hazards, but most of the hazards which are usually faced by those working in the construction industry. Hazardous waste sites are really construction sites with the added hazards involved with the handling, removal, and disposal of hazardous chemicals.

Written Safety and Health Program

As part of paragraph (p), employers must develop and implement a written safety and health program for their employees involved in hazardous waste operations. The program is to be designed to identify, evaluate, and control safety and health hazards, and provide for emergency response for hazardous waste operations. The written safety and health program shall contain the following:

1. An organizational structure.

2. A comprehensive workplan.

3. A site-specific safety and health plan which need not repeat the employer's standard operating procedures.

4. The safety and health training program.

5. The medical surveillance program.

6. The employer's standard operating procedures for safety and health.

7. Any necessary interface between the general program and site specific activities.

An employer, who retains contractor or subcontractor services for work in hazardous waste operations, must inform those contractors, subcontractors, or their representatives of the site emergency response procedures, and any potential fire, explosion, health, safety, or other hazards of the hazardous waste operation that have been identified by the employer, including those identified in the employer's information program. The written safety and health program is to be made available to: any contractor, subcontractor, or their representative, who will be involved with the hazardous waste operation; to employees; to employee designated representatives; to OSHA personnel; and to personnel of other federal, state, or local agencies with regulatory authority over the site.

The organizational structure part of the program must establish the specific chain of command and specify the overall responsibilities of the supervisors and employees. It must include, at a minimum, the following elements: a general supervisor who has the responsibility and authority to direct all hazardous waste operations; a site safety and health supervisor who has the responsibility and authority to develop and implement the site safety and health plan and verify compliance; and all other personnel needed for the hazardous waste site operations and emergency response. They must be informed of their general functions, responsibilities, and the lines of authority, responsibility, and communication. The organizational structure is to be reviewed and updated, as necessary, to reflect the current status of the waste site operations.

The comprehensive workplan part of the program addresses the tasks and objectives of the site operations, and the logistics and resources required to reach those tasks and objectives. The comprehensive workplan addresses anticipated clean-up activities, as well as normal operating procedures, which need not repeat the employer's procedures available elsewhere. The comprehensive workplan defines work tasks and objectives and identifies the methods for accomplishing those tasks and objectives; establishes personnel requirements for implementing the plan; provides for the implementation of the training required; provides for the implementation of the required informational programs; and provides for the implementation of the medical surveillance program.

Site Safety and Health Plan

The site safety and health plan, which must be kept on site, addresses the safety and health hazards of each phase of the site operation, and includes the requirements and procedures for employee protection, as a minimum, and addresses the following:

1. A safety and health risk or hazard analysis for each site task and operation found in the workplan.

2. Employee training assignments to assure compliance.

3. Personal protective equipment to be used by employees for each of the site tasks and operations being conducted, as required by the personal protective equipment program.

4. Medical surveillance requirements in accordance with the program.

5. Frequency and types of air monitoring, personnel monitoring, and environmental sampling techniques and instrumentation to be used, including methods of maintenance and calibration of monitoring and sampling equipment to be used.

6. Site control measures used in accordance with the site control program.

7. Decontamination procedures.

8. An emergency response plan for safe and effective responses to emergencies, including the necessary PPE and other equipment.

9. Confined space entry procedures.

10. A spill containment program.

The site specific safety and health plan must provide for pre-entry briefings to be held prior to initiating any site activity, and at such other times as necessary to ensure that employees are apprised of the site safety and health plan, and that this plan is being followed. The information and data obtained from site characterization and analysis work are to be used to prepare and update the site safety and health plan.

Inspections are to be conducted by the site safety and health supervisor or, in the absence of that individual, another individual who is knowledgeable in occupational safety and health who acts on behalf of the employer, as necessary, to determine the effectiveness of the site safety and health plan. Any deficiencies in the effectiveness of the site safety and health plan shall be corrected by the employer.

Site Evaluation

Hazardous waste sites are to be evaluated to identify specific site hazards and to determine the appropriate safety and health control procedures needed to protect employees from the identified hazards.

A preliminary evaluation of a site's characteristics is to be performed prior to site entry, by a qualified person, in order to aid in the selection of appropriate employee protection methods prior to site entry. Immediately after initial site entry, a more detailed evaluation of the site's specific characteristics is to be performed by a qualified person in order to further identify existing site hazards, and to further aid in the selection of the appropriate engineering controls and personal protective equipment for the tasks to be performed.

All suspected conditions that may pose inhalation or skin absorption hazards that are immediately dangerous to life or health (IDLH), or other conditions that may cause death or serious harm, must be identified during the preliminary survey and evaluated during the detailed survey. Examples of such hazards include, but are not limited to, confined space entry, potentially explosive or flammable situations, visible vapor clouds, or areas where biological indicators, such as dead animals or vegetation, are located.

The following information, to the extent available, shall be obtained by the employer prior to allowing employees to enter a site:

1. Location and approximate size of the site.

2. Description of the response activity and/or the job task to be performed.

3. Duration of the planned employee activity.

4. Site topography and accessibility by air and roads.

5. Safety and health hazards expected at the site.

6. Pathways for hazardous substance dispersion.

7.   Present status and capabilities of emergency response teams that would provide assistance to hazardous waste clean-up site employees at the time of an emergency.

8,   Hazardous substances and health hazards involved or expected at the site, and their chemical and physical properties.

Personal protective equipment (PPE) is to be provided and used during initial site entry. Based upon the results of the preliminary site evaluation, an ensemble of PPE is to be selected and used during the initial site entry, which will provide protection to a level of exposure below permissible exposure limits, and published exposure levels for known or suspected hazardous substances and health hazards, and which will provide protection against other known and suspected hazards identified during the preliminary site evaluation. If there is no permissible exposure limit or published exposure level, the employer may use other published studies and information as a guide to appropriate personal protective equipment. If positive-pressure, self-contained breathing apparatus is not used as part of the entry ensemble, and if respiratory protection is warranted by the potential hazards identified during the preliminary site evaluation, an escape self-contained breathing apparatus of at least five minute's duration is to be carried by employees during the initial site entry. If the preliminary site evaluation does not produce sufficient information to identify the hazards or suspected hazards of the site, an ensemble providing protection equivalent to Level B PPE is to be provided as minimum protection, and direct reading instruments must be used, as appropriate, for identifying IDLH conditions. Once the hazards of the site have been identified, the appropriate PPE is selected and used.

When the site evaluation produces information that shows the potential for ionizing radiation or IDLH conditions, or when the site information is not sufficient to reasonably eliminate these possible conditions, then the following methods are to be used: monitoring with direct reading instruments for hazardous levels of ionizing radiation; monitoring the air with appropriate direct reading test equipment (i.e., combustible gas meters, detector tubes) for IDLH and other conditions that may cause death or serious harm (combustible or explosive atmospheres, oxygen deficiency, toxic substances); monitoring by visually observing for signs of actual or potential IDLH or other dangerous conditions; and monitoring by an ongoing air monitoring program. All these methods are to be implemented after site characterization has determined the site is safe for the start-up of operations.

Once the presence and concentrations of specific hazardous substances and health hazards have been established, the risks associated with these substances shall be identified. Employees who will be working on the site are to be informed of any risks that have been identified, and receive the required training, including hazard communications. The risks to consider include, but are not limited to, exposures exceeding the permissible exposure limits and published exposure levels, IDLH concentrations, potential skin absorption and irritation sources, potential eye irritation sources, explosion sensitivity and flammability ranges, and oxygen deficiency.

Any information that is available to the employer, concerning the chemical, physical, and toxicological properties of each substance known, or expected to be present on the jobsite, and is relevant to the duties an employee who is expected to perform at that jobsite, shall be made available to the affected employees prior to the commencement of their work activities. The employer may utilize information developed for the hazard communication standard for this purpose.

Site Control

Before clean-up work begins, appropriate site control procedures are to be implemented to control employee exposure to hazardous substances. Therefore, a site control program must be developed, for the protection of the employees, during the planning stages of a hazardous waste clean-up operation, and it is to be a part of the employer's site safety and health program. The program shall be modified, as necessary, as new information becomes available.

The site control program shall, as a minimum, include: a site map, site work zones, the use of a "buddy system," site communications (including alerting means for emergencies), standard operating procedures or safe work practices, and identification of the nearest medical assistance.

All employees working on site, such as, but not limited to, equipment operators, general laborers, and others exposed to hazardous substances, health hazards, or safety hazards, and their supervisors and management responsible for the site, shall receive training before they are permitted to engage in hazardous waste operations that could expose them to hazardous substances, safety, or health hazards.

Workers are to be trained to the level required by their job function and responsibility. The training must thoroughly cover the following:

1. Names of personnel and alternates responsible for site safety and health.

2. Safety, health, and other hazards present on the site.

3. Use of personal protective equipment.

4. Work practices by which the employee can minimize risks from hazards.

5. Safe use of engineering controls and equipment on the site.

6. Medical surveillance requirements, including recognition of symptoms and signs which might indicate overexposure to hazards.

7. The contents of the paragraphs of the site safety and health plan.

Training

General site workers (such as equipment operators, general laborers, and supervisory personnel) engaged in hazardous substance removal, or other activities which expose or potentially expose workers to hazardous substances and health hazards, must receive a minimum of 40 hours of instruction off the site, and a minimum of three days actual field experience under the direct supervision of a trained, experienced supervisor. Workers who are only on the site occasionally for a specific limited task (such as, but not limited to, groundwater monitoring, land surveying, or geo-physical surveying), and who are unlikely to be exposed over permissible and published exposure limits are to receive a minimum of 24 hours of instruction off the site, and the minimum of one day actual field experience under the direct supervision of a trained, experienced supervisor. Workers who are regularly on the site, and work in areas which have been monitored, and these areas have been fully characterized and indicate that exposures are under the permissible and published exposure limits and do not require the use of respirators, and that the characterization indicates that there are no health hazards, or the possibility of an emergency developing, must receive a minimum of 24 hours of instruction off the site and the minimum of one day actual field experience under the direct supervision of a trained, experienced supervisor. Workers with 24 hours of training, who are covered by this paragraph, who become general site workers, or who are required to wear respirators, must have the additional 16 hours and two days of training (see Figure 7-57).

On-site management and supervisors who are directly responsible for, or who supervise employees engaged in, hazardous waste operations must receive 40 hours initial training, and three days of supervised field experience. The training may be reduced to 24 hours and one day, and/or at least eight additional hours of specialized training, at the time of job assignment,

Figure 7-57. Trained workers doing hazardous waste work

on such topics as, but not limited to, the employer's safety and health program, the associated employee training program, the personal protective equipment program, the spill containment program, and the health hazard monitoring procedures and techniques.

Trainers are to be qualified to instruct employees in the subject matter that is being presented in training. Such trainers must have satisfactorily completed a training program for teaching the subjects they are expected to teach, or they must have the academic credentials and instructional experience necessary for teaching the subjects. Instructors are to demonstrate competent instructional skills and knowledge of the applicable subject matter.

Employees and supervisors who have received and successfully completed the training and have the field experience are to be certified by their instructor or the head instructor and trained supervisor as having successfully completed the necessary training. A written certificate is given to each person so certified. Any person who has not been so certified, or who does not meet the requirements, is prohibited from engaging in hazardous waste operations.

Employees who respond to hazardous emergency situations at hazardous waste clean-up sites, and may be exposed to hazardous substances, are to be trained how to respond to such expected emergencies.

Employees, managers, and supervisors, specified, are to receive eight hours of refresher training, annually, on specified items. Critiques may be used of incidents that have occurred in the past year; these can serve as training examples for related work and other relevant topics.

Employers, who can show by documentation or certification that an employee's work experience, and/or training, has resulted in training equivalent to that training required, are not required to provide the initial training requirements to these employees; the employer must provide a copy of the certification or documentation to the employee upon request. However, certified employees, or employees who have equivalent training, and who are new to a site, must receive the appropriate site specific training before site entry, and must have the appropriate supervised field experience at the new site. Equivalent training includes any academic training or the training that existing employees might have already received from actual hazardous waste site work experience.

Medical Surveillance

Employers engaged in hazardous waste operations must institute a medical surveillance program which includes employees who are, or may be exposed to hazardous substances or health hazards at, or above the permissible exposure limits, or if there is no permissible exposure limit above the published exposure levels for these substances, without regard to the use of respirators, for 30 days or more a year. This also includes all employees who wear a respirator for 30 days or more a year, or as required by 1926.103; all employees who are injured, become ill, or develop signs or symptoms due to possible overexposure which involves hazardous substances or health hazards from an emergency response or hazardous waste operation; and members of HAZMAT teams.

Medical examinations and consultations are to be made available by the employer to each employee, as follows: prior to an assignment; at least once every twelve months for each employee covered (unless the attending physician believes a longer interval, not greater than biennially, is appropriate); at the termination of employment; when an employee is reassigned to an area and the employee would not be covered if the employee has not had an examination within the last six months; as soon as possible after notification by an employee, that the employee has developed signs or symptoms indicating possible overexposure to hazardous substances or health hazards; when an employee has been injured, or exposed above the permissible exposure limits, or published exposure levels in an emergency situation; and at more frequent times, if the examining physician determines that an increased frequency of examination is medically necessary.

All employees are to be provided medical examinations, including; those who may have been injured; who received a health impairment; those who developed signs or symptoms which may have resulted from exposure to hazardous substances while working at an emergency incident; those who may have been exposed, during an emergency incident, to hazardous substances at concentrations above the permissible exposure limits or published exposure levels without the necessary personal protective equipment being used; as soon as possible following the emergency incident or development of signs or symptoms and at additional times, if the examining physician determines that follow-up examinations or consultations are medically necessary.

Medical examinations must include a medical and work history, or updated history if one is in the employee's file, and should have a special emphasis on the employee's symptoms related to the handling of hazardous substances and health hazards, and the employee's fitness for duty, including the ability to wear any required PPE under conditions (i.e., temperature extremes) that may be expected at the worksite. The content of the medical examinations or consultations made available to employees is to be determined by the attending physician. The guidelines in the *Occupational Safety and Health Guidance Manual for Hazardous Waste Site Activities* should be consulted.

All medical examinations and procedures are to be performed by, or under the supervision of a licensed physician, preferably one knowledgeable in occupational medicine, and are to be provided without cost to the employee, without loss of pay, and at a reasonable time and place. The employer shall provide one copy of hazardous waste standards, and its appendices, to the attending physician. In addition, the following must be supplied for each employee: a description of the employee's duties as they relate to the employee's exposures; the employee's exposure levels, or anticipated exposure levels; a description of any personal protective equipment used, or to be used; information from the employee's previous medical examinations which is not readily available to the examining physician; and information required by 1926.103.

The employer must obtain and furnish the employee with a copy of a written opinion from the attending physician which contains the following:

1. The physician's opinion as to whether the employee has any detected medical conditions which would place the employee at an increased health risk of material impairment when working at hazardous waste operations, when working in emergency response situations, or when using a respirator.

2.   The physician's recommended limitations upon the employee's assigned work.

3.   The results of the medical examination and tests, if requested by the employee.

4.   A statement that the employee has been informed by the physician of the results of the medical examination and any medical conditions which require further examination or treatment.

The written opinion obtained by the employer is not to reveal specific findings or diagnoses unrelated to occupational exposures. An accurate record of the medical surveillance is to be retained for the length of employment, plus 30 years. The required record must include at least the following information:

1.   The name and social security number of the employee.

2.   The physician's written opinions, recommended limitations, and results of examinations and tests.

3.   Any employee medical complaints related to exposure to hazardous substances.

4.   A copy of the information provided to the examining physician by the employer, with the exception of the standard and its appendices.

Hazard Controls

Engineering controls, work practices, personal protective equipment, or a combination of these, are to be implemented to protect employees from exposure to hazardous substances, and safety and health hazards. Engineering controls, when feasible, should include the use of pressurized cabs or control booths on equipment, and/or the use of remote-operated material handling equipment. Also, work practices that should be followed, when feasible, are those of removing all non-essential employees from potential exposure during the opening of drums and the wetting down dusty operations; locate these employees upwind of possible hazards.

Whenever engineering controls and work practices are not feasible, or not required, any reasonable combination of engineering controls, work practices, and PPE are to be used to reduce and maintain employee exposures to, or below the permissible exposure or dose limits for substances. The employer is not to use a schedule of employee rotation as a means of compliance with permissible exposure or dose limits, except when there is no other feasible way of complying with the airborne or dermal dose limits for ionizing radiation. The employer may use the published literature and MSDS as a guide in making a determination as to what level of protection the employer believes is appropriate for hazardous substances and health hazards, when there are no permissible or published exposure limits.

Personal Protective Equipment

Personal protective equipment (PPE) must be selected and used which will protect employees from the hazards and potential hazards that are identified during the site characterization and analysis. Personal protective equipment selection is to be based on an evaluation of the performance characteristics of the PPE relative to the requirements and limitations of the site, the task-specific conditions and duration, and the hazards and potential hazards identified at the site.

Positive pressure, self-contained breathing apparatus, or positive pressure, air-line respirators which are equipped with an escape air supply, are to be used when the chemical exposure levels present at the site create a substantial possibility of immediate death, immediate serious illness or injury, or impair the ability to escape.

Totally encapsulating chemical protective suits are to be used in conditions where skin absorption of a hazardous substance may result in a substantial possibility of immediate death, immediate serious illness or injury, or impair the ability to escape.

The level of protection provided by PPE selection is to be increased when additional information on site conditions indicates that increased protection is necessary in order to reduce employees exposure to below the permissible exposure limits, the published exposure levels for hazardous substances, and health hazards. The level of employee protection provided may be decreased when additional information or site conditions show that decreased protection will not result in hazardous exposure to employees.

A written personal protective equipment program must be established which is part of the employer's safety and health program and also a part of the site-specific safety and health plan. The PPE program must address the elements listed below:

1. PPE selection based upon site hazards.

2. PPE use and limitations of the equipment.

3. Work mission duration.

4. PPE maintenance and storage.

5. PPE decontamination and disposal.

6. PPE training and proper fitting.

7. PPE donning and doffing procedures.

8. PPE inspection procedures prior to, during, and after use.

9. Evaluation of the effectiveness of the PPE program.

10. Limitations during temperature extremes, heat stress, and other appropriate medical considerations.

Monitoring

If there are no permissible exposure limits listed for hazardous substances being used at the jobsite, monitoring is to be performed in order to assure the proper selection of engineering controls, work practices, and personal protective equipment so that employees are not exposed to levels which exceed the permissible exposure limits, or published exposure levels. Air monitoring is used to identify and quantify airborne levels of hazardous substances, and safety and health hazards, in order to determine the appropriate level of employee protection needed on site.

Upon initial entry, representative air monitoring is to be conducted to identify any IDLH condition, exposure over permissible exposure limits or published exposure levels, exposure over a radioactive material's dose limits, or other dangerous conditions such as the presence of flammable atmospheres or oxygen-deficient environments.

Periodic monitoring is to be conducted when the possibility of an IDLH condition or flammable atmosphere has developed, or when there is an indication that exposures may have risen over the permissible exposure limits or published exposure levels after the initial monitoring. These conditions may be: work beginning on a different portion of the site; contaminants, other than those previously identified, being handled; a different type of operation being initiated (e.g., drum opening, as opposed to exploratory well drilling); employees handling leaking drums or containers; or employees working in areas with obvious liquid contamination (e.g., a spill or lagoon).

Handling and Transporting Hazardous Materials

Hazardous substances and contaminated soils, liquids, and other residues are to be handled, transported, labeled, and disposed following accepted guidelines. Drums and containers used during the clean-up must meet the appropriate DOT, OSHA, and EPA regulations for the wastes that they contain. When practical, drums and containers are to be inspected and their integrity assured prior to being moved. Drums or containers that cannot be inspected before being moved, because of storage conditions (i.e., buried beneath the earth, stacked behind other drums, stacked several tiers high in a pile, etc.), are to be moved to an accessible location and inspected prior to further handling. Unlabeled drums and containers are considered to contain hazardous substances and handled accordingly until the contents are positively identified and labeled.

Site operations should be organized to minimize the amount of drum or container movement. Prior to movement of drums or containers, all employees exposed to the transfer operation are to be warned of the potential hazards associated with the contents of the drums or containers. The U.S. Department of Transportation's specified salvage drums or containers, and suitable quantities of proper absorbent, are to be kept available and used in areas where spills, leaks, or ruptures may occur. Where major spills may occur, a spill containment program, which is part of the employer's safety and health program, is to be implemented; it is to contain and isolate the entire volume of the hazardous substance being transferred. Drums and containers that cannot be moved without rupture, leakage, or spillage are to be emptied into a sound container, using a device classified for the material being transferred.

Where drums or containers are being opened and an airline respirator system is used, connections to the source of air supply are to be protected from contamination and the entire system must be protected from physical damage. Employees not actually involved in the opening of drums or containers are to be kept a safe distance. If employees must work near or adjacent to drums or containers being opened, a suitable shield, that does not interfere with the work operation, is to be placed between the employee and the drums or containers being opened; this will protect the employee in case of accidental explosion. Controls for drum or container opening equipment, monitoring equipment, and fire suppression equipment are to be located behind the explosion-resistant barrier.

Drums and containers are to be opened in such a manner that excess interior pressure will be safely relieved. If pressure cannot be relieved from a remote location, appropriate shielding must be placed between the employee and the drums or containers in order to reduce the risk of employee injury. Employees must not stand upon, or work from drums or containers.

Material handling equipment used to transfer drums and containers must be selected, positioned, and operated to minimize sources of ignition, related to the equipment, from igniting vapors released from ruptured drums or containers. Drums and containers containing radioactive wastes are not to be handled until such time as their hazard to employees is properly assessed.

As a minimum, the following special precautions are to be taken when drums and containers containing, or suspected of containing shock-sensitive wastes are handled. All nonessential employees are to be evacuated from the area of transfer. Material handling equipment must be provided with explosive containment devices, or protective shields, to protect equipment operators from exploding containers. An employee alarm system, capable of being perceived above surrounding light and noise conditions, is to be used to signal the commencement and completion of explosive waste handling activities. Continuous communications (i.e., portable radios, hand signals, telephones, as appropriate) are to be maintained between the employee-in-charge of the immediate handling area, both the site safety and health supervisors, and the command post, until such time as the handling operation is completed. Communication equipment, or methods that could cause shock sensitive materials to explode, shall not be used. Drums and containers under pressure, as evidenced by bulging or swelling, are not to be moved until such time as the cause for excess pressure is determined, and appropriate containment procedures have been implemented to protect employees from explosive relief of the drum. Drums and containers containing packaged laboratory wastes are to be considered to contain shock-sensitive or explosive materials until they have been characterized.

Lab Packs

Lab packs are to be opened only when necessary, and then only by an individual knowledgeable in the inspection, classification, and segregation of the containers within the pack, according to the hazards of the wastes. If crystalline material is noted on any container, the contents are to be handled as a shock-sensitive waste until the contents are identified. Sampling of containers and drums is to be done in accordance with a sampling procedure, which is part of the site safety and health plan.

Drum or Container Staging

Drums and containers are to be identified and classified prior to packaging for shipment. Drum or container staging areas are to be kept to the minimum number necessary to identify and classify materials safely and to prepare them for transport. Staging areas are to be provided with adequate access and egress routes.

Bulking of hazardous wastes is permitted only after a thorough characterization of the materials has been completed. Tanks and vaults containing hazardous substances are to be handled in a manner similar to that for drums and containers, taking into consideration the size of the tank or vault. Appropriate tank or vault entry procedures, as described in the employer's safety and health plan, are to be followed whenever employees must enter a tank or vault.

Decontamination

Procedures for all phases of decontamination are to be developed and implemented. A decontamination procedure is to be developed, communicated to employees, and implemented before any employees or equipment may enter areas on site where the potential for exposure to hazardous substances exists. All employees leaving a contaminated area are to be appropriately decontaminated and all contaminated clothing and equipment leaving a contaminated area must be appropriately disposed of, or decontaminated. Decontamination procedures are to be monitored by the site safety and health supervisor to determine their effectiveness.

Decontamination must be performed in geographical areas that will minimize exposure of uncontaminated employees or equipment to contaminated employees or equipment. All equipment and solvents used for decontamination are to be decontaminated or disposed of properly.

Protective clothing and equipment are to be decontaminated, cleaned, laundered, maintained, or replaced, as needed, to maintain their effectiveness. Employees, whose non-impermeable clothing becomes wetted with hazardous substances, must immediately remove that clothing and proceed to the shower. The clothing is to be disposed of, or decontaminated before it is removed from the work zone. Unauthorized employees do not remove protective clothing or equipment from change rooms. Commercial laundries or cleaning establishments that decontaminate the protective clothing or equipment are to be informed of the potentially harmful effects of exposures to hazardous substances.

Emergency Response Plan

To handle anticipated emergencies, and prior to the commencement of hazardous waste operations, an emergency response plan is to be developed and implemented by all employers. The plan must be in writing and be available for inspection and copying by employees, their representatives, OSHA personnel, and other governmental agencies with relevant responsibilities. Employers who will evacuate their employees from the danger area when an emergency occurs, and who do not permit any of their employees to assist in handling the emergency, if they provide an emergency action plan complying with 1926.35 are exempt from the other parts of this paragraph. The employer who develops an emergency response plan for emergencies must address, as a minimum, the following:

1. Pre-emergency planning.

2. Personnel roles, lines of authority, and communication.

3. Emergency recognition and prevention.

4. Safe distances and places of refuge.

5. Site security and control.

6. Evacuation routes and procedures.

7. Decontamination procedures which are not covered by the site safety and health plan.

8. Emergency medical treatment and first aid.

9. Emergency alerting and response procedures.

10. Critique of response and follow-up.

11. PPE and emergency equipment.

Emergency response plans must include site topography, layout, and prevailing weather conditions, and procedures for reporting incidents to local, state, and federal governmental agencies.

The emergency response plan is to be a separate section of the site safety and health plan, and shall be compatible and integrated with the disaster, fire, and/or emergency response plans of local, state, and federal agencies. The emergency response plan is to be rehearsed regularly as part of the overall training program for site operations and is to be reviewed periodically and as necessary, be amended to keep it current with new or changing site conditions or information. An employee alarm system is to be installed in accordance with 29 CFR 1926.159.

Sanitation

All toilets, potable water, showers, washing facilities, change rooms, temporary sleeping quarters, and food service facilities must follow existing construction regulations. When hazardous waste clean-up or removal operations commence on a site, and the duration of the work will require six months or greater time to complete, the employer must provide showers and change rooms for all employees who will be exposed to hazardous substances and health hazards. Showers and change rooms are to be located in areas where exposures are below the permissible exposure limits and published exposure levels. If this cannot be accomplished, then a ventilation system must be provided that will supply air that is below the permissible exposure limits and published exposure levels. Employers need to assure that employees shower at the end of their work shift and when leaving the hazardous waste site.

New Technologies

The employer must develop and implement procedures for the introduction of effective new technologies and equipment developed for the improved protection of employees working with hazardous waste clean-up operations. The same is to be implemented as part of the site safety and health program to assure that employee protection is being maintained. New technologies, equipment, or control measures available to the industry, such as the use of foams, absorbents, neutralizers, or other means used to suppress the level of air contaminates while excavating the site or for spill control, are to be evaluated by employers or their representatives. Such an evaluation is to be done to determine the effectiveness of the new methods, materials, or equipment before implementing their use on a large scale. Information and data from manufacturers or suppliers may be used as part of the employer's evaluation effort. Such evaluations shall be made available to OSHA upon request.

RCRA

Resource Conservation and Recovery Act of 1976 (RCRA). Employers conducting operations at treatment, storage, and disposal (TSD) facilities must develop and implement a written safety and health program for employees involved in hazardous waste operations that shall be available for inspection by employees, their representatives, and OSHA personnel. The program is to be designed to identify, evaluate, and control safety and health hazards at their facilities for the purpose of employee protection, to provide for emergency response, and to address, as appropriate, site analysis, engineering controls, maximum exposure limits, hazardous waste handling procedures, and uses of new technologies. The employer must implement a hazard communication program as part of the employer's safety and health program.

The employer must develop and implement a medical surveillance program, a decontamination procedure, a procedure for introducing new and innovative equipment into the workplace, a program for handling drums or containers, and a training program which is part of the employer's safety and health program for employees exposed to health hazards or hazardous substances at TSD operations to enable the employees to perform their assigned duties and functions in a safe and healthful manner so as not to endanger themselves or other employees. The initial training is for 24 hours and refresher training shall be for 8 hours annually. Employees who have received the initial training required by this paragraph are to be given a written certificate attesting that they have successfully completed the necessary training. Employers who can show by an employee's previous work experience and/or training, can be considered to meet the previous requirements of this paragraph.

Emergency Response

Those emergency response organizations, who have developed and implemented emergency response programs for handling releases of hazardous substances, must follow the previous guidelines.

To avoid duplications, emergency response organizations may use the local emergency response plan, the state emergency response plan, or both, as part of their emergency response plan. Those items of the emergency response plan that are being properly addressed by the SARA Title III plans may be substituted into their emergency plan, or otherwise kept together for the employer and employee's use. The senior emergency response official, responding to an emergency, becomes the individual in charge of a site-specific Incident Command System (ICS). All emergency responders and their communications are to be coordinated and controlled through the individual in charge of the ICS and assisted by the senior official present for each employer.

The individual in charge of the ICS must identify, to the extent possible, all hazardous substances or conditions present, and address, as appropriate, site analysis, use of engineering controls, maximum exposure limits, hazardous substance handling procedures, and any new technologies. Based on the hazardous substances and/or conditions present, the individual in charge of the ICS is to implement the appropriate emergency operations, and assure that the personal protective equipment worn is appropriate for the hazards to be encountered.

Employees engaged in emergency response who are exposed to hazardous substances which present an inhalation hazard, or a potential inhalation hazard, must wear positive pressure self-contained breathing apparatus until such time as the individual in charge of the ICS determines, through the use of air monitoring, that a decreased level of respiratory protection will not result in hazardous exposures to employees.

The individual in charge of the ICS must limit the number of emergency response personnel at the emergency site, in those areas of potential or actual exposure to incident or site hazards, to those who are actively performing emergency operations. However, operations in hazardous areas shall be performed using the buddy system in groups of two or more.

Back-up personnel must stand by with equipment, and be ready to provide assistance or rescue. Advance first aid support personnel, as a minimum, also stand by with medical equipment and transportation capability.

The individual in charge of the ICS may designate a safety official, who is knowledgeable in the operations being implemented at the emergency response site, with the specific responsibility of identifying and evaluating hazards and providing direction with respect to the safety of the operations for the emergency at hand. When activities are judged by the safety official to be an IDLH condition and/or to involve an imminent danger condition, the safety official has the authority to alter, suspend, or terminate those activities. The safety official is to immediately inform the individual in charge of the ICS of any actions needed to be taken to correct these hazards at the emergency scene. After emergency operations have terminated, the individual in charge of the ICS must implement appropriate decontamination procedures.

Personnel, not necessarily an employer's own employees, who are skilled in the operation of certain equipment, such as mechanized earth moving or digging equipment, or crane and hoisting equipment, and who are needed temporarily to perform immediate emergency support work that cannot reasonably be performed in a timely fashion by an employer's own employees, and who will be or may be exposed to the hazards at an emergency response scene, are not required to meet the training required for the employer's regular employees. However, these personnel are to be given an initial briefing at the site prior to their participation in any emergency response. The initial briefing must include instruction in the wearing of appropriate personal protective equipment, what chemical hazards are involved, and what duties are to be performed. All other appropriate safety and health precautions provided to the employer's own employees are to be used to assure the safety and health of these personnel.

Employees who in the course of their regular job duties work with and are trained in the hazards of specific hazardous substances, and who will be called upon to provide technical advice or assistance, at a hazardous substance release incident, to the individual in charge, must, annually, receive training, or demonstrate competency in the area of their specialization.

Training is to be based on the duties and function to be performed by each responder of an emergency response organization. The skill and knowledge levels required for all new responders, those hired after the effective date of this standard, shall be conveyed to them through training before they are permitted to take part in actual emergency operations on an incident.

First responders, at the awareness level, are individuals who are likely to witness or discover a hazardous substance release, and who have been trained to initiate an emergency response sequence by notifying the proper authorities of the release. They would take no further action beyond notifying the authorities of the release. First responders, at the awareness level, shall have sufficient training, or have had sufficient experience to objectively demonstrate competency in the following areas:

1. An understanding of what hazardous substances are, and the risks associated with them in an incident.

2. An understanding of the potential outcomes associated with an emergency created when hazardous substances are present.

3. The ability to recognize the presence of hazardous substances in an emergency.

4. The ability to identify the hazardous substances, if possible.

5. An understanding of the role of the first responder awareness individual in the employer's emergency response plan, including site security and control, and the U.S. Department of Transportation's Emergency Response Guidebook.

6. The ability to realize the need for additional resources, and to make appropriate notifications to the communication center.

First responders, at the operations level, are individuals who respond to releases or potential releases of hazardous substances, as part of the initial response to the site, for the purpose of protecting nearby persons, property, or the environment from the effects of the release. They are trained to respond in a defensive fashion without actually trying to stop the release. Their function is to contain the release from a safe distance, keep it from spreading, and prevent exposures. First responders, at the operational level, must receive at least eight hours of training, or have had sufficient experience to objectively demonstrate competency in the following areas, in addition to those listed for the awareness level, and the employer shall so certify:

1. Knowledge of the basic hazard and risk assessment techniques.

2. Know of how to select and use proper personal protective equipment provided to the first responder, operational level.

3. An understanding of basic hazardous materials terms.

4. Know how to perform basic control, containment, and/or confinement operations within the capabilities of the resources and personal protective equipment available with their unit.

5. Know how to implement basic decontamination procedures.

6. An understanding of the relevant standard operating and termination procedures.

Hazardous materials technicians are individuals who respond to releases, or potential releases, for the purpose of stopping the release. They assume a more aggressive role than a first responder at the operations level in that they will approach the point of release in order to plug, patch, or otherwise stop the release of a hazardous substance. Hazardous materials technicians must have received at least 24 hours of training, equal to the first responder operations level, and in addition, must have competency in the following areas, and the employer shall so certify:

1. Know how to implement the employer's emergency response plan.

2. Know the classification, identification, and verification of known and unknown materials by using field survey instruments and equipment.

3. Be able to function within an assigned role in the Incident Command System.

4.   Know how to select and use proper specialized chemical personal protective equipment provided to the hazardous materials technician.

5.   Understand hazard and risk assessment techniques.

6.   Be able to perform advance control, containment, and/or confinement operations within the capabilities of the resources and personal protective equipment available with the unit.

7.   Understand and implement decontamination procedures.

8.   Understand termination procedures.

9.   Understand basic chemical and toxicological terminology and behavior.

Hazardous materials specialists are to be individuals who respond with, and provide support to hazardous materials technicians. Their duties parallel those of the hazardous materials technician, however, those duties require a more directed or specific knowledge of the various substances they may be called upon to contain. The hazardous materials specialist also acts as the site liaison with federal, state, local, and other government authorities in regard to site activities. Hazardous materials specialists must have received at least 24 hours of training equal to the technician level, and, in addition, must have competency in the following areas, and the employer shall so certify:

1.   Know how to implement the local emergency response plan.

2.   Understand classification, identification, and verification of known and unknown materials by using advanced survey instruments and equipment.

3.   Know of the state emergency response plan.

4.   Be able to select and use proper specialized chemical personal protective equipment provided to the hazardous materials specialist.

5.   Understand in-depth hazard and risk techniques.

6.   Be able to perform specialized control, containment, and/or confinement operations within the capabilities of the resources and personal protective equipment available.

7.   Be able to determine and implement decontamination procedures.

8.   Have the ability to develop a site safety and control plan.

9.   Understand chemical, radiological, and toxicological terminology and behavior.

Incident commanders, who will assume control of the incident scene beyond the first responder awareness level, must receive at least 24 hours of training equal to the first responder operations level, and, in addition, must have competency in the following areas, and the employer shall so certify:

1.   Know and be able to implement the employer's incident command system.

2.   Know how to implement the employer's emergency response plan.

3.   Know and understand the hazards and risks associated with employees working in chemical protective clothing.

4.   Know how to implement the local emergency response plan.

5.   Know of the state emergency response plan and of the Federal Regional Response Team.

6.   Know and understand the importance of decontamination procedures.

Emergency Response Training

Trainers who teach any of the previous training subjects must have satisfactorily completed a training course for teaching the subjects they are expected to teach, such as the courses offered by the U.S. National Fire Academy; or they must have the training and/or academic credentials and instructional experience necessary to demonstrate competent instructional skills. They must also have a good command of the subject matter of the courses they are to teach. Trainers and trainees must receive annual refresher training of sufficient content and duration to maintain their competencies, or demonstrate, annually, their competency in those areas. A statement is to be made of the training or competency, and if a statement of competency is made, the employer must keep a record of the methodology used to demonstrate competency.

Members of an organized and designated HAZMAT team, and hazardous materials specialists, must receive a baseline physical examination and be provided with medical surveillance. Any emergency response employee who exhibits signs or symptoms which may have resulted from exposure to hazardous substances during the course of an emergency incident is to be provided with medical consultation.

Upon completion of the emergency response, if it is determined that it is necessary to remove hazardous substances, health hazards, and contaminated materials (such as contaminated soil or other elements of the natural environment) from the site of the incident, the employer conducting the clean-up must comply with the standard guidelines and rules.

It is very important to refer to the referenced information found in the tables, figures, and appendices which appear in 29 CFR 1916.65, since they are too lengthy to be included in this text.

Some of the information in the appendices include personal protective equipment test methods, general descriptions and discussion of the levels of protection and protective gear, compliance guidelines, references, and training curriculum guidelines.

HEAD PROTECTION (1926.100)

The use of protective hardhats is a must on a construction jobsite. They are not just for protection from falling objects, but for all potential head injuries. The dynamic nature of a construction worksite makes bumping into objects, or being hit by flying objects, a real risk for lateral impact injuries. Other types of potential head injuries can also occur from such things as hot materials or electrical hazards.

Hardhats must meet, according to the OSHA standard, the requirements ANSI standard Z89.1-1969 for impact, and Z89.2-1997 for burns and electricity. Class A hardhats are to protect workers from impact of falling objects and low voltage electricity. Class B hardhats are to protect workers from impact and high voltage electricity. Class C hardhats are to protect workers from impact of falling objects.

There is a new ANSI standard for hardhats, Z89.1-1997, which updates the previous standard and changes the A,B,C classification to G for general, E for electrical, and C for conductive. Also, guidelines on chin straps are incorporated. Chin straps are becoming more acceptable and are used in the United States.

Hardhats are ideal for attaching other types of PPE such as ear muffs, chin straps, face shields, cold weather protection, and welding hoods (see Figure 7-58).

Figure 7-58. Hardhat with hearing protection

HEARING PROTECTION – OCCUPATIONAL NOISE EXPOSURE (1926.52)

Protection against the effects of noise exposure is to be provided when the sound levels exceed those shown in Table 7-11 of this section and are measured on the A-scale of a standard sound level meter at slow response. When employees are subjected to sound levels exceeding those listed in Table 7-11 of this section, feasible administrative or engineering controls are to be utilized. If such controls fail to reduce sound levels within the levels of the table, personal protective equipment is to be provided and used to reduce sound levels to within the levels of the table. If the variations in noise level involves the maxims at intervals of one second or less, it is to be considered continuous. In all cases where the sound levels exceed the values shown as continuing, an effective hearing conservation program is to be administered.

Table 7-11

Permissible Noise Exposure
Courtesy of OSHA

Duration per day (Hours)	Sound level dBA slow response
8	90
6	92
4	95
3	97
2	100
1 1/2	102
1	105
1/2	110
1/4 or less.	115

When the daily noise exposure is composed of two or more periods of noise exposure of different levels, their combined effect should be considered, rather than the individual effect of each. Exposure to different levels of noise for various periods of time is to be computed using the formula in 1926.52. Exposure to impulsive or impact noise should not exceed 140 dB of peak sound pressure level.

HEATING DEVICES (TEMPORARY) (1926.154)

Temporary heating devices often cause carbon monoxide or high levels of carbon dioxide. Thus, the work environment must be supplied with sufficient quantities of fresh air, and the heater itself should be kept ten feet from combustible materials (see Figure 7-59). These heaters should be protected from tipping over, and at no time should the safety cutoff be over ridden. Also, care should be taken to insulate the heater from combustible floors. Solid fuel salamanders are prohibited in buildings and on scaffolds.

Figure 7-59. Example of a temporary heater

HEAVY EQUIPMENT, PREVENTING SLIPS AND FALLS

In preventing slips and falls from heavy equipment, the following precautions should be undertaken:

1. Always dismount equipment while facing ladders or steps.

2. Always make sure that three points of contact are used at all times (Example: two feet and one hand, or two hands and one foot) (see Figure 7-60).

3. Never jump from a piece of equipment.

4, Make sure to have proper foot wear; make sure the foot wear is clear of mud and debris.

5. After operating a piece of equipment for a period of time in the sitting position, muscles may stiffen up and may not be as flexible when dismounting equipment.

6. Use the provided hand holds, ladders, and steps for mounting and dismounting equipment.

7. Be careful when ice or snow exists on surfaces.

8. Cleaning, oiling, fueling, or repairing is not to be done while equipment is operating.

Figure 7-60. Construction worker mounting heavy equipment using three points of contact

HELICOPTERS (1926.551)

The use of helicopters for lifting requires special procedures, an adequately trained ground crew, and an awareness of the unique hazards involved with their use. Helicopter cranes are expected to comply with any applicable regulations of the Federal Aviation Administration. Prior to each day's operation, a briefing is to be conducted. This briefing must set forth the plan of operation for the pilot and ground personnel.

All loads must be properly slung. Taglines should be of a length that will not permit their being drawn up into rotors. Pressed sleeves, swedged eyes, or equivalent means are to be used for all freely suspended loads in order to prevent hand splices from spinning open, or cable clamps from loosening.

All electrically operated cargo hooks must have electrical activating devices so designed and installed as to prevent inadvertent operation. In addition, cargo hooks are to be equipped with an emergency mechanical control for releasing the load. The hooks are to be tested prior to each day's operation to determine that the release functions properly, both electrically and mechanically.

Workers receiving the load must have personal protective equipment which consists of complete eye protection, and hardhats secured by chinstraps. Loose-fitting clothing which is likely to flap in the downwash and be snagged on a hoist line, is not to be worn. Every practical precaution is to be taken to protect the employees from flying objects in the rotor downwash. All loose gear within 100 feet of the place of lifting the load, depositing the load, and all other areas susceptible to rotor downwash, are to be secured or removed. Good housekeeping is to be maintained in all helicopter loading and unloading areas. The helicopter operator/pilot is to be responsible for the size, weight, and manner in which loads are connected to the helicopter. Open fires are not permitted in an area that could result in such fires being spread by the rotor downwash. If, for any reason, the helicopter operator believes the lift cannot be made safely, the lift is not made. The weight of an external load must not exceed the manufacturer's rating.

When employees are required to perform work under hovering craft, a safe means of access is to be provided for employees to reach the hoist line hook, and engage or disengage cargo slings. Employees do not perform work under hovering craft except when necessary to hook or unhook loads.

Static charge on the suspended load is to be dissipated with a grounding device before ground personnel touch the suspended load, or protective rubber gloves are to be worn by all ground personnel touching the suspended load.

Ground lines, hoist wires, or other gear, except for pulling lines or conductors that are allowed to "pay out" from a container or roll off a reel, are not to be attached to any fixed ground structure, or allowed to foul on any fixed structure.

When visibility is reduced by dust or other conditions, ground personnel must exercise special caution to keep clear of main and stabilizing rotors. Precautions are also to be taken by the employer to eliminate, as far as practical, reduced visibility. Signal systems between aircrew and ground personnel must be understood and checked in advance of hoisting the load. This applies to either radio or hand signal systems. Hand signals are as shown in Figure 7-61.

No unauthorized person is allowed to approach within 50 feet of the helicopter when the rotor blades are turning. Whenever approaching or leaving a helicopter with blades rotating, all employees are to remain in full view of the pilot and keep in a crouched position. Employees should avoid the area from the cockpit or cabin rearward unless authorized by the helicopter operator to work there.

For safe helicopter loading and unloading operations, sufficient ground personnel needs to be provided. There must be constant, reliable communication between the pilot and the designated employee, of the ground crew, who is acting as the signalman during the period of loading and unloading. This signalman is to be distinctly recognizable from all other ground personnel.

HOIST, BASE-MOUNTED DRUM (1926.553)

Base-mounted drum hoists, exposed to moving parts, such as gears, projecting screws, setscrews, chains, cables, chain sprockets, and reciprocating or rotating parts, which constitute a hazard, must be guarded. All controls used during the normal operation cycle are to be located within easy reach of the operator's station. Electric motor-operated hoists are to be provided with a device to disconnect all motors from the line, upon power failure, and no motor is permitted to be restarted until the controller handle is brought to the "off" position. And where applicable, an overspeed preventive device is to exist. Also, a means must be present whereby a remotely operated hoist stop can be activated when any control is ineffective. All base-mounted drum hoists in use must meet the applicable requirements for design, construction, installation, testing, inspection, maintenance, and operations, as prescribed by the manufacturer.

HOIST, MATERIALS (1926.552)

All material hoists must comply with the manufacturer's specifications and limitations, which are applicable to the operation of all hoists and elevators. Where manufacturer's specifications are not available, the limitations assigned to the equipment are to be based on the determinations of a professional engineer competent in the field. Rated load capacities, recommended operating speeds, and special hazard warnings or instructions are to be posted on cars and platforms.

Figure 7-61. Hand signals used by helicopter ground personnel. Courtesy of OSHA from Figure N-1 of 29 CFR 1926.551

Hoisting ropes are to be installed in accordance with the wire rope manufacturers' recommendations. Wire rope is to be removed from service when any of the following conditions exist:

1. In hoisting ropes, six randomly distributed broken wires in one rope lay, or three broken wires in one strand in one rope lay.

2. Abrasion, scrubbing, flattening, or peening, causing loss of more than one-third of the original diameter of the outside wires.

3. Evidence of any heat damage resulting from a torch, or any damage caused by contact with electrical wires.

4. Reduction from nominal diameter of more than three sixty-fourths inch for diameters up to and including three-fourths inch; one-sixteenth inch for diameters seven-eights to 1 1/8 inches; and three thirty-seconds inch for diameters 1 1/4 to 1 1/2 inches.

The installation of live booms on hoists is prohibited. The use of endless belt-type manlifts on construction is prohibited. Operating rules for material hoists are to be established and posted at the operator's station of the hoist. Such rules include signal systems and the allowable line speed for various loads. Rules and notices are to be posted on the car frame or crosshead in a conspicuous location, including the statement "No Riders Allowed." No person is allowed to ride on material hoists except for the purposes of inspection and maintenance.

All entrances of the hoistways are to be protected by substantial gates or bars which guard the full width of the landing entrance. All hoistway entrance bars and gates are to be painted with diagonal contrasting colors, such as black and yellow stripes. Bars must not be less than 2- by 4-inch wooden bars, or the equivalent, located two feet from the hoistway line. Bars are to be located not less than 36 inches, nor more than 42 inches above the floor. Gates or bars protecting the entrances to hoistways are to be equipped with a latching device.

An overhead protective covering of 2-inch planking, 3/4-inch plywood, or other solid material of equivalent strength, is to be provided on the top of every material hoist cage or platform. The operator's station of a hoisting machine is to be provided with overhead protection equivalent to tight planking not less than 2 inches thick. The support for the overhead protection is to be of equal strength.

Hoist towers may be used with or without an enclosure on all sides. However, whichever alternative is chosen, the following applicable conditions are to be met:

1. When a hoist tower is enclosed, it is to be enclosed on all sides, for its entire height, with a screen enclosure of 1/2-inch mesh, No. 18 U.S. gauge wire, or equivalent, except for landing access.

2. When a hoist tower is not enclosed, the hoist platform or car is to be totally enclosed (caged) on all sides, for the full height between the floor and the overhead protective covering, with 1/2-inch mesh of No. 14 U.S. gauge wire or equivalent. The hoist platform enclosure must include the required gates for loading and unloading. A 6-foot high enclosure is to be provided on the unused sides of the hoist tower at ground level.

Car arresting devices are to be installed to function in case of rope failure. All material hoist towers are to be designed by a licensed professional engineer. All material hoists must conform to the requirements of ANSI A10.5-1969, Safety Requirements for Material Hoists.

HOIST, OVERHEAD (1926.554)

When operating an overhead hoist, the safe working load, as determined by the manufacturer, is to be indicated on the hoist, and this safe working load shall not be exceeded. Also, the supporting structure to which the hoist is attached must have a safe working load equal to that of the hoist. The support is to be arranged so as to provide for free movement of the hoist, and must not restrict the hoist from lining itself up with the load. Overhead hoists are to be installed only in locations that will permit the operator to stand clear of the load at all times.

An air driven hoist must be connected to an air supply of sufficient capacity and pressure in order to safely operate the hoist. All air hoses supplying air are to be positively connected to prevent their becoming disconnected during use.

All overhead hoists in use must meet the applicable requirements for construction, design, installation, testing, inspection, maintenance, and operation, as prescribed by the manufacturer.

HOIST, PERSONNEL

Personnel hoists/hoist towers outside the structure are to be enclosed for the full height on the side, or the sides used for entrance and exit to the structure. At the lowest landing, the enclosures on the sides which are not being used for exit or entrance to the structure, must be enclosed to a height of at least 10 feet. Other sides of the tower which are adjacent to floors or scaffold platforms are to be enclosed to a height of 10 feet above the level of such floors or scaffolds. Towers inside of structures are to be enclosed on all four sides throughout the full height.

Towers are to be anchored to the structure at intervals not exceeding 25 feet. In addition to tie-ins, a series of guys are installed. Where tie-ins are not practical, the tower is to be anchored by means of guys made of wire rope at least one-half inch in diameter, securely fastened to anchorage to ensure stability.

Hoistway doors or gates are not to be less than 6 feet 6 inches high, and shall be provided with mechanical locks which cannot be operated from the landing side; they are be accessible only to persons on the car.

Cars are to be permanently enclosed on all sides and the top, except sides used for entrance and exit which have car gates or doors. A door or gate is to be provided at each entrance to the car which protects the full width and height of the car entrance opening. An overhead protective covering of 2-inch planking, 3/4-inch plywood, or other solid material or equivalent strength, is to be provided on the top of every personnel hoist. Doors or gates must be provided with electric contacts which do not allow movement of the hoist when the door or gate is open. Safeties are to be capable of stopping and holding the car and rated load when traveling at governor tripping speed. Cars are to be provided with a capacity and data plate, secured in a conspicuous place on the car or crosshead. Internal combustion engines are not permitted for direct drive. Normal and final terminal stopping devices are to be provided, and an emergency stop switch is to be provided in the car and marked "Stop."

The minimum number of hoisting ropes used is: three for traction hoists and two for drum-type hoists. The minimum diameter of hoisting and counterweight wire ropes is 1/2-inch. Minimum factors of safety for suspension wire ropes can be found in 29 CFR 1926.552.

Following assembly and erection of hoists, and before being put in service, inspection and testing of all functions and safety devices are to be made, under the supervision of a competent person. A similar inspection and test is required following a major alteration of an existing installation. All are to be inspected and tested at not more than 3-month intervals. The employer must prepare a certification record which includes the date the inspection and test of all functions and safety devices was performed; the signature of the person who performed the inspection and test; and a serial number, or other identifier, for the hoist that was inspected and tested. The most recent certification record is to be maintained on file.

All personnel hoists used by employees are to be constructed of materials and components which meet the specifications for materials, construction, safety devices, assembly, and structural integrity as stated in the American National Standard A10.4-1963, Safety Requirements for Workmen's Hoists. Personnel hoists used in bridge tower construction must be approved by a registered professional engineer and erected under the supervision of a qualified engineer competent in this field.

When a hoist tower is not enclosed, the hoist platform or car is to be totally enclosed (caged) on all sides, for the full height between the floor and the overhead protective covering, with 3/4-inch mesh of No. 14 U.S. gauge wire or equivalent. The hoist platform enclosure must include the required gates for loading and unloading.

These hoists are to be inspected and maintained on a weekly basis. Whenever the hoisting equipment is exposed to winds exceeding 35 miles per hour, it is to be inspected and put in operable condition before reuse.

Wire rope is to be taken out of service when any of the following conditions exist:

1. In running ropes, six randomly distributed broken wires in one lay, or three broken wires in one strand in one lay.

2. Wear of one-third the original diameter of outside individual wires, kinking, crushing, bird caging, or any other damage resulting in distortion of the rope structure.

3. Evidence of any heat damage from any cause.

4. Reductions from nominal diameter of more than three-sixty-fourths inch for diameters to, and including three-fourths inch, one-sixteenth inch for diameters seven-eights inch to 1 1/8 inches inclusive, and three-thirty-seconds inch for diameters 1 1/4 to 1 1/2 inches inclusive.

5. In standing ropes, more than two broken wires in one lay in sections beyond end connections, or more than one broken wire at an end connection.

Permanent elevators, under the care and custody of the employer, and used by employees for work covered by this Act, shall comply with the requirements of American National Standards Institute A17.1-1965 with addenda A17.1a-1967, A17.1b-1968, A17.1c-1969, A17.1d-1970, and must be inspected in accordance with A17.2-1960 with addenda A17.2a-1965, A17.2b-1967.

HOUSEKEEPING (1926.25)

Housekeeping means keeping everything at work in its proper place and putting things away after they are used. Tools and materials should never be left on the floor, on stairs, in walkways, or aisles. During the course of construction, alteration, or repairs, form and scrap lumber with protruding nails, and all other debris, are to be kept cleared from work areas, passageways, and stairs, in and around buildings or other structures. Combustible scraps and debris are to be removed at regular intervals during the course of construction; a safe means of facilitating such removal must be provided. (See Figure 7-62.)

Containers are to be provided for the collection and separation of waste, trash, oily and used rags, and other refuse. Containers used for garbage and other oily, flammable, or hazardous wastes, such as caustics, acids, harmful dusts, etc. are to be equipped with covers. Garbage and other waste must be disposed of at frequent and regular intervals.

Figure 7-62. Typical housekeeping problems that arise on construction sites

Chapter 8

CONSTRUCTION SAFETY: I THROUGH W

Charles D. Reese

This chapter is a continued discussion of construction safety hazards; many of the most often occurring problem areas are presented from ladder safety to scaffolding safety. It is difficult to summarize the many requirements for construction safety related hazard issues due to the pervasiveness of hazards faced by those in the construction industry, thus you may still need to use the OSHA regulations and other more specialized information sources on some the topics presented within this and the previous chapter.

ILLUMINATION (1926.56)

Determining the adequacy of illumination for an ongoing construction worksite is not always an easy process. Employers should respond to workers' complaints of inadequate lighting; this may be the best indicator of adequate lighting. Although construction areas, ramps, runways, corridors, offices, shops, and storage areas are to meet the minimums found in Table D-3 of 29 CFR 1926.56; actual measurements are seldom done on construction sites. Thus, prudent judgment should be used when evaluating the amount of illumination necessary. This is especially true since it is widely recognized that poorly illuminated areas result in slips and falls which lead to other injuries.

JACKS (1926.305)

When using jacks, the manufacturer's rated capacity must be legibly marked on all jacks and not exceeded. All jacks should have a positive stop to prevent overtravel. When it is necessary to provide a firm foundation, the base of the jack is to be blocked or cribbed. Where there is a possibility of slippage of the metal cap of the jack, a wood block is to be placed between the cap and the load. After the load has been raised, the load must be cribbed, blocked, or otherwise secured at once.

Hydraulic jacks which are exposed to freezing temperatures must be supplied with an adequate antifreeze liquid. All jacks need to be properly lubricated at regular intervals. Each jack must be thoroughly inspected, at times, depending upon the service conditions. Inspections are not to be less frequent than the following:

1. For constant or intermittent use at one locality—once every 6 months.

2. For jacks sent out of shop for special work—when sent out, and when returned.

3. For a jack subjected to abnormal load or shock—immediately before, and immediately thereafter.

Repair or replacement parts are to be examined for possible defects. Jacks which are out of order are to be tagged accordingly, and are not to be used until repairs are made.

LADDERS (1926.1053)

The employer must, as necessary, provide training for each employee using ladders and stairways. The program should enable each employee to recognize hazards related to ladders and stairways, and train each employee in the procedures to be followed to minimize these hazards.

The employer needs to ensure that each employee has been trained by a competent person in the following areas: the nature of fall hazards in the work area; the correct procedures for erecting, maintaining, and disassembling the fall protection systems to be used; the proper construction, use, placement, and care in handling of all stairways and ladders; the maximum intended load-carrying capacities of ladders; and the contents of the OSHA standards regarding ladders. Retraining is to be provided for each employee, as necessary, so that the employee maintains the understanding and knowledge acquired for safe use of ladders and stairways.

Each self-supporting portable ladder must support at least four times the maximum intended load, except that each extra-heavy-duty type 1A metal or plastic ladder shall sustain at least 3.3 times the maximum intended load. Each portable ladder that is not self-supporting must support at least four times the maximum intended load, except that each extra-heavy-duty type 1A metal or plastic ladders shall sustain at least 3.3 times the maximum intended load (see Table 8-1). The ability of a ladder to sustain the loads indicated in this paragraph is to be determined by applying or transmitting the requisite load to the ladder in a downward vertical direction, when the ladder is placed at an angle of 75 1/2 degrees from the horizontal.

Table 8-1

Ladder Types and Load Capacities

Type	Grade	Duty Rating
III	Household	200 lbs.
II	Commercial	225 lbs.
I	Industrial	250 lbs.
IA	Extra Heavy Duty Industrial	300 lbs

Ladder rungs, cleats, and steps shall be parallel, level, and uniformly spaced when the ladder is in position for use. Rungs, cleats, and steps of portable and fixed ladders are to be spaced not less than 10 inches apart, nor more than 14 inches apart, as measured between center lines of the rungs, cleats and steps. Rungs, cleats, and steps of step stools are not to be less than 8 inches apart, nor more than 12 inches apart, as measured between center lines of the rungs, cleats, and steps. Rungs, cleats, and steps of the base section of extension trestle ladders are not to be less than 8 inches, nor more than 18 inches apart, as measured between center lines of the rungs, cleats, and steps. The rung spacing on the extension section of the extension trestle ladder must not be less than 6 inches, nor more than 12 inches, as measured between center lines of the rungs, cleats, and steps. The rungs of individual-rung/step ladders are to be shaped such that the employees' feet cannot slide off the end of the rungs. The rungs and steps of portable metal ladders must be corrugated, knurled, dimpled, coated with skid-resistant material, or otherwise treated to minimize slipping.

The minimum clear distance between the sides of individual-rung/step ladders, and the minimum clear distance between the side rails of other fixed ladders, is to be 16 inches. The minimum clear distance between side rails for all portable ladders is to be 11 1/2 inches.

Ladders are not to be tied or fastened together to provide longer sections, unless they are specifically designed for such use. A metal spreader or locking device is to be provided, on each stepladder, to hold the front and back sections in an open position when the ladder is being used (see Figure 8-1). When splicing is required to obtain a given length of side rail, the resulting side rail must be at least equivalent in strength to a one-piece side rail made of the same material.

Figure 8-1. The safe use of a stepladder is standing no higher than the second step from the top

Except when portable ladders are used to gain access to fixed ladders (such as those on utility towers, billboards, and other structures where the bottom of the fixed ladder is elevated to limit access), or when two or more separate ladders are used to reach an elevated work area, the ladders are to be offset with a platform or landing between the ladders. (The requirements, for having guardrail systems with toeboards for falling objects and overhead protection on platforms and landings, are set forth in Subpart M of this part.)

Ladder components are to be surfaced so as to prevent injury to an employee from punctures or lacerations, and to prevent snagging of clothing. Wood ladders are not to be coated with any opaque covering, except for identification or warning labels which may be placed on one face, only, of a side rail.

The minimum perpendicular clearance between fixed ladder rungs, cleats, and steps, and any obstruction behind the ladder shall be 7 inches, except in the case of an elevator pit ladder for which a minimum perpendicular clearance of 4 1/2 inches is required. The minimum perpendicular clearance between the center line of fixed ladder rungs, cleats, and steps, and any obstruction on the climbing side of the ladder shall be 30 inches, except as provided in paragraph (a)(15) of this section. When unavoidable obstructions are encountered, the minimum perpendicular clearance between the centerline of fixed ladder rungs, cleats, and steps, and the obstruction on the climbing side of the ladder may be reduced to 24 inches, provided that a deflection device is installed to guide employees around the obstruction.

Fixed Ladders

Each fixed ladder must support at least two loads of 250 pounds each, concentrated between any two consecutive attachments (the number and position of additional concentrated loads of 250 pounds each, determined from anticipated usage of the ladder, are also included), plus anticipated loads caused by ice buildup, winds, and rigging, and impact loads resulting from the use of ladder safety devices. Each step or rung shall be capable of supporting a single concentrated load of a least 250 pounds applied in the middle of the step or rung.

Through-fixed ladders at their point of access/egress must have a step-across distance of not less than 7 inches, nor more than 12 inches, as measured from the centerline of the steps or rungs, to the nearest edge of the landing area. If the normal step-across distance exceeds 12 inches, a landing platform shall be provided to reduce the distance to the specified limit. Fixed ladders are to be used at a pitch no greater than 90 degrees from the horizontal, as measured to the back side of the ladder. Fixed ladders, without cages or wells, shall have a clear width of at least 15 inches on each side of the centerline of the ladder, to the nearest permanent object. Fixed ladders are to be provided with cages, wells, ladder safety devices, or self-retracting lifelines where the length of climb is less than 24 feet, but the top of the ladder is to be at a distance greater than 24 feet above lower levels. Where the total length of a climb equals or exceeds 24 feet, fixed ladders shall be equipped with one of the following:

1. Ladder safety devices.

2. Self-retracting lifelines and rest platforms at intervals not to exceed 150 feet.

3. A cage or well and multiple ladder sections and each ladder section must not exceed 50 feet in length. Ladder sections shall be offset from adjacent sections, and landing platforms shall be provided at maximum intervals of 50 feet.

Cages for fixed ladders shall conform to all of the following horizontal bands and shall be fastened to the side rails of rail ladders, or fastened directly to the structure, building, or equipment. For individual-rung ladders, vertical bars are to be on the inside of the horizontal bands and shall be fastened to them. The cages shall not extend less than 27 inches, or more than 30 inches from the centerline of the step or rung (excluding the flare at the bottom of the cage), and must not be less than 27 inches in width. The inside of the cage is to be clear of projections and horizontal bands are to be spaced not more than 4 feet on center vertically. The vertical bars are to be spaced at intervals not more than 9 1/2 inches on center horizontally and the bottom of the cage is to be at a level not less than 7 feet, nor more than 8 feet above the point of access to the bottom of the ladder. The bottom of the cage shall be flared not less than 4 inches all around within the distance between the bottom horizontal band and the next higher band. The top of the cage is to be a minimum of 42 inches above the top of the platform, or the point of access at the top of the ladder, with provision for access to the platform or other point of access.

Wells for fixed ladders must conform to all of the following: they must completely encircle the ladder; they are to be free of projections; the inside face on the climbing side of the ladder must not extend less than 27 inches, nor more than 30 inches from the centerline of the step or rung. The inside clear width is to be at least 30 inches, and the bottom of the wall, on the access side, must start at a level not less than 7 feet, nor more than 8 feet above the point of access to the bottom of the ladder.

Ladder safety devices, and related support systems, for fixed ladders must conform to all of the following: they shall be capable of withstanding, without failure, a drop test consisting of an 18-inch drop of a 500-pound weight. They must permit the employee using the device to ascend or descend without continually having to hold, push, or pull any part of the device, leaving both hands free for climbing. They are to be activated within 2 feet after a fall occurs, and limit the descending velocity of an employee to 7 feet/sec. or less. The connection between the carrier or lifeline, and the point of attachment to the body harness, must not exceed 9 inches in length.

The mounting of ladder safety devices for fixed ladders must conform to the following: mountings for rigid carriers are to be attached at each end of the carrier with intermediate mountings, as necessary, and spaced along the entire length of the carrier in order to provide the strength necessary to stop employees' falls. Mountings for flexible carriers are to be attached at each end of the carrier. When the system is exposed to wind, cable guides for flexible carriers shall be installed at a minimum spacing of 25 feet, and a maximum spacing of 40 feet along the entire length of the carrier in order to prevent wind damage to the system. The design and installation of mountings and cable guides must not reduce the design strength of the ladder.

The side rails of through or side-step fixed ladders must extend 42 inches above the top of the access level or landing platform served by the ladder. For a parapet ladder the access is the top of the roof if the parapet is cut to allow access then the access level shall be the top of the parapet. For through-fixed ladder extensions, the steps or rungs are to be omitted from the extension, and the extension of the side rails is to be flared to provide not less than 24 inches, nor more than 30 inches clearance between side rails. Where ladder safety devices are provided, the maximum clearance between side rails of the extensions must not exceed 36 inches. For side-step fixed ladders, the side rails and the steps or rungs are to be continuous in the extension. Individual-rung/step ladders, except those used where their access openings are covered with manhole covers or hatches, must extend at least 42 inches above an access level or landing platform by either the continuation of the rung spacing as horizontal grab bars, or by providing vertical grab bars that have the same lateral spacing as the vertical legs of the rungs.

Rules for All Ladders

Rules for the use of all ladders, including job-made ladders, are when portable ladders are used for access to an upper landing surface, the ladder's side rails shall extend at least 3 feet above the upper landing surface to which the ladder is used to gain access; or when such an extension is not possible because of the ladder's length, then the ladder shall be secured, at its top, to a rigid support that will not deflect, and a grasping device, such as a grabrail, shall be provided to assist employees in mounting and dismounting the ladder (see Figure 8-2). In no case shall the extension be such that ladder deflection under a load would, by itself, cause the ladder to slip off its support.

Figure 8-2. Extension ladder secured and three rungs above the landing area

Ladders are to be maintained free of oil, grease, and other slipping hazards. Also, ladders are not to be loaded beyond the maximum intended load for which they were built, nor beyond their manufacturer's rated capacity. Ladders must be used only for the purpose for which they were designed.

Non-self-supporting ladders are to be used at an angle such that the horizontal distance from the top support, to the foot of the ladder, is approximately one-quarter of the working length of the ladder (the distance along the ladder between the foot and the top support). Wood job-made ladders, with spliced side rails, are to be used at an angle such that the horizontal distance is one-eighth the working length of the ladder.

Ladders are to be used only on stable and level surfaces unless secured in order to prevent accidental displacement (see Figure 8-3); they are not to be used on slippery surfaces unless secured, or provided with slip-resistant feet which prevents accidental displacement. Slip-resistant feet shall not be used as a substitute for care in placing, lashing, or holding a ladder that is used upon slippery surfaces including, but not limited to, flat metal or concrete surfaces that are constructed so they cannot be prevented from becoming slippery.

Figure 8-3. Safe use of an extension ladder by using an extra person to prevent displacement

Ladders placed in any location where they can be displaced by workplace activities or traffic, such as in passageways, doorways, or driveways, are to be secured in order to prevent accidental displacement, or a barricade must be used to keep the activities or traffic away from the ladder. The area around the top and bottom of ladders is to be kept clear.

The top of a non-self-supporting ladder is to be placed with the two rails supported equally, unless it is equipped with a single support attachment. Ladders are not to be moved, shifted, or extended while occupied. Ladders must have nonconductive siderails if they are used where the employee or the ladder could contact exposed energized electrical equipment.

The top or top step of a stepladder is not to be used as a step. Cross-bracing on the rear section of a stepladder is not to be used for climbing unless the ladder is designed and provided with steps for climbing on both front and rear sections.

Ladders are to be inspected on a periodic basis by a competent person for visible defects, and after any occurrence that could affect their safe use. Portable ladders with structural defects such as, but not limited to, broken or missing rungs, cleats, steps, broken or split rails, corroded components, or other faulty or defective components, are to be either immediately marked in a manner that readily identifies them as defective, or be tagged with "Do Not Use," or similar language, and are to be withdrawn from service until repaired.

Fixed ladders with structural defects such as, but not limited to, broken or missing rungs, cleats, steps, broken or split rails, or corroded components, are to be withdrawn from service until repaired. The requirement to withdraw a defective ladder from service is satisfied if the ladder is

1.   Immediately tagged with "Do Not Use" or similar language, or

2.   Marked in a manner that readily identifies it as defective, or

3.   Blocked (such as with a plywood attachment that spans several rungs).

Ladder repairs must restore the ladder to a condition meeting its original design criteria before the ladder is returned to use. Single-rail ladders are not to be used. When ascending or descending a ladder, the user must face the ladder and keep the user's weight centered between the rails. Each worker should maintain three points of contact when progressing up and/or down the ladder. Employees are not to carry any object or load while on a ladder that could cause the employee to lose balance and fall; also, do not move, shift, or extend the ladder while occupied. Do not use the top two steps of a step ladder, or top three rungs of extension ladder (see Figure 8-4). Do not use a ladder when you need a scaffold.

Figure 8-4. Unsafe use of a stepladder by standing on top step

It is important that job made ladders are substantially built (see Figure 8-5) and that the right size ladder is selected to accomplish the assigned task (see Figure 8-6). No single rail ladders are permitted on construction jobsites.

LIFT-SLAB CONSTRUCTION (1926.705)

Due to the 28 workers who died at L'Ambiance Plaza in April of 1987 while using lift-slab construction procedures, it was deemed that a need existed for a standard regarding this construction process. Thus, lift-slab operations are to be designed and planned by a registered professional engineer who has experience in lift-slab construction. Such plans and designs shall be implemented by the employer and include detailed instructions and sketches indicating the prescribed method of erection. These plans and designs must also include provisions for ensuring the lateral stability of the building/structure during construction.

Jacks/lifting units are to be marked to indicate their rated capacity, as established by the manufacturer. Jacks/lifting units are not to be loaded beyond their rated capacity, as established by the manufacturer. Jacking equipment is to be capable of supporting at least two and one-half times the load being lifted during jacking operations, and the equipment is not to be overloaded. For the purpose of this provision, jacking equipment includes any load bearing component which is used to carry out the lifting operation(s). Such equipment includes, but is not limited, to the following: threaded rods, lifting attachments, lifting nuts, hook-up collars, T-caps, shearheads, columns, and footings. Jacks/lifting units must be designed and installed so that they will neither lift nor continue to lift when they are loaded in excess of their rated capacity. Jacks/lifting units must have a safety device installed which will cause the jacks/lifting units to support the load, in any position, in the event any jacklifting unit malfunctions or loses its lifting ability.

Figure 8-5. Job-made ladder

Figure 8-6. Ladders of different lengths should be available

Jacking operations are to be synchronized in such a manner, as to ensure even and uniform lifting of the slab. During lifting, all points at which the slab is supported are to be kept within 1/2 inch of that needed to maintain the slab in a level position. If leveling is automatically controlled, a device is to be installed that will stop the operation when the 1/2 inch tolerance is exceeded, or where there is a malfunction in the jacking (lifting) system. If leveling is maintained by manual controls, such controls are to be located in a central location and attended by a competent person while the whole lift is in progress.

The maximum number of manually controlled jacks/lifting units on one slab is to be limited to a number that will permit the operator to maintain the slab level within specified tolerances, but in no case shall that number exceed 14.

No workers, except those essential to the jacking operation, are to be permitted in the building/structure while any jacking operation is taking place, unless the building/structure has been reinforced sufficiently to ensure its integrity during erection. The phrase "reinforced sufficiently to ensure its integrity" means that a registered professional engineer, independent of the engineer who designed and planned the lifting operation, has determined, from the plans, that if there is a loss of support at any jack location, that loss will be confined to that location, and the structure as a whole will remain stable. Under no circumstances is any employee, who is not essential to the jacking operation, permitted to be immediately beneath a slab while it is being lifted. A jacking operation begins when a slab, or group of slabs, is lifted, and the operation ends when such slabs are secured (with either temporary connections or permanent connections). When making temporary connections to support slabs, wedges are to be secured by tack welding, or an equivalent method of securing the wedges, in order to prevent them from falling out of position. Lifting rods may not be released until the wedges at that column have been secured. All welding on temporary and permanent connections is to be performed by a certified welder, familiar with the welding requirements, specified in the plans, and specifications for the lift-slab operation. Load transfer from jacks/lifting units to building columns is not to be executed until the welds on the column shear plates (weld blocks) are cooled to air temperature.

Jacks/lifting units are to be positively secured to building columns so that they do not become dislodged or dislocated. Equipment is to be designed and installed so that the lifting rods cannot slip out of position, or the employer must institute other measures, such as the use of locking or blocking devices, which will provide positive connection between the lifting rods and attachments, and will prevent components from disengaging during lifting operations.

LIQUID-FUEL TOOLS (1926.302)

Liquid-fuel tools are usually powered by gasoline. Vapors that can burn or explode and give off dangerous exhaust fumes are the most serious hazards associated with liquid-fuel tools. Only assigned, qualified operators shall operate power, powder-actuated, or air-driven tools. The following safe work procedures must be followed when using liquid-fuel: handle, transport, and store gas or fuel only in approved flammable liquid containers; before refilling the tank for a fuel-powered tool, the user must shut down the engine and allow it to cool to prevent accidental ignition of hazardous vapors; and effective ventilation and/or personal protective equipment must be provided when using a fuel-powered tool inside a closed area. A fire extinguisher must be readily available in the work area.

LIQUID PETROLEUM GAS (1926.153)

The fire hazard involved with LP Gas makes it imperative to follow precautions, as a preventive measure, in the handling of it. LP gas systems must have containers, valves, connectors, manifold valve assemblies, and regulators of an approved type. All cylinders must meet the Department of Transportation's specification identification requirements published in 49 CFR Part 178, Shipping Container Specifications. Welding on LP-Gas containers is prohibited. Container valves, fittings, and accessories connected directly to the container, including primary shut-off valves, must have a rated working pressure of at least 250 psig. and must be of a material and design suitable for LP-Gas service. Every container and every vaporizer is to be provided with one or more approved safety relief valves or devices. Care must be taken when filling the fuel containers of trucks or motor vehicles from bulk storage containers. This must be performed not less than 10 feet from the nearest masonry-walled building, or not less than 25 feet from the nearest building or other construction and, in any event, not less than 25 feet from any building opening. Also, the filling of portable containers or containers mounted on skids from storage containers, must not be performed less than 50 feet from the nearest building. Regulators are to be either directly connected to the container valves or to manifolds connected to the container valves. The regulator is to be suitable for use with LP-Gas. Manifolds and fittings, which connect containers to pressure regulator inlets, are to be designed for at least 250 psig. service pressure.

LP-Gas consuming appliances are to be approved types and in good condition. Aluminum piping or tubing shall not be used. Hose should be designed for a working pressure of at least 250 psig. Design, construction, and performance of hose and hose connections must have their suitability determined by listing by a nationally recognized testing agency.

Portable heaters, including salamanders, should be equipped with an approved automatic device to shut off the flow of gas to the main burner, and pilot, if used, in the event of flame failure. Container valves, connectors, regulators, manifolds, piping, and tubing are not to be used as structural supports for heaters. Containers, regulating equipment, manifolds, pipe, tubing, and hose are to be located where there will be minimized exposure to high temperatures or physical damage. If two or more heater-container units, of either the integral or nonintegral type, are located in an unpartitioned area on the same floor, the container or containers of each unit are to be separated from the container or containers of any other unit by at least 20 feet. Storage of LPG within buildings is prohibited.

Storage outside of buildings, for containers awaiting use, is to be located from the nearest building or group of buildings, in accordance with Table F-3 of 29 CFR 1926.153. Containers are to be placed in a suitable ventilated enclosure or otherwise protected against tampering. Storage locations need to be provided with at least one approved portable fire extinguisher having a rating of not less than 20-B:C. When potential damage to LP-Gas systems from vehicular traffic is a possibility, precautions against such damage are taken.

The minimum separation between a liquefied petroleum gas container, and a flammable or combustible liquid storage tank, is to be 20 feet, except in the case of flammable or combustible liquid tanks operating at pressures exceeding 2.5 psig., or equipped with emergency venting which will permit pressures to exceed 2.5 psig. Suitable means are to be taken to prevent the accumulation of flammable or combustible liquids under adjacent liquefied petroleum gas containers, such as by diversion curbs or grading. When flammable or combustible liquid storage tanks are within a diked area, the liquefied petroleum gas containers are to be outside the diked area and at least 10 feet away from the centerline of the wall of the diked area. The foregoing provisions do not apply when liquefied petroleum gas containers of 125 gallons or less capacity are installed adjacent to fuel oil supply tanks of 550 gallons, or less capacity.

LOCKOUT/TAGOUT (1910.147)

Although there is not a specific regulation for the construction industry regarding lockout/tagout, each construction company should develop a program or policy regarding lockout/tagout which covers the construction, servicing, and maintenance of machines, electrical circuits, piping containing liquids, and equipment in which the "unexpected" energization or start up of the machines or equipment, or release of stored energy, could cause injury to employees.

If a worker is required to remove or bypass a guard or other safety device, is required to place any part of his or her body into an area on a machine or piece of equipment where work is actually performed upon the material being processed (point of operation), or where an associated danger zone exists during a machine operating cycle or where the employee may contact, be engulfed, or inundated by a source of energy, then lockout/tagout procedures should be implemented.

Lockout/tagout should include work on cords and plug connected to electric equipment for which there could be exposure to the hazards of unexpected energizing or start up of the equipment. Lockout/tagout should also include hot tap operations involving transmission and distribution systems for substances such as gas, steam, water, or petroleum products when they are performed on pressurized pipelines. Another example of this is that of a mechanical energy source that needs to be locked or blocked into place, such as when a dump truck's bed is raised. Other sources of energy would include hydraulic, pneumatic, chemical, thermal, or other energy.

When such conditions exist, and in order to prevent injury to employees, employers should establish a program and utilize procedures for affixing appropriate lockout or tagout devices to the energy isolating devices, and to otherwise disabled machines or equipment, in order to prevent unexpected energization, start up, or the release of stored energy.

An energy isolating device is a mechanical device that physically prevents the transmission or release of energy, including, but not limited to, the following: a manually operated electrical circuit breaker, a disconnect switch, a manually operated switch by which the conductors of a circuit can be disconnected from all ungrounded supply conductors and, in addition, no pole can be operated independently; a line valve; a block; and any similar device used to block or isolate energy. Push buttons, selector switches, and other control circuit type devices are not to be used as energy isolating devices.

An energy isolating device must be capable of being locked out, if it has a hasp or other means of attachment to which, or through which, a lock can be affixed, or it has a locking mechanism built into it. Other energy isolating devices are to be capable of being locked out, if lockout can be achieved without the need to dismantle, rebuild, or replace the energy isolating device, or permanently alter its energy control capability.

Energy Control Program

Contractors should establish a program consisting of energy control procedures, employee training, and perform periodic inspections to ensure that before any employee does any servicing, construction, or maintenance on a machine or equipment, where the unexpected energizing, startup, or release of stored energy could occur and cause injury, the machine or equipment must be isolated from the energy source and rendered inoperative.

Although tagout systems have been used when an energy isolating device is not lockable, the only truly safe energy isolating device is the one which is lockable. Although it seems highly unlikely that the contractor can demonstrate that a tagout program will provide a level of safety equivalent to that obtained by using a lockout program, there might be isolated occasions where it might occur. Other means to be considered as part of the demonstration of full employee protection could include the implementation of additional safety measures such as the removal of an isolating circuit element, the blocking of a controlling switch, the opening of an extra disconnecting device, or the removal of a valve handle to reduce the likelihood of inadvertent energization.

Contractors should develop, use, and implement energy control procedures, for the control of potentially hazardous energy, where worker exposure could, or does exist. This may not be necessary if it can be documented that

1. The machine or equipment has no potential for stored or residual energy, or reaccumulation of stored energy after shut down which could endanger employees.

2. The machine or equipment has a single energy source which can be readily identified and isolated.

3. The isolation and locking out of the energy source will completely de-energize and deactivate the machine or equipment.

4. The machine or equipment is isolated from that energy source and locked out during servicing or maintenance.

5. A single lockout device will achieve a locked-out condition.

6. The lockout device is under the exclusive control of the authorized employee performing the servicing or maintenance.

7. The servicing or maintenance does not create hazards for other employees.

8. The employer, in utilizing this exception, has had no accidents involving the unexpected activation or re-energizing of the machine or equipment during servicing or maintenance.

The procedures shall clearly and specifically outline the scope, purpose, authorization, rules, and techniques to be utilized for the control of hazardous energy, and also outline the means to enforce compliance, including, but not limited to, the following:

1. A specific statement of the intended use of the procedure;

2. Specific procedural steps for shutting down, isolating, blocking, and securing machines or equipment to control hazardous energy;

3. Specific procedural steps for the placement, removal, and transfer of lockout devices or tagout devices, and the responsibility for them;

4. Specific requirements for testing a machine or equipment to determine and verify the effectiveness of lockout devices, tagout devices, and other energy control measures.

Lockout/Tagout Devices

Locks, tags, chains, wedges, key blocks, adapter pins, self-locking fasteners, or other hardware shall be provided by the employer for isolating, securing, or blocking of machines or equipment from energy sources. They shall be singularly identified; they shall be the only device(s) used for controlling energy; and they shall not be used for other purposes.

These devices must be durable and capable of withstanding the environment to which they are exposed for the maximum period of time that exposure is expected. Tagout devices shall be constructed and printed so that exposure to weather conditions, or wet and damp locations, will not cause the tag to deteriorate, or the message on the tag to become illegible. Tags and locking devices must not deteriorate when used in corrosive environments, such as areas where acid and alkali chemicals are handled and stored.

Lockout and tagout devices shall be standardized within the worksite in at least one of the following criteria: color, shape, or size, and, additionally, in the case of tagout devices, print and format shall be standardized (see Figure 8-7).

Lockout devices shall be substantial enough to prevent removal without the use of excessive force or unusual techniques, such as with the use of bolt cutters or other metal cutting tools. Tagout devices, including their means of attachment, shall be substantial enough to prevent inadvertent or accidental removal. Tagout device attachment means shall be of a non-reusable type, attachable by hand, self-locking, and non-releasable, with a minimum unlocking strength of no less than 50 pounds and having the general design and basic characteristics of being at least equivalent to a one-piece, all environment-tolerant nylon cable tie.

DO NOT OPERATE	DO NOT OPERATE
DANGER	*DANGER*
DO NOT OPERATE	**DO NOT OPERATE**
This Lock/Tag May Only Be Removed By:	**THIS SERVICE HAS BEEN INTERRUPTED TO PROTECT ME WHILE I AM PERFORMING MAINTENANCE. THIS TAG MUST NOT BE REMOVED EXCEPT BY ME OR AT MY REQUEST**
Name:_____ **Dept.:** _____ **Phone:** _____ **Date:**_____	
See Reverse Side	**Reverse Side**

Figure 8-7. Example of a tag for use in lockout/tagout

Lockout and tagout devices shall indicate the identity of the employee applying the device(s). Tagout devices shall warn against hazardous conditions, if the machine or equipment is energized, and must include a line such as the following: "Do Not Start, Do Not Open, Do Not Close, Do Not Energize, Do Not Operate."

Periodic Inspections

The contractor should conduct a periodic inspection of the energy control procedure, at least annually, to ensure that the procedure and the requirements of this standard are being followed. The periodic inspection needs to be performed by an authorized (competent) employee, utilizing the energy control procedure being inspected. The periodic inspection should be conducted to correct any deviations or inadequacies identified. Where lockout is used for energy control, the periodic inspection shall include a review, between the inspector and each authorized employee, of that employee's responsibilities under the energy control procedure being inspected. Authorized employees are workers who lockout or tagout machines or equipment in order to perform servicing or maintenance on that machine or equipment. An affected employee becomes an authorized employee when that employee's duties include performing servicing or maintenance covered under this section.

Where tagout is used for energy control, the periodic inspection shall include a review, between the inspector and each authorized and affected employee, of that employee's responsibilities under the energy control procedure being inspected. Affected employees are those workers whose job requires him/her to operate or use a machine or equipment on which servicing or maintenance is being performed under lockout or tagout, or whose job requires him/her to work in an area in which such servicing or maintenance is being performed.

The employer should certify that the periodic inspections have been performed. The certification needs to identify the machine or equipment on which the energy control procedure was being utilized; the date of the inspection; the employees included in the inspection; and the person performing the inspection.

Training and Communications

The employer must provide training to ensure that the purpose and function of the energy control program is understood by employees, and that the knowledge and skills required for the safe application, usage, and removal of the energy controls are acquired by employees. The training shall include the following:

1. Each authorized employee shall receive training in the recognition of applicable hazardous energy sources, the type and magnitude of the energy available in the workplace, and the methods and means necessary for energy isolation and control.

2. Each affected employee shall be instructed in the purpose and use of the energy control procedure.

3. All other employees, whose work operations are or may be in an area where energy control procedures may be utilized, shall be instructed about the procedure, and about the prohibition relating to attempts to restart or reenergize machines or equipment which are locked out or tagged out.

When tagout systems are used, employees shall also be trained in the following limitations of those tags: tags are essentially warning devices affixed to energy isolating devices and they do not provide the physical restraint that is provided by a lock; they are attached to an energy isolating means and are not to be removed without authorization of the authorized person responsible for it; they are never to be bypassed, ignored, or otherwise defeated; they must be legible and understandable by all authorized employees, affected employees, and all other employees whose work operations are, or may be, in the area; in order for tags to be effective, they must be made of materials which will withstand the environmental conditions encountered in the workplace, including their means of attachment; they may evoke a false sense of security; their meaning needs to be understood as part of the overall energy control program; and they must be securely attached to energy isolating devices so that they cannot be inadvertently or accidentally detached during use.

Retraining shall be provided for all authorized and affected employees whenever there is a change in their job assignments, or a change in machines, equipment, or processes that presents a new hazard, or when there is a change in the energy control procedures. Additional retraining shall also be conducted whenever a periodic inspection reveals, or whenever the employer has reason to believe that there are deviations from, or inadequacies in the employee's knowledge or use of the energy control procedures. The retraining shall reestablish employee proficiency and introduce new or revised control methods and procedures, as necessary. The employer shall certify that employee training has been accomplished and is being kept up to date. The certification should contain each employee's name and dates of training.

Energy Isolation

Lockout or tagout shall be performed only by the authorized employees who are performing the servicing or maintenance. Affected employees are to be notified by the employer or authorized employee of the application and removal of lockout devices or tagout devices. Notification must be given before the controls are applied, and after they are removed from the machine or equipment.

Established Procedure

The established procedures for the application of energy control (the lockout or tagout procedures) need to cover the following elements and actions, and shall be done in the following sequence:

1. Prior to the preparation for shutdown or turning off a machine or equipment, the authorized or affected employee, who will be performing this task, must have knowledge of the type and magnitude of the energy, the hazards of the energy to be controlled, and the method or means to control the energy.

2. Machine or equipment shutdown must follow the procedures established for the machine or equipment. An orderly shutdown must be utilized to avoid any additional or increased hazard(s) to employees, as a result of the equipment stoppage.

3. There shall be machine or equipment isolation of all energy isolating devices that are needed to control the energy to the machine or equipment, and they shall be physically located and operated in such a manner as to isolate the machine or equipment from the energy source(s).

Lockout or Tagout Device Application

Lockout or tagout devices are to be affixed to each energy isolating device by authorized employees. Lockout devices, where used, shall be affixed in a manner that will hold the energy isolating devices in a "safe" or "off" position. Tagout devices, where used, must be affixed in such a manner that will clearly indicate that the operation or movement of energy isolating devices from the "safe" or "off" position is prohibited. Where tagout devices are used with energy isolating devices designed with the capability of being locked, the tag attachment needs to be fastened at the same point at which the lock would have been attached. Where a tag cannot be affixed directly to the energy isolating device, the tag shall be located as close as safely possible to the device and in a position that will be immediately obvious to anyone attempting to operate the device.

Stored Energy

Following the application of lockout or tagout devices to energy isolating devices, all potentially hazardous stored or residual energy is to be relieved, disconnected, restrained, and otherwise rendered safe. If there is a possibility of reaccumulation of stored energy to a hazardous level, verification of isolation shall be continued until the servicing or maintenance is completed, or until the possibility of such accumulation no longer exists.

The verification of isolation must be done by an authorized employee prior to starting work on machines or equipment that have been locked out or tagged out; the authorized employee must verify that isolation and deenergization of the machine or equipment have been accomplished.

Release from Lockout or Tagout

Before lockout or tagout devices are removed and energy is restored to the machine or equipment, procedures are to be followed and actions taken by the authorized employee(s) to ensure the following:

1.   The work area must be inspected to ensure that nonessential items have been removed and machine or equipment components are operationally intact.

2.   The work area must be checked to ensure that all employees have been safely positioned or removed. Before lockout or tagout devices are removed, and before machines or equipment are energized, affected employees shall be notified that the lockout or tagout devices have been removed. After lockout or tagout devices have been removed, and before a machine or equipment is started, affected employees shall be notified that the lockout or tagout device(s) have been removed.

3.   Each lockout or tagout device shall be removed from each energy isolating device by the employee who applied the device. There may be an exception to this rule when the authorized employee, who applied the lockout or tagout device, is not available to remove it; that device may be removed under the direction of the employer, provided that specific procedures and training for such removal have been developed, documented, and incorporated into the employer's energy control program.

Testing or Positioning

When testing or positioning of machines, equipment, or components in situations in which lockout or tagout devices must be temporarily removed from the energy isolating device, and the machine or equipment energized to test or position the machine, equipment, or component thereof, the following sequence of actions shall be followed:

1.   Clear the machine or equipment of tools and materials.

2.   Remove employees from the machine or equipment area.

3.   Remove the lockout or tagout devices.

4.   Energize and proceed with testing or positioning.

5.   Reenergize all systems and reapply energy control measures to continue the servicing and/or maintenance.

Outside Personnel (Subcontractors, etc.)

Whenever outside personnel are to be engaged in activities covered by the scope and application of the lockout/tagout procedure, the on-site employer and the outside employer shall inform each other of their respective lockout or tagout procedures. The on-site employer shall ensure that his/her employees understand and comply with the restrictions and prohibitions of the outside employer's energy control program.

Group Lockout or Tagout

When servicing and/or maintenance is performed by a crew, craft, department, or other group, they shall utilize a procedure which affords the employees a level of protection equivalent to that provided by the implementation of a personal lockout or tagout device. Group lockout or tagout devices are to be used in accordance, but not necessarily limited to, the following specific requirements:

1.   Primary responsibility is vested in an authorized employee for a set number of employees working under the protection of a group lockout or tagout device (such as an operations lock).

2. Provision for the authorized employee to ascertain the exposure status of individual group members with regard to the lockout or tagout of the machine or equipment.

3. When more than one crew, craft, department, etc. is involved, assignment of the overall job-associated lockout or tagout control should be given to an authorized employee designated to coordinate affected work forces and ensure continuity of protection.

4. Each authorized employee shall affix a personal lockout or tagout device to the group lockout device, group lockbox, or comparable mechanism when he or she begins work, and shall remove those devices when he or she stops working on the machine or equipment being serviced or maintained.

Shift or Personnel Changes

Specific procedures shall be utilized during shift or personnel changes to ensure the continuity of the lockout or tagout protection, including provision for the orderly transfer of lockout or tagout device protection between off-going and oncoming employees in order to minimize exposure to hazards from the unexpected energization or start-up of the machine or equipment, or the release of stored energy.

MARINE EQUIPMENT (1926.605)

The term "longshoring operations" means the loading, unloading, moving, or handling of construction materials, equipment and supplies, etc. into, in, on, or out of any vessel from a fixed structure or shore-to-vessel, vessel-to-shore or fixed structure or vessel-to-vessel. Ramps for access of vehicles to or between barges are to be of adequate strength, provided with side boards, well maintained, and properly secured.

Unless workers can step safely to or from the wharf, float, barge, or river towboat, either a ramp or a safe walkway must be provided. Jacob's ladders are of the double rung or flat tread type. They should be well maintained and properly secured. A Jacob's ladder either hangs without slack from its lashings, or is to be pulled up entirely.

When the upper end of the means of access rests on, or is flush with, the top of the bulwark, substantial steps which are properly secured and equipped with at least one substantial hand rail approximately 33 inches in height, must be provided between the top of the bulwark and the deck. Obstructions must never be laid on or across the gangway, and the means of access is to be adequately illuminated for its full length. Unless the structure makes it impossible, the means of access is to be so located that the load will not pass over workers.

Workers are not permitted to walk along the sides of covered lighters or barges with coamings more than 5 feet high, unless there is a 3-foot clear walkway, or a grab rail, or a taut handline is provided. Decks and other working surfaces are to be maintained in a safe condition. Workers are not permitted to pass fore and aft, over, or around, deckloads, unless there is a safe passage. Workers are not permitted to walk over deckloads from rail to coaming unless there is a safe passage. If it is necessary to stand at the outboard or inboard edge of the deckload where less than 24 inches of bulwark, rail, coaming, or other protection exists, all must be provided with a suitable means of protection against falling from the deckload.

There must be in the vicinity of each barge in use at least one U.S. Coast Guard approved 30-inch life ring with not less than 90 feet of line attached, and at least one portable or permanent ladder which will reach the top of the apron to the surface of the water. If the above equipment is not available at the pier, the employer must furnish it during the time that the worker is working the barge. Workers walking or working on the unguarded decks of barges are to be protected with U.S. Coast Guard-approved work vests or buoyant vests.

MATERIAL HANDLING AND STORAGE (1926.250)

Materials handling accounts for approximately 40 percent of the lost-time incidents that occur in the construction industry. These injuries are often a result of inadequate planning, administrative, and/or engineering approaches. Therefore, in an effort to reduce workplace injuries, the following safe work procedures will be implemented and enforced at all construction projects.

Materials that are stored in tiers must be stacked, racked, blocked, interlocked, or otherwise secured to prevent sliding, falling, or collapse. Noncompatible materials are to be segregated in storage. Bagged materials must be stacked by stepping back the layers and cross-keying the bags at least every 10 bags high. Brick stacks are not to be more than 7 feet in height. When a loose brick stack reaches a height of 4 feet, it is to be tapered back 2 inches in every foot of height above the 4-foot level. When masonry blocks are stacked higher than 6 feet, the stack is to be tapered back one-half block per tier above the 6-foot level.

Also, lumber is to be stacked on a level surface and solidly supported sills, and so stacked as to be stable and self-supporting. Lumber piles must not exceed 20 feet in height, except lumber that is to be handled manually is not to be stacked more than 16 feet high. Any used lumber must have all nails withdrawn before stacking. Structural steel, poles, pipe, bar stock, and other cylindrical materials, unless racked, are to be stacked and blocked so as to prevent spreading or tilting (see Figure 8-8).

Figure 8-8. Securely stacked cylindrical material

Materials stored inside buildings which are under construction are not to be placed within 6 feet of any hoistway or inside floor openings, nor within 10 feet of an exterior wall which does not extend above the top of the material stored. The maximum safe load limits of floors within buildings and structures, in pounds per square foot, must be conspicuously posted in all storage areas, except for floors or slabs on grade. The maximum safe loads are not to be exceeded. Materials are not to be stored on scaffolds or runways in excess of supplies needed for immediate operations.

Also, all aisles and passageways should be kept clear to provide for the free and safe movement of material handling equipment or employees, and these are to be kept in good repair. When a difference in road or working levels exists, means, such as ramps, blocking, or grading, are to be used to ensure the safe movement of vehicles between the two levels.

Any time a worker is required to work on stored material in silos, hoppers, tanks, and similar storage areas, they are to be equipped with personal fall arrest equipment which meets the requirements of Subpart M.

"Housekeeping" storage areas are to be kept free from accumulation of materials that constitute hazards from tripping, fire, explosion, or pest harborage. Vegetation control is to be exercised when necessary.

When portable and powered dockboards are used, they should be strong enough to carry the load imposed on them. Portable dockboards are to be secured in position, either by being anchored or equipped with devices which will prevent their slipping. Handholds, or other effective means, are to be provided on portable dockboards to permit safe handling. Positive protection must be provided to prevent railroad cars from being moved while dockboards or bridge plates are in position.

When handling materials, do not attempt to lift or move a load that is too heavy for one person–get help. Attach handles or holders to the load to reduce the possibility of pinching or smashing fingers.

Wear protective gloves and clothing (i.e., aprons), if necessary, when handling loads with sharp or rough edges. When pulling or prying objects, be sure you are properly positioned.

During the weekly "tool-box" meetings, employees should receive instructions on proper materials handling practices so that they are aware of the following types of injuries associated with manual handling of materials:

1. Strains and sprains from lifting loads improperly, or from carrying loads that are too heavy or large.

2. Fractures and bruises caused by dropping or flying materials, or getting hands caught in pinch points.

3. Cuts and abrasions caused by falling materials which have been improperly stored, or by cutting securing devices incorrectly.

Engineering controls should be used, if feasible, to redesign the job so that the lifting task becomes less hazardous. This includes reducing the size or weight of the object lifted, changing the height of a pallet or shelf, or installing a mechanical lifting aid.

MATERIAL HANDLING EQUIPMENT (1926.602)

Material handling equipment must include the following types of earthmoving equipment: scrapers, loaders, crawlers or wheel tractors, bulldozers, off-highway trucks, graders, agricultural and industrial tractors, and similar equipment (see Figure 8-9).

Seat belts are to be provided on all equipment covered by this section, and meet the requirements of the Society of Automotive Engineers, J386-1969, Seat Belts for Construction Equipment. Seat belts for agricultural and light industrial tractors must meet the seat belt requirements of Society of Automotive Engineers J333a-1970, Operator Protection for Agricultural and Light Industrial Tractors. Seat belts are not needed for equipment which is designed only for standup operation. Seat belts are not needed for equipment which does not have roll-over protective structure (ROPS) or adequate canopy protection.

No construction equipment or vehicles are to be moved upon any access roadway or grade unless the access roadway or grade is constructed and maintained to safely accommodate

Figure 8-9. Example of material handling equipment

the movement of the equipment and vehicles involved. Every emergency access ramp and berm used by an employer is to be constructed to restrain and control runaway vehicles. All earth-moving equipment is to have a service braking system capable of stopping and holding the equipment fully loaded, as specified in Society of Automotive Engineers SAE-J237, Loader Dozer-1971, J236, Graders-1971, and J319b, Scrapers-1971. Brake systems for self-propelled, rubber-tired, off-highway equipment, manufactured after January 1, 1972, must meet the applicable minimum performance criteria set forth in the following Society of Automotive Engineers Recommended Practices:

> Self-Propelled Scrapers .... SAE J319b-1971.
>
> Self-Propelled Graders .... SAE J236-1971.
>
> Trucks and Wagons .... SAE J166-1971.
>
> Front-End Loaders and Dozers .... SAE J237-1971.

Pneumatic-tired, earth-moving haulage equipment (trucks, scrapers, tractors, and trailing units), whose maximum speed exceeds 15 miles per hour, is to be equipped with fenders on all wheels, unless the employer can demonstrate that uncovered wheels present no hazard to personnel from flying materials. Rollover protective structures should be found on all material handling equipment (see Figure 8-10).

Figure 8-10. A scraper with rollover protection

All bidirectional machines, such as rollers, compactors, front-end loaders, bulldozers, and similar equipment, are to be equipped with a horn, distinguishable from the surrounding noise level, and they are to be operated, as needed, when the machine is moving in either direction. The horn shall be maintained in an operative condition. No earth-moving or compacting equipment, which has an obstructed view to the rear, is permitted to be used in the reverse gear unless the equipment has, in operation, a reverse signal alarm which is distinguishable from the surrounding noise level, or there is an authorized employee who signals that it is safe to do so (see Figure 8-11).

Figure 8-11. Front-end loaders are bidirectional and required to have a back-up alarm

Scissor points on all front-end loaders, which constitute a hazard to the operator during normal operation, are to be guarded.

Tractors must have seat belts. They are required for the operators when they are seated in the normal seating arrangement for tractor operation, even though back-hoes, breakers, or other similar attachments are used on these machines for excavating or other work which result in other than normal seating.

Industrial trucks such as lift trucks, forklifts, stackers, etc. must have the rated capacity clearly posted on the vehicle so as to be clearly visible to the operator. When auxiliary removable counterweights are provided by the manufacturer, corresponding alternate rated capacities also must be clearly shown on the vehicle. These ratings shall not be exceeded. No modifications or additions which affect the capacity or safe operation of the equipment shall be made without the manufacturer's written approval. If such modifications or changes are made, the capacity, operation, and maintenance instruction plates, tags, or decals must be changed accordingly. In no case shall the original safety factor of the equipment be reduced. If a load is lifted by two or more trucks working in unison, the proportion of the total load carried by any one truck must not exceed its capacity. Steering or spinner knobs are not to be attached to the steering wheel unless the steering mechanism is of a type that prevents road reactions from causing the steering handwheel to spin. The steering knob shall be mounted within the periphery of the wheel.

All high lift rider industrial trucks are to be equipped with overhead guards which meet the configuration and structural requirements as defined in paragraph 421 of American National Standards Institute B56.1-1969, Safety Standards for Powered Industrial Trucks. All industrial trucks in use shall follow the applicable requirements of design, construction, stability, inspection, testing, maintenance, and operation, as defined in American National Standards Institute B56.1-1969.

Unauthorized personnel are not permitted to ride on powered industrial trucks. A safe place to ride is to be provided when riding on trucks is authorized. Whenever a truck is equipped with vertical only, or vertical and horizontal controls elevatable with the lifting, carriage, or forks for lifting personnel, the following additional precautions are to be taken for the protection of the personnel being elevated. They are to use a safety platform which is firmly secured to the lifting carriage and/or forks, and a means must be provided whereby personnel on the platform can shut off power to the truck. Personnel are also to be protected from falling objects, if falling objects could occur due to the operating conditions.

MEDICAL SERVICES AND FIRST AID (CFR 1926.23 AND .50)

First aid services and provisions for medical care are to be made available, by the employer, for every employee. The employer must insure the availability of medical personnel for advice and consultation on matters of occupational health. In case of serious injury, provisions are to be made for prompt medical attention prior to the commencement of the project. When there is no infirmary, clinic, physician, or hospital that is reasonably located, in terms of time and distance to the worksite, for the treatment of injured employees, a person who has a valid certificate in first aid training from the U.S. Bureau of Mines, the American Red Cross, or equivalent training, and can be verified by documentary evidence, this employee shall be available at the worksite to render first aid. First-aid supplies must be approved by the consulting physician and are to be easily accessible, when required. The first aid kit must consist of materials approved by the consulting physician and must be in a weatherproof container with individual sealed packages for each type of item. The contents of the first aid kit must be checked before being sent out on each job, and must be checked, at least weekly, on each job to ensure that the expended items are replaced. Proper equipment for prompt transportation of the injured person to a physician or hospital, or a communication system for contacting necessary ambulance service, is to be provided. The telephone numbers of the physicians, hospitals, or ambulances are to be conspicuously posted. Where the eyes or body of any person may be exposed to injurious corrosive materials, suitable facilities for quick drenching or flushing of the eyes and body are to be provided within the work area for immediate emergency use.

MOTOR VEHICLES AND MECHANIZED EQUIPMENT (1926.601)

Motor vehicles that operate within an off-highway jobsite, which is not open to public traffic, must have a service brake system, an emergency brake system, and a parking brake system. These systems may use common components, and must be maintained in operable condition. Also, each vehicle must have the appropriate number of seats for occupants, and the seat belts must be properly installed.

Whenever visibility conditions warrant additional light, all vehicles, or combinations of vehicles, in use, are to be equipped with at least two headlights and two taillights in operable condition. All vehicles, or combination of vehicles, must have brake lights, in operable condition, regardless of light conditions. Also, these vehicles are to be equipped with an adequate audible warning device at the operator's station and be in operable condition.

No worker is to use motor vehicle equipment having an obstructed view to the rear, unless the vehicle has a reverse signal alarm, audible above the surrounding noise level, or the vehicle is only backed up when an observer signals that it is safe to do so.

All vehicles with cabs are to be equipped with windshields and powered wipers. Cracked and broken glass is to be replaced. Vehicles operating in areas, or under conditions that cause fogging or frosting of the windshields, are to be equipped with operable defogging or defrosting devices.

All haulage vehicles, whose pay load is loaded by means of cranes, power shovels, loaders, or similar equipment, must have a cab shield and/or canopy which is adequate to protect the operator from shifting or falling materials.

Tools and material are to be secured in order to prevent movement when transported in the same compartment with employees. Vehicles used to transport employees must have seats firmly secured and adequate for the number of employees to be carried and must also include adequate seat belts.

Trucks with dump bodies are to be equipped with positive means of support, permanently attached, and capable of being locked in position to prevent accidental lowering of the body while maintenance or inspection work is being done. Operating levers, controlling hoisting or dumping devices on haulage bodies, are to be equipped with a latch or other device which will prevent accidental starting or tripping of the mechanism. Trip handles for tailgates of dump trucks are to be so arranged that, in dumping, the operator will be in the clear. All rubber-tired motor vehicle equipment manufactured on or after May 1, 1972 is to be equipped with fenders. Mud flaps are to be used in lieu of fenders whenever motor vehicle equipment is not designed for fenders.

All vehicles in use are to be checked at the beginning of each shift to assure that the following parts, equipment, and accessories are in safe operating condition and free of apparent damage that could cause failure while in use: service brakes, including trailer brake connections; parking system (hand brake); emergency stopping system (brakes); tires; horn; steering mechanism; coupling devices; seat belts; operating controls; and safety devices. All defects are to be corrected before the vehicle is placed in service. These requirements also apply to equipment such as lights, reflectors, windshield wipers, defrosters, fire extinguishers, etc., where such equipment is necessary.

All equipment left unattended at night, adjacent to highways or construction areas, must have lights, reflectors, and/or barricades to identify location of the equipment. Supervisory personnel must ensure that all machinery and equipment is inspected prior to each use to verify that it is in safe operating condition. Rated load capacities and recommended rules of operation must be conspicuously posted on all equipment at the operator's station. An accessible fire extinguisher of 5 BC rating or higher must be available at all operator stations. When vehicles or mobile equipment are stopped or parked, the parking brake must be set. Equipment, on inclines, must have wheels chocked, as well as the parking brake set.

NONPOTABLE WATER (1926.51)

Nonpotable water is to be used for industrial or firefighting use only and must be identified by a label. It is not to be used for drinking, washing, or cooking purposes.

PERSONAL PROTECTIVE EQUIPMENT (PPE) (1926.95)

Personal protective equipment (PPE) includes protective equipment for the eyes, face, head, and extremities. This protective equipment may be such items as protective clothing, respiratory devices, protective shields, and barriers. When this equipment is needed because of hazardous processes, environmental hazards, chemical hazards, radiological hazards, or mechanical irritants which may cause injury or impairment to any part of the worker's body, by either absorption, inhalation, or physical contact, personal protective equipment is to be provided, used, and maintained in a sanitary and reliable condition. Where employees provide their own protective equipment, the employer is responsible to assure its adequacy, including proper maintenance and sanitation of such equipment.

All personal protective equipment must be of safe design and construction for the work to be performed. See Table 8-2 for some of the most common guidelines for the wearing of PPE.

Table 8-2

Common Guidelines for Wearing PPE

- Wear all required PPE for the job or task.
- Inspect all PPE for wear or damage prior to use.
- Take care of and clean PPE when necessary.
- Do not use PPE for which worker has not received training.
- Workers working in areas where overhead structures, equipment, or stored materials create a hazard shall wear hard hats and be required to wear them at all times.
- Workers wearing prescription eye glasses should use hardened safety glass lenses.
- Goggles shall be worn during work where flying particles are definite eye hazards.
- Hand protection is required during climbing, lifting, or potential contact with chemicals.
- Approved ear protection shall be worn when required.
- Respirators shall be worn if the concentration of dust, toxic fumes, or other air contaminants exceeds safe exposure levels.
- Safety harnesses and life lines shall be used when working surfaces are above 6 feet.
- Proper work shoes or boots, in good repair, shall be worn on the jobsites.
- Workers shall wear suitable work clothing consisting of at least long pants and a tucked-in short sleeve shirt.
- Rubber boots must be worn when doing concrete work.
- Always wear life jackets when working over, or adjacent to deep water.

PILE DRIVING (1926.603)

Pile driving requires special equipment and unique skills compared with operating other construction equipment and operations. In pile driving there is a need for tremendous amounts of energy. [This energy is in the form of steam in a pressurized vessel which presents a different type of hazard to construction operations] (see Figure 8-12).

Figure 8-12. Example of piles being driven

All pressure vessels which are a part of, or used with pile driving equipment must meet the applicable requirements of the American Society of Mechanical Engineers, Pressure Vessels (Section VIII). Boilers and piping systems which are also part of, or used with pile driving must meet the applicable requirements of the American Society of Mechanical Engineers, Power Boilers (Section I).

Workers are to be provided overhead protection which does not obscure the vision of the operator and which meets the requirements of Subpart N of this part. The protection shall be the equivalent of 2-inch planking or other solid material of equivalent strength.

A stop block is also to be provided for the leads in order to prevent the hammer from being raised against the head block. While employees are working under the hammer, a blocking device, capable of safely supporting the weight of the hammer, is to be provided.

Guards are to be provided across the top of the head block to prevent the cable from jumping out of the sheaves. And, when the leads must be inclined in the driving of batter piles, provisions are to be made to stabilize the leads.

Fixed leads are to be provided with a ladder and adequate rings, or similar attachment points, so that the loft worker may engage his safety belt lanyard to the leads. If the leads are provided with loft platforms(s), such platform(s) shall be protected by standard guardrails.

A steam hose, which leads to a steam hammer or jet pipe, is to be securely attached to the hammer with an adequate length of at least 1/4-inch diameter chain or cable in order to prevent whipping in the event the joint at the hammer is broken. Air hammer hoses are to be provided with the same protection as required for steam lines. Safety chains, or equivalent means, are to be provided for each hose connection to prevent the line from thrashing around in case the coupling becomes disconnected. Steam line controls must consist of two shutoff valves, one of which is a quick-acting lever type which is within easy reach of the hammer operator.

Guys, outriggers, thrustouts, or counterbalances are to be provided, as necessary, to maintain stability of pile driver rigs. Barges or floats supporting pile driving operations must meet the applicable requirements of marine operations.

In order to protect the workers, engineers and winchmen are to only accept signals from the designated signalmen. All employees are to be kept clear when piling is being hoisted into the leads. When piles are being driven in an excavated pit, the walls of the pit are to be sloped to the angle of repose, or sheet-piled and braced. When steel tube piles are being "blown out," employees are to be kept well beyond the range of falling materials. When it is necessary to cut off the tops of driven piles, pile driving operations are to be suspended, except where the cutting operations are located, at least twice the length of the longest pile from the driver. When driving jacked piles, all access pits are to be provided with ladders and bulkheaded curbs to prevent material from falling into the pit.

PNEUMATIC TOOLS (1926.302)

Pneumatic tools are powered by compressed air and include chippers, drills, hammers, and sanders (see Figure 8-13). Only assigned, qualified operators should operate power, powder-actuated, or air driven tools. The following safe work procedures should be implemented and enforced at all company construction projects.

Figure 8-13. Example of pneumatic driven grinder

1. Pneumatic tools that shoot nails, rivets, or staples, and operate at pressures more than 100 pounds per square inch, must be equipped with a special device to keep fasteners from being ejected unless the muzzle is pressed against the work surface.

2. Eye protection is required and face protection recommended for employees working with pneumatic tools.

3. Compressed air guns should never be pointed toward anyone.

4. A safety clip or retainer must be installed to prevent attachments from being unintentionally shot from the barrel of the tool.

5. When using pneumatic tools, check to see that they are fastened securely to the hose to prevent them from becoming disconnected. All hoses exceeding 1/2 inch inside diameter must have a safety device at the supply source, or a branch line to reduce pressure in the event of hose failure.

6. Workers operating a jackhammer are required to wear safety glasses, shoes, and hearing protection.

Compressed air is not to be used for cleaning purposes, except where reduced to less than 30 psi, and then only with effective chip guarding and personal protective equipment.

The 30 psi requirement does not apply for concrete form, mill scale, and similar cleaning purposes.

The manufacturer's safe operating pressure for hoses, pipes, valves, filters, and other fittings is not to be exceeded. The use of hoses for hoisting or lowering tools is not permitted. All hoses exceeding 1/2-inch inside diameter must have a safety device at the source of supply, or branch line to reduce pressure in case of hose failure.

Airless spray guns, which atomize paints and fluids at high pressures (1,000 pounds or more per square inch), are to be equipped with an automatic or visible manual safety devices which prevents pulling the trigger and releasing any paint or fluid until the safety device is manually released.

In lieu of the above, a diffuser nut may be provided which prevents high pressure, high velocity release, when the nozzle tip is removed, plus a nozzle tip guard which prevents the tip from coming into contact with the operator, or other equivalent protection.

POTABLE WATER (1926.51)

An adequate supply of drinking (potable) water, which meets USPHS standard 40 CFR 72, is to be available at each place of employment. Potable water containers are to be clearly marked as drinking water, be fitted with a tap, and be capable of being tightly closed. No dipping from a drinking water container is allowed, nor can a common cup be used. If single service cups are used, a sanitary container for unused cups is to be provided, as well as a receptacle for disposing of the used cups (see Figure 8-14).

Figure 8-14. Potable water container with disposable cups

POWDER-ACTUATED GUNS (1926.302)

Powder-actuated tools operate like a loaded gun and should be treated with the same respect and precautions. Only assigned, qualified operators shall operate power, powder-actu-ated, or air-driven tools (see Figure 8-15). The following safe work procedures must be imple-mented and enforced at all company construction projects.

    1.   All powder-actuated tools must meet American National Standards Institute, A10.3-1970, Safety Requirements for Explosive-Actuated Fastening Tools re-quirements for design, operation, and maintenance.

2.   Never use powder-actuated tools in an explosive or flammable atmosphere.

3.   Before using a powder-actuated tool, the worker should inspect it to determine that it is clean, that all moving parts operate freely, and that the barrel is free from obstructions.

4.   Never point the tool at anyone, or carry a loaded tool from one job to another.

5.   Do not load a tool unless it is to be used immediately. Never leave a loaded tool unattended, especially where it would be available to unauthorized persons.

6.   Suitable eye and face protection are essential when using a powder-actuated tool.

7.   In case of misfire, the operator should hold the tool in the operating position for at least 30 seconds, and then attempt to operate the tool for a second time. If the tool misfires again, wait another 30 seconds (still holding the tool in the operating position) and then proceed to remove the explosive load from the tool, in strict accordance with the manufacturer's instructions.

8.   If the tool develops a defect during use, it should be tagged and taken out of service immediately until it is properly repaired.

9.   Warning signs should be posted within the area of operation of any powder-actuated tool.

10.   Powder-actuated tool operators must be qualified and carry a card certifying this fact at all times. Failure to comply with any or all safety procedures governing the use of powder-actuated tools will be sufficient cause for the immediate revocation of the operator's card.

11.   Always store powder-actuated tools in a container by themselves.

Figure 8-15. Powder-actuated tool being used to drive fasteners into masonry wall

The powder-actuated tool is to be tested each day before loading to see that safety devices are in proper working condition. The method of testing shall be in accordance with the manufacturer's recommended procedure. Fasteners are not to be driven into very hard or brittle materials including, but not limited to cast iron, glazed tile, surface-hardened steel, glass block, live rock, face brick, or hollow tile. Driving into easily penetrated materials is to be avoided unless such materials are backed by a substance that will prevent the pin or fastener from passing completely through and creating a flying missile hazard on the other side. And no fastener is to be driven into a spalled area caused by an unsatisfactory fastening.

POWER TOOLS (1926.300)

The following general guidelines apply to those workers who operate power tools. These workers should remove from service any damaged or worn power tools. They should never use power tools with missing guards, always wear safety shoes, gloves, and safety glasses, and never use electric power outdoors or in wet conditions without a GFCI.

All hand held power tools must be equipped with a deadman switch, and a positive "on-off" control must be provided on all hand-held powered platen sanders, grinders with wheels 2-inch diameter or less, and routers, planers, laminate trimmers, nibblers, shears, scroll saws, and jigsaws with blade shanks one-forth of an inch wide or less. A momentary contact "on-off" control must be provided on all hand-held powered drills, tapers, fasteners drivers, horizontal, vertical, and angle grinders with wheels greater than 2 inches in diameter. Tools designed to accommodate guards must be equipped with such guards when in use. All rotating, reciprocating, or moving parts of equipment (belts, gears, shafts, flyheads, etc.) must be guarded to prevent contact by employees using such equipment. Guarding must meet requirements set forth in ANSI B15.1-1953. All hand-held power tools (e.g., circular saws, chain saws, and percussion tools) without a positive accessory holding means must be equipped with a constant pressure switch that will shut off the power when pressure is released (see Figure 8-16).

Figure 8-16. Circular saw with blade guard and equipped with a constant pressure switch

POWER TOOLS, ELECTRICAL (1926.302)

All electrical power operated tools are to be either of the approved double-insulated type or grounded in accordance with Subpart K. The use of electric cords for hoisting or lowering tools is not permitted.

Electric tools present several dangers to the user; the most serious is the possibility of electrocution (see Figure 8-17). Only assigned, qualified operators shall operate power, powder-actuated, or air driven tools. The following safe work procedures should be implemented and enforced at all company construction projects.

Figure 8-17. Damaged electrical power tool on a construction site which should be removed from service

1. Tools must have either a three-wire cord with ground and be grounded, double insulated, or powered by a low-voltage isolation transformer. A Ground-Fault Circuit Interrupter (GFCI) must be used, or the tool must be double-insulated to prevent the worker from electrical shock hazards.

2. Never remove the third prong from the plug.

3. Electric tools should be operated within their design limitations.

4. Gloves and safety footwear are recommended during use of electric tools.

5. When not in use, tools should be stored in a dry place.

6. Electric tools should not be used in damp or wet locations.

7. Work areas should be well lighted.

POWER TOOLS, FUEL DRIVEN (1926.302)

All fuel powered tools are to be stopped while being refueled, serviced, or maintained, and fuel is to be transported, handled with care, and stored to prevent fires. When fuel powered tools are used in enclosed spaces, the applicable requirements for concentrations of toxic gases and the use of personal protective equipment apply.

POWER TOOLS, HYDRAULIC (1926.302)

The fluid used in hydraulic powered tools is to be fire-resistant fluids approved under Schedule 30 of the U.S. Bureau of Mines, Department of the Interior, and must retain its operating characteristics at the most extreme temperatures to which it will be exposed. The manufacturer's safe operating pressures for hoses, valves, pipes, filters, and other fittings are not to be exceeded.

POWER TRANSMISSION AND DISTRIBUTION (1926.950)

Before beginning power transmission and distribution work, all potential hazardous conditions are to be determined by an inspection or a test. Such conditions include, but are not limited to energized lines and equipment, including power and communications, cable television and fire alarm circuits. All electrical equipment or lines are considered to be energized until determined otherwise by testing, or until grounded. It is important to determine the operating voltage of equipment and lines before working on or near energized parts. Crews working on lines or equipment must assure that the means of disconnect are open or locked out. Workers who work on energized lines are to be trained in emergency and first aid procedures. Workers working near or over water are to be protected from drowning. During night operations, spot lights or portable lights are to be used.

Tools and Protective Equipment (1926.951)

In no case are conductive objects without insulated handles allowed closer than the specifications stated in Table 8-3. All rubber protective equipment must comply with the provisions of ANSI J6-1950 (R-1971) series and must be visually inspected before use and rubber gloves are to be "air tested." Metal or conductive ladders are never to be used near energized lines. All tools are to be inspected, and the defective ones removed from service. Portable electric hand tools are to be double insulated. No metal measuring tapes or ropes are to be used near energized parts. Hydraulic and pneumatic tools which are used around energized lines must have nonconductive hoses. Hydraulic fluids used for insulating sections on equipment must be the insulating type.

Table 8-3

Alternating Current – Minimum Distances
Courtesy of OSHA

Voltage range (phase to phase) (kilovolt)	Minimum working and clear hot stick distance
2.1 to 15	2 ft. 0 in.
15.1 to 35	2 ft. 4 in.
35.1 to 46	2 ft. 6 in.
46.1 to 72.5	3 ft. 0 in.
72.6 to 121	3 ft. 4 in.
138 to 145	3 ft. 6 in.
161 to 169	3 ft. 8 in.
230 to 242	5 ft. 0 in.
345 to 362	(1)7 ft. 0 in.
500 to 552	(1)11 ft. 0 in.
700 to 765	(1)15 ft. 0 in.

NOTE: For 345-362 kv., 500-552 kv., and 700-765 kv., minimum clear hot stick distance may be reduced provided that such distances are not less than the shortest distance between the energized part and the grounded surface.

Mechanical Equipment (1926.952)

All mechanical equipment must be visually inspected each time it is used. Aerial-lift trucks are to be grounded or barricaded since they are considered energized equipment. Except for equipment certified to work on specific voltages, mechanical equipment is not to be operated closer than the distances specified in Table 8-4. Workers are not allowed to pass materials from an aerial-lift to a utility pole or structure unless conductor are protected (insulated).

Table 8-4

Minimum Clearance Distances for Live-Line, Bare-Hand Work (Alternating Current) Courtesy of OSHA

Voltage range (phase to phase) kilovolts	Distance in feet and inches for maximum voltage	
	Phase to ground	Phase to phase
2.1 to 15	2'0"	2'0"
15.1 to 35	2'4"	2'4"
35.1 to 46	2'6"	2'6"
46.1 to 72.5	3'0"	3'0"
72.6 to 121	3'4"	4'6"
138 to 145	3'6"	5'0"
161 to 169	3'8"	5'6"
230 to 242	5'0"	8'4"
345 to 362	(1)7'0"	(1)13'4"
500 to 552	(1)11'0"	(1)20'0"
700 to 765	(1)15'0"	(1)31'0"

For 345-362kv., 500-552kv., and 700-765kv., the minimum clearance distance may be reduced provided the distances are not made less than the shortest distance between the energized part and the grounded surface.

Material Handling (1926.953)

Before unloading steel poles, cross arms, or similar materials, they are to be inspected to assure they have not shifted. Poles being transported are to be secured with a red flag attached to the end of longest pole. Materials are not to be stored under energized buses or conductors. If materials are stored under conductors, then the distance found in Table 8-3 applies, and extreme caution is required. To protect workers, taglines or other suitable devices are to be used during hoisting (see Figure 8-18). Workers are not to work under hoisted or suspended loads unless the loads are properly supported.

Figure 8-18. Worker using a tagline to direct a load

Grounding for Protection of Employees (1926.954)

All conductors and equipment are to be treated as energized until deenergized or grounded. All new lines are considered deenergized when they are grounded, no induced voltage is present, or adequate clearances are maintained. All bare wire communications conductors, on poles or structures, are to be considered energized. Conductors or equipment are to be tested to assure they are deenergized prior to grounding. The ground end must always be attached first, and using insulated tools, the other end is to be attached to the conductors or equipment. Grounds must be placed at the work location, or between the work location and all sources of energy. A ground is to exist at each work location along the conductor on a line section. When making a ground is impractical or too dangerous, then work is considered energized. When grounds are being removed, the ground side is removed first, using insulated tools. During testing, grounds are removed only for testing, and with great caution. Ground electrodes must have a low resistance to ground in order to protect workers. Adequately rated and designed tower clamps are to be used to ground towers. All ground leads for tower are to be driven into the ground and capable of conducting the fault current while having a minimum conductance of a No. 2 AWG Copper.

Overhead Lines (1926.955)

Any structure which cannot support a climber must be guyed, braced, or made safe. Poles, ladders, and elevated structures are to be inspected to assure they can support a climber. When standing on the ground, workers are not to touch equipment or machinery next to energized lines or equipment unless protective equipment exists.

Prior to the removal or installing of wires or cables, the strain on poles or structures is to be determined. Poles being set, removed, or moved by using hoists or other devices are to be protected from contacting energized lines. Pole holes are not to be left unattended or unguarded. Lifting equipment is to be bonded to an effective ground and is to be considered energized when near energized equipment or lines. Taglines shall be of nonconductive type.

Metal Tower Construction

While working in unstable material during metal tower construction, the excavation for pad- or pile-type footings, which are in excess of 5 feet deep, shall be either sloped to the angle of repose, or shored, if entry is required. Ladders shall be provided for access to pad- or pile-type footing excavations in excess of 4 feet.

A designated employee shall be used in directing mobile equipment adjacent to footing excavations. No one shall be permitted to remain in the footing while equipment is being spotted for placement. When it is necessary to assure the stability of mobile equipment, the location to be used for such equipment must be graded and leveled.

When working at two or more levels on a tower assembly, employees are to have a minimum exposure to falling objects. Guy lines are to be used, as necessary, to maintain sections, or parts of sections, in position and to reduce the possibility of tipping. Members and sections being assembled shall be adequately supported. The construction of transmission towers, and the erecting of poles, hoisting machinery, site preparation machinery, and other types of construction machinery shall conform to the normal and required construction requirements.

No one shall be permitted under a tower which is in the process of erection or assembly, except as may be required to guide and secure the section being set. When erecting towers using hoisting equipment adjacent to energized transmission lines, the lines shall be deenergized, when practical. If the lines are not deenergized, extraordinary caution shall be exercised to maintain the minimum clearance distances. Erection cranes are to be set on firm level foundations, and when the cranes are so equipped, outriggers shall be used. Taglines should be utilized to maintain control of tower sections being raised and positioned, except where the use of such lines would create a greater hazard. Loadlines are not to be detached from a tower section until the section is adequately secured.

Except during emergency restoration procedures, erection shall be discontinued in the event of high wind or other adverse weather conditions which would make the work hazardous. Equipment and rigging are to be regularly inspected and maintained in safe operating condition. Adequate traffic control must be maintained when crossing highways and railways with equipment. A designated employee shall be utilized to determine that required clearance is maintained in moving equipment under or near energized lines.

Stringing and Removing Deenergized Conductors

Prior to stringing and removal of deenergized conductors, work procedures are to be reviewed, including equipment needs and precautions. When deenergized conductors can contact an energized conductor, the deenergized conductor is to be grounded and workers are to be insulated or isolated. Any existing line is to be deenergized and each end is to be ground on both sides of the crossover or, if energized, it is to be worked as such.

When crossing over an energized line over 600 volts, the ropes nets or guards structures are to be installed unless insulation or isolation of the live conductors from workers and the automatic reclosing feature of the circuit interrupter is made inoperative.

Tension reels, guard structures, tie lines, or other equivalent means are to be used when conductors are strung or removed. Conductor grips must not be used on wire rope unless designed for that purpose. The clipping crew needs to have a minimum of two structures clipped in between the crew and the conductor being sagged. The crews always work between grounds when working on bare conductors. The ground must remain in place until work is competed. Reliable communications is to be maintained between the reel tender and the pulling-rig operator. No other pulls shall take place until the previous pull is snubbed or dead-ended.

Stringing Adjacent to Energized Lines

The potential for induced voltage is to be assessed when stringing lines parallel to energized lines. When the potential for induced voltage exists, pulling and tension equipment is to be isolated, insulated, or grounded. The ground is to be installed between the tension reel and the first structure to ground each bare conductor, subconductor, or overhead ground. During a stringing operation, the conductor is to be grounded at the first tower adjacent to both the tension and pulling setup, and in increments no more than 2 miles from the ground. These grounds are to be left in place until all work is completed, then removed with a hot stick. These conductors are to be grounded at all dead ends or catch-off points. Conductors, subconductors, and ground conductors are to be bonded to the tower at an isolated tower to complete work on a transmission line.

Live-Line Bare-Hand Work

All deenergized lines are to be grounded when working a dead-end tower. When structures are worked, grounds must exist at all work locations to protect workers on conductors. Live-line and bare-hand work is to be done by trained workers. Workers must know the voltage rating of circuits, clearance to ground line, and voltage limitations of the aerial-lift equipment for live-line or bare-hand work. All equipment is to be designed, tested, and intended for live-line and bare-hand work. This type of work is to be supervised by a trained and qualified person.

The automatic reclosing feature of an interrupting device is to be made inoperative prior to working on energized lines or equipment. A conductive bucket liner, or other suitable conductive devices, is to be used for bonding the insulated aerial device to live-line or equipment. Workers are to be connected to the bucket liner with conductive shoes, leg clips, or other equivalent means. At times, electrostatic shielding or conductive clothing is to be provided. The outrigger on the aerial truck is to be extended prior to elevating the boom. The truck is to be bonded to an effective ground or barricade and is to be considered energized. The truck must be inspected prior to moving into the work position. Only clean and dry tools and equipment, which are intended for live-line and bare-hand work, are to be used.

An arm current test is to be made before the beginning of work each, and any time a higher voltage is to be worked on. The test must last at least 3 minutes on buckets which are in use, and leakage is not to be more than one micro-ampere per kilo-volt of the nominal line to line voltage.

Aerial lifts must have both upper and lower controls. The upper control must have override capabilities and be within easy reach of the worker. Lower controls are to be located at base of the boom and are not to be used without permission of the worker in the bucket. Before workers contact energized parts, the conductive bucket liner is to be bonded to the energized conductor and must not be removed until work is completed. The minimum clearance distances are to be printed on durable, nonconductive material posted inside the bucket and in view of the workers.

Underground Lines (1926.956)

To work on underground lines, workers must post warning signs when a cover of a manhole, handholes, or vault is removed. Before workers enter street openings or vaults, barriers, temporary covers, or other suitable guards are to be in place. Workers do not enter manholes or unvented vaults until forced ventilation is provided, or until the atmosphere is tested for oxygen deficiency, the presence of explosive gases, or toxic atmospheres, and found safe. A safety watch must be in the immediate vicinity to render assistance and may occasionally enter the manhole for short periods other than emergencies. A qualified worker can work alone for brief periods to inspect, perform housekeeping, or take readings.

When open flames are used in manholes, extra precautions are to be taken. When combustible gases or liquids may be present, the atmosphere is to be tested to assure safety.

Dangerous underground facilities are to be located to prevent worker exposure or damage to them prior to excavation. Underground facilities are to be protected to avoid damage when they are exposed. All cables found in an excavation are to be protected from damage. The cable to be worked on, when multiple cables exist, is to be identified electrically, unless it has unique characteristics. Cables are to be identified and verified before being cut or spliced. When buried cable or cable in manholes are worked, metallic sheath continuity is to be maintained by bonding across the opening, or by equivalent means.

Construction in Energized Substations (1926.957)

In order for work to proceed on an energized substation, it must be authorized by an authorized person. Energized facilities are to be identified prior to construction, and appropriate PPE is to be selected, as well as the safety precautions for workers denoted. Extra caution is to be taken when busbars and steel for towers and equipment are handled in the vicinity of energized facilities. At times it is may be necessary to deenergize equipment or lines to protect workers.

Barricades and barriers are to be erected to prevent accidental worker contact with energized lines or equipment. Signs indicating the hazard are to be posted near barricades or barriers.

Work on or near control panels is to be performed by designated workers. Precautions are to be taken to prevent jarring, vibration, or improper wiring causing accidental operation of relays or other protective devices.

Vehicles, gin poles, cranes, and other equipment in restricted or hazardous areas are to be controlled by a designated worker. All mobile cranes and derricks are to be effectively grounded when being moved or operated in close proximity to energized lines or equipment, or equipment worked as energized.

When substation fences are removed, a temporary fence, which affords similar protection, is to be provided when the site is unattended. Adequate interconnection between the temporary fence and permanent fence is to be maintained. All gates to unattended substations are to be locked, except when work is in progress.

External Load Helicopters (1926.958)

When using helicopters, the provisions in 29 CFR 1926.551 must be followed.

Lineman's Body Belts, Safety Straps, and Lanyards (1926.959)

Lineman's body belts, safety straps, and lanyards must meet ASTM standard B-117-64. Workers working at elevated locations are to use body belts, safety straps, and lanyards, except operations where their use creates a greater hazard, or other safeguards are used. Body belts, safety straps, and lanyards are to be inspected prior to each use. Damaged equipment is to be removed from service. Safety lines are never to be shock tested. They are used to lower workers during an emergency. The cushion support of a body belt must not contain any exposed rivets on the inside. Body belts have a maximum of four tool loops, four inches in the center of the back (from D-ring to D-ring) kept free.

PRECAST CONCRETE (1926.704)

Precast concrete construction has found its place in parking garages, bridges, and structures, and can be put together in somewhat of a jig puzzle fashion (see Figure 8-19). The large premade concrete parts of a structure require special handling, lifting, support, and an-

Figure 8-19. Example of construction of a precast parking garage

choring procedures. Precast concrete wall units, structural framing, and tilt-up wall panels are to be adequately supported to prevent overturning and collapse until permanent connections are completed. Lifting inserts which are embedded or otherwise attached to tilt-up precast concrete members must be capable of supporting at least two times the maximum intended load applied or transmitted to them. Lifting inserts which are embedded or otherwise attached to precast concrete members, other than the tilt-up members, must be capable of supporting at least four times the maximum intended load applied or transmitted to them (see Figure 8-20). Lifting hardware must be capable of supporting at least five times the maximum intended load applied or transmitted to the lifting hardware. No employee is permitted under precast concrete members being lifted or tilted into position, except those employees required for the erection of those members.

Figure 8-20. Lifting a piece of precast into place using lifting inserts

PROCESS CHEMICAL SAFETY MANAGEMENT (1926.64)

Over a number of years there has been a series of catastrophic releases of toxic, reactive, flammable, or explosive chemicals. These releases can result in toxic, fire, or explosion hazards and there is a definite need for their prevention. Regulations and guidelines were needed to address processes which involve chemicals at or above the specified threshold quantities, and processes which involve flammable liquids or gases on site and in one location in a quantity of 10,000 pounds or more, except for hydrocarbon fuels which are used solely as a fuel for workplace consumption (e.g., propane used for comfort heating, gasoline for vehicle refueling). As long as such fuels are not a part of a process which contains another highly hazardous chemical, these flammable liquids can be stored in atmospheric tanks or transferred. If they are kept below their normal boiling point without benefit of chilling or refrigeration, then these regulations are to be followed.

Retail facilities, oil or gas well drilling or servicing operations, or normally unoccupied remote facilities are not affected by this standard.

Requirements

Employers are required to develop a written plan of action regarding the implementation of employee participation. Employers must consult with employees and their representatives concerning the conduct and development of the process hazards analyses, and, also, the development of other elements of the process safety management in this standard. Employers must provide to the employees and their representatives access to the process hazard analyses and to all other information required under this standard.

The employer is to complete a compilation of the written process safety information before conducting any process hazard analysis required by the standard. The compilation of the written safety information enables the employer and the employees who are involved in operating the process to identify and understand the hazards involved with highly hazardous chemicals. This process safety information must include information pertaining to the hazards of the highly hazardous chemicals used or produced by the process, information pertaining to the technology of the process, and information pertaining to the equipment in the process.

Information pertaining to the hazards of the highly hazardous chemicals must consist of at least the following:

1. Toxicity information.
2. Permissible exposure limits.
3. Physical data.
4. Reactivity data.
5. Corrosivity data.
6. Thermal and chemical stability data.
7. Hazardous effects of inadvertent mixing of different materials that could foreseeably occur.

Also, information concerning the technology of the process includes at least the following:

1. A block flow diagram or simplified process flow diagram.
2. Process chemistry.
3. Maximum intended inventory.
4. Safe upper and lower limits for such items as temperatures, pressures, flows, or compositions.
5. An evaluation of the consequences of deviations, including those affecting the safety and health of employees.

Where the original technical information no longer exists, such information may be developed in conjunction with the process hazard analysis and it must be in sufficient detail to support the analysis. Information pertaining to the equipment in the process shall include

1. Materials of construction.

2. Piping and instrument diagrams (P&IDs).

3. Electrical classification.

4. Relief system design and design basis.

5. Ventilation system design.

6. Design codes and standards employed.

7. Material and energy balances for processes built after May 26, 1992.

8. Safety systems (e.g., interlocks, detection, or suppression systems).

The employer must document that the equipment complies with the recognized and generally accepted good engineering practices. For existing equipment designed and constructed in accordance with codes, standards, or practices that are no longer in general use, the employer must determine and document that the equipment is designed, maintained, inspected, tested, and operated in a safe manner.

The employer shall perform an initial process hazard analysis (hazard evaluation) on processes covered by this standard. The process hazard analysis is to be appropriate to the complexity of the process, and identify, evaluate, and control the hazards involved in the process. Employers shall determine and document the priority order for conducting process hazard analyses. This is to be based on a rationale which includes such considerations as extent of the process hazards, the number of potentially affected employees, the age of the process, and the operating history of the process. The process hazard analysis shall be conducted as soon as possible, but not later than May 26, 1997.

The employer must use one or more of the following methodologies that are appropriate to determine and evaluate the hazards of the process being analyzed: what-if, checklist, what-if/checklist, hazard and operability study (HAZOP), failure mode and effects analysis (FMEA), fault-tree analysis, or an appropriate equivalent methodology.

Process Hazard Analysis

The process hazard analysis must address the hazards of the process, the identification of any previous incident that had a potential for catastrophic consequences in the workplace, the engineering and administrative controls applicable to the hazards, and their interrelationships, such as the appropriate application of detection methodologies in order to provide early warning of releases (acceptable detection methods might include process monitoring and control instrumentation with alarms and detection hardware such as hydrocarbon sensors), the consequences of the failure of engineering and administrative controls, facility siting, human factors, and a qualitative evaluation of the possible safety and health effects on the employees in the workplace, because of the failure of controls.

The process hazard analysis is to be performed by a team with expertise in engineering and process operations, and the team must include at least one employee who has experience and knowledge specific to the process being evaluated. Also, one member of the team must be knowledgeable in the specific process hazard analysis methodology being used. The employer is to establish a system to promptly address the team's findings and recommendations; assure that the recommendations are resolved in a timely manner and that the resolution is documented; document what actions are to be taken; complete actions as soon as possible; develop a written schedule of when these actions are to be completed; communicate the actions to operating, maintenance, and other employees whose work assignments are in the process, and who may be affected by the recommendations or actions.

At least every five (5) years after the completion of the initial process hazard analysis, the process hazard analysis is to be updated and revalidated by a team which meets to assure that the process hazard analysis is consistent with the current process. Employers are to retain process hazards analyses and updates or revalidations for each process covered by this section, as well as the documented resolution of recommendations described for the life of the process.

Employer Responsibility

The employer is to develop and implement written operating procedures that provide clear instructions for safely conducting activities involved in each covered process. These procedures must be consistent with the process safety information, and must address at least the following steps for each operating phase, initial startup, normal operations, temporary operations, emergency shutdown, including the conditions under which emergency shutdown is required and the assignment of the shutdown responsibility to qualified operators, in order to ensure that emergency shutdown is executed in a safe and timely manner. The written operating procedures must also include the procedures for emergency operations, normal shutdown, startup following a turnaround or after an emergency shutdown, operating limits (including the consequences of deviation), the steps required to correct or avoid deviation, the safety and health considerations (such as properties of, and hazards presented by, the chemicals used in the process), the precautions necessary to prevent exposure (including engineering controls, administrative controls, and personal protective equipment), the control measures to be taken if physical contact or airborne exposure occurs, the quality control for raw materials, the control of hazardous chemical inventory levels, or any special or unique hazards, and safety systems and their functions.

Operating procedures are to be readily accessible to employees who work in, or maintain a process. The operating procedures are to be reviewed as often as necessary to assure that they reflect current operating practices, including changes that result from changes in process chemicals, technology, and equipment, and changes to facilities. The employer must certify annually that these operating procedures are current and accurate and is to develop and implement safe work practices to provide for the control of hazards during operations. These safe work practices include such things as lockout/tagout, confined space entry, opening process equipment or piping, and control over entrance into a facility by maintenance, contractor, laboratory, or other support personnel. These safe work practices apply to both employees and contractor employees.

Each employee presently involved in operating a process, and each employee before being involved in operating a newly assigned process, are to be trained in an overview of the process and in the operating procedures. The training shall include an emphasis on the specific safety and health hazards, emergency operations (including shutdown), and safe work practices applicable to the employee's job tasks. In lieu of the initial training, for those employees already involved in operating a process on May 26, 1992, an employer may certify in writing that the employee has the required knowledge, skills, and abilities to safely carry out the duties and responsibilities as specified in the operating procedures. In order to assure that the employees involved in operating a process understand and adhere to the current operating procedures of the process, refresher training is to be provided at least every three years, and more often if necessary. The employer, in consultation with the employees involved in operating the process, shall determine the appropriate frequency of the refresher training. The employer must ascertain that each employee involved in operating a process has received and understands the training. The employer must prepare a record which contains the identity of the employee, the date of training, and the means used to verify that the employee understood the training.

Contractors performing maintenance or repair, turnaround, major renovation, or specialty work on or adjacent to a covered process are impacted by this regulation. It does not apply to contractors providing incidental services which do not influence process safety, such as janitorial work, food and drink services, laundry, delivery, or other supply services. The employer, when selecting a contractor, must obtain and evaluate information regarding the contract employer's safety performance and programs. The employer must inform the contract employers of the known potential fire, explosion, or toxic release hazards related to the contractor's work and the process. The employer has to explain to contract employers the applicable provisions of the emergency action plan. As part of the responsibility of employers, they must develop and implement safe work practices to control the entrance, exit, and the presence of contract employers and their employees while they are in the covered process areas, and the employer must periodically evaluate the performance of the contract employers. Also, the employer must maintain a contract employee injury and illness log related to the contractor's work in process areas.

Contractor Responsibility

The contract employer must assure that each contract employee is trained in the work practices necessary to safely perform his/her job. He or she must also assure that each contract employee is instructed in the known potential of fire, explosion, or toxic release hazards related to his/her job and the process, and must have applicable provisions for the emergency action plan. The contract employer shall document that each contract employee has received and understood the training and must prepare a record which contains the identity of the contract employee, the date of training, and the means used to verify that the employee understood the training. The contract employer must assure that each contract employee follows the safety rules of the facility, including the safe work practices. The contract employer advises the employer of any unique hazards presented by the contract employer's work, or of any hazards found by the contract employer's work.

The Process

The employer performs a pre-startup safety review for new facilities and for modified facilities, when the modification is significant enough to require a change in the process safety information The pre-startup safety review shall confirm that prior to the introduction of highly hazardous chemicals to a process,

1. Construction and equipment are in accordance with design specifications.

2. Safety, operating, maintenance, and emergency procedures are in place and are adequate.

3. For new facilities, a process hazard analysis has been performed and recommendations have been resolved or implemented before startup; and modified facilities meet the requirements contained in management of change.

4. Training of each employee involved in operating a process has been completed.

The types of process equipment that are most applicable are pressure vessels and storage tanks, piping systems (including piping components such as valves), relief and vent systems and devices, emergency shutdown systems, controls (including monitoring devices and sensors, alarms, and interlocks), and pumps.

The employer shall establish and implement written procedures to maintain the ongoing integrity of process equipment. To assure that the employee can perform the job tasks in a safe manner, the employer shall train each employee, involved in maintaining the on-going integrity of process equipment, in an overview of that process, its hazards, and the procedures applicable to the employee's job tasks. Inspections and tests are to be performed on process equipment. Inspection and testing procedures must follow recognized and generally accepted good engineering practices. The frequency of inspections and tests of the process equipment is to be consistent with applicable manufacturers' recommendations and good engineering practices, and more frequently if determined to be necessary by prior operating experience. The employer must document each inspection and test that has been performed on process equipment. The documentation shall identify the date of the inspection or test, the name of the person who performed the inspection or test, the serial number or other identifier of the equipment on which the inspection or test was performed, a description of the inspection or test performed, and the results of the inspection or test. The employer must correct deficiencies in equipment that are outside acceptable limits before further use, or in a safe and timely manner when other necessary means are taken to assure safe operation.

New Operations

In the construction of new plants and equipment, the employer must assure that the equipment, as it is fabricated, is suitable for the process application for which it will be used. Appropriate checks and inspections are to be performed to assure that the equipment is installed properly and consistent with design specifications and the manufacturer's instructions. The employer must assure that maintenance materials, spare parts, and equipment are suitable for the process application for which they will be used.

Hot Work

The employer shall issue a hot work permit for hot work operations conducted on or near a covered process. The permit shall document that the fire prevention and protection requirements have been implemented prior to beginning the hot work operations and it must indicate the date(s) authorized for hot work and identify the object on which hot work is to be performed. The permit is to be kept on file until completion of the hot work operations.

Management Change

The employer must establish and implement written procedures to manage changes (except for "replacements in kind") to process chemicals, technology, equipment, and procedures, and changes to facilities that affect a covered process. The procedures should assure that the following considerations are addressed prior to any change: the technical basis for the proposed change, the impact of the change on safety and health, the modifications to operating procedures, the necessary time period for the change, and the authorization requirements for the proposed change.

Employees who are involved in the operating and maintenance of a process, and contract employees whose job tasks will be affected by a change in the process, operating procedures, or practices, shall be informed of, and trained in, the change prior to start-up of the process, or the affected part of the process. If a change occurs because of a change in the process safety information, operating procedures, or practices, this information is to be updated accordingly.

The employer shall investigate each incident which resulted in, or could reasonably have resulted in a catastrophic release of a highly hazardous chemical in the workplace. An incident investigation is to be initiated as promptly as possible, but not later than 48 hours following the incident. An incident investigation team is to be established and shall consist of at least one person knowledgeable in the process involved, including a contract employee, if the incident involved work of the contractor, and there must also be other persons with appropriate knowledge and experience who will thoroughly investigate and analyze the incident. A report is to be prepared at the conclusion of the investigation which includes, at a minimum, the date of the incident, the date the investigation began, a description of the incident, the factors that contributed to the incident, and any recommendations resulting from the investigation. The employer must establish a system to promptly address and resolve the incident report findings and recommendations. Resolutions and corrective actions are to be documented. The report is to be reviewed with all affected personnel whose job tasks are relevant to the incident findings, including contract employees, where applicable. All incident investigation reports are to be retained for five years.

Emergency Action Plan

The employer needs to establish and implement an emergency action plan for the entire plant. In addition, the emergency action plan must include procedures for handling small releases. Employers covered under this standard may also be subject to the hazardous waste and emergency response provisions.

Compliance Certification

To verify that the procedures and practices developed under the standard are adequate and are being followed, employers must certify that they have evaluated compliance at least every three years. The compliance audit is to be conducted by at least one person knowledgeable in the process. A report of the findings of the audit is to be developed. The employer must promptly determine and document an appropriate response to each of the findings of the compliance audit, and document that the deficiencies have been corrected. Employers need to retain the two (2) most recent compliance audit reports. They must also make all information that is necessary to achieve compliance available to those persons responsible for compiling the process safety information, to those assisting in the development of the process hazard analysis, to those responsible for developing the operating procedures, and to those involved in incident investigations, emergency planning, and response and compliance audits, without regard to possible trade secret status of such information. Nothing precludes the employer from requiring the persons to whom the information is made available to enter into confidentiality agreements where they cannot disclose information deemed as trade secrets. Employees and their designated representatives have access to trade secret information contained within the process hazard analysis and other documents required to be developed by this standard.

RADIATION, IONIZING (1926.53)

Keep clear of all radioactive materials or areas where work is being done with radioactive material. These areas must be barricaded and posted with a radiation hazard sign. A properly executed permit must be approved prior to bringing radioactive sources onto the worksite. In construction and related activities involving the use of sources of ionizing radiation, the pertinent provisions of the Nuclear Regulatory Commission Standards for Protection Against Radiation (10 CFR Part 20), relating to protection against occupational radiation exposure, shall apply. Any activity which involves the use of radioactive materials or X-rays, whether or not under license from the Nuclear Regulatory Commission, is to be performed by competent persons specially trained in the proper and safe operation of such equipment. In the case of materials used under Commission license, only persons actually licensed, or competent persons under the direction and supervision of the licensee, shall perform such work.

RADIATION, NON-IONIZING (LASERS) (1926.54)

Only qualified and trained employees are to be assigned to install, adjust, and operate laser equipment. Proof of the qualification of the laser equipment operator is to be available and in possession of the operator at all times. Laser equipment must bear a label to indicate maximum output and employees are not to be exposed to light intensities above

1.   Direct staring: 1 micro-watt per square centimeter.

2.   Incidental observing: 1 milliwatt per square centimeter;

3.   Diffused reflected light: 2 1/2 watts per square centimeter.

Laser units in operation should be set up above the heads of the employees, when possible. Employees are not to be exposed to microwave power densities in excess of 10 milliwatts per square centimeter. Employees, when working in areas in which a potential exposure to direct or reflected laser light greater than 0.005 watts (5 milliwatts) exists, are to be provided with antilaser eye protection devices. Areas in which lasers are used are to be posted with standard laser warning placards.

Beam shutters or caps are to be utilized, or the laser turned off when laser transmission is not actually required. When the laser is left unattended for a substantial period of time, such as during lunch hour, overnight, or at change of shifts, the laser is to be turned off.

Only mechanical or electronic means are to be used as a detector for guiding the internal alignment of the laser. The laser beam is never to be directed at employees (see Figure 8-21).

Figure 8-21. Construction worker using a laser for leveling operations

When it is raining or snowing, or when there is dust or fog in the air, the operation of laser systems is prohibited where practicable; in any event, employees are to be kept out of range of the area of source and target during such weather conditions.

RIGGING (1926.251)

Rigging Equipment for Material Handling (1926.251)

OSHA's most common citations related to rigging are those citations that involve no inspection of riggings for defects, no informed competent person, no tags on chain or synthetic slings, the use of defective alloy chains, the company's own manufacturing of their lifting devices, and deaths due to overloading or defective rigging.

The rigging of materials to be handled, moved, or lifted has to be done by individuals who have attained skills in the proper and safe rigging procedures. These skills include not only proper rigging procedures, but the knowledge to select the proper equipment for the task. Rigging equipment for material handling must be inspected prior to use on each shift, and as necessary during its use, to ensure that it is safe. Defective rigging equipment is to be removed from service. Rigging equipment is not to be loaded in excess of its recommended safe working load, which is provided in 29 CFR 1926.251 Tables H-1 through H-20. Rigging equipment, when not in use, is to be removed from the immediate work area so as not to present a hazard to employees.

The old rigging saying of "rig to the center of gravity" is critical; the load must be level. Fouling of a load or rigging gear increases loading and can affect load control. Working conditions such as wind, temperature, and dynamic loading are examples of unusual conditions which could affect load distribution. See The Crosby Lifting Guide in Appendix F for more information on load distribution.

The load is to be placed in a sling and connecting hardware which could be dramatically affected by the sling angle and/or all multiple legs (see Figure 8-22). All working load limits are to be based on inline loading. When side loading is allowed, the capacity is to be reduced by at least 75%. See Appendix F, The Crosby Lifting Guide, for information on sling angles.

Figure 8-22. Using a multiple leg sling for lifting

Special custom design grabs, hooks, clamps, or other lifting accessories, for such units as modular panels, prefabricated structures, and similar materials, are to be marked to indicate the safe working loads, and are to be proof-tested prior to use to 125 percent of their rated load.

Each day before being used, the sling and all fastenings and attachments are to be inspected for damage or defects, by a competent person designated by the employer. Additional inspections are to be performed during sling use, where service conditions warrant. Damaged or defective slings must be immediately removed from service.

Welded Alloy Steel Chains (1926.251)

As part of the material handling and rigging process, chains are often the tool of choice (see Figure 8-23). As with other slings, chains should be inspected upon purchase, before each use, and annually or as dictated by use. These inspections are to be performed by the employer's competent person(s). During these inspections, the competent person should look for wear, nicks, cracks, breaks, gouges, stretch, bends, weld splatter, and temperature damage.

Figure 8-23. Chain using a choker hitch for lifting.

Chains should have a safety factor of four to one. All alloy steel chains should have an identification tag which is durable and permanently affixed and states the size, grade, rated capacity, and reach. Chains should never be used to lift loads which are not equal to or less than the manufacturer's rated capacities. Any hooks, rings, oblong links, pear shaped links, welded or mechanical coupling links, or other attachments are to be rated at least equal to the chain you are using. If this is not the case, then the chain's rated lifting capacity must be reduced to the weakest component.

Welded alloy steel chain slings are to have permanently affixed durable identification stating size, grade, rated capacity, and sling manufacturer. Hooks, rings, oblong links, pear-shaped links, welded or mechanical coupling links, or other attachments, when used with alloy steel chains, must have a rated capacity at least equal to that of the chain. Job or shop hooks and links, or makeshift fasteners, formed from bolts, rods, etc., or other such attachments, are never to be used.

Rated capacity (working load limit) for alloy steel chain slings must conform to the values shown in Appendix F. Whenever wear at any point of any chain link exceeds that shown in Table 1 of Appendix F in the Crosby Lifting Guide, the assembly is to be removed from service.

In addition to the daily inspections, a thorough periodic inspection of alloy steel chain slings in use is to be made on a regular basis. The inspection is to be determined by the frequency of sling use, the severity of service conditions, the nature of the lifts being made, and the experience gained on the service life of slings used in similar circumstances. Such inspections are not to be performed at intervals greater than once every 12 months. The employer must make and maintain a record of the most recent month in which each alloy steel chain sling was thoroughly inspected, and make such records available for examination.

Wire Ropes (1926.251)

When wire rope is used, the safe working loads must be determined for the various sizes and classifications of the improved plow steel wire rope and wire rope slings with various types of terminals. For sizes, classifications, and grades not included in Tables H-3 through H-13 of 29 CFR 1926.251, the safe working load, which is recommended by the manufacturer for specific, identifiable products, is to be followed, provided that a safety factor of not less than 5 is maintained. Some general guidelines for wire ropes can be found in Appendix F, as well as information on wire ropes' sling capacities (see Figure 8-24).

Figure 8-24. Example of wire rope slings being used

Protruding ends of strands in splices on slings and bridles are to be covered or blunted. Wire rope is not to be secured by knots, except on haul back lines on scrapers. An eye splice made in any wire rope is not to be less than three full tucks. However, this requirement does not operate to preclude the use of another form of splice or connection which can be shown to be as efficient and which is not otherwise prohibited.

Except for eye splices at the ends of wires, and for endless rope slings, each wire rope used in hoisting, lowering, or pulling loads, must consist of one continuous piece without a knot or splice. Eyes in wire rope bridles, slings, or bull wires are not to be formed by wire rope clips or knots.

Wire rope is not used if, in any length of eight diameters, the total number of visible broken wires exceeds 10 percent of the total number of wires, or if the rope shows other signs of excessive wear, corrosion, or defect. See Figure 8-25 for an example of a lay.

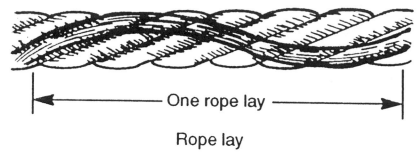

Rope lay

Figure 8-25. Example of rope lay. Courtesy of Department of Energy

When U-bolt wire rope clips are used to form eyes, Appendix F can be used to determine the number and spacing of the clips based upon the wires diameter. When used for eye splices, the U-bolt is to be applied so that the "U" section is in contact with the dead end of the rope (see Figure 8-26).

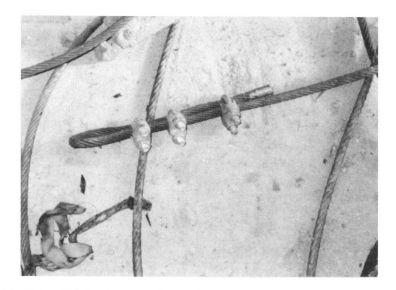

Figure 8-26. Using U-bolt wire rope clips to form an eye

Slings must never be shortened with knots or bolts or other makeshift devices, nor are sling legs to be kinked. Slings used in a basket hitch must have the loads balanced to prevent slippage. Slings must be padded or protected from the sharp edges of their loads. Hands or fingers shall not be placed between the sling and its load while the sling is being tightened around the load. Shock loading is prohibited. Slings are not to be pulled from under a load when the load is resting on the sling.

Cable laid and 6×19 and 6×37 slings must have a minimum clear length of wire rope 10 times the component rope diameter between splices, sleeves, or end fittings. Braided slings must have a minimum clear length of wire rope 40 times the component rope diameter between the loops or end fittings. Cable-laid grommets, strand-laid grommets, and endless slings are to have a minimum circumferential length of 96 times their body diameter.

Fiber core wire rope slings of all grades shall be permanently removed from service if they are exposed to temperatures in excess of 200 degree F. When nonfiber core wire rope slings of any grade are used at temperatures above 400 degree F, or below minus 60 degree F, recommendations by the sling manufacturer regarding use at that temperature shall be followed.

Welding of end attachments, except covers to thimbles, is to be performed prior to the assembly of the sling. All welded end attachments are not to be used unless proof tested by the manufacturer, or equivalent entity at twice their rated capacity, prior to initial use. The employer must retain a certificate of proof of test, and make it available for examination.

Synthetic Rope (1926.251)

When using natural or synthetic fiber rope slings, 29 CFR 1926.251 Tables H-15, 16, 17, and 18 apply. All splices, made to rope slings which are provided by the employer, are to be made in accordance with the manufacturer's fiber rope recommendations. In manila rope, eye splices are to contain at least three full tucks, and short splices must contain at least six full tucks (three on each side of the center line of the splice). In layed synthetic fiber rope, eye splices must contain at least four full tucks, and short splices must contain at least eight full tucks (four on each side of the center line of the splice). Strand end tails are not to be trimmed short (flush with the surface of the rope) but trimmed immediately adjacent to the full tucks. This precaution applies to both eye and short splices and all types of fiber rope. For fiber ropes under 1-inch diameter, the tails must project at least six rope diameters beyond the last full tuck. For fiber ropes 1-inch diameter and larger, the tails must project at least 6 inches beyond the last full tuck. In applications where the projecting tails may be objectionable, the tails are to be tapered and spliced into the body of the rope using at least two additional tucks (this will require a tail length of approximately six rope diameters beyond the last full tuck). For all eye splices, the eye is to be sufficiently large enough to provide an included angle of not greater than 60 degree at the splice, when the eye is placed over the load or support. Knots are never to be used in lieu of splices.

Natural and synthetic fiber rope slings, except for wet frozen slings, may be used in a temperature range from minus 20 degree F to plus 180 degree F without decreasing the working load limit. For operations outside this temperature range, and for wet frozen slings, the manufacturer's recommendations for the sling is to be followed.

Spliced fiber rope slings are not to be used unless they have been spliced in accordance with the minimum requirements, and in accordance with any additional recommendations of the manufacturer. In manila rope, eye splices must consist of at least three full tucks, and short splices must consist of at least six full tucks, three on each side of the splice center line.

For fiber rope under one inch in diameter, the tail shall project at least six rope diameters beyond the last full tuck. For fiber rope one inch in diameter and larger, the tail shall project at least six inches beyond the last full tuck. Where a projecting tail interferes with the use of the sling, the tail shall be tapered and spliced into the body of the rope, using at least two additional tucks (which will require a tail length of approximately six rope diameters beyond the last full tuck).

Fiber rope slings are to have a minimum clear length of rope between eye splices equal to 10 times the rope diameter. Knots are not to be used in lieu of splices. Clamps not designed specifically for fiber ropes are not to be used for splicing. For all eye splices, the eye is to be of such size to provide an included angle of not greater than 60 degrees at the splice, when the eye is placed over the load or support. Fiber rope slings shall not be used if end attachments in contact with the rope have sharp edges or projections.

In synthetic fiber rope, eye splices must consist of at least four full tucks, and short splices must consist of at least eight full tucks, four on each side of the center line. Strand end tails shall not be trimmed flush with the surface of the rope immediately adjacent to the full tucks. This applies to all types of fiber rope, and both eye and short splices. Natural and synthetic fiber rope slings are to be immediately removed from service if any of the following conditions are present: abnormal wear, powdered fiber between strands, broken or cut fibers, variations in the size or roundness of strands, discoloration or rotting, or distortion of hardware in the sling.

Web Slings (1926.251)

Web slings made from synthetic materials with a safety factor of 5 to 1 (ANSI B30.9A-1994) must be proof tested to twice the rated load (see Figure 8-27). Rated capacities are not to be exceeded. Synthetic web slings are to be permanently marked with the manufacturer's name, stock number, type of material, and rate loads for different types of hitches. As with any sling choker, hitch loads are not to exceed 80% of vertical rated load. The rated loads for non-vertical bridle or basket hitches must be adjusted in accordance with the horizontal sling angle. The horizontal angle should never be less than 30 degrees. Guidelines can be found in Appendix F of the Crosby Lifting Guide regarding the use and capacities of web slings.

Figure 8-27. Web sling being used on construction site

Synthetic webbing is to be of uniform thickness and width, and selvage edges are not to be split from the webbing's width. Fittings are to be of a minimum breaking strength equal to that of the sling and must be free of all sharp edges that could in any way damage the webbing. Stitching is to be the only method used to attach end fittings to webbing and to form eyes. The thread must be an even pattern and contain a sufficient number of stitches to develop the full breaking strength of the sling. When synthetic web slings are used, the following precautions are to be taken:

1.   Nylon web slings are not to be used where fumes, vapors, sprays, mists, liquids of acids, or phenolics are present.

2.   Polyester and polypropylene web slings, and slings with aluminum fittings are not to be used where fumes, vapors, sprays, mists, or liquids of caustics are present.

Web slings should be inspected when purchased, prior to each use, and annually. Although OSHA does not require recordkeeping, it is highly recommended. Inspections should be done by your competent person(s) and should be done according to the frequency of use, adverse lifting conditions, or other extenuating circumstances, such as temperature extremes. Synthetic web slings of polyester and nylon are not to be used at temperatures in excess of 180 degree F. Polypropylene web slings are not to be used at temperatures in excess of 200 degree F. Synthetic web slings are to be immediately removed from service if any of the following conditions are present: acid or caustic burns, melting or charring of any part of the sling surface, snags, punctures, tears or cuts, broken or worn stitches, or distortion of fittings.

When inspecting web slings, look for knots/stretching, broken or worn stitches, chemical burns, cuts, snags, holes, and tears or excessive abrasions. The inspection should also include the examination of fittings for cracks, deforming, corrosion/ pitting, breakage, or missing or illegible labels. If any of these are observed, the web sling should be removed from service immediately, and destroyed or tagged out. Remember that ultraviolet or extreme sunlight can also damage web slings; this necessitates storing them out of the sunlight and in a dry area. In all cases, the manufacturer's guidelines, rate loads, and specifications should be followed.

Shackles and Hooks (1926.251)

Appendix F provides information regarding the safe working loads of various sizes of shackles. Higher safe working loads are permissible, when recommended by the manufacturer for specific, identifiable products, provided that a safety factor of not less than five is maintained, as well as hooks. The manufacturer's recommendations are to be followed in determining the safe working loads of the various sizes and types of specific and identifiable hooks. All hooks for which no applicable manufacturer's recommendations are available, are to be tested to twice the intended safe working load, before they are initially put into use (see Figure 8-28). The employer shall maintain a record of the dates and results of such tests.

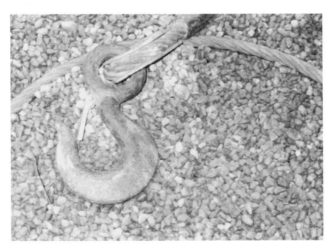

Figure 8-28. Example of a hook with a safety latch

ROLLOVER PROTECTIVE STRUCTURES (1926.1000)

All of the following material handling equipment, rubber-tired self-propelled scrapers, rubber-tired front-end loaders, rubber-tired dozers, wheel-type agricultural and industrial tractors, crawler-type loaders, and motor graders, must have rollover protection structures (ROPS) if manufactured after September 1, 1972. Those manufactured as far back as July 1, 1969 will need to be retrofitted.

ROPS are to be designed to support two times the weight of the equipment at the point of impact; this should protect the operator from being crushed (see Figure 8-29). The ROPS will not protect the operator if he or she is ejected or jumps from the operating cab of the equipment. With ROPS there should be an enforced requirement for operators to wear their seat belts. Many operators feel they can jump clear of the equipment which seldom happens because the direction and momentum of the jump is dictated largely by the direction and speed of the equipment. Suffice to say, most operators who jump are run over by their own equipment which is moving in the same direction as the jumper.

Overhead protection should be provided such that materials cannot fall through the opening in the rollover protection and strike the worker.

Figure 8-29. Dozer with rollover protection and overhead protection

SAFETY HARNESS

As of January 1, 1998, all workers are required to use safety harnesses, or full body harnesses for fall protection; body belts may no longer be used for fall protection. Body belts can still be used for positioning devices. These harnesses' webbing, and their accompanying hardware must posses a tensile strength of 5000 pounds (see Figure 8-30).

SAFETY NETS (1926.105 & 502)

Safety nets are to be used when workplaces are more than 25 feet above the surface, or over water, and when the use of ladders, scaffolds, catch platforms, temporary floors, safety lines, and safety harnesses are not practical. All safety nets are to be inspected daily, and any tools or debris must be removed from them. Safety nets are never used for falls greater than 30 feet. When safety nets are used around the perimeter of a work area, their lateral extension out from the side is to be 8 feet for a potential fall of 5 feet, 10 feet for a potential fall of greater than 5 feet, but less than 10 feet, and 13 feet for a potential fall greater then 10 feet. The maximum size of the net's mesh should be no greater than 36 square inches, or longer than 6 inches. The border ropes for the webbing must have a minimum breaking strength of 5000 pounds. (See Figure 8-31 for example of safety nets.)

When installed safety nets undergo a drop test, this requires a 400 pound bag of sand, approximately 30 inches in diameter, to be dropped into the net from the highest walking/working surface, but not less than 42 inches. This should be repeated after major repairs, and every six months. If a drop test is not practical, the employer must certify and make a record attesting to the fact that it meets the compliance requirements. This activity may be done by a competent person.

Figure 8-30. Example of worker with a full body harness

Figure 8-31. Perimeter safety nets in use to protect steelworkers involved in erection

SCAFFOLDS (1926.450)

The use of scaffolds, as tools for working at varied levels on construction sites, is a fixture in the construction industry (see Figure 8-32). Unfortunately, there have been many accidents involving scaffolding. Some scaffolding has collapsed because of improper erection, some have fallen because the scaffolding could not support the loads placed on it, and some have been improperly used because of the lack of knowledge, or the lack of proper training regarding the use of scaffolds. Scaffolds are one of the leading causes or contributors to construction fall fatalities. They must be designed by a qualified person and be constructed and loaded in accordance with that design.

Figure 8-32. A large scaffold system encompassing an entire building

General Requirements (1926.451)

Each scaffold and scaffold component must be capable of supporting, without failure, its own weight, and at least four times the maximum intended load applied or transmitted to it. Direct connections to the roofs and floors, and counterweights used to balance adjustable suspension scaffolds, are to be capable of resisting at least four times the tipping moment imposed by the scaffold's operating at the rated load of the hoist, or 1.5 (minimum) times the tipping moment imposed by the scaffold operating at the stall load of the hoist, whichever is greater.

Each suspension rope, including connecting hardware, used on non-adjustable or adjustable suspension scaffolds shall be capable of supporting, without failure, at least six times the maximum intended load applied or transmitted to that rope. Adjustable scaffolds must operate at either the rated load of the hoist, or two (minimum) times the stall load of the hoist, whichever is greater. The stall load of any scaffold hoist must not exceed three times its rated load.

Platforms

Each platform on all working levels of scaffolds is to be fully planked or decked between the front uprights and the guardrail supports (see Figure 8-33). The planking is to be installed so that the space between adjacent units, and the space between the platform and the uprights is no more than one inch wide, except where the employer can demonstrate that a wider space is necessary (for example, to fit around uprights when side brackets are used to extend the width of the platform). When the employer can demonstrate this, then the platform is planked or decked as fully as possible and the remaining open space between the platform and the uprights shall not exceed 9 1/2 inches. The requirement to provide full planking or decking does not apply to platforms used solely as walkways, or solely by employees performing scaffold erection or dismantling. In these situations, only the planking that the employer establishes as necessary to provide safe working conditions, is required. Platforms must not deflect more than 1/60 of the span when loaded. Debris is not allowed to accumulate on platforms.

Figure 8-33. A full planked scaffold with guardrails

Each scaffold platform and walkway needs to be at least 18 inches wide. Each ladder jack scaffold, top plate bracket scaffold, roof bracket scaffold, and pump jack scaffold is to be at least 12 inches (30 cm) wide. There is no minimum width requirement for boatswains' chairs. Where scaffolds must be used in areas that the employer can demonstrate are so narrow that platforms and walkways cannot be at least 18 inches wide, such platforms and walkways are to be as wide as feasible, and employees on those platforms and walkways shall be protected from fall hazards by the use of guardrails and/or personal fall arrest systems.

When the front edge of a platform is more than 14 inches from the face of the work, it must have a guardrail system erected along the front edge, and/or personal fall arrest systems to prevent workers from falling. The maximum distance from the face for outrigger scaffolds is 3 inches, and the maximum distance from the face for plastering and lathing operations is 18 inches.

Each end of a platform or its plank, unless cleated or otherwise restrained by hooks or equivalent means, must extend over the centerline of its support at least 6 inches. Each end of a platform or planking 10 feet or less in length is not to extend over its support more than 12 inches, and each platform greater than 10 feet in length must not extend over its support more than 18 inches, unless the platform is designed and installed so that the cantilevered portion of the platform is able to support employees and/or materials without tipping, or has guardrails which block employee access to the cantilevered end. On scaffolds where scaffold planks are abutted to create a long platform, each abutted end must rest on a separate support surface. This does not preclude the use of common support members, such as "T" sections, to support abutting planks, or hooks on platforms designed to rest on common supports. On scaffolds where platforms are overlapped to create a long platform, the overlap must occur only over supports, and is not to be less than 12 inches, unless the platforms are nailed together or otherwise restrained to prevent movement.

At all points of a scaffold where the platform changes direction, such as turning a corner, any platform that rests on a bearer at an angle other than a right angle, is laid first, and a platform which rests at a right angle over the same bearer, is laid second, on top of the first platform. Wood platforms are not to be covered with opaque finishes, except that platform edges may be covered or marked for identification. Platforms may be coated periodically with wood preservatives, fire-retardant finishes, and slip-resistant finishes; however, the coating may not obscure the top or bottom wood surfaces.

Scaffold Components

Scaffold components manufactured by different manufacturers are not to be intermixed unless the components fit together without force, and the scaffold's structural integrity is maintained by the user. Scaffold components manufactured by different manufacturers are not to be modified in order to intermix them, unless a competent person determines the resulting scaffold is structurally sound. Scaffold components made of dissimilar metals are not to be used together unless a competent person has determined that galvanic action will not reduce the strength of any component below accepted support levels.

Supporting Scaffolds

Supported scaffolds with a height to base width (including outrigger supports, if used) ratio of more than four to one (4:1) shall be restrained from tipping, by guying, tieing, bracing, or equivalent means, as follows:

1. Guys, ties, and braces are to be installed at locations where horizontal members support both inner and outer legs.

2. Guys, ties, and braces are to be installed according to the scaffold manufacturer's recommendations, or at the closest horizontal member to the 4:1 height, and to be repeated vertically at locations of horizontal members every 20 feet or less thereafter for scaffolds 3 feet wide or less, and every 26 feet or less thereafter for scaffolds greater than 3 feet wide.

3. The top guy, tie, or brace of completed scaffolds are to be placed no further than the 4:1 height from the top. Such guys, ties, and braces are to be installed at each end of the scaffold and at horizontal intervals not to exceed 30 feet (measured from one end [not both] toward the other).

4. Ties, guys, braces, or outriggers are to be used to prevent the tipping of supported scaffolds in all circumstances where an eccentric load, such as a cantilevered work platform, is applied or is transmitted to the scaffold.

Supported scaffold poles, legs, posts, frames, and uprights are to bear on base plates and mud sills or other adequate firm foundation (see Figure 8-34). Footings are to be level, sound, rigid, and capable of supporting the loaded scaffold without settling or displacement. Unstable objects are not to be used to support scaffolds or platform units, and are not to be used as working platforms. Supported scaffold poles, legs, posts, frames, and uprights are to be plumbed and braced to prevent swaying and displacement.

Figure 8-34. Scaffold legs and base plates supported on mud sills

Front-end loaders and similar pieces of equipment are not to be used to support scaffold platforms unless they have been specifically designed by the manufacturer for such use. Fork-lifts are not to be used to support scaffold platforms unless the entire platform is attached to the fork and the fork-lift is not moved horizontally while the platform is occupied.

Suspension Scaffolds

All suspension scaffold support devices, such as outrigger beams, cornice hooks, parapet clamps, and similar devices, must rest on surfaces capable of supporting at least four times the load imposed on them by the scaffold operating at the rated load of the hoist, or at least 1.5 times the load imposed on them by the scaffold at the stall capacity of the hoist, whichever is greater. Suspension scaffold outrigger beams, when used, are to be made of structural metal or equivalent strength material, and are to be restrained to prevent movement.

The inboard ends of suspension scaffold outrigger beams are to be stabilized to the floor or roof deck by bolts or other direct connections, or they must have their inboard ends stabilized by counterweights; except masons' multipoint adjustable suspension scaffold outrigger beams are not to be stabilized by counterweights. Before the scaffold is used, direct connections are to be evaluated by a competent person who confirms, based on the evaluation, that the supporting surfaces are capable of supporting the loads to be imposed. In addition, masons' multipoint adjustable suspension scaffold connections are to be designed by an engineer experienced in such scaffold design.

Counterweights are to be made of nonflowable material. Sand, gravel, and similar materials that can be easily dislocated are not to be used as counterweights. Only those items specifically designed as counterweights are to be used to counterweight scaffold systems. Construction materials such as, but not limited to, masonry units and rolls of roofing felt, can be used as counterweights. Counterweights are to be secured by mechanical means to the outrigger beams to prevent accidental displacement. Counterweights are not to be removed from an outrigger beam until the scaffold is disassembled.

Outrigger beams, which are not stabilized by bolts or other direct connections to the floor or roof deck, are to be secured by tiebacks and the tiebacks are to be equivalent in strength to the suspension ropes. Outrigger beams are to be placed perpendicular to their bearing supports (usually the face of the building or structure). However, where the employer can demonstrate that it is not possible to place an outrigger beam perpendicular to the face of the building or structure because of obstructions that cannot be moved, the outrigger beam may be placed at some other angle, provided opposing angle tiebacks are used. Tiebacks are to be secured to a structurally sound anchorage on the building or structure. Sound anchorages include structural members, but do not include standpipes, vents, other piping systems, or electrical conduit.

Tiebacks are to be installed perpendicular to the face of the building or structure, or opposing angle tiebacks are to be installed. Single tiebacks which are installed at an angle are prohibited.

Suspension scaffold outrigger beams are to be

1. Provided with stop bolts or shackles at both ends.

2. Securely fastened together with the flanges turned out when channel iron beams are used in place of I-beams.

3. Installed with all bearing supports perpendicular to the beam center line.

4. Set and maintained with the web in a vertical position.

5. When an outrigger beam is used, the shackle or clevis with which the rope is attached to the outrigger beam is to be placed directly over the center line of the stirrup.

Suspension scaffold support devices, such as cornice hooks, roof hooks, roof irons, parapet clamps, or similar devices, are to be

1. Made of steel, wrought iron, or materials of equivalent strength.

2. Supported by bearing blocks.

3. Secured against movement by tiebacks installed at right angles to the face of the building or structure, or opposing angle tiebacks are to be installed and secured to a structurally sound point of anchorage on the building or structure. Sound points of anchorage include structural members, but do not include standpipes, vents, other piping systems, or electrical conduit.

4. Tiebacks are to be equivalent in strength to the hoisting rope.

When winding drum hoists are used on a suspension scaffold, they are to contain not less than four wraps of the suspension rope at the lowest point of scaffold travel. When other types of hoists are used, the suspension rope is to be long enough to allow the scaffold to be lowered to the level below without the rope end passing through the hoist; or, the rope end must be configured or provided with a means to prevent the end from passing through the hoist. The use of repaired wire rope as suspension rope is prohibited. A wire suspension rope is not to be joined together except through the use of eye splice thimbles which are connected with shackles or coverplates and bolts. The load end of wire suspension ropes is to be equipped with proper size thimbles and secured by eyesplicing or equivalent means. Ropes are to be inspected for defects, by a competent person, prior to each workshift and after every occurrence which could affect a rope's integrity. Ropes are to be replaced if any of the following conditions exist:

1. Any physical damage which impairs the function and strength of the rope.

2. Kinks that might impair the tracking or wrapping of rope around the drum(s) or sheave(s).

3. Six randomly distributed broken wires in one rope lay, or three broken wires in one strand in one rope lay.

4. Abrasion, corrosion, scrubbing, flattening, or peening which causes loss of more than one-third of the original diameter of the outside wires.

5. Heat damage caused by a torch, or any damage caused by contact with electrical wires.

6. Evidence that the secondary brake has been activated during an overspeed condition and has engaged the suspension rope.

Swagged attachments, or spliced eyes on wire suspension ropes, are not to be used unless they are made by the wire rope manufacturer or a qualified person. When wire rope clips are used on suspension scaffolds,

1. There must be minimum of three wire rope clips installed, with the clips a minimum of six rope diameters apart.

2. Clips are to be installed according to the manufacturer's recommendations.

3. Clips are to be retightened to the manufacturer's recommendations after the initial loading.

4. Clips are to be inspected and retightened to the manufacturer's recommendations at the start of each workshift thereafter.

5. U-bolt clips are not to be used at the point of suspension for any scaffold hoist.

6. When U-bolt clips are used, the U-bolt is to be placed over the dead end of the rope, and the saddle is to be placed over the live end of the rope.

Suspension scaffold power-operated hoists, and manual hoists, are to be tested by a qualified testing laboratory. Gasoline-powered equipment and hoists are not to be used on suspension scaffolds. Gears and brakes of power-operated hoists which are used on suspension scaffolds are to be enclosed. In addition to the normal operating brake, suspension scaffold power-operated hoists, and manually operated hoists, must have a braking device or locking pawl which engages automatically when a hoist makes either of the following uncontrolled movements: an instantaneous change in momentum, or an accelerated overspeed. Manually operated hoists require a positive crank force to descend.

Two-point and multipoint suspension scaffolds are to be tied or otherwise secured to prevent them from swaying; this is to be determined as necessary, based upon an evaluation by a competent person. Window cleaners' anchors are not to be used for this purpose. When devices are used for the sole purpose of providing emergency escape and rescue, they are not to be used as working platforms. This provision does not preclude the use of systems which are designed to function both as suspension scaffolds and emergency systems.

Accessing Scaffolds

When scaffold platforms are more than two feet above or below a point of access, portable ladders, hook-on ladders, attachable ladders, stair towers (scaffold stairways/towers), stairway-type ladders (such as ladder stands), ramps, walkways, integral prefabricated scaffold access, or direct access from another scaffold, structure, personnel hoist, or similar surface are to be used (see Figure 8-35). Crossbraces are not to be used as a means of access.

Portable, hook-on, and attachable ladders are to be positioned so as not to tip the scaffold. Hook-on and attachable ladders are to be positioned so that their bottom rung is not more than 24 inches above the scaffold supporting level. When hook-on and attachable ladders are used on a supported scaffold more than 35 feet high, they must have rest platforms at 35-foot maximum vertical intervals. Hook on and attachable ladders are ones specifically designed for use with the type of scaffold used. They must have a minimum rung length of 11 1/2

Figure 8-35. Built-in scaffold access ladders

inches and have uniformly spaced rungs with a maximum spacing between rungs of 16 3/4 inches.

Stairway-type ladders shall be positioned such that their bottom step is not more than 24 inches above the scaffold supporting level; shall be provided with rest platforms at 12 foot maximum vertical intervals; shall have a minimum step width of 16 inches (except that mobile scaffold stairway-type ladders must have a minimum step width of 11 1/2 inches); and shall have slip-resistant treads on all steps and landings.

Stairtowers (scaffold stairway/towers) are to be positioned such that their bottom step is not more than 24 inches above the scaffold supporting level.

Stairrails

A stairrail, which consists of a toprail and a midrail, is to be provided on each side of each scaffold stairway. The toprail of each stairrail system is also to be capable of serving as a handrail, unless a separate handrail is provided. Handrails, and toprails that serve as handrails, must provide an adequate handhold for employees grasping them to avoid falling. Stairrail systems and handrails are to be surfaced to prevent injury to employees from punctures or lacerations, and to prevent snagging of clothing. The ends of stairrail systems and handrails are to be constructed so that they do not constitute a projection hazard. Handrails, and toprails that are used as handrails, are to be at least 3 inches from other objects. Stairrails must not be less than 28 inches, nor more than 37 inches from the upper surface of the stairrail to the surface of the tread, and they must be in line with the face of the riser at the forward edge of the tread. A landing platform at least 18 inches wide by at least 18 inches long is to be provided at each level. Each scaffold stairway shall be at least 18 inches wide between stairrails. Guardrails shall be provided on the open sides and ends of each landing.

Stairways and Ramps

Stairways are to be installed between 40 degrees and 60 degrees from the horizontal. The riser height is to be uniform, within 1/4 inch, for each flight of stairs. Greater variations in riser height are allowed for the top and bottom steps of the entire system, not for each flight of stairs. Tread depth is to be uniform, within 1/4 inch, for each flight of stairs. Treads and landings are to have slip-resistant surfaces.

Ramps and walkways 6 feet or more above lower levels must have guardrail systems. No ramp or walkway is to be inclined more than a slope of one (1) vertical to three (3) horizontal (20 degrees above the horizontal.) If the slope of a ramp or a walkway is steeper than one (1) vertical in eight (8) horizontal, the ramp or walkway must have cleats not more than fourteen (14) inches apart, which are securely fastened to the planks to provide footing.

Integrated Scaffold Access

Integral prefabricated scaffold access frames are to be specifically designed and constructed for use as ladder rungs, and must have a rung length of at least 8 inches. They are not to be used as work platforms when rungs are less than 11 1/2 inches in length, unless each affected employee uses fall protection, or a positioning device. They must be uniformly spaced within each frame section and provide rest platforms at 35-foot maximum vertical intervals on all supported scaffolds more than 35 feet high and have a maximum spacing between rungs of 16 3/4 inches. Nonuniform rung spacing, caused by joining end frames together, is allowed provided the resulting spacing does not exceed 16 3/4 inches. Steps and rungs of ladders and stairway type access must line up vertically with each other between rest platforms. Direct access, to or from another surface, is to be used only when the scaffold is not more than 14 inches horizontally, and not more than 24 inches vertically from the other surface.

Access During Erecting or Dismantling

Where the provision of safe access is feasible and does not create a greater hazard, the employer is to provide a safe means of access for each employee erecting or dismantling a scaffold. The employer must have a competent person determine whether it is feasible to provide this access, or if it would pose a greater hazard. This determination is to be based on site conditions and the type of scaffold being erected or dismantled. Hook-on or attachable ladders are to be installed as soon as scaffold erection has progressed to a point that permits safe installation and use. When erecting or dismantling tubular welded frame scaffolds, (end) frames with horizontal members that are parallel, level, and are not more than 22 inches apart vertically, then these may be used as climbing devices for access, provided they are erected in a manner that creates a usable ladder which provides good hand hold and foot space. Cross braces on tubular welded frame scaffolds shall not be used as a means of access or egress.

Other Scaffold Rules

Scaffolds and scaffold components are not to be loaded in excess of their maximum intended loads or rated capacities, whichever is less. The use of shore or lean-to scaffolds is prohibited. Scaffolds and scaffold components are to be inspected for visible defects, by a competent person, before each work shift, and after any occurrence which could affect a scaffold's structural integrity. Any part of a scaffold that is damaged or weakened such that its strength is less than that required is to be immediately repaired or replaced, braced to meet those provisions, or removed from service until repaired.

Scaffolds are not to be moved horizontally while employees are on them, unless they have been designed by a registered professional engineer specifically for such movement.

Scaffolds are not to be erected, used, dismantled, altered, or moved so that they, or any conductive material handled on them, are close to exposed and energized power lines. Table 8-5 provides clearance distances for scaffolds from power lines.

Table 8-5

Scaffold Clearance Distance From Power Lines

*Insulated Lines

Voltage	Minimum distance	Alternatives
Less than 300 volts.	3 feet (0.9 m)	
*300 volts to 50 kv.	10 feet (3.1 m)	
More than 50 kv.	10 feet (3.1 m) plus 0.4 inches (1.0 cm) for each 1 kv over 50 kv.	2 times the length of the line insulator, but never less than 10 feet (3.1 m).

*Uninsulated lines

Voltage	Minimum distance	Alternatives
Less than 50 kv.	10 feet (3.1 m).	
More than 50 kv.	10 feet (3.1 m) plus 0.4 inches (1.0 cm) or each 1 kv over 50 kv.	2 times the length of the line insulator, but never less than 10 feet (3.1 m).

Scaffolds and materials may be closer to power lines than specified above when such clearance is necessary for performance of work, but only after the utility company or electrical system operator has been notified of the need to work closer, and the utility company or electrical system operator has deenergized the lines, relocated the lines, or installed protective coverings to prevent accidental contact with the lines.

Scaffolds are to be erected, moved, dismantled, or altered only under the supervision and direction of a competent person qualified in scaffold erection, moving, dismantling, or alteration. Such activities are to be performed only by experienced and trained employees who are selected, by the competent person, for such work.

Employees are prohibited from working on scaffolds covered with snow, ice, or other slippery material, except as necessary for removal of such materials. Work on or from scaffolds is prohibited during storms or high winds unless a competent person has determined that it is safe for employees to be on the scaffold; those employees are to be protected by a personal fall arrest system or wind screens. Wind screens are not to be used unless the scaffold is secured against the anticipated wind forces imposed.

Where swinging loads are being hoisted onto, or near scaffolds so that the loads might contact the scaffold, tag lines or equivalent measures are to be used to control the loads.

Makeshift devices such as, but not limited to, boxes and barrels are not to be used on top of scaffold platforms to increase the working level height for employees. Ladders are not to be used on scaffolds to increase the working level height for employees, except on large area scaffolds when the ladder can be placed against a structure which is not a part of the scaffold. When this takes place, the scaffold is to be secured against the sideways thrust exerted by the ladder, and the platform units are to be secured to the scaffold to prevent their movement. The ladder legs are to be on the same platform, or other means are to be provided to stabilize the ladder against unequal platform deflection. Also, the ladder legs are to be secured to prevent them from slipping or being pushed off the platform.

To reduce the possibility of the welding current arcing through the suspension wire rope when performing welding from suspended scaffolds, the following precautions are to be taken, as applicable:

1. An insulated thimble is to be used to attach each suspension wire rope to its hanging support (such as cornice hook or outrigger). Excess suspension wire rope, and any additional independent lines from grounding, are to be insulated.

2. The suspension wire rope is to be covered with insulating material extending at least four feet above the hoist. If there is a tail line below the hoist, it is to be insulated to prevent contact with the platform. The portion of the tail line that hangs free below the scaffold shall be guided or retained, or both, so that it does not become grounded.

3. Each hoist is to be covered with insulated protective covers.

4. In addition to a work lead attachment which is required by the welding process, a grounding conductor is to be connected from the scaffold to the structure. The size of this conductor is to be at least the size of the welding process work lead, and this conductor is not to be in series with the welding process or the work piece.

5. If the scaffold grounding lead is disconnected at any time, the welding machine is to be shut off.

6. An active welding rod, or uninsulated welding lead, is not to be allowed to contact the scaffold or its suspension system.

Fall Protection

Each employee on a scaffold, which is more than 10 feet above a lower level, is to be protected from falling to that lower level. The fall protection requirements for employees installing suspension scaffold support systems on floors, roofs, and other elevated surfaces are set forth in Subpart M. Each employee on a boatswains' chair, catenary scaffold, float scaffold, needle beam scaffold, or ladder jack scaffold is to be protected by a personal fall arrest system. Workers on a single-point, or two-point adjustable suspension scaffold, are to be protected by both a personal fall arrest system and a guardrail system. Each employee on a crawling board (chicken ladder) is to be protected by a personal fall arrest system, a guardrail system (with minimum 200 pound toprail capacity), or by a three-fourth inch diameter grabline, or an equivalent handhold which is securely fastened beside each crawling board.

Each employee on a self-contained adjustable scaffold is to be protected by a guardrail system (with minimum 200 pound toprail capacity), when the platform is supported by the frame structure. Each employee is to be protected by both a personal fall arrest system and a guardrail system (with minimum 200 pound toprail capacity), when the platform is supported by ropes.

Each employee who is on a walkway that is located within a scaffold is to be protected by a guardrail system (with minimum 200 pound toprail capacity), and the guardrail system is to be installed within 9 1/2 inches of, and along at least one side of, the walkway.

Workers, who are performing overhand bricklaying operations from a supported scaffold, must be protected from falling from the open sides and ends of the scaffold (except at the side next to the wall being laid) by using a personal fall arrest system, or guardrail system (with minimum 200 pound toprail capacity).

For all other scaffolds, each employee is to use personal fall arrest systems, or guardrail systems for their protection. The employer must have a competent person determine the feasibility and safety of providing fall protection for employees erecting or dismantling supported scaffolds. Employers are required to provide fall protection for employees erecting or dismantling supported scaffolds, where the installation and use of such protection is feasible and does not create a greater hazard. Personal fall arrest systems used on scaffolds are to be attached by lanyard to a vertical lifeline, horizontal lifeline, or scaffold structural member.

When vertical lifelines are used, they are to be fastened to a fixed safe point of anchorage; they are to be independent of the scaffold; and they are to be protected from sharp edges and abrasion. Safe points of anchorage include structural members of buildings, but do not include standpipes, vents, other piping systems, electrical conduit, outrigger beams, or counterweights. Vertical lifelines are not to be used when overhead components, such as overhead protection or additional platform levels, are part of a single-point or two-point adjustable suspension scaffold. Vertical lifelines, independent support lines, and suspension ropes are not to be attached to each other; they are not to be attached to, or used at the same point of anchorage; nor are they to be attached to the same point on the scaffold or personal fall arrest system.

When horizontal lifelines are used, they are to be secured to two or more structural members of the scaffold, or they may be looped around both suspension and independent suspension lines (on scaffolds so equipped) above the hoist and the brake attached to the end of the scaffold. Horizontal lifelines are not to be attached only to the suspension ropes. When lanyards are connected to horizontal lifelines or structural members on a single-point or two-point adjustable suspension scaffold, the scaffold is to be equipped with additional independent support lines and automatic locking devices capable of stopping the fall of the scaffold, in the event one or both of the suspension ropes fail. The independent support lines are to be equal in number and strength to the suspension ropes.

Guardrail systems are to be installed along all open sides and ends of platforms. Guardrail systems are to be installed before the scaffold is released for use by employees other than erection/dismantling crews. The top edge height of toprails, or equivalent members on supported scaffolds which are manufactured or placed in service after January 1, 2000, shall be installed between 38 inches and 45 inches above the platform surface. The top edge height on supported scaffolds manufactured and placed in service before January 1, 2000, and on all suspended scaffolds where both a guardrail and a personal fall arrest system are required, are to be between 36 inches and 45 inches. When conditions warrant, the height of the top edge may exceed the 45-inch height.

When midrails, screens, mesh, intermediate vertical members, solid panels, or equivalent structural members are used, they are to be installed between the top edge of the guardrail system and the scaffold platform. When midrails are used, they are to be installed at a height approximately midway between the top edge of the guardrail system and the platform surface. When screens and mesh are used, they must extend from the top edge of the guardrail system to the scaffold platform, and along the entire opening between the supports. When intermediate members (such as balusters or additional rails) are used, they are not to be more than 19 inches apart. Each toprail, or equivalent member of a guardrail system, is to be capable of withstanding, without failure, a force applied in any downward or horizontal direction, at any point along its top edge, of at least 100 pounds for guardrail systems installed on single-point adjustable suspension scaffolds, or two-point adjustable suspension scaffolds. All other guardrail systems which are installed on all other scaffolds must be capable of withstanding at least 200 pounds. Midrails, screens, mesh, intermediate vertical members, solid panels, and equivalent structural members of a guardrail system are to be capable of withstanding, without failure, a force applied in any downward or horizontal direction, at any point along the midrail or other member, of at least 75 pounds for guardrail systems with a minimum 100 pound toprail capacity, and at least 150 pounds for guardrail systems with a minimum 200 pound toprail capacity. Suspension scaffold hoists and non-walk-through stirrups may be used as end guardrails, if the space between the hoist or stirrup and the side guardrail or structure does not allow passage of an employee to the end of the scaffold.

Guardrails are to be surfaced to prevent injury, to an employee, from punctures or lacerations, and to prevent snagging of clothing. Rails are not to overhang the terminal posts except when such overhang does not constitute a projection hazard to employees. Steel or plastic banding cannot be used as a toprail or midrail. Manila or plastic (or other synthetic) rope, being used for toprails or midrails, is to be inspected by a competent person, as frequently as necessary, to ensure that it continues to meet the strength requirements. Crossbracing is acceptable in place of a midrail, when the crossing point of two braces is between 20 inches and 30 inches above the work platform, or it may be used as a toprail when the crossing point of two braces is between 38 inches and 48 inches above the work platform. The end points at each upright are to be no more than 48 inches apart.

Falling Object Protection

In addition to wearing hardhats, each employee on a scaffold is to be provided with additional protection, from falling hand tools, debris, and other small objects, through the installation of toeboards, screens, or guardrail systems, or through the erection of debris nets, catch platforms, or canopy structures that contain or deflect the falling objects. When the falling objects are too large, heavy, or massive to be contained or deflected by any of the above-listed measures, the employer must place these potential falling objects away from the edge of the surface and must secure those materials, as necessary, to prevent their falling.

Where there is a danger of tools, materials, or equipment falling from a scaffold and striking employees below, the following provisions apply:

1. The area below the scaffold is to be barricaded, and employees are not to be permitted to enter the hazard area.

2. A toeboard is to be erected along the edge of platforms which are more than 10 feet above lower levels, and the distance is to be sufficient to protect employees below. On float (ship) scaffolds, an edging of 3/4 x 1 1/2 inch wood, or the equivalent, may be used in lieu of toeboards.

3. Where tools, materials, or equipment are piled to a height higher than the top edge of the toeboard, paneling or screening which extends from the toeboard or platform to the top of the guardrail is to be erected for a distance sufficient to protect employees below.

4. A guardrail system is to be installed with openings small enough to prevent passage of potential falling objects.

5. A canopy structure, debris net, or catch platform, strong enough to withstand the impact forces of the potential falling objects, is to be erected over the employees below.

Canopies, when used for falling object protection, are to be installed between the falling object hazard and the employees. For falling object protection, canopies are to be used on suspension scaffolds, and the scaffold is to be equipped with additional independent support lines, equal in number to the number of points supported, and equivalent in strength to the strength of the suspension ropes. Independent support lines and suspension ropes are not to be attached to the same points of anchorage.

Where used, toeboards are to be capable of withstanding, without failure, a force of at least 50 pounds applied in any downward or horizontal direction at any point along the toeboard, and at least three and one-half inches high from the top edge of the toeboard to the level of the walking/working surface. Toeboards are to be securely fastened in place at the outermost edge of the platform, and must have no more than 1/4 inch clearance above the walking/working surface. Toeboards are to be solid, or with openings not over one inch in the greatest dimension.

Additional Requirements Applicable to Specific Types of Scaffolds (1926.452)

Pole Scaffolds

When using pole scaffolds and platforms that are being moved to the next level, the existing platform is to be left undisturbed until the new bearers have been set in place and braced prior to receiving the new platforms. Crossbracing is to be installed between the inner and outer sets of poles on double pole scaffolds. Diagonal bracing, in both directions, is to be installed across the entire inside face of double-pole scaffolds, and it is to be used to support loads equivalent to a uniformly distributed load of 50 pounds or more per square foot. Also, diagonal bracing, in both directions, is to be installed across the entire outside face of all double- and single-pole scaffolds. All runners and bearers are to be installed on the edge. Bearers are to extend a minimum of 3 inches over the outside edges of runners, while runners must extend over a minimum of two poles, and are to be supported by bearing blocks which are securely attached to the poles. Braces, bearers, and runners are not to be spliced between poles. Where wooden poles are spliced, the ends are to be squared, and the upper section shall rest squarely on the lower section. Wood splice plates are to be provided on at least two adjacent sides, and they must extend at least 2 feet on either side of the splice, and overlap the abutted ends equally. They must have at least the same cross-sectional areas as the pole. Splice plates, of other materials of equivalent strength, may be used (see Figure 8-36).

Pole scaffolds over 60 feet in height are to be designed by a registered professional engineer, and are to be constructed and loaded in accordance with that design. Nonmandatory Appendix A of 29 CFR 1926.450 to Subpart L contains examples of criteria that will enable an employer to comply with the design and loading requirements for pole scaffolds under 60 feet in height.

Tubular and Coupler Scaffolds

On tublar and coupler scaffold when platforms are being moved to the next level, the existing platform is left undisturbed until the new bearers have been set in place and braced prior to receiving the new platforms. Transverse bracing, which forms an "X" across the width

Figure 8-36. Example of a pole scaffold

of the scaffold, is to be installed at the scaffold ends, at least at every third set of posts horizontally (measured from only one end), and at every fourth runner vertically. Bracing is to extend diagonally from the inner or outer posts or runners, upward to the next outer or inner posts or runners. Building ties are to be installed at the bearer levels between the transverse bracing. On straight run scaffolds, longitudinal bracing is to be installed across the inner and outer rows of posts, diagonally in both directions, and it must extend from the base of the end posts, upward to the top of the scaffold at approximately a 45 degree angle. On scaffolds, where the length is greater than their height, such bracing is to be repeated beginning at least at every fifth post. On scaffolds, where the length is less than their height, bracing is to be installed from the base of the end posts, upward to the opposite end posts, and then in alternating directions until reaching the top of the scaffold. Bracing is to be installed as close as possible to the intersection of the bearer and post or runner and post. Where conditions preclude the attachment of bracing to posts, bracing is to be attached to the runners as close to the post as possible. Scaffolding illustrations of tublar and coupler scaffolds can be found in Appendix O.

Bearers are to be installed transversely between posts, and when coupled to the posts, they must have the inboard coupler bear directly on the runner coupler. When the bearers are coupled to the runners, the couplers are to be as close to the posts as possible. Bearers must extend beyond the posts and runners, and provide full contact with the coupler. Runners are to be installed along the length of the scaffold, and they are to be located on both the inside and outside posts at level heights (when tube and coupler guardrails and midrails are used on outside posts, they may be used in lieu of outside runners). Runners are to be interlocked on straight runs in order to form continuous lengths, and they are to be coupled to each post. The bottom runners and bearers are to be located as close to the base as possible.

Couplers are to be of a structural metal, such as drop-forged steel, malleable iron, or structural grade aluminum. The use of gray cast iron is prohibited. Tube and coupler scaffold, over 125 feet in height are to be designed by a registered professional engineer, and are to be constructed and loaded in accordance with such design. Nonmandatory Appendix A of 29 CFR 1926.450 to this subpart contains examples of criteria that will enable an employer to comply with design and loading requirements for tube and coupler scaffolds under 125 feet in height.

Fabricated Frame Scaffolds

On tubular welded frame scaffolds (see Appendix O), when moving platforms to the next level, the existing platform is to be left undisturbed until the new end frames have been set in place and braced prior to receiving the new platforms. Frames and panels are to be braced by cross, horizontal, or diagonal braces, or combination thereof, which secure vertical members together laterally. Frames and panels are to be joined together vertically, by coupling or stacking pins, or by equivalent means. Where uplift can occur which would displace scaffold end frames or panels, the frames or panels are to be locked together vertically by pins or equivalent means.

The cross braces are to be of a length that will automatically square and align vertical members so that the erected scaffold is always plumb, level, and square. All brace connections are to be secured.

Brackets used to support cantilevered loads shall

1.   Be seated with side-brackets parallel to the frames, and end-brackets at 90 degrees to the frames.

2.   Not be bent or twisted from these positions.

3.   Be used only to support personnel, unless the scaffold has been designed for other loads, by qualified engineer, and has been built to withstand the tipping forces caused by those other loads being placed on the bracket-supported section of the scaffold.

Scaffolds over 125 feet (38.0 m) in height above their base plates are to be designed by a registered professional engineer, and are to be constructed and loaded in accordance with such design.

Plasterers', Decorators', and Large Area Scaffolds

Plasterers', decorators', and large area scaffolds are to follow the construction guidelines and use which have been set forth for pole, tubular, coupler, and fabricated frame scaffolds. Illustrations of these scaffolds can be found in Appendix O.

Bricklayers' Square Scaffolds

Bricklayers' scaffolds are usually made of wood and are to be reinforced with gussets on both sides of each corner, and diagonal braces are to be installed on all sides of each square. Diagonal braces are to be installed between squares on the rear and front sides of the scaffold, and they must extend from the bottom of each square to the top of the next square. The scaffolds are not to exceed three tiers in height, and are to be so constructed and arranged that one square rests directly above the other. The upper tiers must stand on a continuous row of planks which are laid across the next lower tier, and they are to be nailed down, or otherwise secured, to prevent displacement.

Horse Scaffolds

Horse scaffolds are not to be constructed or arranged with more than two tiers, or be more than 10 feet in height, whichever is less (see Appendix O). When horses are arranged in tiers, each horse is to be placed directly over the horse in the tier below. Thus, when the horses are arranged in tiers, the legs of each horse are to be nailed down, or otherwise secured to prevent displacement. If the horses are arranged in tiers, each tier is to be crossbraced.

Form Scaffolds and Carpenters' Bracket Scaffolds

All form or bracket scaffolds, except those for wooden bracket-form scaffolds, are to be attached to the supporting formwork or structure by means of one or more of the following:

nails, a metal stud attachment device, welding, and by hooking over a secured structural supporting member with the form wales either bolted to the form, or secured by snap ties or tie bolts which extend through the form and are securely anchored. For carpenters' bracket scaffolds, only, they may be attached to the supporting formwork or structure by a bolt which extends through to the opposite side of the structure's wall.

Wooden bracket-form scaffolds are an integral part of the form panel. Folding type metal brackets, when extended for use, are either to be bolted or secured with a locking-type pin. Examples of these scaffolds can be found in Appendix O.

Roof Bracket Scaffolds

Scaffold brackets are to be constructed to fit the pitch of the roof and shall provide a level support for the platform. Brackets (including those provided with pointed metal projections) are to be anchored in place by nails, unless it is impractical to use nails. When nails are not used, brackets are to be secured in place with first-grade manila rope of at least three-fourth inch diameter, or equivalent (see Figure 8-37).

Figure 8-37. Example of roof bracket scaffolds

Outrigger Scaffolds

The inboard end of outrigger beams of an outrigger scaffold, measured from the fulcrum point to the extreme point of anchorage, is not to be less than one and one-half times the outboard end in length. Outrigger beams are to be fabricated in the shape of an I-beam or channel, and are to be placed so that the web section is vertical. The fulcrum point of outrigger beams must rest on secure bearings at least 6 inches in each horizontal dimension. Outrigger beams are to be secured in place against movement, and are to be securely braced at the fulcrum point against tipping. The inboard ends of outrigger beams are to be securely anchored, either by means of braced struts bearing against sills which are in contact with the overhead beams or ceiling, or by means of tension members which are secured to the floor joists underfoot, or by both. The entire supporting structure must be securely braced to prevent any horizontal movement. To prevent their displacement, platform units are to be nailed, bolted, or otherwise secured to outriggers. The scaffolds and scaffold components are to be designed by a registered professional engineer, and are to be constructed and loaded in accordance with such design.

Pump Jack Scaffolds

Pump jack brackets, braces, and accessories are to be fabricated from metal plates and angles. Each pump jack bracket must have two positive gripping mechanisms to prevent any failure or slippage. Poles are to be secured to the structure by rigid triangular bracing or equivalent at the bottom, top, and other points, as necessary. When the pump jack has to pass bracing already installed, an additional brace is to be installed approximately 4 feet above the brace to be passed, and it is to be left in place until the pump jack has been moved and the original brace reinstalled. When guardrails are used for fall protection, a workbench may be used as the toprail (see Figure 8-38). Work benches are not to be used as scaffold platforms.

Figure 8-38. Pump jack scaffold

When poles are made of wood, the pole lumber is to be straight-grained, free of shakes, large loose or dead knots, and other defects which might impair strength. If the wood poles are constructed of two continuous lengths, they are to be joined together with the seam parallel to the bracket. When two by fours are spliced to make a pole, mending plates are to be installed at all splices in order to develop the full strength of the member.

Ladder Jack Scaffolds

On ladder jack scaffold platforms, do not exceed a height of 20 feet (see Appendix O). All ladders used to support ladder jack scaffolds must meet the requirements of Subpart X of this part – Stairways and Ladders, except job-made ladders are not to be used to support ladder jack scaffolds. The ladder jack is to be so designed and constructed that it will bear on the side rails and ladder rungs, or on the ladder rungs alone. If bearing on rungs only, the bearing area must include a length of at least 10 inches on each rung. Ladders used to support ladder jacks are to be placed, fastened, or equipped with devices to prevent slipping. Scaffold platforms are never to be bridged one to another.

Window Jack Scaffolds

Window jack scaffolds are to be securely attached to the window opening. These scaffolds are to be used only for the purpose of working at the window opening through which the jack is placed. Window jacks are not to be used to support planks placed between one window jack and another, or for other elements of scaffolding (see Appendix O).

Crawling Boards (Chicken Ladders)

Crawling boards must extend from the roof peak to the eaves, when used in connection with roof construction, repair, or maintenance. Crawling boards are to be secured to the roof by ridge hooks, or by a means that meets equivalent criteria (e.g., strength and durability) (see Appendix O).

Step, Platform, and Trestle Ladder Scaffolds

Step, platform, and trestle ladder scaffolds are not to be placed any higher than the second highest rung or step of the ladder supporting the platform. All ladders used in conjunction with step, platform, and trestle ladder scaffolds must meet the pertinent requirements of Subpart X of this part – Stairways and Ladders, except that job-made ladders are never to be used to support such scaffolds. Ladders used to support step, platform, and trestle ladder scaffolds are to be placed, fastened, or equipped with devices to prevent slipping, and these scaffolds shall not be bridged one to another. Examples of this scaffold can be found in Appendix O.

Single-Point Adjustable Suspension Scaffolds

When two single-point adjustable suspension scaffolds are combined to form a two-point adjustable suspension scaffold, the resulting two-point scaffold must comply with the requirements for two-point adjustable suspension scaffold (see Appendix O).

On a single-point adjustable suspension scaffold, the supporting rope between the scaffold and the suspension device is to be kept vertical unless all of the following conditions are met:

1. The rigging has been designed by a qualified person.

2. The scaffold is accessible to rescuers.

3. The supporting rope is protected to ensure that it will not chafe at any point where a change in direction occurs.

4. The scaffold is positioned so that swinging cannot bring the scaffold into contact with another surface.

Boatswains' Chair

Boatswains' chair tackle must consist of the correct size ball bearings or bushed blocks which contain safety hooks, and must be properly "eye-spliced" with a minimum five-eighth (5/8) inch diameter first-grade manila rope, or other rope which will satisfy the criteria (e.g., strength and durability) of manila rope. Boatswains' chair seat slings are to be reeved through four corner holes in the seat; cross each other on the underside of the seat; and are to be rigged so as to prevent slippage that could cause an out-of-level condition. Boatswains' chair seat slings are to be a minimum of five-eight (5/8) inch diameter fiber, synthetic, or other rope which will satisfy the criteria (e.g., strength, slip resistance, durability, etc.) of first grade manila rope. When a heat-producing process such as gas or arc welding is being conducted, boatswains' chair seat slings are to be a minimum of three-eight (3/8) inch wire rope. Non-cross-laminated wood boatswains' chairs are to be reinforced on their underside by cleats securely fastened to prevent the board from splitting (see Appendix O).

Two-Point Adjustable Suspension Scaffolds (Swing Stages)

To prevent unstable conditions, two-point adjustable suspension scaffolds' (swing stages) platforms are not to be more than 36 inches wide, unless designed by a qualified person. The platforms are to be securely fastened to hangers (stirrups) by U-bolts, or by other satisfactory means. The blocks for fiber or synthetic ropes must consist of at least one double and one single block. The sheaves of all blocks shall fit the size of the rope to be used.

Platforms are to be of the ladder-type, plank-type, beam-type, or light-metal type. Light metal-type platforms having a rated capacity of 750 pounds or less, and platforms 40 feet or less in length which are tested and listed by a nationally recognized testing laboratory, are to be used.

Two-point scaffolds are not to be bridged or otherwise connected one to another during raising and lowering operations, unless the bridge connections are articulated (attached), and the hoists are properly sized. Passage may be made from one platform to another only when the platforms are at the same height, are abutting, and walk-through stirrups, which are specifically designed, are used for this purpose (see Appendix O).

Multipoint Adjustable Suspension Scaffolds, Stonesetters' Multipoint Adjustable Suspension Scaffolds, and Masons' Multipoint Adjustable Suspension Scaffolds

Multipoint adjustable suspension scaffolds, stonesetters' multipoint adjustable suspension scaffolds, and masons' multipoint adjustable suspension scaffolds which use two or more scaffolds, are not to be bridged one to another unless they are designed to be bridged. Examples of these types of scaffolds can be found in Appendix O. The bridge connections are to be articulated, and the hoists are to be properly sized. If bridges are not used, passage may be made from one platform to another only when the platforms are at the same height and are abutting. Scaffolds are to be suspended from metal outriggers, brackets, wire rope slings, hooks, or by means that meet equivalent criteria (e.g., strength, durability).

Catenary Scaffolds

No more than one platform is to be placed between consecutive vertical pickups, and no more than two platforms are to be used on a catenary scaffold (see Appendix O). Platforms, supported by wire ropes, must have hook-shaped stops on each end of the platforms to prevent them from slipping off the wire ropes. These hooks are to be placed so that they will prevent the platform from falling if one of the horizontal wire ropes breaks. Wire ropes must not be tightened to the extent that the application of a scaffold load will overstress them. Wire ropes are to be continuous and without splices between anchors.

Float (Ship) Scaffolds

On a float scaffold, the platform is to be supported by a minimum of two bearers, each of which must project a minimum of 6 inches beyond the platform on both sides (see Appendix O). Each bearer is to be securely fastened to the platform. Any rope connections are to be such that the platform cannot shift or slip. When only two ropes are used with each float, they are to be arranged so as to provide four ends which are securely fastened to overhead supports. Each supporting rope is to be hitched around one end of the bearer and pass under the platform to the other end of the bearer, where it is hitched again, leaving sufficient rope at each end for the supporting ties.

Interior Hung Scaffolds

Interior hung scaffolds are to be suspended only from the roof structure or other structural member, such as ceiling beams. An example of an interior hung scaffold can be found in Appendix O. Overhead supporting members (roof structure, ceiling beams, or other structural members) are to be inspected and checked for strength before the scaffold is erected. Suspension ropes and cables are to be connected to the overhead supporting members by shackles, clips, thimbles, or other means that meet equivalent criteria (e.g., strength, durability).

Needle Beam Scaffolds

These scaffold's support beams are to be installed on edge (see Appendix O). Ropes or hangers are to be used for supports, except that one end of a needle beam scaffold may be supported by a permanent structural member. The ropes are to be securely attached to the needle beams. The support connection is to be arranged so as to prevent the needle beam from rolling or becoming displaced. Platform units are to be securely attached to the needle beams by bolts or equivalent means. Cleats and overhang are not considered to be adequate means of attachment.

Multi-Level Suspended Scaffolds

Multi-level suspended scaffolds are to be equipped with additional independent support lines, equal in number to the number of points supported, and of equivalent strength to the suspension ropes (see Appendix O). They must be rigged to support the scaffold in the event the suspension rope(s) fail. Independent support lines and suspension ropes are not to be attached to the same points of anchorage. Supports for platforms are to be attached directly to the support stirrup and not to any other platform.

Mobile Scaffolds

Mobile scaffolds must be braced by cross, horizontal, or diagonal braces, or a combination thereof, to prevent racking or collapse of the scaffold, and to secure vertical members together laterally so as to automatically square and align the vertical members (See Figure 8-39). Scaffolds are to be plumb, level, and squared. All brace connections are to be secured.

Scaffolds constructed of tube and coupler or fabricated frame components must comply with the previous requirements for these types of scaffolds. When scaffolds are equipped with casters and wheels, they are to be the locking type with positive wheel and/or wheel and swivel locks, or equivalent means, to prevent movement of the scaffold while the scaffold is used in a stationary manner.

The use of manual force to move the scaffold is to be applied as close to the base as practicable, but not more than 5 feet above the supporting surface. Power systems used to propel mobile scaffolds are to be designed for such use. Forklifts, trucks, similar motor vehicles, or add-on motors are not to be used to propel scaffolds unless the scaffold is designed for such propulsion systems.

Scaffolds are to be stabilized to prevent tipping during movement. Workers are not allowed to ride on scaffolds unless the following conditions exist:

1.　The surface on which the scaffold is being moved is within 3 degrees of level, and free of pits, holes, and obstructions.

2.　The height-to-base width ratio of the scaffold during movement is two to one or less, unless the scaffold is designed and constructed to meet or exceed nationally recognized stability test requirements.

Figure 8-39. Example of a mobile scaffold

3. Outrigger frames, when used, are to be installed on both sides of the scaffold.

4. When power systems are used, the propelling force is to be applied directly to the wheels, and it must not produce a speed in excess of one foot per second.

5. No employee is to be on any part of the scaffold which extends outward beyond the wheels, casters, or other supports.

Platforms must not extend outward beyond the base supports of the scaffold, unless outrigger frames or equivalent devices are used to ensure stability. Where leveling of the scaffold is necessary, screw jacks or equivalent means are to be used. Caster stems and wheel stems are to be pinned or otherwise secured in scaffold legs or adjustment screws. Before a scaffold is moved, each employee on the scaffold is to be made aware of the move.

Repair Bracket Scaffolds

Brackets are to be secured in place by at least one wire rope at least 1/2 inch in diameter. Each bracket is to be attached to the securing wire rope (or ropes) by a positive locking device capable of preventing the unintentional detachment of the bracket from the rope, or by equivalent means. At the contact point between the supporting structure and the bottom of the bracket, a shoe (heel block or foot) must be provided which is capable of preventing the lateral movement of the bracket. Platforms are to be secured to the brackets in a manner that will prevent the separation of the platforms from the brackets and the movement of the platforms or the brackets on a completed scaffold.

When a wire rope is placed around the structure in order to provide a safe anchorage for personal fall arrest systems used by employees erecting or dismantling scaffolds, the wire rope is to be at least 5/16 inch in diameter. Each wire rope used for securing brackets in place, or used as an anchorage for personal fall arrest systems, is to be protected from damage due to contact with edges, corners, protrusions, or other discontinuities of the supporting structure or scaffold components. Tensioning of each wire rope used for securing brackets in place, or used as an anchorage for personal fall arrest systems, is to be by means of a turnbuckle at least 1 inch in diameter, or by equivalent means. Each turnbuckle is to be connected to the other end of its rope by use of an eyesplice thimble of a size appropriate to the turnbuckle to which it is attached. U-bolt wire rope clips are not to be used on any wire rope used to secure brackets, or used to serve as an anchor for personal fall arrest systems. The employer must ensure that materials are not dropped to the outside of the supporting structure. Scaffold erection shall progress in only one direction around any structure.

Stilts

An employee may wear stilts on a scaffold only if it is a large area scaffold. When an employee is using stilts on a large area scaffold where a guardrail system is used to provide fall protection, the guardrail system is to be increased in height by an amount equal to the height of the stilts being used by the employee. Surfaces on which stilts are used are to be flat and free of pits, holes, and obstructions such as debris, as well as other tripping and falling hazards. Stilts are to be properly maintained and any alteration of the original equipment is to be approved by the manufacturer.

Training Requirements (1926.454)

The employer must assure that each employee who performs work while on a scaffold is trained by a person qualified in the subject matter to recognize the hazards associated with the type of scaffold being used, and to understand the procedures to control or minimize those hazards. The training shall include the following areas, as applicable:

1. The nature of any electrical hazards, fall hazards, and falling object hazards in the work area.

2. The correct procedures for dealing with electrical hazards, and the procedures for erecting, maintaining, and disassembling the fall protection and falling object protection systems being used.

3. The proper use of the scaffold, and the proper handling of materials on the scaffold.

4. The maximum intended load, and the load-carrying capacities of the scaffolds used.

5. Any other pertinent requirements or information.

The employer must have each employee, who is involved in erecting, disassembling, moving, operating, repairing, maintaining, or inspecting a scaffold, trained, by a competent person, to recognize any hazards associated with the work in question. The training shall include the following topics, as applicable:

1. The nature of scaffold hazards.

2. The correct procedures for erecting, disassembling, moving, operating, repairing, inspecting, and maintaining the type of scaffold in question.

3. The design criteria, maximum intended load-carrying capacity, and intended use of the scaffold.

4. Any other pertinent requirements or information.

When the employer has reason to believe that an employee lacks the skill or understanding needed to work safely with the erection, use, or dismantling of scaffolds, the employer must retrain the employee so that the requisite proficiency is regained. Retraining is to be required in at least the following situations: where changes at the worksite present a hazard;

when an employee has not been previously trained in the types of scaffolds, fall protection, falling object protection, or other equipment that presents a hazard; and when an employer notices inadequacies in an employee who is working on scaffolds and there is an indication that the employee has not retained the requisite proficiency.

SIGNS, SIGNALS, AND BARRICADES

Construction activities at the jobsite may present several potential hazards to workers. The use of signs, signals, and barricades is essential in order to make employees aware that an immediate or potential hazard exists (see Figure 8-40). Therefore, the following safe work procedures for signs, signals, and barricades must be implemented and enforced on each construction project.

Figure 8-40. Warning sign for site requirements

Accident Prevention Signs/Tags

1. **Danger Signs** must be used wherever an immediate hazard (i.e., electrical conductor) exists. The danger signs must have red as the predominant color, in the upper panel, and a white lower panel, for additional sign wording (see Figure 8-41).

2. **Caution Signs** must be used to warn against potential hazards, or to caution against unsafe practices. The caution signs must have yellow as the predominant color, with a black upper panel (yellow lettering of "caution" on the upper panel), and a yellow lower panel, for additional sign wording.

3. **Exit Signs,** when required, shall be in legible red letters, not less than 6 inches high, on a white field.

4. **Safety Instruction Signs**, when used, must be white with a green upper panel, and white lettering to convey the principal message. Any additional wording must be in black lettering on the white background.

Figure 8-41. An example of an electrical danger sign

5. **Directional Signals** must be white with a black panel, and a white directional symbol. Any additional wording must be in black lettering on a white background.

6. **Traffic Signs** must be posted, at points of hazards, in all construction areas. All traffic control signs or devices must conform to ANSI D6.1-1971, *Manual on Uniform Traffic Control Devices for Streets and Highways.*

7. **Accident Prevention Tags** will be used as a temporary means of warning employees of an existing hazard, such as defective tools, equipment, etc.

8. **Out of Order Tags** will be used to designate equipment which requires repair or maintenance. Equipment with such a tag may not be used until the tag is removed.

9. Additional rules, not specifically prescribed in this section, are contained in ANSI Z35.1 1968, Specifications for Accident Prevention Signs and Z35.2-1968, Specifications for Accident Prevention Tags.

SITE CLEARANCE (1926.604)

When workers are engaged in clearing a site, they must be protected from hazards of irritant and toxic plants, and they must be suitably instructed in the first aid treatment available. All equipment used in site clearing operations is to be equipped with rollover protection. In addition, rider-operated equipment is to be equipped with an overhead and rear canopy guard which meets the following requirements:

1. The overhead covering on this canopy structure is not to be less than an 1/8-inch steel plate, or a 1/4-inch woven wire mesh, with openings no greater than 1 inch, or equivalent.

2. The opening in the rear of the canopy structure is to be covered with not less than 1/4-inch woven wire mesh, with openings no greater than 1 inch.

SLIPS, TRIPS, AND FALLS

Slips, trips, and falls result in approximately 20% of workers' compensation claims, 15% of all disabling injuries, 69% of fall-to-same-level injuries, and 15% of occupational deaths due to falls.

The causes for slips, trips, and falls have been linked to an unsafe mindset, unsafe conditions, and unsafe behaviors. Many of these types of accidents could be prevented by becoming aware of state of mind, identifying unsafe conditions or behaviors, selecting the right tools for the job, and using the correct body mechanics. Some examples of using good body mechanics are

1. Don't tilt head.

2. Use all your fingers to grip.

3. Shorten your stride and point feet outward slightly.

4. Walk with knees slightly bent which will help to avoid falling forward.

5. Balance any load you carry.

6. Avoid reaching too far.

Maintain a good center of balance by using eyes, ears, and muscles; correct vision problems; stay in shape; maintain a normal weight; and do not use alcohol or drugs.

Also, some time must be spent looking for potential hazards, such as loose or bent boards, broken floor tiles, unsecured mats or carpets, floor surfaces that change elevation, broken concrete, uncovered manholes, uncovered drains, unsafe ladders or stairs, slippery surfaces, obstructions in walkways, poor lighting, the use of improper shoes, and running or moving too fast. Some hazards and remedies can be found in Table 8-6.

Table 8-6

Making the Work Area Safe

HAZARD	REMEDY
Wet/uneven surfaces	Clean up/report spills
Improper lighting	Keep areas well lit
Winter weather	Wear the right shoes
Wrong tools	Use job-appropriate tools
Poor housekeeping	Clean clutter
Big loads	Carry smaller loads
Not enough time	Slow down

Some shoes are better than others for certain sets of circumstances. Soles which are better for wet surfaces are those that are synthetic rubber soles (neoprene); they should not be worn when oil exists on the surface. Crepe soles (crinkle rubber) are good on rough concrete, but not on tile, wood, or smooth concrete; for these surfaces, aluminum oxide bonded soles are appropriate. For dry surfaces, the use of neoprene soles, crepe soles, or soft rubber soles work well. Hard rubber soles are good for oily surfaces, but are not good for tile, wood surfaces, or wet or dry concrete.

Stairways are often culprits in causing slips, trips, and falls. Some of the stairways' most common conditions and hazards are: stairways without handrails; tools, equipment, litter, and spills left on stairways; stairways covered with ice or snow; and stairways not fully constructed.

Most workers often trip or stumble over unexpected objects in their way; thus, the importance of housekeeping is a key strategy in the prevention of slips, trips, and falls. (see Table 8-7).

Table 8-7

Housekeeping

HOUSEKEEPING DO'S

- Keep everything at work in its proper place.

- Put things away after use.

- Have adequate lighting, or use a flashlight.

- Walk and change directions slowly, especially when carrying anything.

- Make sure the teeth or head on a wrench is in good shape and won't slip when you pull on it.

HOUSEKEEPING DON'TS

- Don't leave machines, tools, or other materials on the floor.

- Don't block walkways or aisles with machines or equipment.

- Don't use a "cheater" on a wrench; get a larger wrench with a larger head or handle, if you need it.

- Don't leave cords, power cables, or air hoses in walkways.

- Don't place anything on stairs.

- Don't leave drawers open.

- Don't carry or push loads that block your vision.

When falls to the same level happen, it usually occurs because someone is running or walking too fast; dangerous surfaces exist, such as icy or wet surfaces; poor visibility exists due to dust, glare, or smoke; or because a worker is carrying a load that blocks the worker's vision.

Clothing can be, at times, a contributing factor to an accident. A fall can occur when a worker stumbles over a loose pant cuff, or when the shoes the worker is wearing aren't appropriate for the job or activity. A worker who is not watching where he or she is going has a greater potential of having an accident, as well as does a worker who does not keep all four legs on the floor when sitting on a chair.

In order to prevent slips, the following guidelines can be followed:

1. Clean up spills, drips, and leaks immediately.

2. Sand icy spots immediately — and tread carefully.

3. Use slip-resistant floor waxes and polishes in offices.

4.   Use steel drains, grates, and splash guards.

5.   Use non-slip paint, mats, treads, and abrasive surfaces to roughen plates, grills, and concrete walking surfaces.

6.   Use rough or grained steel surfaces in areas where there are often spills.

7.   Put up signs or barriers to keep workers away from temporary slip hazards.

8.   Wear shoes with anti-skid soles and materials that resist oils and acids; ask your supervisor about the right shoes for the job.

9.   Avoid turning sharply when walking on a slippery surface.

10.   Keep hands at your sides, not in your pockets.

11.   Walk slowly and slide feet when on wet, slippery, or uneven surfaces.

12.   Don't count on the other workers to report the hazard.

13.   Most important, be careful and take responsibility for your own actions.

Workers often slip, trip, or fall because they are busy performing other tasks and do not pay attention to the risks and the existing hazards. Although there are no foolproof ways to prevent slips, trips, and falls, there is much that can be done, by everyone on a construction site, to mitigate the hazards and foster a mindset which can prevent these types of accidents.

STAIRWAYS (1926.1052)

A stairway or ladder is to be provided, at all personnel points of access, where there is a break in elevation of 19 inches or more, and where there is no ramp, runway, sloped embankment, or personnel hoist provided. Workers are not to use any spiral stairways that will not be a permanent part of the structure on which construction work is being performed.

Stairways that will not be a permanent part of the structure on which construction work is being performed must have landings of not less than 30 inches in the direction of travel, and must extend at least 22 inches in width at every 12 feet or less of vertical rise. Stairs are to be installed between 30 degrees and 50 degrees from the horizontal. Riser height and tread depth must be uniform within each flight of stairs, including any foundation structure used as one or more treads of the stairs. Variations in riser height or tread depth shall not be over 1/4-inch in any stairway system.

Where doors or gates open directly on a stairway, a platform is to be provided, and the swing of the door is not to reduce the effective width of the platform to less than 20 inches.

Metal pan landings and metal pan treads, when used, are to be secured in place before filling with concrete or other material (see Figure 8-42). Except during stairway construction, foot traffic is prohibited on stairways with pan stairs, where the treads and/or landings are to be filled in with concrete or other material at a later date, unless the stairs are temporarily fitted with wood or other solid material at least to the top edge of each pan. Such temporary treads and landings are to be replaced when worn below the level of the top edge of the pan. Metal pan landings and treads should not be left empty since this results in a severe tripping hazard (see Figure 8-43).

All parts of stairways are to be free of hazardous projections, such as protruding nails. Slippery conditions on stairways are to be eliminated before the stairways are used to reach other levels. Except during stairway construction, foot traffic is prohibited on skeleton metal stairs where permanent treads and/or landings are to be installed at a later date, unless the stairs are fitted with secured temporary treads and the landings are long enough to cover the entire

Figure 8-42. Stair pans filled with wood

tread and/or landing area. Treads for temporary service can be made of wood or other solid material, and are to be installed the full width and depth of the stair.

Stairways having four or more risers, or rising more than 30 inches, whichever is less, are to be equipped with at least one handrail and one stairrail system, along each unprotected side or edge. Handrails and the top rails of stairrail systems are to be capable of withstanding, without failure, a force of at least 200 pounds applied within 2 inches of the top edge, in any downward or outward direction, at any point along the top edge.

Figure 8-43. Unfilled stair pans create a real slip/trip/fall hazard

The height of handrails are not to be more than 37 inches, nor less than 30 inches from the upper surface of the handrail to the surface of the tread, and are to be in line with the face of the riser at the forward edge of the tread. Handrails must provide an adequate handhold for employees to grasp them to avoid falling. Handrails that will not be a permanent part of the structure being built must have a minimum clearance of 3 inches between the handrail and walls, stairrail systems, and other objects. Winding and spiral stairways are to be equipped with a handrail, offset sufficiently, to prevent walking on those portions of the stairways where the tread width is less than 6 inches.

The height of stairrails are not to be less than 36 inches from the upper surface of the stairrail system to the surface of the tread, and they must be in line with the face of the riser at the forward edge of the tread. Stairrails installed before March 15, 1991 are not to be less than 30 inches, nor more than 34 inches from the upper surface of the stairrail system to the surface of the tread, and they must be in line with the face of the riser at the forward edge of the tread. Midrails, screens, mesh, intermediate vertical members, or equivalent intermediate structural members, are to be provided between the top rail of the stairrail system and the stairway steps. Midrails, when used, are located at a height midway between the top edge of the stairrail system and the stairway steps.

Screens or mesh, when used, must extend from the top rail to the stairway step, and along the entire opening between top rail supports. When intermediate vertical members, such as balusters, are used between posts, they are not to be more than 19 inches apart. Other structural members, when used, are to be installed such that there are no openings in the stairrail system that are more than 19 inches wide.

The ends of stairrail systems and handrails are to be constructed so as not to constitute a projection hazard. Stairrail systems and handrails are to be surfaced to prevent injury to employees from punctures or lacerations, and to prevent snagging of clothing.

When the top edge of a stairrail system also serves as a handrail, the height of the top edge is not to be more than 37 inches, nor less than 36 inches from the upper surface of the stairrail system to the surface of the tread, and it must be in line with the face of the riser at the forward edge of the tread.

Unprotected sides and edges of stairway landings must be provided with guardrail systems.

When using stairways, hold the handrails when ascending or descending, maintain a good path for vision, if carrying objects while on stairs, and watch your footing in poorly lighted stairwells, or where there are spills, ice, or snow on the stairs (see Table 8-8 and Table 8-9).

Table 8-8

Unsafe Acts on Stairways

- Climbing or descending stairs without holding onto the handrail.
- Carrying a load, especially one that blocks visibility.
- Not cleaning known slippery surfaces.
- Lack of concentration.
- Not keeping stairs free of clutter.
- Forgetting or ignoring safe work practices.
- Slow physical reaction, dizziness, or vision problems.

Table 8-9

Stairs Checklist

_____	1.	Do all stairs have handrails?
_____	2.	Are stairs constructed of sound materials — free of loose steps, splits, cracks, warping, rust, corrosion, or other defects?
_____	3.	Are spills cleaned up immediately?
_____	4.	Are stairs free of ice or snow?
_____	5.	Is the jobsite equipped with lifting devices that keep load carrying on stairs to a minimum?
_____	6.	Do workers go up and down stairs without using handrails?
_____	7.	Do workers carry heavy loads while going up and down stairs?
_____	8.	Do workers carry loads which block their vision while going up and down stairs?
_____	9.	Do workers run up or down stairs?
_____	10.	Do workers go up or down stairs cluttered with tools, equipment, litter, or personal belongings?
_____	11.	Do workers go up or down stairs covered with ice or snow?

STEEL ERECTION (1926.750)

Although erecting steel is one of the unique skills that steel erectors or ironworkers must have, it is still one of the most dangerous activities for construction workers (see Figure 8-44). OSHA has been actively working at coming up with a revised regulation regarding steel erection. See Table 8-10 for a review of the highlights of the proposed changes. The performing of steel erections has resulted in an average of 28 deaths and 1,800 lost-workday injuries each year. The existing hazards include working under loads, hoisting, landing and placing decking, column stability, double connections, landing and placing steel joists, and falling to lower levels.

Figure 8-44. A steel worker's precarious perch

Table 8-10

Highlights of SENRAC Consensus Proposal for Steel Erection

Site Layout and Construction Sequence
- Requires certification of proper curing of concrete in footings, piers, etc., for steel columns.
- Requires controlling contractor to provide erector with a safe site layout, including preplanning routes for hoisting loads.

Site-Specific Erection Plan
- Requires preplanning of key erection elements, including coordination with controlling contractor before erection begins, in certain circumstances.

Hoisting and Rigging
- Provides additional crane safety for steel erection by including elements of American National Standards Institute (ANSI) B30.5 - 1994 in pre-shift inspection.
- Minimizes employee exposure to overhead loads through preplanning and work practice requirements.
- Prescribes proper procedures for multiple lifts ("Christmas-treeing") that reduces overhead exposure, operator fatigue, and eliminates improper work practices.

Structural Steel Assembly
- Provides safer walking/working surfaces by eliminating tripping hazards, such as shear connectors, and minimizes slips through new slip resistance requirements.
- Provides specific work practices regarding landing deck bundles safely, and promotes prompt protection from fall hazards in interior openings.

Anchor Bolts
- Eliminates a major cause of collapse by requiring four anchor bolts per column, along with other column stability requirements.
- Requires procedures that assure the adequacy of anchor bolts which have been modified in the field.

Beams and Columns
- Eliminates the extremely dangerous collapse hazards associated with making double connections at columns by prescribing a safe procedure.

Open Web Steel Joists
- Minimizes the hazards associated with collapse of lightweight steel joists by addressing the need for erection bridging, and a method of attachment.
- Requires bridging terminus anchors as a performance standard, with illustrations and drawings provided in a non-mandatory appendix.
- Minimizes collapse hazards, while placing loads on steel joists, with new requirements that reflect proper industry procedures.

Pre-Engineered Metal Buildings
- Minimizes collapse hazards in the erection of these specialized structures, which account for the majority of steel erection in the U.S.

Falling Object Protection
- Provides protection against all hazards of falling objects in steel erection.

Table 8-10

Highlights of SENRAC Consensus Proposal for Steel Erection
(Continued)

Fall Protection
- Requires fall protection at heights greater than 15 feet, with exceptions for connectors and deckers working in a controlled decking zone. A controlled decking zone is an area where work, such as the initial installation and placement of the metal deck, may be performed without the use of guardrail systems, personal fall arrest systems, or safety net systems, and where access to the zone is controlled.
- Connectors and deckers in the controlled decking zone must be protected at heights greater than two stories, or 30 feet. Connectors, working between 15 and 30 feet, must wear fall arrest or restraint equipment, and those workers be able to be tied off or be provided with another means of fall protection.
- Requires additional provisions for controlled decking zones, because of the number of decking fatalities which have been reported.

Training
- Requires a qualified person to train exposed workers in fall protection.
- Requires a qualified person to train exposed workers engaged in special activities, such as "Christmas treeing," connecting, and decking.

Floor Requirements (1926.750)

During steel erection the installation of permanent floors, as the erection of structural members progresses, and is not to be more than eight stories between the erection floor and the uppermost permanent floor, except where the structural integrity is maintained as a result of the design (see Figure 8-45). There is not to be more than four floors, or 48 feet of unfinished bolting or welding above the foundation, or uppermost permanently secured floor.

Figure 8-45. Flooring installed during steel erection

All derrick or erection floor is to be solidly planked or decked over its entire surface, except for access openings. Planking or decking, of equivalent strength, must be of the proper thickness to carry the working load. Planking is not to be less than 2 inches thick full size undressed, and is to be laid tight, and secured to prevent movement.

If a building or structure is not adaptable to temporary floors, and where scaffolds are not used, safety nets are to be installed and maintained, whenever the potential fall distance exceeds two stories, or 25 feet. The nets are to be hung with sufficient clearance to prevent contacts with the surface of structures below.

Floor periphery-safety railing must be of 1/2-inch wire rope, or its equal, and it is to be installed approximately 42 inches high around the periphery of all temporary-planked floors, or temporary metal-decked floors of tier buildings, and other multi-floored structures, during structural steel assembly.

Where skeleton steel erection is being done, a tightly planked and substantial floor is to be maintained within two stories, or 30 feet, whichever is less, below and directly under that portion of each tier of beams on which any work is being performed, except when gathering and stacking temporary floor planks on a lower floor, in preparation for transferring such planks for use on an upper floor. When gathering and stacking temporary floor planks, the planks are to be removed successively, working toward the last panel of the temporary floor so that the work is always done from the planked floor. Also, when gathering and stacking temporary floor planks from the last panel, the employees assigned to such work are to be protected by safety harnesses with safety lines attached to a catenary line, or other substantial anchorage.

In the erection of a building having double wood floor construction, the rough flooring is to be completed as the building progresses, including the tier below the one on which floor joists are being installed. For single wood floors, or other flooring systems, the floor immediately below the story where the floor joists are being installed, is to be kept planked or decked over.

Structural Steel Assembly (1926.751)

During the final placing of solid web structural members, the load is not to be released from the hoisting line until the members are secured with not less than two bolts, or the equivalent, at each connection, and drawn up wrench tight (see Figure 8-46).

Figure 8-46. Steel worker trying to align the members prior to bolting

Open web steel joists are not to be placed on any structural steel framework unless such framework is safely bolted or welded. No load is to be placed on open web steel joists until security requirements are met.

In steel framing, where bar joists are utilized and columns are not framed in at least two directions with structural steel members, a bar joist is to be field-bolted at the columns to provide lateral stability during construction.

Where longspan joists or trusses 40 feet or longer are used, a center row of bolted bridging must be installed in order to provide lateral stability during construction, and this must be done prior to the slacking of the hoisting line. Taglines are to be used for controlling the loads.

Bolting, Riveting, Fitting-Up, and Plumbing-Up (1926.752)

Containers are to be provided for storing or carrying rivets, bolts, and drift pins, and they must be secured against accidental displacement when aloft. When bolts or drift pins are being knocked out, means are to be provided to keep them from falling. If rivet heads are knocked off or backed out, a means shall be provided to keep them from falling. Riveting is not to be done in the vicinity of combustible material unless precautions are taken to prevent fire.

Pneumatic hand tools are to be disconnected from the power source, and the pressure in hose lines is to be released, before any adjustments or repairs are made. Air line hose sections are to be tied together, except when quick disconnect couplers are used to join sections. A safety wire is to be properly installed on the snap and handle of the pneumatic riveting hammer, and it is to be used at all times. The wire size is not to be less than No. 9 (B&S gauge), leaving the handle and annealed No. 14 on the snap, or equivalent. Impact wrenches are to be provided with a locking device for retaining the socket, and eye protection is to be provided.

Connections of the equipment used in plumbing-up are to be properly secured. The turnbuckles are to be secured to prevent unwinding while under stress. Plumbing-up guys, and related equipment are to be placed so that employees can get at the connection points. Plumbing-up guys are to be removed only under the supervision of a competent person.

Wood planking is to be of the proper thickness so it can carry the working load, but it is not to be less than 2 inches thick full size undressed, exterior grade plywood; it must be at least 3/4-inch thick, or equivalent material. Metal decking of sufficient strength is to be laid tight, and secured to prevent movement. Planks must overlap the bearing on each end by a minimum of 12 inches. Wire mesh, exterior plywood, or equivalent, is to be used around columns where planks do not fit tightly. Provisions are to be made to secure temporary flooring against displacement. All unused openings in floors, temporary or permanent, are to be completely planked over or guarded. Workers are to be provided with safety harnesses when they are working on float scaffolds.

TEMPORARY SLEEPING QUARTERS (1926.51)

Any temporary sleeping quarters must be heated, ventilated, and lighted.

TIRE CAGES (1926.600)

Split rims, or rims with locking rings, have been the cause of serious injuries and deaths. Thus, a safety tire rack, cage, or equivalent, to protect workers during tire maintenance and changing, must be provided when inflating, mounting, and dismounting tires on this type of rim.

TOEBOARDS

Generally speaking, toeboards are to be provided when guard rails are required, or when tools and other materials could be accidentally knocked off of a landing, work platform, scaffold, or other raised area. It is also a reminder to workers that they are near the edge. The acceptable height for a toeboard is usually between 3 1/2 inches and 4 inches. The most common height of a toeboard in the construction industry is the height of the width (3 1/2 inches) of a standard 2×4 board (see Figure 8-47).

Figure 8-47. 3 1/2 inch toeboard on a guardrail system

TOILETS (1926.51)

Toilets are to be provided according to the number of employees. For 20 or less workers, one toilet is provided; for 20 or more workers, one toilet and one urinal are provided per 40 workers; and for 200 or more workers, one toilet and one urinal are provided per 50 workers. Even under temporary field conditions, at least one toilet is to be available. When no sewer is available, privies, chemical, recirculating, or combustible toilets are permissible (see Figure 8-48). The previous requirements do not apply to mobile crews who have transportation to a nearby toilet.

Figure 8-48. Typical toilets found on construction jobsites

TRANSPORTATION

When you have moving vehicles going and coming on a construction site, the potential for errant movement, changes in traffic patterns, and workers getting in the pathway can always occur. All vehicle or heavy equipment operators must obey all traffic signs and posted speeds (See Figure 8-49). All loads are to be secured properly, and hauled material that overhangs the sides or ends of a truck are to be marked. Workers are never to ride running boards, fenders, siderails, tailgates, or tops of vehicles. Workers are not to extend any part of their body outside of a truck bed, or stand in it while moving. Passengers can only be transported when adequate seats, and/or safety provisions are made. With the movement of vehicular traffic by those who are delivering and removing materials to and from the site, the heavy equipment operators may experience blind spots; therefore, workers themselves must be alert and take precautions to protect themselves from being struck by vehicles and equipment.

Figure 8-49. Dangers due to loading and movement of truck traffic frequently occur on construction

TUNNELS/SHAFT (UNDERGROUND CONSTRUCTION) (1926.800)

Construction tunnels, and shafts underground, pose the same construction hazards as aboveground. There are also some added problems, the least of which are: a more confining environment with less space, and a lack of stability of the rocks or materials on all sides. Workers must be alert for loose soil, rock, or fractured materials; these should be removed or properly supported. Also, workers are to be alert to moving equipment, especially around loading and hauling equipment.

During underground, construction workers should never work alone. They must be alert to hazards, and get help to correct them. Workers need to wear their personal protective equipment, such as eye protection from dust, flying objects, lasers, or other eye hazards.

Since the underground operations do not have adequate ventilation, a special effort must be made to ensure that the ventilation is adequate at all times. Dust should be controlled during such operations as drilling, and the lighting must be adequate in underground operations.

With space at a premium, waste materials must not be allowed to accumulate in work the areas or passageways. All hoses, lines, and cords from work operations must be protected from damage.

Underground hazards, such as reduced natural ventilation and light, difficult and limited access and egress, exposure to air contaminants, fire, and explosion, can frequently exist. Therefore, in an effort to minimize injuries and illness associated with underground construction, the following safe work procedures are be implemented and enforced during all underground construction operations on company projects.

Underground Construction (1926.800)

The construction of underground tunnels, shafts, chambers, and passageways, as well as cut-and-cover excavations which are both physically connected to ongoing underground construction operations, and covered in such a manner as to create conditions characteristic of underground construction, are all viewed as underground construction.

Egress and Access

The employer's primary responsibility is to provide and maintain a safe means of access and egress to all work stations. This access and egress is to be designed in such a manner that employees are protected from being struck by excavators, haulage machines, trains, and other mobile equipment.

The employer is to control access to all openings in order to prevent unauthorized entry underground. Unused chutes, manways, or other openings are to be tightly covered, bulkheaded, or fenced off, and are to be posted with warning signs indicating "Keep Out" or similar language. Completed or unused sections of the underground facility are to be barricaded.

Check-In/Check-Out

The employer must maintain a check-in/check-out procedure that will ensure that aboveground personnel can determine an accurate count of the number of persons underground in the event of an emergency. However, this procedure is not required when the construction of underground facilities designed for human occupancy has been sufficiently completed so that the permanent environmental controls are effective and no structural failure could occur within the facilities.

All employees are to be instructed in the recognition and avoidance of hazards associated with underground construction activities including, where appropriate, the following subjects: air monitoring, ventilation, illumination, communications, flood control, mechanical equipment, personal protective equipment, explosives, fire prevention and protection, and emergency procedures, including evacuation plans and check-in/check-out systems.

Communications

Any oncoming shifts are to be informed of any hazardous occurrences or conditions that have affected, or might affect employee safety, including liberation of gas, equipment failures, earth or rock slides, cave-ins, floodings, fires, or explosions.

The employer is to establish and maintain direct communications, for the coordination of activities, with other employers whose operations at the jobsite affect, or may affect the safety of employees underground. When natural, unassisted voice communication is ineffective, a power-assisted means of voice communication is to be used to provide communication between the work face, the bottom of the shaft, and the surface. Two effective means of communication, at least one of which is voice communication, are to be provided in all shafts which are being developed, or are used for either personnel access or hoisting.

Powered communication systems must operate on an independent power supply, and are to be installed so that the use of or disruption of any one phone or signal location will not disrupt the operation of the system from any other location. Communication systems are to be tested upon initial entry of each shift to the underground, and as often as necessary at later times, to ensure that they are in working order.

Emergencies

Any employee working alone in a hazardous underground location, who is both out of the range of natural unassisted voice communication, and who is not under observation of other persons, is to be provided with an effective means of obtaining assistance in an emergency.

When a shaft is used as a means of egress, the employer must make advance arrangements for power-assisted hoisting capability to be readily available in an emergency, unless the regular hoisting means can continue to function in the event of an electrical power failure at the jobsite. Such hoisting means are to be designed so that the load hoist drum is powered in both directions of rotation, and so that the brake is automatically applied upon power release or failure.

The employer is to provide self-rescuers, who have current approval from the National Institute for Occupational Safety and Health, and the Mine Safety and Health Administration, to be immediately available to all employees who are at underground work station areas and might be trapped by smoke or gas.

At least one designated person needs to be on duty, aboveground, whenever any employee is working underground. This designated person is to be responsible for securing immediate aid, and must keep an accurate count of the employees underground, in case of emergency. The designated person must not be so busy that the counting function is encumbered.

Each employee underground shall have an acceptable portable hand lamp, or cap lamp, in his or her work area for emergency use, unless natural light or an emergency lighting system provides adequate illumination for escape.

On jobsites where 25 or more employees work underground at one time, the employer must provide (or make arrangements in advance with locally available rescue services to provide) at least two 5-person rescue teams – one on the jobsite, or within one-half hour travel time from the entry point, and the other one within a 2 hour travel time. On jobsites where less than 25 employees work underground at one time, the employer must provide (or make arrangements in advance with locally available rescue services to provide) at least one 5-person rescue team to be either on the jobsite, or within one-half hour travel time from the entry point.

Rescue team members are to be qualified in rescue procedures, the use and limitations of breathing apparatus, and the use of firefighting equipment. Qualifications shall be reviewed not less than annually. On jobsites, where flammable or noxious gases are encountered or anticipated in hazardous quantities, rescue team members must practice donning and using self-contained breathing apparatus, monthly. The employer must ensure that rescue teams are familiar with conditions at the jobsite.

Gassy Operations

Underground construction operations are to be classified as potentially gassy: if either air monitoring discloses 10 percent or more of the lower explosive limit for methane; if other flammable gases measured at 12 inches to + or – 0.25 inch from the roof, face, floor, or walls, in any underground work area, for more than a 24-hour period on three consecutive days; if the history of the geographical area or geological formation indicates that 10 percent or more of the lower explosive limit for methane, or other flammable gases are likely to be encountered in such underground operations; if there has been an ignition of methane, or of other flammable gases emanating from the strata which indicates the presence of such gases; and if the underground construction operations are both connected to an underground work area, which is currently classified as gassy, and is also subject to a continuous course of air containing the flammable gas concentration.

Underground construction gassy operations may be declassified to Potentially Gassy, when air monitoring results remain under 10 percent of the lower explosive limit for methane or other flammable gases, for three consecutive days. Only acceptable equipment, maintained in suitable condition, shall be used in gassy operations. Mobile diesel-powered equipment, used in gassy operations, is to be approved or verified to be equivalent to requirements of 30 CFR Part 36. Each entrance to a gassy operation is to be prominently posted with signs notifying all entrants of the gassy classification, and smoking is to be prohibited in all gassy operations. The employer is responsible for collecting all personal sources of ignition. A fire watch is to be maintained when hot work is performed.

Once an operation has been classified as gassy, all operations in the affected area, except the following, are to be discontinued until the operation is either in compliance with all of the gassy operation requirements, or has been declassified.

1. Operations related to the control of the gas concentration.

2. Installation of new equipment, or conversion of existing equipment to comply with requirements.

3. Installation of above-ground controls for reversing the air flow.

Air Quality

The employer must assign a competent person to perform all air monitoring. The competent person shall make a reasonable determination as to which substances to monitor, and how frequently to monitor. This is to be based on the location of the jobsite, including the proximity to fuel tanks, sewers, gas lines, old landfills, coal deposits, swamps, and geology of the jobsite, particularly those involving the soil type and its permeability presence of air contaminants in nearby jobsites, as well as the changes in the levels of substances which were monitored on the prior shift. Also the following work practices and jobsite conditions should be taken into account: the use of diesel engines, the use of explosives, the use of fuel gas, the volume and flow of ventilation, the visible atmospheric conditions, the decompression of the atmosphere, welding, cutting and hot work, and the employees' physical reactions to working underground.

The atmosphere in all underground work areas is to be tested, as often as necessary, to ensure that the atmosphere, at normal atmospheric pressure, contains at least 19.5 percent oxygen, and no more than 22 percent oxygen. The atmosphere in all underground work areas is to be tested quantitatively for carbon monoxide, nitrogen dioxide, hydrogen sulfide, and other toxic gases, dusts, vapors, mists, and fumes, as often as necessary, to ensure that the permissible exposure limits prescribed are not exceeded. The atmosphere in all underground work areas is to be tested quantitatively for methane and flammable gases as often as necessary.

If diesel-engine, or gasoline-engine driven ventilating fans or compressors are used, an initial test is to be made of the inlet air of the fan or compressor, with the engines operating, to ensure that the air supply is not contaminated by engine exhaust.

When rapid excavation machines are used, a continuous flammable gas monitor is to be operated at the face, with the sensor(s) placed as high and close to the front of the machine's cutter head as practicable.

Hydrogen Sulfide

Whenever air monitoring indicates the presence of 5 ppm or more of hydrogen sulfide, a test is to be conducted in the affected underground work area(s), at least at the beginning and midpoint of each shift, until the concentration of hydrogen sulfide has been less than 5 ppm for 3 consecutive days. Whenever hydrogen sulfide is detected in an amount exceeding 10 ppm, a continuous sampling and indicating hydrogen sulfide monitor is to be used to monitor the affected work area. Employees are to be informed when a concentration of 10 ppm hydrogen sulfide is exceeded. The continuous sampling and indicating hydrogen sulfide monitor is to be designed, installed, and maintained to provide a visual and aural alarm, when the hydrogen sulfide concentration reaches 20 ppm, to signal that additional measures, such as respirator use, increased ventilation, or evacuation, might be necessary to maintain hydrogen sulfide exposure below the permissible exposure limit.

When the competent person determines, on the basis of air monitoring results or other information, that air contaminants may be present in sufficient quantity to be dangerous to life, the employer must prominently post a notice at all entrances to the underground jobsite to inform all entrants of the hazardous condition, and must ensure that the necessary precautions are taken.

Flammable Gases

Whenever five percent or more of the lower explosive limit for methane or other flammable gases is detected in any underground work area(s), or in the air return, steps are to be taken to increase ventilation air volume, or to otherwise control the gas concentration, unless the employer is operating in accordance with the potentially gassy, or gassy operation requirements. Such additional ventilation controls may be discontinued when gas concentrations are reduced below five percent of the lower explosive limit, but are reinstituted whenever the five percent level is exceeded. Whenever 10 percent or more of the lower explosive limit for methane or other flammable gases is detected in the vicinity of welding, cutting, or other hot work, such work is to be suspended until the concentration of such flammable gas is reduced to less than 10 percent of the lower explosive limit.

Whenever 20 percent or more of the lower explosive limit for methane or other flammable gases is detected in any underground work area(s), or in the air return, all employees, except those necessary to eliminate the hazard, are to be immediately withdrawn to a safe location aboveground. Electrical power, except for acceptable pumping and ventilation equipment, is to be cut off to the area endangered by the flammable gas, until the concentration of such gas is reduced to less than 20 percent of the lower explosive limit.

Operations which meet the criteria for potentially gassy and gassy operations are subject to the additional monitoring. A test for oxygen content is to be conducted in the affected underground work areas, and the work areas immediately adjacent to such areas, at least at the beginning and midpoint of each shift. When using rapid excavation machines, continuous automatic flammable gas monitoring equipment is to be used to monitor the air at the heading, on the rib, and in the return air duct. The continuous monitor signals the heading and shuts down electric power in the affected underground work area, except for acceptable pumping and ventilation equipment, when 20 percent or more of the lower explosive limit for methane or other flammable gases are encountered. A manual flammable gas monitor is to be used, as needed, and at least at the beginning and midpoint of each shift, to ensure that the limits are not exceeded. In addition, a manual electrical shutdown control is to be provided near the heading. Local gas tests are to be made prior to, and continuously during any welding, cutting, or other hot work.

In underground operations driven by drill-and-blast methods, the air in the affected area is to be tested, after blasting, and prior to re-entry, for flammable gas, and it is to be tested continuously when employees are working underground.

A record of all air quality tests is to be maintained aboveground at the worksite, and it is to be made available upon request. The record must include the location, date, time, substance, and amount monitored. Records of exposures to toxic substances are to be retained. All other air quality test records shall be retained until completion of the project.

Ventilation

Fresh air is to be supplied to all underground work areas in sufficient quantities to prevent dangerous or harmful accumulation of dusts, fumes, mists, vapors, or gases. Mechanical ventilation is to be provided in all underground work areas, except when the employer can demonstrate that natural ventilation provides the necessary air quality through sufficient air volume and air flow. A minimum of 200 cubic feet of fresh air per minute is to be supplied for each employee underground. The linear velocity of air flow in the tunnel bore, in shafts, and in all other underground work areas, must be at least 30 feet per minute where blasting or rock drilling is conducted, or where other conditions which are likely to produce dust, fumes, mists, vapors, or gases in harmful or explosive quantities, are present. The direction of the mechanical air flow must be reversible. Following blasting, ventilation systems should exhaust smoke and fumes to the outside atmosphere before work is resumed in affected areas.

Ventilation doors are to be designed and installed so that they remain closed when in use, regardless of the direction of the air flow. When ventilation has been reduced to the extent that hazardous levels of methane or flammable gas may have accumulated, a competent person must test all affected areas after ventilation has been restored, and the competent person must determine whether the atmosphere is within flammable limits, before any power, other than for acceptable equipment, is restored, or before work is resumed. Whenever the ventilation system has been shut down with all employees out of the underground area, only competent persons, who are authorized to test for air contaminants, are allowed to be underground until the ventilation has been restored and all affected areas have been tested for air contaminants and declared safe.

When drilling rocks or concrete, appropriate dust control measures are to be taken to maintain dust levels within limits. Such measures may include, but are not limited to, wet drilling, the use of vacuum collectors, and water mix spray systems. Internal combustion engines, except diesel-powered engines on mobile equipment, are prohibited underground. Mobile diesel-powered equipment, used underground in atmospheres other than gassy operations, is either to be approved or certified by the employer as equivalent. (Each brake horsepower of a diesel engine requires at least 100 cubic feet of air per minute for suitable operation in addition to the air requirements for personnel. Some engines may require a greater amount of air to ensure that the allowable levels of carbon monoxide, nitric oxide, and nitrogen dioxide are not exceeded.)

Potentially gassy or gassy operations need to have ventilation systems installed which are constructed of fire-resistant materials and must have acceptable electrical systems, including fan motors.

Gassy operations are to be provided with controls, located aboveground, for reversing the air flow of ventilation systems. In potentially gassy or gassy operations, where mine-type ventilation systems are used which have an offset main fan installed on the surface, they are to be equipped with explosion-doors, or have a weak-wall which has an area at least equivalent to the cross-sectional area of the airway.

Illumination

Illumination requirements, applicable to underground construction operations, must be employed. Only acceptable portable lighting equipment is to be used within 50 feet of any underground heading during explosives handling.

Fire Prevention

Fire prevention and protection requirements, applicable to underground construction operations, are to be followed. Open flames and fires are to be prohibited in all underground construction operations, except as permitted for welding, cutting, and other hot work operations. Smoking may be allowed only in areas free of fire and explosion hazards. Readily visible signs prohibiting smoking and open flames are to be posted in areas having fire explosion hazards. The employer is not permitted to store underground more than a 24-hour supply of diesel fuel for the underground equipment used at the worksite. The piping of diesel fuel, from the surface to an underground location, is permitted only if the diesel fuel is contained at the surface. This fuel must be in a tank which has a maximum capacity of no more than the amount of fuel required to supply, for a 24-hour period, the equipment serviced by the underground fueling station. The surface tank is to be connected to the underground fueling station by an acceptable pipe or hose system, and this system is to be controlled at the surface by a valve, and at the shaft bottom by a hose nozzle. The pipe is to be empty at all times, except when transferring diesel fuel from the surface tank to a piece of equipment in use underground. In the shaft, hoisting operations are to be suspended during refueling operations, if the supply piping, in the shaft, is not protected from damage. Gasoline is not to be carried, stored, or used underground. Acetylene, liquefied petroleum gas, and methylacetylene propadiene stabilized gas may be used underground only for welding, cutting, and other hot work.

Oil, grease, and diesel fuel, stored underground, are to be kept in tightly sealed containers and in fire-resistant areas at least 300 feet from underground explosive magazines. They are to be at least 100 feet from shaft stations and steeply inclined passageways. Storage areas are to be positioned or diked so that the contents of ruptured or overturned containers will not flow from the storage area.

Flammable or combustible materials are not to be stored aboveground within 100 feet of any access opening to any underground operation. Where this is not feasible because of space limitations at the jobsite, such materials may be located within the 100-foot limit, provided that they are located as far as practicable from the opening. Either a fire-resistant barrier, of not less than one-hour rating, is to be placed between the stored material and the opening, or additional precautions are to be taken which will protect the materials from ignition sources. Fire-resistant hydraulic fluids are to be used in hydraulically actuated underground machinery and equipment unless such equipment is protected by a fire suppression system, or by multi-purpose fire extinguisher(s), rated at sufficient capacity for the type and size of hydraulic equipment involved, but rated at least 4A:40B:C.

Electrical installations in underground areas where oil, grease, or diesel fuel are stored, are to be used only for lighting fixtures. Lighting fixtures in storage areas, or within 25 feet of underground areas where oil, grease, or diesel fuel are stored, are to be approved for Class I, Division 2 locations.

Leaks and spills of flammable or combustible fluids are to be cleaned up immediately. A fire extinguisher of at least 4A:40B:C rating, or other equivalent extinguishing means, is to be provided at the head pulley, and at the tail pulley of underground belt conveyers. Any structure located underground, or within 100 feet of an opening to the underground, is to be constructed of material having a fire-resistance rating of at least one hour.

When underground welding, cutting, and other hot work is occurring, no more than the amount of fuel gas and oxygen cylinders necessary to perform welding, cutting, or other hot work during the next 24-hour period, is permitted underground, and noncombustible barriers are to be installed below welding, cutting, or other hot work being done in, or over a shaft or raise.

Unstable Formations

Portal openings and access areas are to be guarded by shoring, fencing, head walls, shotcreting, or other equivalent protection, to ensure safe access of employees and equipment. Adjacent areas are to be scaled, or otherwise secured to prevent loose soil, rock, or fractured materials from endangering the portal and access area. The employer must ensure ground stability in hazardous subsidence areas by shoring, by filling in, or by erecting barricades and posting warning signs to prevent entry. A competent person is to inspect the roof, face, and walls of the work area at the start of each shift, and as often as necessary to determine ground stability. Competent persons, who conduct such inspections, must be protected from loose ground by location, by ground support, or by equivalent means. Ground conditions along haulageways and travelways are to be inspected as frequently as necessary to ensure safe passage. Loose ground that might be hazardous to employees is to be taken down, scaled, or supported.

Torque wrenches are to be used wherever bolts, used for ground support, depend on torsionally applied force. A competent person must determine whether rock bolts meet the necessary torque, and must also determine the testing frequency in light of the bolt system, ground conditions, and the distance from vibration sources.

Suitable protection is to be provided for employees exposed to the hazard of loose ground while installing ground support systems. Support sets are to be installed so that the bottoms have sufficient anchorage to prevent ground pressures from dislodging the support base of the sets. Lateral bracing (collar bracing, tie rods, or spreaders) is to be provided between immediately adjacent sets in order to ensure added stability. Damaged or dislodged ground supports that create a hazardous condition are to be promptly repaired or replaced. When replacing supports, the new supports are to be installed before the damaged supports are removed.

A shield, or other type of support, is to be used to maintain a safe travelway for employees who are working in dead-end areas and are ahead of any support replacement operation. Where employees must enter shafts and wells over 5 feet in depth, the shafts and wells are to be supported by a steel casing, concrete pipe, timber, solid rock, or other suitable material.

The full depth of the shaft is to be supported by casing, or bracing, except where the shaft penetrates into solid rock having characteristics that will not change as a result of exposure. Where the shaft passes through earth into solid rock, or through solid rock into earth, and where there is a potential for shear, the casing or bracing must extend at least 5 feet into the solid rock. When the shaft terminates in solid rock, the casing or bracing must extend to the end of the shaft, or 5 feet into the solid rock, whichever is less. The casing or bracing must extend 42 inches, plus or minus 3 inches, aboveground level, except that the minimum casing height may be reduced to 12 inches, provided that a standard railing is installed; that the ground adjacent to the top of the shaft is sloped away from the shaft collar to prevent entry of liquids; and that effective barriers are used to prevent mobile equipment operating near the shaft from jumping over the 12 inch barrier.

Explosives and Blasting

No explosives or blasting agents are to be permanently stored in any underground operation until the operation has been developed to the point where at least two modes of exit have been provided. Permanent underground storage magazines are to be at least 300 feet from any shaft, adit, or active underground working area. Permanent underground magazines, containing detonators, are not to be located closer than 50 feet to any magazine containing other explosives or blasting agents.

When using and transporting explosives while underground, special precautions are to be taken. All explosives or blasting agents in transit underground are to be taken to the place of use, or storage, without delay. The quantity of explosives or blasting agents taken to an underground loading area must not exceed the amount estimated to be necessary for the blast. Again, explosives in transit are not to be left unattended. The hoist operator is to be notified before explosives or blasting agents are transported in a shaft conveyance. Trucks used for the transportation of explosives underground must have the electrical system checked weekly to detect any failures which may constitute an electrical hazard. A certification record, which includes the date of the inspection, the signature of the person who performed the inspection, and a serial number, or other identifier of the truck inspected, is to be prepared, and the most recent certification record is to be maintained on file. The installation of auxiliary lights on truck beds, which are powered by the truck's electrical system, is prohibited.

When firing from a power circuit in underground operations, a safety switch is to be placed, at intervals, in the permanent firing line. This switch is to be made so it can be locked only in the "Off" position, and it is to be provided with a short-circuiting arrangement of the firing lines to the cap circuit. In underground operations, there is to be a "lightning" gap of at least 5 feet in the firing system, ahead of the main firing switch that is between this switch and the source of power. This gap is to be bridged by a flexible jumper cord just before firing the blast.

During underground operations when explosives and blasting agents are hoisted, lowered, or conveyed in a powder car, no other materials, supplies, or equipment are to be transported in the same conveyance, at the same time. No one, except the operator, his helper, and the powderman, is permitted to ride on a conveyance transporting explosives and blasting agents. No person is to ride in any shaft conveyance transporting explosives and blasting agents. No explosives or blasting agents are to be transported on any locomotive. At least two car lengths must separate the locomotive from the powder car. No explosives or blasting agents are to be transported on a man haul trip. The car or conveyance containing explosives or blasting agents is to be pulled, not pushed, whenever possible. The powder car, or conveyance, should be especially built for the purpose of transporting explosives or blasting agents. It must bear a reflectorized sign on each side, with the word "Explosives" in letters not less than 4 inches in height, and these letters are to be placed upon a background of sharply contrasting color. Compartments, for transporting detonators and explosives in the same car or conveyance, are to be physically separated by a distance of 24 inches, or by a solid partition at least 6 inches thick. Detonators, and other explosives, are not to be transported at the same time in any shaft conveyance. Explosives, blasting agents, or blasting supplies are not to be transported with other materials. Explosives or blasting agents, not in original containers, are to be placed in a suitable container when transported manually. Detonators, primers, and other explosives are to be carried in separate containers when transported manually.

When blasting under compressed air during excavation work, detonators and explosives are not to be stored or kept in tunnels, shafts, or caissons. Detonators and explosives, for each round, are to be taken directly from the magazines to the blasting zone, and immediately loaded. Detonators and explosives left over after loading a round are to be removed from the working chamber before the connecting wires are connected up. When detonators or explosives are brought into an air lock, no employee, except the powderman, blaster, lock tender, and the employees necessary for carrying, are permitted to enter the air lock. No other material, supplies, or equipment are to be locked through with the explosives. Detonators and explosives are to be taken separately into pressure working chambers. The blaster or powderman is responsible for the receipt, unloading, storage, and on-site transportation of explosives and detonators. All metal pipes, rails, air locks, and steel tunnel lining are to be electrically bonded together and grounded at or near the portal or shaft, and such pipes and rails are to be cross-bonded together at not less than 1,000-foot intervals throughout the length of the tunnel. In addition, each low air supply pipe shall be grounded at its delivery end. No explosives are to be loaded or used underground in the presence of combustible gases or combustible dusts.

The explosives suitable for use in wet holes are to be water resistant and shall be Fume Class 1. When tunnel excavation in rock face is approaching a mixed face, or when tunnel excavation is in a mixed face, blasting is to be performed with light charges and with light burden on each hole. Advance drilling is to be performed, as tunnel excavation in rock face approaches mixed face, to determine the general nature and extent of the rock cover, and the remaining distance ahead to soft ground as the excavation advances.

After blasting operations in shafts, a competent person must determine if the walls, ladders, timbers, blocking, or wedges have loosened. If so, necessary repairs are to be made before employees, other than those assigned to make the repairs, are allowed in or below the affected areas. Further requirements for blasting and explosives operations, including handling of misfires, are found in 29 CFR 1926 Subpart U.

Blasting wires are to be kept clear of electrical lines, pipes, rails, and other conductive material, excluding earth, to prevent explosives' initiation, or employee exposure to electric current. Following blasting, workers should not enter a work area until the air quality meets all the requirements.

Drilling

A competent person must inspect all drilling and associated equipment prior to each use. Equipment defects, which affect safety, are to be corrected before the equipment is used. The drilling area is to be inspected for hazards before the drilling operation is started. Employees are not allowed on a drill mast while the drill bit is in operation, or the drill machine is being moved. When a drill machine is being moved from one drilling area to another, drill steel, tools, and other equipment must be secured and the mast is to be placed in a safe position. Receptacles or racks need to be provided for storing drill steel located on jumbos.

Employees working below jumbo decks are to be warned whenever drilling is about to begin. Drills on columns are to be anchored firmly before starting drilling, and are to be retightened, as necessary, thereafter. The employer must provide a mechanical means on the top deck of a jumbo for lifting unwieldy or heavy material. When jumbo decks are over 10 feet in height, the employer shall install stairs wide enough for two persons. Jumbo decks more than 10 feet in height are to be equipped with guardrails on all open sides, excluding access openings of platforms, unless an adjacent surface provides equivalent fall protection.

Only employees assisting the operator are to be allowed to ride on jumbos, unless the jumbo meets the specific requirements. Jumbos are to be chocked to prevent movement while employees are working on them. Walking and working surfaces of jumbos are to be maintained to prevent the hazards of slipping, tripping, and falling. Jumbo decks and stair treads are to be designed to be slip resistant, and they are to be secured to prevent accidental displacement.

General Guidelines

Scaling bars are to be available at scaling operations, and are to be maintained in good condition at all times. Blunted, or severely worn bars, are not to be used. Blasting holes are not to be drilled through blasted rock (muck) or water. Employees in a shaft are to be protected either by location, or by suitable barrier(s), if powered mechanical loading equipment is used to remove muck which contains unfired explosives. A caution sign which reads, "Buried Line," or similar wording, is to be posted where air lines are buried, or otherwise hidden by water or debris.

Power Haulage

A competent person is to inspect haulage equipment before each shift. Equipment defects which affect safety and health are to be corrected before the equipment is used. Powered mobile haulage equipment must have a suitable means of stopping. Power mobile haulage equipment, including trains, are to be equipped with audible warning devices to warn employees to stay clear.

The operator must sound the warning device before moving the equipment, and whenever necessary during travel. The operator must assure that lights, which are visible to employees at both ends of any mobile equipment, including a train, are turned on whenever the equipment is operating. In those cabs where glazing is used, the glass is to be safety glass, or its equivalent, and is to be maintained and cleaned so that vision is not obstructed.

Anti-roll back devices or brakes are to be installed on inclined conveyer drive units to prevent conveyers from inadvertently running in reverse. Employees are not permitted to ride a power-driven chain, belt, or bucket conveyer unless the conveyer is specifically designed for the transportation of persons. Endless belt type manlifts are prohibited in underground construction. The usual rules for conveyors apply.

No employee shall ride haulage equipment unless it is equipped with seating for each passenger, and it protects passengers from being struck, crushed, or caught between other equipment or surfaces. Members of train crews may ride on a locomotive, if it is equipped with handholds, and nonslip steps or footboards.

Powered mobile haulage equipment, including trains, are not to be left unattended unless the master switch or motor is turned off. The operating controls are to be in neutral or the park position, and the brakes are to be set, or equivalent precautions are to be taken to prevent rolling. Whenever rails serve as a return for a trolley circuit, both rails are to be bonded at every joint, and crossbonded every 200 feet.

When dumping cars by hand, the car dumps must have tiedown chains, bumper blocks, or other locking or holding devices to prevent the cars from overturning. Rocker-bottom or bottom-dump cars are to be equipped with positive locking devices to prevent the cars from overturning.

Equipment to be hauled is to be loaded and secured to prevent sliding or dislodgment. Mobile equipment, including rail-mounted equipment, is to be stopped for manual connecting or service work. Employees are not to reach between moving cars during coupling operations. Couplings are not to be aligned, shifted, or cleaned on moving cars or locomotives. Safety chains or other connections are to be used, in addition to couplers, to connect man cars, or powder cars, whenever the locomotive is uphill of the cars.

When the grade exceeds one percent and there is a potential for runaway cars, safety chains or other connections are to be used, in addition to couplers, to connect haulage cars or, as an alternative, the locomotive must be downhill of the train. Such safety chains or other connections shall be capable of maintaining connections between cars in the event of coupler disconnect, failure, or breakage. Parked rail equipment is to be chocked, blocked, or have the brakes act to prevent inadvertent movement.

Berms, bumper blocks, safety hooks, or equivalent means, are to be provided to prevent overtravel and overturning of haulage equipment at dumping locations. Bumper blocks, or equivalent stopping devices, are to be provided at all track dead ends.

Only small handtools, lunch pails, or similar small items may be transported with employees in mancars, or on top of a locomotive. When small hand tools or other small items are carried on top of a locomotive, the top is to be designed or modified to retain them while traveling.

Where switching facilities are available, occupied personnel-cars are to be pulled, not pushed. If personnel-cars must be pushed and visibility of the track ahead is hampered, a qualified person is to be stationed in the lead car to give signals to the locomotive operator. Crew trips are to consist of personnel-loads only.

Electrical Safety

In addition to the normal electrical construction safety requirements, electric power lines are to be insulated, or located away from water lines, telephone lines, air lines, or other conductive materials so that a damaged circuit will not energize the other systems. Lighting circuits are to be located so that movement of personnel or equipment will not damage the circuits or disrupt service. Oil-filled transformers are not to be used underground, unless they are located in a fire-resistant enclosure which is suitably vented to the outside, and surrounded by a dike to retain the contents of the transformers in the event of rupture.

Cranes

Although the usual crane requirements apply, there are other general requirements for underground cranes and hoists which deal with materials, tools, and supplies that are being raised or lowered. Whether these materials, tools, or supplies are within a cage or otherwise, they are to be secured or stacked in a manner to prevent the load from shifting, snagging, or falling into the shaft. A warning light, suitably located, is to flash and warn employees at the shaft bottom, and subsurface shaft entrances, whenever a load is above the shaft bottom or subsurface entrances, or the load is being moved in the shaft. This does not apply to fully enclosed hoistways.

Whenever a hoistway is not fully enclosed and employees are at the shaft bottom, conveyances or equipment are to be stopped at least 15 feet above the bottom of the shaft and held there until the signalman at the bottom of the shaft directs the operator to continue lowering the load, except that the load may be lowered without stopping, if the load or conveyance is within full view of a bottom signalman who is in constant voice communication with the operator.

Before maintenance, repairs, or other work commence in a shaft served by a cage, skip, or bucket, the operator and other employees in the area are to be informed and given suitable instructions. A sign warning that work is being done in the shaft is to be installed at the shaft collar, at the operator's station, and at each underground landing.

Any connection between the hoisting rope and the cage or skip is to be compatible with the type of wire rope used for hoisting. Spin-type connections, where used, are to be maintained in a clean condition, and protected from foreign matter that could affect their operation. Cage, skip, and load connections to the hoist rope are to be made so that the force of the hoist pull, vibration, misalignment, release of lift force, or impact will not disengage the connection. Moused or latched open-throat hooks do not meet this requirement. When using wire rope wedge sockets, means are to be provided to prevent wedge escapement, and to ensure that the wedge is properly seated.

Cranes are to be equipped with a limit switch to prevent overtravel at the boom tip. Limit switches are to be used only to limit travel of loads when operational controls malfunction, and are not used as a substitute for other operational controls. Hoists are to be designed so that the load hoist drum is powered in both directions of rotation, and so that brakes are automatically applied upon power release or failure.

Control levers are to be of the "deadman type" which return automatically to their center (neutral) position upon release. When a hoist is used for both personnel hoisting and material hoisting, load and speed ratings for personnel and for materials are to be assigned to the equipment. Material hoisting may be performed at speeds higher than the rated speed for personnel hoisting, if the hoist and components have been designed for such higher speeds, and if shaft conditions permit.

Employees do not ride on top of any cage, skip, or bucket, except when necessary to perform inspection or maintenance of the hoisting system, in which case they are to be protected by a body harness system to prevent falling.

Personnel and materials (other than small tools and supplies which are secured in a manner that will not create a hazard to employees) are not to be hoisted together in the same conveyance. However, if the operator is protected from the shifting of materials, then the operator may ride with materials in cages or skips which are designed to be controlled by an operator within the cage or skip.

Line speed must not exceed the design limitations of the systems. Hoists are to be equipped with landing level indicators at the operator's station. Marking the hoist rope does not satisfy this requirement.

Whenever glazing is used in the hoist house, it is to be safety glass, or its equivalent, and must be free of distortions and obstructions. A fire extinguisher that is rated at least 2A:10B:C (multi-purpose, dry chemical) is to be mounted in each hoist house. Hoist controls are to be arranged so that the operator can perform all operating cycle functions, and reach the emergency power cutoff without having to reach beyond the operator's normal operating position.

Hoists are to be equipped with limit switches to prevent overtravel at the top and bottom of the hoistway. Limit switches are to be used only to limit travel of loads when operational controls malfunction, and they are not to be used as a substitute for other operational controls. Hoist operators are to be provided with a closed-circuit voice communication system to each landing station, with speaker microphones so located that the operator can communicate with individual landing stations during hoist use.

When sinking shafts 75 feet or less in depth, cages, skips, and buckets that may swing, bump, or snag against shaft sides or other structural protrusions, are to be guided by fenders, rails, ropes, or a combination of those means. When sinking shafts more than 75 feet in depth, all cages, skips, and buckets are to be rope or rail guided to within a rail length from the sinking operation. Cages, skips, and buckets in all completed shafts, or in all shafts being used as completed shafts, are to be rope or rail-guided for the full length of their travel.

Wire rope used in load lines of material hoists is to be capable of supporting, without failure, at least five times the maximum intended load, or the factor recommended by the rope manufacturer, whichever is greater. The design factor is to be calculated by dividing the breaking strength of wire rope, as reported in the manufacturer's rating tables, by the total static load, including the weight of the wire rope in the shaft when fully extended.

A competent person is to visually check all hoisting machinery, equipment, anchorages, and hoisting rope at the beginning of each shift, and during hoist use, as necessary. Each safety device is to be checked by a competent person, at least weekly during hoist use, to ensure suitable operation and safe condition.

In order to ensure suitable operation and the safe condition of all functions and safety devices, each hoist assembly is to be inspected and load-tested to 100 percent of its rated capacity at the time of installation; after any repairs or alterations affecting its structural integrity; after the operation of any safety device; and annually when in use. The employer must prepare a certification record, which includes the date each inspection and load-test was performed; the signature of the person who performed the inspection and test; and a serial number or other identifier for the hoist that was inspected and tested. The most recent certification record is to be maintained on file until completion of the project.

Before hoisting personnel or material, the operator must perform a test run of any cage or skip, whenever it has been out of service for one complete shift, and whenever the assembly or components have been repaired or adjusted. Unsafe conditions are to be corrected before using the equipment.

Hoist drum systems are to be equipped with at least two means of stopping the load, each of which are capable of stopping and holding 150 percent of the hoist's rated line pull. A broken-rope safety, safety catch, or arrestment device is not a permissible means for stopping. The operation must remain within sight and sound of the signals at the operator's station.

All sides of personnel cages are to be enclosed by one-half inch wire mesh (not less than No. 14 gauge or equivalent) to a height of not less than 6 feet. However, when the cage or skip is being used as a work platform, its sides may be reduced in height to 42 inches, when the conveyance is not in motion.

All personnel cages are to be provided with positive locking doors that do not open outward. All personnel cages are to be provided with a protective canopy. The canopy is to be made of steel plate, at least 8/16-inch in thickness, or be material of equivalent strength and impact resistance. The canopy is to be sloped to the outside, and be so designed that a section may be readily pushed upward to afford emergency egress. The canopy must cover the top in such a manner as to protect those inside from objects falling in the shaft.

Personnel platforms, operating on guide rails or guide ropes, are to be equipped with broken-rope safety devices, safety catches, or arrestment devices that will stop and hold 150 percent of the weight of the personnel platform and its maximum rated load. During sinking operations in shafts where guides and safeties are not yet used, the travel speed of the personnel platform must not exceed 200 feet per minute. Governor controls set for 200 feet per minute are to be installed in the control system and shall be used during personnel hoisting. The personnel platform may travel over the controlled length of the hoistway at rated speeds up to 600 feet per minute, during sinking operations, in shafts where guides and safeties are used. The personnel platform may travel at rated speeds greater than 600 feet per minute in completed shafts.

Caissons (1926.801)

In caisson work in which compressed air is used, and the working chamber is less than 11 feet in length, and when such caissons are at any time suspended or hung while work is in progress, so that the bottom of the excavation is more than 9 feet below the deck of the working chamber, a shield is to be erected therein for the protection of the employees. Shafts are to be subjected to a hydrostatic or air-pressure test, at the pressure at which they shall be air tight. The shaft is to be stamped on the outside shell, about 12 inches from each flange, to show the pressure to which they have been subjected. Whenever a shaft is used, it is to be provided, where space permits, with a safe, proper, and suitable staircase for its entire length, including landing platforms, not more than 20 feet apart. Where this is impracticable, suitable ladders are to be installed, with landing platforms located about 20 feet apart to break the climb.

All caissons, having a diameter or side greater than 10 feet, are to be provided with a man lock and shaft for the exclusive use of employees. In addition to the gauge in the locks, an accurate gauge is to be maintained on the outer and inner side of each bulkhead. These gauges are to be accessible at all times, and are to be kept in accurate working order. In caisson operations, where employees are exposed to compressed air working environments, standard compressed air guidelines are to be followed.

Cofferdams (1926.802)

If overtopping of the cofferdam by high waters is possible, means are to be provided for controlled flooding of the work area. Warning signals, for the evacuation of employees in case of emergency, are to be developed and posted. Cofferdam walkways, bridges, or ramps with at least two means of rapid exit, are to be provided with guardrails. Cofferdams located close to navigable shipping channels are to be protected from vessels in transit, where possible.

Compressed Air (1926.803)

When work is occurring under compressed air, there must be present, at all times, at least one competent person, designated by and representing the employer, who is familiar with all aspects, and responsible for full compliance of working under compressed air.

Medical Requirements

Every employee is to be instructed in the rules and regulations which concern their safety, or the safety of others. There must be at least one or more licensed physicians retained who is familiar with, and experienced in the physical requirements and the medical aspects of compressed air work, and the treatment of decompression illness. The physician must be available at all times while work is in progress, in order to provide medical supervision for employees employed in compressed air work. The physician must be physically qualified, and must be willing to enter a pressurized environment.

No employee is to be permitted to enter a compressed air environment until examined by the physician, and reported to be physically qualified to engage in such work. In the event an employee is absent from work for 10 days, or is absent due to sickness or injury, that employee shall not resume work until reexamined by the physician. After the examination, the employee's physical condition is to be reported, by the physician, to be such as to permit the worker to work in compressed air. After an employee has been continuously employed in compressed air work for a period, designated by the physician, but not to exceed one year, the worker is to be reexamined by the physician to determine if the employee is still physically qualified to engage in compressed air work.

Such physicians must, at all times, keep a complete and full record of the examinations made, and must keep accurate records of any decompression illness, or other illness or injury, which incapacitates any employee for work. The physicians must also report all loss of life that occurs in the operation of a tunnel, caisson, or other compartment in which compressed air is used. Records are to be available for the inspection of the Secretary or his representatives, and a copy is to be forwarded to OSHA within 48 hours following the occurrence of the accident, death, injury, or decompression illness. It is to state, as fully as possible, the cause of said death or decompression illness, the place where the injured or sick employee was taken, and such other relative information as may be required by the Secretary.

A fully equipped first aid station is provided at each tunnel project regardless of the number of persons employed. An ambulance or transportation suitable for a litter case is to be at each project. Where tunnels are being excavated from portals more than 5 road miles apart, a first aid station and transportation facilities are to be provided at each portal.

Medical Lock

A medical lock is to be established and maintained in the immediate working order, whenever air pressure in the working chamber is increased above the normal atmosphere. The medical lock has to have at least 6 feet of clear headroom at the center, be subdivided into not less than two compartments, and is to be readily accessible to employees working under compressed air. The medical lock is to be kept ready for immediate use for at least 5 hours subsequent to the emergence of any employee from the working chamber. It is to be properly heated, lighted, ventilated, and maintained in a sanitary condition. A non-shatterable port shall be present through which the occupant(s) may be kept under constant observation. The medical lock is to be designed for a working pressure of 75 psig.; equipped with internal controls which may be overridden by external controls; provided with air pressure gauges to show the air pressure within each compartment to observers inside and outside the medical lock; equipped with a manual type sprinkler system that can be activated inside the lock, or by the outside lock tender; and provided with oxygen lines and fittings leading into external tanks. The lines are to be fitted with check valves to prevent reverse flow. The oxygen system inside the chamber is to be of a closed circuit design, and to be so designed as to automatically shut off the oxygen supply whenever the fire system is activated. An attendant is to be in constant charge, and under the direct control of the retained physician. The attendant is to be trained in the use of the lock; to be suitably instructed regarding the steps to be taken in the treatment of employee exhibiting symptoms compatible with a diagnosis of decompression illness; to be adjacent to an adequate emergency medical facility, which is equipped with demand-type oxygen inhalation equipment approved by the U.S. Bureau of Mines that is capable of being maintained at a temperature not to exceed 90 degree F., nor be less than 70 degree F; and to be provided with sources of air, free of oil and carbon monoxide, for normal and emergency use, which are capable of raising the air pressure in the lock from 0 to 75 psig. in 5 minutes.

Identifying Workers

Identification badges are to be furnished to all employees, indicating that the wearer is a compressed air worker. A permanent record is to be kept of all identification badges issued. The badge must give the employee's name, the address of the medical lock, the telephone number of the licensed physician for the compressed air project, and instructions in case there is an emergency of an unknown or doubtful cause, or illness, where the wearer may need to be rushed to the medical lock. The badge is to be worn at all times – off the job, as well as on the job.

Communications

Effective and reliable means of communication, such as bells, whistles, or telephones, are to be maintained, at all times, between all the following locations: the working chamber face, the working chamber side of the man lock near the door, the interior of the man lock, the lock attendant's station, the compressor plant, the first aid station, the emergency lock (if one is required), and the special decompression chamber, if one is required.

Signs and Records

The time of decompression is to posted in each manlock, using the format in Table 8-11. This form is to be posted in the manlock at all times.

Any code of signals used is to be conspicuously posted near workplace entrances, and other locations, as may be necessary, to bring them to the attention of all employees concerned. For each 8-hour shift, a record of the employees, who are employed under air pressure, is to be kept by an employee who remains outside the lock, near the entrance. This record must show the period each employee spends in the air chamber, and the time taken from decompression. A copy is to be submitted to the appointed physician after each shift.

Table 8-11

Time of Decompression Form

Time of Decompression for This Lock

___ pounds to ___ pounds in ___ minutes.

___ pounds to ___ pounds in ___ minutes.

(Signed by) _____(Superintendent)

Compression

Every employee going under air pressure for the first time is to be instructed on how to avoid excessive discomfort. During the compression of employees, the pressure is not to be increased to more than 3 psig. within the first minute. The pressure is to be held at 3 psig., and again at 7 psig. sufficiently long enough to determine if any employees are experiencing discomfort. After the first minute, the pressure is to be raised uniformly, and at a rate not to exceed 10 psi per minute. If any employee complains of discomfort, the pressure is to be held to determine if the symptoms are relieved. If after 5 minutes the discomfort does not disappear, the lock attendant is to gradually reduce the pressure until the employee signals that the discomfort has ceased. If the worker does not indicate that the discomfort has disappeared, the lock attendant is to reduce the pressure to atmospheric, and the employee is to be released from the lock. No employee is to be subjected to pressure exceeding 50 pounds per square inch, except in emergency.

Decompression

Decompression to normal condition is to be in accordance with the Standard Decompression Tables. In the event it is necessary for an employee to be in compressed air more than once in a 24-hour period, the appointed physician is responsible for the establishment of methods and procedures of decompression applicable to repetitive exposures. If decanting is necessary, the appointed physician must establish procedures before any employee is permitted to be decompressed by decanting methods. The period of time that an employee spends at atmospheric pressure, when the employee is between the decompression (following the shift) and recompression, is not to exceed 5 minutes.

Manlocks

Except in emergency, no employees employed in compressed air, are to be permitted to pass from the working chamber to atmospheric pressure until after decompression. The lock attendant in charge of a manlock is to be under the direct supervision of the appointed physician. He/she is to be stationed at the lock controls, on the free air side, during the period of compression and decompression, and must remain at the lock control station whenever there are individuals in the working chamber, or in the manlock. Except where air pressure in the working chamber is below 12 psig., each manlock is to be equipped with automatic controls which, through taped programs, cams, or similar apparatus, automatically regulates decompressions. It is also to be equipped with manual controls which permits the lock attendant to override the automatic mechanism in the event of an emergency. A manual control, which can be used in the event of an emergency, is to be placed inside the manlock. A clock, thermometer, and continuous recording pressure gauge, with a 4-hour graph, are to be installed outside of each manlock, and are to be changed prior to each shift's decompression. The chart is to be of sufficient size to register a legible record of variations in pressure within the manlock, and is to be visible to the lock attendant. A copy of each graph is to be submitted to the appointed physician after each shift. In addition, a pressure gauge clock and thermometer are also to be installed in each manlock. Additional fittings are to be provided so that test gauges may be attached whenever necessary. Except where air pressure is below 12 psig. and there is no danger of rapid flooding, all caissons having a working area greater than 150 square feet, and each bulkhead in tunnels of 14 feet or more in diameter, or equivalent area, must have at least two locks in perfect working condition, one of which is used exclusively as a manlock, the other, as a materials lock.

Where only a combination man-and-materials lock is required, this single lock is to be of sufficient capacity to hold the employees of two successive shifts.

Emergency locks are to be large enough to hold an entire heading shift and a limit maintained of 12 psig. There is to be a chamber available for oxygen decompression therapy to 28 psig. The manlock is to be large enough so that those using it are not compelled to be in a cramped position; it must not have less than 5 feet clear head room at the center; and it must have a minimum of 30 cubic feet of air space per occupant.

Locks on caissons are to be located so that the bottom door is not less than three feet above the water level surrounding the caisson on the outside. (The water level, where it is affected by tides, is construed to mean high tide.) In addition to the pressure gauge in the locks, an accurate pressure gauge is to be maintained on the outer and inner side of each bulkhead. These gauges are to be accessible at all times and are to be kept in accurate working order. Manlocks must have an observation port at least four inches in diameter, and they must be located in such a position that all occupants of the manlock may be observed from the working chamber, and from the free air side of the lock. Adequate ventilation in the lock is to be provided. Manlocks are to be maintained at a minimum temperature of 70 degree F. When locks are not in use and employees are in the working chamber, lock doors are to be kept open to the working chamber, where practicable. Provision is to be made to allow for rescue parties to enter the tunnel, if the working force is disabled.

Special Decompression Chamber

A special decompression chamber, of sufficient size to accommodate the entire force of employees being decompressed at the end of a shift, is to be provided whenever the regularly established working period requires a total time of decompression exceeding 75 minutes. The headroom in the special decompression chamber is not to be less than a minimum of 7 feet, and the cubical content must provide at least 50 cubic feet of airspace for each employee. For each occupant, there is to be provided 4 square feet of free walking area, and 3 square feet of seating space, exclusive of the area required for lavatory and toilet facilities. The rated capacity is to be based on the stated minimum space per employee, and is to be posted at the chamber entrance. The posted capacity is not to be exceeded, except in case of emergency. Each special decompression chamber is to be equipped with: a clock, or clocks, suitably placed so that the attendant and the chamber occupants can readily ascertain the time; pressure gauges which will indicate to the attendants, and to the chamber occupants, the pressure in the chamber; valves which will enable the attendant to control the supply and discharge of compressed air into and from the chamber. Valves and pipes, which are in connection with the air supply and exhaust, are to be arranged so that the chamber pressure can be controlled from within and without; an effective means of oral intercommunication between the attendant, occupants of the chamber, and the air compressor plant is to be supplied; and an observation port, at the entrance, is to be used to permit observation of the chamber occupants.

Seating facilities, in special decompression chambers, are to be so arranged as to permit a normal sitting posture without cramping. Seating space, not less than 18 inches by 24 inches wide, is to be provided, per occupant. Adequate toilet and washing facilities, in a screened or enclosed recess, are to be provided. Toilet bowls must have built-in protectors on the rim, so that an air space is created when the seat lid is closed. Fresh pure drinking water is to be available. This may be accomplished by either piping water into the special decompression chamber and providing drinking fountains, by providing individual canteens, or by some other sanitary means. Community drinking vessels are prohibited. No refuse or discarded material of any kind shall be permitted to accumulate, and the chamber is to be kept clean. Unless the special decompression chamber is serving as the manlock to atmospheric pressure, the special decompression chamber is to be situated, where practicable, adjacent to the manlock on the atmospheric pressure side of the bulkhead. A passageway is to be provided, connecting the special chamber with the manlock in order to permit employees, in the process of decompression, to move from the manlock to the special chamber, without a reduction in the ambient pressure from that designated for the next stage of decompression. The passageway is to be so arranged as to not interfere with the normal operation of the manlock, nor with the release of the occupants, of the special chamber, to atmospheric pressure upon the completion of the decompression procedure.

Compressor Plant and Air Supply

The compressor plant and air supply is to be manned, at all times, by a thoroughly experienced, competent, and reliable person; the competent person is to be at the air control valves as a gauge tender who regulates the pressure in the working areas. During tunneling operations, a gauge tender may regulate the pressure in two headings, but only if the gauges and controls are all in one location. In caisson work, there is to be a gauge tender for each caisson.

The low air compressor plant shall be of sufficient capacity to not only permit the work to be done safely, but also must provide a margin to meet emergencies and repairs. Low air compressor units must have at least two independent and separate sources of power supply, and each is to be capable of operating the entire low air plant, and its accessory systems. The capacity, arrangement, and number of compressors are to be sufficient enough to maintain the necessary pressure, without overloading the equipment, and they must also be sufficient enough to assure the maintenance of such pressure in the working chamber during periods of breakdown, repair, or an emergency.

Switching from one independent source of power supply to the other is to be done periodically, to ensure the workability of the apparatus in an emergency. Duplicate low-pressure air feedlines and regulating valves are to be provided between the source of air supply, and a point beyond the locks, with one of the lines extending to within 100 feet of the working face. All high- and low-pressure air supply lines are to be equipped with check valves. Low-pressure air is to be regulated automatically. In addition, manually operated valves are to be provided for emergency conditions. The air intakes for all air compressors are to be located at a place where fumes, exhaust, gases, and other air contaminants will be at a minimum. Gauges, indicating the pressure in the working chamber, are to be installed in the compressor building, the lock attendant's station, and at the employer's field office.

Compressed Air Ventilation and Air Quality

Exhaust valves and exhaust pipes are to be provided and operated so that the working chamber is well ventilated, and there are no pockets of dead air. Outlets may be required at intermediate points along the main low-pressure air supply line, to the heading, to eliminate such pockets of dead air. Ventilating air is not to be less than 30 cubic feet per minute. The air in the workplace is to be analyzed by the employer, not less than once each shift, and records of such tests are to be kept on file at the place where the work is in progress. The test results are to be within the threshold limit values for hazardous gases, and within 10 percent of the lower explosive limit of flammable gases. If these limits are not met, immediate action to correct the situation must be taken by the employer. The temperature of all working chambers which are subjected to air pressure shall, by means of after-coolers or other suitable devices, be maintained at a temperature not to exceed 85 degree F.

Forced ventilation is to be provided during decompression. During the entire decompression period, forced ventilation, through chemical or mechanical air purifying devices that will ensure a source of fresh air, is to be provided. Whenever heat-producing machines (moles, shields) are used in compressed air tunnel operations, a positive means of removing the heat build-up at the heading shall be provided.

All lighting in compressed-air chambers is to be by electricity, exclusively, and two independent electric-lighting systems, with independent sources of supply, are to be used. The emergency source is to be arranged to become automatically operative in the event of failure of the regularly used source. The minimum intensity of the light on any walkway, ladder, stairway, or working level is not to be less than 10 foot-candles, and in all workplaces the lighting is, at all times, to be such as to enable employees to see clearly. All electrical equipment and wiring for light and power circuits must comply with the requirements of construction electrical safety for use in damp, hazardous, high temperature, and compressed air environments. External parts of lighting fixtures and all other electrical equipment, when within 8 feet of the floor, are to be constructed of noncombustible, non-absorptive, insulating materials, except that metal may be used if it is effectively grounded. Portable lamps are to be equipped with noncombustible, non-absorptive, insulating sockets, approved handles, basket guards, and approved cords. The use of worn or defective portable and pendant conductors is prohibited.

Sanitation

Sanitary, heated, lighted, and ventilated dressing rooms and drying rooms are to be provided for all employees engaged in compressed air work. Such rooms must contain suitable benches and lockers. Bathing accommodations (showers at the ratio of one to 10 employees per shift), are to be equipped with running hot and cold water, and suitable and adequate toilet accommodations are to be provided. One toilet for each 15 employees, or fractional part thereof, is to be provided. When the toilet bowl is shut by a cover, there should be an air space so that the bowl or bucket does not implode when pressure is increased. All parts of caissons, and other working compartments, are to be kept in a sanitary condition.

Fire Prevention

Firefighting equipment is to be available at all times, and it shall be maintained in working condition. While welding or flame-cutting is being done in compressed air, a firewatch, with a fire hose or approved extinguisher, must stand by until such operation is completed. Shafts and caissons containing flammable material of any kind, either above or below ground, are to be provided with a waterline. A fire hose is to be connected so that all points of the shaft or caisson are within reach of the hose stream, and the fire hose must be at least 1 1/2 inches in nominal diameter. The water pressure must be, at all times, adequate for efficient operation of the type of nozzle used, and the water supply must be sufficient enough to ensure an uninterrupted flow. The fire hose, when not in use, is to be located in an area where it is guarded from damage.

The power house, compressor house, and all buildings which house ventilating equipment, are to be provided with at least one hose connection in the water line, with a fire hose connected thereto. A fire hose is to be maintained within reach of structures of wood over or near shafts.

Tunnels are to be provided with a 2-inch minimum diameter water line which extends into the working chamber, and it must be within 100 feet of the working face. Such lines must have hose outlets with 100 feet of fire hose attached, and one must be maintained at each of the following locations: at the working face, inside the bulkhead of the working chamber, and one immediately outside the bulkhead. In addition, hose outlets are to be provided at 200-foot intervals throughout the length of the tunnel, and 100 feet of fire hose is to be attached to the outlet nearest any location where flammable material is being kept or stored, or where any flame is being used.

In addition to the required fire hose protection, every floor of every building which is not under compressed air work, but is used in connection with the compressed air work, is to be provided with at least one approved fire extinguisher of the proper type for the hazard involved. At least two approved fire extinguishers are to be provided in the working chamber, with one at the working face and one immediately inside the bulkhead (pressure side). Extinguishers which are used in the working chamber, and use water as the primary extinguishing agent, must not use any extinguishing agent which could be harmful to the employees in the working chamber. The fire extinguisher is to be protected from damage.

Highly combustible materials are not to be used or stored in the working chamber. Wood, paper, and similar combustible material, are not to be used in the working chamber in quantities which could cause a fire hazard. The compressor building must be constructed of noncombustible material.

Manlocks are to be equipped with a manual type fire extinguisher system that can be activated inside the manlock, and also outside by the lock attendant. In addition, a fire hose and portable fire extinguisher are to be provided inside and outside the manlock. The portable fire extinguisher is to be the dry chemical type.

Equipment, fixtures, and furniture in manlocks and special decompression chambers are to be constructed of noncombustible materials. Bedding, etc. is to be chemically treated so as to be fire resistant. Head frames are to be constructed of structural steel, or open frame-work fireproofed timber. Head houses and other temporary surface buildings or structures that are within 100 feet of the shaft, caisson, or tunnel opening are to be built of fire-resistant materials.

No oil, gasoline, or other combustible material is to be stored within 100 feet of any shaft, caisson, or tunnel opening, except oils may be stored in suitable tanks, in isolated fire-proof buildings, provided the buildings are not less than 50 feet from any shaft, caisson, or tunnel opening, or any building directly connected to underground operations. Positive means are to be taken to prevent leaking flammable liquids from flowing into the areas previously mentioned. All explosives used in connection with compressed air work are to be selected, stored, and transported according to the standard rules for underground use and storage.

Bulkheads and Safety Screens

Intermediate bulkheads, with locks or intermediate safety screens, or both, are to be required where there is the danger of rapid flooding. In tunnels 16 feet or more in diameter, hanging walkways are to be provided from the face to the manlock, and as high in the tunnel as practicable, with at least 6 feet of head room. Walkways are to be constructed of noncombustible material. Standard railings are to be securely installed, on open side, throughout the length of all walkways. Where walkways are ramped under safety screens, the walkway surface is to be skidproofed by cleats, or by equivalent means. Bulkheads used to contain compressed air, shall be tested, where practicable, to prove their ability to resist the highest air pressure which may be expected to be used.

VERMIN CONTROL (1926.51)

In workplaces where there are rodents, insects, and other vermin, a continuous extermination program is to be used.

WASHING FACILITIES (1926.51)

The employer must provide adequate washing facilities for employees engaged in the application of paints, coating, herbicides, or insecticides, or in other operations where contaminants may be harmful to the employees. Such facilities are to be in near proximity to the worksite, and are to be equipped to enable employees to remove such substances. Washing facilities shall be maintained in a sanitary condition.

Lavatories are to be made available in all places of employment. These requirements do not apply to mobile crews, or to normally unattended work locations, if employees working at these locations have transportation readily available to them to nearby washing facilities, and these facilities meet the other requirements of this paragraph. Each lavatory is to be provided with hot and cold running water, or tepid running water. Hand soap, or similar cleansing agents, are to be provided; as well as individual hand towels, sections of cloth or paper towels, clean, individual sections of continuous cloth toweling, or warm air blowers, convenient to the lavatories, must be provided.

Whenever showers are required by a particular standard, one shower is to be provided for each 10 employees of each sex, or numerical fraction thereof, who are required to shower during the same shift.

Body soap, or other appropriate cleansing agents, convenient to the showers, must be provided. Showers are to be provided with hot and cold water which feeds a common discharge line. Employees who use showers are to be provided with individual clean towels.

WELDING (1926.350)

Only experienced persons are allowed to do electrical or acetylene welding or cutting. No welding or burning is to be done in hazardous areas without a permit. Warning signs, signaling overhead welding and cutting, must be posted. No welding or cutting should be done on barrels or tanks. Special precautions need to be taken while welding or cutting in a confined space. When welding, workers must wear the proper eye and face protection. Welders may be subjected to the inhalation of toxic fumes, which can cause illness; and they may also be subject to safety hazards, such as fire, which could result in a fatality, serious injury, and/or property damage; therefore, special precautions need to be followed when welding operations are in progress. See Table 8-12 for a summary of welding requirements.

Fire Prevention (1926.352)

When practical, objects to be welded, cut, or heated are to be moved to a designated safe location or, if the objects to be welded, cut, or heated cannot be readily moved, all movable fire hazards in the vicinity are to be taken to a safe place, or otherwise protected. If the object to be welded, cut, or heated cannot be moved, and if all the fire hazards cannot be removed, positive means must be taken to confine the heat, sparks, and slag, and to protect the immovable fire hazards from them. No welding, cutting, or heating is to be done where the application of flammable paints, or the presence of other flammable compounds, or heavy dust concentrations create a hazard. Suitable fire extinguishing equipment is to be immediately available in the work area, and is to be maintained in a state of readiness for instant use.

When the welding, cutting, or heating operation is such that normal fire prevention precautions are not sufficient, additional personnel must be assigned to guard against fire while the actual welding, cutting, or heating operation is being performed, and the additional personnel must remain for a sufficient period of time, after the completion of the work, to ensure that no possibility of fire exists. Such personnel are to be instructed as to the specific anticipated fire hazards, and how the firefighting equipment provided is to be used.

When welding, cutting, or heating is performed on walls, floors, and ceilings, and since direct penetration of sparks or heat transfer may introduce a fire hazard to an adjacent area, the same precautions are to be taken on the opposite side, as are taken on the side on which the welding is being performed.

For the elimination of possible fire in enclosed spaces, that are a result of gas escaping through leaking or improperly closed torch valves, the gas supply to the torch is to be positively shut off at a point which is outside the enclosed space; this should be done whenever the torch is not to be used, or whenever the torch is left unattended for a substantial period of time, such as during the lunch period. Overnight, and at the change of shifts, the torch and hose are to be removed from the confined space. Open-end fuel gas and oxygen hoses are to be immediately removed from enclosed spaces when they are disconnected from the torch, or other gas-consuming device.

Except when the contents are being removed or transferred, drums, pails, and other containers, which contain or have contained flammable liquids, are to be kept closed. Empty containers are to be removed to a safe area apart from hot work operations or open flames.

Before welding, cutting, or heating is undertaken on drum containers, or hollow structures which have contained toxic or flammable substances, they are to either be filled with water, or thoroughly cleaned of such substances, and ventilated and tested. For welding, cutting, and heating on steel pipelines containing natural gas, the pertinent portions of regulations issued by the Department of Transportation, Office of Pipeline Safety, 49 CFR Part 192, Minimum Federal Safety Standards for Gas Pipelines, apply. Before heat is applied to a drum, container, or hollow structure, a vent or opening is to be provided for the release of any built-up pressure during the application of heat.

Ventilation and Protection in Welding, Cutting, and Heating (1926.353)

Mechanical ventilation must meet the following requirements:

1.   Mechanical ventilation must consist of either general mechanical ventilation systems or local exhaust systems.

Table 8-12

Summary of Welding Requirements

General Requirements

1. Only qualified welders are to be authorized to do any welding, heating, or cutting.

2. Inspect your work area for fire hazards and proper ventilation before welding or cutting.

3. Avoid welding or cutting sparks, and hot slag. Be alert to hot surfaces, and avoid touching metal surfaces until they have cooled.

4. Place compressed gas cylinders in an upright position, and secure them in place to prevent dropping or falling. Handle them with extreme care, and do not store them near any sources of heat.

5. Remove any combustibles when welding or cutting must be done. If removal is not feasible, cover combustibles with a noncombustible material. When welding near any combustible material, another employee must be posted to serve as a fire watch. Make sure this person has a fire extinguisher available, and keep him/her in the area after welding/cutting is completed, and until all danger of fire is past.

6. When working in the vicinity of welding operations, wear approved goggles and avoid looking directly at the flash, as serious flash burns could result.

7. When opening valves on tanks that have regulators installed, be sure the pressure adjustment screw is all the way out, and do not stand in front of the regulator. An internal failure could rupture the regulator and cause the adjustment screw to become a missile.

Gas Welding and Cutting

1. When transporting, moving, and storing compressed gas cylinders, always ensure that the valve protection cap is in place and secured.

2. Secure cylinders on a cradle, slingboard, or pallet when hoisting. Never hoist or transport by means of magnet or choker slings.

3. Move cylinders by tilting and rolling them on their bottom edges. Do not allow cylinders to be dropped, struck, or come into contact with other cylinders, violently.

4. Secure cylinders in an upright (vertical) position, when transporting by powered vehicles.

5. Do not hoist cylinders by lifting on the valve protection caps.

6. Do not use bars under valves, or valve protection caps, to pry cylinders loose when frozen. Use warm, <u>not boiling</u>, water to thaw cylinders loose.

7. Remove regulators and secure valve protection caps prior to moving cylinders, unless cylinders are firmly secured on a special carrier intended for transport.

8. Close the cylinder valve when work is finished, when cylinders are empty, or when cylinders are moved at any time.

Table 8-12

Summary of Welding Requirements (*Continued*)

9. Secure compressed gas cylinders in an upright position (vertical), except when cylinders are actually being hoisted or carried.

Arc Welding and Cutting

1. Use only manual electrode holders which are specifically designed for arc welding and cutting.

2. All current-carrying parts, passing through the portion of the holder, must be fully insulated against the maximum voltage encountered to ground.

3. All arc welding and cutting cables must be completely insulated, flexible type, and capable of handling the maximum current requirements of the work in progress.

4. Report any defective equipment to your supervisor, immediately, and refrain from using such equipment.

5. Shield all arc welding and cutting operations, whenever feasible, by noncombustible or flameproof screens, to protect employees and other persons working in the vicinity from the direct rays of the arc.

Fire Prevention

1. Locate the nearest fire extinguisher in your work area in case of future need for an emergency. Fire extinguishing equipment must be immediately available in the work area.

2. Never use matches or cigarette lighters. Use only friction lighters to light torches.

3. Never strike an arc on gas cylinders.

4. Move objects to be welded, cut, or heated to a designated safe location. If the objects cannot readily moved, then all movable fire hazards, in the vicinity, must be taken to a safe place, or otherwise protected.

5. Do not weld, cut, or heat where the application of flammable paints, or the presence of other flammable compounds, or heavy dust concentrations creates a hazard.

6. Additional employees must be assigned to guard against fire, while the actual welding, cutting, or heating is being performed, when the operation is such that normal fire prevention precautions are not sufficient.

7. Prior to applying heat to a drum, container, or hollow structure, provide a vent or opening to release any built-up pressure during the application of heat.

8. Never cut, weld, or heat on drums, tanks, or containers that have contained flammable liquids, until they have been cleaned.

2. General mechanical ventilation is to be of sufficient capacity and so arranged as to produce the number of air changes necessary to maintain welding fumes and smoke within safe limits.

3. Local exhaust ventilation must consist of freely movable hoods, intended to be placed by the welder or burner, as close as practicable to the work. This system must be of sufficient capacity and so arranged as to remove fumes and smoke at the source, and keep the concentration of them, in the breathing zone, within safe limits.

4. Contaminated air, exhausted from a working space, is to be discharged into the open air, or otherwise clear of the source of intake air.

5. All air replacing shall be clean and respirable.

6. Oxygen is not to be used for ventilation purposes, comfort cooling, blowing dust from clothing, or for cleaning the work area.

General mechanical, or local exhaust ventilation is to be provided whenever welding, cutting, or heating is performed in a confined space. When sufficient ventilation cannot be obtained without blocking the means of access, employees in the confined space are to be protected by air line respirators. An employee, on the outside of such a confined space, is to be assigned to maintain communication with those working within it, and the employee is to aid them in an emergency. When a welder must enter a confined space through a manhole or other small opening, means are to be provided for quickly removing him in case of emergency. When safety harnesses and lifelines are used for this purpose, they are to be attached to the welder's body so that his body cannot be jammed in a small exit opening. An attendant, with a preplanned rescue procedure, is to be stationed outside to observe the welder, at all times, and is to be capable of putting rescue operations into effect.

Welding, cutting, or heating, in any enclosed space which involves metals that have zinc-bearing bases, filler metals, metals coated with zinc-bearing materials, lead base metals, cadmium-bearing filler materials, chromium-bearing metals, or metals coated with chromium-bearing materials, is to be performed with either general mechanical or local exhaust ventilation.

Any welding, cutting, or heating, in any enclosed space involving the following metals: metals containing lead, other than as an impurity, or metals coated with lead-bearing materials, cadmium-bearing or cadmium-coated base metals, metals coated with mercury-bearing metals, beryllium-containing base or filler metals shall be done with ventilation and air-line respirators. Because of its high toxicity, work involving beryllium is done with both local exhaust ventilation and air line respirators. All other welding, cutting, or heating of these metals is to be performed with local exhaust ventilation, or the employees shall be protected by air line respirators.

Employees performing such operations in the open air are to be protected by filter-type respirators, except that employees performing such operations on beryllium-containing base, or filler metals are to be protected by air line respirators. Other employees, who are exposed to the same atmosphere as the welders or burners, are to be protected in the same manner as the welders or burners.

Since the inert-gas metal-arc welding process involves the production of ultra-violet radiation of intensities of 5 to 30 times that produced during shielded metal-arc welding, the decomposition of chlorinated solvents by ultraviolet rays, and the liberation of toxic fumes and gases, the employees are not to be permitted to engage in, or be exposed to the process until the following special precautions have been taken. The use of chlorinated solvents are to be kept at least 200 feet, unless shielded, from the exposed arc, and the surfaces prepared with chlorinated solvents, are to be thoroughly dry before welding is permitted on such surfaces. Employees in the area, not protected from the arc by screening, shall be protected by filter lenses. When two or more welders are exposed to each other's arc, filter lens goggles of a suitable type, are to be worn under welding helmets. Hand shields, to protect the welder against flashes and radiant energy, are to be used when either the helmet is lifted, or the shield is removed. Welders, and other employees who are exposed to radiation, are to be suitably protected so that the skin is completely covered, to prevent burns and other damage by ultraviolet rays. Welding helmets, and hand shields, are to be free of leaks and openings, and free of highly reflective surfaces.

When inert-gas metal-arc welding is being performed on stainless steel, workers are to be protected against dangerous concentrations of nitrogen dioxide.

Welding, cutting, and heating, involving normal conditions or materials, may be done without mechanical ventilation or respiratory protective equipment, but where, because of unusual physical or atmospheric conditions, an unsafe accumulation of contaminants exists, suitable mechanical ventilation or respiratory protective equipment shall be provided. Employees performing any type of welding, cutting, or heating are to be protected by suitable eye protective equipment.

Welding, Cutting, and Heating of Preservative Coatings (1926.354)

Before welding, cutting, or heating commences on any surface covered by a preservative coating and the flammability is not known, a test is to be made by a competent person to determine its flammability. Preservative coatings are to be considered highly flammable when scrapings burn with extreme rapidity. Precautions are to be taken to prevent ignition of highly flammable, hardened preservative coatings. When coatings are determined to be highly flammable, they are to be stripped from the area to be heated to prevent ignition. In enclosed spaces, all surfaces covered with toxic preservatives are to be stripped of all toxic coatings for a distance of at least 4 inches from the area of heat application, or the employees are to be protected by air line respirators. When the previous task is conducted in the open air, employees are to be protected by a respirator. To ensure that the temperature of the unstripped metal will not be appreciably raised, the preservative coatings are to be removed at a sufficient distance from the area to be heated. Artificial cooling of the metal surrounding the heating area may be used to limit the size of the area required to be cleaned.

WOODWORKING TOOLS (1926.304)

All woodworking tools and machinery must meet other applicable requirements of the American National Standards Institute, 01.1-1961, Safety Code for Woodworking Machinery. All fixed power-driven woodworking tools are to be provided with a disconnect switch that can either be locked or tagged in the off position. The operating speed is to be etched, or otherwise permanently marked on all circular saws over 20 inches in diameter, or operating at over 10,000 peripheral feet per minute. Any saw so marked is not to be operated at a speed other than that marked on the blade. When a marked saw is retensioned for a different speed, the marking is to be corrected to show the new speed.

Automatic feeding devices are to be installed on machines, whenever the nature of the work will permit. Feeder attachments must have the feed rolls, or other moving parts covered or guarded so as to protect the operator from hazardous points.

All portable, power-driven circular saws are to be equipped with guards above and below the base plate or shoe. The upper guard must cover the saw to the depth of the teeth, except for the minimum arc required to permit the base to be tilted for bevel cuts. The lower guard must cover the saw to the depth of the teeth, except for the minimum arc required to allow proper retraction and contact with the work. When the tool is withdrawn from the work, the lower guard must be automatically and instantly returned to the covering position.

On radial saws, the upper hood must completely enclose the upper portion of the blade, down to a point that will include the end of the saw arbor. The upper hood is to be constructed in such a manner, and of such material, that it will protect the operator from flying splinters, broken saw teeth, etc. and will defect sawdust away from the operator. The sides of the lower exposed portion of the blade are to be guarded to the full diameter of the blade, by a device that will automatically adjust itself to the thickness of the stock, and it must remain in contact with the stock being cut, to give the maximum protection possible for the operation being performed.

Each circular crosscut table saw is to be guarded by a hood which meets all the requirements for the hoods for circular ripsaws. Each circular hand-fed ripsaw is to be guarded by a hood which completely encloses the portion of the saw above the table, and the portion of the saw above the material being cut. The hood and mounting are to be arranged so that the hood will automatically adjust itself to the thickness of, and remain in contact with, the material being cut, but it must not offer any considerable resistance to insertion of the material to the saw, or to passage of the material being sawed. The hood is to be made of adequate strength to resist blows and strains incidental to reasonable operation, adjusting, and handling, and is to be so designed as to protect the operator from flying splinters and broken saw teeth. It is to be made of material that is soft enough so that it will be unlikely to cause tooth breakage. The hood is to be so mounted as to ensure that its operation will be positive, reliable, and in true alignment with the saw; and the mounting shall be adequate in strength to resist any reasonable side thrust, or other force tending to throw it out of line.

Workers should wear eye protection, hearing protection, and other personal protective equipment, as is appropriate to protect them from injury.

WORKING OVER WATER (1926.106)

Many of the deaths which occur while working around or over water are due to the failure to wear a U.S. Coast Guard approved life jacket. The life jacket is often found near the scene of the accident, or a life jacket was not supplied by the employer. Even when the victim is a swimmer and falls into the water, the victim may be injured by the fall, may strike another object, may find the current too strong, or may find the water too cold, making survival difficult.

Precautions, prior to an occurrence, are the easiest prevention. The simple solution to this problem is to require life jackets to always be worn when working over or around water. It is also important to provide a life boat for rescue purposes, as well as ring buoys with 90 feet of line, spaced 200 feet apart.

Chapter 9

INDUSTRIAL HYGIENE ACTIVITIES IN CONSTRUCTION

James V. Eidson

Although many of the health hazards associated with general industry are the same as those found in construction, there are unique differences that are addressed in this chapter. As noted in previous chapters, the constant changing of the workplace where each work day is different from the next, leads to hazard exposure problems that are constantly changing as well. (See Figure 9-1.) Workers in the construction industry are exposed to a wide range of chemical and biological hazards that lead to numerous diseases, cancers, and/or disorders. These health hazards come in all forms – dusts, mists, vapors, gasses – and enter the body by inhalation, ingestion, and absorption. Studying body systems to understand the physical and chemical origin and development is called physiology. Every conceivable physical hazard is found in construction.

Figure 9-1. Today's construction site

BACKGROUND

With the down-sizing and out-sourcing of jobs that were once the responsibility of

441

general industry employees, there has been a tremendous jump in construction work over the past twenty years. Often, even entire maintenance responsibilities for a company are handled by contract staff. Many companies found it much cheaper to contract work out so they weren't responsible for insurance and all the other benefits given to their employees. And often, jobs that were not wanted by company employees, such as cleaning hazardous tanks, clearing fugitive lead or arsenic dust from miscellaneous surfaces, or confined space entry, etc. were the type of jobs contracted out. Although many outside contractors specialized in hazardous clean-up type of work, their work force turns over frequently making it very difficult to keep up with hazard communication and other required worker training programs that help reduce worker injuries.

Construction work doesn't normally provide steady employment the year round. Workers are always working themselves out of a job. Some projects may last for years but most are short lived. Bad weather and layoffs compound the problem of maintaining a well-trained work force. Work assignments change with the changing project. For these reasons, workers are often not aware of hazards that are outside their trade responsibilities. Construction workers constitute about 6% of the labor force but have 15% of the fatalities and over 9% of the "lost work day" injuries according to the Bureau of Labor Statistics.

Figure 9-2. The construction site is constantly changing

In some ways it is good for health's sake that construction is constantly changing. Many times worker exposures are short in duration and have no health effects, or at most, short-term effects. This is likely to reduce the number of construction workers with chronic health effects whereas those in general industry are often exposed every day for long periods of time. On the other hand, construction workers are more likely to experience short-term acute effects from chemicals found in the workplace. For example, it is likely that a construction worker would check the integrity of a chemical tank being repaired, sand blasted, or moved. In such cases the construction worker may not be as familiar with or as well trained about hazards associated with the job as employees of the company where work will be performed. In this scenario, a construction worker may enter a tank without knowing the toxic effects of chemicals that had been stored in the tank and without knowing that a health hazard could still remain in the tank.

Construction work will normally fall into three broad categories, new, repair or maintenance, and demolition. Each category has its own unique set of health hazards. New construction will have hazards associated with materials and chemicals transported to the site that the contractor is familiar with and most often has a material safety data sheet on hand. Maintenance or repair contract work will often involve unknown hazards and chemicals that have been previously used at the site or that are being used adjacent to work to be performed. Demolition involves many hidden or unknown hazards. Even if hazards such as asbestos or lead have been removed prior to demolition of a building, once walls start coming down, new surfaces are exposed that may not have been abated. Hazardous waste abatement work would also fit into this category even though most sites are characterized to the nth degree. Of course, in all categories, hazards are often not communicated between different contractors working at the same site.

CONSTRUCTION INDUSTRIAL HYGIENE

An industrial hygienist concerned about exposure hazards associated with construction work must be familiar with various activities of different trades. The classic recognition, evaluation, and control strategy of an industrial hygienist applies in construction as well. Sometimes exposures can be attributed to the job. For example, for a worker sand blasting a piece of mechanical equipment, the industrial hygienist may need to investigate silica exposure, correct personal protective equipment used in the correct way, surrounding environment, and more than likely, lead exposure from the paint blasted off the equipment. With the exception of confined space, construction hazards involving normal activities can usually be predicted by a trained industrial hygienist.

It is, however, very unpredictable how much airborne exposure a worker is getting from a particular source. Many times the same type of work conducted at one site is much different from an exposure condition at another. Inside exposures will remain more constant than outside where wind and weather conditions play a major role. For example, asbestos abatement work that is conducted in a controlled atmosphere inside should remain fairly constant if work practices such as negative air filtration and surfaces are properly wetted. Conversely, work on an asbestos roof on the outside, even though there is a difference in the type of asbestos, will depend more on weather conditions. Work practices such as location of the worker in relationship to the wind, (up stream or down stream), and how "intact" the shingles are as they are removed also play an important part in overall exposure. The more broken up they are, the more likely an asbestos exposure will result, although inside exposures sometimes can vary vastly with the size of an area and individual work practices.

If the airborne exposure is to be determined for a particular job, the industrial hygienist must be prepared to monitor quickly. The next day may be too late. Concentrations usually need to be high to find time-weighted averages (TWAs) that exceed OSHA permissible exposure limits. More often than not the construction worker is not conducting the same job for an eight-hour period. Many tasks are usually required to accomplish a day's work, which also makes it difficult to evaluate a particular hazard. A construction worker cutting and burning all day on a bridge demolition project may have no exposure or wind up in the hospital undergoing chelation therapy with a blood lead level in the hundreds. All depends upon work habits, weather, type of paint on the steel, and personal protective equipment used.

PHYSICAL HAZARDS

As noted in the American Conference of Governmental Hygienist (ACGIH) TLV booklet, physical hazards are characterized separately from the health hazards. Physical hazards are defined as those type of hazards that can harm to a worker from an external source. Types of physical hazards are loud noise (equipment), temperature extremes (working in Tyvek coveralls in an asbestos containment area), radiation (exposures to the sun or welding flash), chemical burn (acids or caustics), fire and/or explosions. Other physical hazards include, but are not limited to, slips and falls, trench or excavation cave-ins, exposed machinery because of improper guarding, equipment moving about on site, confined spaces, and falling objects. The physical hazards listed below are some of the frequent concerns on construction jobs.

Ergonomic Hazards

Ergonomic hazards, also called repetitive motion disorders, are encountered when workers use vibrating tools like chain saws and powered hammers, or continuously move their hand and arm in the same motion throughout the work shift. According to the National Institute for Occupational Safety and Health, an estimated 1.45 million workers who use vibrating tools may experience injuries to their fingers and hands. Symptoms of vibration-induced health problems include numbness, pain, whiteness of the fingers, and loss of finger movement and coordination. This syndrome is often called "white finger disease" or Raynaud's disease. Stonecutters that used their hands to guide the cutting tool called this injury "dead fingers." Whether a worker develops a hand-arm vibration problem depends on several factors.

- The amount of vibration the tool causes.
- The length of time a worker uses the tool per day as well as the cumulative amount of time per month or year.
- Environmental conditions (cold or hot weather).
- The workers vibration "tolerance."
- Whether the worker uses tobacco, alcohol, and drugs.

Figure 9-3. Positioning of tools may cause ergonomic hazards

Another ergonomic injury that some workers experience is called Carpal Tunnel Syndrome. It occurs when repetitive and stressful wrist motion causes irritation, fluid build up, or thickening of the carpal ligaments in the wrists. This puts pressure on a nerve, sending a message of pain to the brain. Continued use under these conditions results in permanent damage to the nerve. The typical symptoms of carpal tunnel syndrome are tingling of the thumb and fingers, and night pain. The pain awakens the worker from sleep and is often relieved by shaking, hanging, or massaging the hand. Pain may also occur in the arm and the shoulder. Numbness and loss of movement may occur in more advanced cases. Weakness of the hand also occurs, causing difficulty with pinch and grasp motions. The patient may drop objects, be unable to use keys, or count change with the affected hand. Surgical treatment may be necessary if the symptoms are severe and if other measures do not provide relief.

Other ergonomic hazards include manual handling of objects and materials where lifting and carrying is done. Lifting is so much a part of many everyday jobs that most of us do not think about it. But it is often done wrong, with unfortunate results such as pulled muscles, disk injuries, or painful hernias.

Noise

Noise is a serious hazard when it results in temporary or permanent hearing loss, physical or mental disturbance, any interference with voice communications, or the disruption of a job, rest, relaxation, or sleep. Noise is any undesired sound and is usually a sound that bears no information with varying intensity. It interferes with the perception of wanted sound, and is likely to be harmful, cause annoyance, and/or interfere with speech.

The noise created by circular saws, planers, or high speed grinders and similar power tools is narrow band noise. This high frequency type of noise is very damaging to the inner ear. Impulse type noise is generated by energy bursts occurring repetitively or one at a time. Noise from a jack hammer is an example of repetitive impulse noise. The firing of a gun is an example of a singular impulse noise. All types of noise can harm you if it is high intensity and/or the exposure time is prolonged or repeated over and over.

Frequency is the number of times the sound vibrates in air. It is usually expressed in Hertz (Hz) units or in cycles per second (cps). If it vibrates fast, we say it has a high frequency

Figure 9-4. Saws are normally generators of high frequency noise

and it sounds high pitched to us. The keys on the right side of a piano keyboard produce a high pitch or frequency. If the frequency is low, the air is vibrating more slowly and we hear a low frequency sound. Piano keys on the left side of the keyboard produce low frequency sounds. A healthy young person can detect sounds in the 20 to 20,000 cycles-per-second range. As aging takes place, some hearing is lost. Higher frequencies cause the most damage to our ears and most people who have hearing loss have high frequency losses first.

Loudness or softness is determined by the intensity or sound pressure. The more power driving the sound, the higher the pressure. This is measured with an instrument called a sound level meter (SLM) in units called decibels (dB). Sounds that can just be heard by a person with very good hearing in an extremely quiet location is assigned the value of 0 dB. Ordinary speech is around 50 to 60 dB. At about 120 dB, the threshold of pain is reached. This would be like hearing a jet engine about 50 feet away.

Exposure to intense noise causes hearing losses which may be temporary, permanent, or a combination of both. Temporary hearing losses are recoverable after a period of time away from the noise. Such losses may occur after only a few minutes of exposure to intense noise. The greatest portion of temporary hearing losses occur within the first two hours of exposure. Recovery from such losses is usually within one or two hours after being removed from the exposure.

Permanent hearing loss, seen in workers who have been exposed to noise daily for a period of many years, is very similar to the pattern of temporary hearing loss. You do not recover from permanent loss and it does not respond to any known treatment or cure. To measure hearing loss, you must first establish a baseline of your hearing ability by conducting hearing tests, called audiometric tests, then retest at a later time (usually a year). Hearing loss may occur in one or both ears with varying degree of loss.

Noise dose limits are now required for workplaces to minimize hearing loss from occupational exposure. Although louder noise is allowed for brief periods during the workday, the mandatory noise level limit, (set by OSHA), is 90 dBA, time weighed average (TWA) over 8 hours. An employer must make hearing protection available, provide training, and provide hearing tests, when the noise level exceeds 85 dBA, time-weighted average. As a basic rule, if you cannot hear the snap of your fingers at arms length, you should be using hearing protection. Over 90 dBA, employers must assure that protection is being used. Studies have shown that in most individuals injuries are likely to occur at this level or greater.

The Ear

Airborne sound is rapid fluctuation of normal atmospheric pressure caused by a vibrating source. These air pressure fluctuations, or waves, can vary in intensity, harmonic content, frequency, and directionality. The word "sound" is also used to indicate the sensation

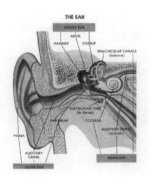

Figure 9-5. How we hear

experienced when these waves hit the ear. The ear enables an individual to detect sound waves within a range from about 20 to 20,000 cycles per minute and translate them into electrical impulses that are transferred to the brain. The ear can be divided into three parts, the outer, middle, and inner ear. Each of these parts plays a different role in transmitting sound to the brain.

The external ear acts as a funnel to channel sound waves to the middle ear. This outermost portion of the ear is also referred to as the external auditory canal. It is a skin-lined pouch about one and a half inches long supported by cartilage and bone. At its innermost end lies the tympanic membrane or eardrum. This eardrum separates the external from the middle ear.

There are small hairs that secrete a waxy substance located in the skin of the outer third of the ear canal. The function of these hairs is to filter out particulate matter thereby protecting the canal and eardrum. The waxy secretion is both sticky and bactericidal. The stickiness helps prevent small particles from entering the ear canal and the bactericidal quality keeps a healthy canal free from infection.

A very tiny space or cavity, about 1/4 inch square and lying between the eardrum and the bony wall of the inner ear, is called the middle ear. The middle ear is lined with a mucous membrane about the same as that which lines the mouth. Within the middle ear space are located the smallest bones in the body. These bones connect the eardrum to an opening in the wall of the inner ear called the oval window.

The sound conducting equipment is housed in the middle ear and includes the eardrum and three bones (ossicles) that are supported by ligaments and moved by two muscles. The ossicles, which together are called the ossicular chain, are the malleus, incus, and stapes, also often referred to as the hammer, anvil, and stirrup.

A tube running from the inner ear to the back of the throat (the eustachian tube), serves to equalize the pressure in the middle ear with the external atmospheric pressure. To accomplish this, it opens during swallowing and yawning. It normally tends to remain closed but when the pressure is increasing or decreasing such as during the rapid decent or take off of an airplane, the ears pop. Whenever the pressure is unequal on the two sides of the eardrum, it is not free to vibrate in response to sound waves.

The mucous membrane in the middle ear is connected to that of the pharynx via the eustachian tube. Thus, it is very easy for infection to travel along the mucous membrane from the nose or the throat to the middle ear. The inner ear contains the nerve receptors for hearing and for balance or position sense.

Biological Effects of Noise Exposure

Although the human ear is subject to a number of disorders that can cause hearing loss, the major cause of damage is excessive occupational noise. There are, however, a number of off-the-job non-noise induced hearing impairments such as

- Physical blockage of the auditory canals with excessive wax, etc.

- Traumatic damage, such as punctured eardrums or displacement of the ossicles.

- Disease damage such as childhood smallpox, infections, tumors, etc.

- Hereditary damages.

- Drug-induced damages such as from use of streptomycin or quinine.

- A natural reduction of hearing due to aging.

Other off-the-job noise exposures such as to motorcycles, snowmobiles, airplanes, etc. can also cause hearing loss. The outer and middle ear are rarely damaged by exposure to intense sound energy, although explosive sounds or blasts can rupture the eardrum and possibly dislodge the ossicular chain. Work-related hearing loss is most often caused by excessive exposure that involves injury to the hair cells of the inner ear.

When a worker is first exposed to hazardous noise, the initial change is usually a loss of hearing in the higher frequency range – about 4000 Hz. Initially this may be a temporary threshold shift (TTS) with recovery in about 14 hours. After prolonged exposure it may result in permanent damage referred to as noise-induced hearing loss. The hearing loss does not necessarily stop at the 4000 Hz range. Further exposure may result in a deepening and widening of the loss. This hearing loss will involve the speech frequency range resulting in considerable difficulty in hearing conversational speech.

The communication problem of a worker with a hearing loss is very frustrating to him and is easily misunderstood by coworkers, family, and friends. The person may appear to hear very well at times and poorly at other times. It is important to understand this kind of communication problem and assure that a coworker hears everything you have said.

Heat Stress

Heat stress is a serious physical hazard that should always be considered on a construction jobsite especially during the summer months. The chance of developing heat stress increases with increased humidity, hot environments, and the use of personal protective equipment. Sweating is the most effective means of losing excess heat, as long as adequate fluids are taken in to replace the sweat. When individuals are severely stressed by the heat, they may stop sweating with the most severe consequences of heat stress occurring. Adequate rest periods, availability of large amounts of replacement fluids, and frequent monitoring are essential to prevent the consequences of heat stress which may occur without warning symptoms. The body maintains a normal temperature (98.6° F) in a hot environment by two methods,

- Sending more blood to the skin.

- Sweating.

As the temperature in the air increases, sweat production allows more heat to be carried away through evaporation from your skin. As the humidity in the air increases or if the sweat can't be removed from the skin because of protective clothing, the body has more difficulty keeping an acceptable safe temperature. The body also cools by sending more blood to the skin. This reduces the blood available for the brain and muscles. Therefore, people who work in hot environments may feel tired sooner and less mentally alert. Both these factors, plus the awkwardness of protective equipment may contribute to increased accidents on a construction site.

Heat stress can be more critical than some chemical hazards on a construction site. Early stages can cause rash (prickly heat), cramps or muscle spasms, irritability, discomfort, and drowsiness. Advanced stages can cause nausea, dizziness or weakness, confusion, collapse, convulsions, coma, and death. Individuals who survive the most advanced stages (heat stroke) are likely to have damage to the brain and other organs. Heat stroke is a serious, life threatening medical emergency. It can happen in just a few minutes of extreme heat conditions. Individuals at particular risk of heat stress may be workers

- Wearing protective clothing.

- Who are dehydrated from blood pressure medications or infections such as diarrhea or fever.

- Who are not physically fit or have not worked in a hot environment in the preceding week (become acclimated).

- With chronic disease such as heart disease or diabetes.

- Who drink excessive alcohol or use drugs.

- Who are very overweight.

- Who regularly take certain medications for depression, nervous conditions, high blood pressure, diabetes, or heart disease.

Figure 9-6. Protective clothing that prohibits body cooling may be a source of heat stress

The following steps may be taken by a contractor to eliminate or reduce the risk of heat related injuries when working under hot conditions:

- The work schedule should be adjusted with adequate rest periods.
- Rest areas should be shaded and, if possible, air-conditioned.
- Provide cool fluids to drink.
- Provide medical screening including vital signs.
- Restrict activities.
- Provide adequate first aid facilities for quick treatment of heat stress illness.

Taking salt tablets to replace salt lost when sweating is not recommended. A normal diet contains more than adequate amounts of salt. Individuals working in hot environments may take the following steps to reduce the possibility of injury from heat stress:

- Drink adequate amounts of fluids (water or juices) throughout the day, even if you are not thirsty.
- The amount should be 4–6 liters (1 to 1.5 gallons). Alcohol, coffee, soda, and tea are not good fluids to replace water lost during sweating.
- Maintain good physical fitness. Individuals should work cautiously until their body has adjusted (become acclimatized) to the heat.
- Recognize the signs and symptoms of heat stress.
- Monitor pulse, temperature, and weight.
- Check with your doctor if you have chronic health problems or are taking medication.

Checking heart rate during periods workers are in hot environments is a valuable tool for controlling heat stress. Check the heart rate by locating the radial artery.

1. Placing the back of your hand down with your palm up on the table in front of you, then on the upper part of your wrist (thumb side) locate the spot 1/2"–1" down from upper edge of your wrist and 1" down from where the wrist meets the thumb part of your hand. This is between the hard chords (tendons) in the middle of the wrist and upper edge of the wrist.

2. Feel the pulse by gently placing the first two fingers of your other hand on the spot located in Step 1. You should feel the pulsation.

3. Taking the pulse by counting the number of times your fingers feel the pulsation during a one minute period. This number is your heart rate. The pulse may be taken for 10 seconds and multiplied by 6 to get your heart rate, but this will not be as accurate as taking your pulse for a full minute.

If your heart rate reaches 127 per minute, you should work for no longer than 1 hour without a rest period during which time your heart rate returns to between 60 and 80 per minute. If your heart rate reaches 145, you should work no longer than 15 minutes without a break. A healthy worker should not have a pulse rate above 150–180 for more than 4–6 minutes. The Table below gives the amount and length of rest periods recommended during hot weather for acclimated workers.

Table 9-1

Heart Beat Rates

60–80	90–100	105–110	110–120	120–130	130–140	140–150
at rest	greater than 8 hr	8 hours	2 hours	1 hour	30 min.	15 min.

Workers in hot environments should protect themselves from the symptoms of heat stress by

1. Checking heart rate during rest breaks. If it is greater than 120, work time should be reduced and rest time increased.

2. Check temperature at end of the work period before drinking fluids. If it is greater than 99.6° F (37.6° C), work time needs to be reduced and rest time increased. If it is greater than 100.6°F (38.1° C), protective work clothing should be removed.

3. Check weight (in the nude) before and at the end of work. If the loss of weight is greater than 1.5% of the total weight, take in more fluids during work. As an example, if normal weight is 200 pounds and the loss is more than 3 pounds from the start to the end of the shift, more fluids are needed ($.015 \times 200$ lbs = 3 lbs).

4. Check for symptoms of heat stress and seek treatment.

The three primary injuries that may occur from working in hot environments are

1. **Heat Cramps** – Heat cramps are caused by heavy sweating with inadequate electrolyte replacement. Signs and symptoms include muscle spasms, pain in the hands, arms, feet, legs, and abdomen.

2. **Heat Exhaustion** – Heat exhaustion occurs from increased stress on various body organs and the blood circulation system and is due to the inability of the heart to work properly and/or dehydration. Signs and symptoms include

- • Pale, cool, moist skin.
- • Heavy sweating.

- • Dizziness.
- • Nausea.

- • Fainting.
- • Rapid, shallow breathing.

3. **Heat stroke –** Heat stroke is the most serious form of heat stress. Temperature regulation fails and the body temperature rises to critical levels. Immediate action must be taken to cool the body before serious injury and death occur. Competent medical help must be obtained. Signs and symptoms are

- • Dizziness and confusion.
- • Nausea.

- • Strong, rapid pulse.
- • Coma.

- • Red, hot, usually dry skin.
- • Lack of or reduced sweating.

Heat stress potential is monitored by using the Wet Bulb Globe Temperature Index (WBGT) developed by the American Conference of Governmental Industrial Hygienists (ACGIH). A work site competent person, safety or industrial hygiene technician, or medical professional may monitor heat stress conditions using the WBGT index. The formula for this index is

Outdoors with solar load: $WBGT = 0.7\,NWB + 0.2\,GT + 0.1\,DB$
Indoors or outdoors without solar load: $WBGT = 0.7\,NWB + 0.3\,GT$

Where: WBGT = Wet Bulb Globe Temperature Index
         NWB = Natural Wet Bulb Temperature
         DB   = Dry Bulb Temperature
         GT   = Globe Temperature

Suggested permissible heat exposure threshold limit values using the WBGT index according to the above calculations are given below in degrees Fahrenheit.

Table 9-2
Work Load Index

Work – Rest Regiment	Light	Moderate	Heavy
Continuous Work	86.0	80.0	77.0
75% Work/25% Rest Each Hour	87.0	82.4	78.6
50% Work/50% Rest Each Hour	88.5	85.0	82.2
25% Work/75% Rest Each Hour	90.0	88.0	86.0

NOTE: The temperatures listed above are a combination of all temperature measurements. These measurements are usually obtained with a direct reading instrument. One type is called a Reuter-Stokes WiBGeT that integrates all the temperature measurements into one number that is compared with the chart above.

There are several factors to consider when establishing a work-rest schedule. These factors will be explained in some detail because it is important that workers have the skills to assess whether the rest break schedule established by employers is frequent enough to protect their health. As previously stated, the factors to consider include temperature and humidity, work load, clothing, individual reactions to heat as measured by heart rate, weight loss and body temperature, and individual sensitivity due to lack of acclimatization, physical conditioning, etc.

The environmental "adjusted temperature" can be determined by an industrial hygienist or safety specialist using the Wet Bulb Glove Temperature Index recommended by NIOSH. The Wet Bulb Globe test can be easily performed and requires simple instrumentation. The WBGT reflects the combined effects of factors which cause heat stress including humidity, air movement, air temperature, and radiation.

A somewhat cruder measurement can also be made using a regular thermometer while considering the percentage of sunshine in the sky. To get the adjusted temperature, measure the air temperature with the bulb shielded from direct sunlight. Next, determine the percent sunshine by judging what percent time the sun is not covered by clouds. (100 percent sunshine = no clouds and a sharp distinct shadow. Zero percent sunshine = cloudy, no shadows. The formula for calculating adjusted temperature is air temperature + (13 × percent sunshine) = adjusted temperature. For example, if the air temperature is 80° F and the percent sunshine is 75%, then the Adjusted Temperature = 80 + (13 × .75) = 89.75 F (90° F).

Cold Stress

When temperatures go down, the body maintains its temperature by reducing blood flow to the skin. This causes a marked decrease in skin temperature. The most extreme effect is on the extremities (fingers, toes, earlobes, and nose). When hands and fingers become cold, they become numb and insensitive, and there is an increased possibility of accidents. If the restriction of blood flow to the skin is not adequate to maintain temperature, then shivering occurs. If this is not adequate to warm the body, then a marked decrease in temperature (hypothermia) may occur. Workers that may be at increased risk are

- Doing hard labor who become fatigued and/or wet either from sweating or contact with water.

- Taking sedatives or drinking alcohol before or during work.

- Workers with chronic diseases that affect the heart and/or blood vessels of the hands or feet.

- Not physically fit or have not worked in a cold environment recently. Those who use pavement breakers or other vibrating equipment.

Harmful effects of exposure to cold stress include

1. **Frostbite**: Freezing parts of the body, particularly the fingers, toes, earlobes, and nose. The first warning is a sharp, pricking sensation. However, the numbness caused by the cold increases the chance of frostbite occurring without warning. Injuries vary from redness of the skin, numbness, loss of skin (as with a sunburn), to the loss of the body part.

2. **Immersion Foot:** Injury to the skin without freezing after long exposure to cold combines with dampness or contact with water. Injuries vary from swelling, tingling, itching, and pain to the loss of skin and skin ulcers.

3. **Hypothermia:** Inability of the body to maintain core temperature can lead to hallucinations, sleepiness, irregular heart beat, unconsciousness, and death.

Preventing cold stress may be accomplished by

- Wearing several layers of loosely fitted dry clothes that can be adjusted to match changing temperatures. Keep extremities warm. A top layer of wind-proof clothing is useful if there is a wind.

- Making sure you are not becoming overheated and sweaty, and that your extremities are not becoming numb.

- If you become chilled, sleepy, or develop pain and cold in your extremities, you need to seek warm shelter.

- Be sure you don't have any chronic heart or blood vessel disease, use sedatives, or drink excessive alcohol. All make you more susceptible to cold stress.

Radiation

Another type of physical hazard often encountered during construction activities is various forms of radiation. Radiation is energy which is emitted, transmitted, or absorbed in wave or energy form. It is a form of energy that each of us encounters every day. The radiation from the sun, visible light, and heat are examples and are familiar to everyone. Radiation is energy that, for the most part, cannot be seen, heard, felt, smelled, or tasted. Although we cannot sense radiation, it is easily detected with instrumentation. Radiation is measurable and through research, we have gained an understanding of its behavior. Today, different types of radiation are used as a valuable tool for medicine, construction, manufacturing, and agriculture.

Radiation is divided into two major categories, based on its effect on living tissue, 1) non-ionizing and 2) ionizing radiation. Ionizing radiation has the ability to change or destroy the atomic structure of cells, non-ionizing radiation does not. Some types of non-ionizing radiation that we are exposed to every day include microwave energy used for cooking and radio waves used in broadcasting radio and television. Types of ionizing radiation we are exposed to are cosmic rays from the sun and stars, terrestrial radiation from the earth, and medical radiation from X-rays and lasers.

Although non-ionizing radiation is not as hazardous as ionizing, there are exposures that can cause severe injuries. Non-ionizing radiation is generated by such things as the sun, lamps, welding arcs, lasers, plastic sealers, and radio or radar broadcast equipment. Since the eye is the primary organ at risk to all types of non-ionizing radiation, eye protection is very important. Protective glasses should be selected based on the type of radiation exposure, for example, sun light or welding flashes. Ionizing radiation is so named because it has enough energy to change (ionize) atoms and molecules, the building blocks of all matter. There are four natural types of ionizing radiation – alpha, beta, gamma, and neutron particles.

Alpha particles are the largest and most highly charged radiation energy. Due to their size, alpha particles are easily stopped. A sheet of paper will block the flow of alpha particles. An alpha particle striking you is normally stopped by the dead skin cells of your body. Alpha emitting radioactive material is primarily a concern if it is retained in the body. It is because of the alpha particle's large positive charge that it is easily shielded, but it is this same characteristic that makes the alpha the most damaging of all types of radiation when inside the body. This internal hazard exists if alpha particles have been swallowed, inhaled, or absorbed through broken skin. Protective clothing should be used to prevent contamination of the skin.

Beta particles are commonly emitted by the sun. Coleman lantern mantels and smoke detectors in your home contain and emit small amounts of beta particles. Beta particles are much smaller than alpha particles and have a velocity approaching the speed of light. Beta particles are stopped by a quarter inch piece of plastic or a light sheet of aluminum. The need for denser shielding indicates that the beta particle is more penetrating than the alpha particle. It will travel about 10 to 12 feet in air or penetrate the skin. It will usually be stopped by a wall unless the energy is extremely high. Beta particles are both an internal and external hazard. Externally they can damage the skin and eyes. Internally they can damage any tissue they come in contact with. Skin burns may result from excessively high doses of beta radiation.

Gamma energy is a type of radiation that is similar to X-radiation in that it is also an electromagnetic and ionizing radiation. Unlike light rays which we can see, gamma rays are invisible. Because gamma radiation has no electrical charge, it is very penetrating. With such penetrating ability, very dense materials such as lead or concrete are needed for a protective shield. Gamma radiation has the ability to penetrate completely through your body. It normally takes about 3 inches of lead or 3 to 4 feet of concrete to stop its energy.

Neutron radiation is another ionizing radiation hazard developed during the fission process normally found in nuclear reactions. When the fission process stops, the radiation hazard from neutrons also disappears. This means that when a nuclear plant is shut down, no neutrons are present to create a hazard. During nuclear plant operation, however, it takes about 10 feet of concrete and 8 inches of steel to stop this high energy radiation. Contractors working around nuclear power plants should take extra precaution to protect workers from this added hazard. Neutron radiation travels hundreds of feet in the air.

The exposure of living tissue to ionizing radiation has harmful effects to cells ranging from inhibition of cell function to impairment or death. The extent of damage is dependent upon the radiation dosage and the organs exposed. Some organs are more susceptible to radiation than others. The effects of low dosage radiation may be so delayed that they can only be detected by medical observation.

Material emitting radiation may be in the form of a gas, liquid, or solid. It may be in the air we breath, a water puddle on the ground, or dust on piping and equipment. Any time radioactive material is where we don't want it, we call it contamination. Once contamination is identified, steps must be taken to ensure that it is not spread around. With available instruments, we can determine where the radioactive contamination is, identify the type of radiation, identify material that is emitting the radiation, and calculate the potential biological effects of exposure.

Commonly used radiation detection equipment include film badges, pocket ionization chambers, dosimeters, Geiger counters, and portable air samplers. This equipment is also used when conducting radiation surveys. Radiation surveys should be done before contract employees are required to work in hazardous areas. Contractors often have the responsibility of obtaining survey readings on a radioactive hazardous waste site where their employees are working. Only qualified technicians that have the appropriate on-site training and the proper personal protective equipment should have this responsibility.

Figure 9-7. Radiation areas must be labeled

Due to risks that may be associated with radiation exposure, OSHA has developed regulations under 29 CFR 1910.96, that limit the amount of radiation you may receive. The forearm, hands, ankle, and feet are allowed the highest exposure at 18.75 rem per quarter (every three months). There are no vital organs in the extreme parts of the body; therefore, higher limits are permitted. Beta radiation can penetrate the skin tissue. For this reason, the limit set for radiation of the skin is 7.5 rem per quarter. Though the limit refers to the skin of the whole body, the organ which we are most concerned about is the eyes.

Other limits include those concerned with penetrating gamma rays. There are two limits set for gamma radiation. Three rem per quarter is the limit established for workers who have an available, known exposure history. Workers whose previous exposure has not been documented are restricted to 1.25 rem per quarter or 5 rem per year. The total amount of radiation any worker is allowed to accumulate is controlled by the formula 5(N–18), where N is the age of the individual. For this limit you would subtract eighteen from your age then multiply the result by five rem.

For example, a worker who is 33 years old would have a life-time exposure limit of 5(33–18) = 75 REM. It is important to note that all radiation exposure should be reduced to levels As Low As Reasonably Achievable (ALARA). This may be done by control measures such as engineering, administrative, or personal protective equipment.

HEALTH HAZARDS

Health hazards are caused by any chemical or biological exposure that interacts adversely with organs within our body causing illnesses or injuries. The majority of chemical exposures result from inhaling chemical contaminants in the form of vapors, gases, dusts, fumes, and mists, or by skin absorption of these materials. The degree of the hazard depends on the length of exposure time and the amount or quantity of the chemical agent. This is considered to be the dose of a substance. A chemical is considered a poison when it causes harmful effects or interferes with biological reactions in the body. Only those chemicals that are associated with a great risk of harmful effects are designated as poisons.

Dose is the most important factor determining whether or not you will have an adverse effect from a chemical exposure. The longer you work at a job and the more chemical agent gets into the air or on your skin, the higher the dose potential. Two components that make up dose are:

- The length of exposure, or how long you are exposed – 1 hour, 1 day, 1 year, 10 years, etc.

- The quantity of substance in the air (concentration), how much you get on your skin, and/or the amount eaten or ingested.

Another important factor to consider about the dose is the relationship of two or more chemicals acting together that cause an increased risk to the body. This interaction of chemicals that multiply the chance of harmful effects is called a synergistic effect. Many chemicals can interact and although the dose of any one chemical may be too low to affect you, the combination of doses from different chemicals may be harmful. For example, the combination of chemical exposures and a personal habit such as cigarette smoking may be more harmful than just an exposure to one chemical. Smoking and exposure to asbestos increases the chance of lung cancer by as much as 50 times.

The type and severity of the bodies response is related to dose and the nature of specific contaminant present. Air that looks dirty or has an offensive odor may, in fact, pose no threat whatsoever to the tissues of the respiratory system. In contrast, some gases that are odorless or at least not offensive, can cause severe tissue damage. Particles that normally cause lung damage can't even be seen. Many times, however, large visible clouds of dust are a good indicator that smaller particles may also be present.

Figure 9-8. Smoking contributes to a synergistic effect

The body is a complicated collection of cells, tissue, and organs having special ways of protecting themselves against harm. We call these the body's defense systems. The body's defense system can be broken down, overcome, or bypassed. When this happens, injury or illness can result. Sometimes job-related injuries or illness are temporary, and you can recover completely. Other times, as in the case of chronic lung diseases like silicosis or cancer, permanent changes may lead to death.

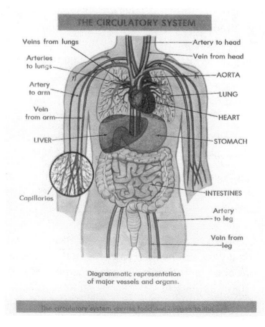

Figure 9-9. Blood carries contaminants absorbed in the lungs to all parts of the body

When it is impossible to prevent exposures to harmful materials, the next best protection is to reduce exposures through the use of various controls. Controlling hazards and informing employees of any danger are contractor responsibilities. In some cases employees may find that a material could cause harmful effects and must inform the employer so that steps can be taken to eliminate the hazard. Controlling hazards are implemented in the following sequence depending upon circumstances:

- Substitution or elimination.

- Engineering (ventilation, etc.).

- Administrative (rotating workers) (not a valid control for asbestos).

- Personal protective equipment.

The last control method that should be implemented is "personal protective equipment" which, unfortunately, is sometimes the first control used in construction.

Acute Health Effects

Chemicals can cause acute (short-term) or chronic (long-term) effects. Whether or not a chemical causes an acute or chronic reaction depends both on the chemical and the dose you are exposed to. Acute effects are seen quickly, usually after exposures to high concentrations of a hazardous material. For example, the dry cleaning solvent perchloroethylene can immediately cause dizziness, nausea, and at higher levels, coma and death. Most acute effects are temporary and reverse shortly after being removed from the exposure. But at high enough exposures, permanent damage may occur. For most substances neither the presence nor absence of acute effects can be used to predict whether chronic effects will occur. Dose is the determining factor. Exposures to cancer-causing substances (carcinogens) and sensitizers may lead to both acute and chronic effects.

An acute exposure may occur, for example, when we are exposed to ammonia while using another cleaning agent. Acute exposure may have both immediate and delayed effects on the body. Nitrogen dioxide poisoning can be followed by signs of brain impairment (such as confusion, lack of coordination, and behavioral changes) days or weeks after "recovery."

Chemicals can cause acute effects on breathing. Some chemicals irritate the lungs and some sensitize the lungs. Fluorides, sulfides, and chlorides are all found in various welding and soldering fluxes. During welding and soldering, these materials combine with the moisture in the air to form hydrofluoric, sulfuric, and hydrochloric acid. All three can severely burn the skin, eyes, and respiratory tract. High levels can overwhelm the lungs, burning and blistering them, and causing pulmonary edema. (Fluid building up in the lungs that will cause shortness of breath and if severe enough, can cause death.)

In addition, chemicals can have acute effects on the brain. When inhaled, solvent vapors enter the blood stream and travel to other parts of the body, particularly the nervous system. Most solvents have a narcotic effect. This means they affect the nervous system by causing dizziness, headaches, feelings of "drunkenness," and tiredness. One result of these symptoms may be poor coordination which can contribute to falls and other accidents on a construction site. Exposure to some solvents may increase the effects of alcoholic beverages.

Chronic Health Effects

A chronic exposure occurs during longer and/or repeated periods of contact, sometimes over years and often at relatively low concentrations of exposure. Perchlorethylene or alcohol, for example, may cause liver damage or other cancers 10 to 40 years after first exposure. This period between first exposure and the development of the disease is called the latency period. An exposure to a substance may cause adverse health effects many years from now with little or no effects at the time of exposure. It is important to avoid or eliminate all exposures to chemicals that are not part of normal ambient breathing air. For many chemical agents, the toxic effects following a single exposure are quite different from those produced by repeated exposures. For example, the primary acute toxic effect of benzene is central nervous system damage, while chronic exposures can result in leukemia.

There are two ways to determine if a chemical causes cancer – studies conducted on people and studies on animals. Studies of humans are expensive, hard to do, and very often not even possible. This type of long-term research is called epidemiology. Studies on animals are less expensive and easier to do. This type of research is sometimes referred to as toxicology. Results showing increased occurrences of cancer in animals are generally accepted to indicate that the same chemical causes cancer in humans. The alternative to not accepting animal studies means we would have a lot less knowledge about the health effects of chemicals. We would never be able to determine the health effects of the more than 100,000 chemicals used by industry.

There is no level of exposure to cancer-causing chemicals that is safe. Lower levels are considered safer. One procedure for setting health standard limits is called Risk Assessment. Risk assessment on the surface appears very scientific yet the actual results are based on many assumptions. It is differences in these assumptions that allow scientists to come up with very different results when determining an acceptable exposure standard. Following are major questions that assumptions are based on:

- Is there a level of exposure below which a substance won't cause cancer or other chronic diseases? (Is there a threshold level?)

- Can the body's defense mechanisms inactivate or breakdown chemicals?

- Does the chemical need to be at a high enough level to cause damage to a body organ before it will cause cancer?

- How much cancer should we allow? (One case of cancer among one million people, or one case of cancer among one hundred thousand people, or one case of cancer among ten people?)

For exposures at the current permissible exposure limit (PEL), the risk of developing cancer from vinyl chloride is about 700 cases of cancer for each million workers exposed. The risk for asbestos is about 6,400 cases of cancer for each million workers exposed. The risk for coal tar pitch is about 13,000 cases for each million workers exposed. Permissible exposure limits set for current Federal standards differ because of these different risks.

The dose of a chemical causing cancer in human or animal studies is then used to set a standard PEL below which only a certain number of people will develop illness or cancer. This standard is not an absolute safe level of exposure to cancer-causing agents, so exposure should always be minimized even when levels of exposure are below the standard. Just as the asbestos standard has been lowered in the past from 5 fibers/cubic centimeter to .2 fibers/cubic centimeter, and now to .1f/cc (50 times lower). It is possible that other standards will be lowered in the future as new technology for analysis is discovered and public outrage insists on fewer deaths for a particular type of exposure. If a chemical is suspected of causing cancer, it's best to minimize exposure, even if the exposure is below accepted levels.

Chronic Disease

Chronic disease is not always cancer. There are many other types of chronic diseases which can be as serious as cancer. These chronic diseases affect the function of different organs of the body. For example, chronic exposure to asbestos or silica dust (fine sand) causes scarring of the lung. Exposure to gases such as nitrogen oxides or ozone may lead to destruction of parts of the lung. No matter what the cause, chronic disease of the lungs will make the individual feel short of breath and limit their activity. Depending on the extent of disease, chronic lung disease can kill. In fact, it is one of the top ten causes of death in the United States.

Scarring of the liver (cirrhosis) is another example of chronic disease. It is also one of the top ten causes of death in the United States. The liver is important in making certain essential substances in the body and cleaning certain waste products. Chronic liver disease can cause an individual to be tired all the time, have their muscles waste away, and cause swelling of their stomach from fluid accumulation. Many chemicals such as carbon tetrachloride, chloroform, and alcohol can cause cirrhosis of the liver.

The brain is also affected by chronic exposure. Chemicals such as lead can decrease IQ, decrease ability to remember things, and/or make someone more irritable. Many times these changes are small and can be found only with special medical tests. Workers exposed to solvents, such as toluene or xylene in oil-based paints, may develop neurological changes over a period of time.

Scarring of the kidney is another example of a chronic disease. Individuals with severe scarring must be placed on dialysis to remove the harmful waste products or have a kidney transplant. Chronic kidney disease can cause an individual to be tired all the time, have high blood pressure and swollen feet, as well as many other symptoms. Lead, mercury, and solvents are suspect causes of chronic kidney disease.

Birth Defects/Infertility

The ability to have a healthy child can be affected by chemicals in many different ways. A woman may be unable to conceive because a man is infertile. The production of sperm may be abnormal, reduced, or stopped by chemicals that enter the body. Men working in a plant that manufactured batteries were exposed to lead fumes and realized after talking among themselves that none of their wives had been able to become pregnant. When tested, all the men were found to have marked reduction in the number of sperm.

Figure 9-10. The first few days after conception are most critical to the fetus

A woman may be unable to conceive or may have frequent early miscarriages because of mutagenic or embryotoxic effects. Changes in genes in the woman's ovaries or man's sperm from exposure to chemicals may cause the developing embryo to die. A woman may give birth to a child with a birth defect because of a chemical with mutagenic or teratogenic effects. When a chemical causes a teratogenic effect, the damage is caused by the woman's direct exposure to the chemical. When a chemical causes a mutagenic effect, changes in genes from either the man or woman have occurred.

Many chemicals used in the work place can damage the body. Effects range from skin irritation and dermatitis to chronic lung diseases such as silicosis and asbestosis or even cancer. The body may be harmed at the point where a chemical touches or enters it. This is called a local effect. When the solvent benzene touches the skin, it can cause drying and irritation (local effect).

A systemic effect develops at some place other than the point of contact. Benzene can be absorbed through the skin, breathed into the lungs, or ingested. Once in the body, benzene can affect the bone marrow, leading to anemia and leukemia. (Leukemia is a kind of cancer affecting the bone marrow and blood.) Adverse health effects may take years to develop from a small exposure or may occur very quickly with large concentrations.

Biological Hazards

Biological agents may be a part of the total environment or may be associated with certain occupations such as agriculture. Biological agents in the workplace include viruses, rickettsiae (organisms that cause diseases), bacteria, and parasites of various types. Diseases transmitted from animal to man are common. Infections and parasitic diseases may also result from exposure to insects or by drinking contaminated water. Exposure to biohazards may seem obvious in occupations such as nursing, medical research, laboratory work, farming, and handling of animal products (slaughterhouses and meat packing operations). Workers on a construction site exposed to fire ants may not be so obvious.

Figure 9-11. Poison ivy may be a serious biological hazard for some workers

Biohazards may be transmitted to a person through inhalation, injection, ingestion, or physical contact. Many plants and animals produce irritating, toxic, or allergenic (causing allergic reactions) substances. Dusts may contain many kinds of allergenic materials, including insect scale, hairs, and fecal dust, sawdust, plant pollens, and fungal spores. Other hazards include bites or attacks by domestic and wild animals. Workers on hazardous waste sites may risk exposure to bites from venomous snakes or poisonous spiders.

ROUTES OF ENTRY AND MODES OF ACTION

The Cell

The basic living unit of our body is the cell. Each organ is a collection of many different cells held together by a supporting structure. Each cell performs one particular function. For example, red blood cells, the most abundant type of cell in the body with twenty-five trillion out of a total of 75 trillion cells in the body, transport oxygen from the lungs to body tissues.

About 56% of the adult body is fluid. Fluid inside the cells is called intercellular and outside the cell it's called extracellular. The extracellular fluid is in constant motion throughout the body and is rapidly mixed by blood circulation and by crossing between the blood and tissue fluids. Therefore, all cells live in essentially the same environment. Each time the blood passes through the body it also passes through the lungs. The blood picks up oxygen in the alveoli, thus acquiring the oxygen needed for cells throughout the body. Cells are capable of living, growing, and providing their special functions so long as they get the proper amounts of oxygen, glucose, the different amino acids, and fatty substances, which are normally available in the internal environment.

Eyes

The importance of the human visual system in life is clear. There is a necessity for intact healthy eyes for the performance of tasks including those where man and machine interact. Of all the major body organs prone to work-site injuries, the eye is probably the most vulnerable. Consequently, protection against eye and face injuries is of major concern and importance for workers. The eye is an organ of sight and is not designed for the demands of prolonged viewing at close distances as is commonplace in today's workplace. Although the eye does have some natural defenses, it has none to compare with the healing ability of the skin, the automatic cleansing abilities of the lungs, or the recuperative powers of the ear. This is why an eye injury may be described as the most traumatic loss to the human body.

The eyeball is housed in a case of cushioning fatty tissue that insulates it from the skull's bony eye socket. The skull, brow, and cheek ridges serve to help protect the eyeball which is comprised of several highly specialized tissues.

The front of the eyeball is protected by a smooth, transparent layer of tissue called the conjunctiva. A similar membrane covers the inner surface of the eyelids. The eyelids also contain dozens of tiny glands that secrete oil to lubricate the surfaces of the eyelids and the eyeball. Another gland located at the outer edge of the eye socket secretes tears to clean the protective membrane and keep it moist.

Eye Hazards

Of physical injuries, foreign materials are by far the most common. Effects that can be expected from foreign bodies entering the eye are

- Pain, because the cornea is heavily covered with nerves and an object sitting on the surface of the cornea will constantly stimulate the nerves.

- Infection, because a foreign particle may carry bacteria or fungi, or may be carried by fingers used to rub the eye.

- Scarring from tissue that has healed and that may obscure vision.

- Damage, depending on the angle and point of entry and speed of a particle.

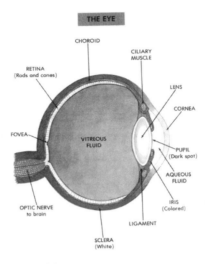

Figure 9-12. Schematic drawing of the eye

Heat can destroy eye and eyelid tissue just as it does other body tissue. High intensity light may have sufficient energy to damage the eye tissue. Exposure to ultraviolet light from welding operations (know as welder's flash) may severely damage the eye. Also, the effects of accidental exposure of the eye to chemicals can vary from mild irritation to complete loss of vision. In some cases, a chemical which does not actually damage the eye may be absorbed through the eye tissue in sufficient quantities so as to cause systemic poisoning.

Exposure to caustics are much more injurious to the eyes than acids. The appearance of an eye that has been exposed to a caustic on the first day after exposure may not look too bad. It may, however, deteriorate markedly on succeeding days. This is in contrast to acid burns where the initial appearance is a good indication of the ultimate damage.

Lungs and Inhalation

The respiratory system consists of all the organs of the body that contribute to normal breathing. This includes the nose, mouth, upper throat, larynx, trachea, and bronchi, all airways that lead to the lungs. It is in these airways that the first defense against contaminants exists. The adult human lung has an enormous area (75 square yards total surface area) where the body exchanges waste carbon dioxide for needed oxygen. This large surface, together with the blood vessel network (117 square yards total surface area) and continuous blood flow, makes it possible for an extremely rapid rate of absorption of oxygen from the air in the lungs to the blood stream. Some highly soluble substances such as gases, may pass through the lungs and into the blood stream so fast that it is not detected by the worker until ill effects set in. On the other hand there are substances, such as asbestos that by reason of being insoluble in body fluids, remains in our lungs for extended periods of time. Bodily attempts to destroy or remove these substances may result in irritation, inflammation, edema, emphysema, fibrosis, cancer, or allergic reactions and sensitization. Impairment of the lungs will not be noticed in the day-to-day activities of a worker. It does, however, reduce a workers ability to withstand future exposures.

Air enters through the nostrils and passes through a web of nasal hairs. Air is warmed and moistened as some particles are removed by compacting on the nasal hairs and at the bends in the air path. Interior walls of the nose are covered with membranes that secrete a fluid called mucus. The mucus drains slowly into the throat and serves as a trap for bacteria and dust in the air. It also helps dilute toxic substances that enter the airway.

Cilia, another important air cleaner, are hair-like filaments that wave back and forth a dozen times per second. Millions of cilia lining the nose and nasal airway help the mucus clean, moisten, and heat the air before it reaches the lungs. As the air moves into the bronchi it is divided and subdivided into smaller, finer, and more numerous tubes, much like those of the branches of a tree. There are two main branches, each getting smaller until they reach the lungs located on each side of the chest cavity. The respiratory tract branches from the trachea to some 25 to 100 million branches. These branches terminate in about 300 million air sacs called alveoli.

The lungs are suspended within the chest by the trachea, arteries, veins running to and from the heart, and by the pulmonary ligaments. They extend from the collarbone to the diaphragm, one on the right and the other on the left of the body, which fills most of the chest cavity. The right lung is slightly larger and is divided partially into three lobes, whereas the left is divided into two lobes.

The lungs are covered by a double membrane. The pleural membrane lies over the lungs and the other membrane lies in the chest cavity. Separating the two is a thin layer of fluid, which prevents the membranes from rubbing against each other during breathing. Negative pressure in the space between membranes prevents the lungs from collapsing. This negative pressure acts like a suction cup to pull the lungs against the chest wall and keep them expanded.

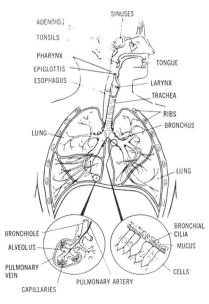

Figure 9-13. The respiratory passages

The ability of the lungs to function properly can be adversely affected in many ways. There may be blocked or restricted passageways, reduced elasticity, and/or damaged membranes. The first line of defense is the nose. It filters the air and prevents many contaminants from reaching lower portions. Many times however, we bypass this filtering defense system by breathing through our mouth. Coughing is another mechanism that expels foreign particles from the trachea and bronchi. Hair cells (called cilia) serve as a continuous cleaning mechanism for the nose, trachea, bronchi, and bronchioles. These hair-like extensions move like an escalator to sweep foreign particles back to the trachea where it is swallowed or spat out. Macrophages also help reduce particle levels by engulfing or digesting bacteria and viruses.

Respiration

The process by which the body combines oxygen with food nutrients to produce energy is called metabolism. To produce energy the body must exchange oxygen for carbon dioxide, which is accomplished in the lungs and is called respiration. While the other structures of the respiratory system are necessary to move air in and out of the lungs, it is at the alveoli that oxygen is delivered to blood for circulation throughout the body. Blood contains a chemical that is part protein and part iron pigment called hemoglobin. The hemoglobin attracts oxygen when it flows through the alveoli. Similarly, the carbon dioxide produced by the body cells during metabolism is attracted by the hemoglobin as it flows through the tissues and is then carried back to the lungs where it is released. Thus, blood acts as a tank car, unloading carbon dioxide and loading oxygen for distribution throughout the body. Carbon dioxide is always present in air, but the amount of carbon dioxide in exhaled air from the lungs is 100 times greater.

Often, gases are not blocked or restricted by the filtering defense system. One type of inhalation hazard found on construction sites is carbon monoxide which is present in exhaust from heavy equipment, generators, or compressors. It is also produced as a by-product of welding and soldering operations. Carbon monoxide's main effect is to rob the body of its oxygen supply. After being breathed, carbon monoxide combines more readily with the blood's oxygen carrier, hemoglobin, than oxygen. So exposures to high levels of carbon monoxide can prevent the body from getting enough oxygen, severely affecting the heart and brain. First symptoms may be headache, dizziness, and nausea. Higher exposures can result in fainting, coma, or even death. Persons with existing heart conditions, if exposed to carbon monoxide, are more likely to suffer additional heart damage as a result. And if you smoke, you already have higher than normal levels in your blood stream. A burning cigarette produces fairly high carbon monoxide levels.

The fate of substances that reaches the lungs depends on its solubility and reactivity. The more soluble the contaminant, the more likely it will be an upper respiratory irritant, such as sulfur dioxide (SO_2). Soluble reactive particles may cause acute inflammatory reactions and build-up of fluid (pulmonary edema). The less soluble gases and materials reach the lower lungs causing lung dysfunction or the particles that stick in the alveoli are engulfed by macrophages that move them back to the mouth, where they are expectorated or swallowed. Some chemicals that reach the digestive tract by this method are then absorbed and may still cause adverse health effects. The size of the particle greatly influences where it will be deposited in the air passages.

The normal atmosphere consists of 78 percent nitrogen, 21 percent oxygen, 0.9 percent inert or non-reactive gases, and 0.04 percent carbon dioxide. An atmosphere containing toxic contaminants, even at very low concentrations, could be a hazard to the lungs and the body. A concentration large enough to decrease the percentage of oxygen in the air can lead to asphyxiation or suffocation, even if the contaminant is an inert gas.

Inhaled contaminants that adversely affect the lungs or body fall into three categories.

- Aerosols and dusts that, when deposited in the lungs, may produce either tissue damage, tissue reaction, disease, or physical plugging.

- Toxic gases that produce adverse reaction in the tissue of the lungs themselves. For example, hydrogen fluoride is a gas that causes chemical burns.

- Toxic aerosols or gases that do not affect the lung tissue, but are passed from the lung into the blood stream. From there they are carried to other organs, or have adverse affects on the oxygen carrying capacity of the bloodstream itself.

For example, carbon monoxide gas, after being inhaled through the lungs, binds tightly to the red blood cells in our bloodstream and prevents the cells from getting oxygen and carrying it to the other parts in the body that need oxygen.

Four things must be known about inhaled contaminants before the toxic affects can be determined.

- Identity of the contaminant. (What chemical or material?)

- The concentration inhaled. (How much?)

- Duration of exposure. (How long?)

- The frequency of exposure. (How often?)

One way to gauge damage to our lungs is to measure the amount of air that we can get into them (vital capacity). If we know the vital capacity of our lungs before they are exposed to contaminants, and know what it is after exposure, then sometimes we may be able to determine the damage. This type of measurement is called pulmonary function testing. Normal activity such as aging is taken into consideration for an accurate assessment of lung functions.

Natural Defenses

Both rate and depth of respiration increase with heavy work, resulting in an increase in the amount of air and possibly toxicants in the lung. Certain toxicants, such as the nicotine in cigarettes, also disables the normal protective mechanisms of the lungs.This increases the damage caused by exposure to other toxicants, such as tar in cigarettes. Some of the bodies natural defense systems include the following responses:

- Coughing is one of the first signs of air passage irritation. This may occur at the time of an adverse exposure or may be delayed. Delayed effects often occur first thing in the morning.

- Sputum production, especially at night while lying down, helps clear toxic particles. This is normally greater for smokers due to the paralyzing effect of smoke on the cilia. During an extended period of time without smoking some of the cilia will recover and move contaminants out of the airways.

- Macrophage cell production is another method the body uses to eliminate particles from the air passages. This often is the result of an allergic or sensitization reaction in the respiratory tract.

Skin Absorption

The skin is the largest organ of the body, covering about 19 square feet of surface area. It is often the first barrier to come in contact with hazardous contaminants. The skin must protect the worker from heat, cold, moisture, radiation, bacteria, fungus, and penetrating objects. The skin is the organ that senses touch or hurt for the central nervous system. One square inch of skin contains about 72 feet of nerves. Contact with a substance may initiate the following actions:

- The skin and its associated layer of fat (lipid) cells can act as an effective barrier against penetration, injury, or other forms of irritation.

- The substance can react with the skin surface and cause a primary irritation (dermatitis).

- The substance can penetrate the skin and accumulate in the tissue, resulting in allergic reactions (skin sensitization).

- The substance can penetrate the skin, enter the bloodstream, and act as a poison to other body organs (systemic action).

- The substance can penetrate the skin, dissolve the fatty tissues, and allow other substances to penetrate skin layers.

Most job-related skin conditions are caused by repeated contact with irritants such as solvents, soaps detergents, particulate dusts, oils, grease, and metal working fluids. This is called contact dermatitis, and the symptoms are red, itchy skin, swelling ulcers, and blisters. The length of exposure and the strength of the irritant will affect the severity of the reaction as well as abrasions, sores, and cuts, which open a pathway through the skin and into the body.

The skin performs a number of important functions.

- Against invasion by bacteria.

- Against injury to other organs that are more sensitive.

- Against radiation such as from the sun.

- Against loss of moisture.

- Providing a media for the nervous system.

The nervous system in the skin contains hundred of pain receptors, plus pressure, heat, and cold receptors. Temperature regulation is an important function of the skin. One square inch of the skin contains about 15 feet of blood vessels which expand (dilate) when the body needs to lose heat, or shrink (constrict) when the body must maintain heat. There are about 2 million sweat glands over the surface of the body that are controlled by a heat regulator in the brain. Individual's skin reacts in different ways with the same chemical, physical, and biological exposures.

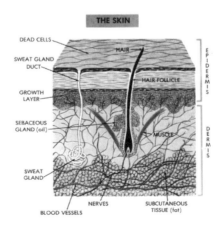

Figure 9-14. Diagram of the skin's protective layers

The skin provides protection, sensory reception, and regulation of temperature. Three layers of tissue, (epidermis, dermis, and subcutaneous), make up the skin. The dermis, middle layer, contains blood vessels, nerves, hair follicles, sweet glands, and oil glands. The skin thickness ranges from about 0.5 mm on the eyelid to 3 or 4 mm on the palm and sole of the foot. There may be as many as 60 layers of cells on the palm or sole of the foot. Beneath the dermis is the subcutaneous layer of skin that helps to cushion the outer layers and serve as an insulator. This layer also contains fat lobules, blood vessels, and nerves. It connects the skin with the tissue covering the muscles and bones.

Serious and even fatal poisoning has occurred from brief skin exposures to highly toxic substances such as parathion or other related organic phosphates (weed and insect killers), phenol, and hydrocyanic acid. Compounds that are good solvents for grease or oil, such as toluene and xylene, may cause problems by being readily absorbed through the skin. Abrasions, lacerations, and cuts may greatly increase the absorption, thus increasing the exposure to toxic chemicals.

Ingestion

Workers on the jobsite may unknowingly eat or drink harmful toxic chemicals. These toxic chemicals, in turn, are then capable of being absorbed from the gastrointestinal tract into the blood. Lead oxide, found in red paint on I beams, can cause serious problems if workers eat or smoke on the jobsite. Good personal hygiene habits, such as thoroughly washing face and hands before eating or smoking, are essential to preventing exposure.

Figure 9-15. Workers often eat contaminants due to inadequate washing facilities

Inhaled toxic dusts can also be swallowed and ingested in amounts large enough to cause poisoning. Toxic materials that are easily dissolved in digestive fluids may speed absorption into the blood stream. Ingestion toxicity is normally lower than inhalation toxicity for the same material, due to relatively poor absorption of many chemicals from the intestines into the blood stream.

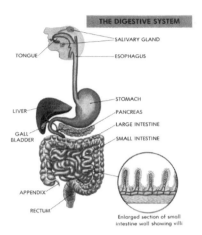

Figure 9-16. The digestive system

After absorption from the intestinal tract into the blood stream, the toxic material generally targets the liver which may alter or break down the material. This detoxification process is an important body defense mechanism. It involves a sequence of reactions.

- Deposition in the liver.

- Conversion to a non-toxic substance.

- Transportation to the kidney via the bloodstream.

- Excretion through the kidney and urinary tract.

Sometimes this process will have a reverse effect by breaking down a chemical into components that are much more toxic than the original compound. These components may stay in the liver causing adverse effects, or they may be transported to other body organs damaging them.

Personal Exposure Guides

A variety of hazard guidelines exist to evaluate worker exposure to chemical or other hazardous conditions at work sites. Most of these guidelines can be used to evaluate the dangers present at sites and determine the appropriate level of protection to be worn or other action necessary to protect workers health. Personal exposure guides are indications that hazardous conditions may exist. Workers should watch for the following personal signs of exposure to toxic chemicals or work stress. If any of these occur, they should leave the site and report the problem immediately. Return should not occur until the cause of the symptoms have been checked by a qualified person. Warning signs of chemical exposure may be

1. Breathing difficulties – breathing faster or deeper, soreness and a lump in the throat.

2. Dizziness, drowsiness, disorientation, difficulty in concentration.

3. Burning sensation in the eyes or on the skin, redness, or soreness.

4. Weakness, fatigue, lack of energy.

5. Chills, upset stomach.

6. Odors and/or a strange taste in your mouth.

CHEMICAL EXPOSURE GUIDELINES

Exposure guidelines are set by reviewing previous experience with hazards from several sources, including actual experience in dealing with hazards, results of studies of human exposure to toxic chemicals, and laboratory studies on animals. Because we do not have absolute knowledge about most hazards and opinions vary about the degree of hazards posed by different chemicals, guidelines will vary, even for the same chemical. Guidelines can and do change as new information is discovered. The goal is to minimize any worker exposure to hazardous conditions.

OSHA regulations require the employee to know about chemicals to which they are being exposed. General guidelines do not require that you know the amount of chemical present or its concentrations in the air. These are often found on labels or placards on chemicals containers. General guidelines often use short phrases, a word, numbers, or symbols to communicate hazards such as "Avoid skin contact" or "Avoid breathing vapors."

Specific OSHA regulations also require the employer to know both the identity and air concentration of the chemicals that may be present at the worksite. The results of air monitoring are compared to specific permissible levels to make decisions about worker exposure. Three different organizations have developed specific chemical exposure levels that are widely used at work sites to reduce worker exposures to levels thought to be safe.

1. Permissible Exposure Limit (PEL) (Set by the Occupational Safety and Health Administration, OSHA) PELs are legal enforceable standards. PELs are meant to be minimum levels of protection. Employers may use more protective exposure levels for chemicals. In many cases, current PELs are derived from threshold limit values published in the 1998 ACGIH TLV list. Many PELs are not set to protect workers from chronic effects such as cancer. In addition, most PELs that apply to the construction industry were established in 1969 and are rather outdated.

2. Recommended Exposure Limits (REL) (Set by the National Institute for Occupational Safety and Health, NIOSH) These are advisory levels and are not legally enforceable. RELs are sometimes more protective than PELs. Long-term or chronic health effects are considered when setting the RELs.

3. Immediately Dangerous to Life and Health (IDLH) (Set by the National Institute for Occupational Safety and Health) These values are established to recognize serious exposure levels that could cause death and serve as a blueprint for selecting specific types of respiratory protection.

4. Threshold Limit Value (TLV) (Set by the American Conference of Governmental Industrial Hygienists, ACGIH) TLVs are advisory and are not legally enforceable. A revised list of TLVs is published each year making them more current than PELs. However, chronic effects such as cancer are not always given consideration when setting TLVs. Ways to list chemical hazard guidelines are

   - Time-weighted average, (TWA).
   - Short-Term Exposure Limit, (STEL).
   - Ceiling Values.
   - Skin Absorption Hazard.

Time-Weighted Average

Time-weighted average is the average concentration of a material over a full work shift (set as 8 hours per day and 40 hours per week). The changes in exposure that occur during the work shift are averaged out. Exposures may be collected throughout the work shift with varying levels. They must be averaged for the 8 hours to compare the results with OSHA PEL lists. For example, a worker's exposure to toluene is 90 parts per million (ppm) for 2 hours, 120 ppm for 1 hour, 20 ppm for 5 hours. The TWA is calculated,

$$TWA = \frac{(90 \text{ ppm}) (2 \text{ hours}) + (120 \text{ ppm}) (1 \text{ hour}) + (20 \text{ ppm}) (5 \text{ hours})}{8 \text{ hours}}$$

$$TWA = \frac{180 \text{ ppm hours} + 120 \text{ ppm hours} + 100 \text{ ppm hours}}{8 \text{ hours}}$$

$$TWA = \frac{400 \text{ ppm hours}}{8 \text{ hours}} \quad\quad OR \quad\quad TWA = 50 \text{ ppm}$$

The actual TWA exposure to toluene for this worker is 50 ppm. The allowable TWA exposure for toluene is 100 ppm. This indicates that on this particular day this worker was not over-exposed according to OSHA limits. If the employee works longer than eight hours, over-time calculations must be conducted. In addition, if the worker is exposed to more than one substance or a mixture of substances, mixture calculations must be conducted.

Short-Term Exposure Limits (STELs)

Short-Term Exposure Limits are the maximum concentration level that workers can be exposed to for a short period of time (usually 15 to 30 minutes) without suffering from irritation, chronic or irreversible tissue damage, dizziness sufficient to increase the risk of accidents, impair self-rescue, or reduce work efficiency. Exposures above the PEL should not occur more than four (4) times per shift with at least sixty (60) minutes between exposures. The daily TWA PEL must not be exceeded without appropriate personal protective equipment. Not all chemicals have been assigned STELs. For substances without STELs, it is generally recommended that exposure should not exceed three (3) times the TWA for a short term (10 to 30 minutes). OSHA's STEL for toluene is 300 ppm over 10 minutes.

Ceiling Limit

Often workers can experience acute health effects if a ceiling limit level listed in OSHA's PEL values are exceeded. If a ceiling limit is not assigned to a substance or chemical, it is generally recommended that exposures never exceed five times their PEL.

Skin Absorption Notation

The notation "skin" listed in OSHA's PELs indicate that the chemical can be absorbed through the skin as a route of entry into the body. Remember that PELs, RELs and TLVs refer only to inhalation exposure! No concentration guidelines for skin exposure exist. Steps should be taken to avoid skin contact with chemicals. Even if the PEL, REL, or TLV are within the standard, you may be overexposed to a chemical by skin absorption. Again, if the notation "skin" is listed, there should be no skin exposure allowed. Chemical recommended exposure lists are guidelines and not absolute safe levels. Exposure should be minimized as much as possible. What is thought to be safe today may not be tomorrow. The following table gives some of the common hazards on a construction site and the action that should be taken at certain levels.

Table 9-3

Common Hazard Guidelines

Hazard	Level	Action
Explosive atmosphere	Less than 10% LEL	Seek advise
	10–25% LEL	Stop work
	Greater than 25% LEL	Explosion hazard Leave area immediately
Oxygen	Less than 19.5%	SCBA or other SAR must be worn. Combustible gas readings are invalid.
	19.5–23%	Continue work. Variations may be due to other gasses.
	Greater than 23%	Fire hazard. Leave area immediately.
Radiation	Less than 1 m rem/hr.	Continue work. Any reading may be background.
	Greater than 2 m rem/hr.	Radiation hazard. Leave area immediately.

Many of the worker exposures are the result of airborne contaminants such as dusts, fumes, gases. mists, or vapors. Each of these contaminants has different actions and physical properties which will be covered in the following paragraphs. These contaminants are instrumental in creating respiratory hazards such as asbestosis or silicosis.

TYPES OF AIRBORNE CONTAMINANTS

Dusts

Dusts are solid particles suspended in air. They may be produced by crushing, grinding, sanding, sawing, or the impact of materials against each other. Some dusts have no effect on the body. They don't seem to harm the body or be changed by the body's chemistry into other harmful substances. Most harmful dusts cause damage after being breathed. Some, such as cement and arsenic, can also directly affect the skin.

When considering health effects from inhaled dust, we must be concerned about a solid material that is small enough to reach the air sacks in our lungs where oxygen and carbon dioxide exchange takes place. This area is called the alveoli. Only particles smaller than about 5 micrometers (um) or microns (about 1/100 the size of a speck of pepper) are likely to reach this area of the lung. Particles in the range from 5–10 um will be deposited in the upper respiratory tract airways, (nose, throat, trachea, and major bronchial tubes) and cause bronchitis. Particles larger than 10 um, like wood dusts, can deposit in the nasal airways with the possibility of causing nasal ulcerations and cancer. Particles smaller than about one micrometer are likely to be exhaled during normal breathing.

The body has defenses against dusts. The most damaging dusts to the alveoli are small enough to get past these defenses and are too small to be seen with the naked eye. We call these "respirable" or breathable dusts. Your body's defenses against large-sized dusts that get

lodged in the trachea are mucous and the hair-like cells called cilia. Cilia work like an escalator, moving particles back to the throat where it is swallowed or coughed up. Cigarette smoke
paralyzes the cilia, which is the likely reason that smokers have about 50 times the normal
cancer rate.

Figure 9-17. Dust containing silica is generated by concrete finishing

Special white blood cells or macrophages can capture and remove damaging dust.
Often these cells can be overwhelmed with a large or chronic exposures. Some chemicals can
dry out the mucous. A result of such a breakdown is more dust particles reaching the lungs
where contaminant is eventually absorbed by the blood and carried to target organs. A broad
term used to describe lung injury developed from chronic effects of breathing dusts is called
pneumoconiosis (pronounced, new-mo-cone-e-o-sis). This tongue-twisting Greek word means
"lung" and "dust" or "dusty lung."

Figure 9-18. Cutting pipe could generate dust containing asbestos

Fumes

Fumes, like dust, are also solid particles in the air. They are usually formed when metals are heated to their melting points, especially during welding or soldering. Fumes are produced when metal is welded. Solder, electrode, welding rod, or metallic coating on materials may be vaporized generating additional fumes. Chromium and nickel exposures are possible when fumes are generated from stainless steel during arc welding. Sometimes plumbers generate lead fumes when molten lead is used for joining black pipe. Lead fumes are also generated by melting lead to make fishing sinkers or burning lead paint off surfaces.

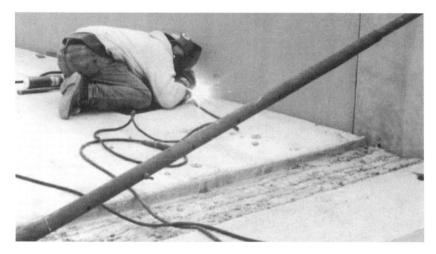

Figure 9-19. Welding fumes' toxic effect depends on rod, metal, and amount inhaled

Although many fumes can irritate the skin and eyes, these fine particles primarily affect the body when they are inhaled. This type of exposure sometimes results in an acute health effect, referred to as "metal fume fever" especially if the fumes are from metals such as zinc, cadmium, or magnesium. Workers often generate a lot of lead and metal fumes during demolition projects when using torches to cut and burn I beams. Dangerous fumes may also be produced by heating asphalt during hot tar roofing or road paving. An ingredient used in this process is called "coal tar pitch." These hazardous fumes are regarded as a serious cancer threat.

Fume particles' small size allows many of them to get past the body's natural defenses. They can then reach and irritate the lungs. Their small size and ability to spread out in the lung fluids allows fumes to pass easily from the lungs into the blood stream, thereby damaging other parts of the body. Many fumes, such as lead, affect the liver, kidneys, and nervous system and are called systemic poisons.

Gases

Gases are formless at room temperature and always expand to fill their containers. They can be changed into liquids or solids by increasing the pressure and/or decreasing their temperature. It is in these changed forms that gases are normally stored and/or transported. Toxic gases can directly irritate the skin, throat, eyes, or lungs, or they may pass from the lungs into the blood stream to damage other parts of the body. Some gases such as methane can also cause a worker to suffocate by displacing oxygen in the air. Many fatalities have occurred due to the improper entry of confined spaces such as under ground silos containing manure. As the manure decays, it generates methane gas displacing the oxygen.

The body's defenses against some gases include smelling, tearing eyes, and coughing. Ammonia's irritating effects and odor warn workers of exposure. However, workers may be exposed to some gases without knowing it. Carbon monoxide is the most widespread gas risk. It can be found whenever heavy equipment or motors are being used. It is a colorless, odorless gas formed by burning carbon-containing materials such as coal, oil, gasoline, wood, or paper. The chief source of carbon monoxide in the environment is the automobile. Carbon monoxide has no warning properties. You can't see or smell it, and it doesn't irritate the nose, eyes, throat, or lungs. Special respirators must be used due to the lack of warning properties.

Other gases such as hydrogen sulfide with its characteristic rotten egg odor may dull the sense of smell after awhile, then the natural warning sign of smell no longer works. Therefore, the sense of smell is a poor way of detecting any type of exposure to hazardous substances and workers should not rely on it.

Figure 9-20. Confined spaces may contain toxic gases

Steel cylinders of oxygen, hydrogen, nitrogen, and air are among the many gases that are routinely used on construction sites. These cylinders should be stored in an upright position, strapped securely to a permanent structure, and protected from high temperature. Gases in cylinders are generally safe under such conditions as long as the temperature of the gas does not exceed 125° Fahrenheit. Beyond this temperature, the cylinder might burst.

Mists

Mists and fogs are drops of liquid suspended in the air. Fogs may be created by vapors condensing to the liquid state, while mists are droplets being splashed or sprayed. Examples of mists used in construction include oil mist sprayed onto concrete forms, paint spray mists, and acid mists produced by fluxes used in soldering. Many mists and fogs can damage the body if they are breathed or if they make direct contact with skin or eyes. Like fumes, mists are small enough to by-pass the respiratory system's defenses and get deep inside the lungs. There they pass easily into the blood stream, then to other parts of the body.

Figure 9-21. Cleaning buildings with high pressure chemicals may generate hazardous mists or vapors

Vapors

Vapors are gaseous forms of certain materials that are usually solid or liquid at room temperatures. Vapors may be formed when liquids or solids are heated. Some materials, such as solvents, form vapors without being heated. Solvent vapors are one of the most common exposures at a hazardous waste and/or construction site. Mercury is an example of a metal that vaporizes at room temperature and can be a serious health hazard. Mercury was used by our ancestors during felt-hat manufacturing. Hence the name "mad hatters" due to central nervous system damage. Both vapors and the materials from which they evaporate can harm the body. Many directly affect the skin causing dermatitis, while some can be absorbed through the skin. As with gases and fumes, most vapors when breathed pass to the blood stream and damage other parts of the body. Some of these materials can damage the liver, kidneys, blood, or cause cancer.

TYPICAL HAZARDOUS CHEMICALS IN CONSTRUCTION

There are many different types of hazardous chemicals used in construction that you may be exposed to. Many of these chemicals can be grouped into a set of general categories because they pose the same types of hazards. In this way it simplifies the general hazards that maybe encountered on the work site. Hazards associated with some common materials found on construction sites are reviewed in paragraphs below.

• Solvents	• Acids, Bases, and Alkalines	• Fuels	• Wood
• Cleaners	• Adhesives and Sealants	• Concrete	

Solvents

A solvent is a liquid that dissolves another substance without changing the basic characteristic of either material. When the solvent evaporates, the original material is the same. In construction, we most often see them as cleaners, degreasers, thinners, fuels, and glues. Solvents are lumped into three main types or classes: those containing water (aqueous solutions) such as acids, alkalines, and detergents, and those containing carbon (organic solvents) like acetone, toluene, and gasoline. The third group contains chlorine in their chemical makeup and are called chlorinated solvents like methylenechloride and trichloroethylene.

Solvents can enter into your body in two ways, by inhalation or by absorption through the skin. Any solvent inhaled may cause dizziness or headaches as it affects the central nervous system. If breathing solvent vapors continues over time, the development of nose, throat, eye and lung irritation and even damage to the liver, blood, kidneys, and digestive system, may result. Most solvents in contact with skin can be absorbed into the body. Because solvents dissolve oils and greases, contact with skin can also dry it out producing irritation, cracking, and skin rashes. Once a solvent penetrates through the skin, it enters into the bloodstream and can attack the central nervous system or other body organs.

Figure 9-22. Some organic solvents will readily absorb through the skin

Like all chemicals, the effect on the body will depend on a number of factors; how toxic it is, how long the exposure, the body's sensitivity, and how concentrated or strong the solvent is. Solvent hazards may be minimized by following a few simple rules:

- Know what chemicals you are working with.
- Use protective equipment like gloves, safety glasses, and proper respirators to prevent contact with skin, eyes, and lungs.
- Make sure the work area has plenty of fresh air.
- Avoid skin contact with solvents.
- Wash with plenty of soap and water if contact with skin occurs.
- If a solvent splashes into eyes, flush with running water for a minimum of 15 minutes and get medical help. Remember, gasoline should never be used as a solvent or cleaning agent.

Cleaners

Cleaners contain acids, alkalizes, aromatics, surfactants, petroleum products, ammonia, and hypochlorite. Because of these ingredients, cleaners are considered to be irritants, and can be harmful if swallowed or inhaled. Many can cause eye, nose, throat, skin, and lung irritation. Some cleaners are flammable and burn easily. Others may be caustic or corrosive and cause severe skin damage. Because many cleaners used in industrial situations are consumer products commonly found in our homes, you may underestimate the hazard they pose. Close review of precautions listed in the "material safety data sheet" (MSDS) is needed to protect workers from these chemicals. Often, gloves and eye protection are required. Respirators may be needed to avoid inhaling the vapors and mists. The lack of worker personal hygiene is one of the greatest exposure problems. Hands and face should be washed thoroughly before eating, drinking, or smoking.

Figure 9-23. Many cleaners may be incompatible if mixed

Mixing of cleaning chemicals should be avoided unless specifically instructed to do so. For example, a dangerous gas, chlorine, will be created if you mix bleach and ammonia, or bleach and drain cleaner together. Because of the variety of cleaning materials in use, there are many signs and symptoms of overexposure. Contractors should rely on the MSDS for the particular product being used.

Acids and Bases

Acids and bases (caustics) can easily damage the skin and eyes. How serious the damage is depends on how strong the chemical is, how long contact is maintained, and what actions are taken after an exposure. Acids and bases can be in the form of liquids, solid granules, powders, vapors, and gases. A few commonly used acids include sulfuric acid, hydrochloric acid, muriatic acid, and nitric acid. Some common bases (caustics) are lye (sodium hydroxide) and potash (potassium hydroxide). Both acids and bases can be corrosive, causing damage to whatever they contact. The more concentrated the chemical, the more dangerous it can be. Vinegar is a mild form of acetic acid and as such it can be swallowed or rubbed on the skin with no damage, but a concentrated solution of acetic acid can cause serious burns.

Various acids react differently when they contact the skin. Sulfuric acid mixes with water to produce heat, so when it contacts the skin, it reacts with moisture and causes burns. Hydrofluoric acid may not even be noticed if it spills on the skin, but hours later as the acid is absorbed into the muscle tissue, can cause deep burns that are very painful and take a long time to heal. Most acids in a gas or vapor form when breathed react with the moisture in the nose and throat causing irritation or damage. Acetic and nitric acids do not react as readily with water but when these vapors are inhaled, they quickly penetrate into the lungs causing serious damage.

Figure 9-24. Acids and bases should not be stored together

Bases, as a class of chemicals, feel slippery or soapy. In fact, soap is made from a mixture of a base (lye) and animal fat. Concentrated bases easily dissolve tissue and, therefore, can cause severe skin damage on contact. Concentrated caustic gases like ammonia vapors can damage the skin, eyes, nose, mouth, and lungs. Even dry powder forms of bases can damage tissue when breathed because they react with the moisture in your skin, eyes, and respiratory tract. Cement and mortar are alkali compounds in their wet or dry form. As dust and powder they can cause damage to the skin and eyes when they react with moisture in the body. Concrete and mortar can also cause an allergic reaction in people who become sensitive to them. Contractors should follow these rules when working with acids and bases:

- Know what chemicals you are working with and how strong (concentrated) they are.
- Use Personal Protective Equipment as noted on the MSDS.
- In case of skin or eye contact, flush with cool water for at least 15 minutes but do not rub the skin or eyes.
- Always add acid to the water to prevent splatter.
- Keep acids and bases apart, store separately, and clean up spills promptly. Acid and bases react, often violently, when mixed together.

Adhesives and Sealants

Most adhesives and sealants have some type of hazard warning on the label. Because of their common usage at home and on the job, these warnings are sometimes taken lightly or ignored altogether. Many adhesives and sealants are toxic because of their chemically reactive ingredients, or because of the solvent base which permits them to be more easily applied.

Adhesives or sealants that contain solvents may be flammable. Other types of adhesives, such as wood glue, may be eye and skin irritants. When working with any glue, care should be taken to avoid eye and skin contact. If the label indicates the adhesive is flammable, use and store away from sources of ignition. Epoxies contain epoxy amine resins and polyamide hardeners, which cause skin sensitization and respiratory tract irritation. Overexposure to epoxies can result in dizziness, drowsiness, nausea, and vomiting. In instances of extreme or prolonged exposure, kidney and liver damage may occur.

Figure 9-25. Floor adhesives may contain coal tar pitch volatiles

Floor adhesives may contain acrylics that can be irritating to the skin, may cause nausea, vomiting, headache, weakness, asphyxia, and death. Other adhesives or sealants may contain coal tar derivatives which are suspected carcinogens. Prolonged breathing of vapors and skin contact should be avoided.

Fuels

The primary hazard posed by fuels is, obviously, fire. Fuels are either flammable or combustible. Whether flammable (a material that easily ignites and burns with a vapor pressure below 100°F), or combustible (a material that ignites with a vapor pressure over 100°F), they should be handled with care. Gasoline is a flammable liquid and diesel fuel is an example of a combustible liquid.

Proper storage and transport of fuels in approved, self-closing, safety containers is extremely important, and should be strictly adhered to at all times. When filling portable containers with flammable materials, proper grounding and bonding is a must to prevent ignition caused by static electricity. Store gasoline in containers marked or labeled "Gasoline." Store kerosene in containers marked "Kerosene." Never use kerosene containers for the transport or storage of gasoline.

Figure 9-26. Gasoline and flammables should be stored in approved cans

Excessive skin contact with fuels can result in dermatitis. Fuels entering the body through the skin and over a long period of time can break down the fatty tissues and possibly build up in the body. Excessive inhalation of fuels may cause central nervous system depression and aggravation of any existing respiratory disease. Leukemia is a potential side effect of chronic exposure to some fuels and may lead to death. Ingestion of fuels may cause poisoning and possible lung damage if aspirated into the lungs when ingested. Short exposures to fuel may cause skin, lung, and respiratory tract irritation.

When using portable containers for flammable liquids, spark-arrestors must be in place. When dispensing or using fuels, be aware of the location of fire extinguishers, fire alarms, and evacuation procedures. Fuels are flammable, so do not store, use, or dispense near arc welding or open flame. Use a bonding clamp to bond and ground containers when dispensing fuels. Remember, if there is a spill of fuel, the vapors may travel some distance to an ignition source resulting in fire or an explosion. Do not pour waste fuel and flammable liquids down the drain. See the MSDS for proper waste disposal procedures. Specific emergency procedures will be detailed on the Material Safety Data Sheet. In general, if a fuel gets into eyes, flush the affected eye with running water for at least 15 minutes, then seek medical attention. If it gets on the skin, wash the area of contact with soap and water.

Wood

Wood which has been preserved by pressure treatment with a pesticide, to protect it from insect attack or decay, can be dangerous. Inorganic arsenic, copper, zinc, a pesticide, or some combination of these, sometimes called CCA (chromate copper arsenate), is used in the pressure treatment process. These chemicals are forced deeply into the wood where they remain for long periods of time. As a result treated wood, whether it is from the lumber yard or found in an existing structure, can pose health hazards if not properly handled.

Inhalation of sawdust from treated wood must be avoided. In addition, cutting, routing, sanding, or working treated wood, may be hazardous unless approved respirators are used or engineering controls are implemented. Whenever possible, these operations should be performed outdoors to avoid indoor accumulations of airborne sawdust from treated wood. Bystanders, children, and pets should be refrained from walking in the accumulated sawdust.

Some treated wood may appear damp and have chemical residue on the surface. Gloves should be used when handling treated lumber and especially the sawdust from treated wood. After working with treated wood, workers should wash hands and face thoroughly. Meals and breaks should be in a location away from the work area. Work clothing should be laundered after each use and separate from other clothing. Treated wood should not be used in circumstances where it will come in direct contact with food or with public drinking water sources. Do not burn in home stoves, fireplaces, or open fires because the chemicals may become part of the smoke and ashes. Disposal of treated wood should be at approved waste disposal sites. Treated wood may be burned in commercial or industrial incinerators or boilers in accordance with state and federal regulations. Clean up all construction debris.

Acute allergic reactions (hives, respiratory tract irritation, general swelling) have been reported following contact with mahogany, birch, beech, and other untreated woods. To reduce the likelihood of such a reaction, personal hygiene is important. Upon completion of work, clothing should be laundered separately from non-work clothing. Workers must shower thoroughly to remove any material on their skin.

Figure 9-27. Wood processed inside without adequate ventilation may be hazardous

EXPOSURE MONITORING

The role of monitoring is to tell you what contaminants are present, and at what levels. Yet the limitations of many instruments means that you can't be sure of the readings unless all perimeters are taken into consideration or you already know what is in the air. This seems to be a contradiction. After all, how can you know what is present if the instruments can't tell you? Often, determining contaminant levels are possible only after extensive diagnostic work with a variety of sampling strategies. Air sampling instruments can provide very important information to clarify the hazards at a construction site. Monitoring surveys can help answer questions like

- What types of air contaminants are present?
- What are the levels of these contaminants?
- How far does the contamination range?
- What type of protective gear is needed for the workers?

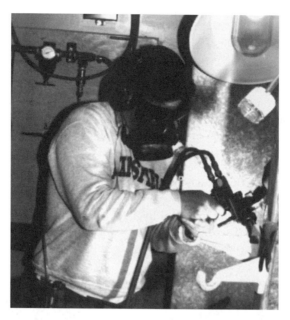

Figure 9-28. Worker "personal" exposure monitoring and "area" monitoring

Effective monitoring can be difficult work. It is much more than pushing buttons on a "high-tech" gadget. As you will see, it is more like detective work. The issues fall into three major categories.

- What are the limitations of instruments used?

- What strategy should be used to get useful information?

- How do you evaluate results that you get?

There are two types of air monitoring methods, (1) direct reading and (2) laboratory sampling. Direct reading instruments have built-in detectors to give "on the spot" results. However, there is a trade-off between sophistication and the weight of the unit. The instruments must be truly portable to be useful. Because of this, it is important to be aware that there are limits to any given instrument.

Laboratory sampling emphasis is on collecting a sample in the field, then conducting the actual analysis later back at the lab. The disadvantage is the delay in obtaining results. An advantage is that the instruments in the lab do not have to be portable. They can be large and more sophisticated for more precise analysis. For example, labs can utilize an instrument known as a gas chromatography and analyze a mixture of 5 different chemicals and separate them so they each can be separately examined. Lab procedures can also use computers to compare analytical results to known chemical properties. These unique properties serve to allow identification – similar to fingerprinting.

It is common to use both types of procedures to investigate exposures in the workplace. Direct-reading methods are ideal for quick checks especially when the contaminants are known or suspected. However, they are limited in accuracy. No instrument can read every contaminant. Two common instruments, Organic Vapor Analyzer (OVA) and the Photoionizer (HNU), can detect hundreds of compounds but can't detect important toxic chemicals such as phosgene, cyanides, arsenic, or chlorine.

Another example of a direct-reading instrument is an air pump and detector tubes, which are simple but important direct-reading instruments. There is a wide range of detector tubes for gases but accuracy is only about + or – 25%. The detectable range for each type of tubes must be carefully reviewed in addition to the number of strokes or the amount of sample needed. It is important to be aware of monitoring limits of any instrument.

Figure 9-29. Direct-reading instrument

Most direct-reading instruments respond to several chemicals. For example, benzene detector tubes give the same response for the related chemicals toluene, xylene, and ethyl benzene. As an example, the response on an OVA meter to 100 ppm of a given chemical may read as shown below.

Chemical	Instrument Response
Methane	100
Benzene	235
Vinyl Chloride	35
Methyl Isobutyl Ketone	100

If you know that only Methyl Isobutyl Ketone is present, then you can be fairly confident about the meaning of an instrument reading of 100. If you are not sure about what is in the air, then you are in a much different position. The nature of construction work is that there are a variety of types of contaminants that may be present. This complicates the use of direct-reading instruments. The values given above for the organic vapor analyzer reflect this. The instrument is less sensitive to vinyl chloride (100 ppm gives a response of 35). On the other hand, it is more sensitive to benzene (100 ppm gives a response of 235). When you add in the fact that a mixture of chemicals is usually present, you can see that it can be difficult to precisely determine the levels at an early stage in monitoring.

You can begin to see that while direct-reading instruments can give you numbers "on the spot," it takes longer to determine the actual amount of a substance present and determine the hazard to workers. You have to go through several steps to identify the chemical, then additional steps can be taken to determine the actual level of contaminants.

Calibrating direct-reading instruments is also an important step in getting accurate measurements. Calibration is the term used to describe checking the instrument response against a known source. This check is critical to insure accuracy. Instruments can "drift" caused by low batteries, rough handling, and several other factors. An uncalibrated instrument is like a clock that is 20 minutes too slow. It still works, it just isn't accurate. An instrument is generally calibrated to see if it reads zero with no contaminant present and the correct amount with a known level of gas. For example, an organic vapor analyzer is calibrated with 0 and 100 ppm of methane. Sometimes special calibration is needed. For example, an oxygen meter must be calibrated for air pressure, due to different readings at extremes such as sea level or high elevations. Sometimes instruments are calibrated with different chemicals to aid in determining the level of a given chemical.

Care must be taken using monitoring data for decision making about personal protective equipment. Higher levels of protection are needed on the jobsite at the early stages when only general information about exposures is known. Only when contaminants are further identified and exposure levels are more precisely known can the level of protective equipment be confidently lowered and the jobsite classified into various hazard areas. More accurate monitoring usually requires samples to be collected for laboratory analysis. Even when monitoring seems to have validity, it is important to realize that this is no guarantee that exposures will stay the same day after day especially on a construction site.

Worker exposures are influenced by several factors: (1) Change in Location – Contaminants are not evenly distributed at most work sites. One area may have more solvents and less metals than another area. Monitoring often must be done when work is initiated due to the rapid change of conditions on a construction site. (2) Change in Operation – Exposures tend to vary with jobs. Bagging-out asbestos material will have a different exposure potential than removing the asbestos from the ceiling. (3) Site and Environmental Conditions – Construction work outside will have exposures that are variable with the wind and immediate weather conditions, while inside work will be more consistent with ventilation systems and type of enclosure. Temperature and season, and even rainfall can affect contaminant exposures. (4) Mishaps – Leaks and spills can have obvious effects on exposure levels.

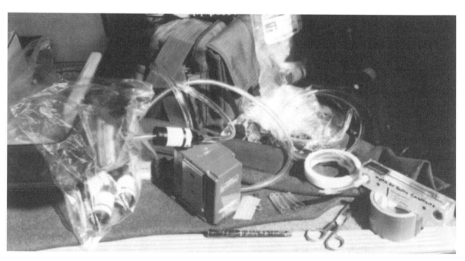

Figure 9-30. Monitoring apparatus for field asbestos survey

Because of these factors, the worker exposure monitoring job is never done. It must be done on a periodic basis over the course of the entire job. Such is the case with asbestos and lead abatement work where continuous monitoring is required. Two major categories of samples collected to draw an exposure profile are normally analyzed by laboratories and are called "area" and "personal." In general, direct-reading instruments are used to obtain area or background samples, and personnel samples are obtained with lab based analysis methods.

- Area samples are obtained in a given location. For example, a confined space might be checked for contaminants or oxygen level. Area samples are collected to verify background levels such as asbestos outside a regulated area. Sometimes high background levels prevent achieving clearance levels for reuse of space. Area samples are also a valuable tool in locating contaminant movement and documenting worse case scenarios.

- Personal samples are obtained to determine a worker exposure level without regard to respiratory equipment. It gives the most accurate profile of the worker's daily exposure level. An air monitoring pump drawing the same amount of air as we would normally breath is typically worn for the work shift, and the results compared with an 8-hour time-weighted average "permissible exposure limit" established by OSHA. It is not necessary to monitor every worker to obtain a valid exposure profile, although, each type of job in an exposure area should be monitored.

Instead, workers who are "representative" of a typical job are usually sampled. It is best to choose those who are expected to have the highest exposure. Since occupational exposures are affected the most by worker activity, this type of sampling is typically done after work begins. Personal monitoring samples are typically taken in the worker's "breathing zone" which is the area directly outside the respirator face piece within one foot of the nose.

Immediately dangerous to life or health (IDLH) sampling is done at the beginning of a hazardous job, and at appropriate periods throughout the job, such as in a confined space entry with an oxygen level under 19.5 percent. This sampling is conducted to answer the question, (are dangerous conditions present?). Personnel performing this type of work must wear appropriate personal protective equipment. Sampling should include "worst case" conditions and should be conducted on the actual approach to worst case conditions to give the monitoring person a degree of warning.

Worker exposure monitoring produces numbers. These numbers must be evaluated to be useful in decision making. The skill and judgment used by an industrial hygienist or safety professional is critical. Correctly interpreting the numbers directly affects the health of workers and the profitability of a project. As previously noted, several groups have generated exposure limits.

- PELs (Permissible Exposure limits) – These are the legal enforceable limits set in standards and regulations drafted by OSHA.

- NIOSH RELs (Recommended Exposure Limits) – These levels are not strictly enforceable in most cases but serve as a guideline for toxic chemicals.

- ACGIH TLVs (Threshold Limit Values) – These also are not enforceable. In general these are more up to date than PELs because of the ongoing committees that look at changes in technology and toxicological effects of various chemicals. It is good industrial hygiene practice to list all relevant exposure limits when discussing sampling results especially if there are differences in levels. The OSHA Hazardous Waste Standard (29 CFR 1926.65) that applies to construction workers does require NIOSH and ACGIH exposure limits be considered.

BIOLOGICAL MONITORING

Biological monitoring is the analysis of body systems such as blood, urine, finger-nails, teeth etc. that provide a baseline level of contaminants in the body. Medical testing can have several different purposes, depending on why the worker is visiting a doctor. If it is a pre-employment examination, it is usually considered a baseline to use as a reference for future medical testing. Baselines are a valuable tool to measure the amount of toxic substances in the body and often give an indication of the effectiveness of personal protective equipment.

OSHA regulations allow the examining physician to determine most of the content reviewed in the examination. Benefits received from an examination will vary with content of the examination. No matter what tests are included in the examination, there are certain impor-tant limitations of medical testing.

- Medical testing cannot prevent cancer. Cancer from exposure to chemicals or as-bestos can only be prevented by reducing or eliminating an exposure.

- For many conditions, there are no medical tests for early diagnosis. For example, the routine blood tests conducted by doctors for kidney functions do not become abnormal until half the kidney function is lost. Nine of 10 people with lung cancer die within five years because chest X-rays do not diagnose lung cancer in time to save the individual.

- No medical test is perfect. Some tests are falsely abnormal and some falsely normal.

Medical Questionnaire

A medical and work history, despite common perceptions, are probably the most im-portant part of an examination. Most diagnoses of disease in medicine are made by the history. Laboratory tests are used to confirm past illnesses and injuries. Doctors are interested in the history of lung, heart, kidney, liver, and other chronic diseases for the individual and family. The doctor will also be concerned about symptoms indicating heart or lung disease and smok-ing habits.

A physical examination is very beneficial for routine screening. Good results are im-portant, but an individual may have a serious medical problem while physical examination results seem perfectly normal. Blood is taken to check for blood cell production (anemia), liver function, kidney function, and if taken while fasting, for increased sugar, cholesterol, and fat in the blood. Urine tests are obtained to check for kidney function and diabetes (sugar in the urine). It is possible to measure in the blood and urine chemicals that get into the body from exposures on a jobsite. This type of testing is called biological monitoring.

Pulmonary Function Tests

When an individual breathes into a spirometer it measures how much air volume is in their lungs and how quickly they can breathe in and out. This is called pulmonary function testing. This is useful for diagnosing diseases that cause scarring of the lungs that affects expandability (asbestosis). Emphysema or asthma may also be diagnosed with pulmonary func-tion testing. It is vital for evaluating the ability of an individual to wear a respirator without additional health risk.

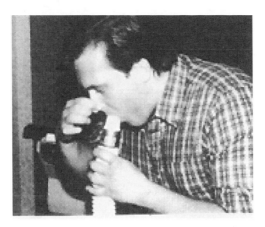

Figure 9-31. Respirator use is dependent on passing a pulmonary function test

Electrocardiogram

An electrocardiogram is a test used to measure heart injury or irregular heat beats. Construction work can be extremely strenuous, particularly when wearing protective equipment in hot environments. A stress test utilizing an electrocardiogram while exercising is sometimes a help in determining fitness, especially if there are indications from the questionnaire that an individual has a high risk of heart disease.

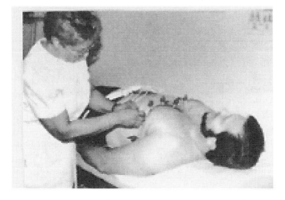

Figure 9-32. Electrocardiogram can often detect damage to the heart

Chest X-rays

X-rays are useful in determining the cause of breathing problems or to use as a baseline to determine future problems. A chest X-ray is used to screen for scarring of the lungs from exposure to asbestos or silica. It should not be routinely performed, unless the history indicates a potential lung or heart problem and the physician thinks a chest X-ray would be necessary. Some OSHA regulations require chest X-rays as part of the medical surveillance program. Unnecessary X-ray screening should be eliminated. Five-year intervals are sufficient for work-related biological monitoring.

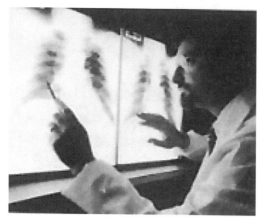

Figure 9-33. Chest X-rays for screening should be at least 5 years apart

KNOWN CANCER-CAUSING CHEMICALS AND HAZARDOUS CONSTRUCTION MATERIALS

Substances or groups of substances and technological or manufacturing processes that are known to be carcinogenic.

4-aminobiphenyl	Conjugated estrogens
Analgesic mixtures containing pheacetin	Cyclophosphamide
Arsenic and certain arsenic compounds	Diethylstilbestrol
Asbestos	Hematite underground mining
Azathioprine	Isopropyl alcohol manufacturing (strong-acid process)
Benzene	
Benzidine	Manufacture of auramine
N,N-bis (2-chloroethyl)-2-napthylamnine (chlornaphazine)	Melphalan
Bis(chloromethyl)ether and technical grade chloromethyl methyl ether	Methoxsalen with ultra-violet A therapy (PUVA)
1,4-Butanediol dimethylsulfonate (myleran)	Mustard gas
Certain combined chemotherapy for lymphomas	2-Naphthylamine
	Nickel refining
Chlorambucil	Rubber industry (certain occupations)
Chromium and certain chromium compounds	Soots, tars, and mineral oils
Coke oven emissions	Thorium dioxide
	Vinyl chloride

* Based on human evidence

A listing of hazardous construction materials which provides route of entry, physical hazards, health hazards, target organs, how to detect them, and types of protection needed is found in Table 9-4. Note: Construction materials listed in the following table are only a fraction of items used around a jobsite on an everyday basis. Hazards and appropriate types of protection may not be valid for all substances listed. Material safety data sheets should be reviewed for all chemicals used.

Table 9-4
Hazardous Construction Materials

ITEM	EXAMPLES	ENTRY ROUTES	PHYSICAL HAZARDS	HEALTH HAZARDS	TARGET ORGAN	HOW DETECTED	TYPES OF PROTECTION
Abrasives	Abrasive belts, disks, wheels, silica, sandblasting	Inhalation skin or eye contact	Strike by, injury to skin or eyes	Damage to skin, eyes, and lungs, chronic lung disease cancer if removing nickel or chrome alloy metals or asbestos coatings	Lungs, eyes, or skin	Airborne dust sampling	Ventilation, respirator, safety glasses, face-shields, gloves
Adhesives and Solvents	Caulking, epoxy, plastic cements, flooring adhesives, super glue, urethane sealant, white glue	Inhalation, skin or eye absorption,	Flammable	Skin dermatis, sensitizer, nervous system effects, mucous membrane irritant	Lungs, kidneys, liver,	Odor, labeling, monitoring	Ventilation, respirators, gloves, eye protection, good personal hygiene
Asbestos	Insulation, tiles, roofing, mastics	Inhalation, ingestion	None	Damage to lungs, asbestosis, mesothelioma	Lungs, gut	X-rays, lab analysis	Ventilation, HEPA vacuums, decon. with shower, approved respirators

Table 9-4

Hazardous Construction Materials (*continued*)

ITEM	EXAMPLES	ENTRY ROUTES	PHYSICAL HAZARDS	HEALTH HAZARDS	TARGET ORGAN	HOW DETECTED	TYPES OF PROTECTION
Asphalt	Coal tar, creosote, pitch, cutback, treated wood, roofing, pavement	Inhalation, skin or eye absorption, ingestion	Flammable, may be solid or liquid	Damage to skin, eyes, and lungs, chronic lung disease, cancer	Lungs, liver, kidneys, eyes, skin, CNS, bladder	Odor, air sampling, lab analysis	Ventilation, respirators, gloves, eye protection, good personal hygiene, PPE
Biological agents	Sewage, rabies, ticks, ants, bees, flies, poison ivy-oak, spiders	Inhalation, skin, skin or eye absorption, ingestion	Bites, stings	Skin infections, diseases, rashes, allergic reactions, damage to skin	Skin, entire body	Visual, irritation ill feeling	Immunization, personal hygiene, insecticides, PPE
Cleaners	Bleach, drain cleaners, germi-cides, metal polish and cleaners, stain removers	Inhalation, skin or eye absorption, ingestion	Flammable, reactive, solid or liquids	Skin dermatitis, sensitizer, nervous system effects, mucuous membrane irritant, some carcinogens, damage to eyes and lungs, corrosive	Lungs, skin, eyes	Odor, labeling, monitoring	Ventilation, respirators, gloves, eye protection, good personal hygiene, PPE
Coatings	Waterproof, anti-corrosion, epoxies, polish, varnishes, waxes, wood treatments	Inhalation, skin or eye absorption, ingestion	Flammable, lipophilic, hydrophilic	Skin dermatitis, sensitizer, nervous system effects, mucuous membrane irritant, some carcinogens, damage to eyes and lungs, reproductive hazard, bone damage	Skin, eyes, lungs, nervous system, reproductive organs, blood	Odor, container labeling, lab analysis	Ventilation, respirators, gloves, eye protection, good personal hygiene, PPE
Fuels	Diesel, gasoline, kerosene, propane	Inhalation, skin or eye irritation, absorption, ingestion	Flammable liquids	Damage to skin, eyes, and lungs, Chronic lung disease, cancer if removing nickel or or chrome alloy metals or asbestos coatings	Lungs, eyes, or skin	Sampling, odor, labels	Ventilation, respirator, safety glasses, face-shields, gloves, grounding and bonding, approved containers

Table 9-4

Hazardous Construction Materials *(continued)*

ITEM	EXAMPLES	ENTRY ROUTES	PHYSICAL HAZARDS	HEALTH HAZARDS	TARGET ORGAN	HOW DETECTED	TYPES OF PROTECTION
Gases – compressed	Acetylene, oxygen hydrogen, freon, ammonia, propane, LPG, nitrogen	Inhalation	Flammable, oxidizers, compressed, explosive, caustics	Asphyxiant, corrosive, damage to skin, eyes, and lungs, freon – heart attacks	Lungs, kidneys, liver, CNS, skin, eyes, heart, bones	Odor, labeling, monitoring	Ventilation, respirators, gloves, eye protection, Sample confined spaces prior to entering
Gases – non-compressed	Carbon monoxide, hydrogen sulfide, hydrogen fluoride, chlorine, methane	Inhalation	Flammable, caustic, explosive	Acute lung damage, chronic cancers, hydrogen fluoride – bone seeker causing tissue burns, asphyxiant, central nervous system damage, IDLH conditions – death	Lungs, liver, kidneys, eyes, skin, CNS blood	Odor, air sampling, lab analysis, hydrogen sulfide – rotten egg smell	Ventilation, respirators, gloves, eye protection. Sample confined spaces prior to entering, PPE
Insulation – non-asbestos	Graphite, fiber-glass, rockwool, vermiculite, kaowool, inswool,	Inhalation, skin or eye irritation	None	Skin infections – irritation, rashes, allergic reactions, upper respiratory tract irritation, sensitizer	Skin, respiratory tract	Labeling, sampling, fibrous materials or foam	Ventilation, respirators, gloves, eye protection, good personal hygiene, PPE
Lubricant	Oils, greases, cutting oils, CRC, WD-40, trans-mission fluid	Skin or eye irritation, absorption, inhalation	Flammable, solid or liquids	Skin dermatitis, sensitizer, nervous system effects, damage to eyes and lungs	Lungs, skin, eyes	Container, labeling, lab analysis, thick viscous liquids	Ventilation, gloves, eye protection, good person-al hygiene, PPE
Masonry	Concrete, brick, lime, mortar, muriatic acid, sand-silica, gunite refractory	Inhalation, skin or eye irritant ingestion	Corrosive	Skin dermatitis, sensitizer, nervous system effects, mucuous membrane irritant – muriatic acid, some carcinogens, silicosis, damage to eyes and lungs	Skin, eyes, lungs	Container labeling, lab analysis, dust, sand	Ventilation, respirators, gloves, eye protection, good personal hygiene, PPE, goggles

Table 9-4
Hazardous Construction Materials *(continued)*

ITEM	EXAMPLES	ENTRY ROUTES	PHYSICAL HAZARDS	HEALTH HAZARDS	TARGET ORGAN	HOW DETECTED	TYPES OF PROTECTION
Metals	Lead, nickel, maganese, zinc, chromium, cadmium, iron, mercury, welding rod	Ingestion, inhalation	Dusts may be explosive	Skin dermatitis, sensitizer, nervous system effects, mucous membrane irritant – muriatic acid, some carcinogens, damage to eyes and lungs, reproductive damage	Lungs, blood, CNS, bones, liver, kidneys	Air sampling, labeling, solids that become liquid and fumes when heated	Ventilation, respirators, gloves, eye protection, good personal hygiene and other PPE
Paints – oil-based products	Enamel, thinners, lacquers, primers, stains, turpentine, linseed oil, epoxies	Inhalation, ingestion, absorption	Flammable, explosive caustic	Asphyxiant, damage to skin, eyes, and lungs, CNS damage, cancers	Lungs, kidneys, liver, CNS, skin, eyes	Odor, labeling, air monitoring	Ventilation, respirators, gloves, eye protection, full body clothing
Pesticide, herbicide, fungicide	Parathion diazinon, DDVP, round-up organo-derivatives	Inhalation, ingestion, absorption	Flammable, explosive, liquids and powders	Highly toxic with damage to all crgans possible, readily absorbed through the skin	Lungs, blood, CNS, bones, liver, kidneys	Warning labels, sampling for lab analysis	Ventilation, respirators, gloves, eye protection, good personal hygiene and other PPE
Radiation – ionizing	Lasers, XRF analyzers for lead paint	Penetration through eyes and skin	Burns	Lasers, damage to eyes, soil contaminated with radon can cause internal organ damage if inhaled	Lungs, liver, kidneys, eyes, skin, CNS, blood	Warning labels, sampling for analysis	Eye protection with appropriate lenses, avoid laser beam. Keep distance from sources

Table 9-4
Hazardous Construction Materials *(continued)*

ITEM	EXAMPLES	ENTRY ROUTES	PHYSICAL HAZARDS	HEALTH HAZARDS	TARGET ORGAN	HOW DETECTED	TYPES OF PROTECTION
Solvents	Acetone, alcohol, toluene, xylene, MEK, chlorinated solvents, spirits, benzene	Inhalation, skin or eye irritation, absorption	Flammable, explosive	Toxic, skin dermatitis, sensitizer, nervous system effects, mucous membrane irritant, some carcinogens, damage to eyes and lungs, reproductive damage	Lungs, kidneys, liver, CNS, skin, eyes	Labeling, sampling materials for lab analysis	Ventilation, respirators, gloves, eye protection, good personal hygiene, PPE
Welding, soldering, cutting, brazing	Rod, fluxes, lead solder, metals, gases	Skin or eye damage, absorption, inhalation	Burns, electrical	Skin dermatitis, sensitizer, nervous system effects, damage to eyes from flash, lungs, highly toxic fumes, lead fumes readily absorbed into the blood stream	Lungs, liver, kidneys, eyes, skin, CNS, blood	Container labeling, lab analysis	Ventilation, gloves, eye protection, good personal hygiene, other PPE
Wood Products	Treated lumber, hardwood and softwood dust	Inhalation, skin absorption ingestion	Corrosive or acidic	Skin dermatitis, nervous system effects, some carcinogens, damage to eyes and lungs, some treated woods contain arsenic	Skin, eyes, lungs	Container labeling, lab analysis, dust, sand	Gloves, eye protection, good personal hygiene, PPE

OSHA's Standards on Hazard Communication, Hazardous Waste Operations, and Emergency Response require all employees on a construction site to receive general classroom training. In addition, Section 1926.65(e)(2) also requires a minimum of three days of actual field experience under the direct supervision of a trained, experienced supervisor. This training is essential to provide the employees with information regarding specific hazards, equipment, and other characteristics unique to the jobsite. Worker training on the components of material safety data sheets (MSDSs) is necessary to appraise them of chemical hazards and appropriate personal protective equipment. Tables 9-5 and 9-6 contain material safety data sheets for lead and a solvent

CONSTRUCTION TRAINING REQUIREMENTS

OSHA has mandated that worker training be provided for conditions such as but not limited to respirator use, hazard communication where chemicals may be present, hazardous waste work, and asbestos work. Keep in mind that some training requirements for construction such as hazardous waste work may be found in the General Industry Standards. Hazards on site that require contractor training programs are

1. Special chemical hazards.

    • Flammable hazards.

    • Shock-sensitive wastes.

    • Incompatibles.

    • Radioactive wastes.

    • IDLH conditions.

2. Description of safety hazards on site.

Figure 9-34. Toolbox meeting is often used for on-the-job training

Hands-on review of personal protective equipment including respirators to be used should be part of a contractor training program. Some concerns for personal protective use are

- Selection issues with respect to site hazards.
- Inspection – review condition of gear.
- Maintenance and storage.
- Proper fitting – conduct fit testing and provide eyeglass fittings, if required.
- In-use monitoring.

Employees must have a good working knowledge of what specific conditions require special protection on the jobsite. It is suggested that supervisors review specific needs prior to starting a job, then train workers including inspection and maintenance of equipment necessary for conducting safe work activities.

Table 9-5

Sample Lead Material Safety Data Sheet

IDENTITY NOTE:	Blank spaces are not permitted. The space must be marked if an (as used on label/list) item is not applicable or no information is available, the space must be LEAD marked.

SECTION I

Manufacturer's Name and Address(s)	Emergency Telephone Number		1-314-000-0000
The ABC Lead Company	Telephone Number for Information		N/A
Suite 4CABC	Lead Company Smelter	ABC Lead Company Smelter	
1266 Metal Road	230 Main Street	RR 5	
Leadland, MO 00000-0000	Galena, IL 00000	Leaddust, MT 00000	
Chemical Name and Synonyms	Lead, Lead alloys, and lead strip.		
Chemical Family N/A			
Signature of Preparer (Optional)	Date Prepared		March 1, 1998

SECTION II – Hazardous Ingredients/Identity Information

Hazardous Components (Specific)	OSHA PEL	ACGIH TWA	MSHA Limit	Sarah 313	%
Chemical Identity, Common Name)					
Lead (Pb) CAS # 7439-2-1.	05 mg/m^3	.15 mg/m^3	.15 mg/m^3	X	99-99.99
This product contains toxic chemicals which are subject to the reporting requirements of Section 313 of Title III of the Superfund Amendments and Reauthorization Act of 1986, and 40 CFR Part 372. No other hazardous material is present in concentration greater than 1% (0.1% for carcinogens).					

SECTION III – Physical/Chemical Characteristics

Boiling Point 3164° F Specific Gravity (H$_2$0 = 1)		11.3
Vapor Pressure 0 mm Hg. (approx.)	Melting Point	621° F
Vapor Density N/A	Evaporation Rate	N/A
(Air = 1) (Butyl Acetate = 1)		
Solubility in Water	Insoluble	
Appearance and Odor	Bluish gray soft metal, no odor.	

SECTION IV – Fire and Explosion Hazard Data

Flash Point (Method Used)	N/A				
Flammable Limits N/A		LEL	N/A	UEL	N/A

Extinguishing Media Lead should not burn. However, if area where lead is stored or used is burning then water spray, fog, or standard foam is recommended.

Special Fire-Fighting Procedures When this material is heated to the point of vaporization then toxic fumes are emitted. Respiratory protection in the form of NIOSH/MSHA approved dust/fume respirators with high efficiency filters or if in conditions where unknown levels/types of contaminants are present then Self Contained Breathing Apparatus (SCBA) should be worn. Other appropriate protective equipment such as clothing and face protection should also be worn to protect the individual from splash or metal burns.

Unusual Fire/Explosion Hazards Fire and explosion hazards are moderate when this material is in the form of dust and is exposed to heat or flames. When heated, lead emits highly toxic fumes which can react vigorously with oxidizing materials. (See reactivity data.)

LEAD MATERIAL SAFETY DATA SHEET *(Continued)*

SECTION V – Reactivity Data

Stationary	Unstable	Stable	Conditions to Avoid
		XX	Lead can react violently with oxidizing materials. Water may become trapped within surface cracks which may cause an explosion when the metal is molten.

Incompatibility (Materials to Avoid) Strong oxidizing materials, e.g., chlorine triflouride, hydrogen peroxide, sodium oxide, sodium and potassium.

Hazardous Decomposition/Byproducts Release of toxic materials at high temperatures.

Hazardous Polymerization		
May Occur	Conditions to Avoid	N/A
Will Not Occur	XX	

SECTION VI – Health and Hazard Data   *HAZARD RATING : DANGER*

Routes of Entry	Inhalation X	Skin (only in organic form) Ingestion X

Health Hazards (Acute and Chronic)   Acute effects include flu like symptoms, insomnia, weakness, constipation, nausea, and abdominal pain. It is unlikely that occupationally related exposure to this material would result in an acute illness. However, if symptoms are present, the individual should be removed from exposure and a physician consulted. For hot metal burns, the skin should be cooled with water and medical attention sought. Chronic effects may include kidney damage, nervous system involvement, hypertension, and reproductive effects.

Target Organs   Eyes, skin, CNS, and respiratory system.

Carcinogenicity: NIP No IARC Monographs Yes   OSHA Regulated N/A

IARC classifies lead and some lead compounds as Group 2B carcinogens (possibly carcinogenic to humans).   This classification based primarily on the Carcinogenicity of certain soluble lead salts in lab animals. Neither lead nor its insoluble salts appear to be carcinogenic to humans or lab animals.

Signs & Symptoms of Exposure Possibly a metallic taste in the mouth. Rise in blood lead levels.

Medical Conditions Aggravated by Exposure Diseases of the blood and blood-forming organs, kidneys, central nervous system, and possibly reproductive system.

Emergency & First Aid Procedures If symptoms indicate possible acute effects of lead, the individual should be removed from exposure and a physician consulted. Good hygiene practices should be followed. It is unlikely that occupational exposure will result in an acute illness; however, injury may result may result from hot metal burns if employees work with molten metal.

SECTION VII – Precautions for Safe Handling and Use

Procedures for Material Release or Spill Material should be recovered for use or recycled by vacuuming, shoveling (if in pieces too large to vacuum), bobcat, or wash down procedures. Dry sweeping methods should never be used because they may result in increased concentrations of airborne lead.

Waste Disposal Method   Scrap or waste should be recycled if at all possible. If recycling is not possible, dispose of in accordance with Federal, State, and Local regulations.

Handling and Storage Precautions   Avoid inhalation of dust or fumes. Avoid accidental ingestion by using good personal hygiene practices. If lead has been handled, wash thoroughly before eating or smoking.

Other Precautions Food and drink should not be consumed, tobacco products should not be used, nor cosmetics applied in areas where exposures exceed applicable limits.

LEAD MATERIAL SAFETY DATA SHEET *(Continued)*

SECTION VIII – Control Measures

Respiratory Protection (Specify Type) A NIOSH/MSHA approved dust/fume respirator with high efficiency filters should be worn where applicable limits may be exceeded.
Ventilation Local Exhaust Local exhaust ventilation should be used where possible. Adequate ventilation should be used when material is in molten or dusty state.
Mechanical (General) Adequate ventilation should be used when material is in molten or dusty state.
Special   N/A
Other   N/A
Protective Gloves Should be worn to reduce burn exposure. May be needed to prevent excessive skin contact which could result in inadvertent ingestion. Inorganic lead is not absorbed through intact skin.
Eye Protection   Eye and/or face protection recommended for exposure to dust and potential hot metal splash.
Other Protective Clothing or Equipment   Where applicable limits are exceeded, work clothes should be worn and laundered in accordance with the OSHA lead standard (see 29 CFR 1910.1025). Vacuums with HEPA filtration abilities should be used in cleaning up change houses, control rooms, and offices.
Work/Hygienic Practices   Do not eat, drink, or smoke in the area.

Table 9-6

Sample Solvent Material Safety Data Sheet

May be used to comply with OSHA's Hazard
Communication Standard 29 CFR 1926.59
Standard must be consulted for specific requirements.
(Non-Mandatory Form)

U.S. Department of Labor
Occupational Safety and
Health Administration

IDENTITY (as used on label and list)   NOTE: Blank spaces are not permitted. If any item is not applicable or no information is available, the space must be marked.

SECTION I

Manufacturer's Name	Emergency Telephone Number
ABC SOLVENT Company, Inc.	1-800-111-2222
Address	Telephone Number for Information
Canary Street	1-999-333-4445
Somewhere, U.S. 12345	Date Prepared 2/1/98
May 1, 1989	
Signature of Preparer (Optional)	

SECTION II – Hazardous Ingredients/Identity Information

Hazardous Components	OSHA PEL (STEL)	ACGIH	Other	%
(Specific Chemical Identity, Common Name)			NIOSH	
Acetone	750 ppm (1000 ppm)	Same as	Same as	20
Ethylene Glycol	50 ppm Ceiling	OSHA	OSHA	3
Methyl Ethyl Ketone (2-Butanone)	200 ppm (300 ppm)	PEL/	PEL/	2.5
Xylene	100 ppm (150 ppm)	STEL	STEL	5.8
(Inert Ingredients - nonhazardous)				68.7

SECTION III – Physical/Chemical Characteristics

Boiling Point 125° F	Specific Gravity ($H_2O = 1$) 0.75
Vapor Pressure 135 mm Hg.	Melting Point -150° F
Vapor Density 1.1	Evaporation Rate N/A
(Air = 1)	(Butyl Acetate = 1) 4.5
Solubility in Water 27 %	
Appearance and Odor Clear liquid, fragrant mint-like odor.	

SECTION IV – Fire and Explosion Hazard Data

Flash Point (Method Used)	5° F (open cup)
Flammable Limits	LEL 1.5% UEL 10%
Extinguishing Media	NFPA Class B Extinguishers.
Special Fire-Fighting Procedures	Water spray may be ineffective. Use Class B extinguisher, Carbon Dioxide, dry chemical or alcohol foam.
Unusual Fire/Explosion Hazards	Closed container may explode if exposed to heat.

SAMPLE SOLVENT MATERIAL SAFETY DATA SHEET *(Continued)*

SECTION V – Reactivity Data

Stationary Unstable Stable	Conditions to Avoid
Explosive	Heat, open flames, flame, electrical equipment.
Incompatibility (Materials to Avoid)	Unknown.
Hazardous Decomposition/Byproducts	Carbon Monoxide, Carbon Dioxide.
Hazardous Polymerization	
May Occur	Conditions to Avoid
Will Not Occur	X

SECTION VI – Health and Hazard Data   *HAZARD RATING : DANGER*

Routes of Entry	Inhalation X	Skin X	Ingestion X
Health Hazards (Acute and Chronic) Central	ACUTE: Eye, nose, throat, lung irritation. nervous system. Skin irritant. CHRONIC: No chronic effects reported.		
Target Organs	Eyes, Skin, CNS, and Respiratory System.		
Carcinogenicity: NIPIARC Monographs OSHA Regulated			
	Not Reported		
Signs & Symptoms of Exposure	Eye, nose, throat, lung, skin irritation, headache, dizziness.		
Medical Conditions Aggravated by Exposure	Asthma		
Emergency & First Aid Procedures	Fumes - Remove to fresh air and notify a physician. Eye splash - Wash with running water for at least 15 min. Skin splash - Wash with water.		

SECTION VII – Precautions for Safe Handling and Use

Procedures for Material Release or Spill	Remove all sources of ignition, avoid breathing vapors, ventilate (non-sparking equipment).
Waste Disposal Method	Dispose in accordance with local/state/federal regulations.
Handling and Storage Precautions	Do not store over 110°F. Store as Class 1A NFPA flammable liquid.
Other Precautions	Group containers when pouring – WITH ADEQUATE VENTILATION

SECTION VIII – Control Measures

Respiratory Protection (Specify Type)	NIOSH/MSHA approved positive, pressure/pressure demand airline or SCBA.
Ventilation	Local Exhaust In use area.
Mechanical (General)	Area ventilation.
Special Spills or confined areas.	
Other	Use explosion proof electrical.
Protective Gloves	Butyl Gloves
Eye Protection	Chemical Splash Goggles
Other Protective	Clothing or Equipment Butyl splash apron when required to avoid splash exposure.
Work/Hygienic Practices	Do not eat, drink, or smoke in the area.

Chapter 10

PERSONAL PROTECTIVE EQUIPMENT

James V. Eidson

INTRODUCTION

It is not necessary for contractors to become medical professionals, but it is necessary to know enough about the body to anticipate hazardous conditions that could cause severe or even permanent injury. This section will provide construction personnel with an overview of appropriate personal protective equipment. Each chemical hazard has its own set of characteristics that impact the human body. It is the understanding of these characteristics that is necessary to make an informed decision on personal protective equipment for each specific hazard.

Information presented on personal protective equipment merely presents pertinent recommendations from OSHA's *Field Operations Manual,* organized into a form convenient for quick reference during a contractor project. It is designed to deal with both the human body and equipment to protect it. Therefore, only those recommendations directed primarily toward control of toxic exposures are cited in this section. The recommendations from "Chemical Protective Clothing" have been paraphrased, combined, or otherwise reorganized, and headings have been added. However, their intent has not been changed.

Protective Equipment for Noise

Hearing loss results from long-term exposures to sound levels at or above 85 decibels. Equipment such as back hoes, vacuums, and street sweepers produce noise which can damage hearing over time. Ear plugs are available in many shapes and sizes. Disposable foam ear plugs reduce noise by 15 dBA, conform to the ear, and are washable. Soft rubber ear plugs are available in three sizes (large, medium, and small) and reduce noise up to 13 dBA. These are washable and reusable. A hearing conservation program must be implemented when noise exposures are at or above an 8-hour average of 85 dBA (29 CFR 1926.101 for Construction).

When it is not feasible to eliminate noise that exceeds 85 dBA, it may be necessary to protect the worker from the environment with hearing protection (see Figure 10-1). This is considered to be a last resort if other controls are not possible, for example, enclosing or isolating the noise source.

Protection against noise-induced hearing impairment may be accomplished with the use of ear plugs or ear muffs. Ear plugs usually provide about a 15 dBA attenuation or reduction in sound intensity, depending on the type and manufacturer of the plugs. For example, if the noise source was 93 dBA, with ear plugs the actual noise would be reduced to about 78 dBA. Ear muffs surprisingly reduce the noise intensity about the same amount. Use of ear plugs and ear muffs increase the noise reduction by only about 3 dBA over the level achieved separately by either one. One drawback to hearing protection is that it must be worn properly to get maximum protection. Ear plugs must be firmly inserted into the ear canal (see Figure 10-2).

Figure 10-1. Hearing protection may be needed to assist communication

Figure 10-2. Ear plugs must be firmly seated in the ear canal

Figure 10-3. Safety glasses must have side shields

Care must be exercised to assure that dirt and grease do not contaminate the ear from handling plugs prior to washing hands. Ear muffs are heavier and hotter to wear and must be securely positioned over the ears without interference from hair or glasses. Communication is not drastically reduced by hearing protection. Speech is transmitted at a frequency that is not eliminated by hearing protection as much as higher frequencies.

Eye Protection

General types of equipment used to protect eyes from flying particles encountered in such jobs as chipping, grinding and overall construction work, are glasses with impact resistant lenses and side shields, cushion fitting goggles, with face shields as added protection (see Figure 10-3).

To protect the eyes from damaging radiation, it is necessary to wear glasses that are designed to filter out the harmful rays. Ultraviolet, visible, and infrared light are all able to produce harmful effects. It is necessary to choose the lenses based on the type of radiation exposure. The eyes and face can be protected from chemical splashes and dust by wearing face shields and eye goggles. Workers who wear prescription glasses may be more comfortable using a full-face mask (see Figure 10-4).

Figure 10-4. Welding is a classic generator of ultraviolet rays

Head Protection

Hard hats may be used in work areas where there is risk from falling objects or where heavy equipment is used. Generally, hard hats are less likely to be used at residential abatement sites unless required by the employer or requested by the worker. Head coverings are available to keep hair and scalp clean.

Foot Protection

Abatement workers are as subject to foot injuries as workers in most construction sites. Steel-toed boots or shoes may prevent fractures from falling objects and punctures from nails or other debris. Shoe/boot covers (plastic, Tyvek R, rubber, etc.) may be worn over shoes or boots to prevent slips and falls when working on wet surfaces.

Disposable foot covers (booties) may also prevent the spread of lead from the work site. The booties must be removed when workers or visitors leave the abatement site. Contaminated booties can be properly disposed of at the end of the work shift or whenever leaving the work area. However, disposable foot covers should not be worn on ladders or scaffolding. An alternative would be to use work boots which never leave the jobsite.

Hand Protection

Hands may be exposed to both physical and chemical hazards. In paint abatement projects, removal and replacement techniques will expose workers to nails, wood, splinters, and other sharp objects. Similarly, soil and dust abatement projects may expose workers to sharp objects. Leather gloves are recommended as protection from these hazards.

When strippers are used as the abatement strategy, workers are exposed to the chemicals in the strippers. Appropriate rubber gloves should be used for worker protection. Since the chemicals in strippers vary, the type of rubber glove may also vary. The environmental professional responsible for worker safety must verify that the gloves specified are adequate for the chemicals being used.

CHEMICAL PROTECTIVE EQUIPMENT STANDARDS

The protective equipment buyers may wish to specify equipment that meets specific standards, such as 1910.120, 1910.132, or the NFPA standards. Keep in mind, some of the NFPA Standards do not apply to all forms of protective clothing and applications. The following OSHA regulations are applicable for PPE compliance:

1910.132(a) — Personal Protective Equipment (see Figure 10-5).

Figure 10-5. Personal protective equipment

(a) Application.

Protective equipment, including personal protective equipment for eyes, face, head, and extremities, protective clothing, respiratory devices, and protective shields and barriers, shall be provided, used, and maintained in a sanitary and reliable condition wherever it is necessary by reason of hazards of processes or environment, chemical hazards, radiological hazards, or mechanical irritants encountered in a manner capable of causing injury or impairment in the function of any part of the body through absorption, inhalation, or physical contact.

(b) Employee-owned equipment.

Where employees provide their own protective equipment, the employer shall be responsible to assure its adequacy, including proper maintenance and sanitation of such equipment.

(c) Design.

All personal protective equipment shall be of safe design and construction for the work to be performed.

(d) Hazard assessment and equipment selection.

(1) The employer shall assess the workplace to determine if hazards are present, or are likely to be present, which necessitate the use of personal protective equipment (PPE). If such hazards are present, or likely to be present, the employer shall

(d)(1)(i) Select, and have each affected employee use, the types of PPE that will protect the affected employee from the hazards identified in the hazard assessment;

(d)(1)(ii) Communicate selection decisions to each affected employee; and,

(d)(1)(iii) Select PPE that properly fits each affected employee.

(2) The employer shall verify that the required workplace hazard assessment has been performed through a written certification that identifies the workplace evaluated; the person certifying that the evaluation has been performed; the date(s) of the hazard assessment; and, which identifies the document as a certification of hazard assessment.

(e) Defective and damaged equipment.

Defective or damaged personal protective equipment shall not be used.

(f) Training.

(1) The employer shall provide training to each employee who is required by this section to use PPE. Each such employee shall be trained to know at least the following:

(f)(1)(i) When PPE is necessary;

(f)(1)(ii) What PPE is necessary;

(f)(1)(iii) How to properly don, doff, adjust, and wear PPE;

(f)(1)(iv) The limitations of the PPE; and,

(f)(1)(v) The proper care, maintenance, useful life and disposal of the PPE.

(2) Each affected employee shall demonstrate an understanding of the training specified in paragraph (f)(1) of this section, and the ability to use

PPE properly, before being allowed to perform work requiring the use of PPE.

(3)   When the employer has reason to believe that any affected employee who has already been trained does not have the understanding and skill required by paragraph (f)(2) of this section, the employer shall retrain each such employee. Circumstances where retraining is required include, but are not limited to, situations where

(f)(3)(i)   Changes in the workplace render previous training obsolete; or

(f)(3)(ii)   Changes in the types of PPE to be used render previous training obsolete;

(f)(3)(iii)   Inadequacies in an affected employee's knowledge or use of assigned PPE indicate that the employee has not retained the requisite understanding or skill.

(4)   The employer shall verify that each affected employee has received and understood the required training through a written certification that contains the name of each employee trained, the date(s) of training, and that identifies the subject of the certification.

(g)   Paragraphs (d) and (f) of this section apply only to 1910.133, 1910.135, 1910.136, and 1910.138. Paragraphs (d) and (f) of this section do not apply to 1910.134 and 1910.137. [59 FR 16360, April 6, 1994; 59 FR 33910, July 1, 1994; 59 FR 34580, July 6, 1994].

1910.133(a) — Eye and face protection (general requirements)

(1)   The employer shall ensure that each affected employee uses appropriate eye or face protection when exposed to eye or face hazards from flying particles, molten metal, liquid chemicals, acids or caustic liquids, chemical gases or vapors, or potentially injurious light radiation.

Figure 10-6. Eye and face protection

(2) The employer shall ensure that each affected employee uses eye protection that provides side protection when there is a hazard from flying objects (see Figure 10-6). Detachable side protectors (e.g., clip-on or slide-on side shields) meeting the pertinent requirements of this section are acceptable.

(3) The employer shall ensure that each affected employee who wears prescription lenses while engaged in operations that involve eye hazards wears eye protection that incorporates the prescription in its design, or wears eye protection that can be worn over the prescription lenses without disturbing the proper position of the prescription lenses or the protective lenses (see Figure 10-7).

Figure 10-7. Special personal protective equipment

(4) Eye and face PPE shall be distinctly marked to facilitate identification of the manufacturer.

(5) The employer shall ensure that each affected employee uses equipment with filter lenses that have a shade number appropriate for the work being performed for protection from injurious light radiation. Table 10-1 indicates the appropriate shade numbers for various operations (see Figure 10-8).

1910.133(b) — Criteria for protective eye and face devices.

(1) Protective eye and face devices purchased after July 5, 1994 shall comply with ANSI Z87.1-1989, "American National Standard Practice for Occupational and Educational Eye and Face Protection," which is incorporated by reference as specified in Sec. 1910.6.

(2) Eye and face protective devices purchased before July 5, 1994 shall comply with the ANSI "USA standard for Occupational and Educational Eye and Face Protection," Z87.1-1968, which is incorporated by reference as specified in Sec. 1910.6, or shall be demonstrated by the employer to be equally effective.

Table 10-1

Filter Lenses for Protection Against Radiant Energy

Filter Lenses for Protection Against Radiant Energy

Operations Electrode Size 1/32 in.	Minimum(*) Arc Current	Protective Shade
Shielded metal arc welding	Less than 3 .... Less than 60	7
	3-5 60-160	8
	5-8 160-250	10
	More than 8 250-550	11
Gas metal arc welding and flux cored arc welding	Less than 60 7	
	60-160 10	
	160-250 10	
	250-500 10	
Gas Tungsten arc welding	Less than 50 8	
	50-150 8	
	150-500 10	
Air carbon Arc cutting	(Light) .... Less than 500 10	
	(Heavy) 500-1000 11	
Plasma arc welding	Less than 20 6	
	20-100 8	
	100-400 10	
	400-800 11	
Plasma arc cutting	(light)(**) .... less than 300 8	
	(medium)(**) 300-400 9	
	(heavy)(**) 400-800 10	
Torch brazing.	 3	
Torch soldering	 2	
Carbon arc welding	 14	

Filter Lenses for Protection Against Radiant Energy

Minimum(*) Operations	Plate thickness-inches	Plate thickness-mm	Protective Shade
Gas Welding:			
Light	Under 1/8	Under 3.2 4	
Medium	1/8 to 1/2	3.2 to 12.7 5	
Heavy	Over 1/2	Over 12.7 6	
Oxygen cutting:			
Light	Under 1	Under 25 3	
Medium	1 to 6	25 to 150 4	
Heavy	Over 6	Over 150 5	

Footnote(*) As a rule of thumb, start with a shade that is too dark to see the weld zone. Then go to a lighter shade which gives sufficient view of the weld zone without going below the minimum. In oxyfuel gas welding or cutting where the torch produces a high yellow light, it is desirable to use a filter lens that absorbs the yellow or sodium line in the visible light of the (spectrum) operation.

Footnote(**) These values apply where the actual arc is clearly seen. Experience has shown that lighter filters may be used when the arc is hidden by the workpiece.

Figure 10-8. Eye and face protection for welding

CHEMICAL PROTECTIVE CLOTHING

The purpose of chemical protective clothing and equipment is to shield or isolate individuals from the chemical, physical, and biological hazards that may be encountered during hazardous materials operations. During chemical operations, it is not always apparent when exposure occurs. Many chemicals pose invisible hazards and offer no warning properties. These guidelines describe the various types of clothing that are appropriate for use in various chemical operations, and provides recommendations in their selection and use. The final section discusses heat stress and other key physiological factors that must be considered in connection with protective clothing use (see Figure 10-9).

Figure 10-9. Protective clothing and equipment specific for the job

It is important that protective clothing users realize that no single combination of protective equipment and clothing is capable of protecting against all hazards. Thus protective clothing should be used in conjunction with other protective methods. For example, engineering or administrative controls to limit chemical contact with personnel should always be considered as alternative measures for preventing chemical exposure.

The use of protective clothing can itself create significant wearer hazards, such as heat stress, physical and psychological stress, in addition to impaired vision, mobility, and communication. In general, the greater the level of chemical protective clothing, the greater the associated risks. For any given situation, equipment and clothing should be selected that provide an adequate level of protection. Overprotection as well as underprotection can be hazardous and should be avoided.

Workers should change into work clothes only at the worksite. Street clothes should be stored separately from work clothes in a clean area provided by the employer so that they are not contaminated. Workers should change back into their street clothes after washing or showering and before leaving the worksite to prevent the accumulation of lead dust in the workers' cars and homes, and thereby protect family members from exposure to lead.

Appropriate disposable or washable work clothes should be provided by the employer. To reduce the potential for heat stress, breathable clothing should be used for all methods except for chemical removal, where chemical-resistant clothing is necessary. Worker shoes or disposable booties should have non-skid soles.

Employers should arrange for the laundering of protective clothing; or if disposable protective clothing is used, the employer should maintain an adequate supply at the worksite and arrange for its safe disposal according to applicable Federal and State regulations. The launderer of lead-contaminated clothing should be advised (in writing) of the lead contamination and of the potentially harmful effects of lead.

Facilities for worker personal hygiene should be improved to minimize workers exposure to lead and other contaminants through ingestion, and carry-home of lead contamination. Adequate washing facilities including running hot and cold water and wherever feasible, showers should be provided at the worksite so that workers can remove lead particles from skin and hair.

Contractors should arrange for collection and disposal of the waste water in accordance with local and state requirements. Wherever feasible, contractors should supply a portable trailer to contain storage, washing facilities, and clean areas.

All workers exposed to hazardous chemicals should wash their hands and faces before eating, drinking, or smoking, and they should not eat, drink, or use tobacco products in the work area or other potentially contaminated areas on site. Tobacco and food products should never be permitted in the work area. Contaminated work clothes should be vacuumed or removed before eating.

Arms and legs may be exposed to chemicals from paint stripping agents, dusts, and other skin irritants at the work site. Disposable suits such as TYVEK provide limited protection from chemicals but do protect against dust and minor injuries such as scrapes and cuts. When disposable suits are not used, long-sleeved shirts and pants should be worn provided they are not taken home and are cleaned by the employer. When working near heat sources, clothing should not be loose and should be non-flammable and should not be polyester. Shirts should be tucked in and buttoned to prevent catching fire or getting caught on other objects.

Protective clothing and equipment guard against injury to the body including the head, eyes, face, and feet. Generally, abatement site workers wear coveralls, safety glasses, ear plugs, and steel-toed boots or shoes. At other times equipment such as boot covers, hard hats, and face shields are either required or recommended, depending on the job conditions and requirements. The following are types of equipment commonly used.

PROTECTIVE CLOTHING APPLICATIONS

Protective clothing must be worn whenever the wearer faces potential hazards arising from chemical or biological exposure (see Figure 10-10). Some examples include:

- Emergency response.

- Chemical manufacturing and process industries.

- Hazardous waste site clean up and disposal,

- Asbestos removal and other particulate operations, and

- Agricultural application of pesticides.

- Environmental hazards such as poison ivy.

Figure 10-10. Insects and poisonous plants may require special clothing

Within each application, there are several operations which require chemical protective clothing. For example, in emergency response, the following activities dictate chemical protective clothing use:

- Site Survey – The initial investigation of a hazardous material incident; these situations are usually characterized by a large degree of uncertainty and mandate the highest levels of protection.

- Rescue – Entering a hazardous materials area for the purpose of removing an exposure victim; special considerations must be given to how the selected protective clothing may affect the ability of the wearer to carry out rescue and to the contamination of the victim.

- Spill Mitigation – Entering a hazardous materials area to prevent a potential spill or to reduce the hazards from an existing spill (i.e., applying a chlorine kit on railroad tank car). Protective clothing must accommodate the required tasks without sacrificing adequate protection.

- Emergency Monitoring – Outfitting personnel in protective clothing for the primary purpose of observing a hazardous materials incident without entry into the spill site; this may be applied to monitoring contract activity for spill clean up.

- Decontamination – Applying decontamination procedures to personnel or equipment leaving the site; in general, a lower level of protective clothing is used by personnel involved in decontamination.

THE CLOTHING ENSEMBLE

The approach in selecting personal protective clothing must encompass an "ensemble" of clothing and equipment items which are easily integrated to provide both an appropriate level of protection and still allow one to carry out activities involving chemicals. In many cases, simple protective clothing by itself may be sufficient to prevent chemical exposure such as the wearing of gloves in combination with a splash apron and faceshield (or safety goggles).

The following is a checklist of components that may form the chemical protective ensemble:

- Protective clothing (suit, coveralls, hoods, gloves, boots).
- Respiratory equipment (SCBA, combination SCBA/SAR, air purifying respirators).
- Cooling system (ice vest, air circulation, water circulation).
- Communications device.
- Head protection.
- Eye protection.
- Ear protection.
- Inner garment.
- Outer protection (overgloves, overboots, flashcover).

Factors that affect the selection of ensemble components include

- How each item accommodates the integration of other ensemble components. Some ensemble components may be incompatible due to how they are worn (e.g., some SCBAs may not fit within a particular chemical protective suit or allow acceptable mobility when worn).

- The ease of interfacing ensemble components without sacrificing required performance (e.g., a poorly fitting overglove that greatly reduces wearer dexterity).

- Limiting the number of equipment items to reduce donning time and complexity (e.g., some communication devices are built into SCBAs which, as a unit, are NIOSH certified).

LEVEL OF PROTECTION

Ensemble components based on the widely used EPA Levels of Protection are Levels A, B, C, and D. These lists can be used as the starting point for ensemble creation; however, each ensemble must be tailored to the specific situation in order to provide the most appropriate level of protection. For example, if an emergency response activity involves a highly contaminated area or if the potential of contamination is high, it may be advisable to wear a disposable covering such as Tyvek coveralls or PVC splash suits over the protective ensemble.

EPA Levels of Protection

LEVEL A:

Vapor protective suit (meets NFPA 1991). Pressure-demand, full-face SCBA, Inner chemical-resistant gloves, Chemical-resistant safety boots, Two-way radio communication (see Figure 10-11).

Figure 10-11. Level A ensemble. Courtesy of Operating Engineers, Local #478, Meriden, CT

OPTIONAL: Cooling System, Outer Gloves, Hard Hat

Protection Provided: Highest available level of respiratory, skin, and eye protection from solid, liquid, and gaseous chemicals.

Used When: The chemical(s) have been identified and have high level of hazards to respiratory system, skin, and eyes. Substances are present with known or suspected skin toxicity or carcinogenity. Operations must be conducted in confined or poorly ventilated areas.

Limitations: Protective clothing must resist permeation by the chemical or mixtures present. Ensemble items must allow integration without loss of performance.

LEVEL B:

Liquid splash-protective suit (meets NFPA 1992). Pressure-demand, full-facepiece SCBA. Inner chemical-resistant gloves, Chemical-resistant safety boots, Two-way radio communications, hard hat (see Figure 10-12).

Figure 10-12. Level B ensemble. Courtesy of Operating Engineers, Local #478, Meriden, CT

OPTIONAL: Cooling System, Outer Gloves

Protection Provided: Provides same level of respiratory protection as Level A, but less skin protection. Liquid splash protection, but no protection against chemical vapors or gases.

Used When: The chemical(s) have been identified but do not require a high level of skin protection. Initial site surveys are required until higher levels of hazards are identified. The primary hazards associated with site entry are from liquid and not vapor contact.

Limitations: Protective clothing items must resist penetration by the chemicals or mixtures present. Ensemble items must allow integration without loss of performance.

LEVEL C:

Support Function Protective Garment (meets NFPA 1993). Full-facepiece, air-purifying, canister-equipped respirator, chemical resistant gloves, chemical resistant safety boots, two-way communications system, and hard hat (see Figure 10-13).

OPTIONAL: Faceshield, Escape SCBA.

Protection Provided: The same level of skin protection as Level B, but a lower level of respiratory protection. Liquid splash protection but no protection to chemical vapors or gases.

Used When: Contact with site chemical(s) will not affect the skin. Air contaminants have been identified and concentrations measured. A canister is available which can remove the contaminant. The site and its hazards have been completely characterized.

Limitations: Protective clothing items must resist penetration by the chemical or mixtures present. Chemical airborne concentration must be less than IDLH levels. The atmosphere must contain at least 19.5% oxygen.

NOTE: NOT ACCEPTABLE FOR CHEMICAL EMERGENCY RESPONSE.

LEVEL D:

Coveralls, safety boots/shoes, safety glasses or chemical splash goggles (see Figure 10-14).

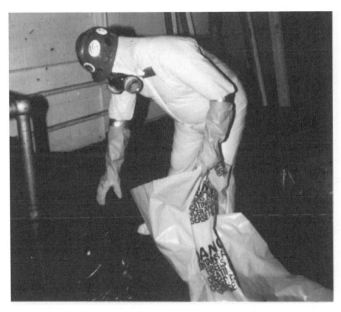

Figure 10-13. Level C ensemble. Courtesy of Professional Health & Safety Consultants, Storrs, CT

OPTIONAL – Gloves, Escape SCBA, Face-shield

Protection Provided: No respiratory protection, minimal skin protection.

Used When: The atmosphere contains no known hazard. Work functions preclude splashes, immersion, potential for inhalation, or direct contact with hazard chemicals.

Limitations: This level should not be worn in the Hot Zone. The atmosphere must contain at least 19.5% oxygen.

NOTE: NOT ACCEPTABLE FOR CHEMICAL EMERGENCY RESPONSE.

Figure 10-14. Level D ensemble

The type of equipment used and the overall level of protection should be periodically reevaluated as the amount of information about the chemical situation or process increases, and when workers are required to perform different tasks. Personnel should upgrade or downgrade their level of protection only with the concurrence of the site supervisor, safety officer, or plant industrial hygienist.

It is important to realize that selecting items only by how they are designed or configured is not sufficient to ensure adequate protection. In other words, just having the right components to form an ensemble is not enough. The EPA levels of protection do not define what performance the selected clothing or equipment must offer. Many of these considerations are described in the "limiting criteria" column in this chapter. Additional factors relevant to the various clothing and equipment items are also described in subsequent sections.

ENSEMBLE SELECTION FACTORS

Chemical Hazards

Chemicals present a variety of hazards such as toxicity, corrosiveness, flammability, reactivity, and oxygen deficiency. Depending on the chemicals present, any combination of hazards may exist (see Figure 10-15).

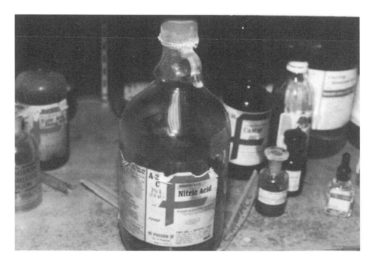

Figure 10-15. Chemical hazards

Physical Environment

Chemical exposure can happen anywhere: in industrial settings, on the highways, in residential areas; they may be either indoors or outdoors; the environment may be extremely hot, cold, or moderate; the exposure site may be relatively uncluttered or rugged presenting a number of physical hazards; chemical handling activities may involve entering confined spaces, heavy lifting, climbing a ladder, or crawling on the ground. The choice of ensemble components must account for these conditions.

Duration of Exposure

The protective qualities of ensemble components may be limited to certain exposure levels (e.g., material chemical resistance, air supply). The decision for ensemble use time must be made assuming the worst-case exposure so that safety margins can be applied to increase the protection available to the worker.

Protective Clothing

Hopefully, an array of different clothing or equipment are available to workers to meet all intended applications. Reliance on one particular clothing or equipment item may severely limit a facility's ability to handle a broad range of chemical exposures. In its acquisition of equipment and clothing, the safety department or other responsible authority should attempt to provide a high degree of flexibility while choosing protective clothing and equipment which is easily integrated and provides protection against each conceivable hazard.

Classification of Protective Clothing

Personal protective clothing includes

- Fully encapsulating suits.
- Non-encapsulating suits.
- Gloves, boots, and hoods.
- Firefighter's protective clothing.
- Proximity or approach clothing.
- Blast or fragmentation suits.
- Radiation-protective suits.

Firefighter turnout clothing, proximity gear, blast suits, and radiation suits by themselves are not acceptable for providing adequate protection from hazardous chemicals contractor employees may encounter.

Performance

The National Fire Protection Association (NFPA) has classified suits by their performance as

- Vapor-protective suits (NFPA Standard 1991). Vapor-protective suits provide "gas-tight" integrity and are intended for response situations where no chemical contact is permissible. This type of suit would be equivalent to the clothing required in EPA's Level A.

- Liquid splash-protective suits (NFPA Standard 1992). Liquid splash-protective suits offer protection against liquid chemicals in the form of splashes, but not against continuous liquid contact or chemical vapors or gases. Essentially, the type of clothing would meet the EPA Level B needs. It is important to note, however, that by wearing liquid splash-protective clothing, the wearer accepts exposure to chemical vapors or gases because this clothing does not offer gas-tight performance. The use of duct tape to seal clothing interfaces does not provide the type of wearer encapsulation necessary for protection against vapors or gases.

- Support function protective garments (NFPA Standard 1993). Support function protective garments must also provide liquid splash protection but offer limited physical protection. These garments may comprise several separate protective clothing components (i.e., coveralls, hoods, gloves, and boots). They are intended for use in nonemergency, nonflammable situations where the chemical hazards have been completely characterized. Examples of support functions include proximity to chemical processes, decontamination, hazardous waste clean up, and training.

Support function protective garments should not be used in chemical emergency response or in situations where chemical hazards remain uncharacterized.

- These NFPA standards define minimum performance requirements for the manufacture of chemical protective suits. Each standard requires rigorous testing of the suit and the materials that comprise the suit in terms of overall protection, chemical resistance, and physical properties. Suits which are found compliant by an independent certification and testing organization may be labeled by the manufacturer as meeting the requirements of the respective NFPA standard. Manufacturers also have to supply documentation showing all test results and characteristics of their protective suits.

- Protective clothing should completely cover both the wearer and his or her breathing apparatus. In general, respiratory protective equipment is not designed to resist chemical contamination. Level A protection (vapor-protective suits) require this configuration. Level B ensembles may be configured either with the SCBA on the outside or inside. However, it is strongly recommended that the wearer's respiratory equipment be worn inside the ensemble to prevent its failure and to reduce decontamination problems. Level C ensembles use cartridge or canister type respirators which are generally worn outside the clothing (see Figure 10-16).

Table 10-2 describes various types of protection clothing available, details the type of protection they offer, and lists factors to consider in their selection and use.

Design

Categorizing clothing by design is mainly a means for describing what areas of the body the clothing item is intended to protect. In emergency response, hazardous waste site clean up, and dangerous chemical operations, the only acceptable types of protective clothing include fully or totally encapsulating suits and nonencapsulating or "splash" suits plus accessory clothing items such as chemically resistant gloves or boots. These descriptions apply to how the clothing is designed and not to its performance.

Service Life

Clothing item service life is an end-user decision depending upon the costs and risks associated with clothing decontamination and reuse. For example, a Saranex/Tyvek garment may be designed to be a coverall (covering the wearer's torso, arms, and legs) intended for liquid splash protection, which is disposable after a single use.

Protective clothing may be labeled as

- Reusable, for multiple wearing.

- Disposable, for one-time use.

The distinctions between these types of clothing are both vague and complicated. Disposable clothing is generally lightweight and inexpensive. Reusable clothing is often more rugged and costly. Nevertheless, extensive contamination of any garment may render it disposable. The basis of this classification really depends upon the costs involved in purchasing, maintaining, and reusing protective clothing versus the alternative of disposal following exposure. If an end-user can anticipate obtaining several uses out of a garment while still maintaining adequate protection from that garment at a lower cost than its disposal, the suit becomes reusable. The key assumption in this determination is the viability of the garment following exposure. This issue is further discussed in the section on decontamination.

Figure 10-16. Chemical splash suit. Courtesy of Operating Engineers, Local #478, Meriden, CT

PROTECTIVE CLOTHING SELECTION FACTORS

Clothing Design

Manufacturers sell clothing in a variety of styles and configurations.

Design Considerations

- Clothing configuration.
- Components and options.
- Sizes.
- Ease of donning on and off.
- Clothing construction.
- Accommodation of other selected ensemble equipment.
- Comfort.
- Restriction of mobility.

Material Chemical Resistance

Ideally, the chosen material(s) must resist permeation, degradation, and penetration by the respective chemicals.

Permeation

This is the process by which a chemical dissolves in or moves through a material on a molecular basis. In most cases, there will be no visible evidence of chemicals permeating a material.

Permeation breakthrough time is the most common result used to assess material chemical compatibility. The rate of permeation is a function of several factors such as chemi-

Table 10-2

Types of Protective Clothing for Full Body Protection

DESCRIPTION	TYPE OF PROTECTION	USE CONSIDERATIONS
Fully encapsulating suit One-piece garment. Boots and gloves may be integral, attached and replaceable, or separate.	Protects against splashes, dust, gases, and vapors.	Does not allow body heat to escape. May contribute heat stress in wearer, particularly if worn in conjunction with a closed-circuit SCBA; a cooling garment may be needed. Impairs worker mobility, vision, and communication.
Non-encapsulating suit Jacket, hood, pants or bib overalls, and one-piece coveralls.	Protects against splashes, dust, and other materials but not against gases and vapors. Does not protect parts of head or neck.	Do not use where gas-tight or pervasive splashing protection is required. May contribute to heat stress in wearer. Tape-seal connection between pant cuffs and boots and between gloves and sleeves.
Aprons, leggings, and sleeve protectors Fully sleeved and gloved apron. Separate coverings for arms and legs. Commonly worn over non-encapsulating suit.	Provides additional splash protection of chest, forearms, and legs.	Whenever possible should be used over a non-encapsulating suit to minimize potential heat stress. Useful for sampling, labeling, and analysis operations. Should be used only when there is a low probability of total body contact with contaminants.
Firefighters' protective clothing Gloves, helmet, running or bunker coat, running or bunker pants (NFPA No. 1971, 1972, 1973, and boots (1974).	Protects against heat, hot water, and some particles. Does not protect against gases and vapors, or chemical permeation or degradation. NFPA Standard No. 1971 specifies that a garment consists of an outer shell, an inner liner, and a vapor barrier with a minimum water penetration of 25 lbs /in(2) (1.8 kg/cm (2) to prevent passage of hot water.	Decontamination is difficult. Should not be worn in areas where protection against gases, vapors, chemical splashes, or permeation is required.
Proximity garment (approach suit) One- or two-piece over garment with boot covers, gloves, and hood of aluminized nylon or cotton fabric. Normally worn over other protective clothing, firefighters' bunker gear, or flame-retardant coveralls.	Protects against splashes, dust, gases, and vapors.	Does not allow body heat to escape. May contribute to heat stress in wearer, particularly if worn in conjunction with a closed-circuit SCBA; a cooling garment may be needed. Impairs worker mobility, vision, and communication.
Blast and fragmentation suit Blast and fragmentation vests and clothing, bomb blankets, and bomb carriers.	Provides some protection against very small detonations. Bomb blankets and baskets can help redirect a blast.	Does not provide for hearing protection.

Table 10-2

Types of Protective Clothing for Full Body Protection (Continued)

DESCRIPTION	TYPE OF PROTECTION	USE CONSIDERATIONS
Radiation-contamination protective suit		
Various types of protective clothing designed to prevent contamination of the body by radioactive particles.	Protects against alpha and beta particles. Does NOT protect against gamma radiation.	Designed to prevent skin contamination. If radiation is detected on site, consult an experienced radiation expert and evacuate personnel until the radiation hazard has been evaluated.
Flame/fire-retardant coveralls		
Normally worn as an undergarment.	Provides protection from flash fires.	Adds bulk and may exacerbate heat stress problems and impair mobility.

The following table provides a listing of clothing classifications. Clothing can be classified by

Design	Performance	Service Life
Gloves, boots, aprons, jackets, coveralls, full body suits.	Particulate protection, liquid-splash protection, vapor protection.	Single use, limited use, reusable.

cal concentration, material thickness, humidity, temperature, and pressure. Most material testing is done with 100% chemical over an extended exposure period. The time it takes a chemical to permeate through the material is the breakthrough time. An acceptable material is one where the breakthrough time exceeds the expected period of garment use. However, temperature and pressure effects may enhance permeation and reduce the magnitude of this safety factor (see Figures 10-17 and 18). For example, small increases in ambient temperature can significantly reduce breakthrough time and the protective barrier properties of a protective clothing material.

Figure 10-17. Site clean-up clothing is dependent on the hazards anticipated.
Courtesy of the Operating Engineers, Local #478, Meriden, CT

*Figure 10-18. Chemical permeation must be considered for hazardous waste work
Courtesy of the Operating Engineers, Local #478, Meriden, CT*

Degradation

Degradation involves physical changes in a material as the result of a chemical exposure, use, or ambient conditions (e.g., sunlight). The most common observations of material degradation are discoloration, swelling, loss of physical strength, or deterioration.

Penetration

Penetration is the movement of chemicals through zippers, seams, or imperfections in a protective clothing material. It is important to note that no material protects against all chemicals and combinations of chemicals, and that no currently available material is an effective barrier to any prolonged chemical exposure.

Sources of Information

- Guidelines for the Selection of Chemical Protective Clothing, 3rd Edition. This reference provides a matrix of clothing material recommendations for approximately 500 chemicals based on an evaluation of chemical resistance test data, vendor literature, and raw material suppliers. The major limitation for these guidelines are their presentation of recommendations by generic material class. Numerous test results have shown that similar materials from different manufacturers may give widely different performances. That is to say manufacturer A's butyl rubber glove may protect against chemical X, but a butyl glove made by manufacturer B may not.

- Quick Selection Guide to Chemical Protective Clothing. Pocket size guide that provides chemical resistance data and recommendations for 11 generic materials against over 400 chemicals. The guide is color-coded by material-chemical recommendation. As with the "Guidelines..." above, the major limitation of this reference is its dependence on generic data.

- Vendor data or recommendations. The best source of current information on material compatibility should be available from the manufacturer of the selected clothing. Many vendors supply charts which show actual test data or their own recommendations for specific chemicals. However, unless vendor data or the recommendations are well documented, end-users must approach this information with caution. Material recommendations must be based on data obtained from tests performed to standard ASTM methods. Simple ratings of "poor," "good," or "excellent" give no indication of how the material may perform against various chemicals.

Mixtures of chemicals can be significantly more aggressive toward protective clothing materials than any single chemical alone. One permeating chemical may pull another with it through the material. Very little data are available for chemical mixtures. Other situations may involve unidentified substances. In both the case of mixtures and unknowns, serious consideration must be given to deciding which protective clothing is selected. If clothing must be used without test data, garments with materials having the broadest chemical resistance should be worn, i.e., materials which demonstrate the best chemical resistance against the widest range of chemicals.

Physical Properties

As with chemical resistance, manufacturer materials offer wide ranges of physical qualities in terms of strength, resistance to physical hazards, and operation in extreme environmental conditions. Comprehensive manufacturing standards such as the NFPA Standards set specific limits on these material properties, but only for limited applications, i.e., emergency response. End-users in other applications may assess material physical properties by posing the following questions:

- Does the material have sufficient strength to withstand the physical strength of the tasks at hand?

- Will the material resist tears, punctures, cuts, and abrasions?

- Will the material withstand repeated use after contamination and decontamination?

- Is the material flexible or pliable to allow end-users to perform needed tasks?

- Will the material maintain its protective integrity and flexibility under hot and cold extremes?

- Is the material flame resistant or self-extinguishing (if these hazards are present)?

- Are garment seams in the clothing constructed so they provide the same physical integrity as the garment material?

Ease of Decontamination

The degree of difficulty in decontaminating protective clothing may dictate whether disposable or reusable clothing is used, or a combination of both.

Cost

Protective clothing end-users must endeavor to obtain the broadest protective equipment they can buy with available resources to meet their specific application.

GENERAL GUIDELINES

Decide if the Clothing Item is Intended to Provide Vapor, Liquid-Splash, or Particulate Protection

Vapor protective suits also provide liquid splash and particulate protection. Liquid splash protective garments also provide particulate protection.

Many garments may be labeled as totally encapsulating but do not provide gas-tight integrity due to inadequate seams or closures. Gas-tight integrity can be determined only by performing a pressure or inflation test and a leak detection test of the respective protective suit. This test involves

- Closing off suit exhalation valves,

- Inflating the suit to a prespecified pressure, and

- Observing whether the suit holds the above pressure for a designated period of time.

ASTM Standard Practice F1052 (1987 Edition) offers a procedure for conducting this test. Splash suits must still cover the entire body when combined with the respirator, gloves, and boots. Applying duct tape to a splash suit does not make it protect against vapors.

Particulate protective suits may not need to cover the entire body, this is dependent on the hazards posed by the particulate. In general, gloves, boots and some form of face protection are required. Clothing items may be needed to cover only a limited area of the body such as gloves on hands. The nature of the hazards and the expected exposure will determine if clothing should provide partial or full body protection.

Determine if the Clothing Item Provides Full Body Protection

- Vapor-protective or totally encapsulating suit will meet this requirement by passing gas-tight integrity tests.

- Liquid splash-protective suits are generally sold incomplete (i.e., less gloves and boots).

- Missing clothing items must be obtained separately and match or exceed the performance of the garment.

- Buying a PVC glove for a PVC splash suit does not mean that you obtain the same level of protection. This determination must be made by comparing chemical resistance data.

Evaluate Manufacturer Chemical Resistance Data Provided with the Clothing

Manufacturers of vapor-protective suits should provide permeation resistance data for their products, while liquid and particulate penetration resistance data should accompany liquid splash and particulate protective garments, respectively. Ideally, data should be provided for every primary material in the suit or clothing item. For suits, this includes the garment, visor, gloves, boots, and seams (see Figure 10-19).

Permeation data should include the following:

- Chemical name,

- Breakthrough time (shows how soon the chemical permeates),

- Permeation rate (shows the rate that the chemical comes through),

- System sensitivity (allows comparison of test results from different laboratories), and

Figure 10-19. Check with the manufacturer for protection suit will provide
Courtesy of Operating Engineers, Local #478, Meriden, CT

- A citation that the data were obtained in accordance with ASTM Standard Test Method F739-85.

If no data are provided or if the data lack any one of the above items, the manufacturer should be asked to supply the missing data. Manufacturers that provide only numerical or qualitative ratings must support their recommendations with complete test data.

Liquid penetration data should include a pass or fail determination for each chemical listed, and a citation that testing was conducted in accordance with ASTM Standard Test Method F903-86. Protective suits which are certified to NFPA 1991 or NFPA 1992 will meet all of the above requirements.

Particulate penetration data should show some measure of material efficiency in preventing particulate penetration in terms of particulate type or size and percentage held out. Unfortunately, no standard tests are available in this area and end-users may have little basis for company products.

Suit materials which show no breakthrough or no penetration to a large number of chemicals are likely to have a broad range of chemical resistance. (Breakthrough times greater than one hour are usually considered to be an indication of acceptable performance.) Manufacturers should provide data on the ASTM Standard Guide F1001-86 chemicals. These 15 liquid and 6 gaseous chemicals listed in Table 10-3 represent a cross-section of different chemical classes and challenges for protective clothing materials. Manufacturers should also provide test data on other chemicals as well. If there are specific chemicals within your operating area which have not been tested, ask the manufacturer for test data on these chemicals.

Obtain and Examine the Manufacturer's Instruction or Technical Manual

This manual should document all the features of the clothing, particularly suits, and describe what material(s) are used in its construction. It should cite specific limitations for the clothing and what restrictions apply to its use. Procedures and recommendations should be supplied for at least the following:

- Donning and doffing.
- Inspection, maintenance, and storage.
- Decontamination.
- Use.

The manufacturer's instructions should be thorough enough to allow the end-users to wear and use the clothing without a large number of questions.

Obtain and Inspect Sample Clothing Item Garments

Examine the quality of clothing construction and other features that will impact its wearing. The questions listed under Protective Clothing Selection Factors, Clothing Design should be considered. If possible, representative clothing items should be obtained in advance and inspected prior to purchase, and discussed with someone who has experience in their use. It is also helpful to try out representative garments prior to purchase by suiting personnel in the garment and having them run through exercises to simulate expected activities.

Field Selection of Chemical Protective Clothing

Even when end-users have gone through a very careful selection process, a number of situations will arise when no information is available to judge whether or not their protective clothing will provide adequate protection (see Table 10-3). These situations include

- Chemicals which have not been tested with the garment materials.
- Mixtures of two or more different chemicals.
- Chemicals which cannot be readily identified.
- Extreme environmental conditions (hot temperatures).
- Lack of data on all clothing components (e.g., seams, visors).

Testing material specimens using newly developed field test kits may offer one means for making on-site clothing selection. A portable test kit has been developed by the EPA using a simple weight loss method that allows field qualification of protective clothing materials within one hour. Use of this kit may overcome the absence of data and provide additional criteria for clothing selection.

Selection of chemical protective clothing is a complex task and should be performed by personnel with both extensive training and experience. Under all conditions, clothing should be selected by evaluating its performance characteristics against the requirements and limitations imposed by the application.

MANAGEMENT PROGRAM

Written Management Program

A written Chemical Protective Clothing Management Program should be established by all end-users who routinely select and use protective clothing. Reference should be made to [1910.120 Hazardous Waste] and [1910.132 (d)(1) Hazard Assessment] for those covered.

The written management program should include policy statements, procedures, and guidelines. Copies should be made available to all personnel who may use protective clothing in the course of their duties or job. Technical data on clothing, maintenance manuals, relevant regulations, and other essential information should also be made available.

Table 10-3

**Recommended Chemicals to Evaluate the
Performance of Protective Clothing Materials**

CHEMICAL	CLASS
Acetone	Ketone
Acetonitrile	Nitrile
Ammonia	Strong base (gas)
1,3-Butadiene	Olefin (gas)
Carbon Disulfide	Sulfur-containing organic
Chlorine	Inorganic gas
Dichloromethane	Chlorinated hydrocarbon
Diethylamine	Amine
Dimethyl formamide	Amide
Ethyl Acetate	Ester
Ethylene Oxide	Oxygen heterocyclic gas
Hexane	Alihatic hydrocarbon
Hydrogen Chloride	Acid gas
Methanol	Alcohol
Methyl Chloride	Chlorinated hydrocarbon (gas)
Nitrobenzene	Nitrogen-containing organic
Sodium Hydroxide	Inorganic base
Suylfuric Acid	Inorganic acid
Tetrachloroethylene	Chlorinated hydrocarbon
Tetrahydrofuran	Oxygen heterocyclic
Toluene	Aromatic hydrocarbon

The two basic objectives of any management program should be to protect the wearer from safety and health hazards, and to prevent injury to the wearer from incorrect use and/or malfunction of the chemical protective clothing. To accomplish these goals, a comprehensive management program should include hazard identification; medical monitoring; environmental surveillance; selection, use, maintenance, and decontamination of chemical protective clothing; and training.

Program Review and Evaluation

The management program should be reviewed at least annually. Elements which should be considered in the review include

- The number of person-hours that personnel wear various forms of chemical protective clothing and other equipment.

- Accident and illness experience.

- Levels of exposure.

- Adequacy of equipment selection.

- Adequacy of the operational guidelines.

- Adequacy of decontamination, cleaning, inspection, maintenance, and storage programs.

- Adequacy and effectiveness of training and fitting programs.

- Coordination with overall safety and health program.

- The degree of fulfillment of program objectives.

- The adequacy of program records.

- Recommendations for program improvement and modification.

- Program costs.

The results of the program evaluation should be made available to all end-users and presented to top management so that program changes may be implemented.

Types of Standard Operating Procedures

Personal protective clothing and equipment can offer a high degree of protection only if it is properly used. Standard Operating Procedures (SOPs) should be established for all workers involved in handling hazardous chemicals. Areas that should be addressed include

- Selection of protective ensemble components.

- Protective clothing and equipment donning, doffing, and use.

- Decontamination procedures.

- Inspection, storage, and maintenance of protective clothing/equipment.

- Training.

Selection of Protective Clothing Components

Protective clothing and equipment SOPs must take into consideration the factors presented in Ensemble and Clothing sections. All clothing and equipment selections should provide a decision tree that relates chemical hazards and information to levels of protection and performance needed.

Responsibility in selecting appropriate protective clothing should be vested in a specific individual who is trained in both chemical hazards and protective clothing use such as a safety officer or industrial hygienist. Only chemical protective suits labeled as compliant with the appropriate performance requirements should be used. In cases where the chemical hazards are known in advance or encountered routinely, clothing selection should be predetermined. That is, specific clothing items should be identified in specific chemical operations without the opportunity for individual selection of other clothing items.

CLOTHING DONNING, DOFFING, AND USE

The procedures below are given for vapor protective or liquid-splash protective suit ensembles and should be included in the training program.

Donning the Ensemble

A routine should be established and periodically practiced for donning the various ensemble configurations that a facility or team may use. Assistance should be provided for donning and doffing since these operations are difficult to perform alone, and solo efforts may increase the possibility of ensemble damage (see Figure 10-20).

Figure 10-20. Donning and doffing ensemble should be accomplished with haz-waste partner. Courtesy of Operating Engineers, Local #478, Meriden, CT

Once the equipment has been donned, its fit should be evaluated. If the clothing is too small, it will restrict movement, increase the likelihood of tearing the suit material, and accelerate wearer fatigue. If the clothing is too large, the possibility of snagging the material is increased, and the dexterity and coordination of the wearer may be compromised. In either case, the wearer should be recalled and better fitting clothing provided.

Doffing an Ensemble

Exact procedures for removing a totally encapsulating suit/SCBA ensemble must be established and followed in order to prevent contaminant migration from the response scene and transfer of contaminants to the wearer's body, the doffing assistant, and others.

Doffing procedures should be performed only after decontamination of the suited end-user. They require a suitably attired assistance. Throughout the procedures, both wearer and assistant should avoid any direct contact with the outside surface of the suit.

The following are sample procedures for donning a totally encapsulating suit/SCBA ensemble. These procedures should be modified depending upon the suit and accessory equipment used. The procedures assume the wearer has previous training in respirator use and decontamination procedures (see Figure 10-21).

Sample Donning Procedures

1.   Inspect clothing and respiratory equipment before donning.

2.   Adjust hard hat or headpiece, if worn, to fit user's head.

Figure 10-21. Donning and doffing require step-by-step procedures
Courtesy of Operating Engineers, Local #478, Meriden, CT

3. Open back closure used to change air tank (if suit has one) before donning suit.

4. Standing or sitting, step into the legs of the suit; ensure proper placement of the feet within the suit; then gather the suit around the waist.

5. Put on chemical-resistant safety boots over the feet of the suit. Tape the leg cuff over the tops of the boots. If additional chemical-resistant safety boots are required, put these on now. Some one-piece suits have heavy-soled protective feet. With these suits, wear short, chemical resistant safety boots inside the suit.

6. Put on air tank and harness assembly of the SCBA. Don the facepiece and adjust it to be secure, but comfortable. Do not connect the breathing hose. Open valve on air tank.

7. Perform negative and positive respirator facepiece seal test procedures. To conduct a negative-pressure test, close the inlet part with the palm of the hand or squeeze the breathing tube so it does not pass air, and gently inhale for about 10 seconds. Any inward rushing of air indicates a poor fit. Note that a leaking facepiece may be drawn tightly to the face to form a good seal, giving a false indication of adequate fit. To conduct a positive-pressure test, gently exhale while covering the exhalation valve to ensure that a positive pressure can be built up. Failure to build a positive pressure indicates a poor fit.

8. Depending on type of suit, put on long-sleeved inner gloves (similar to surgical gloves). Secure gloves to sleeves, for suits with detachable gloves (if not done prior to entering the suit). Additional overgloves, worn over attached suit gloves, may be donned later.

9. Put sleeves of suit over arms as assistant pulls suit up and over the SCBA. Have assistant adjust suit around SCBA and shoulders to ensure unrestricted motion.

10. Put on hard hat, if needed.

11. Raise hood over head carefully so as not to disrupt face seal of SCBA mask. Adjust hood to give satisfactory comfort.

12. Begin to secure the suit by closing all fasteners on opening until there is only adequate room to connect the breathing hose. Secure all belts and/or adjustable leg, head, and waistbands.

13. Connect the breathing hose while opening the main valve.

14. Have assistant first ensure that wearer is breathing properly and then make final closure of the suit.

15. Have assistant check all closures.

16. Have assistant observe the wearer for a period of time to ensure that the wearer is comfortable, psychologically stable, and that the equipment is functioning properly.

Sample Doffing Procedures

If sufficient air supply is available to allow appropriate decontamination before removal,

1. Remove any extraneous or disposable clothing, boot covers, outer gloves, and tape.

2. Have assistant loosen and remove the wearer's safety shoes or boots.

3. Have assistant open the suit completely and lift the hood over the head of the wearer and rest it on top of the SCBA tank.

4. Remove arms, one at a time, from suit. Once arms are free, have assistant lift the suit up and away from the SCBA backpack—avoiding any contact between the outside surface of the suit and the wearer's body – and lay the suit out flat behind the wearer. Leave internal gloves on, if any.

5. Sitting, if possible, remove both legs from the suit.

6. Follow procedure for doffing SCBA.

7. After suit is removed, remove internal gloves by rolling them off the hand, inside out.

8. Remove internal clothing and thoroughly cleanse the body.

Sample doffing procedures, if the low-pressure warning alarm has sounded, signifying that approximately 5 minutes of air remain,

1. Remove disposable clothing.

2. Quickly scrub and hose off, especially around the entrance/exit zipper.

3. Open the zipper enough to allow access to the regulator and breathing hose.

4. Immediately attach an appropriate canister to the breathing hose (the type and fittings should be predetermined). Although this provides some protection against any contamination still present, it voids the certification of the unit.

5. Follow Steps 1 through 8 of the regular doffing procedure above. Take extra care to avoid contaminating the assistant and wearer.

User Monitoring and Training

The wearer must understand all aspects of clothing/equipment operation and their limitations; this is especially important for fully encapsulating ensembles where misuse could potentially result in suffocation. During protective clothing use, end-users should be encouraged to report any perceived problems or difficulties to their supervisor. These malfunctions include, but are not limited to

- Degradation of the protection ensemble.

- Perception of odors.

- Skin irritation.

- Unusual residues on clothing material.

- Discomfort.

- Resistance to breathing.

- Fatigue due to respirator use.

- Interference with vision or communication.

- Restriction of movement.

- Physiological responses such as rapid pulse, nausea, or chest pain.

Work Mission Duration

Before end-users undertake any activity in their chemical protective ensembles, the anticipated duration of use should be established. Several factors limit the length of a mission. These include

- Air supply consumption as affected by wearer work rate, fitness, body size, and breathing patterns;

- Suit ensemble permeation, degradation, and penetration by chemical contaminants, included expected leakage through suit or respirator exhaust valves (ensemble protection factor);

- Ambient temperature as it influences material chemical resistance and flexibility, suit and respirator exhaust valve performance, and wearer heat stress; and

- Coolant supply (if necessary).

DECONTAMINATION PROCEDURES

Definition and Types

Decontamination is the process of removing or neutralizing contaminants that have accumulated on personnel and equipment. This process is critical to health and safety at hazardous material response sites. Decontamination protects end-users from hazardous substances that may contaminate and eventually permeate the protective clothing, respiratory equipment, tools, vehicles, and other equipment used in the vicinity of the chemical hazard; it protects all plant or site personnel by minimizing the transfer of harmful materials into clean areas; it helps prevent mixing of incompatible chemicals; and it protects the community by preventing uncontrolled transportation of contaminants from the site.

There are two types of decontamination

- Gross decontamination – To allow end-user to safely exit or doff their chemical protective clothing, and

- Decontamination for reuse of chemical protective clothing.

Prevention of Contamination

The first step in decontamination is to establish Standard Operating Procedures that minimize contact with chemicals and thus the potential for contamination. For example,

- Stress work practices that minimize contact with hazardous substances (e.g., do not walk through areas of obvious contamination, do not directly touch potentially hazardous substances).

- Use remote sampling, handling, and container-opening techniques (e.g., drum grapples, pneumatic impact wrenches).

- Protect monitoring and sampling instruments by bagging. Make openings in the bags for sample ports and sensors that must contact site materials.

- Wear disposable outer garments and use disposable equipment where appropriate.

- Cover equipment and tools with a strippable coating which can be removed during decontamination.

- Encase the source of contaminants, e.g., with plastic sheeting or overpacks.

- Ensure all closures and ensemble component interfaces are completely secured; that no open pockets are present which could serve to collect contaminant.

Types of Contamination

Surface contaminants may be easy to detect and remove. Contaminants that have permeated a material are difficult or impossible to detect and remove. If contaminants that have permeated a material are not removed by decontamination, they may continue to permeate to either surface of the material where they can cause an unexpected exposure. Four major factors that affect the extent of permeation are

- Contact time. The longer a contaminant is in contact with an object, the greater the probability and extent of permeation. For this reason, minimizing contact time is one of the most important objectives of a decontamination program.

- Concentration. Molecules flow from areas of high concentration to areas of low concentration. As concentrations of chemicals increase, the potential for permeation of personal protective clothing increases.

- Temperature. An increase in temperature generally increases the permeation rate of contaminants.

- Physical state of chemicals. As a rule, gases, vapors, and low-viscosity liquids tend to permeate more readily than high-viscosity liquids or solids.

Decontamination Methods

Decontamination methods (1) physically remove contaminants, (2) inactivate contaminants by chemical detoxification or disinfecting/sterilization, or (3) remove contaminants by a combination of both physical and chemical means.

In general, gross decontamination is accomplished using detergents (surfactants) in water combined with a physical scrubbing action. This process will remove most forms of surface contamination including dusts, many inorganic chemicals, and some organic chemicals. Soapy water scrubbing of protective suits may not be effective in removing oily or tacky organic substances (e.g., PCBs in transformer oil). Furthermore, this form of decontamination is unlikely to remove any contamination that has permeated or penetrated the suit materials. Using organic solvents such as petroleum distillates may allow easier removal of heavy organic contamination but may result in other problems, including

- Permeation into clothing components, pulling contaminant with it;
- Spreading localized contaminant into other areas of the clothing; and
- Generating large volumes of contaminated solvents which require disposal.

One promising method for removing internal or matrix contamination is the forced circulation of heated air over clothing items for extended periods of time. This allows many organic chemicals to migrate out of the materials and evaporate into the heated air. The process does require, however, that the contaminating chemicals be volatile. Additionally, low level heat may accelerate the removal of plasticizer from garment materials and affect the adhesives involved in garment seams.

Unfortunately, both manufacturers and protective clothing authorities provide few specific recommendations for decontamination. There is no definitive list with specific methods recommended for specific chemicals and materials. Much depends on the individual chemical-material combination involved.

Testing the Effectiveness of Decontamination

Protective clothing or equipment reuse depends on demonstrating that adequate decontamination has taken place. Decontamination methods vary in their effectiveness and unfortunately, there are no completely accurate methods for non-destructively evaluating clothing or equipment contamination levels.

Methods which may assist in a determination include

- Visual examination of protective clothing for signs of discoloration, corrosive effects, or any degradation of external materials. However, many contaminants do not leave any visible evidence.

- Wipe sampling of external surfaces for subsequent analysis; this may or may not be effective for determining levels of surface contamination and depends heavily on the material-chemical combination. These methods will not detect permeated contamination.

- Evaluation of the cleaning solution; this method cannot quantify clean method effectiveness. Since the original contamination levels are unknown, the method can only show if chemical has been removed by the cleaning solution. If a number of garments have been contaminated, it may be advisable to sacrifice one garment for destructive testing by a qualified laboratory with analysis of contamination levels on and inside the garment.

Decontamination Plan

A decontamination plan should be developed and set up before any personnel or equipment may enter areas where the potential for exposure to hazardous substances exists. The decontamination plan should

- Determine the number and layout of decontamination stations.

- Determine the decontamination equipment needed.

- Determine appropriate decontamination methods.

- Establish procedures to prevent contamination of clean areas.

- Establish methods and procedures to minimize wearer contact with contaminants during removal of personal protective clothing.

- Establish methods for disposing of clothing and equipment that are not completely decontaminated.

The plan should be revised whenever the type of personal protective clothing or equipment changes, the use conditions change, or the chemical hazards are reassessed based on new information.

The decontamination process should consist of a series of procedures performed in a specific sequence. For chemical protective ensembles, outer, more heavily contaminated items (e.g., outer boots and gloves) should be decontaminated and removed first, followed by decontamination and removal of inner, less contaminated items (e.g., jackets and pants). Each procedure should be performed at a separate station in order to prevent cross contamination. The sequence of stations is called the decontamination line.

Stations should be physically separated to prevent cross contamination and should be arranged in order of decreasing contamination, preferably in a straight line. Separate flow patterns and stations should be provided to isolate workers from different contamination zones containing incompatible wastes. Entry and exit points to exposed areas should be conspicuously marked. Dressing stations for entry to the decontamination area should be separate from redressing areas for exit from the decontamination area. Personnel who wish to enter clean areas of the decontamination facility, such as locker rooms, should be completely decontaminated.

All equipment used for decontamination must be decontaminated and/or disposed of properly. Buckets, brushes, clothing, tools, and other contaminated equipment should be collected, placed in containers, and labeled. Also, all spent solutions and wash water should be collected and disposed of properly. Clothing that is not completely decontaminated should be placed in plastic bags, pending further decontamination and/or disposal.

Decontamination of workers who initially come in contact with personnel and equipment leaving exposure or contamination areas will require more protection from contaminants than decontamination workers who are assigned to the last station in the decontamination line. In some cases, decontamination personnel should wear the same levels of protective clothing as workers in the exposure or contaminated areas. In other cases, decontamination personnel may be sufficiently protected by wearing one level lower protection (e.g., wearing Level B protection while decontaminating workers who are wearing Level A).

Decontamination for Protective Clothing Reuse

Due to the difficulty in assessing contamination levels in chemical protective clothing before and after exposure, the responsible supervisor or safety professional must determine if the respective clothing can be reused. This decision involves considerable risk in determining clothing to be contaminant free. Reuse can be considered, if in the estimation of the supervisor,

- No significant exposures have occurred; and

- Decontamination methods have been successful in reducing contamination levels to safe or acceptable concentrations.

Contamination by known or suspected carcinogens should warrant automatic disposal. Use of disposable suits is highly recommended when extensive contamination is expected.

Emergency Decontamination

In addition to routine decontamination procedures, emergency decontamination procedures must be established.

In an emergency, the primary concern is to prevent the loss of life or severe injury to personnel. If immediate medical treatment is required to save a life, decontamination should be delayed until the victim is stabilized. If decontamination can be performed without interfering with essential life-saving techniques or first aid, or if a worker has been contaminated with an extremely toxic or corrosive material that could cause severe injury or loss of life, decontamination should be continued.

If an emergency due to a heat-related illness develops, protective clothing should be removed from the victim as soon as possible to reduce the heat stress. During an emergency, provisions must also be made for protecting medical personnel and disposing of contaminated clothing and equipment.

INSPECTION, STORAGE, AND MAINTENANCE

The end-user in donning protective clothing and equipment must take all necessary steps to ensure that the protective ensemble will perform as expected. During emergencies is not the right time to discover discrepancies in the protective clothing. Teach end-user care for his clothing and other protective equipment in the same manner as parachutists care for parachutes. Following a standard program for inspection, proper storage, and maintenance together with realizing protective clothing/equipment limitations is the best way to avoid chemical exposure during emergency response.

Inspection

An effective chemical protective clothing inspection program should feature five different inspections.

- Inspection and operational testing of equipment as received from the factory or distributor.

- Inspection of equipment as it is selected for a particular chemical operation.

- Inspection of equipment after use or training and prior to maintenance.

- Periodic inspection of stored equipment.

- Periodic inspection when a question arises concerning the appropriateness of selected equipment, or when problems with similar equipment are discovered.

Each inspection will cover different areas with varying degrees of depth. Those personnel responsible for clothing inspection should follow manufacturer directions; many vendors provide detailed inspection procedures. The generic inspection checklist provided below may serve as an initial guide for developing more extensive procedures (see Figure 10-22).

SAMPLE PPE INSPECTION CHECKLIST

CLOTHING

Before use
Determine that the clothing material is correct for the specific task at hand.

Visually inspect for
- Imperfect seams
- Nonuniform coatings
- Tears
- Malfunctioning closures
Hold up to light and check for pinholes

Flex product
- Observe for cracks
- Observe for other signs of shelf deterioration

If the product has been used previously, inspect inside and out for signs of chemical attack
- Discoloration
- Swelling
- Stiffness

During the work task, periodically inspect
- Evidence of chemical attack such as discoloration, swelling, stiffening, and softening. Keep in mind, however, that chemical permeation can occur without any visible effects.
- Closure failure
- Tears
- Punctures
- Seam discontinuities

GLOVES

Before use
Pressurize glove to check for pinholes. Either blow into glove, then roll gauntlet towards fingers or inflate glove and hold under water. In either case, no air should escape.

FULLY ENCAPSULATING SUITS

Before use
- Check the operation of pressure relief valves
- Inspect the fitting of wrists, ankles, and neck
- Check faceshield, if equipped, for cracks and fogginess

Figure 10-22. Sample PPE inspection checklist

Records

Records must be kept of all inspection procedures. Individual identification numbers should be assigned to all reusable pieces of equipment (many clothing and equipment items may already have serial numbers), and records should be maintained by that number.

At a minimum, each inspection should record

- Clothing/equipment item ID number.

- Date of the inspection.

- Person making the inspection.

- Results of the inspection.

- Any unusual conditions noted.

Periodic review of these records can provide an indication of protective clothing which requires excessive maintenance and can also serve to identify potentially failing clothing.

Storage

Clothing must be stored properly to prevent damage or malfunction from exposure to dust, moisture, sunlight, damaging chemicals, extreme temperatures, and impact. Procedures are needed for both initial receipt of equipment and after use or exposure of that equipment. Many manufacturers specify recommended procedures for storing their products. These should be followed to avoid equipment failure resulting from improper storage.

Some guidelines for general storage of chemical protective clothing include

- Potentially contaminated clothing should be stored in an area separate from street clothing or unused protective clothing.

- Potentially contaminated clothing should be stored in a well-ventilated area, with good air flow around each item, if possible.

- Different types and materials of clothing and gloves should be stored separately to prevent issuing the wrong material by mistake (e.g., many glove materials are black and cannot be identified by appearance alone).

- Protective clothing should be folded or hung in accordance with manufacturer instructions.

Maintenance

Manufacturers frequently restrict the sale of certain protective suit parts to individuals or groups who are specially trained, equipped, or authorized by the manufacturer to purchase them. Explicit procedures should be adopted to ensure that the appropriate level of maintenance is performed only by those individuals having this specialized training and equipment. In no case should you attempt to repair equipment without checking with the person in your facility responsible for chemical protective clothing maintenance.

The following classification scheme is recommended to divide the types of permissible or non-permissible repairs:

- Level 1 – User or wearer maintenance, requiring a few common tools or no tools at all.

- Level 2 – Maintenance that can be performed by the response team's maintenance shop, if adequately equipped and trained.

- Level 3 – Specialized maintenance that can be performed only by the factory or an authorized repair person.

Each facility should adopt the above scheme and list which repairs fall into each category for each type of protective clothing and equipment. Many manufacturers will also indicate which repairs, if performed in the field, void the warranty of their products. All repairs made must be recorded on the records for the specific clothing along with appropriate inspection results.

TRAINING

Training in the use of protective clothing

- Allows the user to become familiar with the equipment in a non-hazardous, nonemergency condition.
- Instills confidence of the user in his/her equipment.
- Makes the user aware of the limitations and capabilities of the equipment.
- Increases worker efficiency in performing various tasks.
- Reduces the likelihood of accidents during chemical operations.

Training should be completed prior to actual clothing use in a nonhazardous environment and should be repeated at the frequency required by OSHA SARA III legislation. As a minimum, the training should point out the user's responsibilities and explain the following, using both classroom and field training when necessary:

- The proper use and maintenance of selected protective clothing, including capabilities and limitations.
- The nature of the hazards and the consequences of not using the protective clothing.
- The human factors influencing protective clothing performance.
- Instructions in inspecting, donning, checking, fitting, and using protective clothing.
- Use of protective clothing in normal air for a long familiarity period.
- The user's responsibility (if any) for decontamination, cleaning, maintenance, and repair of protective clothing.
- Emergency procedures and self-rescue in the event of protective clothing /equipment failure.
- The buddy system.

The discomfort and inconvenience of wearing chemical protective clothing and equipment can create resistance to its conscientious use. One essential aspect of training is to make the user aware of the need for protective clothing and to instill motivation for the proper use and maintenance of that protective clothing.

HEAT STRESS DUE TO CLOTHING

Wearing full-body chemical protective clothing puts the wearer at considerable risk of developing heat stress. This can result in health effects ranging from transient heat fatigue to serious illness or death (see Figure 10-23).

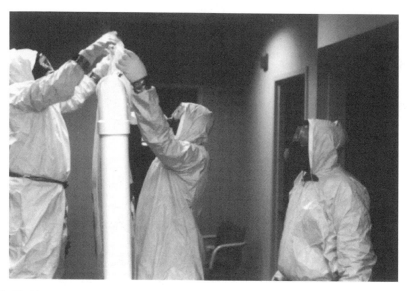

Figure 10-23. Protective clothing may cause severe heat stress problems

Heat stress is caused by a number of interacting factors, including

- Environmental conditions.
- Type of protective ensemble worn.
- The work activity required.
- The individual characteristics of the responder.

When selecting chemical protective clothing and equipment, each item's benefit should be carefully evaluated to its potential to increase the risk of heat stress. For example, if a lighter, less insulating suit can be worn without a sacrifice of protection, then the lighter suit should be worn. Because the incidence of heat stress depends on a variety of factors, all workers wearing full-body chemical protective ensembles should be monitored. The following physiological factors should be monitored.

- Heart Rate
- Oral Temperature

Do not permit an end-user to wear protective clothing and engage in work when his or her oral temperature exceeds 100.6 degrees F (38.1 degrees C).

Use a clinical thermometer (3 minutes under the tongue) or similar device to measure oral temperature at the end of the work period (before drinking).

- If the oral temperature exceeds 99.6 degrees F (37.6 degrees C), shorten the next work period by at least one third the time.
- If the oral temperature exceeds 99.6 degrees F (37.6 degrees C) at the beginning of a response period, shorten the mission time by one third.

Body Water Loss

Measure the end-user's weight on a scale accurate to plus or minus 0.25 pounds prior to any response activity. Compare this weight with his or her normal body weight to determine if enough fluids have been consumed to prevent dehydration. Weights should be taken while the end-user wears similar clothing, or ideally, in the nude. The body water loss should not exceed 1.5% total body weight loss from a response.

RESPIRATORY PROTECTION

Respiratory Protection Physical Limitations

Workers must be evaluated by competent medical personal to ensure that they are physically and mentally able to wear respirators under simulated and actual working conditions. In general, the added breathing resistance of respiratory cartridges and the dead space inside the facepiece causes additional stress to the wearer. The dead space contains air that must be rebreathed before fresh air is obtained. In some asthmatics, an attack may be induced by a variety of factors including exercise, cold and hot air, and additional stress, all of which may be experienced when wearing a respirator. A typical respirator might double the work of breathing. Some workers may experience claustrophobia when wearing a respirator. Claustrophobic reactions might not be detected when a device is first tried on or during the fitting phase (see Figure 10-24).

Whenever the potential for exposure is present and cannot be addressed through engineering controls, administrative controls, or work practices on a lead-abatement site, respiratory protection is required. Only NIOSH/MSHA approved respirators should be used when respiratory protection is needed. It must be provided to employees at no cost. Employers are responsible for the selection of respiratory protection. The concentration of lead and its physical state (vapor or solid) determine what type of respirator should be used.

Figure 10-24. Powered air-purifying respirator used by a window contractor for potential lead dust exposure

Abatement site workers generally wear air-purifying respirators when a respirator is required. Some workers prefer to use the powered air purifying respirators (PAPR). In abrasive blasting work, such as is common for removing paint from steel structures, a supplied air respirator (SAR) is usually required. A self-contained breathing apparatus (SCBA) would provide a sufficient level of protection, but the weight of the unit may make it impracticable.

Air-Purifying Respirators

Air-purifying respirators (APR) remove toxic particles such as dusts, gases, and vapor from the air. Their effectiveness depends on the respirator fit and the type of filter or cartridge used. APRs consist of a facepiece with inhalation and exhalation value. Most commonly used APRs are the full-faced mask and half-mask.

The two types of elements used with the APR are particulate filters and chemical cartridges. These filters and cartridges are selected based upon their ability to filter hazardous materials and prevent them from being inhaled. A color code system helps identify the type of filter or cartridge (see table on following page).

As a guide to the selection of respirators, NIOSH has prepared a manual on Respirator Decision Logic (NIOSH, May 1997) which should be consulted by those responsible for establishing a respiratory protection program.

Limitations of Air Purifying Respirators (APRs)

APRs are not for use in atmosphere containing less than 19.5% oxygen. They are not for use in atmospheres immediately dangerous to life or health. They are also not for use against gases and vapors with poor warning protection unless equipped with an end-of-service life indicator.

How do you know if the cartridge needs to be changed? This is difficult to answer. For dust, a wearer may notice that it is more difficult to breathe as the filter becomes loaded with dust. For gas/vapors, a common guideline is to get a new cartridge when the wearer smells or tastes the gas or vapor or to change the cartridge periodically. This is not a good guideline if a person has a poor sense of smell/taste, or when the wearer has a cold or other respiratory infection. However, current technology has not developed an objective indicator that the cartridge must be changed. Generally, for dusts, replace cartridges whenever the wearer notices difficulty breathing through the cartridges or, for gas/vapor, when the odor of the gas/vapor is detected.

Respirator Fit

A respirator will only be effective if there is a good seal between the facepiece and the wearer's face. Since different people have faces of different shapes and sizes, respirators are available in a variety of sizes and models. Each individual who wears a respirator must be fit tested. A wearer will have to be retested after weight loss or gain, changes in dentures, dental work, or facial injury. These factors may change the size or shape of the face and cause the person to require a different size or model respirator.

Fit testing must be done to determine the correct size and model respirator for each individual. Two types of fit testing, qualitative and quantitative, are used to determine these factors as well as the integrity of the face-to-facepiece seal. These tests should be repeated periodically to document the respirator's effectiveness. Quantitative fit testing must be used for full facepieces unless used at the levels permissible for a half mask. Half masks may also be quantitatively fit tested or a qualitative test may be used.

This is usually done by introducing an irritant smoke around the mask to verify a good face-to-facepiece fit. Caution the worker to close their eyes before smoking the respirator to prevent irritation. When using isoamyl acetate (banana oil), the respirator must be equipped with organic vapor cartridges. Fit testing should be performed every six months according to OSHA regulations.

Qualitative Fit Testing

Purpose: To check the effectiveness of a respirator in preventing substances from entering the facepiece. Method: An individual, while wearing a respirator, enters an atmosphere where a test substance has been released (banana oil or an irritant special smoke tube). The wearer should not be able to detect the substance.

There are several important precautions for qualitative fit testing. Some test substances may irritate the eyes or cause coughing. Since the test relies on the wearer stating she/he cannot smell the test agent, there is some concern that test results are not always reliable. Some people have a poor sense of smell.

Quantitative Fit Testing

Purpose: To measure the effectiveness of a respirator in preventing substances from entering the facepiece. Methods: An individual, while wearing a respirator modified with a probe, enters a chamber. A test substance is released and air inside and outside of the respirator are tested for the substance.

This test provides an objective assessment of the effectiveness of a respirator. It measures the protection factor (PF), which is a comparison of the concentration of the substance outside the mask to the concentration of the substance inside the mask. This PF is useful in determining the maximum concentration in which a cartridge type of respirator may be used.

It does not determine if respirator will effectively protect the wearer from a specific chemical. A disadvantage of this test is that special equipment (microcomputer programs and hardware accessories) and trained personnel are needed. Also, fit testing is often done under "ideal" conditions and not after the respirator has been worn for several hours or during strenuous activity. Therefore, test results may not be accurate.

Routine Personal Fit Tests

Two types of testing, positive and negative pressure tests, should be done each time a respirator is donned to check the face seal. They do not replace yearly fitting or provide a routine assessment as to whether or not the fit is still adequate. For a negative pressure test, the worker places hand over cartridges and inhales. No outside air should be felt leaking into the facepiece and the facepiece should slightly collapse. This test should be done before each use.

The purpose of the positive pressure test is to test the face seal. A slight positive pressure should be built up inside the facepiece with no evidence of outward leakage. The wearer covers the exhalation valve with hand and blows out. Air should escape only from the seal around the facepiece. This test should be done before each use. Positive and negative pressure tests can be done quickly and easily in the field. They do, however, rely on the wearer's ability to detect leaks.

Medical Fitness to Wear a Respirator

Before an employee receives clearance to wear a respirator, a medical examination must be performed by a licensed physician. Medical examinations are required for all employees who wear respirators as outlined by the OSHA Respiratory Protection Standard (1910.134) located in the appendices.

Some medical conditions that may prevent an individual from wearing a respirator include:

- lung disease
- severe high blood pressure
- heart disease

Facial Hair and Respiratory Protection

Any facial hair growth between the facepiece sealing surface and the skin that prevents a good sealing surface is a direct violation of OSHA's respirator standard for workers (see Figure 10-25). Quantitative respirator fit test results demonstrate that workers cannot get adequate protection with facial hair. It's the extremely small toxic particles that can cause the greatest health damage to the lungs and respiratory system.

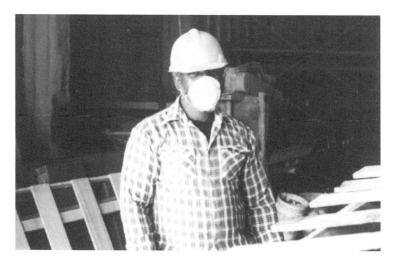

Figure 10-25. Even disposable dust respirators cannot be worn with facial hair

Care and Cleaning of Respirators

Any abatement project which requires the routine use of respirators should provide a respirator care and cleaning program. The purpose of this program is to assure that all respirators are maintained for optimal effectiveness. Generally, one person in an organization is trained to inspect, clean, repair, and store respirators. Respirators should not be shared. All workers should have their own respirator. Respirator programs should be based on the types of respirators, working conditions, and hazards involved. In general, the program should include

- Inspection (including a leak check).
- Cleaning and disinfecting.
- Repair.
- Storage.

Inspection

Inspect respirators before and after each use. For air-purifying respirators, thoroughly check all connections for gaskets and "O" rings and for proper tightness. Check the condition of the facepiece and all its parts, connecting air tube, and headbands. Inspect rubber or elastic parts for flexibility and signs of deterioration. Replacement parts should be readily available. Maintain a record for each respirator inspection, including date, name of inspector, and any unusual conditions or findings.

Cleaning and Disinfecting

Respirators should be cleaned and disinfected after each use to prevent worker exposure to chemical hazards and to prevent deterioration of respirator parts. Clean and disinfect respirators as follows:

- Remove all cartridges, canisters, and filters, plus loose gaskets.

- Remove elastic headbands.

- Remove exhalation cover.

- Remove speaking diaphragm or speaking diaphragm-exhalation valve assembly.

- Remove inhalation valves.

- Wash facepiece and breathing tube in cleaner/sanitizer powder mixed with warm water, preferably at 120°C or 140°F. Wash components separately from the facemask, as necessary. Remove heavy soil from surfaces with a hand brush.

- Remove all parts from the wash water and rinse twice in clean, warm water.

- Air-dry parts in a designated clean area.

- Wipe facepieces, valves, and seats with a damp lint-free cloth to remove any remaining soap or other foreign materials.

NOTE: Most respirator manufacturers market their own cleaners/sanitizers as dry mixtures in one-ounce packets. Respirators should not be cleaned in dish washers.

Repairs

Only a trained person with proper tools and replacement parts should repair respirators. No one should ever attempt to replace respirator parts, make adjustments, or repairs beyond the manufacturer's recommendations. Replace all faulty or questionable parts or assemblies. Only use parts specifically designed and approved for the particular respirator. Reassemble the entire respirator and visually inspect the completed assembly and tightly seal.

Storage

Manufacturers provide cleaning, maintenance, and storage instructions with new respirators. The following instructions may be helpful as well:

- After respirators have been inspected, cleaned, and repaired, store them in a location which will protect them against dust, excessive moisture, damaging chemicals, extreme temperatures, and direct sunlight.

- Do not store respirators in clothes lockers, bench drawers, or tool boxes. Place them in wall compartments at work stations or in a work area designated for emergency equipment. Store respirators in the original carton or carrying case when possible.

Respirator Program

Employers who are required by OSHA to provide respirators must have a written respirator program. This program should be evaluated at least annually or as requirements change. A respirator program must include the following as minimum requirements according to the Respiratory Protection Standard (29 CFR 1910.134):

- Written standard operating procedures for selection and use of respirators.

- Respirators must be certified and selected on the basis of hazards to which the worker is exposed.

- The wearer must be instructed and trained in the proper use of respirators and their limitations.

- Where practical, respirators should be assigned to individual workers for their exclusive use.

- Respirators must be regularly cleaned and disinfected after each use or more often if necessary. Those used by more than one worker must be cleaned and disinfected after each use.

- Respirators must be stored in a convenient, clean, and sanitary location.

- Respirators routinely used must be inspected during cleaning. Worn, deteriorated parts must be replaced.

- Survey employee work areas, conditions, and degree of employee exposures.

- Medical review.

- Evaluation of respiratory protection program.

NIOSH Recommendation

"Respiratory protection may be necessary for certain operations or methods such as paint removal by chemicals, heat gun, or abrasive techniques, and some set-up, and cleaning operations. However, respirators are the least preferred method of controlling airborne lead exposure, and they should not be used as the only means of preventing or minimizing exposures.

Respiratory protection requirements are not an acceptable substitute for adequate training, supervision, appropriate engineering controls, and environmental or medical monitoring. Initial respiratory protection requirements for abatement work (which may be based on conservative assumptions) should be modified with appropriate job-specific requirements based on air monitoring results. Respirator selection for each job category at every worksite should be determined by an industrial hygienist or other qualified individual, based on maximum airborne exposures measured."

Chapter 11

ERGONOMICS IN CONSTRUCTION

Charles D. Reese

Only in recent years has the construction industry become interested in ergonomics or at least acknowledged the existence of this new field. Because of the uniqueness of construction work and the lack of information which accurately describes the incidence of injuries and illnesses related to ergonomic factors, only small steps have been made to address ergonomic hazards on construction worksites. Although most construction operations recognize the large numbers of back-related injuries, few, if any, associate construction work activities with the other common cumulative trauma disorders (CTDs).

This chapter discusses back injuries and their prevention, as well as the nature and causes of other CTDs. These types of injuries and disorders will be linked to the types of job-related tasks which construction workers perform. Also, the chapter presents a discussion of the tools, equipment, and ergonomic devices available for intervention and prevention of cumulative trauma disorders, as well as providing investigative tools used for identifying the risk factors and causes of CTDs.

BACK INJURY PREVENTION

Back Injury Background

In a letter to the Secretary of Labor, Lynn Martin, the AFL-CIO's Building and Construction Trades Department President, Mr. Robert A. Georgine, stated: *"Back injuries are reaching epidemic proportions in the construction industry. American workers sustain more than a million such injuries a year. Low back injuries are common among carpenters, bricklayers, roofers, plasterers, cement masons, laborers, truck drivers, and others who are involved in manual materials handling. A recent survey of accident reports revealed that construction electricians suffer from an excessive rate of strains and sprains, most of which were connected with materials handling and unsafe worksites. A large percentage of these lower back injuries can be avoided through simple and inexpensive changes in worksite design and materials packaging.*

"Other back injuries are also traceable to ergonomic hazards in the construction industry. Operating engineers and truck drivers are exposed to whole body vibration, which causes progressive deterioration of the lower back region. Many worksites require tradespeople to adopt strained or awkward work postures to complete their tasks. Plumbers, electricians, boilermakers, laborers, and insulators incur upper back injuries as a result. Many of these workers are also at risk for thoracic outlet syndrome, which is caused by constant work with the arms in an elevated posture."

How serious is the problem of low back injuries in the United States? Back problems will be encountered by 80 million people in their adult life, which means that eight out of ten adults will be affected. About 7 million new cases of back pain occur each year. Twenty-five percent of all on-the-job lost work time is due to back injuries, which results in 93 million lost work days a year. Back injuries result in thirty three percent of all compensation costs. The economic cost of back injuries is about $14 billion each year. The average back injury results in $9,000 in direct medical expenses for each injury. Back injuries are the second most frequent cause for visits to a physician's office. (The first is the common cold and respiratory disorders.) We need to prevent low back injuries and, if we cannot prevent all injuries, we must have treatment interventions that are effective and efficient.

The lower back is the body site most frequently injured by construction workers. These back injuries most often occur in the form of sprains and strains due to overexertion. This information is not new to most who work in or are involved with the construction industry. Most construction trades involve handling materials or equipment as part of the normal work process. The work itself requires extreme force which is often combined with repetitive work in awkward positions – positions that are often prolonged (static exertion). Many construction tasks involve rapid twisting and bending of the back. Whole-body vibration transmitted through the seats in heavy equipment poses further cumulative risk of back injury.

The risk factors for back injuries are well known and have been evaluated over some period of time. Below are some of the factors.

1. Sudden or frequent twisting of the back.

2. Sudden or frequent bending of the back.

3. Strain of unused muscles.

4. Strain occurring from longer periods of sitting or standing.

5. Smoking is an aggravating factor.

6. Muscles of the trunk and back lose their strength and flexibility due to lack of exercise.

7. Poorly designed work areas and stations.

8. Work resulting in a jerking activity.

9. Material handling such as lifting, lowering, pulling, pushing, and carrying.

10. Using improper lifting techniques.

11. Carrying loads too heavy for the individual.

12. Stress or other types of illness.

13. Lifting or holding objects above the shoulders.

All of the above risk factors are found in the construction industry; thus, it is understandable that the construction industry is plagued with so many back injuries. How can the construction industry address this problem? Since back injuries cost time and dollars, it makes sense to develop a back injury control program. Most construction companies need a program which emphasizes training, materials handling, safe lifting practices, load lifting limits, buddy lifting, mechanical lifting devices, physical conditioning in the form of strength and flexibility exercises, vibration control, and the proper use of personal protective devices.

If you ask a group of construction workers if they have ever had a sore back, the predominant answer is "yes." There are large numbers of workers who have had injuries that have caused them to miss work, but only few have needed surgery on their backs. This seems to say that most back injuries are sprains or strains that will heal themselves, given proper care and time and not the very serious injuries which result in permanent back problems.

Most back injuries are strains or sprains of the muscles, ligaments, or tendons. These tissues are capable of healing (although ligaments and tendons heal slowly). In 80 percent of the cases, the worker will get better in a few days, or it may take as long as a few weeks. If you injure your back, the key to effective and efficient treatment is early activity. Keep as active as you can within the discomfort level. What this means is that you need to move as much as possible without re-injuring your back. Each of us have different levels of pain tolerance, thus some individuals will be able to get out of bed more frequently or walk further than others. Reasonable movement will facilitate the healing process.

Most people know how to lift. Perhaps the best prevention strategy is to plan the lift before carrying it out. Safe lifting techniques, which are applicable to field situations, are important. Remember to exercise, warm up, and use safe lifting techniques and, when lifting is required, follow these simple steps.

STOP!

Think about what will be lifted. Is it heavy, light, bulky? What lifting technique will be used? The lifting technique chosen will determine the risk of injuring the back. Lifting should be done with a straight back and, as the lift occurs, keep the curve in the small of the back. Use the leg muscles instead of the back muscles. Although back muscles are generally extremely strong, the anatomy of the back does not allow them to use this strength in an efficient manner; thus, back muscles and the associated tissues can be easily overloaded. The anatomy of the leg allows the thigh muscles to more efficiently convert the muscle contraction into lifting force. Always rely on the leg muscles and protect the back while keeping the load close to the body. The important thing is to **STOP** before lifting and think about the proper lifting techniques. In the following section there are pictures of the proper lifting techniques.

LOOK!

Look around the area. Is the ground uneven? Is the load in a cramped spot? Can walking be accomplished without tripping? Is it necessary to twist the body in order to lift? Look for all these things when planning the lift. A few seconds of attention to the items noted in STOP and LOOK could save a few weeks of back discomfort.

LISTEN!

Listen to your body, to coworkers, and to your knowledge of lifting techniques. If you feel awkward when lifting, it is probably not a safe lift. If help is needed from a co-worker, get it! If a coworker needs help with a lift, please offer it. Listen to what your brain is saying about the proper lifting techniques. Whenever possible, remember to use lifting aids. A few minutes of preparation could save a few days of back pain.

- Use proper lifting techniques. (see Figure 11-1.)

A

As you approach the load determine its weight, size and shape. Consider your physical ability.

B

Stand close to the object with feet 8 to 12 inches apart for good balance.

C

Bend the knees and get a firm grasp on the object.

D

Using both leg and back muscles lift the load straight up. Keep the object close to the body.

E

Do not twist or turn until the object is in carrying position.

F

Rotate body by turning your feet and make sure path of travel is clear.

G

To set the object down, use leg and back muscles and lower the object by bending the knees.

Figure 11-1.   Good Lifting Practices – Courtesy of the Bureau of Mines

- Lift naturally: bend your knees and use your leg muscles. (See Figure 11-2.)

Figure 11-2. Laborer setting down a load using his legs

- Position the beginning and end points of the lift as close as possible to waist height.
- Keep your back straight and keep the load close to your body. (See Figure 11-3.)

Figure 11-3. Carpenter carrying a board with his back straight and load close to his body

- Do not twist your back when you lift and move the load. Move your feet to position your body. (See Figure 11-4.)

Figure 11-4. The movement of a worker's feet is especially critical during shoveling

- Move slowly. Rapid twisting and bending are as dangerous as a heavy load.
- Make sure you have a solid, comfortable grip on the load.
- Minimize the distance the load is lifted and carried.
- Make sure the surface on which you stand and move the load is unobstructed and not slippery.
- Avoid repetitive lifting.
- Avoid lifting heavy or awkward loads. (See Figure 11-5.)

Figure 11-5. An example of heavy awkward lifting

- Get help if you need it! Ask for the help of a fellow worker, or use a mechanical device. (See Figure 11-6.)

Figure 11-6. A typical two-person lifting task

Preventive Actions

At times excessive force is used to accomplish the lifting of loads during material handling. This is often when back injuries occur. When a worker extends himself or herself over a load, the lower back is called upon to support the weight of the body and the load. Also, great force is utilized by construction workers when pushing and pulling loads. Some of the actions which can be taken to prevent back injuries and correct these types of situation are

- Use buddy lifting or additional workers.
- Use mechanical lifting devices.
- Make sure lifting devices are conveniently located.
- Add handles or lifting eyelets to such things as boxes, parts, and equipment which are to be lifted.
- Place materials in lighter or less bulky lots or packages.
- Redesign equipment so that the maintenance access and process will be easier.
- Make sure that the right tools are available to prevent slipping or the lack of leverage.
- Do not jerk on loads or tools.

There are times when the redesigning of equipment and processes can reduce the possibility of back injuries. These may include redesigning for a better fit for the worker or substituting human power with mechanical power such as conveyors, hoists, and hand trucks. Some actions which could be taken to mitigate back injuries are

- Improve seats on vehicles.
- Attach lifting hands or eyelets to panels or equipment.

- Place hinges on the side of covers rather than the top.
- Use nonskid materials on surfaces when possible.
- Develop lifting aids which are easily used and convenient.

Often the actual work practices are the culprits which cause or exacerbate back injuries. By reviewing your injuries you should be able to identify work procedures which have led to injuries. By addressing these work practices, you may be able to prevent further injuries by

- Making sure supplies are delivered as close to the point of use as possible.
- Not storing supplies on the ground which will require low lifts.
- Having good housekeeping as a priority on construction sites.
- Using tool pouches and carrying devices for small tools and parts.
- Minimizing the amount of stretching, reaching, and twisting in a work activity.
- Using enough workers for lifting and moving large items.
- Using mechanical lifting aids.
- Assuring material handling and lifting training are job specific.
- Encouraging workers to discourage fellow workers from making unsafe lifts.

Sudden movement often results or contributes to back injuries. Common examples of this are: pulling on a door that is stuck, trying to catch a shifting load, or slipping. These sudden movements can be anticipated and, thus, prevented. Some actions which can be undertaken to prevent sudden movement are

- Walk slowly and carefully in slippery, slushy, or muddy conditions.
- Avoid yanking on tools or loads.
- Provide both operator and passenger restraints (i.e., seat belts).
- Use the right tools for the job in order to avoid sudden slippage or releases.
- Mark areas where bumps or low spots exist in roadways and travelways.
- Identify and prevent the release of large forces or energy.
- Anchor vehicles and loads to assure no sudden movement occurs.

Good Health and a Healthy Back

Back injuries occurring among construction workers are increasing at alarming rates. The medical management of these injuries is, at times, not efficient and the medical costs are skyrocketing. The purpose of this section is to present material on the aspects of general health, the management of back discomfort and acute back injuries, the prevention of back injuries, and lifting techniques.

It is important to understand the relationship between good health and a healthy back. Good general health is directly related to having and maintaining a healthy back. General health is usually defined as having good eating habits, regular exercise (walking is great!), keeping off excess weight, and maintaining good posture. Back injuries generally happen in the adult years and are usually not permanent; most back injuries will heal. But healing depends on general health, as well as the nature of the injury. If one is healthy, recovery occurs much sooner than if one is unhealthy.

Health is frequently thought to be a state of mind. If an individual feels healthy, looks healthy, and maintains a positive attitude, that individual is usually considered healthy by him/herself and others.

Stress is an important contributor to back injury and many cumulative trauma disorders. It is important to recognize the stress that occurs both on and off the worksite. The first step in controlling stress is to recognize and acknowledge when we are stressed. If we are stressed, we must work on ways to decrease or eliminate as much stress as possible from our life and/or work situation. Stress management techniques are often helpful in reducing work and non-work stress. But many stressful work and non-work situations require action in order to change these situations. For instance, jobs that are characterized by high demands, by the worker having minimal control over the job, or by low levels of social support from coworkers, supervisors, and/or the company, can significantly elevate the risk of back injury. When these factors are present, changes are necessary in the work organization in order to reduce these causes of stress.

Good eating habits also contribute to your general health. FDA guidelines suggest most of us could benefit by increasing our intake of fruits, vegetables, and grains, while reducing the amount of red meat in our diet. But one does not need to be a vegetarian or a "health nut" to have and maintain good, sensible eating habits. Maintain a diet with the "basic four" food groups as part of your daily life: dairy, bread and grains, fruit/vegetables, and meat and fish. Reduce the amount of fats consumed, particularly "saturated" fats and fats of animal origin.

General exercise also helps to maintain a healthy back.

- On-the-job exercises are different from the general exercises one may perform at home or at a health club. Work activity at a jobsite is often in short bursts, sometimes very strenuous and repetitive, and sometimes easy. Generally, you should perform short periods of exercise or warm up before attempting work at the jobsite.

- Devote a sustained period of time (at least twenty minutes or more) to a general exercise program, in addition to on-the-job exercises or activity.

- Walking is a great general exercise. Walking uses every muscle in the body, including your heart. If the average individual can take a twenty-minute walk three times a week, he/she will feel better and be healthier.

Keeping off excess weight will also help keep the back free from injury. Excessive weight is similar to walking around all day with a ten or twenty pound bag of potatoes strapped to your stomach area. If that weight is lost, the result is feeling better and being able to move and lift things more easily.

Good posture maintains a healthy back. Keep shoulders directly above your hips. Even a slight stoop will increase the risk of having back problems. Your back has a natural inward curve just above the hips and an outward curve below the shoulders. This S-shape serves as a spring to absorb the normal shocks of walking, running, and jumping. As much as possible, try to maintain these spinal curves as you work.

Anatomy (Back Structure)

New Terms:

Muscle – Body tissue that moves the bones of the body.

Ligament – Body tissue that holds bones together.

Tendon – Body tissue that attaches muscle to bones.

Joint — The area where bones meet. Joints allow for motion and movement.

The anatomic structure of the back requires guidelines and bonds to support the back in an upright position. Muscles are the guidelines which help maintain stability and support; they have the ability to contract and are attached to the bones by tendons. Ligaments are the bonds which tie the bones together; these ligaments are not very flexible. The spine is made up of many bony structures (vertebrae) which are stacked one on top of another, held together by ligaments, and stabilized by several different contracting muscle systems. The joints between each vertebra allow for movement. Each joint has a very limited movement but, when combined, allows for significant motion of the spine. At each joint the vertebrae are separated by shock-absorbing discs; these discs are like small hockey pucks with a tough outer layer and a jelly-like center. Under the intense pressure of an improper load, this center material can push through the tough outer layer and press on the spinal cord (causing intense pain). These discs get much attention in the news and chiropractic offices, but it is very rare that the discs give way or rupture. This means that most back injuries do not result in spinal cord pressure (which may require surgery). Much more common, back injuries are muscle, ligament, and tendon disorders that can heal much like any injury.

The spine is surrounded by muscles that support, stabilize, and move the spine. These muscles are the abdominal muscles that are in front of the stomach, the muscles of the hips and buttock area, and the muscles along the side of the spine. They provide for the twisting of the trunk, as well as pulling it upright.

The muscles of the upper leg also help to support the back. These muscles control the hip motion and, to a certain extent, the back motion. The leg muscles take considerable strain off the back by allowing the legs to do the lifting. Strength and flexibility are required for the muscles that support and control the activity of the back. If the muscles are strong and flexible, they will help to keep the back in proper alignment.

Back Injury Management

When you experience an episode of serious back pain, first call a doctor to discuss treatment. When the back is injured, many workers feel as though they will never be free of pain. Take heart, over eighty percent of back injuries get better in a short period of time (i.e., from several days to a few weeks).

- Most back injuries involve the ligaments, tendons, or muscles. These tissues can heal, although sometimes healing may be slow.

- Remember the times when you banged your elbow or knee and you rubbed it and moved it in order to make it feel better. Back injuries are similar to this except the back has many more joints, muscles, tendons, and ligaments. With moderate and careful activity, this usually promotes healing.

- If the back is injured, the injured individual may move, but move gently and within the tolerable pain level.

Of course, some back injuries are more serious. A physician should be called if you experience any of the following symptoms:

- Constant pain—pain that does not improve somewhat over a period of days.

- Pain and/or numbness in the back of the legs or below the knee.

- Bowel or bladder trouble (for example, a bowel movement or urination occurs without your knowing).

- Any worsening in the discomfort level.

See the doctor whenever you have any questions about your health or back concerns. If the doctor recommends that you stay in bed, discuss with him/her how long you should remain at rest. If possible, stay just long enough for the back discomfort to be significantly reduced (i.e., a few hours at a time at the most), and then continue moving (carefully) within the discomfort level that can be tolerated.

First Aid: Taking Care of Your Own Back

Here are some things that each of us can do if we suffer a back injury.

- When in bed, lie on either side with a pillow under the head and another pillow between the knees. Bring the knees toward the chest, causing a bend at the hips and knees. Adjust this position until you feel reasonably comfortable.

- During the first three days of the injury, consider putting an ice compress on the area of the back that hurts. Take a thin plastic bag filled with some crushed ice, and place the bag over the affected area. Use the ice three to four times a day. Discuss with a physician the appropriate time to begin the application of heat to the area.

- If the doctor says that aspirin or other mild medications can be taken to reduce the discomfort and possible inflammation, take them according to the physician's suggestions.

- As the discomfort subsides, try to move the legs and hips in bed. Move slowly at first, then with slightly increased speed as you feel better. If you start to feel more discomfort, **STOP!**

- As you feel the muscles and joints relax and loosen up in the area of discomfort, try to get out of bed.

- To get out of bed, first roll onto your side so that your face is close to the edge of the bed. Bend the hips and let the bend in your knees reach the outside edge of the bed. Slowly let the legs slide off the edge of the bed, and use one arm to help push yourself to a sitting position. Now stand up.

- Try to stand as straight as possible; try to arch the small of your back (look at the ceiling). Now start to walk a little. Do not sit in an easy chair. Sitting can be harmful at the very beginning of a back injury. To eat, stand up or partially sit at the kitchen counter on a high type stool.

- Walk around the house, using as straight a posture as possible. Try to measure the distance walked (once through the living room and kitchen). Each time a walk is taken, try to match or exceed the previous distance walked.

- When you get tired, lie back down on your side and put ice on your back. Continue this ice treatment for the first three days after the injury. After that, standing in the shower with warm water running on the area of discomfort should help. A heating pad can also provide the same effect. Again, check with a doctor for the proper mix of ice and heat treatments.

- You should experience gradual improvement (i.e., less discomfort and more movement) as the days go on. If improvement does not occur, call a physician. As the back begins to feel better, start to walk longer distances. Walk outside, when possible, to provide a change in the environment.

- As much as possible, continue to avoid sitting. Use a reclining chair or lie on the sofa when watching television; this reduces the pressure on the spinal discs. If driving, use a lumbar support. A good thing to use as a lumbar support is a rounded piece of material, soft but not too soft, that is approximately five inches in diameter (e.g., a bath towel tightly rolled up.) A lumbar support helps keep the small of the back in the normal inward curved position. When sitting in the car, place this roll in the small of your back. Numerous other lumbar supports are commercially available. Many of them tie to the car seat or attach with Velcro strips.

- When going back to work, stop, look, and listen before lifting anything. Remember to get help from a fellow worker when lifting large bulky items.

- Continue walking as much as possible. Try to get up to twenty minutes of walking without stopping. Remember, walking at work is a stop-and-go type of walking and is not as useful for healing and exercise as continuous walking.

- Keep the physician informed of all that is done when taking self-care of the back.

Warm-Up and Stretching Exercises

Scientific research has not yet determined whether exercise programs are an effective prevention technique. However, many companies that have implemented exercise programs report that warm-up exercises have helped them reduce back injuries and other cumulative trauma disorders. The experience of athletes suggests that the body and its muscle groups have to be warmed up, just as an automobile engine needs warmed up before going down the road. Warm-up exercises prepare the muscles and tendons for activity by increasing blood flow and gently stretching tissues that will be required to do the work. In recent years, some construction companies have been requiring that all employees do stretching, flexibility, and warm-up exercises prior to beginning work. Jumping jacks, walking in place, standing knee lifts, and back stretches are all good warm-up exercises. This activity is not widely practiced in the industry, but certainly merits further consideration and implementation. Remember, workers are just like professional athletes; it is important to warm up their bodies before attempting any strenuous work activity.

STRETCHING exercises help keep the muscles from tightening up too much. Sometimes the muscles may tighten up a little at a time and workers may not even be aware of it. Stretching exercises are done to prevent major muscle groups from tightening up too much and should be done every day. Some companies encourage short stretching sessions (1-3 minutes) several times a day.

STRENGTHENING of the major muscle groups is very important for a healthy back. The stretching and warm-up exercises, noted previously, are designed to strengthen the muscle groups that support the back and keep it straight.

Use of Back Belts

The use of back belts to support the back during lifting has been both maligned as useless and touted as a panacea. The truth probably lies somewhere in between. The National Institute for Occupational Safety and Health (NIOSH) does not support the use of back belts at this time, but many companies that use these belts have seen not only a decrease in their back injuries rates but a decrease in the cost per back injury. When injuries do occur, they are not as severe as they were in the past. Now, do not get the idea that back belts are the answer to back injuries. It is possible that the studies which show back belt use to be effective may actually be measuring the effect of other factors. The companies which had success also had good safety and health programs, provided training in lifting and back injury prevention, and provided instruction on how to properly use the back belts. Many companies have instituted voluntary back belt use; this may be more effective, but no conclusive studies have been done (see Figure 11-7).

Figure 11-7. Mason tender wearing a back belt

Medical problems such as high blood pressure, previous back problems, and other conditions may make the use of back belts a greater risk for some individuals. These conditions should be considered prior to prescribing back belts for everyone. Also, back belts may only remind a person to lift properly and keep the back muscles warm; this seems to be an "ok" purpose, but it may also give workers a false sense that they can lift more than they should. In addition, some researchers worry that wearing the back belts all the time will cause the muscles to atrophy or "decondition;" this is a natural process when muscles are underutilized. It is the opposite of the process by which muscles increase bulk and strength during carefully performed weight training. If they are used, the belts should only be tight or tightened when actually lifting. Great care should be taken not to look to back belts as a cure or band-aid for a problem that should be solved by redesigning lifting tasks, decreasing the weight being lifted, or providing lifting devices to assist workers.

Preventing Back Injuries

A good safety program, combined with proper lifting aids and procedures, can prevent many back injuries. But sometimes back injuries can be caused by poor technique, inadequate lifting aids, loads that are too heavy, excessive production pressures, etc. If the back is injured, follow the recommendations found in the previous sections of this chapter.

Back Injury Prevention Program

When developing a back injury prevention program, analyze the previous back injuries and their causes, the activities at time of injury, the job classification, the locations, and any other contributing factors. Approach management for a commitment and policy statement regarding their concern and involvement in back injury prevention. Supervisors must be involved in analyzing the material handling tasks, redesigning jobs to fit the workers, training workers in safe lifting procedures, and monitoring the effectiveness of the program.

Evaluate the capabilities of workers and place them in work assignments that they can physically handle. Eliminate as many manual lifting and carrying tasks as possible and replace them with mechanical lifting aids. Establish safety standards that establish weight limits for manual lifting.

Conduct training sessions which include the company's history and the risk of back injuries from handling materials. Training should include the basic principles of manual material handling, as well as the effects of it on the human body. Workers need to know how to evaluate their physical capability for the weight they are required to lift, and make safe decisions when they know the load they are going to lift is beyond their strength.

Workers need to be taught how to avoid accidents. They need to be taught to ask the following questions of themselves prior to beginning a lift:

- Is the load free to move or is it stuck?

- Is it a weight that can be safely handled by one person?

- Are lifting aids available?

- Does the load have handles to grasp or can they be provided?

- Is protective clothing needed?

- Is the work area clear of obstructions?

- Is the floor clean, dry, and nonslippery?

- Is the area clear where the load will be set down?

Following the training, supervised observations of the workers should take place while they are performing hands-on lifting of the materials most often used by the company. At that time, the supervisor should make sure that the proper lifting techniques are being used by the newly trained workers; hands-on training is a way to verify that the classroom training has been transferred to the worksite. It should not be assumed that workers know how to use the mechanical materials handling devices; they should undergo classroom training, as well as be able to demonstrate the proper use of the devices used in the work environment.

The back injury prevention program (BIPP) needs to provide prompt medical care to insure early treatment and rehabilitation of any injured worker. The BIPP's results should be monitored and changed as the need arises. The BIPP should encourage wellness and personal physical fitness to strengthen the muscles and joint areas. It is appropriate to encourage warm-up exercises prior to a difficult lifting task; this can reduce the risk of overexertion injuries.

Summary

General health is important for a healthy back. Health is affected by many environmental and genetic factors, but each individual can take an important degree of control over his or her own health. To keep muscles healthy, consistent exercise is important. Walk at least twenty minutes three times a week; it is good for your heart and other muscles. Eat right and try to keep your body weight down!

If you injure your back, see your physician! Know the warning signs of potentially serious back injuries. Generally, back injuries involve the muscles, ligaments, and tendons. These structures heal, although healing may be slow in ligaments and tendons. Keep as active as possible during a back injury; stay in bed for short periods of time only to decrease the pain. Prevent the occurrence of a back injury at work by following company lifting procedures and practices.

CUMULATIVE TRAUMA DISORDERS IN CONSTRUCTION

It is almost certain that cumulative trauma disorders (CTDs) are under reported within the construction industry. The reason for this assumption is because of the known activities and risk factors involved. Construction work involves

- Constantly repeated motions,
- High levels of force.
- Unnatural or awkward body positions.
- Body positions that are held for long periods of time (static postures).
- Compression of body tissues against hard and sharp corners.
- Rapid movement of the body and body segments.
- Whole body vibration, as well as hand and arm vibration.
- Mental stress due to time pressure, low control of job processes, and problems with social support on the job.
- Fatigue due to inadequate recovery times; having days off or having frequent breaks are not the normal mode of operation on construction sites.

CTDs are best understood as diseases caused by repeated and accumulated exposure to the risk factors noted above. The "micro-traumas" (tiny breaks and tears) caused by these risk factors, if insufficient recovery time is not available, can result in serious damage to the musculoskeletal system. Thus, these risk factors affect the bones, joints, muscles, tendons, ligaments, nerves, and blood vessels. Due to overuse over a period of time, these injuries become very painful and debilitating; they often result in permanent damage.

Construction workers are required to frequently utilize their hands, arms, and shoulders in the performance of their jobs. It is no wonder that most of the CTDs suffered by construction workers are of these "upper extremities." Some of the most common CTDs which affect construction workers are tendonitis, thoracic outlet syndrome, tennis elbow (sometimes called "carpenters' elbow"), rotator cuff tendonitis, Raynaud's syndrome, golfer's elbow, trigger finger, ganglion cyst, and carpal tunnel syndrome. Certain construction workers are also at risk for disorders of the lower extremities (hips, legs, knees, ankles, and feet). Perhaps the most well-known disorder is "carpet layer's knee" which occurs when there is repetitive use of the carpet stretcher. But, anybody who works on their knees can also have this problem. Likewise, overloading on the legs, knees, ankles, or feet with repeated heavy loads can also be problematic.

The following are descriptions of the most common upper extremity CTDs which affect construction workers. Also included are summaries of the work activities which cause or exacerbate them.

Bursitis – When the fluid-filled sacs found in the joints of the shoulder, elbow, or knee (which help tendons and muscles move over boney areas) become irritated, these bursas can swell and become very painful. This disorder can occur when a carpet layer is using a knee kicker to stretch a carpet in place, or the shoulder is injured by prolonged overhead assembly work by sheet metal workers.

Tendonitis – This is the inflammation of the tendon(s) which attach muscle(s) to the bones and allow for movement of body parts. This inflammation occurs with continued use or the constant tensing of the muscle which pulls on the tendon. Small injuries to the tendons heal rather quickly, but the overuse and lack of recovery time can result in torn and frayed tendons; these can become a serious medical problem. These types of inflammation usually occur in the wrist, elbows, and shoulders but also can occur in the knees and ankles. Examples of work activities that can result in tendonitis are working with the hands above the shoulder or over the head (rotator cuff tendonitis); forceful extension, flexing, and twisting of the wrist; carrying heavy loads or repeated throwing of heavy loads which affect the shoulders; and repeated extension, flexing, and adduction (pulling towards the chest) of the arms.

Thoracic Outlet Syndrome – This occurs when the muscular and connective tissue structures of the shoulder and upper arm entrap and compress the blood vessels from the heart and nerves from the spine which serve the arms. The effects can be felt as numbness in the fingers and a weakened wrist pulse; this causes the arm to feel like it is "falling asleep." Thus, this condition is often misdiagnosed as Carpal Tunnel Syndrome (see below). Many unnecessary wrist surgeries have been performed because of this confusion. The risk for Thoracic Outlet Syndrome is increased by a number of activities: working above one's shoulders/head (see Figure 11-8); carrying heavy items, such as a concrete blocks; pulling the shoulders outward/back and up; and carrying heavy loads while using a shoulder strap. Research indicates that the greatest risk may come from these awkward shoulder postures, if they are held for some time (static postures) without moving. A few of the most common trades affected by this particular CTD are painters, sheet metal workers, laborers, welders, and truck drivers.

Figure 11-8. Mason laying block above shoulder height

Tennis Elbow (Lateral Epicondylitis) – This is a painful inflammation of the forearm tendons that attach to the outer side of the elbow. The inflammation is caused by a jerky throwing motion, strong wrist/finger extension (pulling the wrist and fingers away from a curled position), and is often in combination with the forearm/wrist twisting. Construction work often involves repetitive throwing of heavy loads which can cause or exacerbate this disorder.

Golfer's Elbow (Medial Epicondylitis also called "Carpenter's Elbow") – This is a similar and painful inflammation of the forearm tendons that attach to the inside of the elbow. It is caused or exacerbated by the repeated rotating of the forearm and bending of the wrist and fingers at the same time. Higher rates of incidence have been found among individuals doing wiring operations. Although tennis and golfer's elbow are named for sports injuries, most individuals who suffer these disorders have never played golf or tennis. Both of these often occur to workers who perform repetitive and awkward tasks and use poorly designed tools.

Rotator Cuff Tendonitis – This is the most common shoulder tendon disorder and is usually caused by repeated elevation of the arm above the shoulder. The condition may progress to what is called "frozen shoulder," which is a very painful loss of shoulder function. Workers performing sheet metal work, plumbing, painting, and drywall installation may have a higher than normal incidence of this disorder. (See Figure 11-9.)

Figure 11-9. Electrician in a work position which causes him to have to raise his arms
above his shoulders

Raynaud's Syndrome – This disorder is often called "vibration white finger" or "hand-arm vibration syndrome." It is caused by the forceful gripping and/or prolonged use of vibrating tools driven by electrical or pneumatic sources. There is a clear increase of this disorder when these vibrating power tools are used in cold weather conditions. The fingers become numb and turn pale and cold; the eventual result may be loss of sensation and control of the fingers and hands. Extreme cases also involve visible tissue damage. In construction, the use of powered vibrating hand tools, as well as working in cold weather conditions, is a very common practice (see Figure 11-10). Poorly designed pneumatic tools often contribute to the risk by directing cold exhaust air over the hands of the operator.

Figure 11-10. Example of a vibrating tamping tool

Tenosynovitis – The tendons that transmit force from the forearm muscles to the fingers (both closing and opening motions – flexion and extension) run through a special protective and lubricating sheath – the synovial sheath. With overuse the tendon sheath can become irritated and inflamed, which causes the increased production of synovial fluid. The sheath becomes swollen and painful and is sometimes accompanied by cracking sounds and some loss of function. Forceful hand motions, awkward hand postures, and poor tool design – often present in construction work – are risk factors contributing to this disorder.

Trigger Finger – This is a variant of tenosynovitis. With overuse, the synovial sheath can become so swollen that the tendon becomes temporarily locked inside. When a worker attempts to move the finger, it will often snap or move in a jerking manner. (This is actually a form of "stenosing tendonitis," which also involves the formation of nodes, or swollen portions of the tendon that catch when the finger is at certain positions.) This type of disorder is often caused by using tools with hard edges or ridged handles and/or repeated bending of the finger while maintaining a forceful grip.

Ganglion Cyst – There is yet another possible response to the overuse of the synovial sheath. The sheath can become irritated and filled with fluid (synovial fluid) to the point that the sheath swells and becomes a bump under the skin. This often occurs at the wrist and may be due to either a blunt force injury or repetitive and forceful gripping or opening of the hand. Construction workers are often pounding objects with their hands or accidentally striking body parts against hard objects.

Carpal Tunnel Syndrome – A major nerve (the median nerve) and nine tendons in their synovial sheaths pass through a narrow tunnel between the wrist bones and the strong ligament that stabilizes the underside of the wrist. If the tendons in their sheaths become irritated through overuse and awkward postures, they can become swollen and cause a narrowing of this tunnel (the Carpal Tunnel). The compression and crowding can cause pain, swelling, numbness, tingling, or burning sensations in the thumb and first three fingers (due to pressure on the median nerve). This, in turn, results in a loss of finger sensation and grip strength and may cause permanent disability of the hand and fingers. This is one of the most serious disorders, if allowed to progress to the point of disability; but, it is often over diagnosed (see Thoracic Outlet Syndrome, above). Repetitive and forceful gripping motions (flexing) of the fingers, combined with awkward wrist postures, are risk factors in this disorder. Construction tasks which may increase the risk of carpal tunnel syndrome are performing electrical work, caulking windows, and using a hammer. All of these tasks require bending, twisting, and straining of the wrist, hands, and fingers.

De Quervain's Disease – This type of tendonitis is an inflammation of the two thumb tendons that share a common sheath. The tendons become swollen and restrict the movement of the thumb. This disorder is often caused when a worker grips an object firmly and performs a wringing motion – hence the old name "washerwoman's thumb." Construction activities such as screwing pipe together or forcefully twisting are tasks which present risk of this disorder.

Any feeling of soreness could be a beginning symptom of a CTD. If these initial symptoms are disregarded, the injury could progress into a full blown case of CTD. When this transpires, much suffering, lost of productivity, and excessive medical costs may result. This is why it makes good business sense today to conduct a pain survey and encourage workers to report signs and symptoms as early as possible. This will allow for early intervention and prevention activities to be carried out. Even more important, construction employers and safety and health personnel need to take an proactive approach in dealing with potential CTD problems. Ergonomic committees should carry out job analyses throughout the company to identify areas of risk and develop solutions before problems occur.

Preventing CTDs

Some fairly simple measures can be taken to decrease or prevent CTDs. These include

- Use tools with smooth handles, rounded edges, long handles, and the proper diameter.
- Reengineer tools or insulating/isolating handles to decrease vibration.
- Set up the work area to minimize excessive stretching, bending, or other awkward postures.
- Rotate between tasks so that different muscles and body parts are being used.
- Decrease repetition rates, reduce force requirements, and reduce awkward postures, wherever possible.
- Take regular mini-breaks (about five minutes), to allow for some recovery time.
- Take longer breaks between shifts to allow for natural healing periods.
- Use insulated/vibration isolating tool handles to decrease impact shock and vibration.
- Workers should get up, move, and stretch whenever stress, strain, or pain is felt (micro-breaks of thirty seconds to one minute).
- The use of gloves can provide some protection from some forms of CTDs. For example, there are antivibration gloves which can reduce chilling of the hands. However, gloves should be used with caution (see above).

CONSTRUCTION TASKS AND ERGONOMIC RISKS

Heavy equipment operators (earth movers, cranes, etc.) are subjected to prolonged static work and vibration. Seat designs that support the lumbar and neck regions of the spine and are adjustable present significant improvement for operators. In addition, seats are now available that dampen most whole-body vibration from the equipment. Noise has always been another issue faced by heavy equipment operators and has been decreased somewhat by cab designs.

Excavation work requires lifting, walking on uneven surfaces, and heavy exertion as when soil is moved by hand and the worker repeatedly climbs into and out of trenches. Once the concert footers and retaining walls are poured, the forms are removed by hand and the soil is back-filled and compacted by motorized tampers. These transmit extreme vibration to the hands and arms. Seldom are ergonomically designed shovels used when digging or moving soil. Little attention is paid to the vibrating tools in use (see Figure 11-11).

Figure 11-11. Worker using a pneumatic jack hammer

Masonry work usually involves hand carrying and the lifting of blocks, bricks, and mortar, as well as the changing height of the work area. This type of work requires much lifting and stooping, which puts stress on the lower back (see Figure 11-12). Many practical solutions are available, including the use of adjustable scaffolds, the pumping of mortar to work platforms, the use of cranes or mechanical lifts to get materials to work areas, and the use of lift tables for materials.

Figure 11-12. Typical bricklayer work and work area

Those workers who erect structural steel are at high risk for strains and sprains. Steel erection workers are very much like circus performers. While they erect these steel structures, they are forced to be in awkward working postures and also have to use force when bolting the steel members together as well as placing themselves at risk due to performing their jobs at heights. In this process there is a lot of squatting and kneeling as they weld, use pneumatic tools, and secure beams (see Figure 11-13). All of these factors can contribute to sprains and strains, but little has been done over the years to mitigate these risks in either tool design or work practices.

Figure 11-13. Steel worker bolting a steel member into place

Erecting structural steel is usually followed by the installation of steel decking and preparation for pouring the concrete floors. Also, the reinforcing rebar has to be tied together with wire. This is an operation that requires continuous bending and squatting while cutting and twisting wire to make the ties. Not only are awkward and static postures a problem, but the continuous use of cutting pliers and the twisting of wire around the rebar causes strain on the hands, wrists, and arms (see Figure 11-14). Other trades suffer similar work conditions.

Figure 11-14. Construction workers tieing rebar

Heavy use of hand tools increases the injury risks to sheet metal workers. Work positions for these workers may range from below the knees to over the head. Through forceful and awkward postures, the use of metal cutting tools has the potential to harm the hands, wrists, and arms. Also, powder-actuated tools (Hilti guns) can cause vibration and impact-related disorders.

When pouring concrete floors, the heavy concrete must be moved by using a rake. Furthermore, pumping concrete, using mechanical finishers, and using hand-held concrete vibrators, all present risks associated with vibration (see Figure 11-15). Much of the finishing work, such as troweling, is done by hand. These all result in awkward postures and the forceful use of hand-held tools.

Figure 11-15. Workers performing hand troweling while worker in the background uses a mechanized trowel

Roofing is usually a highly manual task which involves heavy materials and the risk of back injuries are quite high. Many times the workers are required to carry and lift the roof materials. The use of buddy lifting should be fostered, as well as the use of dollies and material handling devices. Another area of this work where risk of injury may occur is when the worker is standing on slanted surfaces; this places an uneven strain on the muscles of the legs. (See Figure 11-16.)

Figure 11-16. Roofer working on slanted surface

Plumbing and the installation of pipe for heating, cooling, water, and the fire suppression systems is usually done in cramped spaces and often over head. Thus, for plumbers, neck and shoulder strain is often a risk. Much of the pipe must be welded; this requires the worker to bend over for long periods of time. These strained work positions can be partly mitigated by better architectural construction designs which take into account the work positions required of builders.

Similar to plumbing, electrical work is often done in cramped spaces. It also requires the continuous pulling of wires which places force and posture stresses on the arms and shoulders. Electricians are constantly making connections which require fine motor skills and extensive use of hand tools. This also produces risks because of awkward postures, high force requirements, and compression of hand tissues.

Although the installers of floors, walls, and ceiling materials are not usually cramped for work space, they often handle heavy materials and struggle to get them into position. They may work on their knees, as well as over their heads. Each task has its own impact and stress on the various body parts. The floor installers are constantly on their knees, while the ceiling installers use their shoulders and arms as they work on scaffolds and support heavy guns to anchor ceiling materials in place.

Installing the exterior covering of buildings is done from fixed or suspended scaffolds. This type of activity, whether it be bricklaying, placing metal cladding in place, or facing a building with granite cladding, results in very awkward material movement, storage, and installation. Much of this activity requires extensive heavy lifting and placement of materials on less than stable work platforms. Window installation poses similar problems.

Workers doing the finishing work on the interior of a building have similar risks of CTDs, due to the unique tasks of this job. Examples are using poorly designed knives to cut dry wall, using knee kickers to tighten carpet, using stilts to plaster, using rollers and brushes to paint, installing ceiling fixtures, and hanging heavy doors or sealing joints (see Figure 11-17).

Figure 11-17. A sealer working on his knee in a strained posture

These risks are related to the use of hand tools, awkward and static work postures, the change of working levels, high force requirements, rapid motions, vibrations, and lifting. The expected injuries are to the back, hands, arms, shoulders, and knees. These types of ergonomically related injuries can be addressed by correcting workplace design, changing work practices, using newly designed tools, and using material-handling devices. By taking these steps, you may alleviate and, ideally, prevent many of these injuries.

TOOLS AND THEIR DESIGN

Hand Tool Design to Reduce CTDs

With pounding, twisting, static postures, vibrations, and the force involved in using construction hand tools, it is not surprising that soft tissue injuries occur. Since prevention of CTDs is critical, contractors and construction companies should begin to utilize new tools that help prevent some of these injuries. Tools should be designed so that the body parts are kept in a neutral position and grip forces are minimized. The most neutral position for the arm and wrist is the upper arm in a position down the side of the body with the forearm at right angle and the wrist/hand in a handshake position. This can be accomplished by bending the tool handle (not the body part, such as the wrist) and modifying the grip characteristics.

When considering tool design, remember the basic risk factors that interact and cause or exacerbate Cumulative Trauma Disorders.

- Repetition.

- High force requirements.

- Awkward and/or static postures.

- Rapid movement of body parts.

- Compression of sensitive tissues.

- Vibration, especially in cold conditions.

- Stress.

- Fatigue; lack of sufficient recovery time.

Obviously, in construction work the risks depend to some extent on work practices and techniques. Proper tool maintenance is also a crucial part of an ergonomic prevention program; properly sharpened tools accomplish the task with fewer repetitions and less force. Also, well-maintained power tools reduce exposure to vibration. But it is also possible and important to build features into tool design that will reduce these risks.

Repetition can be reduced by designing a tool that is well balanced and makes the most efficient use of the operator's input force without increasing the other risk factors. A good example of this is a well balanced hammer that drives a nail home in fewer strokes. The development of the hammer drill (if the handles are well isolated to reduce vibration transmitted to the hands) allows drilling through concrete with fewer repetitive hand motions.

Awkward postures can be reduced by designing tools that allow the body to adopt a "neutral posture" during as much of the work cycle as possible (see Figure 11-18). The "ergonomic" rake and snow shovel are good examples of this. The bent handles of these tools reduce the need for severe bending of the back. Likewise, if a hand tool requires constant bending of the wrist, it is possible to design a tool that is bent. The downward bend of some hammer handles allows the wrist to maintain the most comfortable and neutral hand position: that of the "handshake." Also, overhead arm postures have been reduced by the development of drills and drivers with extra long shafts. This allows the motorized part of the tool to be held at waist to chest level while drilling or driving overhead.

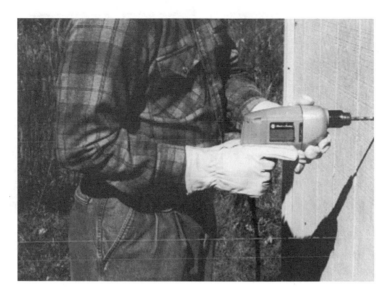

Figure 11-18. A worker with a 90-degree bend of arm and a straight wrist is in a neutral position

Static postures (body parts held in one position for a long time) can be reduced by designing tools that reduce holding and positioning demands. Tool designers have developed a wide variety of holding and clamping devices; these devices allow the worker to stop using one hand as a clamp or vise that requires them to hold the work piece while the other hand performs the work. The development of self-feeding screw guns frees one hand from holding the screw in place while bringing the screw gun into position.

Force demands can be reduced in a number of ways. The primary method used to reduce grip and holding force is by reducing the weight of the tool. The maximum tool weight for one hand should be maintained below four pounds. If the tool is very heavy, many devices are available to support or suspend the tool, thus removing a portion of the weight from the operator. It is better if the tool has handles which provide for two-handed use. This reduces the grip force required from each individual hand. Also, a second handle for supporting power tools should be reversible; this allows for both left- and right-hand use. Triggers/actuation switches on power tools should be long enough to allow for the use of two or three fingers in holding the trigger/switch closed for prolonged periods of time during use. Compact and well-balanced tools reduce the need for high grip forces; for instance, a drill with a pistol grip close to the tool's center of gravity requires less force to keep the head from dropping down.

Handle diameter is also a crucial aspect of tool design. The hand exerts maximum grip force on a handle with a diameter around $1\frac{1}{4}$ to $1\frac{1}{2}$ inches, depending on the worker's hand size. Handles that are fatter or thinner require greater muscle force to grip. Ideally, the use of interchangeable handles or grips allows each worker to select the diameter that gives maximum efficiency of grip force. Handles should be long enough to prevent undue pressure upon the soft tissues of the hand or other susceptible body parts (see Figure 11-19). Slip-resistant handle materials reduce the grip force needed to stabilize the tool. If this material has good insulating properties, it reduces the need for wearing gloves. The wearing of gloves can reduce the efficiency of the grip by up to 30% and means the worker must exert a higher grip force to prevent slipping, thus increasing the risk of CTD. Finally, handle contour can also reduce grip force requirements; hence the flare at the end of many modern hammer handles.

Figure 11-19. A handle on a screwdriver which fits the individual's hand and prevents soft tissue injuries

Operating force requirements can also be reduced by tool design. A well-designed drill bit can drill a hole faster and with less pushing than a poorly designed bit. The shape and angle of saw blade teeth dramatically affects the force required to push a hand saw or power saw through the work. Nail removal and board-bending tools should be designed to provide the best leverage (mechanical advantage) to reduce operating force. To distribute the force requirements between arms, it is helpful to have a tool designed for left- or right-handed use. This also reduces the extra force and postural stress on left-handed workers. It is easier for a worker to close a pliers-type tool since the muscles which close the hand are stronger. It is, therefore, a good idea to have springs which assist in opening hand tools. Finally, the addition of torque absorbing fixtures to driving and drilling tools can significantly reduce requirements for both grip force and operating force (see Figure 11-20).

Figure 11-20. Typical power tool which can be used with either hand by changing the handle and also using both hands to prevent torquing

Mechanical compression risks can be reduced by handle design. In order to prevent hand injuries, sharp corners and edges should be rounded. Also, tools with ridges running the length of the handle may not slip as easily but expose the hand nerves and vessels to possible compression damage. Likewise, contoured handles with ridges that fit between the fingers may be easier for some workers to grip, but may pose danger of compression to workers with smaller or larger hands. The length of the handles should be at least four inches without gloves and five inches when gloves are worn (see Figure 11-21). This keeps the end of the handles from damaging hand tissues.

Figure 11-21. The use of gloves requires longer handles since grip strength is also reduced by gloves

Many tool manufacturers now produce power tools that dampen vibration. Vibration-absorbing couplings on handles reduce vibration transmission to the hands. Well-designed tool handles (see above) that reduce grip force requirements also reduce the transmission of vibration from tool to hand. Handles of impact tools (such as hammers) can be designed to significantly dampen the transmission of impact shock from the handle to the hand and arm.

Serious consideration should be given to purchasing ergonomically designed hand tools when replacement tools are purchased. Properly designed hand tools can go a long way toward reducing the occurrence of CTDs.

Existing Equipment/Tools and Their Applications

In construction, materials and tools are always in motion. Since much of this material is heavy and/or awkward to carry, the use of cranes, mechanical lifts, forklifts, lift tables, and other material-moving devices is extremely important.

The use of battery powered screw drivers and pneumatic or electrical hammers can have the effect of reducing stress on the hands, wrists, arms, and shoulders, but some of the powered tools may cause other types of ergonomic stress, such as vibration. This places the worker in the "damned if you don't and damned if you do" predicament. These new risk factors need to be weighed against the risk factors of the nonpowered tool. For different applications, you might select the hand tool or the powered equivalent, depending on the balance of risk factors in that particular task.

Hand tools are not the only aids which are in use on construction sites. There are handles which are designed to aid in the carrying of 4×8 pieces of plywood or other such materials. There are also handles with straps which can be attached to equipment or to materials which do not have handles.

When surfaces permit, wheeled carts for transporting dry wall or other materials should be used. The faithful hand truck also finds a multitude of uses in the construction business (see Figure 11-22).

Figure 11-22. Typical hand truck

Using tool boxes with wheels and using rolling scaffolds (not only as work platforms, but also to transport tools and materials) will eliminate the need to manually transport items. Wheeled carts can be used for the transport of hot tar for roofing and for compressed gas cylinders.

Some of the portable lifting devices are designed to lift materials (see Figure 11-23) and improvements have been undertaken to use such devices for the handling of drums.

Figure 11-23. An example of a material lift

Some other lifting aids are portable gantry cranes, roof derricks, and boom floor cranes (see Figure 11-24). These types of lifting devices decrease the need to manually handle materials and equipment.

Figure 11-24. A floor crane

When possible, the palletizing of materials makes the use of industrial lift trucks (forklifts) or pallet trucks possible (see Figure 11-25) for the safe lifting and movement of materials. Although lift tables are not in wide use in the construction industry, there are times when they could be appropriately used. Their most logical use would be for concrete blocks and bricks during masonry work or for unloading bagged material onto a conveyor. Even though to this point conveyors have not been mentioned, they decrease the volume of materials which the worker needs to handle.

Figure 11-25. Palletized concrete blocks being moved by a pallet truck

CONDUCTING ERGONOMICS ASSESSMENTS

At times it is necessary to take a systematic look at the workplace, work pieces, and work process to determine if risk factors exist for cumulative trauma disorders. These assessments are a formalized approach to problem solving and usually generate a written record of the evaluation findings. There are many different assessment instruments available; they are the symptom survey, risk analysis, hazard identification, workstation feature checklist, task analysis checklist, tool analysis checklist, and material handling checklist. Examples of each of these assessment instruments are found in the NIOSH Guide on "Elements of Ergonomics Programs" and in Appendix H.

ERGONOMICS REGULATIONS

At present no federal ergonomic regulations exist. Thus, few employers feel a strong motivation to address ergonomic issues since they already have a large number of existing regulations with which to comply. However, it is becoming clearer to a larger number of employers that there are significant costs associated with CTDs: Workers' Compensation costs, reduced productivity, increased turnover, recruitment, and retraining costs. Many large employers have voluntarily introduced ergonomic programs into their organizations. As a result of these practical "experiments," a large amount of information is now available. This information demonstrates the effectiveness of ergonomic programs and the financial savings involved, as well as the companies improvement of employee safety and health.

ERGONOMICS PROGRAMS

If you need to formalize an ergonomic program, that program should contain the same key elements found in other safety and health programs discussed in this text. The very first item you should consider is obtaining management commitment in supporting the program, both in deeds and dollars. Once this is accomplished, you will need to collect data on previous ergonomically related diseases and injuries. Also, you will need to conduct job and task analyses to identify high risks or potential hazards for workers performing the job/task. This sets a baseline, as well as identifies targets. This ergonomic hazard identification effort has been most successful for many operations when the ergonomic team they use is composed of a combination of ergonomic experts, safety and health personnel, supervisors, and workers who have been trained.

Next, identify the possible solutions and controls to the problems. Once the feasibility of the solution and controls has been determined, the program should be implemented, monitored, and adjusted as needed; then, evaluate the effectiveness of the program.

Of course, any safety and health program initiative often results in at least some basic hazard avoidance training and at times, the ergonomic team and supervisors may require more detailed training. Implementation and follow through will be the cornerstone of the program's success and this success can be measured by the number of ergonomically related disease and injuries that are prevented.

SUMMARY

If your employees are experiencing back injuries or cumulative trauma disorders, it is to your benefit to take action. When these types of injuries and diseases are not addressed quickly, they often develop into serious and costly medical problems. Thus, prevention and early intervention are the keys to curtailing a potentially costly situation which not only affects medical costs but is a detriment to production as well.

REFERENCES

U.S. Department of the Interior, Bureau of Mines, *Back Injuries (IC 8948),* Pittsburgh, PA, 1983.

U.S. Department of Labor/MSHA/National Mine Health and Safety Academy, *Back Injuries in the Mining Industry (Safety Manual No. 24)*, Beckley, WV, 1992.

U.S. Department of Health and Human Services/PHS/CDC/NIOSH, *Elements of Ergonomics Programs,* Cincinnati, OH, March, 1997.

U.S. Department of Health and Human Services/PHS/HCPR, *Understanding Acute Low Back Problems (Pub. No. 95-0644),* Rockville, MD, December, 1994.

U.S. Department of Health and Human Services/PHS/HCPR, *Acute Low Back Problems in Adults: Assessment and Treatment (Pub. No. 95-0643),* Rockville, MD, December, 1994.

McKenzie, Robin, *Treat Your Own Back,* Spinal Publications, Ltd., Waikanaw, NZ, 1985.

National Safety Council, *Ergonomics: A Practical Guide (2nd Edition),* Itasca, IL, 1993.

Putz-Anderson, Vern, *Cumulative Trauma Disorders: A Manual for Musculoskeletal Diseases of the Upper Limbs,* Taylor & Francis, Inc., Bristol, PA, 1988.

Chapter 12

CONSTRUCTION SAFETY AND HEALTH MANUAL

Charles D. Reese

Many construction contractors have developed a construction safety and health manual which is distributed to their workers. These manuals have been used to set the company's safety and health (S&H) policy and procedures, define requirements, spell out specific rules regarding hazards, and convey specific company rules and expectations.

The S&H manual is of no use unless it is a living document that is used for training or tool box sessions and reinforcing safe operating procedures. These manuals should not just be issued and then expected to be used; rather, they need to be an integral part of a company's program.

These types of manuals should not try to cover all the bases, but should be company or site specific. The manual should address only those hazards which will be faced by your workers. Brevity, conciseness, and clarity are essential to the effectiveness of these types of manuals or they will not be used or read.

Contractors should realize that all of the OSHA construction safety and health regulations (29 CFR 1926) will not apply to their type of construction work since highway paving contractors' and residential building contractors' workers are not exposed to the same hazards. Thus, only those safety and health rules, regulations, or safe operating procedures which are applicable to the tasks and type of work being performed should be included within their S&H manual.

When developing a S&H manual the following contents should be included:

1. The cover.

2. A statement of the company's safety and health policy.

3. Either a table of contents or an index at the end of the manual.

4. The responsibilities regarding safety and health.

5. Specific policies and procedures of the company.

6. The cardinal safety and health rules of the company.

7. Special programs, high hazard initiatives, or unique permit required activities.

8. The rules or SOPs governing the specific hazards faced by the company's employees.

9. Any reference materials, diagrams, or drawings needed.

10. A certification that the employee has read the S&H manual and received a copy.

For a construction company, a S&H manual can be the foundation or visible portion of the jobsite safety program. This is possible only if the manual is kept updated, used for training, used for tool box meetings, and accepted as the company's stance on job safety and health.

COVER

The cover should include the name of the company, the title of the manual, date produced, a logo or graphic, and the company address. An example is found in Figure 12-1.

DIG EXCAVATION CONTRACTORS, INC.

SAFETY GUIDE AND JOB RULE MANUAL

JANUARY, 1999

5 FT. TRENCH DRIVE
SOIL, EX 63450
(680) 820-5133

Figure 12-1.   Sample cover for a safety and health manual

SAFETY POLICY

DIG EXCAVATION CONTRACTORS, INC. has made a serious commitment to safety and health. In addition to humanitarian reasons for preventing personal injury, increased awareness and attention to safety by everyone is essential to the economic health of this company.

This manual is issued to identify certain safety and work rules that are mandatory for <u>all</u> individuals on our jobsites. Following these rules will contribute significantly to your well-being and a successful completion of our construction projects. These rules will be enforced and violation of these rules can result in termination.

Accidents can be prevented through use of good judgment and common sense. By following safe practices along with all of our safety regulations and applicable OSHA standards, you can help us in making our projects a safe place of employment for yourself and co-workers.

Remember, everyone loses when an accident happens. We are concerned with your safety. Help us keep you safe! If you don't know the safe way, we'll teach you. If you won't work safely, you won't work here.

Mr. R. T. Safety
President
XYZ Construction Company

RESPONSIBILITY AND ACCOUNTABILITY

The responsibility for safety and health should be assigned. This does not relieve other company officials, managers, supervisors, forepersons, or workers of their responsibility to the company's safety and health rules. Everyone should know who has accountability, responsibility, and authority for safety and health on the jobsite. A statement to that effect should be included within this manual.

A possible statement of responsibility might be written as follows:

The construction manager is responsible for the safety of all employees. The manager will require his forepersons or supervisors to comply with and enforce all safety rules and regulations. It is a basic policy of the company that appropriate disciplinary measures be taken for the violation of safety rules or use of unsafe practices regardless of the employee involved. Disciplinary action for the infraction of safety regulations will be decided by the construction manager whether it be a foreperson or the recommendation of a foreperson.

The forepersons are the key to promoting safety on the job. They are directly and personally responsible to the construction manager for the safety of all employees within their jurisdiction. They must be fully conversant with all safety rules and must set the example for all their personnel. They must make clear to the employees the company's requirements that they work safely at all times and that the violations of this requirement may be cause for termination. Where safety rule violations occur, the foreperson may exercise independent judgment as to whether the employee involved will be removed from the job temporarily until a decision can be made as to what action is warranted.

The forepersons's principal duties are to

1.   *Train new workers on the safe operating procedures on the job, thoroughly familiarize them with the specific job and the company's safety rules, and perform follow-up at periodical intervals.*

2.   *Reinforce the principals of safe operation to all workers in the crew and coach them that to think safety is to act safely.*

3.   *Be constantly watchful for unsafe conditions, unsafe acts, and upon detection, take immediate, positive corrective action.*

4.   *Request the necessary safety equipment and protective devices and insist that the workers use them regularly and assure the equipment is maintained in good condition.*

5.   *Keep tools and machines in proper repair and adjustment.*

6.   *Participate in the company's safety and health program.*

7.   *Conduct safety meetings as needed.*

All employees are required, not requested, to work safely. Each must know the safe way to do their job and the safe way to conduct themselves during working hours. The employees can become a valuable part of the company's safety program through safety suggestions and assisting their coworkers to work safely. Safety can only be obtained through complete participation and cooperation of each employee.

TABLE OF CONTENTS OR INDEX

Handbooks or manuals should have a table of contents or an index to act as a quick way of finding information. Since there is no standard for organizing manuals like this, some mechanism must exist to identify the location of materials within the manual.

COMPANY POLICIES AND PROCEDURES

This manual may address such topics as equal employment opportunity, sexual harassment, Americans with Disabilities Act, and integrity. It will often include job rules not specific to safety and health but critical to the total jobsite and ones that may have interrelated affects on job safety. The job rules often impact identification and security, hiring or referrals, parking policies, hours of work, checking in or out, pay periods, a full day's work, absence from work, remaining in the work area, visits to the doctor, coffee, smoking, vending machines, drinking water, sanitary facilities, and the issuing, care, and use of tools and of material and equipment.

Some examples, as well as representative language for some of the previously mentioned job rules, are as follows.

Parking Facilities and Site Access

At most projects, parking facilities are provided for your use and at your own risk. Please park in designated areas and respect your coworkers and their automobiles in the parking lot. Enter the jobsite through designated entry gates and proceed to your work area by the most direct route.

Identification

Badge numbers are often required for gate clearance, payroll, and time keeping identification. Workers should wear their company badge in full view at all times.

Driving Rules

Safe driving rules are strictly enforced. Two violations of the driving rules will result in loss of driving privileges or termination.

Security

Security rules on the job may require that a lunch box or electronic detection check may be conducted at the gate. Refusal of any employee to submit to such checks should be cause for termination.

Employment

The project managers or their authorized representatives will do all hiring for the job They will hire individuals who, in their judgment, are able to perform a full day's work and who will work at all times as directed. The company will maintain personnel records and will hire in accordance with these records.

Hours of Work

The regular workday and lunch period will be determined by the contract in effect for each project. The workday should start and end at a time designated in advance by the company. Generally, the normal workday and work week will be announced at the pre-job conference subject to later change by management.

Most projects are scheduled on a non-overtime basis. This does not mean employees are guaranteed their regular hours of work, as there may be time loss due to weather, etc. There may be occasional overtime; however, this will be on an "as needed" basis for the portion of the project and only for the number of workers of a craft or work team that is required. No overtime work is anticipated for most projects; however, if it becomes necessary, the trades involved will be notified prior to the overtime being worked and proper arrangements made.

Workers must be at their appointed work locations ready to work at the regular starting time and remain at work until the normal quitting time. Sufficient time will be allowed for workers to pick up their tools on the company's time. Loitering in the change room or other late starting of work and early quitting will not be permitted.

Checking In and Out

Based upon the project, a badge may be used for check in and out at starting and quitting time. Each worker is responsible for their own check in and out.

Workers arriving late must sign in at the gate prior to entering the jobsite. Workers authorized to leave the jobsite during working hours must sign out at the gate prior to leaving.

Payday

The payroll week and payday will be determined by the contract and agreement in effect for each project. Payment will be by check. Employees absent from work on their regular payday will be paid on the first day they work following such absence.

Request for checks to be direct-deposited or mailed to any worker must be in writing and signed by the worker. Checks for workers who are terminated for unexcused absenteeism will be mailed to the address shown on the employee's personnel file at the time of termination.

A Full Day's Work

Each worker is expected to perform a full day's work. Workers should not restrict production or interfere with others in their job.

Workers with poor performance, as indicated in the rules of conduct, may be subject to termination.

Issuing, Care, and Use of Materials and Equipment

Company tools will be issued on a receipt system. Tools must be properly cared for. A tool room clearance will be required upon termination. Loss of or damage to tools will be noted on the worker's record.

Visits to the Doctor

Workers who are injured and require a doctor's treatment may return to work with a written doctor's release. Light duty may be arranged with a doctor's permission and as determined by the responsible Project Manager or their representative.

Coffee, Smoking, and Vending Machines

Coffee making or cooking will not be permitted on the job. Smoking will be permitted in designated areas. Vending machines will not be provided.

Any of the client's facilities, such as vending machines and sanitary facilities, are not to be used unless written permission has been received from the client.

Drinking Water

Cool, clean drinking water will be available at all times. Used cups will be disposed of in the containers provided for this purpose and are not to be discarded on the jobsite.

Sanitary Facilities

Adequate sanitary facilities are provided on the jobsite for all workers. It is requested that workers assist in maintaining these facilities in a clean and orderly condition.

Remaining in the Work Area

Workers must remain on the jobsite at their work location at all times during regular working hours, unless authorized to leave by their supervisor.

Absence from Work

The company performs work on large construction and plant sites. The work is periodic and must be performed on a very prompt basis. Employees must be ready and able to work at all times. Unexcused absences are not permitted. A worker with repeated or unexcused absences will be subject to termination. Unexcused tardiness is also not permitted and will subject a worker to possible termination.

This section may also include rules and policies regarding enforcement and consequences of violation of job or safety rules. An example of the content of job conduct and outcomes of violating these rules is as follows:

A Category-One Violation

A worker terminated for any of the following reasons will not be eligible for rehire. These workers will not be referred to other contractors for employment. The category one rules are

1. Willful or gross negligence of any safety, fire, security, sanitary, or medical rules or practices or engaging in any conduct which tends to create a safety hazard.

2. Theft of company's, client's, or other worker's property.

3. Possession of or being under the apparent influence of alcohol or any illegal or nonprescription drugs while on the project.

4. Possession of firearms or other weapons.

5. Insubordination or refusal to accept a work assignment.

6. Assault on supervisory personnel.

7. Making false claims or falsifying any reports or records.

8. Fighting on the project.

9. Acts of sabotage or defacing buildings, facilities, or equipment.

10. Picking up or using another worker's time card, employee number, or other identification.

11. Leaving the company premises without permission.

A Category-Two Violation

A worker terminated for violation of a category-two violation will be ineligible for rehire for a period of six months. If another category-two violation occurs within one year, the worker will not be eligible for rehire by the company. The category-two rules are

1. Engaging in horseplay.

2. Sleeping on the jobsite.

3. Eating, drinking, or smoking in any area where prohibited.

4. Gambling, selling items, or holding raffles at any time on project's site.

5. Loafing or unsatisfactory work on the project.

6. Late starts and early quits.

7. Failure to use or wear personal protective equipment.

8. Violation of a safety or security rule.

9. Driving a vehicle during a period of driving suspension.

10. Violation of the non-discrimination or sexual harassment policy.

These preceding rules are only examples and should be tailored to meet the needs of the company. Construction companies should assess their own work-rule requirements and either add to or delete inappropriate rules from the previous list.

SOME CARDINAL SAFETY AND HEALTH RULES

Some safety and health practices should be followed by everyone. These might be called cardinal safety and health rules. Some of them are as follows:

1. Wear and take care of personal protective equipment.

2. Do not wear jewelry in hazardous areas (e.g., moving machines, electrical) which can hook or catch on protrusions.

3. Don't use drugs and alcohol on the job.

4. Avoid activities which create a hazard (e.g., horseplay, scuffling, or practical jokes).

5. Report to your supervisor any unsafe work conditions or practices, any accidents, injuries or illnesses, and toxic chemical exposures.

6. Make sure all safeguards are always in place and never removed.

7. Do not use any damaged hand or power tools and remove them from service.

8. Follow all regulatory requirements as well as employer's safety and health rules.

9. Keep work areas free of debris and organize and maintain good housekeeping practices.

10. Never work on electrical circuits or equipment unless properly trained and qualified. Never touch loose or broken wires.

11. Follow good lifting practices when handling materials.

12. Do not operate any equipment or machinery or perform any work task unless trained and qualified.

13. Stay alert to hazards which exist around moving vehicles. Do not hitch rides on moving equipment, know the traffic patterns, and listen for warning signals.

The above list can be added to or deleted from to tailor the manual for a specific company (also see Appendix G). Some other basic rules which could be incorporated into the manual are

1. Pay special attention to instructions given by your supervisor/foreperson.

2. Perform only the specific work you have received instruction to do.

3. Follow the requirements as specified on all permits.

4. Follow lockout and isolation procedures required for your work. Personally verify systems are deenergized.

5. Stay alert for changes in conditions while you work.

6. Report any unexpected or unusual liquids, gases, or vapors that appear.

7. Respond to any emergency alarms and respond using evacuation routes.

These are the types of rules which everyone on the construction site must follow at all times and are not the task-specific rules related to operating a crane or building a form.

SPECIAL EMPHASIS PROGRAMS AND PERMIT-REQUIRED PROCEDURES/ INITIATIVES

Some companies initiate programs that target specific problems such as developing a back-injury prevention program or reducing first aid cases. Other procedures are so dangerous that permits or special procedures are required. Some of these types of activities are confined space entry or lockout/tagout. These unique programs or requirements are often included within the S&H Manual.

SPECIAL SECTION ON FIRST AID

A section with common first aid practices and procedures may be included. This section may include the proper response for bleeding, burns, broken bones, shock, suffocation, and movement of injured workers. A decision needs to be made as to how much first aid is expected to be provided by a coworker. Based on the extent of the first aider's training, this may range from calling for help to assisting the injured worker.

OTHER SECTIONS

Some companies may want to include the introduction, the company's equal employment opportunity policy, or their sexual harassment policy in this section..

RULES AND SOPs FOR CONSTRUCTION-SPECIFIC TASKS

This section should incorporate the rules and safe operating procedures (SOPs) for those construction activities performed by a specific type of contractor. The list of activities and rules are not the same for all contractors. In order to use the following listings, you will need to review and determine their appropriateness to meet the needs of your company/contractor. While some of these items under each heading could be used verbatim, others will need to be deleted or revised. These were not written in a standard format but were written to be used as examples to assist you in developing your own rules or SOPs. This listing is not meant to be all-inclusive but to act as a guide for your own use.

<u>SAFE OPERATING PROCEDURES</u>

SAFETY IN THE WORK AREA

GENERAL

- Unsafe conditions in a work area are to be immediately reported to a supervisor.
- Stress alertness and awareness of stepping or walking hazards. Comply with caution and danger signs.
- Good housekeeping will be practiced at all times. There should be no materials in aisles, walkways, stairways, roads, or other points of entry or exit.
- No one should be permitted to enter a work area where there may be exposure to electrically energized circuits and conductors until the area has been tagged as deenergized.
- No employee should be permitted to work in such proximity to an electric power circuit that the employee may come in contact with the circuit during the course of work.
- Consult the safety inspection checklist for that project for safety requirements in a specific work area.

BARRICADES AND HOLE COVERS

- Excavations and openings on working surfaces must be protected with barricades or hole covers.
- Barricades or signs should always be provided as warning of hazards such as overhead work, crane swing, and excavations.
- When a hole or floor opening is created during the performance of a work activity, a cover or a barricade must be installed immediately.

FIRE PREVENTION

- When using heat-producing equipment, make sure the area is clear of all fire hazards and all sources of potential fires are eliminated.
- Do not use a salamander or other open-flamed device in confined or enclosed structures. Vent heaters to the atmosphere and make sure they are located an adequate distance from walls, ceilings, and floors.
- Have fire extinguishers available at all times when using heat-producing equipment.

EXCAVATIONS

- Have supervisory personnel determine whether excavations, trenches, or cuts more than four feet in depth require shoring or some other hold-back means.
- Excavations must be checked daily for cracks, slides, and scaling. During rain, snow, and other hazardous weather conditions, checks should be performed more often.

- Heavy equipment must be kept back from edges of all excavations. The access for excavations should be ladders or steps and should be located within twenty-five feet of any worker.

HOUSEKEEPING

- Proper housekeeping is the foundation for a safe work environment. Good house-keeping definitely prevents accidents and fires, as well as it creates a business-like work area.

- Pile or store material in a stable manner so that it will not be subject to falling.

- Rubbish, scraps, and debris should be removed from the work areas as soon as practical.

- It is not permissible to leave materials and supplies in stairways, walkways, near floor openings, or at the edge of the building when exterior walls are not built.

INDUSTRIAL HYGIENE AND OCCUPATIONAL HEALTH

- Employees must be protected against exposure to injurious sound levels by controlling exposure or by using the proper personal protective equipment.

- Employees must be protected against exposure to ionizing (X-ray, radioactive) and non-ionizing (laser beam) radiation.

- Protection against exposure to harmful gasses, fumes, dust, and similar airborne hazards must be furnished through proper ventilation or personal respiratory equipment.

WELDING AND CUTTING

GENERAL

- A suitable fire extinguisher or other fire control device should be ready for instant use in any location where welding is done. Where welding must be carried on near combustible materials, all proper precautions should be taken.

- Screens, shields, or other safeguards should be provided for the protection of workers from combustible materials below or sparks or falling objects. When others must work nearby, they should be protected from the arc rays by screens or by other adequate individual protection.

- When welding or cutting lead, zinc, cadmium coated, lead bearings, chrome, or other toxic materials, provisions should be made for removal of fumes, or the use of proper personal respiratory protection enforced.

WELDING EQUIPMENT

- Only standard electric arc welding equipment such as generators, motor-generator units, transformers, rectifiers, etc. conforming to the requirements of the National

Electrical Manufactures' Association or the Underwriters' Laboratories, Inc., or both, should be used.

- Power circuits should be installed and maintained in accordance with the National Electrical Code. Check to see what voltage the machine is wired for before connecting.

- All electric welding machines should be effectively grounded.

- Electrode and ground cables should be supported, not off equipment, so as to avoid obstruction interfering with the safe passage of workers. The ground lead for the welding circuit should be mechanically strong and electrically adequate for the service required.

- Where it is necessary to couple several lengths of cable for use such as a welding lead circuit, insulated connectors should be used on both the ground and electrode holder lines if occasional coupling or uncoupling is necessary.

- An electrode holder of adequate rated current capacity, insulated against shock and shorting or flashing when laid on grounded material, should be used.

- Adequate exhaust and adequate ventilation to the outside should be provided where internal combustion engines are used to operate gas welding machines in enclosed spaces.

CHIPPING, CLEANING

- When removing excess weld metal, faulty weldments, or slag, the welder removes or raises his helmet in order to see. The chips flying from the cleaning hammer are dangerous, especially to the eyes. Safety goggles and a protective face shield will be used whenever chipping and cleaning is being performed.

- Gloves should be worn to protect the hands and wrists. Also, flying chips are liable to travel a considerable distance and create danger for other personnel. In this instance screening and shielding may be required.

- Gloves will be worn when wire brushing.

- When cleaning and brushing surfaces to be welded, use caution to avoid metal slivers and sharp edges.

- Gauntlet gloves are advisable.

GRINDING

- Serious injuries result from unsafe practices with grinding wheels. Many of the most serious injuries result from exploding wheels.

- Most wheel exploding accidents are caused by one or more of the following:

    1. <u>Overspeeding the Wheel</u>: All wheels should be checked to see that the RPM rating is sufficient for the machine.

    2. <u>Cracks</u>: All wheels should be ring tested prior to installation. To do this, suspend the wheel from a string and tap lightly with a wooden hammer handle. <u>Good wheels will ring; cracked wheels will thud.</u>

3. <u>Mounting</u>: Never force a wheel over the spindle. Don't overtighten a holding nut as this may cause the wheel to crack. Resinoid wheels show less tendency to crack.

ELECTRODE STUBS

- A fire-resistant bucket or container should be provided for the disposal of electrode stubs.

- Careless disposal of hot stubs may result in: 1) injury to other workmen, especially those working at lower levers; 2) a falling hazard, if dropped on the floor, walkway, or other surfaces; and 3) a fire hazard, if dropped into combustible material.

FIRE PREVENTION IN WELDING AND CUTTING

- In welding and cutting operations, suitable fire extinguishing equipment should be maintained in a state of readiness for instant use. Such equipment may consist of pails of water, buckets of sand, or portable extinguishers, depending upon the nature and quantity of the combustible material exposed.

- All fire prevention procedures should apply when welding.

- Cutting and welding are major producers of fires on construction projects because of molten metal and sparks. Sparks from cutting and welding operations may be showered 25 to 30 feet, and may retain heat for several seconds which is sufficient to ignite combustible material.

PRECAUTIONS RECOMMENDED

- Practice good housekeeping to remove all loose, easily combustible materials such as wood shavings, wood scraps, sawdust, paper, rags, and especially oil and grease soaked materials.

- Remove all highly volatile materials, such as gasoline and solvents.

- Shield wood planking, scaffolds, wooden forms, and other combustible material that cannot be removed with sheet metal or other suitable material.

- Take precautions when welding or cutting work must be done on an automatic sprinkler system, or where the sprinklers are shut off for any reason.

EXPLOSION HAZARDS

- Cutting or repair welding of closed containers which have contained flammable liquids requires extreme caution. Containers should be thoroughly steam cleaned. If removal and handling for steam cleaning is impractical, containers may be filled with water or an inert gas. Frequent checking with an explosive vapor meter is recommended. Detailed recommendations are contained in "Safety in Welding and Cutting," Z491, published by American National Standards Institute.

GAS WELDING AND CUTTING

GENERAL

- Only standard oxyacetylene welding and cutting equipment, in first-class operating condition, should be used. Most insurance companies and contracting authorities approve equipment listed by the Underwriter's Laboratories, Inc. or approved and listed by Factory Mutuals Laboratories. Avoid oil contamination of gauge connections. If they become contaminated, do not use them.

- All gas cylinders should have their contents clearly marked on the outside of each cylinder.

STORAGE OF CYLINDERS

- Keep cylinders away from sources of heat. If stored in buildings, keep away from highly combustible materials, stoves, radiators, etc.

- Store securely. Cylinders should be securely placed to prevent tipping over and should not be near elevators, gangways, or other places where they are likely to be knocked over.

- Cylinders of oxygen should not be stored closer than twenty feet to cylinders of acetylene or other fuel gas.

- Cylinders stored in the open should be protected from accumulations of ice and snow and should be shielded from the direct rays of the sun.

- Full and empty cylinders are to be stored separately. Close valves on empty cylinders.

- Valve protection caps should always be in place when cylinders are not connected.

- Cylinders should be stored so as to avoid possible destruction or obliteration of coloring, tags, and other means of identifying the contents.

- While in use, valve key wrenches should be kept in place on valve spindle.

USE OF CYLINDERS

- Gas cylinders are exposed to many dangers at the construction site. Select a location for setting up cylinders which will be exposed to as little contact as possible from moving equipment, materials, and the like.

- Cylinders should be placed in a rack, chained, or otherwise positively secured against tipping over.

- Cylinders should be used in the order received from the supplier. When empty, valve should be closed and cylinder marked accordingly.

- Keep cylinders from contact with electric wires.

- Shield from sparks or flames from welding and cutting.

- Do not allow storing, temporary or otherwise, of tools, materials, or anything else on top of cylinders.

HANDLING CYLINDERS

- Whenever a cylinder is being moved, be sure valve protection cap is in place and closed.
- Never use valves or caps for lifting.
- For raising or lowering, use suitable sling, boat, cradle, or platform.
- Always handle carefully. Do not drop or jar.
- Do not lift with electromagnets.
- Cylinders may be moved by tilting and rolling on bottom edge; avoid dragging and sliding.
- When moving with hand truck or motor vehicle, be sure cylinders are securely held in place in a vertical position.

TOOLS AND EQUIPMENT

GENERAL

- It is imperative that the right tool be used for the job and that it be used in a correct manner. Consult the safety inspection checklist for information.
- Keep tools in good working condition. Damaged, worn, or defective tools can cause injuries and should not be used. Report defective tools to your supervisor immediately.
- Do not use tools until you have been properly instructed and authorized to do so. Only employees who have been trained in the operation of the particular tool in use should be allowed to operate a power actuated tool.
- Never remove machinery or equipment guards.
- Never make repairs to tools or equipment unless authorized by your supervisor.
- Inspect all chainfalls and hoisting lines daily. Report any defects to your supervisor immediately.

ELECTRICAL TOOLS AND EQUIPMENT

- All power tools should be inspected prior to delivery to the jobsite.
- Inspect electrical extension cords and other wiring to be certain they are properly insulated. Do not use frayed or damaged cords.
- Take special precautions when using power tools on a scaffold or other locations with limited movement area.
- Get a good footing, use both hands, keep cords clear of obstructions and do not overreach.
- Be sure that a power tool is off and motion stopped before setting tool down.
- Disconnect tool from power source before changing drills, blades, or bits, or attempting repairs or adjustments.

- Never leave a running tool unattended.
- Where Ground Fault Circuit Interrupters are in use, disconnecting is prohibited.

SCAFFOLDING/GENERAL

- Scaffolds should be capable of supporting at least four times the maximum intended load without failure.
- Maintain scaffolds in safe condition and do not overload.
- Scaffolds should be inspected by the user, daily before use, for damaged or weakened components, loosened or pulled out nails, unsafe guardrails, and other defects. Inspection is particularly important after high winds or job shutdowns occur. No scaffold should be erected, moved, dismantled, or altered except under the supervision of competent persons.
- Construct scaffolds on firm footing and use only for the purpose for which they were designated.
- Brace, guy, or tie scaffolds into structure whenever possible. Tie and brace scaffolds at intervals not to exceed thirty feet horizontally and twenty-six feet vertically.
- Provide safe guardrails and tow boards for all scaffolds over ten feet high and at lesser heights where warranted, except float and ship scaffolds. All personnel working on scaffolds are required to wear safety belts and to tie off. Guardrails should be constructed of two-by-four-inch boards or the equivalent, approximately 42 inches high with a midrail when required. Toe boards should be constructed using two-by-four-inch boards. Any material must meet plant fire protection requirements.
- Never permit climbing on cross-bracing. An access ladder or the equivalent safe access should be provided.
- Tubular metal scaffolds should be carefully inspected for broken joints or bends. All cross-bracing should be secured tightly with screws and sections kept plumb.
- Where hoists are used in conjunction with scaffolds, platforms from hoist framing should be tied in with scaffold and anchored. Provide guardrails on platforms.
- When concrete buggies, wheelbarrows, or extra materials are to be used on scaffolding, additional bracing should be added to carry the extra load.
- Rope used for needlebeams should be a minimum of one inch in thickness. Care should be taken to use softeners over rough and sharp edges.

ROLLING SCAFFOLDS

- Lock all caster brakes when a tower is in working position.
- Height of working platform should not exceed four times smallest base dimension.
- Do not fully extend adjusting screws.
- Provide sufficient horizontal bracing.
- Never allow workers to stay on a tower when it is in motion.

TUBE AND COUPLER SCAFFOLDS

- Keep scaffold plumb and level. Tighten all coupler bolts before using scaffold.

- Tie fixed scaffolds to building or other firm support every 20 to 30 feet horizontally and every 26 feet vertically.

- Always use guardrails and tow boards.

- Use sills when erecting on soft ground. Make sure planking is of good quality.

SUSPENDED AND SWINGING SCAFFOLDS

- Keep all bearing points well lubricated.

- Inspect ratchet and locking pawl springs to see that they are in good working condition. Keep pawls engaged at all times, except when unreeling wire ropes or lowering scaffolds. Any motor-driven power units or manually operated hoisting devices must be approved by Underwriters' Laboratories or Factory Mutual Engineering Corporation.

- Keep weight of workers and materials evenly distributed. Never overload. Where electric resistance welding is being done from wire rope suspended scaffolds, a means for preventing electrical contact is needed; rubber hoses eight feet in length should be used on cables. In areas where burning and welding hazards exist, a five pound ABC portable fire extinguisher should be placed on staging for personnel safety.

- Reinforce outriggers if the overhang exceeds specified distance. Outriggers should not project more than six feet from bearing point.

- Properly install toe boards and guardrails. Overlap planks.

- Use independent life lines and safety belts when using a suspended scaffold.

TUBULAR WELDED FRAME

- Keep scaffolds plumb and level at all times. Use adjusting screws rather than blocking.

- To prevent movement, tie into the structure every 26 feet of height and 30 feet of length.

- A landing platform must be provided at intervals not to exceed 30 feet.

- Use guardrailings and toe boards at all heights. Never climb on cross braces.

- Set base plates on sills to spread the load on soft ground.

- Check all wing nuts and locks for tightness. Use horizontal cross braces in addition to diagonal braces.

- Make sure planking meets load requirements. Never use even slightly damaged material.

- Drawings and specifications for all frame scaffolds over 125 feet in height above the base plates should be designed by a registered professional engineer.

FORMWORK SCAFFOLD

- Formwork scaffolds supported by partially cured concrete should not be moved except in accordance with design engineer specifications.

- For more specific information on scaffolds and scaffolding, consult the OSHA Standard for the Construction Industry 1926.451, Scaffolding.

- When scaffold planks are not pretested and certified before purchased, or when planks of doubtful strength are to be used, the following procedures should be employed for testing:

    1. Examine plank for large knots, excessive grain slope, shakes, decay, or other disqualifying defects.

    2. Determine allowable load from standard tables.

    3. Place plank on supports about one foot high, spaced the same as span to be used.

    4. Load plank as near center as possible, with twice the allowable load, leaving load on plank for not more than five minutes. Do not jump on plank.

- Discard plank if there are any signs of weakness, or if obvious deflection remains after load is removed.

- Mark accepted planks on top side as tested. Use them with that side upward.

- While in use do not load plank in excess of allowable load.

- Planks should have corners cut and be banded across each end to prevent splitting and to prolong life of planks.

LADDERS

- All purchased ladders should meet with federal, state, or municipal requirements. All ladders, "job built," should have a safety factor equal to the requirements of similarly purchased ladders.

- Portable ladders should be used at such a pitch that the horizontal distance from the top support to the foot is not greater than one-fourth the vertical distance between these points.

- Ladders with broken or missing rungs or steps, broken or split side rails, or other faulty or defective construction should not be used. Ladders with a single rail should not be used.

- The side rails of ladders should extend three feet beyond the platform or floor served.

- Climb or descend ladder with both hands holding on and facing the ladder. As a general practice, portable metal ladders should be prohibited in order to eliminate the possibility of contacting live electrical equipment in "hot" areas.

- Ladders should not be placed in doorways, etc. unless precautions are taken to prevent displacement.

- Single cleat ladders should not be less than 15 inches nor more than 20 inches wide between the rails.

- Cleats should be uniformly spaced 12 inches top to top.

- Cleats should be inset one-half inch or filler blocks placed between cleats and they should be secured in place with not less than three 10D common nails in each rail.

- Ladders, when in use, should be secured in place at the top or bottom to prevent displacement. See OSHA 1926.450, Ladders.

MOTOR VEHICLES AND MECHANIZED EQUIPMENT

- At the operator's station, rated load capacities and recommended rules of operation should be conspicuously posted on all equipment.

- An accessible fire extinguisher of five BC rating or higher should be available at all operator stations.

- When vehicles or mobile equipment are stopped or parked, parking brakes should be set. Equipment on inclines should have wheels locked, as well as having parking brakes set.

- All vehicles, or combinations of vehicles, should have in operable condition at least:

  1. Two (2) headlights.

  2. Two (2) taillights.

  3. Brake lights.

  4. Audible warning device at operator's station.

  5. Seat belts properly installed.

  6. Seats, firmly secured, for the number of persons carried.

  7. Service, parking, and emergency brake system.

- Operators should not use motor equipment that has an obstructed rear view unless

  1. Vehicle has an audible reverse signal alarm.

  2. Vehicle is backed up only when observer says it is safe to do so.

MATERIAL HANDLING AND STORAGE

GENERAL

- When storing materials, do not leave materials in aisles, walkways, stairways, roads, or other points of entries or exits.

- When moving or lifting materials, avoid stress or strain. Use the correct lifting technique and utilize hoisting equipment or engage the help of a coworker whenever heavy or unwieldy objects are to be moved.

- Used lumber should have nails removed before being stacked.

- Be sure that ground floors or scaffolds are capable of supporting the load that will be stacked or stored there.

- Review storage requirements to ensure compliance with manufacturers recommendations.

- Clearance requirements (to roadways, buildings, railroad tracks, and fences) should be known and observed.

- Keep fire hydrants readily visible and accessible.

- Flammable liquids and grease should be stored in a "No Smoking" area and properly separated from other stored material. Each container should be identified as to its contents.

- "Hitching a ride" on tractors, cranes, forklifts, and other vehicles is dangerous. Ride in cab of truck or where a seat is provided. Jumping on or off a moving vehicle is prohibited.

STACKING

- Materials should be segregated as to kind, size, and length, and placed in neat and orderly stacks that are racked, blocked, or interlocked to prevent falling, collapsing, or tripping hazards.

- Material stacks will be stepped back as the height increases and should be secured.

- Stacks of materials should be arranged to allow passageways between them and be well marked and visible at night.

- Lumber stacks should be kept as level as possible and should not exceed 16 feet.

- Bagged materials should be stacked by stepping back the layers and crosskeying the bags at least every ten bags high.

- Brick stacks should not be more than seven feet high. When loose bricks reach a height of four feet, they should be tapered back 2 inches for every foot above four feet.

- Masonry blocks stacked over six feet high should be tapered back one-half block on each tier.

RIGGING

- Good rigging is essential for moving construction materials and equipment and, at the same time, keeping them under control.

- Loads should never be swung over the heads of workers in the area.

- Only qualified flagpersons and signalpersons should direct the operation, using hand signals established as standard for the industry.

- Tag lines must be used to control loads and keep workers away.

- Rigging should not be over-loaded. When loads are just off the ground, check them for stability before hoisting.

- A suspended load should never be left unattended until safely landed.

- Loads, booms, or rigging should never be allowed to approach within ten feet of energized electrical lines rated 50 KV or lower unless the lines are deenergized. For lines rated greater than 50 KV, when moving loads in close proximity, follow OSHA Regulations 29 CFR.550(a) (15) (ii).

- Cranes should always be operated on firm, level ground or use mats, particularly for near-capacity lifts with outriggers down.

- A space of 360 degrees should be roped off or barricaded around all cranes operating on the jobsite. This 360 degrees should be established according to the extent of the swing radius of the rear of the rotating structure.

CRANES, DERRICKS, AND HOISTS

- All material and/or passenger hoists and derricks should be erected and operated according to federal, state, or local governing codes.

- All modifications, extensions, replacement parts, or repairs of equipment should maintain at least the factor of safety as the original designed equipment. No modifications should be made without the manufacturer's written approval and load testing, prior to use.

- All drums on local hoisting equipment should be equipped with proper dogs or pawls.

- Braking equipment capable of effectively braking and holding a load of at least one and one half times the full rated load should be provided.

- Capacity plates should be attached to all load hoisting equipment. On boom cranes and derricks, they should clearly indicate the safe load for maximum and minimum positions of the boom and for at least two intermediate stations. These indications should be for loads both with and without outriggers, where so equipped.

- There should be at least two full wraps of cable on the drum of the hoisting equipment at all times of operation.

- Riding on loads, hooks, hammers, buckets, or material hoists is prohibited.

- While hoisting equipment is in operation, the operator should not be permitted to perform any other work. The operator should not leave his/her position at the controls until the load has been safely landed or returned to the ground level.

- Tag lines for controlling loads should be used at all times.

- A standard signal system should be used on all hoisting equipment.

- The operator must be able to read and understand signs, notices, operating instructions, and signal codes used.

- Unless the hoist is provided with the necessary protection equipment for passenger transportation, no one should ride the hoist.

- When a project is located in the vicinity of an airport, it should be ascertained from the airport personnel whether a derrick, hoist, etc. on the job can constitute a hazard. This can be true when a derrick is "jumped" to the next tier.

- Outside steel hoisting towers should be suitably grounded, mechanically secure, and electrically safe.

- Whenever electric power is used, the motor, panel boards, sources of power, brakes, and other devices should be installed and operated in accordance with the National Electrical Code.

- Emergency controls should be readily accessible.

- Engines should be securely anchored to prevent movement.

- Adequate lighting should be provided at landing platforms, in and around machine housing, and at switches.

- Shaftways of outside hoists should be enclosed for the entire height, facing open sides of the car with heavy mesh wire (not greater than 2 inch openings), except at loading or unloading platforms. All entrances to shaftways must be protected by substantial gates.

- Loading platform should have rails and toe boards, and should be built to withstand the loads to be imposed on them.

- Where any platform entrance to or at the top of a shaftway is exposed to falling objects, a substantially large and strong overhead catch platform should be provided. It is understood that the floor of each landing should be considered the roof of the platform immediately below.

- A substantially large and strong overhead roof should be provided over the hoisting engineer and equipment. (See OSHA 1926.552 for additional requirements.)

REFERENCE MATERIALS, DIAGRAMS, AND ILLUSTRATIONS

In this area of the S&H manual you often find illustrations or visual stimuli for workers regarding special or standardized procedures. Some examples of these types of items are

- Varied rigging diagrams and procedures.

- Standard hand signals for crane operations.

- Trenching and shoring requirements.

- Scaffolding and ladder placements.

- Respirator selection charts or other specialized use of PPE.

- Charts on strength of materials.

ACKNOWLEDGMENT OF RECEIPT OF S&H HANDBOOK/MANUAL

A form that acknowledges receipt of the S&H handbook and verification that the worker has read the manual might look something like the following:

This is to acknowledge that I have received and read my copy of the Company *Safety and Health Handbook* and received training regarding its contents (only if training occurred). I have read and will abide by all the rules and regulations within this handbook and any additional safety and health rules and regulations required which pertain to my job.

_____

Signed

_____

Date

_____

Craft

This is to acknowledge that I delivered the *Safety and Health Handbook* to the worker who signed the above receipt and that I provided training regarding its contents and other safety and health rules and regulations of the workers' job.

_____

Signed

_____

Date

_____

Title

CHAPTER 13

OSHA COMPLIANCE

Charles D. Reese

Workers should expect to go to work each day and return home uninjured and in good health. There is no logical reason that a worker should be part of workplace carnage. Workers do not have to become yearly workplace statistics.

Construction employers who enforce the occupational safety and health rules and safe work procedures are less likely to have themselves or their workers become one of the 7,500 occupational trauma deaths, one of 90,000 occupational illness deaths, or even one of the 6.8 million nonfatal occupational injuries and illnesses that occur each year in the United States.

OSHA (Occupational Safety and Health Administration) and their regulations should not be the driving force that ensures workplace safety and health. Since OSHA has limited resources and inspectors, enforcement is usually based upon serious complaints, catastrophic events, and workplace fatalities. The essence of workplace safety and the strongest driving catalyst should first be the protection of the workforce, followed by economic incentives for the employer. A contractor having a good safety and health program and record will reap the benefits: a better opportunity to win more contracts, lower insurance premiums for workers' compensation, decrease liability, and increase employee morale and efficiency. Usually, safety and health are linked to the bottom line (company's income), an action seldom perceived as humanitarian.

This chapter will provide answers to many of the questions asked regarding OSHA, and workplace safety and health, and suggest how workers and employers can work together to provide a safe and healthy workplace. This information is a guide to understanding OSHA, OSHA compliance, and ensuring safer and healthier construction sites.

During the many years preceding OSHA, it became apparent that construction contractors needed guidance and incentives to insure safety and health on the jobsite. The contractors needed to realize that workers had a reasonable right to expect a safe and healthy workplace. This guidance and the guarantee of a safe and healthy workplace came to fruition with the enactment of the Occupational Safety and Health Act of 1970 (OSHAct). The Occupational Safety and Health Administration was created by the Act to

- Encourage employers and employees to reduce workplace hazards and to improve existing safety and health programs or implement new programs.

- Provide for research in occupational safety and health in order to develop innovative ways of dealing with occupational safety and health problems.

- Establish "separate-but-dependent" responsibilities and rights for employers and employees for the achievement of better safety and health conditions.

- Maintain a reporting and record keeping system to monitor job-related injuries and illnesses.

- Establish training programs to increase the numbers and competence of occupational safety and health personnel.

- Develop mandatory job safety and health standards and enforce them effectively.

- Provide for the development, analysis, evaluation, and approval of state occupational safety and health programs.

Thus, the purpose of OSHA is to insure, as much as possible, a healthy and safe workplace free of hazardous conditions for workers in the United States.

OSHA STANDARDS

OSHA standards found in the *Code of Federal Regulations* [*CFR*], (see Figure 13-1) include the standards for the following industry groups. OSHA standards cover construction, maritime, agriculture, and the general industry which includes manufacturing, transportation and public utilities, wholesale and retail trades, finance, insurance, and services. Some of the specific areas covered by regulations are found in Table 13-1.

An employer can seek relief (variance) from an OSHA standard. The reasons for variances approved by OSHA are

- The employer may not be able to comply with the standard by its effective date.

- The employer may not be able to obtain the materials, equipment, or professional or technical assistance needed to comply.

- The employer already has processes or methods in place which provide protection to workers and are "at least as effective as" the standard's requirements.

A "temporary" variance that meets the criteria listed above may be issued until compliance is achieved or for one year, whichever is shorter. It can also be extended or renewed for six months (twice).

Employers may obtain a "permanent variance" if the employer can document with a preponderance of evidence that existing or proposed methods, conditions, processes, procedures, or practices provide workers with protections equivalent or better than the OSHA standard. Employers are required to post a copy of the variance in a visible area in the workplace, as well as make workers aware of a request for a variance.

PROTECTIONS UNDER THE OSHAct

Usually all employers and their employees are considered to be protected under the OSHAct, with the exception of

- Self-employed persons.

- Farms where only immediate family members are employed.

- Workplaces already protected by other federal agencies under federal statutes such as the Department of Energy and the Mine Safety and Health Administration.

- State and local employees.

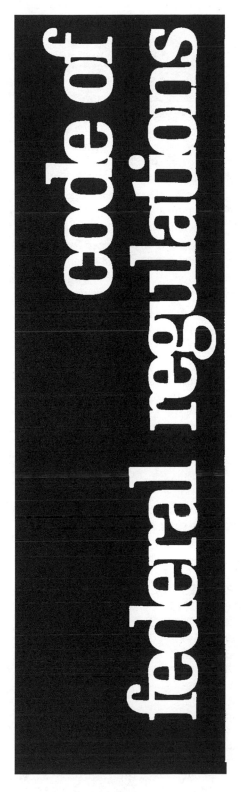

Labor

29

PART 1926
Revised as of July 1, 1996

Figure 13-1. Code of federal regulations. Courtesy of USGPO

Table 13-1

Standard Topics

• Lockout/Tagout	• Electrical Safety
• Housekeeping	• Training Requirements
• Noise Exposure	• Fire Prevention
• Hazard Communication	• Confined Spaces
• Personal Protective Equipment	• Ventilation Requirements
• Sanitation	• Medical and First Aid
• Fall Protection	• Working with
• Emergency Planning	Hazardous Substances
• Record Keeping	• Guarding
• Use of Hand Tools	• Machine and Equipment Safety
• Ladders and Scaffolds Safety	• Radiation
• Explosives/Blasting	• Blood-Borne Pathogens
• Compressed Gases	

OSHA standards and regulations for occupational safety and health are found in Title 29 of the *Code of Federal Regulations* (*CFR*) and can be obtained through the Government Printing Office (GPO). The standards for specific industries are found in Title 29 of the *Code of Federal Regulations* (Table 13-2).

Table 13-2

CFRs for Industry Specific Regulations

- General Industry - 29 CFR PART 1910
- Ship Repairing - 29 CFR PART 1915
- Ship Building - 29 CFR PART 1916
- Ship Breaking - 29 CFR PART 1917
- Longshoring - 29 CFR PART 1918
- Gear Certification - 29 CFR PART 1919
- Construction - 29 CFR PART 1926
- Agriculture - 29 CFR PART 1928
- Federal Agencies - 29 CFR 1960

NATIONAL INSTITUTE FOR OCCUPATIONAL SAFETY AND HEALTH (NIOSH)

Although the formation of NIOSH was a requirement of the OSHAct of 1970, NIOSH is not part of OSHA. NIOSH is one of the Centers for Disease Control and Prevention, head-quartered in Atlanta, Georgia. NIOSH reports to the Department of Health and Human Services (DHHS) and not to the Department of Labor (DOL) as OSHA does. Its functions are to

- Recommend new safety and health standards to OSHA.
- Conduct research on various safety and health problems.
- Conduct Health Hazard Evaluations (HHEs) of the workplace when called upon.
- Publish an annual listing of all known toxic substances and recommend exposure limits (RELs).
- Conduct training that will provide qualified personnel under the OSHAct.

An employer, worker's representative, or worker can request a Health Hazard Evaluation from NIOSH to have a potential health problem invstigated. It is best to use the NIOSH standard form (see Figure 13-2). It can be obtained by calling **1-800-35-NIOSH.**

The Health Hazard Evaluation request should include the following information:

- A description of the problem.
- The symptoms being exhibited by the worker(s).
- Name of the suspected substance (trade or chemical name).
- The process in which the problem is occurring.
- The hazard warning from the label or Material Safety Data Sheet (MSDS) of the substance.
- The length of time worker(s) are exposed to it.
- When the symptoms were first noticed.
- Is this a new or old process or material being used?
- Has this problem occurred previously?
- Has the complaint been registered with OSHA or another government agency?

OCCUPATIONAL SAFETY AND HEALTH REVIEW COMMISSION (OSHRC)

The Occupational Safety and Health Review Commission (OSHRC) was established, under the OSHAct, to conduct hearings when OSHA citations and penalties are contested by employers or by their employees. As with NIOSH, the OSHRC formation was a requirement of the OSHAct, but is a separate entity apart from OSHA.

EMPLOYER RESPONSIBILITIES UNDER THE OSHAct

The employer is held accountable and responsible under the OSHAct. The "General Duty Clause," Section 5(a)(1) of the OSHAct states that employers are obligated to provide a workplace free of recognized hazards that are likely to cause death or serious physical harm to employees. Employers must

Form Approved
OMB No. 0920-0102
Expries Nov. 30, 1995

U. S. DEPARTMENT OF HEALTH AND HUMAN SERVICES
U. S. PUBLIC HEALTH SERVICE
CENTERS FOR DISEASE CONTROL AND PREVENTION
NATIONAL INSTITUTE FOR OCCUPATIONAL SAFETY AND HEALTH

REQUEST FOR HEALTH HAZARD EVALUATION

ESTABLISHMENT WHERE POSSIBLE HAZARD EXISTS

Company name: _____

Address: _____

City: _____ State: _____ Zip Code: _____

What product or service is provided at this workplace? _____
_____

Specify the particular worksite, such as building or department, where the possible hazard exists: _____
_____

How many people are exposed? _____ Duration of exposure (hrs/day)? _____

What are the occupations of the exposed employees; what is the process/task?

Occupation: _____

Process/task: _____

To your knowledge, has NIOSH, OSHA, MSHA, or any other government agency previously evaluated this workplace?
YES _____   NO _____

Is a similar request currently being filed with, or is the problem under investigation by any other local, state, or federal
agency? YES _____   NO _____

If either question is answered yes, give the name and location of each agency. _____
_____
_____

Which company official is responsible for employee health and safety?

Name: _____ Title: _____ Phone: _____

[Send completed form to address listed on the reverse side.]

This form is provided to assist in requesting a health hazard evaluation from the U.S. Department of Health and Human Services. Public reporting burden for this collection of information is estimated to average 12 minutes per response. Send comments regarding this burden estimate or any other aspect of this collection of information, including suggestions for reducing this burden to PHS Reports Clearance Officer; ATTN: PRA, Hubert H. Humphrey Bldg. Rm 721-B; 200 Independence Ave., SW; Washington, DC 20201, and to the Office of Management and Budget; Paperwork Reduction Project (0920-0102); Washington, DC 20503. (See Statement of Authority on reverse.)

Figure 13-2. Request for health hazard evaluation form. Courtesy of NIOSH

DESCRIPTION OF THE POSSIBLE HAZARD OR PROBLEM

Please list all substances, agents, or work conditions which you believe may contribute to the possible health hazard. (Include chemical name, trade name, manufacturer or other identifying information, as appropriate.)

_____

_____

In what physical form(s) do(es) the substance(s) exist? _____ Dust _____ Gas _____ Liquid _____ Mist _____ Other

How are the affected employees exposed (route of exposure)? _____ Breathing _____ Skin contact _____ Swallowing _____ Other (please list) _____

What health problem(s) do employees have as a result of these exposures? _____

_____

_____

Use the space below to supply any additional relevant information.

_____

_____

Requester's Signature: _____ Date: _____

Typed or printed name: _____

Address: _____

City: _____ State: _____ Zip Code: _____

Business Phone: _____ Home phone: _____ Best time of day to call: _____

CHECK ONLY ONE OF THE FOLLOWING:

_____ I am an employer representative.

_____ I am an authorized representative of, or an officer of the union or other organization representing the employees for collective bargaining purpose. Name and address of this organization:

_____

_____ I am a current employee of the employer, and an authorized representative of two or more other current employees in the workplace where the exposures are found. Signatures of the authorizing employees are below:

Signature: _____ Phone: _____

Signature: _____ Phone: _____

_____ I am one of three or fewer employees in the workplace where the substance, hazard, or health problem exists.

Please indicate your desire: _____ I do not want my name revealed to the employer.
_____ My name may be revealed to the employer.

STATEMENT OF AUTHORITY:
Sections 20(a)(3-6) of the Occupational Safety and Health Act (29 USC 669(a)(6-9)). and Section 501(a)(11) of the Federal Mine Safety and Health Act (30 USC 951(a)(11)). Confidentiality of the respondent requester will be maintained in accordance with the provisions of the Privacy Act (5 USC 552a). The voluntary cooperation of the respondent requester is required to initiate the Health Hazard Evaluation.

SEND COMPLETED FORM TO: **National Institute for Occupational Safety and Health**
Hazard Evaluations and Technical Assistance Branch
4676 Columbia Parkway, Mail Stop R-9
Cincinnati, Ohio 45226

Phone: (513) 841-4382   FAX: (513) 841-4488

Figure 13-2. Request for health hazard evaluation form. Courtesy of NIOSH (Contd.)

- Abide and comply with the OSHA standards.

- Maintain records of all occupational injuries and illnesses.

- Maintain records of workers' exposure to toxic materials and harmful physical agents.

- Make workers aware of their rights under the OSHAct.

- Provide, at a convenient location and at no cost, medical examinations to workers when the OSHA standards require them.

- Report within eight hours to the nearest OSHA office all occupational fatalities or catastrophes where three or more employees are hospitalized.

- Abate cited violations of the OSHA standard within the prescribed time period.

- Provide training on hazardous materials and make MSDSs available to workers upon request, such as hazard communication training.

- Assure workers are adequately trained under the regulations.

- Post information required by OSHA such as citations, hazard warnings, and injury/illness records.

WORKERS' RIGHTS AND RESPONSIBILITIES UNDER THE OSHAct

Workers have many rights under the OSHAct. These rights include the right to

- Review copies of appropriate standards, rules, regulations, and requirements that the employer should have available at the workplace.

- Request information from the employer on safety and health hazards in the workplace, precautions that may be taken, and procedures to be followed if an employee is involved in an accident or is exposed to toxic substances.

- Access relevant worker exposure and medical records.

- Be provided Personal Protective Equipment (PPE).

- File a complaint with OSHA regarding unsafe or unhealthy workplace conditions and request an inspection.

- Not be identified to the employer as the source of the complaint.

- Not be discharged or discriminated against in any manner for exercising rights under the OSHAct related to safety and health.

- Have an authorized employee representative accompany the OSHA inspector and point out hazards.

- Observe the monitoring and measuring of hazardous materials and see the results of the sampling, as specified under the OSHAct and as required by OSHA Standards.

- Review the occupational injury and illness records (OSHA No. 200 or equivalent) at a reasonable time and in a reasonable manner.

- Have safety and health standards established and enforced by law.

- Submit a request to NIOSH for a Health Hazard Evaluation (HHE) of the workplace.

- Be advised of OSHA actions regarding a complaint and request an informal review of any decision not to inspect or issue a citation.

- Participate in the development of standards.

- Speak with the OSHA inspector regarding hazards and violations, during the inspection.

- File a complaint and to receive a copy of any citations issued and the time allotted for abatement.

- Be notified by the employer if the employer applies for a variance from an OSHA standard and testify at a variance hearing and appeal the final decision.

- Be notified if the employer intends to contest a citation, abatement period, or penalty.

- File a Notice of Contest with OSHA if the time period granted to the company for correcting the violation is unreasonable, provided it is contested within fifteen working days of the employer's notice.

- Participate at any hearing before the OSHA Review Commission or at any informal meeting with OSHA when the employer or a worker has contested an abatement date.

- Appeal the OSHRC's decisions in the U.S. Court of Appeals.

- Obtain a copy of the OSHA file on a facility or workplace.

Along with rights go responsibilities and workers should be expected to conform to these responsibilities. Workers are expected to

- Comply with the OSHA Regulations and Standards.

- Not remove, displace, or interfere with the use of any safeguards.

- Comply with the employer's safety and health rules and regulations.

- Report any hazardous condition to the supervisor or employer.

- Report any job-related injuries and illness to the supervisor or employer.

- Cooperate with the OSHA Inspector during inspections when requested to do so.

DISCRIMINATION AGAINST WORKERS

Workers have the right to expect safety and health on the job without fear of punishment. This is spelled out in Section 11(c) of the OSHAct and under 49 U.S.C. 31105 (formerly Section 405) for the trucking industry. The law states that employers shall not punish or discriminate against workers for exercising rights such as

- Complaining to an employer, union, or OSHA (or other government agency) about job safety and health.

- Filing a safety and health grievance.

- Participating in OSHA inspections, conferences, hearings, or OSHA-related safety and health activities.

If workers believe they are being discriminated against, they should contact the nearest OSHA office *within 30 days* of the time they sense that discriminatory activity has started.

To file a formal complaint a worker should visit, call, or write their nearest OSHA office or state OSHA office, if a state program exists there. If a worker calls or visits, then a written follow-up letter should be sent. This may be the only documentation of a complaint. Complaints should be filed only when the following is occurring:

- Discrimination has been continuing.

- The employer has been devious, misleading, or concealing information regarding the grounds for the worker's discriminatory treatment.

- The worker has attempted to use the grievance or arbitration procedures under the collective bargaining agreement during the 30 days.

When OSHA receives a Worker's Discrimination Complaint, OSHA will review the facts of the complaint and decide whether to conduct an investigation. If an investigation ensues, the worker and the employer will be notified of the results within 90 days.

If the investigation indicates the worker's case has merit to process the case through the courts, OSHA or the state agency will attempt to negotiate with the employer. The settlement might include reinstatement of the worker's job, full back pay, and purging of the worker's personnel records. The employer might also be required to post a notice on the jobsite warning about any further workplace safety and health discrimination.

At times, employers may not decide to settle. In this instance, OSHA or the State Agency will submit the case to the U.S. District Court. The court can order the employer to reinstate the employee, pay lost wages, purge the worker's personnel records, and protect him/her from further discrimination.

If the investigation determines the worker does not have a case, the worker may feel the decision was in error and still has the right to appeal the decision of OSHA or the State Agency. The worker will need to provide a detailed explanation, as well as documentation, for contesting the prior decision.

Workers can file a discrimination complaint with federal OSHA if the worker's State Program and its courts do not offer protection from discrimination.

RIGHT TO INFORMATION

Workers have a "Right-to-Know." This means that the employer must establish a written, comprehensive hazard communication program that includes provisions for container labeling, materials safety data sheets, and an employee training program. The program must include

- A list of the hazardous chemicals in the workplace.

- The means the employer uses to inform employees of the hazards of non-routine tasks.

- The way the employer will inform other employers of the hazards to which their employees may be exposed.

Workers have the right to information regarding the hazards to which they are or will be exposed. They have the right to review plans such as the hazard communication plan. They have a right to see a copy of a MSDS during their shift and receive a copy of a MSDS when requested. Also, information on hazards which may be brought to the workplace by another employer should be available to workers. Other forms of information such as exposure records, medical records, etc. are to be made available to workers upon request.

ASSURING A SAFE AND HEALTHY WORKPLACE

Workers have the right to refuse hazardous work. This is not a right which is free of stipulations. Workers must assure that three criteria are met.

1. Workers have a reasonable belief, based on what they know at the time, that there is a real likelihood they could be killed or suffer serious injury (imminent danger).

2. When the employer or supervisor has been asked to eliminate the danger and does not take action, the worker should ask for another assignment while the previous one is made safe. The worker should not return to the jobsite unless ordered to do so.

3. If the employer does not respond, workers should call the closest OSHA office to explain the circumstances. If OSHA cannot respond in a timely manner, because of time constraints, the worker has no other alternative but to refuse the work.

Workers have the right to receive the results of any OSHA test for vapors, noise, dusts, mists, fumes, radiation, etc. This includes observation of any measurement of hazardous materials in the workplace. If the hazards of the workplace are such that personal protective equipment is required for workers, workers are to be provided, at no cost, the proper and well-maintain personal protective equipment appropriate for the job.

WORKERS' COMPLAINTS

Workers have the right to complain to OSHA regarding workplace safety and health concerns. Workers will need to contact the nearest federal OSHA office or state OSHA office if the state has a state plan. A listing of the OSHA offices for each state can be found in Appendix J or use the **National OSHA Hotline (after hours) 1-800-321-OSHA.** An OSHA complaint should include the following information:

- A description of the problem, work process, or job.
- Type of inspection being requested.
- Location of the hazard.
- Identification of the problem as a health or a safety hazard.
- Number of workers endangered or exposed.
- Identification of the intensity of the hazard and if there is an immediate danger to life and health.
- Identification of the standard being violated.
- Record of a previous violation for this hazard.
- If the work has been shut down by the employer or by a federal or state agency.
- Time when the hazard was first noticed.
- Notification of employer.
- Request for confidentiality.
- Request for an employee representative to accompany the inspector.
- Request for a closing conference.
- Notification of any worker reprimanded or discriminated against for complaining about the hazard.
- Notification if the complaint is going through any internal grievance procedure.
- Proof of any written documentation regarding the hazard.

OSHA INSPECTIONS

OSHA has the right to conduct workplace inspections as part of their enforcement mandate. OSHA can routinely initiate an unannounced inspection of a business. Other inspections occur due to fatalities/catastrophes, routine program inspections, or by referrals and complaints. These occur during normal working hours.

Workers have the right to request an inspection. The request should be in writing, either by letter or by using the OSHA Complaint Form to identify the employer and the alleged violations. Send the letter or form to the area director or state OSHA director. If workers receive no response, they should contact the OSHA regional administrator. It is beneficial to call the OSHA office to verify its normal operating procedures. If workers allege an imminent danger, they should call the nearest OSHA office.

These inspections include checking company records, reviewing the compliance with the hazard communication standard, fire protection, personal protective equipment, and review of the company's health and safety plan. This inspection will include conditions, structures, equipment, machinery, materials, chemicals, procedures, and processes. OSHA's priorities for scheduling an inspection are rank ordered as follows:

- Situations involving imminent danger.
- Catastrophes or fatal accidents.
- Complaints by workers or their representatives.
- Referral from other state/federal agencies or media.
- Regular inspections targeted at high-hazard industries.
- Follow-up inspections.

Usually no advance notice is given to an employer prior to an inspector appearing at a jobsite. But, there are times when advance notice is an acceptable practice. They are

- In case of an imminent danger.
- When it would be effective to conduct an inspection after normal working hours.
- When it is necessary to assure the presence of the employer or a specific employer or employee representatives.
- When the area director determines that an advance notice would enhance the probability of a more thorough and effective inspection.

No inspection will occur during a strike, work stoppage, or picketing action unless the area director approves such action. Usually this type of inspection would be due to extenuating circumstance such as an occupational death inside the facility. The steps of an OSHA inspection encompass the following:

- The inspector becoming familiar with the operation including previous citations, accident history and business demographics, and gaining entry to the operation. OSHA is forbidden to make a warrantless inspection without the employer's consent. Thus, the inspector may have to obtain a search warrant if reasonable grounds for an inspection exist, and entry has been denied by the employer.

- The inspector holding an opening conference with the employer or a representative of the company. It is required that a representative of the company be with the inspector during the walkaround and a representative of the workers be given the opportunity to accompany the inspector.

- An inspection tour taking hours or possibly days, depending on the size of the operation. The inspector usually covers every area within the operation while assuring compliance with OSHA Regulations.

A closing conference is conducted that gives the employer an opportunity to review the inspectors findings. The inspector will request from the employer an abatement time for the violation(s) to be corrected. An employee representative (union) will also be afforded an opportunity to have separate opening and closing conferences.

The area director will issue, to the employer, the written citations with proposed penalties and abatement dates. This document is called "Notification of Proposed Penalty."

WORKERS' COMPLAINTS AND REQUESTS FOR INSPECTIONS

Requesting an OSHA Inspection is a right which should be used in a prudent and responsible manner and only after all other options have been exhausted. Workers' complaints are the most frequent reason for OSHA inspections.

Requests for these types of inspections should be in writing, using the OSHA complaint form or letter. The complaint should include information about the ongoing work process, the number of workers affected, the nature of the problem, a safety or health hazard, and an indication that the worker has tried to get the employer to fix the problem or remove the hazard.

A written complaint guarantees a written record. This means that OSHA has to keep the worker informed of the results and it will protect a worker against employer discrimination.

OSHA will not inspect if the complaint does not indicate adequate cause or if the complaint was aimed at harassing the employer.

When OSHA receives a complaint, OSHA gathers information and decides whether or not the complaint warrants sending a compliance officer (inspector) to the site. In the case of a non-inspection, the complainant will be notified and a copy of the complaint sent to the employer. Workers requesting an inspection have the right to know of any actions OSHA takes concerning their request and may have an informal review when OSHA decides not to inspect. If OSHA decides to conduct an inspection, workers should do the following during the inspection:

- Cooperate with the OSHA compliance inspector.
- Have a worker representative accompany the inspector.

After a formal complaint is made, OSHA's normal time constraints for conducting an inspection are based upon the seriousness of the complaint. The usual times are

- Within 24 hours if the complaint alleges an imminent danger.
- Within three days if the complaint is serious.
- Within 20 days for all other complaints.

Upon completion of the inspection, the workers' representative can request the inspector to conduct a closing conference for labor. There will be a separate closing conference for the employer, and the workers' information and the employer must post all citations issued. During the closing conference workers should

- Ask the inspector to describe all hazards discovered and standard violations found.
- Make sure the inspector has all the information, as well as information on other complaints.

- Keep a written record and specific notes of the closing conference.
- Ask about the procedures and results which will occur from the inspection.

CITATIONS, PENALTIES, AND OTHER ENFORCEMENT MEASURES

If violations of OSHA Standards are detected during an inspection, the citations will include the following information:

- The violation.
- The workplace affected by the violation.
- Specific control measures to be taken.
- The abatement period or time allotted to correct the hazard.

Upon receipt of the penalty notification, the employer has 15 working days to submit a Notice of Contest which must be given to the workers' authorized representative or, if no representative exists, it must be posted in a prominent location in the workplace. During the 15 days, it is recommended that the employer first request an informal conference with the area director. During the informal conference, the issues concerning the citations and penalties can be discussed. If the employer is not satisfied, a Notice of Contest can be filed. An employer who has filed a Notice of Contest may withdraw it prior to the hearing date by

- Showing that the alleged violation has been abated or will be abated.
- Informing the affected employees or their designated representative of the withdrawal of the contest.
- Paying the assessed fine for the violation.

Workers can contest the length of the time period for abatement of a citation. They may also contest the employer's petition for an extension of time for correcting the hazard. Workers must do this within 10 working days of posting. Workers cannot contest the

- Employer's citations.
- Employer's amendments to citations.
- Penalties for the employer's citations.
- Lack of penalties.

Copies of the citation should be posted near the violation's location for at least three days or until the violation is abated, whichever is longer. Violations are categorized in the following manner (see Table 13-3).

In describing these violations, the *De Minimis* is the least serious and carries no penalty since it violates a standard which has no direct or immediate relationship to safety and health. An *Other-than-Serious* violation would probably not cause death or serious harm, but could have a direct affect on the safety or health of employees. *Serious* violations are those violations where a substantial probability of death or serious physical harm could result. The *Willful* violations are violations where an employer has deliberately, voluntarily, or intentionally violated a standard. And, *Repeat* violations are ones which occur within three years of an original citation.

The values or penalties applied to citations are based upon four criteria.

- The seriousness or gravity of the alleged violation.
- The size of the business.

- The employer's good faith in genuinely and effectively trying to comply with the OSHAct before the inspection and, during and after the inspection, a genuine effort is made to abate and comply.

- The employer's history of previous violations.

Table 12-3

OSHA Violations and Penalties

De Minimis	No Penalty
Other than serious	Up to $7,000 per violation
Serious	$1,500 - $7,000 per violation
Willful, No Death	Up to $70,000 per violation Minimum of $5,000
Willful, Repeat Violations	Same as Willful, No Death
Willful, Death Results	Up to $250,000, or $500,000 for a corporation, and six months in jail
Willful, Death Results, Second Violation	$250,000 and one year in jail
Failure to Correct a Cited Violation	$7,000/day till abated
Failure to post official documents	$7,000 per poster
Falsification of Documents	$10,000 and six months in jail
Assaulting a Compliance Officer	Not more than $5,000 and not more than three years imprisonment

Employers can contest either the citation or the penalty by requesting an informal hearing with the area director to discuss these issues and the area director can enter into a settlement agreement if the situation merits it. But, if a settlement cannot be reached, the employer must notify the area director, in writing, a Notice of Contest of the citation, penalties, or abatement period within 15 days of receipt of the citation.

Workers can challenge an OSHA decision on employer appeals, but have limited challenge rights regarding OSHA's decision. They are

- The time element in the citation for abatement of the hazard.

- An employer's Petition for Modification of Abatement (PMA). Workers have 10 days to contest the PMA.

STATE OSHA PLANS

Most state plans provide for the state to take over the enforcement of workplace safety and health rather than have Federal OSHA perform this service within the state. Many states have opted to take on this responsibility. They are denoted in Appendix J.

The states of Connecticut and New York have unique plans in which they cover only state and local employees (public sector), while federal OSHA covers the general and construction industries. If a state has a federally approved plan or program, the following conditions must exist:

- The state must create an agency to carry out the plan.

- The state's plan must include safety and health standards and regulations. The enforcement of these standards must be at least as effective as the federal plan.

- The state plan must include provisions for right of entry and inspection of the workplace, including a prohibition on advance notice of inspections.

- The state's plan must also cover state and local government employees.

If a state has a plan, are there state specific standards and regulations? The answer is " yes," and they must be at least as stringent as the federal standards and regulations. Some states have standards and regulations which go beyond the requirements of the existing federal standards and regulations, while others simply adopt the federal standards and regulations verbatim.

Anyone who feels their state program has not responded to requests for inspections, complaints of discrimination, or appeals on citations or variances, can file a complaint with federal OSHA. Federal OSHA is responsible to monitor state programs and make evaluations on their effectiveness. A written Complaint Against State Program Administration (CASPA) should contain the

- Description of the attempts to get action from the state and the justification.

- State's response(s) or action(s) which demonstrated poor administration of the state OSHA program.

- Date the incident occurred.

- Exact location where the incident occurred.

- Name of the employer.

- Name and occupations of those involved in the incident.

- Notification to the state agency that a CASPA has been filed.

- Statement requesting confidentiality during the investigation.

A CASPA should be filed when the state plan agency has not

- Conducted an inspection in a timely and effective way.

- Enforced state OSHA standards and regulations.

- Responded to a request for an inspection.

- Protected workers' rights against discrimination.

- Issued citations for violations discovered.

- Complied with proper procedures for granting variances.

Federal OSHA will evaluate the complaint and then notify the worker in writing of its decision. If the filer of a CASPA is not happy with OSHA's response, a written request for a re-evaluation should be sent to the nearest OSHA area office.

WORKER TRAINING

Many standards promulgated by OSHA specifically require the employer to train employees in the safety and health aspects of their jobs. Other OSHA Standards make it the employer's responsibility to limit certain job assignments to employees who are "certified," "competent," or "qualified" — meaning that employees have had special previous training, in or out of the workplace. OSHA regulations imply that an employer has assured that a worker has been trained prior to being designated as the individual to perform a certain task.

In order to make a complete determination of the OSHA requirement for training, one would have to go directly to the regulation that applies to the specific type of activity. The regulation may mandate hazard training, task training, and length of the training, as well as specifics to be covered by the training.

It is always a good idea for the employer, as well as the worker, to keep records of training. These records may be used by a compliance inspector during an inspection, after an accident resulting in injury or illness, as a proof of good intentions, by the employer, to comply with training requirement for workers including new workers and those assigned new tasks.

OCCUPATIONAL INJURIES AND ILLNESSES

The recording and reporting of occupational injuries and illness requirements can be found in 29 CFR 1910 part 1904 – Recording and Reporting Occupational Injuries and Illnesses. These requirements are summarized in the following paragraphs.

Any illness that has been caused by exposure to environmental factors such as inhalation, absorption, ingestion, or direct contact with toxic substances or harmful agents and has resulted in an abnormal condition or disorder that is acute or chronic is classified as an occupational disease. Repetitive motion injuries are also included in this category. All illnesses are recordable, regardless of severity. Injuries are recordable when

- An on-the-job death occurs regardless of length of time between injury and death.
- One or more lost workdays occurs.
- Restriction of work or motion transpires.
- Loss of consciousness occurs.
- Worker is transferred to another job.
- Worker receives medical treatment beyond first aid.

Construction employers with more than ten employees are required to complete and maintain occupational injury and illness records. The OSHA 101 (see Figure 13-3) or equivalent First Report of Injury must be completed within six days of the occurrence of an injury at the worksite and the OSHA 101 must be retained five years. Also, the OSHA 200 log is to be completed within six days when a recordable injury or illness occurs, maintained five years, and posted yearly from February 1 to March 1 (see Figure 13-4). Following is a list of those industries that are required to maintain occupational injury and illness records (Table 13-4).

Bureau of Labor Statistics
Supplementary Record of
Occupational Injuries and Illnesses

U.S. Department of Labor

This form is required by Public Law 91-596 and must be kept in the establishment for 5 years. Failure to maintain can result in the issuance of citations and assessment of penalties.	Case or File No.	Form Approved O.M.B. No. 1220-0029

Employer

1. Name

2. Mail address *(No. and street, city or town, State, and zip code)*

3. Location, if different from mail address

Injured or Ill Employee

4. Name *(First, middle, and last)*   Social Security No.

5. Home address *(No. and street, city or town, State, and zip code)*

6. Age   7. Sex: *(Check one)*   Male ☐   Female ☐

8. Occupation *(Enter regular job title, not the specific activity he was performing at time of injury.)*

9. Department *(Enter name of department or division in which the injured person is regularly employed, even though he may have been temporarily working in another department at the time of injury.)*

The Accident or Exposure to Occupational Illness

If accident or exposure occurred on employer's premises, give address of plant or establishment in which it occurred. Do not indicate department or division within the plant or establishment. If accident occurred outside employer's premises at an identifiable address, give that address. If it occurred on a public highway or at any other place which cannot be identified by number and street, please provide place references locating the place of injury as accurately as possible.

10. Place of accident or exposure *(No. and street, city or town, State, and zip code)*

11. Was place of accident or exposure on employer's premises?   Yes ☐   No ☐

12. What was the employee doing when injured? *(Be specific. If he was using tools or equipment or handling material, name them and tell what he was doing with them.)*

13. How did the accident occur? *(Describe fully the events which resulted in the injury or occupational illness. Tell what happened and how it happened. Name any objects or substances involved and tell how they were involved. Give full details on all factors which led or contributed to the accident. Use separate sheet for additional space.)*

Occupational Injury or Occupational Illness

14. Describe the injury or illness in detail and indicate the part of body affected. *(E.g., amputation of right index finger at second joint; fracture of ribs; lead poisoning; dermatitis of left hand, etc.)*

15. Name the object or substance which directly injured the employee. *(For example, the machine or thing he struck against or which struck him; the vapor or poison he inhaled or swallowed; the chemical or radiation which irritated his skin; or in cases of strains, hernias, etc., the thing he was lifting, pulling, etc.)*

16. Date of injury or initial diagnosis of occupational illness   17. Did employee die? *(Check one)*   Yes ☐   No ☐

Other

18. Name and address of physician

19. If hospitalized, name and address of hospital

Date of report	Prepared by	Official position

OSHA No. 101 (Feb. 1981)

Figure 13-3. OSHA 101. Courtesy of OSHA

Figure 13-4. OSHA 200 Log. Courtesy of OSHA

Table 13-4

Industries Required to Record Injuries and Illnesses

- Agriculture, forestry, and fishing
- Oil and gas extraction
- Construction
- Manufacturing
- Transportation
- Wholesale trade

- Building materials and garden supplies
- General merchandise and food stores
- Hotels and other lodging places
- Repair services
- Amusement and recreation services
- Health services

Some employers are normally not required to keep OSHA records and are included in Table 13-5.

Table 13-5

Employers Not Required to Keep OSHA Injury and Illness Records

- All employers with no more than 10 full- or part-time employees at any one time in the previous calendar year.
- Employers in the following retail trades: finance, insurance, real estate, and services industries.
- Automotive dealers and gasoline service stations.
- Apparel and accessory stores.
- Furniture, home furnishings, and equipment stores.
- Eating and drinking places.
- Miscellaneous retail (SIC 59).
- Banking.
- Credit agencies other than banks.
- Security, commodity brokers, and services.

- Insurance.
- Insurance agents, brokers, and services.
- Real estate.
- Combined real estate, insurance, etc.
- Holding and investment offices.
- Personal services.
- Business services.
- Motion pictures.
- Legal services.
- Educational services.
- Social services.
- Museums, botanical and zoological gardens.
- Membership organizations.
- Private households.
- Miscellaneous services (SIC 89).

Some employers and individuals who never keep OSHA records are

- Self-employed Individuals
- Partners with No Employees
- Employers of Domestics
- Employers Engaged in Religious Activities

MEDICAL AND EXPOSURE RECORDS

Medical examinations are required by OSHA regulations for workers before they can perform certain types of work. This work includes, at the present,

- Asbestos abatement.
- Lead abatement.
- Hazardous waste remediation.
- When workers are required to wear a respirator for 30 days during a year.

Exposure records (monitoring records) are to be maintained by the employer for 30 years. These records include personal sampling, air sampling, and other industrial hygiene sampling records. Medical records are to be maintained by the employer for the length of employment plus 30 years.

In order to access copies of medical records, a worker must make a written request to obtain a copy of their medical records or to make them available to their representative or physician. (See a sample of medical record request letter in Figure 13-5.) A worker's medical record is considered confidential and a request, in writing, from the worker to the physician is required for the records to be released.

I _____, *hereby authorize* _____
(full name of worker/patient)   *(individual or organization holding*
_____ *to release to* _____
the medical records)   *(individual or organization authorized*
_____, *the following medical information from my personal*
to receive the medical information)
medical records: _____
            (Describe generally the information desired to be released)
I give permission for this medical information to be used for the following pur-
pose: _____
_____, *but I do not give*
permission for any other use or re-disclosure of this information.

            _____
            (Full Name of Employee or Legal Representative)

            _____
            (Signature of Employee or Legal Representative)

            _____
            (Date of Signature)

Figure 13-5. Sample authorization letter for the release of employee medical information to a designated representative

POSTING

Employers are required to post in a prominent location the following:

- <u>Job Safety and Health Protection</u> workplace poster (OSHA Form 2203) or state equivalent. (See Figure 13-6.)
- Copies of any OSHA citations of violations of the OSHA standard are to be posted at or near the location of the violation for at least three days or until the violation is abated, whichever is longer.
- Copies of summaries of petitions for variances from any standard, including recordkeeping procedures.
- The summary portion of the "Log and Summary of Occupational Injuries and Illnesses." (OSHA 200 Log or equivalent is to be posted annually from February 1 to March 1.)

WHAT TO DO WHEN OSHA COMES KNOCKING

When an inspector from the Occupational Safety and Health Administration or a corporate or insurance company's safety and health professional shows up at a project or jobsite, there is nothing to worry about if a safety and health program has been implemented and its mandates are being enforced.

To start, the following items should be in place:

1. A job safety and health protection poster (OSHA 2203) is posted on a bulletin board which is visible to all workers.

2. Summaries are available of any petitions for variances.

3. Copies exist of any new or unabated citations.

4. A summary of the OSHA 200 Log which has been posted during the month of February.

5. The following are available for the workers' and inspector's examination:

    - Any exposure records for hazardous materials.

    - The results of medical surveillance.

    - All NIOSH research records for exposure to potentially harmful substances.

6. Verification that workers have been told

    - If exposures have exceeded the levels set by the standard and if corrective measures are being taken.
    - If there are hazardous chemicals in their work area.

7. Training records are available at the time of inspection.

Inspection Process

The inspection process should be handled in a professional manner and mutual respect between the inspector and the employer or representative needs to be developed in a short period of time. It is appropriate to

JOB SAFETY & HEALTH PROTECTION

The Occupational Safety and Health Act of 1970 provides job safety and health protection for workers by promoting safe and healthful working conditions throughout the Nation. Provisions of the Act include the following:

Employers

All employers must furnish to employees employment and a place of employment free from recognized hazards that are causing or are likely to cause death or serious harm to employees. Employers must comply with occupational safety and health standards issued under the Act.

Employees

Employees must comply with all occupational safety and health standards, rules, regulations and orders issued under the Act that apply to their own actions and conduct on the job.

The Occupational Safety and Health Administration (OSHA) of the U.S. Department of Labor has the primary responsibility for administering the Act. OSHA issues occupational safety and health standards, and its Compliance Safety and Health Officers conduct jobsite inspections to help ensure compliance with the Act.

Inspection

The Act requires that a representative of the employer and a representative authorized by the employees be given an opportunity to accompany the OSHA inspector for the purpose of aiding the inspection.

Where there is no authorized employee representative, the OSHA Compliance Officer must consult with a reasonable number of employees concerning safety and health conditions in the workplace.

Complaint

Employees or their representatives have the right to file a complaint with the nearest OSHA office requesting an inspection if they believe unsafe or unhealthful conditions exist in their workplace. OSHA will withhold, on request, names of employees complaining.

The Act provides that employees may not be discharged or discriminated against in any way for filing safety and health complaints or for otherwise exercising their rights under the Act.

Employees who believe they have been discriminated against may file a complaint with their nearest OSHA office within 30 days of the alleged discriminatory action.

Citation

If upon inspection OSHA believes an employer has violated the Act, a citation alleging such violations will be issued to the employer. Each citation will specify a time period within which the alleged violation must be corrected.

The OSHA citation must be prominently displayed at or near the place of alleged violation for three days, or until it is corrected, whichever is later, to warn employees of dangers that may exist there.

Proposed Penalty

The Act provides for mandatory civil penalties against employers of up to $7,000 for each serious violation and for optional penalties of up to $7,000 for each nonserious violation. Penalties of up to $7,000 per day may be proposed for failure to correct violations within the proposed time period and for each day the violation continues beyond the prescribed abatement date. Also, any employer who willfully or repeatedly violates the Act may be assessed penalties of up to $70,000 for each such violation. A minimum penalty of $5,000 may be imposed for each willful violation. A violation of posting requirements can bring a penalty of up to $7,000.

There are also provisions for criminal penalties. Any willful violation resulting in the death of any employee, upon conviction is punishable by a fine of up to $250,000 (or $500,000 if the employer is a corporation), or by imprisonment for up to six months, or both. A second conviction of an employer doubles the possible term of imprisonment. Falsifying records, reports, or applications is punishable by a fine of $10,000 or up to six months in jail or both.

Voluntary Activity

While providing penalties for violations, the Act also encourages efforts by labor and management, before an OSHA inspection, to reduce workplace hazards voluntarily and to develop and improve safety and health programs in all workplaces and industries. OSHA's Voluntary Protection Programs recognize outstanding efforts of this nature.

OSHA has published Safety and Health Program Management Guidelines to assist employers in establishing or perfecting programs to prevent or control employee exposure to workplace hazards. There are many public and private organizations that can provide information and assistance in this effort, if requested. Also, your local OSHA office can provide considerable help and advice on solving safety and health problems or can refer you to other sources for help such as training.

Consultation

Free assistance in identifying and correcting hazards and in improving safety and health management is available to employers, without citation or penalty, through OSHA-supported programs in each State. These programs are usually administered by the State Labor or Health department or a State university.

Posting Instructions

Employers in States operating OSHA approved State Plans should obtain and post the State's equivalent poster.

Under provisions of Title 29, Code of Federal Regulations, Part 1903.2(a)(1) employers must post this notice (or facsimile) in a conspicuous place where notices to employees are customarily posted.

More Information

Additional information and copies of the Act, specific OSHA safety and health standards, and other applicable regulations may be obtained from your employer or from the nearest OSHA Regional Office in the following locations:

Atlanta, GA	(404) 347-3573
Boston, MA	(617) 565-7164
Chicago, IL	(312) 353-2220
Dallas, TX	(214) 767-4731
Denver, CO	(303) 391-5858
Kansas City, MO	(816) 426-5861
New York, NY	(212) 337-2378
Philadelphia, PA	(215) 596-1201
San Francisco, CA	(415) 744-6670
Seattle, WA	(206) 553-5930

Robert B. Reich, Secretary of Labor

U.S. Department of Labor

Occupational Safety and Health Administration

Washington, DC
1995 (Reprinted)
OSHA 2203

This information will be made available to sensory impaired individuals upon request. Voice phone: (202) 219-8615. TDD message referral phone: 1-800-326-2577

Figure 13-6. OSHA form 2203. Courtesy of OSHA

1. Check the compliance officer's credentials and security clearance, if required.

2. Discuss company's safety and health program and its implementation.

3. Delineate activities and initiatives taken to improve safety and health on the job, as well as worker protection.

4. Ask for recommendations and advice that will improve what is being done.

5. Discuss any consultation programs or voluntary participation programs and pursue any inspection exemptions.

6. Ask the purpose, scope, and applicable standards for the inspection and obtain a copy of the employee complaint, if that is what triggered the inspection.

7. Make sure the employer's representative who accompanies the inspector is knowledgeable.

8. Have, if possible, an employees' representative.

9. The employer's representative must be familiar with the project and should try to choose an appropriate route for the inspection. However, the inspector's route cannot be dictated; they can choose their own route for the inspection, if they desire.

10. Make sure all observations, conversations, photographs, readings, and records examined are duplicated. Take good notes and ask appropriate questions.

11. Have records available for the inspector such as the OSHA 200 Log, OSHA 101, exposure records, and training records.

12. Pay close attention to unsafe or unhealthy conditions that are observed. Discuss how to correct them with the inspector and take corrective actions immediately, if possible.

13. Never at any time interfere with employee interviews with the inspector.

Mitigating the Damage

There is no turning back; the inspection will occur. It is imperative that the inspection's outcomes result in as little damage as possible. This can be accomplished in many ways during the inspection process. Some actions may seem redundant, but they need to be reinforced.

1. Ask for an OSHA consultation service or pursue an exemption if the inspector cannot tell you how to abate or correct a violation.

2. Know the jobsite and be familiar with all the processes and equipment.

3. Try to select the inspector's route, if possible. Save the known or suspect problem areas for last.

4. Take good notes and document the inspection process completely. Photograph anything that the inspector does.

5. Many benefits are gained from good recordkeeping.

6. Correct apparent violations immediately, if possible.

7. Maintain updated copies of any required written programs.

Closing Conference

During the closing conference, when the culmination of the inspection process occurs, adhere to the following items to maintain the overall continuity of the process:

1. Listen actively and carefully to the discussion of unsafe or unhealthy conditions and apparent violations.

2. Ask questions for clarification so as to avoid confrontation. Confrontation will accomplish nothing.

3. Make sure the inspector discusses the appeal rights, informal conference procedure rights, and procedures for contesting a citation.

4. Produce documentation to support the company's compliance efforts or special emphasis programs.

5. Provide information that will guide the inspector in setting the times for abatement of citations.

After the Inspector Leaves

Citations and notices will arrive by certified mail and will need to be posted at or near the area where the violation occurred for at least 3 days or until abated, whichever is longer. Any notice of contest or objection must be received by the OSHA area director in writing within 15 days of the receipt of any citations. The area director will forward the notice of contest to the Occupational Safety and Health Review Commission (OSHRC). It is also a good idea to request an informal meeting with the area director during the fifteen day period.

The notice of contest will be assigned to an administrative law judge by the OSHRC. Once the judge rules on the contest notice, further review by OSHRC may be requested. If necessary, the OSHRC ruling can be appealed to the U.S. Court of Appeals.

Remember that all citations or violations must be corrected or abated by the prescribed date unless the citation or abatement date is formally contested. If the response to a citation or violation cannot be abated in the time allotted, due to factors which are beyond reasonable control, a petition to modify the time for abatement must be filed with the area director to extend the date.

Make the Inspection a Positive Experience

A proactive safety and health preparation can make for a quality safety program. This is a safeguard for property, equipment, profits, and liability, as well as for workers. Working with employees to correct deficiencies can foster better safety attitudes. A safer workplace is also a more productive workplace. This also safeguards a very important asset, "the worker." Remember, OSHA has a great deal of expertise within its ranks. Use OSHA as a resource to improve your safety and health program.

Prior to the knock by OSHA, it is necessary to implement a safety and health program. This includes, as iterated earlier, the following:

1. Formal written program or safety manual.

2. Standard operating procedures (SOPs) that incorporate OSHA standards.

3. Worker and supervisor training.

4. Standard recordkeeping procedures.

5. Workplace inspections, audits, job observations, and job safety analyses.

6. Safety and health committees, if possible.

7. Accident or incident investigation procedures.

8. Hazard recognition or reporting procedures.

9. First aid and medical facility availability.

10. Employee medical surveillance or examinations.

11. Consultation services available.

With these as prerequisites, an OSHA inspection quickly becomes a positive learning experience from which many benefits will be reaped. It can produce higher morale, better production, a safer workplace, and a better bottom line since many negatives have been avoided by good pre-activity and planning.

MULTI-EMPLOYER WORKSITES

On multi-employer worksites, both construction (Figure 13-7) and non-construction citations are normally issued to the employer whose employees are exposed to workplace hazards (The Exposing Employer). In addition, the following employers normally shall be cited, whether or not their own employees are exposed:

• The employer who actually creates the hazard (The Creating Employer).

• The employer who is responsible, by contract or through actual practice, for safety and health conditions on the worksite: i.e., the employer who has the authority for ensuring that the hazardous condition is corrected (The Controlling Employer).

• The employer who has the responsibility for actually correcting the hazard (The Correcting Employer).

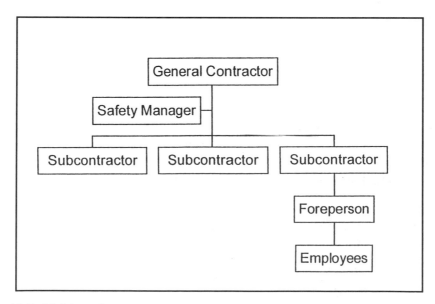

Figure 13-7. Multi-employer construction worksites. Courtesy of OSHA

Prior to issuing citations to an exposing employer, it must first be determined whether or not the available facts indicate the employer has a legitimate defense to the citation. This is accomplished by answering the following questions:

- Did the employer create the hazard?
- Did the employer have the responsibility or authority to have the hazard corrected?
- Did the employer have the ability to correct or remove the hazard?
- Did the employer demonstrate that the creating, the controlling, or the correcting employers, as appropriate, have been specifically notified of the hazard to which their employees are exposed?
- Did the employer instruct employees to recognize the hazard?

Where feasible, an exposing employer must have taken appropriate, alternative means of protecting employees from the hazard; and when extreme circumstances justify it, to avoid a citation, the exposing employer shall remove employees from the job. If an exposing employer has met all of the previous criteria, then the employer shall not be cited.

If all employers on a worksite who have employees exposed to a hazard meet the previous criteria, the citation shall be issued only to the employers who are responsible for creating the hazard or are in the best position to correct or ensure correction of the hazard. In such circumstances, the controlling employer and/or the hazard-creating employer shall be cited even though none of their employees are exposed to the condition that resulted in the violation. Penalties for such citations shall be appropriately calculated by using the exposed employees of all employers as the number of employees for probability assessment.

SUMMARY

It is envisioned that this chapter will be an asset to both contractors and workers. The knowledge will place both entities, labor and management, on a "so-called" level playing field. Knowledge has been shown to fix accountability, as well as responsibility, upon those who claim ignorance of it. The workplace is where both labor and management spend the bulk of their waking hours. With this in mind, the safety and health of those in the workplace should be everyone's concern and responsibility.

Contractors and safety and health professionals need to know how OSHA provides for worker safety and health on construction worksites. This will also assist in assuring that the workers rights are protected and give them the knowledge to help mitigate health and safety issues and problems which may arise. This type of knowledge should ensure a safer and more productive worksite. Respect for the efficient, effective, and proper use of the health and safety rules will have a positive effect upon those in the workplace.

Although it is the ultimate responsibility of the employer to provide for workplace safety and health, adherence to OSHA occupational safety and health rules is the foundation upon which a good safety and health program can be built. The program should hold everyone responsible for the well-being of those in the workplace, including the employer, managers, supervisors, and workers. All should abide by the safety and health rules and the OSHA standards. Together, and through cooperation, all parties can assure a safe and healthy workplace.. A safe and healthy home away from home is and should be the ultimate goal.

FURTHER READING

Access to Medical Records and Exposure Records (OSHA 3110), U.S. Department of Labor/
    OSHA, Washington, D.C., 1988.

All About OSHA (OSHA 2056), U.S. Department of Labor/OSHA, Washington, D.C., 1985.

Anton, Thomas J., *Occupational Safety and Health Management*, Second Edition, McGraw-
    Hill, Inc., New York, NY, 1989.

Bertinuson, Janet and Weinstein, Sidney, *Occupational Hazards of Construction: A Manual
    For Building Trades Apprentices*, Labor Occupational Health Program, University of
    California, Berkeley, CA 1978.

Blosser, Fred, *Primer on Occupational Safety and Health*, The Bureau of National Affairs,
    Inc., Washington, D.C., 1992.

Murphy, W.C. and Hanson, J.R., *A Maine Guide to Employment Law,* The University of Maine,
    Orono, ME, 1995.

OSHA: Employee Workplace Rights (OSHA 3021), U.S. Department of Labor/OSHA,
    Washington, D.C., 1991.

Protecting Workers Lives: A Safety and Health Guide for Unions, Second Edition, National
    Safety Council, Itasca, IL, 1992.

Recordkeeping Guidelines for Occupational Injuries and Illnesses (OMB No. 1220-0029),
    U.S. Department of Labor/OSHA. Washington, D.C., 1986.

Training Course in OSHA for the Construction Industry(Course #500), OSHA Training Institute,
    Des Plaines, IL, 1997.

Chapter 14

CONSTRUCTION STANDARDS

Charles D. Reese

FEDERAL LAWS

Congress establishes federal laws (legislation or acts) and the President signs them into law. These laws often require that regulations (standards) be developed by the federal agencies who are responsible for the intent of the law.

OSHAct

The Occupational Safety and Health Act (OSHAct) of 1970 is such a law and is also called the Williams-Steiger Act. It was signed by President Richard Nixon on December 29, 1970 and became effective April 29, 1971. [The OSHAct was not amended until November 5, 1990 by Public Law 101-552.] The OSHAct assigned the responsibility of implementing and enforcing the law to a newly created agency, the Occupational Safety and Health Administration (OSHA), located in the Department of Labor (DOL).

Most such federal laws (acts) contain the following content or elements:

1. The reason for the law.

2. A statement of the national policy related to the law.

3. Objectives/goals/outcomes expected of the law.

4. Authorization of the agency responsible for implementation.

5. Requirements and structure of the regulations to be developed.

6. Time frames for regulation implementation or deadlines.

7. Enforcement guidelines to be followed.

8. Fines or assessments available to the enforcing agency.

9. Specific actions required by the law.

THE CONTENT OF THE OSHAct

Prior to the OSHAct, there were some state laws, a few pieces of federal regulations, and a small number of voluntary programs by employers. Most of the state programs were limited in scope and the federal laws only partially covered workers.

631

Another important reason for the OSHAct was the increasing number of injuries and illnesses within the workplace. Thus, the OSHAct was passed with the express purpose of assuring that every working man and woman in the nation would be provided safe and healthful work conditions while preserving this national human resource, the American worker. The OSHAct is divided into sections with each having a specific purpose. The full text of the OSHAct, all thirty-one pages, can be found in Appendix I. As a quick reference to the OSHAct, the following paragraphs summarize what each section includes.

The OSHAct starts in Section 2 and contains congressional findings. Due to excessive injuries and illnesses, employers now have specific responsibilities regarding occupational safety and health (OS&H). It is the responsibility of the Secretary of Labor to institute OSHA. He or she will oversee the development and implementation of workplace health and safety standards, including any research and training required, as well as assure the enforcement of OSHA standards, entice states to become involved, develop reporting requirements for injuries and illnesses, and foster joint labor/management efforts regarding OS&H.

Section 3 of the OSHAct defines the employer as a person whose company is engaged in a business that affects commerce. This definition does not include the United States or other government entities. Also, the definition of employees are those employed by an employer who affects commerce. There are also other definitions in this section that are pertinent to the OSHAct.

Section 4 explains the applicability of the Act. In this section the OSHAct is described as not applicable to other federal agencies that exercise their own authority over OS&H. The Act supersedes other existing federal laws and regulations related to OS&H and will not have a similar affect on any workers' compensation laws which already exist.

Section 5 includes the "General Duty Clause" Section (5)(a)(1) which states that each employer shall furnish employment free from recognized hazards. This allows OSHA inspectors to cite an employer even if no OSHA regulation exists for an observed/known workplace hazard. Also, it requires employers to comply with the OSHA standards and employees to comply with rules and regulations.

Section 6 provides OSHA the authority to promulgate start-up standards without following a formal rule-making procedure. This section addresses rule-making procedures, emergency temporary standards, variances from standards, the use of the *Federal Register* for publishing the required public notices during the standard development process, as well as the final standard. Many other issues are also addressed: medical examinations, toxic materials, PPE, labels, etc. The main intent of this section is the promulgation of OS&H standards.

Section 7 delineates the responsibility of the Secretary of Labor to establish an advisory committee on OS&H and provides the resources for the mission and intent of the advisory committee. The procedures and resources available to the committee are explained. This section authorizes OSHA to make use of the services and personnel of state and federal agencies and to provide OS&H consultative services.

Section 8 deals with inspections, investigations, and recordkeeping. It gives the OSHA representative the authority to enter workplaces without delay, at reasonable times, and inspect during regular working hours. During the inspection, the OSHA inspector may be accompanied by an employer representative and an employee representative, if they so desire. The OSHA inspector has the authority to question, privately, employers and employees. **(Note: The Marshall versus Barlow decision {1978} requires a warrant if denied entry.)** Section 8 also provides OSHA subpoena power. Employers are required to maintain and post injury and illness records as well as exposure records. Workers can file a complaint with OSHA if they believe that their workplace is subject to physical hazards or imminent danger. OSHA will make the determination on whether or not the complaint merits a formal inspection.

Section 9 states that employers who have violated Section 5 of the Act or any standard, regulation, rule, or order related to Section 6 of the Act, shall be issued a citation. The citation will be in writing, describing the particular violation, and reference the standard, rule, regulation, or order location in the Act. These citations are to be posted by the employer. Citations must be issued within six months following a detected violation.

Section 10 sets forth the enforcement procedures. The employer has the right to contest any citation, procedure, and time for abatement, and to receive information concerning how the contested citation will be handled. The employees' rights are limited to contesting the abatement time for a hazard only.

Section 11 provides for the appeal and review of any orders issued by the Occupational Safety and Health Review Commission. This section also addresses discrimination by the employer against workers who decide to exercise their right to complain formally or informally regarding safety and health issues.

Section 12 mandates the formation of the Occupational Safety and Health Review Commission (OSHCR) which is composed of three members appointed by the President for a six year term. The commission conducts hearings, when necessary, relevant to the OSHAct or reviews processes, violations, and concerns.

Section 13 requires the Secretary of Labor to take action to protect workers from imminent danger. The Secretary can be held liable for arbitrary or capricious disregard of an imminent danger which is brought to his/her attention.

Section 14 provides for the Solicitor of Labor to represent the Secretary during litigation.

Section 15 protects the trade secrets of a company by requiring that any information gathered during performance of an inspection, by the either the Secretary or his/her representative, be confidential.

Section 16 provides the Secretary with the power to make variations, tolerances, and exemptions from any or all provision of the OSHAct when the impairment of the national defense is threatened. This can take place for a period of six months without notifying employees or having a hearing.

Section 17 deals with the issuance of citations and their accompanying penalties. The types of violations and the amounts of the penalties, as well as the reason for such penalties are discussed.

Section 18 allows for states to assume responsibility from the federal authorities for the safety and health program, but federal OSHA must approve the plan. If no federal standards are in effect, the states may issue their own standards. Federal OSHA will monitor, support, and evaluate the approved states' plans.

Executive Order 12196, Section 19, states the responsibilities of federal agencies regarding safety and health and requires these agencies to have effective occupational safety and health programs.

Section 20 mandates that the Department of Health and Human Services (DHHS) be responsible for the research functions under the Act, and that the National Institute for Occupational Safety and Health (NIOSH) carry out most of these functions.

Section 21 requires DHHS to carry out training and employee education by utilizing grants, contracts, and short-term training.

Section 22 mandates the establishment of the National Institute for Occupational Safety and Health to conduct research and training relevant to occupational safety and health.

Section 23 authorizes the Department of Labor to make grants available to the states in order to assist them in the operation of their occupational safety and health programs.

Section 24 provides for the collection and analysis of statistics concerning occupational fatalities, injuries, and illnesses. This data is to be collected and compiled by the Bureau of Labor Statistics (BLS).

Section 25 requires the recipients of grants to maintain records. It also gives authority to the Secretaries of DHHS and DOL to conduct audits when deemed appropriate and necessary.

Section 26 requires the Secretaries of DOL and DHHS to provide an annual report within 120 days of the convening of each regular session of Congress. This section also dictates the required content of these reports.

Section 27 establishes a National Committee on State Workmen's Compensation Laws to study and evaluate the fairness and adequacy of the present laws.

Section 28 amends the Small Business Act and allows for loans to be given to small businesses in order for them to comply with the OSHAct.

Section 29 adds an Assistant Secretary of Labor for Occupation Safety and Health.

Section 30 allows for an additional 25 DOL and 10 DHHS administrative positions to aid in the implementation of the Act.

Section 31 amends the Federal Aviation Act of 1958 to require fixed-wing powered aircraft that are used in air commerce to have an emergency locator beacon.

Section 32 states, if any provision or application of the Act is invalid for any person, then the remainder of the Act or its application or provisions are held invalid for that person.

Section 33 gives authorization to OSHA to receive funding to carry out the mandate of the Act. This is based upon Congress's approval of necessary funding levels.

Section 34 specifies the effective date of this Act was 120 days after the date of its enactment.

THE REGULATORY PROCESS

The Occupational Safety and Health Administration (OSHA) was mandated to develop, implement, and enforce regulations relevant to workplace safety and health and the protection of workers. Time constraints prevented the newly formed OSHA from developing brand new regulations. Therefore, OSHA adopted previously existing regulations from other government regulations, consensus standards, proprietary standards, professional groups standards, and accepted industry standards. This is the reason that today the hazardous chemical exposure levels, with a few exceptions, are the same as the existing Threshold Limit Values (TLVs) published by the American Congress of Government Industrial Hygienist in 1968. Once these TLVs were adopted, it became very difficult to revise them. Even though research and knowledge in the past 30 years have fostered newer and safer TLVs, they have not been adopted by OSHA.

As stated previously, the original OSHA standards and regulations have come from three main sources: consensus standards, proprietary standards, and federal laws that existed when the Occupational Safety and Health Act became law.

Consensus standards are developed by industry-wide standard-developing organizations and are discussed and substantially agreed upon through industry consensus. OSHA has incorporated into its standards the standards of two primary groups: the American National Standards Institute (ANSI) and the National Fire Protection Association (NFPA).

As an example, ANSI A10.33, Safety and Health Program Requirements for Multi-Employer Projects, covers minimum elements and activities of a program. It also defines the duties and responsibilities of the individual construction employers who will be working on a construction project.

Another consensus standard source was the NFPA standards. NFPA No. 30-1969, Flammable and Combustible Liquids Code, was the source standard for Part 1910, Section 106. It covers the storage and use of flammable and combustible liquids that have flash points below 200°F.

Proprietary standards are prepared by professional experts within specific industries, professional societies, and associations. The proprietary standards are determined by a straight membership vote, not by consensus.

An example of these standards can be found in the "Compressed Gas Association, Pamphlet P-1, Safe Handling of Compressed Gases." This proprietary standard covers requirements for safe handling, storage, and use of compressed gas cylinders.

Some of the pre-existing federal laws that are enforced by OSHA include the Federal Supply Contracts Act (Walsh-Healy), the Federal Service Contracts Act (McNamara-O'Hara), the Contract Work Hours and Safety Standard Act (Construction Safety Act), and the National Foundation on the Arts and Humanities Act. Standards issued under these Acts are now enforced in all industries where they apply.

When OSHA needs to develop a new regulation or even revise an existing one, it becomes a lengthy and arduous process. This is why it took so long to get the following regulations passed:

- Process Chemical Safety Standard – 7 years
- Hazard Communications Standard – 10 years
- Lockout/Tagout Standard – 12 years (Still does not apply to construction.)
- Confined Spaces – 17 years (Still does not apply to construction.)

But, it took only 3 years to get a new regulation passed covering lift-slab construction after the collapse of L'Ambience Plaza in Bridgeport, CT where 28 workers died. Also, only a short period of time lapsed in getting a bloodborne pathogen standard when people were scared to death of HIV (AIDS) and Hepatitis B virus (HBV).

Standards are sometimes referred to as being either "horizontal" or "vertical" in their application. Most standards are "horizontal" or "general." This means they apply to any employer in any industry. Fire protection, working surfaces, and first aid standards are examples of "horizontal" standards.

Some standards are relevant to only a particular industry and are called "vertical" or "particular" standards. Examples of these standards applying to the construction industry, the longshoring industry, and the special industries are covered in Subpart R of 29 CFR 1910.

Through the newspapers and conversations, it certainly sounds as if OSHA is producing new standards each day which will impact the workplace. This simply is not true. The regulatory process is very slow. Why in some cases is the time so long and others so short? Aren't the same steps followed for each regulation? The answer is yes, the process is the same, but at each step the time and the stumbling blocks may not be the same. The steps are as follows:

1. The Agency (OSHA) opens a Regulatory Development Docket for a new or revised regulation.

2. This indicates that OSHA believes a need for a regulation exists.

3. An Advanced Notice of Proposed Rulemaking (ANPRM) is published in the *Federal Register* and written comments are requested to be submitted within 30-60 days.

4. The comments are analyzed.

5. A Notice of Proposed Rulemaking (NPRM) is published in the *Federal Register* with a copy of the proposed regulation.

6. Another public comment period transpires usually for 30–60 days.

7. If no additional major issues are raised by the comments, the process continues to step 10.

8. If someone raises some serious issues, the process goes back to step 4 for review and possible revision of the NPRM.

9. Once the concerns have been addressed, it continues forward to steps 5 and 6 again.

10. If no major issues are raised, a Final Rule (FR) will be published in the *Federal Register*, along with the date when the regulation will be effective (usually 30–120 days).

11. There can still be a Petition of Reconsideration of the Final Rule. There are times when an individual or industry may take legal action to bar the regulation's promulgation.

12. If the agency does not follow the correct procedures or acts arbitrarily or capriciously, the court may void the regulation and the whole process will need to be repeated.

If you desire to comment on a regulation during the development process, feel free to do so; your comments are important. You should comment on the areas where you agree or disagree. This is your opportunity to speak up. If no one comments, it is assumed that nobody cares one way or the other. You must be specific. Give examples, be precise, give alternatives, and provide any data or specific information which can back up your opinion. Federal agencies always welcome good data which substantiates your case. Cost/benefit data are always important in the regulatory process and any valid cost data that you are able to provide may be very beneficial. But, make sure that your comments are based upon what is published in the *Federal Register* and not upon hearsay information. Remember that the agency proposing the regulation may be working under specific restraints. Make sure you understand these constraints. Due to restrictions, the agency may not have the power to do what you think ought to be done.

Sometimes the agency feels that there is not a need for the proposed regulation, but it has been mandated to develop it. Your comments could be useful in stopping the development of this regulation. Just be sure your comments are polite, not demeaning or combative. Remember an individual has worked on this proposed regulation and is looking for constructive and helpful comments. Even if you are against this regulation, do not let your comments degenerate to a personal level. Focus on the regulation, not individuals.

THE FEDERAL REGISTER

The *Federal Register* is the official publication of the United States Government. If you are involved in regulatory compliance, you should obtain a subscription to the *Federal Register*. The reasons for obtaining this publication are clear. It is official, comprehensive, and not a summary done by someone else. It is published daily and provides immediate accurate information. The *Federal Register* provides early notices of forthcoming regulations, informs you of comment periods, and gives the preamble and responses to questions raised about a final regulation. It provides notices of meetings, gives information on obtaining guidance documents, and supplies guidance on findings, on cross references, and gives the yearly regulatory development agenda. It is the "Bible" for regulatory development. It is published daily and is recognizable by brown paper and newsprint quality printing (see Figure 14-1).

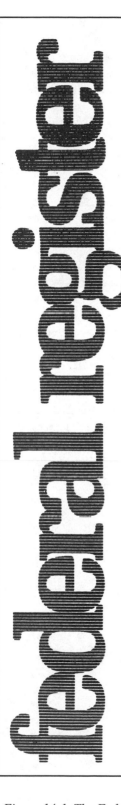

**Friday
December 6, 1991**

Part II (Excerpts)
Pages 64175 thru 64182

Department of Labor

Occupational Safety and Health
Administration

29 CFR Part 1910.1030
Occupational Exposure to Bloodborne
Pathogens; Final Rule

Figure 14-1. The Federal Register

THE CODE OF FEDERAL REGULATIONS

Probably one of the most common complaints from people who use the *U.S. Code Of Federal Regulations* is, "How do you wade through hundreds of pages of standards and make sense out of them?" From time to time you may have experienced this frustration and been tempted to throw the standards into the "round file."

The *Code of Federal Regulations* (CFR) is a codification of the general and permanent rules published in the *Federal Register* by the Executive departments and agencies of the Federal Government. The code is divided into 50 titles which represent broad areas that are subject to federal regulations. Each title is divided into chapters which usually bear the name of the issuing agency. Each chapter is further subdivided into parts covering specific regulatory areas. Based on this breakdown, the Occupational Safety and Health Administration is designated Title 29 – Labor, Chapter XVII (Occupational Safety and Health Administration) and Part 1926 for the Construction Industry Sector.

Each volume of the *Code of Federal Regulations* is revised at least once each calendar year and issued on a quarterly basis. OSHA issues regulations at the beginning of the fourth quarter, or July 1 of each year (the approximate revision date is printed on the cover of each volume).

The Code of Federal Regulations is kept "up-to-date" by individual revisions issued in the Federal Register. These two publications (The CFR and The Federal Register) must be used together to determine the latest version of any given rule.

To determine whether there have been any amendments since the revision date of the U.S. Code volume in which you are interested, the following two lists must be consulted: The "Cumulative List of CFR Sections Affected," issued monthly; and the "Cumulative List of Parts Affected," appearing daily in the Federal Register. These two lists refer to the *Federal Register* page where you may find the latest amendment of any given rule. The pages of the *Federal Register* are numbered sequentially from January 1 to January 1 of the next year.

As stated previously, Title 29, Chapter XVII has been set aside for the Occupational Safety and Health Administration. Chapter XVII is broken down into parts. Part 1926 contains the Construction Industry Standards. The Construction Industry Standards are further broken down into subparts, sections, and paragraphs.

REGULATION PARAGRAPH NUMBERING SYSTEM

In order to use the *Code of Federal Regulations,* you need an understanding of the hierarchy of the paragraph numbering system. The numbering system is a mixture of letters and numbers. Prior to 1979, italicized small case letters and small case roman numerals were used. A change was made after 1979.

CFR Numbering Hierarchy

<1979	1980
(a)	(a)
(1)	(1)
(i)	(i)
Italicized (a)	(A)
Italicized (1)	(1)
Italicized (i)	(*i*)

When trying to make use of the regulations, having knowledge of the regulatory numbering system will help remove a lot of the "headaches." This should make them easier to comprehend and more user friendly. The following illustrates and explains the numbering system using an example from the

29 CFR 1926.59 (h)(3)(ii)

Employee training shall include at least the physical and health hazards of the chemical in the work area.

Title	Code of Fed. Reg.	Part	Subpart	Section	Paragraph
29	CFR	1926	D	.59	

As can be seen from this example, the first number (29) stands for the Title. Next comes CFR which, of course, stands for the Code of Federal Regulations, followed by 1926 which is the Part. Finally, there is a period which is followed by an arabic number. This will always be the Section number. In this case Section .59 is the hazard communication standard. If the number had been .146, the section would pertain to permit-required confined spaces.

29 CFR 1926.59 (h)(3)(ii)

Employee training shall include at least the physical and health hazards of the chemical in the work area.

Title	Code of Fed. Reg.	Part	Subpart	Section	Paragraph
29	CFR	1926	D	.59	(h)

This means that the next breakdown of paragraphs will be sequenced by using small case letters in parentheses (a), (b), (c), etc. If you had three major paragraphs of information under a section, they would be lettered .59(a), .59(b), and .59(c).

29 CFR 1926.59 (h)(3)(ii)

Employee training shall include at least the physical and health hazards of the chemical in the work area.

Title	Code of Fed. Reg.	Part	Subpart	Section	Paragraph and Subparagraph
29	CFR	1926	D	.59	(h)(3)

The next level of sequencing involves the use of arabic numbers. As illustrated, if there were three paragraphs of information between subheadings (a) and (b), they would be numbered (a)(1), (a)(2), and (a)(3).

29 CFR 1926.59 (h)(3)(ii)

Employee training shall include at least the physical and health hazards of the chemical in the work area.

Title	Code of Fed. Reg.	Part	Subpart	Section	Paragraph and Subparagraphs
29	**CFR**	**1926**	**D**	**.5**	**(h)(3)(ii)**

The next level uses the lower case roman numerals. An example would be between paragraphs (2) and (3). If there were five paragraphs of information pertaining to arabic (2) they would be numbered (2)(i), (2)(ii), (2)(iii), (2)(iv), and (2)(v).

If there are subparagraphs to the lower case roman numerals, then a capital or upper case letter is used such as (A), (B),...(F). Any other subparagraph falling under an upper case letter is numbered using brackets for example {*1*}, {*5*}...{*23*}, and any subparagraph to the bracketed numbers would be denoted by an italicized roman numeral as follows: *(i), (iv)...(ix)*.

The construction standard 29 CFR 1926 is divided into 26 subparts lettered A through Z. In the following section of this chapter, each subpart will be highlighted with an overview paragraph, a listing of all the sections in it, a listing of the most frequent citation issued by OSHA related to the subpart (number of times violation was cited is found in parentheses), and a short checklist to assist in deciding which subparts of this regulation applies to your type of construction work. If you check an entry on a subpart, then you will need to comply with part or all of that subpart.

29 CFR 1926 – SAFETY AND HEALTH REGULATIONS FOR CONSTRUCTION

Subpart A – General

Subpart A's first section explains the purpose, scope, and applicable policies. This subpart continues with a discussion of variances, inspections, and rules or guidelines used through the construction regulation. Right of entry is discussed in some detail. The Secretary of Labor or his/her representative must be allowed the right of entry at each worksite which is subject to the Contract Work Hours and Safety Standards Act (Section 107). The Secretary has the right to conduct inspections to ensure compliance. Also, the rules of practice for administration adjudications for the enforcement of safety and health is discussed.

Sections:

1926.1   Purpose and scope.
1926.2   Variances from safety and health standards.
1926.3   Inspections - right of entry.
1926.4   Rules of practice for administrative adjudications for enforcement of safety and health standards.

<u>Checklist:</u>

_____ Does your company have a contract subject to Section 107?
_____ Does your company have to comply with the construction standards?
_____ Has your company had a need for a variance from a safety or health standard?
_____ Has your company been previously inspected by OSHA?

Subpart B – General Interpretations

Subpart B applies to construction work conducted under contract to government agencies and the application and interpretation of how construction health and safety standards are applicable. Section 107 (Contract Work Hours and Safety Standards Act) is related to construction contracts and states that a contract is one which is entered into under the auspices and subject to Reorganization Plan Number 14. This plan relates to the prescribing by the Secretary of "appropriate standards, regulations, and procedures" with respect to enforcement of labor standards under federal and federally assisted contracts. These standards are subject to various statutes listed in this subpart including, for example, the Davis-Bacon Act, Federal-Aid Highway Acts, National Housing Act, etc. This subpart includes the requirements of the Walsh-Healy Public Contracts Act which calls for worksite safety and health requirements for the construction industry.

Also, the aim of Subpart B is to assure that no contractor requires a worker to work in conditions which are unsanitary, hazardous, or dangerous to the worker's safety and health. This subpart ultimately holds the prime contractor responsible for safety and health requirements related to such issues as toilet facilities and first aid. Subcontractors may by agreement let the prime contractor provide certain requirements but are not relieved of legal responsibility if the prime contractor fails to perform. Thus, all regulations are to be followed while performing under these contracts.

<u>Sections:</u>

1926.10 Scope of subpart.
1926.11 Coverage under Section 103 of the act distinguished.
1926.12 Reorganization Plan No. 14 of 1950.
1926.13 Interpretation of statutory terms.
1926.14 Federal contracts for "mixed" types of performance.
1926.15 Relationship to the Service Contract Act; Walsh-Healey Public Contracts Act.
1926.16 Rules of construction.

<u>Checklist:</u>

_____ Does your company perform work under contracts let by federal agencies?
_____ Does your company subcontract parts of the work under these contracts?
_____ Do any of the 58 statutes under the Reorganization Plan Number 14 apply to any of your contracts?
_____ Do you hold subcontractors accountable or delegate responsibilities to them?

Subpart C – General Safety and Health Provisions

Subpart C covers a myriad of general safety and health provisions and again iterates that no contractor or subcontractor can require a worker to perform in hazardous, unsafe, or unhealthy conditions. This subpart delineates the requirement for a safety and health program (accident prevention program). It also requires inspections of jobsites by competent persons. Other requirements include first aid, housekeeping at the site, illumination, fire protection and prevention, sanitation, egress, personal protective equipment, and emergency plans.

Included in this subpart is mandated safety training and education for workers. Also, the medical and exposure record access and preservations stipulations are listed as well as the agreement of OSHA to maintain the confidentiality of trade secrets which are to be used only in emergency situations by doctors or inspectors.

Subpart C addresses the need for exits (egress) stating that all exits must be kept free of obstructions and impediments. Emergency action plans are required which can assure adequate escape procedures, evacuation routes, alarm systems, and other emergency actions.

Sections:

1926.20 General safety and health provisions.
1926.21 Safety training and education.
1926.22 Recording and reporting of injuries. [Reserved]
1926.23 First aid and medical attention.
1926.24 Fire protection and prevention.
1926.25 Housekeeping.
1926.26 Illumination.
1926.27 Sanitation.
1926.28 Personal protective equipment.
1926.29 Acceptable certifications.
1926.30 Shipbuilding and ship repairing.
1926.31 Incorporation by reference.
1926.32 Definitions.
1926.33 Access to employee exposure and medical records.
1926.34 Means of egress.
1926.35 Employee emergency action plans.

Violations/Citations:

1926.20(b)(1)	–	Employer responsibility to initiate and maintain safety and health programs. (5345)
1926.20(b)	–	Employer responsibility to provide for frequent and regular inspections by designated competent persons. (4385)
1926.21(b)	–	Employer responsibility to instruct each employee in the recognition and avoidance of unsafe conditions. (7975)
1926.25(a)	–	Employer responsibility for debris and scrap lumber with protruding nails, not cleared from work areas, stairs, and around structures. (3478)
1926.28(a)	–	Employer responsibility for requiring the wearing of appropriate personal protective clothing. (3240)

Checklist:

_____ Should you have an accident prevention program in place?
_____ Do you provide training and education to construction workers regarding hazardous conditions, safe work practices, and accident prevention?
_____ Does your company have medical or exposure records?
_____ Should you have a emergency action or escape plan or procedure?
_____ Are all exits unlocked and free from impediments?
_____ Do your workers wear personal protective equipment?
_____ Do you take an initiative to prevent fires?
_____ Is housekeeping a part of jobsite procedures?

Subpart D – Occupational Health and Environmental Controls

Subpart D encompasses a host of topics of which the sections are somewhat unique and at times seem unrelated. The first section addresses medical services and making medical professionals available to workers for consultation and advice regarding issues of occupational health. The requirement exists for first aid supplies to be available to workers.

Sanitation deals with the availability of adequate toilet facilities for workers and requires a certain number based upon the number of workers at the site. This also includes the requirement for washing facilities when working with harmful substances and the requirements for drinking water (both the quality of the water and the use of single service drinking cups).

Harmful substances are delineated in this subpart either as individual chemicals such as lead, cadmium, and methylenedianiline or in the form of exposure limits (threshold limit values - TLVs) for some 600 different chemicals. Since most of these chemicals present an airborne hazard, a rather extensive section is given on ventilation which includes the use of exhaust fans and airborne hazard removal from operating equipment such as grinders to airborne contaminants to achieve compliance.

Other sections in subpart D contain requirements regarding other hazards such as occupational noise exposure and noise levels and maximum exposure times for those levels. Also, precautions involving both ionizing and nonionizing radiation sources are included, as well as the minimum illumination requirements.

The remaining sections deal with chemical and physical hazards. The hazard communication standard conveys information concerning workplace hazards to workers and requires information on hazardous chemicals be made available to workers, as well as appropriate training on these hazards. The hazard communication standard also requires the use of DOT marking, placards, and labels for chemicals being used in construction. In cases where workers are required to deal with high hazard chemical processes on a regular basis, either for construction or maintenance purposes, specific procedures and plans are required. Very stringent requirements also exist for the remediation of hazardous waste and emergency response with regard to site planning, training of workers, safe operating procedures, medical surveillance, and the use of personal protective equipment.

Sections:

1926.50 Medical services and first aid.
1926.51 Sanitation.
1926.52 Occupational noise exposure.
1926.53 Ionizing radiation.
1926.54 Nonionizing radiation.
1926.55 Gases, vapors, fumes, dusts, and mists.
1926.56 Illumination.
1926.57 Ventilation.
1926.58 [Reserved]
1926.59 Hazard communication.
1926.60 Methylenedianiline.
1926.61 Retention of DOT markings, placards, and labels.
1926.62 Lead.
1926.63 Cadmium (this standard has been redesignated as 1926.1127).
1926.64 Process safety management of highly hazardous chemicals.
1926.65 Hazardous waste operations and emergency response.
1926.66 Criteria for design and construction for spray booths.

Violations/Citations:

1926.50 — First aid shall be available in the absence of an infirmary, or other access that is reasonable. First aid supplies shall be accessible and telephone number posted. (3358)

1926.59 — Hazard Communication Program, list of chemicals, training, and MSDSs. (66008)

Checklist:

_____ Does your company use any chemicals which could be considered hazardous?
_____ Does your company have a medical officer for examinations, advice, or consultation?
_____ Have you had injuries or illnesses which require first aid?
_____ Have you had to do environmental or air monitoring?
_____ Do you provide drinking water to workers?
_____ Do you provide for toilets and washing facilities?
_____ Do you have high noise worksites or tasks?
_____ Do you have sources of ionizing or nonionizing (lasers) radiation at your worksites?
_____ Do you do contracting jobs where chemical processes involving high hazardous chemicals take place?
_____ Do you use some form of ventilation to remove airborne contaminants?
_____ Do you do hazardous waste remediation work?
_____ Do you do night work or work in areas with limited light?

Subpart E – Personal Protective and Life Saving Equipment

Subpart E requires employers to provide employees with the proper personal protective equipment (PPE) for the construction work being performed. As part of this requirement, the employer must conduct a hazard survey of the work to determine the control measures to use where hazards cannot be eliminated.

This serves as a resource in guiding the selection of the appropriate PPE. This includes PPE for eyes, face, head, and extremities. Other types of equipment which may be required are protective clothing, respiratory devices, protective shields, safety nets, life jackets, and fall protection (Subpart M). All PPE is to be maintained and in a sanitary condition.

Not only are the employers required to provided needed PPE, but they are required to train workers how to use and wear their PPE. Equipment for emergency use should be stored and accessible in a location known to all workers.

This subpart provides the standard for quality, as well as selection of PPE, for hearing protection, eye/face protection, head protection, respiratory protection, foot protection, safety harnesses, lifelines, lanyards, and safety net. The protections include requirements for working over water.

Sections:

1926.95   Criteria for personal protective equipment.
1926.96   Occupational foot protection.
1926.97   Protective clothing for fire brigades.
1926.98   Respiratory protection for fire brigades.
1926.99   [Reserved]
1926.100 Head protection.
1926.101 Hearing protection.
1926.102 Eye and face protection.
1926.103 Respiratory protection.
1926.104 Safety belts, lifelines, and lanyards.
1926.105 Safety nets.
1926.106 Working over or near water.
1926.107 Definitions applicable to this subpart.

Violations/Citations:

1926.95(a)   — Personal protective equipment shall be provided, used, and maintained in a sanitary and reliable condition. (921)

1926.100(a)   — Head protection, where there is a possible danger of head injury. (6912)

1926.102(a)(1) — Eye and face protection shall be provided. (2131)

1926.105(a)   — Workplaces more than 25 feet aboveground or water shall have safety nets when ladders, safety lines/belts, temporary floors, scaffolds, and catch platforms are not practical. (1527)

Checklist:

_____ Do you require personal protective equipment to be used?
_____ Do you have potential for falling, flying, or electrical hazards?
_____ Do you require head protection?
_____ Is there the opportunity for heavy material to fall onto the workers' feet?
_____ Do you have an area above 85 dbA of noise?
_____ Do you have the potential at any time for workers to suffer eye injuries?
_____ Are workers working at heights?
_____ Does your company perform work over water?
_____ Do environment or air contaminants require the use of respirators?

Subpart F – Fire Protection and Prevention

Subpart F is concerned with fire protection, fire prevention, flammable and combustible liquids, liquid petroleum, and temporary heating devices. Employers are to develop and implement a fire protection program, which is part of their safety and health program, when potential for fires exist.

Firefighting equipment is to be available and readily accessible. Portable fire extinguishers are to be inspected and maintained on each floor of a structure and are not to exceed 100 feet from a worker. Workers are to be trained annually on the use of fire extinguishers. Automatic sprinkler protection should be installed as soon as possible during construction and removed as late as possible during demolition work.

This subpart discusses the requirements for handling and storage of flammable and combustible liquids both inside and outside, as well as liquid petroleum gas. The requirements include the specifications for approved container and storage areas.

The requirements for different types of storage tanks are contained within this subpart, as well as the specifications for building them. Each tank is required to meet certain criteria regarding pressure regulations. These tanks are to be correctly labeled and inspected on a regular basis to prevent fires or accidents.

The last sections of the subpart discuss the use of fire detection systems and conducting inspections to assure that they are in proper working condition and are in OSHA compliance. The fire detection system should be in a specific labeled location. Last, a unique alarm system must be established at the worksite which will alert employees to a fire.

Sections:

1926.150 Fire protection.
1926.151 Fire prevention.
1926.152 Flammable and combustible liquids.
1926.153 Liquified petroleum gas (LP-Gas).
1926.154 Temporary heating devices.
1926.155 Definitions applicable to this subpart.

FIXED FIRE SUPPRESSION EQUIPMENT
1926.156 Fixed extinguishing systems, general.
1926.157 Fixed extinguishing systems, gaseous agent.

OTHER FIRE PROTECTION SYSTEMS
1926.158 Fire detection systems.
1926.159 Employee alarm systems.

Violations/Citations:

1926.150(c)(1) — Portable firefighting equipment shall be provided and extinguishers shall be inspected periodically. (2136)

1926.152(a)(1) — Electrical wiring shall be installed with the requirement of Subpart K. (3618)

Checklist:

_____ Does your worksite have a fire hazard potential?
_____ Do you have a fire prevention program?
_____ Do you use fire extinguishers at your site?
_____ Do you train workers in fire prevention and firefighting?
_____ Is there flammable or combustible liquids used or stored on your site?
_____ Do you use LP-gas on your site?
_____ Do you have or build flammable or combustible liquid storage tanks?
_____ Do you have a fire detection system or fire alarm system?
_____ Do you use temporary heating devices?

Subpart G – Signs, Signals, and Barricades

Subpart G provides the requirements for signs when hazards or specific information is needed by workers and others when they are in close proximity to hazards. The signs most commonly used on most construction sites are those designating danger, caution, traffic, directional, exit, and safety precautions. Specifications for the color and size of signs are also found in this subpart as well as barricades and the use of flagmen, when needed, that must conform to the Manual on Uniform Traffic Control Devices for Streets and Highways. Also located in this section is the requirement for the use of temporary accident prevention tags to warn workers of hazards such as defective tools or equipment.

Sections:

1926.200 Accident prevention signs and tags.
1926.201 Signaling.
1926.202 Barricades.
1926.203 Definitions applicable to this subpart.

Checklist:

_____ Does your jobsite have hazards which require the use of signs?
_____ Do you use flagmen to direct traffic?
_____ Do you do work where barricades are needed around highway traffic areas?
_____ Do your workers need to recognize hazard signs and accident prevention tags?
_____ Do you use signs on your jobsite or project?

Subpart H – Materials Handling, Storage, Use, and Disposal

Subpart H details the storage of materials and how to stack, rack, and secure them against falling or sliding. Materials should not create a hazard due to storage in aisles or passageways. Housekeeping is an important component of handling and storing of construction materials.

The rigging of materials for handling is a critical component of Subpart H. This includes the safe use of slings made from wire rope, chains, synthetic fibers ropes or webs, and natural fiber ropes. Specifications of the use of rigging are found in this subpart regarding carrying capacity, inspection for defects, and safe operating procedures.

This subpart also addresses disposal of materials such as scrap wood, etc. All waste material is to be removed from the work area immediately to prevent accidents. Flammable materials and liquids are to be kept in covered containers until they can be removed from the jobsite.

Sections:

1926.250 General requirements for storage.
1926.251 Rigging equipment for material handling.
1926.252 Disposal of waste materials.

Checklist:

_____ Do you have materials stored on the worksite?
_____ Do you have waste materials on the jobsite?
_____ Do your workers have to handle materials on your projects?
_____ Do your workers use rigging to handle materials?
_____ Do your workers know the limitations for the use of wire ropes, chains, etc.?
_____ Does your company have responsibility for housekeeping?

Subpart I – Tools – Hand and Power

The Subpart I regulation is dedicated to the safe use of both power and hand tools including employer and employee owned tools. The subpart requires that hand tools be safe and free from defects. It also cautions against misuse of tools.

This subpart addresses the need for properly guarded power tools. It discusses the areas where guarding is required and the types of guards which should be used, as well as the proper protective equipment to be used, when tools create such hazards as flying materials. The power tools which are covered by the regulation include electrical, pneumatic, fuel, hydraulic, and powder-actuated power tools. These tools are to be secured if maintained in a fixed place and all electrically powered equipment must be effectively grounded. Special attention is given to abrasive wheels and tools. Some special requirements exist for powder-actuated tools.

Because of the nature of construction work and the use of woodworking tools, Subpart I specifically addresses woodworking tools. Also this subpart gives guidance on the use of air receivers and mechanical power-transmission apparatus. Jacks and their use are covered regarding the blocking and securing of objects being lifted. This includes jack maintenance and inspection.

<u>Sections:</u>

1926.300 General requirements.
1926.301 Hand tools.
1926.302 Power operated hand tools.
1926.303 Abrasive wheels and tools.
1926.304 Woodworking tools.
1926.305 Jacks – lever and ratchet, screw and hydraulic.
1926.306 Air Receivers.
1926.307 Mechanical power-transmission apparatus.

<u>Violations/Citations:</u>

1626.300(b)(2) — Guards for power tools shall be used and moving parts of equipment shall be guarded. (2037)

1926.304(f) — Woodworking tools/machinery shall meet the requirements of ANSI 01.1-1961. (1501)

<u>Checklist:</u>

_____ Do your workers use hand or power tools?
_____ Do your workers use woodworking tools?
_____ Do your workers use abrasive wheels or tools?
_____ Do you supply tools to workers?
_____ Do your workers use jacks?
_____ Are there air receivers on your jobsite?
_____ Are there mechanical power-transmission apparatuses on your worksites?

Subpart J – Welding and Cutting

Subpart J covers the procedures and precautions associated with gas welding, cutting, arc welding, fire prevention, compressed gas cylinders, and welding materials. Special attention is given to the transporting, moving, and storing of compressed gas cylinders, as well as the apparatuses such as hoses, torches, and regulators used for welding. Defective gas cylinders should not be used. All cylinders should be marked and labeled with one inch letters. Hoses should be identifiable and designed such that they cannot be misconnected to the wrong cylinder regulators. Prework inspections are an important component of this subpart.

Arc welding and its unique precautions are covered by this regulation. This includes grounding, care of cables, and care of electrode holders. As with all welding and cutting operations, appropriate personal protective equipment and safety are addressed in this subpart.

Fire prevention is an important part of welding and cutting and such work is not to be performed near flammable vapors, fumes, or heavy dust concentrations. Firefighting equipment must be readily accessible and in good working order.

Last, the need for adequate ventilation is delineated. This is especially important for confined spaces and when welding or cutting on toxic materials or coatings such as lead, cadmium, chrome, or beryllium.

<u>Sections:</u>

1926.350 Gas welding and cutting.
1926.351 Arc welding and cutting.
1926.352 Fire prevention.
1926.353 Ventilation and protection in welding, cutting, and heating.
1926.354 Welding, cutting, and heating in way of preservative coatings.

<u>Violations/Citations:</u>

1926.350(a)(9) — Oxygen cylinders in storage shall be separated from fuel gas cylinders by at least 20 feet or a 5-foot high fire resistant barrier. (1919)

1926.350(j)   — Technical portions of ANSI Z49.1-1967. (1777)

<u>Checklist:</u>

_____ Do your workers perform welding and cutting tasks?
_____ Do you have compressed gas cylinders on your jobsite?
_____ Do you have adequate firefighting equipment?
_____ Is there a need for ventilation?
_____ Do your welders wear personal protective equipment?
_____ Does your company weld or cut in confined spaces?
_____ Do your workers have to weld or cut on toxic materials?

Subpart K – Electrical

Subpart K relates to the installation and use of electrical power on construction worksites, including both permanent and temporary. These standards do not apply to existing permanent installations that were on site prior to the construction project. The four areas of emphasis within this subpart are installation safety requirements, safety related work practices, safety related maintenance and environmental considerations, and safety requirements for special equipment.

Installation safety requirements sections of subpart K require that all electrical parts be inspected for durability, quality, and appropriateness. Installation which follows the National Electric Code is considered in compliance with OSHA. Grounding is an important part of this regulation and the use of GFCIs or assured grounding is required. Emphasis is placed upon temporary and portable lighting, as well as the use of extension cords. All listed, labeled, and certified equipment must be installed according to instructions from the manufacturer. This subpart includes special purpose equipment installation such as cranes and monorail hoist, electric welders, and X-ray equipment. It discusses work in high hazard locations (see Appendix E) as well as special systems such as remote control and power limited circuits.

Safety-related work practices include not working on energized circuits, and precautions for working on hidden underground power sources. This subpart addresses the use of barriers to protect workers from electrical sources. Also, working around electrically energized equipment and power lines is explained as well as the procedures for lockout/tagout of energized circuits to protect workers.

Safety related maintenance and environmental considerations require that all hazardous locations be maintained in a dust tight/explosive proof condition. It also specifies that wet, flammable, or excessive temperature should be avoided and electrical installations should be protected from corrosion.

Special requirements for special equipment address batteries and battery recharging. Subpart K also deals with power sources for elevators, escalators, and moving walkways; only electrical equipment designed for high hazard areas should be installed.

Violations/Citations:

1926.403(b)(2) — Employer shall ensure electrical equipment is free from recognized hazards, is suitable, is used in accordance with the listing, labeling or certification. (3149)

1926.403(i)(2) — Electrical live parts shall be guarded. (1402)

1926.404(b) — Ground fault circuit interrupters or an assured equipment ground conductor program are in use. (8974)

1926.404(f)(6) — Grounding paths not permanent and continuous. (5850)

1926.404(f)(7) — Electrical equipment connected by cord and plug shall be grounded except if there is an isolated transformer or the tools are double insulated. (1064)

1926.405(a)(2) (ii)(e&f) — Temporary lights shall be protected from breakage and shall not be suspended by their cords or extension cord. (1815)

1926.405(a)(2)—   Extension cords used with portable electric tools shall be of the
(ii)(j)   three wire type and designed for hard or extra hard use. (1672)

1926.405(a)   —   Temporary wiring over 600 volts shall have barriers to prevent
(2)(iii) access or unauthorized and unqualified personnel. (1702)

1926.405(b)   —   Electrical boxes and fittings shall have covers, faceplates or canopies, and holes shall be smooth where cords pass through. Un-
used openings in cabinets/boxes shall be closed. (2471)

1926.405(g)(2)—   Flexible cords shall be used without splices or tape; strain relief
shall be provided. (2731)

1926.416(e)(1) —   Electrical cords or cables shall not be used when worn or frayed.
(1339)

Checklist:

____   Do you employee electricians?
____   Do your employees perform electrical installations?
____   Do your workers work around energized electrical circuits?
____   Do you follow a lockout/tagout procedure?
____   Do you use temporary lighting and extension cords?
____   Do your workers use GFCIs?
____   Does your operation include battery recharging?
____   Do you have workers working in hazard environments?
____   Do your workers use electrically powered tools?
____   Are there energized power lines on your jobsite?
____   Do your workers work around energized power lines?
____   Is there special electrically powered equipment on your worksite?

Subpart L – Scaffolding

Subpart L provides the guidance for scaffolds which are used as elevated work platforms. This includes strength capacities for scaffolds and components, platform specifications, guardrails and toeboards requirements, access ladders and stairway specifications, and
provisions for scaffolding supporting structures. This subpart includes requirements for erected
scaffold to suspended scaffolds. The stipulations for fall protection, which starts at 10 feet, is
an integral part of this subpart.

Each type of scaffold used in construction operations is described and its unique safety
requirements are provided in this section of Subpart L. This encompasses needle beam scaffolds, ladder jack scaffolds, mobile scaffolds, etc. as well as provisions for the safe use of
aerial scaffolds.

A competent or qualified person must train workers and oversee the erection, moving, disassembling, operation, repair, inspecting, and maintenance of the scaffolds. If procedures change or a new type of scaffolding is used, workers need to be retrained.

Sections:

1926.450   Scope, application, and definitions.
1926.451   General requirements.
1926.452   Additional requirements applicable to specific types of scaffolds.
1926.453   Aerial Lifts.
1926.454   Training requirements.

Violations/Citations:

1926.451(a)(2) — Scaffold footing or anchorage shall be sound, rigid, and capable of carrying the maximum intended load. (1183)

1926.451(a)(3) — Scaffolding shall be erected, moved, dismantled, or altered under the supervision of a competent person. (922)

1926.451(a)(4) — Scaffolding shall have guardrails and toeboards when more than 10 feet, and 4–10 feet when there is less than 45 inches of workspace. (2811)

1926.451(a)(13) — Scaffolding safe access not provided by ladder or equivalent. (3306)

1926.451(a)(14) — Scaffold planking shall extend over the end supports not less than 6 inches and not more than 12 inches. (940)

1926.451(d) — Tubular welded scaffolds shall be properly braced so that they are plumb, square, and rigid; legs on plumb, adjustable, mud sills, etc. to support the maximum load; guardrails and toeboards shall installed. (7178).

1926.451(e) — Manually propelled scaffolds shall have tight planking for the full width, platforms secured, ladder or stairways provided, suitable footing, stand plumb, wheels locked, guardrails and toeboards. (3800)

Checklist:

____ Do your workers use scaffolds in the performance of their work?
____ Do you own scaffolds?
____ Do you erect, tear down, or maintain scaffolds for other contractors or subcontractors?
____ Are you responsible for training workers regarding scaffolds and their safety?
____ Are there scaffolds on your worksite?

Subpart M – Fall Protection

Subpart M states that fall protection is to be provided for workers who are working above the ground except if making an inspection, investigation, or assessment of a workplace prior to actual construction activities or after all work has been completed. This regulation does not cover work related to scaffolds, cranes, derricks, steel work, tunneling, power transmission, or stairways and ladders. It does include leading edges, hoist areas, holes, formwork and reinforced steel, ramps/runways/walkways, excavation, dangerous equipment, overhand bricklaying, roofing, wall opening, and other walking working surfaces. Also, the need for overhead protection from falling objects is mandated.

This subpart details the installation, construction, and proper use of fall protection. The specifications for each form of fall protection are provided in this subpart including guardrail systems, safety net systems, personal fall arrest systems, positioning devices, warning line systems, controlled access zones, and safety monitoring systems. Each fall protection has its limits and criteria explained in this subpart and it also delineates the components of a fall protection plan.

The remaining section provides the requirements for training and retraining regarding fall hazards and the use of fall protection, as well as the requirements for certification of the training.

Sections:

1926.500 Scope, application, and definitions applicable to this subpart.
1926.501 Duty to have fall protection.
1926.502 Fall protection systems criteria and practices.
1926.503 Training requirements.
APPENDIX A TO SUBPART M – DETERMINING ROOF WIDTHS
APPENDIX B TO SUBPART M – GUARDRAIL SYSTEMS
APPENDIX C TO SUBPART M – PERSONAL FALL ARREST SYSTEMS
APPENDIX D TO SUBPART M – POSITIONING DEVICE SYSTEMS
APPENDIX E TO SUBPART M – SAMPLE FALL PROTECTION PLANS

Violations/Citations:

1926.500(b&d)— Guardrails on open sided floors, floor holes, and runways. (9837)

1926.500(c)(1) — Wall openings shall be guarded. (1120)

1926.501(b)(1) — Guardrails, safety nets, or personal fall arrest systems shall be used at 6 feet or more. (985)

Checklist:

_____ Do you have workers working six feet aboveground level?
_____ Do you have some form of fall protection?
_____ Are workers required to wear fall protection?
_____ Do you have any workers exposed to a fall hazard?
_____ Are you required to have a fall protection plan?
_____ Are your workers trained in fall hazards and protection?
_____ Is any of the type of work that your workers do covered by Subpart M – fall protection?

Subpart N – Cranes, Derricks, Hoists, Elevators, and Conveyors

Subpart N provides provisions for the construction industry for cranes, derricks, hoists, helicopters, conveyors, and aerial lifts. This subpart delimits many common safety requirements for construction vehicles and reinforces the need to follow the manufacturer's requirements regarding load capacities, speed limits, special hazards, and unique equipment characteristics. A competent person must inspect all cranes and derricks prior to daily use and a thorough inspection must be accomplished annually by an OSHA recognized qualified person. A record must be maintained of that inspection for each piece of hoisting equipment.

Requirements regarding wire rope, guarding of gears, etc. area of operation, and operation near energized power lines is found in this subpart. The subpart contains specific stipulations for varying types of cranes and derricks including crawlers, locomotives, trucks, overheads, gantries, and tower cranes. If cranes or derricks are to support personnel platforms, then they are to follow defined criteria within this subpart, which includes a safety factor of seven (seven times expected load), trial lifts, and an inspection by a competent person after the trial lift. The rule of thumb is that usually as a last resort or when they are the safest method, cranes and derricks are used to hoist personnel. The path of travel of crane or derricks is to be kept free of possible obstructions and should not exceed their lifting capacity. Workers are to be protected by a means of fall protection at all times to prevent falling from heights or from a personnel platform.

This subpart applies to the use of helicopters for lifting purposes. Helicopters must comply with the Federal Aviation Administration regulations. The pilot of the helicopter has the primary responsibility for the load's weight, size, and rigging. Static charge must be eliminated prior to workers touching the load. Visibility is critical to the pilot in maintaining visual contact with ground crew members so that constant communications can be maintained.

All hoists are to comply with the manufacturer's specifications. If these do not exist, then as with cranes and derricks, the limitations are based upon the determination of a professional engineer. In the operation of a hoist there should be a signaling system, specified line speed, and a sign stating "No Riders." Permanently enclosed hoist cars are to be used to hoist personnel and these cars must be able to stop at any time by safety breaks or a similar system. All hoists are to be tested, inspected, and maintained on an ongoing basis and at least every three months. Also, requirements exist in this subpart for base-mounted drum hoists and overhead hoists.

Conveyors are to have an alarm that sounds prior to start of operation to warn workers and the operator must be capable of stopping the conveyor at any time. Workers must be protected from contact with conveyors. Finally, aerial lifts are addressed, including ladder trucks, tower trucks, and articulating boom platforms.

Sections:

1926.550 Cranes and derricks.
1926.551 Helicopters.
1926.552 Material hoists, personnel hoists, and elevators.
1926.553 Base-mounted drum hoists.
1926.554 Overhead hoists.
1926.555 Conveyors.
1926.556 Aerial lifts.

Violations/Citations:

1926.550(b)(2) — Cranes, crawlers, trucks, or locomotives shall meet design, testing, maintenance, and operation per ANSI B30.5-1968. The most recent certification shall be on file until a new one is prepared. (921)

1926.556(b)(2) — When working from an aerial lift, a body belt and lanyard must be attached to the boom or basket. (1009)

Checklist:

____ Does your company own or use cranes or derricks?
____ Does your company employ helicopters for lifting purposes?
____ Do you use material hoists on your worksite?
____ Do your workers work around cranes, derricks, helicopters, or hoists?
____ Does your company use cranes or hoists for lifting personnel?
____ Do you have conveyors on your jobsite and do workers have exposure to them?
____ Do you rent cranes, derricks, hoists, or other lifting devices?
____ Do you use aerial lifts such as ladder trucks?

Subpart O – Motor Vehicles, Mechanized Equipment, and Marine Operations

Subpart O addresses those vehicles which operate within off-highway jobsites. These vehicles are required to have an operating service, emergency, and parking brake system. Also, they must have two headlights and taillights, operating brake lights, an audible warning device for operator use, and an audible back-up alarm or observer when view to the rear is hampered. Vehicles with cabs are to be equipped with safety glass, windshield wipers, and defogging systems. Special precautions are required when parking vehicles overnight on construction sites. Blocking or cribbing is required to support raised parts of vehicles to be worked under. All controls are to be in a neutral position. Special protections are to be taken when working on tires such as racks, cages, etc. Any vehicle used to transport workers must have seats and safety belts for each worker. Seat belts are to be provided for all construction equipment except stand-up operated and on equipment with no ROPS.

Industrial truck's capacity will be posted on the vehicle in clear view of the operator. No changes or modifications are to be made to equipment without approval of the manufacturer. All industrial trucks must be designed to meet the safety requirements for ANSI standard B56.1-1969.

All boilers, piping systems, and pressure vessels used with pile-driving equipment must meet the requirements of ASME standards. Pile-driving equipment is required to have overhead protection, guards, and stop blocks. A blocking device, which will support the weight of the hammer to protect workers, is to be provided. The regulation delineates other protections for workers and procedures to be followed when driving piles.

This subpart describes the requirements for site clearing, such as worker protection from toxic plants, to the need for rollover protection on equipment. And last, marine operations are discussed regarding the stipulations for material handling, access to barges, work surfaces of barges, and first aid and life saving equipment for work around water going vessels.

Sections:

1926.600 Equipment.
1926.601 Motor vehicles.
1926.602 Material handling equipment.
1926.603 Pile driving equipment.
1926.604 Site clearing.
1926.605 Marine operations and equipment.
1926.606 Definitions applicable to this subpart.

Violations/Citations:

1926.602(a)(9) — Bi-directional earth moving equipment shall have audible alarms. (932)

Checklist:

_____ Does your company use motor vehicles or industrial trucks on your jobsites?
_____ Does your company use mechanized equipment on your jobsite?
_____ Do any of your workers operate motor vehicles or mechanized equipment on your projects?

_____ Does your company own motor vehicles, industrials trucks, or mechanized equipment?

_____ Do your workers operate or work around or with pile-driving equipment?

_____ Does your company own pile-driving equipment?

_____ Do you contract for work or have workers conducting on or around water-going vessels?

_____ Do you contract to perform cite clearing work?

Subpart P – Excavations

Subpart P concerns open excavations made in the earth's surface, including trenches. Specific procedures for protecting workers at depths greater than five feet are provided. These requirements include protective systems for workers, locating and protecting underground utilities, inspections by a competent person, the potential for hazardous atmospheres in excavation, the danger of water, and the location of equipment, materials, and other structures at the excavation.

The regulation describes how to determine class A, B, and C soil types. It also delineates the requirements for sloping and benching based on soil types. The subpart describes the common types of shoring, from timber and aluminum hydraulic shoring to the use of trench boxes.

This subpart requires the location and protection of underground utilities. Materials and equipment are to be kept two feet from the edge of any excavation; and adjacent structures are not to be undercut and are to be supported to maintain stability. When the lack of oxygen, or flammable gas or toxic chemicals exists, the excavation is treated as a confined space which requires atmospheric sampling and emergency rescue equipment.

All excavations deeper than four feet must have a means of egress, either a structural ramp or a ladder within 25 feet of workers. Daily inspections are required, as needed, and after adverse weather, especially rain.

Sections:

1926.650 Scope, application, and definitions applicable to this subpart.

1926.651 Specific Excavation Requirements.

1926.652 Requirements for protective systems.

APPENDIX A TO SUBPART P – SOIL CLASSIFICATION.

APPENDIX B TO SUBPART P – SLOPING AND BENCHING.

APPENDIX C TO SUBPART P – TIMBER SHORING FOR TRENCHES.

APPENDIX D TO SUBPART P – ALUMINUM HYDRAULIC SHORING FOR TRENCHES.

APPENDIX E TO SUBPART P – ALTERNATIVES TO TIMBER SHORING.

APPENDIX F TO SUBPART P – SELECTION OF PROTECTIVE SYSTEMS.

Violations/ Citations:

1926.651(j)(2) — Excavations shall have materials or equipment placed at least two feet from the edge. (2235)

1926.651(c)(2) — Excavations shall have a safe means of egress, such as ladders, ramps, etc. (2157)

1926.651(k)(1) — Excavations and protective systems shall be inspected daily by a competent person and as needed. (3252)

1926.652(a)(1) — Excavations protective systems shall be examined by a competent person when greater than 5 feet. ((5529)

Checklist:

____ Are there going to be any forms of excavations on your projects?
____ Is excavating a primary operation of your company?
____ Do you slope, bench, or shore to remove excavation hazards?
____ Do you own or use any type of shoring or trench shields?
____ Do your workers work around excavations?
____ Do your workers operate equipment around excavations?

Subpart Q – Concrete and Masonry Construction

Subpart Q sets forth the requirements that protect construction workers from hazards related to masonry and concrete work. Workers are to be protected from impalement on rebar by capping or covering them, by bracing walls over eight feet high, and by having limited access zones around masonry wall construction. Workers are to be kept out of areas until concrete has sufficient strength to support itself.

Plans are needed for all formwork and jacks used on the construction site. This subpart addresses riding concrete buckets, working under loads, and the need for personal protective equipment. Also, it addresses bulk storage of concrete, concrete mixers, power concrete trowels, concrete buggies, concrete pump systems, concrete buckets, tremies, masonry saws, lockout/tagout procedures, and bull floats.

Requirements for each type of concrete work are found in this subpart, including cast-in-place, precast, lift-slab, and masonry concrete work. It is apparent from the regulation that special precautions are to be taken with all masonry and concrete work.

Sections:

1926.700 Scope, application, and definitions applicable to this subpart.
1926.701 General requirements.
1926.702 Requirements for equipment and tools.
1926.703 Requirements for cast-in-place concrete.
1926.704 Requirements for precast concrete.
1926.705 Requirements for lift-slab construction operations.
1926.706 Requirements of masonry construction.
APPENDIX TO SUBPART Q – References to subpart Q of Part 1926

Violations/Citations:

1926.701(b)   — Reinforcing steel onto which employees could fall shall be guarded. (2434)

Checklist:

____ Does your company erect forms for concrete work?
____ Does your company perform concrete work?
____ Do your employees have exposure to masonry and concrete work on your projects?
____ Does your company perform either lift-slab or precast construction?
____ Does your company have equipment for performing concrete or masonry work?

Subpart R – Steel Erection

Subpart R describes the provisions for structural steel assembly, bolting, riveting, fitting-up, and plumbing-up. This subpart also discusses the use of fall protection in steel erection. Permanent floors are to be installed as erection progresses. Temporary flooring should be fully planked with full sized, undressed two inch planks and periphery-safety railing of one-half inch wire rope is to be installed 42 inches high. All flooring is to be secured and of adequate strength to support intended loads. Fall protection in the form of nets is to be provided when fall distance exceeds two stories.

In structural steel assembly, the load (member) should not be released from the rigging until it has been bolted with not less than two bolts and/or made laterally stable. Taglines are to be used in controlling loads.

All containers for storing rivets, bolts, or drift pins are to be secured from falling. Pneumatic tools are to be disconnected from power sources prior to adjusting or repairing them and eye protection should be worn. Anytime the potential exists for tools, drift pins, or rivets to fall, a means to prevent them from falling needs to be employed. All equipment used for plumbing-up should be secured. During bolting, riveting, fitting-up, and plumbing-up, fall protection in the form of safety harnesses and lanyards may be required when using floating scaffolds.

Sections:

1926.750 Flooring requirements.
1926.751 Structural steel assembly.
1926.752 Bolting, riveting, fitting-up, and plumbing-up.
1926.753 Safety nets.

Checklist:

____ Do your workers do steel erection work?
____ Does your company contract for steel erection work?
____ Do your workers work around or with those doing steel erection?
____ Does your company have equipment for doing steel erection?

Subpart S – Tunnels, Shafts, Caissons, Cofferdams, and Compressed Air

Subpart S applies to the construction of underground tunnels, shafts, chambers, and passageways, but does not apply to excavation/trenching operations or underground electrical transmission and distribution lines. This subpart does apply to the provision of access and egress, as well as controlling access to underground working to authorized persons. Each person who enters an underground operation must use the check-in/ check-out procedures. Safety instructions are to be provided on air monitoring, ventilation, illumination, communications, flood control, mechanical equipment, personal protective equipment, explosives, fire prevention/protection, and emergency procedures. Since the tunneling environment continuously changes, communication regarding arising hazard must be conveyed to oncoming shifts; thus, this subpart emphasizes both voice and power communication.

The importance of emergency planning is critical in underground operations, and the subpart contents include provisions for hoisting capability, self-rescuers, designated persons, emergency lighting, and rescue teams.

Air monitoring and air quality are integral to this subpart. The potential for explosion in any gassy tunneling operation requires continuous monitoring to maintain levels below the explosive limits. This subpart gives the criteria for gassy operations, as well as air quality and air monitoring procedures. Ventilation does much to mitigate potential air and explosive hazards in under ground operations; thus, this subpart stipulates the minimum criteria concerning the velocity and quantity of air to be provided.

Fire is always a distinct possibility in tunneling operations. This is especially true during welding and other hot work. Open flames for any other reason are prohibited and adequate fire fighting equipment is required to be available.

The person who conducts inspections and reports unsafe conditions is responsible for the control of roof, face, and wall conditions which could fracture and fall. Workers are to have suitable protection from the hazards of loose ground whether in a tunnel or shaft.

Blasting and explosives follow the requirements of Subpart U, while the criteria for both materials and personnel hoists are found in Subpart S. These provisions include the securing and safe movement of materials, as well as special protection for workers who must be transported on a hoist. The load capacity of the hoist must have a safety factor of five.

The remainder of this subpart provides requirements for caisson work and the hazards associated with cofferdams. Also discussed is the importance of having medical surveillance for workers who are working under compressed air.

Sections:

    1926.800 Underground construction.
    1926.801 Caissons.
    1926.802 Cofferdams.
    1926.803 Compressed air.
    1926.804 Definitions applicable to this subpart.
    APPENDIX A TO SUBPART S – DECOMPRESSION TABLES

Checklist:

_____ Does your company perform underground construction?
_____ Are your workers exposed to the hazards of underground construction?
_____ Does your company do caisson work?
_____ Are there, or do you build cofferdams within your worksite?
_____ Are any of your workers exposed to compressed air conditions?
_____ Do you have equipment for underground construction?
_____ Do you have medical facilities or do you monitor the medical condition of workers working in compressed air?

Subpart T – Demolition

It is important to obtain a written engineering survey of structures to be demolished prior to starting demolition, so as to preclude an unplanned collapse of the structure. In buildings which have been damaged by fire, flood, or explosions, supports need to be erected for walls and floors. All utilities, such as water, gas, sewer, steam, or electricity, are to be cut off prior to demolition activities. Any materials which pose a hazard, such as glass, should be removed. The standard demands that any holes, whether used to drop debris or where employees could fall, need to be protected or covered and overhead protection needs to be provided.

No stairways, ladders, or passageways which are not approved for the demolition process should be used. Even these stairways, ladders, or passageways are to be inspected, illuminated, and have overhead protection.

Chutes are to be guarded at the top and bottom and enclosed when angles are steeper than 45 degrees. They need to be substantial enough to handle the falling debris.

The process becomes more hazardous as demolition progresses. Holes cut in floors or walls should not weaken the structure. Any loose materials, such as masonry, must be removed. Ladders and walkways must be available to access scaffolds and walls. Once the floor is removed, substantial walkways at least 18 inches wide are to be provided.

Floors must have the capacity to support any equipment and stored materials and should have barriers to prevent equipment or materials from falling over an edge or into a hole. This subpart provides requirements for removal of materials through floors and the removal of walls, masonry sections, chimneys, floors, and of steel construction.

Only designated workers are to be near the area where mechanical demolition is occurring. All steel members should be cut prior to using the demolition ball. Structural integrity is to be continuously monitored by the demolition process and workers are to removed when hazards exist. Selective demolition by explosives is a very complicated process and should be done in accordance with Subpart U.

Sections:

1926.850 Preparatory operations.
1926.851 Stairs, passageways, and ladders.
1926.852 Chutes.
1926.853 Removal of materials through floor openings.
1926.854 Removal of walls, masonry sections, and chimneys.
1926.855 Manual removal of floors.
1926.856 Removal of walls, floors, and material with equipment.
1926.857 Storage.
1926.858 Removal of steel construction.
1926.859 Mechanical demolition.
1926.860 Selective demolition by explosives.

Checkist:

_____ Does your company do any demolition work?
_____ Do your workers go onto sites where demolition work is transpiring?
_____ Are your workers expected to do demolition work?
_____ Do you have equipment used for demolition work?

Subpart U – Blasting and Use of Explosives

Subpart U states that only authorized and qualified persons can handle and use explosives. No person using explosives is allowed to be under the influence of alcohol or drugs. Nothing which could be an ignition source, such as matches, open flames, or smokers, is to be around explosives. Accountability is required to assure that explosives are under the care of a qualified person. All blasting aboveground is done between sunup and sundown and, when blasting is done, blasters are to take special precaution near public utilities, around transportation conveyances, and near public areas to assure safety and mitigate any damage. Care must be taken to assure that accidental premature ignition does not occur from stray electrical sources or radio transmitters. The blaster is to be considered a competent person in the use and care of explosives, and have experience with the type of blasting method being used.

The transportation of explosives and blasting materials must conform to the Department of Transportation regulatory provisions. Drivers of trucks containing explosives and blasting equipment must shall be licensed and in good physical and mental condition. No blasting materials are to be transported with other cargo and blasting caps are not to be transported in the same vehicle as other explosives. These vehicles should be marked with a placard signifying "Explosives" and have a fully charged fire extinguisher.

The requirements for underground transportation of explosives have many similarities to the aboveground transportation. The storage of any blasting materials must meet the stipulations of the Bureaus of Alcohol, Tobacco, and Firearms regulations.

Subpart U delineates the provisions for the loading of blasting agents into drill holes and the use of electric blasting techniques, safety fuses, detonating cords, and the procedures for detonating an explosive. It also includes the post-blasting inspection and the handling of misfires. Two unique environmental blasting procedures are part of Subpart U; they are underwater blasting and blasting under compressed air.

Sections:

    1926.900 General provisions.
    1926.901 Blaster qualifications.
    1926.902 Surface transportation of explosives.
    1926.903 Underground transportation of explosives.
    1926.904 Storage of explosives and blasting agents.
    1926.905 Loading of explosives or blasting agents.
    1926.906 Initiation of explosive charges – electric blasting.
    1926.907 Use of safety fuse.
    1926.908 Use of detonating cord.
    1926.909 Firing the blast.
    1926.910 Inspection after blasting.
    1926.911 Misfires.
    1926.912 Underwater blasting.
    1926.913 Blasting in excavation work under compressed air.
    1926.914 Definitions applicable to this subpart.

Checklist:

    _____ Does your company have equipment used for explosives and blasting?
    _____ Do any of your workers perform explosive handling and blasting operations?
    _____ Do you have blasting materials on your jobsites?

_____ Do you have individuals who are qualified blasters and employed by you?
_____ Do you have a contract blaster doing your blasting operations?
_____ Does your company contract to do blasting activities?
_____ Does your company or workers transport explosives or blasting materials?
_____ Do blasting activities occur on your jobsites or projects?

Subpart V – Power Transmission and Distribution

Subpart V applies to the construction of electrical transmission and distribution lines and equipment. The initial inspection and testing are to be performed prior to starting any work activity and include determining the operating voltage and whether electrical equipment and lines are energized.

Also, care is to be taken to maintain safe distances from energized lines and equipment. This also includes providing specialized personal protective equipment such as gloves, sleeves, insulated handles, guarding, insulating, hot stick work, lineman's belts, safety straps, and lanyards.

Subpart V provides the regulatory requirements for de-energizing lines and equipment in excess of 600 volts. Emergency procedures for first aid, night work, working over water, and hydraulic fluids are described and provisions are made for climbing, ladders, live line tools, and other tools. This subpart also describes the regulation for mechanical equipment such as aerial lifts, derrick trucks, and cranes which are used for personnel and material lifting. Material handling is specifically delineated regarding pole hauling and rigging procedures.

This regulation discusses the requirements and importance of using grounding as a protection for workers. Much of the construction of power transmission lines and facilities deals specifically with overhead lines, underground lines, and energized substation construction. The activities related to overhead transmission lines, which have regulatory requirements, include metal tower construction, stringing and removing deenergized conductors, stringing near energized lines, and live line, bare-hand work. The provision for underground lines deals with guarding and ventilating street openings, working in manholes, and trenching and excavation work. Subpart V addresses precautions concerning energized substations, barricades, energized control panels, fencing, and mechanized equipment.

Sections:

1926.950 General requirements.
1926.951 Tools and protective equipment.
1926.952 Mechanical equipment.
1926.953 Material handling.
1926.954 Grounding for protection of employees.
1926.955 Overhead lines.
1926.956 Underground lines.
1926.957 Construction in energized substations.
1926.958 External load helicopters.
1926.959 Lineman's body belts, safety straps, and lanyards.
1926.960 Definitions applicable to this subpart.

Checklist:

_____ Does your company bid on construction work for power transmission lines and equipment?
_____ Do your workers conduct power line or substation construction activities?

_____ Is there power transmission and distribution work taking place on your jobsites?

_____ Does your company have equipment for doing power transmission and distribution work?

_____ Does your company bid on power distribution or transmission work?

Subpart W – Rollover Protective Structures; Overhead Protection

Subpart W covers rollover protection structures (ROPS) and overhead protection. All material handling equipment manufactured after September 1, 1972 shall be equipped with rollover protection to meet the requirements within this subpart. Equipment built prior to this date (1969-1970) will need to be retrofitted. ROPS clearance must be 52 inches from the work deck to the ROPS. The regulation provides the minimum performance standards for ROPS for scrapers, loaders, dozers, graders, and crawler tractors. The criteria for the testing procedures include energy absorbing capacity, support capacity, and impact strength.

This subpart includes protective frame ROPS test procedures and performance requirements for wheel-type agriculture and industrial tractors used in construction. Subpart W details overhead protection for operators of agriculture and industrial tractors required by Subpart W. The overhead protection must be of solid material and should not create a hazard. Also, overhead protection frames must pass a drop and crushing test.

Sections:

1926.1000 Rollover protective structures (ROPS) for material handling equipment.
1926.1001 Minimum performance criteria for rollover protective structures for designated scrapers, loaders, dozers, graders, and crawler tractors.
1926.1002 Protective frame (ROPS) test procedures and performance requirements for wheel-type agricultural and industrial tractors used in construction.
1926.1003 Overhead protection for operators of agricultural and industrial tractors.

Checklist:

_____ Does your company have material handling equipment?

_____ Does any of your equipment have ROPS?

_____ Do your workers operate equipment with ROPS?

_____ Do you or have your ever removed ROPS?

_____ Have you ever manufactured you own ROPS or overhead protection for equipment?

_____ Did you test the ROPS or overhead protection which you manufactured?

_____ Do you have equipment which should have ROPS or overhead protection?

_____ Do you have or own any agricultural or industrial tractors?

Subpart X – Stairways and Ladders

Subpart X is concerned with stairways and ladders used in construction workplaces. The scope includes the construction, demolition, alterations, and repairs in building and on jobsites. Stairways are required when a break in elevation of 19 inches or more exists and no ramp, runway, sloped embankment, or personnel hoist is provided. Employers are required to provide and install stairways and ladder fall protection systems before workers begin work.

Stairways are to be at least 22 inches wide and between 30 and 50 degrees of incline. Riser or tread depth is not to vary more than one-fourth inch and must be uniform. Stairways which have more than four risers or are more than 30 inches high must have at least one handrail and one stairrail along each unprotected side or edge. Handrails and the top rails of stairrails should withstand at lease 200 pounds without failure. Height and other requirements are part of this subpart. Unprotected sides and edges of stairway landings are to be provided guardrails according to Subpart M requirements.

Ladders should be capable of supporting four times the maximum intended load. Other load requirements are provided for extra heavy duty, portable, and fixed ladders. Ladder rungs, cleats, and steps are to be parallel, level, and uniformly spaced. Specifications for distance between rungs and cleats are usually 12 inches including widths between side rails. Ladders are never to be spliced together. Ladders are to be inspected for damage or defects prior to use and, if defects are found, the ladders are to be tagged and removed from service. Selecting the proper ladder for the work is important.

Fixed ladders have unique and specific requirements regarding strength. Fixed ladders with a total length of more than a 24 foot climb must be provided with cages, wells, ladder safety devices, or self-retracting lifelines. When ladder safety devices or self-retracting lifelines are used, rest platforms must exist every 150 feet. If a cage, well, or multiple ladder sections exist, rest platforms are to be at 50 feet.

Subpart X describes the major safety provisions when using ladders on jobsites. It also states that employers must train workers to use ladders and stairways. This training is to be done by a competent person and must cover fall hazards, as well as how to use, maintain, and remove a fall protection system. Workers must also know the proper construction procedures regarding ladders and stairways, maximum load capacities, and the standards of Subpart X.

Sections:

1926.1050 Scope, application, and definitions applicable to this subpart.
1926.1051 General requirements.
1926.1052 Stairways.
1926.1053 Ladders.
1926.1054   1926.1059 [Reserved]
1926.1060 Training requirements
APPENDIX A TO SUBPART X – Ladders

Violations/Citations:

1926.1051(a)   — Stairway or ladder shall be provided to all access points where there is a break in elevation of 19 inches or more. (1486)

1926.1052(c)(1) — Stairrails and handrails along each unprotected edge. (3614)

1926.1053(b)(1) — Portable ladder side rail extend at least 3 feet or be secured at top. (2396)

1926.1060(a)   — Ladder and stairway training shall be provided. (1127)

Checklist:

_____ Do your workers use ladders in performing their work?
_____ Do your workers have to construct or use stairways on your jobsites?
_____ Does your company own ladders?

_____ Does your company construct stairways or fixed ladders on jobsites?
_____ Do your workers have to climb fixed ladders?
_____ Do you have a person competent in ladder and stairway safety?

Subpart Y – Commercial Diving Operations

Subpart Y applies to dives and diving support operations which take place within all waters in the U.S., trust territories, D.C., Commonwealth of Puerto Rico, other U.S. protected islands, etc. It does not apply to instructional diving and search and rescue. This subpart describes requirements, qualifications, and training certifications for divers and dive teams, as well as the need to use specific safe practices for pre, during, and post dives. It also includes emergency care procedures such as recompression and evacuation.

This subpart delineates the criteria and procedures for different types of diving operations such as SCUBA, surface supplied air, and mixed gas diving. The margins for error and risk are high; thus, all diving procedures within this regulation are very precise and require more than superficial knowledge and experience with diving operations.

The care and maintenance of all equipment involved, whether cylinders, decompression chambers, oxygen safety, or other diving equipment, require a unique expertise. This subpart makes all diving and diving operation procedures very exacting and requires recordkeeping of all dives and injuries.

Sections:

GENERAL
1926.1071 Scope and application.
1926.1072 Definitions.

PERSONNEL REQUIREMENTS
1926.1076 Qualifications of dive team.

GENERAL OPERATIONS PROCEDURES
1926.1080 Safe practices manual.
1926.1081 Pre-dive procedures.
1926.1082 Procedures during dive.
1926.1083 Post-dive procedures.

SPECIFIC OPERATIONS PROCEDURES
1926.1084 SCUBA diving.
1926.1085 Surface-supplied air diving.
1926.1086 Mixed-gas diving.
1926.1087 Liveboating.

EQUIPMENT PROCEDURES AND REQUIREMENTS
1926.1090 Equipment

RECORDKEEPING
1926.1091 Recordkeeping requirements.
1926.1092 Effective date.
APPENDIX A TO SUBPART Y – Examples of Conditions Which May Restrict or
    Limit Exposure to Hyperbaric Conditions
APPENDIX B TO SUBPART Y – Guidelines for Scientific Diving

Checklist:

_____ Does your company employ any divers?
_____ Does your company oversee any diving operations?
_____ Does your company own any diving equipment?
_____ Do you have divers or diving operations on your project which belong to other contractors?

Subpart Z – Toxic and Hazardous Substances

Subpart Z provides specific regulations for a select group of 24 toxic or hazardous chemicals. The regulations set specific exposure limits, detail acceptable work procedures, delineate workplace/environmental sampling requirements, set specific personal protective equipment requirements, and denote the need for regulated work areas. Subpart Z discusses, in some detail, working with and around potential cancer-causing chemicals. With many of the chemicals unique training requirements exist, as well as medical monitoring and surveillance. Requirements exist for posting and labels which warn of the dangers from exposure to specific chemicals. In many cases precise decontamination is required, along with hygiene procedures to minimize potential contamination to workers or the spread of contamination. These regulations communicate the hazards involved and discuss the target organs, signs, and symptoms which accompany an occupational illness from one of these hazardous or toxic chemicals.

Because each of these chemicals has unique properties, adverse affects, handling procedures, signs and symptoms of overexposure, and regulatory requirements, the regulation specific to each chemical must be consulted and complied with.

Sections:

1926.1100 [Reserved]
1926.1101 Asbestos
1926.1102 Coal tar pitch volatiles; interpretation of term.
1926.1103 4-Nitrobiphenyl.
1926.1104 alpha-Naphthylamine.
1926.1105 [Reserved]
1926.1106 Methyl chloromethyl ether.
1926.1107 3.3'-Dichlorobenzidine (and its salts).
1926.1108 bis-Chloromethyl ether.
1926.1109 beta-Naphthylamine.
1926.1110 Benzidine.
1926.1111 4-Aminodiphenyl.
1926.1112 Ethyleneimine.
1926.1113 beta-Propiolactone.
1926.1114 2-Acetylaminofluorene.
1926.1115 4-Dimethylaminoazobenzene.
1926.1116 N-Nitrosodimethylamine.
1926.1117 Vinyl chloride.
1926.1118 Inorganic arsenic.
1926.1127 Cadmium.
1926.1128 Benzene.

1926.1129 Coke oven emissions.

1926.1144 1,2-dibromo-3-chloropropane.

1926.1145 Acrylonitrile.

1926.1147 Ethylene oxide.

1926.1148 Formaldehyde.

APPENDIX A TO PART 1926 – Designations for General Industry Standards Incorporated into Body of Construction Standards.

Checklist:

_____ Does your company use any of the chemicals listed in Sections 1100 through 1148?

_____ Do any of the chemical mixtures which you use on your jobsites contain any chemicals in Sections 1100 through 1148.

_____ Do your workers do asbestos or lead abatement work?

_____ Do your workers perform hazardous waste remediation work?

_____ Do other contractors use any of the chemical in Sections 1100 through 1148 that might inadvertently expose your own workers?

_____ Do you provide training to your workers on any of the chemicals listed in Sections 1100 through 1148?

_____ Does any of your work take your workers onto or into worksites where exposure to any of the chemicals in Sections 1100 through 1148 could occur?

MORE DETAILED AND OTHER SOURCES OF REGULATORY INFORMATION

In Appendices K and L, you will find key words and their location within the 29 CFR 1926, as well as regulatory definitions for the Subparts of 29 CFR 1926. This information will assist in using and understanding regulations.

Remember that every regulation probably will not apply to your operation. It is doubtful that many of you would need to comply with Subpart Y – diving, especially if you have no divers and no diving operations. Even some sections of a subpart may not apply to you. For instance, Subpart O – Motor Vehicles, Mechanized Equipment, and Marine Operations, Section 1926.603 (Pile-Driving Equipment), would not apply if you do not have pile-driving equipment or perform this type of work. Thus, you can disregard that portion of the regulations which is not applicable to your type of construction activity. As can be seen, it is possible to pick and choose, based upon your construction activity, what subparts of the 29 CFR 1926 you need to use to evaluate your degree of compliance.

Chapter 15

WORKERS' COMPENSATION

Rodney Allen

EMPLOYERS' LIABILITY

Since 1911, the financial liability of employers for the work-related debilities of their employees has been codified in state and federal workers' compensation laws. There exist within the United States 50 separate state acts which provide financial assistance to workers who are injured in the course of employment as well as compensation laws governing American Samoa, Guam, Puerto Rico, and the U.S. Virgin Islands. Federal statutes govern compensation for the District of Columbia, federal employees, and persons engaged in longshoring and harbor work. The Longshore and Harbor Workers' Compensation Act provides for both public and private employees engaged in activities directly related to maritime functions, such as dock construction, but does not normally include workers involved in facilities construction away from the water. In general, the construction industry is almost always ruled by the compensation law for the state or other political unit, such as a territory, in which the construction work is being done.

Prior to the institution of workers' compensation legislation, the financial liability of American employers for worker injury was judged by principles established by English common law. These principles virtually assured that employers would not be held liable for work-related harms to employees. Under the common law, employees did not automatically receive payments when injured on the job and were required to sue the employer in order to obtain compensation for their injuries. If the employee did sue his employer for damages, the employer had four legal defenses which could be asserted and, if any one of these defenses was accepted by the court, the employer did not have to pay. These defenses were

1. The employee contributed to the cause of the accident.

2. Another employee contributed to the accident cause.

3. The employee knew of the hazards involved in the accident before the injury, and still agreed to work in the condition for pay.

4. There was no employer negligence.

Through the institution of a no-fault system for benefits, workers' compensation laws eliminated the requirement for employees to sue employers to obtain financial assistance for an injury suffered on the job. Employers were also protected by the no-fault system through relief of liability from common-law suits asserting negligence. Because of due process concerns, early workers' compensation laws were held to be unconstitutional by the United States Supreme Court. However, the sustainment by the high court of the second Federal Employers' Liability Act of 1908 opened the way for the approval of state laws. There is some controversy as to which state had the first viable workers' compensation law, with New York,

New Jersey, and Wisconsin most often mentioned; all states now provide this protection but with varying coverage and amount of benefits.

Although there is considerable variation in the laws among the states, The U.S. Chamber of Commerce notes six basic objectives for workers' compensation laws. These objectives are:

1. to provide sure, prompt, and reasonable income and medical benefits to work-accident victims, or income benefits to their dependents, regardless of fault;

2. to provide a single remedy and reduce court delays, costs, and workloads arising out of personal injury litigation;

3. to relieve public and private charities of financial drains incident to uncompensated industrial accidents;

4. to eliminate payment of fees to lawyers and witnesses as well as time-consuming trials and appeals;

5. to encourage maximum employer interest in safety and rehabilitation through appropriate experience-rating mechanisms;

6. to promote frank study of causes of accidents (rather than concealment of fault) – reducing preventable accidents and human suffering.

To some extent, most state and federal laws address the objectives set forth above. Nevertheless, the number of workers' compensation laws is great and variation in the laws often significant. Therefore, the bulk of this chapter will be dedicated to a general discussion of workers' compensation laws noting common aspects and identifying differences when appropriate. The control of costs related to workers' compensation will be discussed in the last part of the chapter.

WORKERS' COMPENSATION AS AN EXCLUSIVE REMEDY

Although more than 50 workers' compensation laws exist within the United States, these laws can be classified as either compulsory or elective. Under an elective law, an employer may either accept or reject coverage. However, if an employee sues for damages because of an injury or illness arising out of employment, rejecting employers lose the common law defenses discussed above. Texas, New Jersey, and South Carolina are the only states which have a form of elective coverage. In New Jersey employers or employees may elect to defer coverage by the state's worker's compensation law. Texas allows only employees to choose to forgo protection by worker's compensation and in South Carolina employers of less than four persons, including partners and sole proprietors, may choose not to provide coverage. Compulsory laws require all employers of covered employees to provide benefits according to the provisions of the applicable act, which are for the most part no-fault in nature and therefore disallow employee suits against employers for damages due to injury or illness as a result of work activities. In those jurisdictions where an employee can reject coverage by the act and sue an employer, the employer is normally allowed to assert some of the common law defenses in court.

Even though most states provide for workers' compensation as the exclusive remedy for covered employees impacted by injury or illness arising out of work, a few jurisdictions now allow workers to sue employers in tort and also receive workers' compensation benefits. An example of such a state is West Virginia where the courts have held that some employers' conduct resulting in injury or illness to employees was so egregious that the employer no longer deserved protection by the exclusive remedy provision of the workers' compensation law. In some jurisdictions, employers may be included as a third party in a law suit brought by an employee against another firm, such as an equipment manufacturer, and found financially liable for a portion of the award. In these cases, the firm sued by the employee will sue the employer for any negligence of the employer that they can demonstrate contributed to the injury or illness. Employer equipment maintenance is often a point of issue in these suits.

Since it is impossible to document the exclusive remedy status of all jurisdictions in this text, construction employers should be certain to check the law and its administration in the jurisdictions in which they work to fully understand their potential implications. It is important to remember that the best way of avoiding these legal quagmires is to provide a safe and healthy work environment and thus avoid employee injury or illness.

COVERED EMPLOYMENT

Most workers' compensation laws cover both public and private employment; however, in every jurisdiction there are classes of employees not covered by the law. Not all jurisdictions exempt the same categories of workers. As noted above, in Texas and New Jersey coverage in some circumstances is elective and in South Carolina coverage is required only after four persons are employed. Most other states have much broader coverage requirements primarily exempting employees based upon the type of employment. Many jurisdictions specifically exempt farm workers, domestic help, and casual employees, while some laws do not cover employees engaged in non-hazardous work or employers with a payroll of less than a specified sum. Except in those jurisdictions where an employer may be excepted from coverage because of a requirement to employ a certain number of employees or have a specified amount of payroll, construction employment is universally covered. If you have a question concerning your status as to workers' compensation coverage requirements, contact the workers' compensation administrator for the appropriate jurisdiction.

Employers, who elect to voluntarily provide coverage for exempted employees, are normally protected by the exclusive remedy doctrine from employee law suits for the injuries and illness of voluntarily covered employees. This is an important consideration when an employer is determining whether to voluntarily provide coverage and whether to classify a worker as an independent contractor or an employee. Independent contractors, other non-employees, and employees not covered by workers' compensation, are at liberty to sue an employer in tort for damages due to work injuries and illness. Although the process is adversarial and fault is determined, awards to damaged employees are often much larger than would have been exacted through the workers' compensation system.

COVERED EVENTS

Originally workers' compensation legislation provided benefits for harms to employees due to injuries only, ignoring the adverse consequences of occupational exposures which result in illness. Now all jurisdictions recognize that illnesses as well as injuries arise out of employment and have provided for benefits in their compensation statutes for both. Although most states do not provide coverage for ordinary illnesses of living which are unconnected to employment, both injuries and illness arising out of and in the course of employment are normally covered.

Workers' compensation statutes normally limit benefits to injuries or occupational illness which arise out of and occur in the course of employment. This test, which is almost universally accepted, is designed to assure a relationship between the work activity and the injury or illness. Although the goals of this bifurcated test are relatively straightforward, the facts surrounding individual claims are often fraught with uncertainty. This uncertainty is most pronounced with occupational illness claims. Although there are sometimes controversies as to whether an employee is entitled to benefits following an injury, it is normally not as difficult to determine if an injured employee fulfills the requirements for benefits as it is an employee claiming an occupational illness. There is often a long latency period between an employee's exposure to a disease agent and the onset of illness. During this period employees frequently change jobs or even employers, moving away from the disease agent and making it difficult to substantiate exposure.

The test aspect, in the course of employment, most often refers to the time frame and location of the event producing the injury or illness. Most employees are deemed to be in the course of employment upon entering their employer's property at the start of the work day until they leave at their shift's end. There are some obvious problems with this definition, particularly with respect to employees who do not work at a fixed location, such as traveling salespeople and truck drivers. Normally, the adjudication of disputes surrounding this test have turned around the relationship of the activity when injured to employment. For example, a worker, following the instructions of the supervisor, injured away from a construction site while shopping at a local store to purchase a product necessary to complete a job would, in most jurisdictions, receive compensation. On the other hand, employees are normally not compensated for events occurring while traveling to and from work unless an unusual employee relationship can be demonstrated for the travel, such as being recalled to the work site after leaving.

The arising-out-of component of the test for workers' compensation benefits examines the causal relationship between the employment and the injury or illness. In general, employees are considered to be fulfilling this test if they are involved in activities related to their employment. Employees performing work not related to their employment, such as running personal errands, doing work for themselves on company equipment, or because of intoxication, horseplay, or serious and willful misconduct may not receive benefits.

The following example illustrates the arising-out-of test with respect to illness. Asbestosis is a disease held to be occupational related to older shipyard workers because of their work in close proximity to asbestos fibers during ship construction. However, if a shipyard worker were to contract tuberculosis from a fellow worker, in many jurisdictions the disease would not be a compensable work-related illness because it did not arise out of a particular hazard of employment. On the other hand, if a construction worker employed in a hospital contracted tuberculosis after exposure to a patient, then that case of TB would be compensable in most jurisdictions because TB is a common hazard of work in hospitals if not a normal hazard of construction trades.

SELECTING THE PHYSICIAN

The permanence and severity of an injury or illness is initially determined by the treating physician. In jurisdictions which provide injured employees the opportunity to select the initial treating physician, employers can require employees to also be evaluated by a physician of the employers choosing. In some jurisdictions, employers may require employees to first be evaluated by a physician selected by the employer or require the employee to choose from a group of physicians established by the employer.

If the employer is allowed to choose the first treating physician, employees are often allowed to consult a physician of their choosing at no cost. Jurisdictions, which allow employers to establish a stable of physicians from which employees are encouraged to select, usually allow employees, who do not desire to be treated by a physician from the employers list, to select a practitioner not on the list. However, employers are required to pay only for medical services provided by non-listed physicians at the same fee schedule as that for the accepted list.

If the employee's and employer's physicians disagree as to the medical consequences of an injury or illness, in many jurisdictions, a third physician, agreeable to both the employee and employer, will be selected as an impartial medical evaluator. In selecting a physician, the most important consideration should be that person's ability to treat the injury or illness and help the employee get well. However, and unfortunately, some physicians have acquired a reputation for being biased as an employees' physician witness or an employers' witness with respect to workers' compensation issues and care must therefore be exercised in this selection process.

After an employee accepts treatment beyond that provided in response to an emergency, there are normally restrictions impeding the employees ability to change doctors unless by referral from the treating physician, by agreement with the employer and the insurance company paying the medical bills, or with the approval of the workers' compensation regulatory body of the jurisdiction in which the employee was injured. It is therefore very important to select medical doctors with appropriate skills and for employers to assist employees in selecting appropriate medical practitioners.

BENEFITS

Workers' compensation benefits are provided; 1) as replacement income for wages lost by employees because of their inability to work due to an occupational injury or illness and as financial remuneration for permanent loss of all or part of a bodily function, 2) for medical costs associated with an injury or illness from a job-related exposure, and 3) for rehabilitation following a job-related debility. Most jurisdictions attempt to structure benefits so as to prevent workers from suffering economic hardship because of occupational injury or illness.

CASH BENEFITS

Replacement income and payment for loss of function benefits are often referred to as cash benefits. Cash benefits are normally further categorized according to the severity and permanence of the disability associated with the injury or illness as follows; temporary total disability, permanent total disability, temporary partial disability, permanent partial disability, and death. Cash benefits normally comprise a greater percentage of workers' compensation benefits costs than costs associated with medical care and rehabilitation.

Temporary total disability is an often misunderstood term which does not imply that a worker is confined to bed. It means that the treating physician has determined that the employee is unable to perform any work. This is of course a legal definition and the reality may be somewhat different.

A more thorough analysis often leads to the realization that in many instances employees are considered totally disabled to perform work only because of their employer's policy. For example, if an employer's policy requires that an injured employee be competent to perform all function of a job before returning to work following an injury, an employee who can perform most of the job tasks will still be considered totally disabled by most treating physicians until capable of performing all functions related to the work. Such employees may be able to perform many other functions of normal life unimpeded. This seeming contradiction often results in misunderstandings and accusations of "featherbedding." Permanent total disability implies that an employee's infirmities are such that the worker cannot for the foreseeable future perform any regular work in the existing job market.

Temporary partial disabilities limit employees capacity to do their regular job. These workers are often able to perform many tasks related to their job but are in some ways limited in the performance of the total job or are unable to do the job for the normal length of time. These employees are also often able to perform other jobs which do not require the abilities for which the employee is temporarily deficit. Partially disabled employees can often participate in "return to work" programs which allow employees to perform tasks within their limitations while restricting other activities. Well-run, ethical, return-to-work programs enhance employee recovery and limit workers' compensation costs. Such programs also allow many workers, who would otherwise be classified as temporarily totally disabled, to return to useful work. Permanent partial disability relates to the permanent loss of a body part or functional limitations to the body as a result of an injury or illness. The extent of this type of disability is determinable only after the treating physician concludes that additional physical improvement will not occur.

Death is the final category for which income replacement benefits are paid. Although the death of an employee is the most catastrophic event provided for in workers' compensation legislation, it is often not the most expensive.

Benefits for Total Disability

In all jurisdictions, income replacement cash benefits for total disability are calculated according to formulas which use the injured employees average weekly wage, normally including base pay, overtime pay, all differentials such as shift or task, commissions, tips, and income from other jobs, as the base against which benefits are determined. The number of past weeks used to calculate the average weekly wage may vary, although many jurisdictions use either the 26 or 52 weeks immediately proceeding the injury for this purpose. The average weekly wage of employees working less than the designated number of weeks for calculating benefits is usually determined based upon the weeks actually worked. However, if the employee worked too few weeks to determine the average weekly wage accurately, other methods may be used. The alternative methods include multiplying the hourly salary the person would have received by the hours of expected employment per week, or using the wage history of another person similarly employed.

The preponderance of workers' compensation laws provide cash benefits of 66% of the average pre-tax weekly wage for workers who are totally disabled; however, some jurisdictions have established another formula which provides for the payment of a percentage of after tax income. Connecticut pays benefits amounting to 75% of after-tax income, gross pay minus federal income tax and SSI, for injuries occurring after October 1, 1991. Alaska, Iowa, Maine, and Michigan provide 80% of spendable income, while Rhode Island pays 75% of spendable income.

The formula for determining weekly benefit payments is tempered in all jurisdictions by a maximum weekly payment amount. This is a ceiling amount for benefit payments which will not be exceeded irrespective of an employee's average weekly wage and which varies by jurisdiction. Normally the ceiling payment is some percentage of the state average weekly wage. Many but not all jurisdictions have a minimum weekly payment amount which may also be calculated as a percent of the state average weekly wage or may be only a token, arbitrarily determined, amount such as the $40 established by New York.

Some workers' compensation laws also restrict benefit payments by instituting time limits, beyond which benefits will not be paid, or gross payment amount limits, or both. Although most states provide for temporary total disability payments for the length of the disability and permanent total for life without restrictions as to the amount of such payments, 16 states have instituted time limits expressed in calendar concepts or monetary limits in dollars, or both in concert. In those states which provide for such limits, persons judged to be permanently and totally disabled usually receive more generous treatment than those temporarily disabled.

More common than time or monetary limits for reducing workers' compensation benefit payments are offsets, which are considered when benefits are determined. Twenty-four states provide for workers' compensation benefit reduction if the injured is receiving income from other named social welfare programs, such as social security or unemployment compensation payments. On the positive side, at least from the worker's point of view, 14 states have a built-in cost of living adjustment provision in their workers' compensation payment structure.

Benefits for Partial Disability

Partial disability benefit payments are made to compensate workers who, following an occupational injury or illness, are able to work but are unable to earn comparable wages to those earned prior to the debilitating event. This wage adjustment benefit is normally temporary and often provides only a percentage of the difference between the two salary schedules. Partial disability benefit payments are also provided to compensate injured workers for a permanent disability which results in the partial or complete loss of the use of a body part or function.

Permanent partial disability benefit payments, payments for the loss of a body part, are divided into two categories, scheduled and non-scheduled injuries. Scheduled injuries are those which are specifically addressed in the workers' compensation law and for which a value has been determined by law. The method of expressing these values may differ according to the jurisdiction; however, in most instances, the award is granted as a specified number of weeks of benefits at the injured worker's average weekly wage rate. Some states have no schedule for injuries, but determine the benefit paid for permanent partial disability based upon such factors as the degree of disability in relation to the person as a whole; the degree of impairment; wage loss, age, education, and physical loss.

The values ascribed to scheduled injuries in the workers' compensation laws are predicated on the assumption that the use of the body part has been totally lost to the employee. Since this assumption does not always square with reality, partial disability awards are often granted based on something less than full loss of use. These awards are generally determined by taking into account the treating physician's judgment as to the extent of the loss of use, how that degree of loss effects the person as a whole, and, hopefully, the AMA Guide to the Evaluation of Permanent Impairment. The amount of permanent partial disability payments for non-scheduled injuries, such as back and head injuries, are frequently based on similar criteria.

Survivor Benefits

Survivor benefits are designed to provide replacement income for the families of workers whose deaths are related to work and to pay a portion of the burial expenses. All jurisdictions provide for such benefits but, as with all other workers' compensation benefits, with considerable variation.

As a general rule, the same calculations and factors used to determine the amount of benefit payments a worker would have received for total disability are used in determining weekly benefits to be paid to a surviving spouse with children. This includes a percentage, usually 66% but this varies by jurisdiction, of the worker's average weekly wage with maximum and minimum thresholds similar to those discussed for total disability. In 21 states the percent of the average weekly wage paid is reduced if the worker is survived only by a wife or by only a child. In some instances, these reductions are very significant. Pennsylvania, for example, reduces the percentage for a surviving spouse without children to 51% of the deceased worker's average weekly wage and 32% for a child who is the deceased's sole survivor, from 66% if survived by both a spouse and child.

Most states have established time limits for benefit payments which provide for continuation of payments for dependent children until they reach a designated age, normally 18. Benefit payments for children may be extended beyond the established age if the dependent has not finished school. Spouses usually receive benefits for life or until they remarry. In many states when a spouse remarries, they receive two years' benefits payment as a lump sum award. Some jurisdictions limit payments to a designated number of weeks; the most common is 500 weeks, or to a specified monetary sum.

MEDICAL BENEFITS

Every jurisdiction requires the employer to provide medical care for the injured worker without cost to the employee. Care includes first aid treatment, emergency transportation, the services of a physician and surgeon, hospital and nursing services, drugs, supplies, and prosthetic devices. Some states have introduced fee schedules to control the cost of providing medical care for injured employees or allow employers to contract with specific providers of medical services to reduce expenses.

As previously discussed, the selection of the treating physician is a very important and sometimes emotional issue. In about half of the states, employees may choose their physician while about the same number of states allow employers to make this selection. Although the main concern in selecting physicians should be their competency in helping the employee get well, in fact many other variables are usually considered. The employee may desire that a known and trusted practitioner, often the family doctor, be the treating physician. In such instances, trust is often more important than the physicians specialization. Employees also normally prefer to be in hospitals close to their homes, if possible, so as not to inconvenience their families. Employers usually prefer the treating physician to be a specialist in the area of medicine of concern to the employees condition.

REHABILITATION

Background

Not all injured workers respond to treatment and return to gainful employment. A 1989 review of workers' compensation data by the National Council on Compensation Insurance found that 90% of injured workers recovered and returned to work within 18 months of injury. However, the 10% that did not return accounted for more than 75% of the expenditures for medical care, lost wages, and rehabilitation. Moreover, only 2% of this group had severely disabling injuries or illnesses.

Both the Occupational Safety and Health Administration and the California Workers' Compensation Institute report that less than 2% of all injuries to workers are catastrophic. Catastrophic injuries include spinal cord damage, brain injury, amputation of major limbs, and blindness. The California Institute also found that 8% of claims involving minor disabilities such as strains, sprains, and muscle or ligament tears resulted in no return to gainful employment and were classified as permanent total disability.

Acknowledging the difficulty of overcoming some injuries without significant assistance, most jurisdictions now require employers to provide rehabilitative services for employees. The goal of such services is to return the injured employee to productive work and gainful employment. In general, these services are classified as either medical or vocational. Jurisdictions differ significantly in the way rehabilitative services are provided and funded. It is therefore necessary for interested persons to understand the requirements of the jurisdiction in which they operate.

Medical Rehabilitation

Medical rehabilitation is usually dictated by the accepted course of treatment for a specific injury and can often not be distinguished from medical treatment. Rehabilitation is viewed as an integral part of standard medical treatment and often begins immediately after the patient is stable. Services such as physical and occupational therapy are rehabilitative in nature and are often prescribed to assist employees regain function, mobility, and strength in injured extremities. Medical rehabilitation frequently requires many months to complete, with the rehabilitative services accessed once or more weekly.

Most workers' compensation laws obligate employers and insurers to pay for all medical care including rehabilitation. This can be accomplished by providing the services in-house, contracting with outside providers, or through local medical facilities as part of the medical treatment process.

Vocational Rehabilitation

Vocational rehabilitation is the next step in the process of returning injured workers to gainful employment. Injured persons, who do not adequately recover during medical treatment and medical rehabilitation to continue their former profession in the same manner as before injury, are prepared for either a new career or taught to perform their former job in a new manner during vocational rehabilitation.

Tested methods of eliminating obstacles to employability through vocational rehabilitation include

1. Job modification. After performing a job analysis, the workplace is modified to allow an employee to return to work with the residual disabilities.

2. Adaptive devices. These include grasping aids, orthopedic supports, or special tools.

3. Work hardening. An individualized work-oriented treatment program to improve the employee's physical capacity by involving the person in supervised simulated or actual work.

4. Retraining. This should be done only after careful consideration of other options as it requires the most adjustment and therefore risk.

Injured workers are provided several options for vocational rehabilitation. All states now have agencies dedicated to providing vocational rehabilitation services for injured workers. In many states the insurance company will compensate the state for this service through the worker's compensation law. The federal government also provides funding to aid states in providing rehabilitative services for injured workers through the Federal Vocational Rehabilitation Act.

Vocational rehabilitation is an important step in the return to employment process for many injured persons. Although some employers hesitate to participate in this process because of an ill-founded concern over the initial cost of vocational rehabilitation, the benefits of rehabilitation are great. The employee benefits both psychologically and physically by becoming a productive citizen engaged in meaningful work and the employer benefits by reducing the worker's compensation costs of the firm.

ADMINISTRATION

All jurisdictions have an agency or department responsible for overseeing and administering the worker's compensation law for that jurisdiction. The primary goal of the administrative organizations is to ensure that the laws are complied with and that workers receive appropriate compensation under the statutes. The responsibilities and functions of the agencies are set forth in the statutes of each jurisdiction and vary significantly.

In some jurisdictions the law requires the administrative organization to closely monitor each case to insure that all is performed in accordance with the statute. Others only serve to adjudicate contested claims and issues of law. In many states employees are left to their own devices in obtaining benefits, as the state assumes that employees are knowledgeable regarding workers' compensation benefits.

Workers' compensation claims may be either contested or uncontested. If uncontested, the employer, or the insurance company of the employer, pays the injured employee according to the system set forth in the law. The direct payment system and the agreement system are the most usual methods employed for this purpose. In the direct payment system the employer automatically begins paying compensation without an agreement being in place. The laws set forth the amount of the benefit. In the agreement system, the parties agree on the amount of benefit before payment is made. The agreement system is the most commonly used.

The administrative agency is usually responsible for providing a system by which contested claims can be judged. This system usually begins with a hearing before a referee or hearing officer. The findings of this hearing can usually be appealed to a larger commission or board of appeals by either the employee or the employer. The next appeal is normally to the courts, who are interested in issues of law rather than fact.

RISK MANAGEMENT

Insurance

The risks associated with workers' compensation exposures are normally managed by either purchasing insurance or self-insurance. In most jurisdictions, companies that wish to self-insure their worker's compensation risk must demonstrate that they have sufficient financial resources to fund losses. Many self-insuring companies also purchase insurance for catastrophic risks. This insurance is designed to protect the company's financial stability should a large, unexpected loss occur. Self-insuring companies fund normal worker's compensation costs through internal capital and contract for insurance coverage only for unexpected, significant losses. Companies which elect to self-insure worker's compensation risks frequently employ the services of a third party to administer the compensation program. This service normally includes following up on claims with physicians and writing checks to employees and providers.

Most companies provide workers' compensation coverage through traditional insurance methods in which the insurance company pays the costs associated with worker's compensation claims for the period insured. Insurance companies also provide many services such as third party administration, loss control, and injury analysis depending on the contract and state law.

Insurance companies are compensated for the services they provide through premium payments made by the insured company to the insurer. The amount of the annual premium is principally based on the type of work the insured performs, the classification code, and the insured firm's history of injuries and illness over a given past period, referred to as the experience modification rating.

Classification codes were developed by The National Council on Compensation Insurance (NCCI) to rationalize insurance premiums so that firms engaged in work which is relatively more dangerous will pay more in premiums than those performing less dangerous tasks. For example, working as a coal miner is inherently more dangerous than working as a banker and the class code for coal miners requires a larger premium.

Whereas classification codes are based on the risks associated with a type of work without regard to how safely an individual firm performs, experience modification codes (EMR) are concerned with the safety performance of the firm to be insured. Each firm begins with an EMR of 1.0 (100%), which means it must pay the average premium for the appropriate classification code. Adjustments of this percentage are made by the NCCI based on the firm's injury and illness losses over a specified period, normally three years, and in compliance with jurisdictional requirements. If a firm's safety performance is better than expected, the EMR can be reduced so that the firm pays less than 100% of the average premium. The percentage reduction is predicated on the firm's success in avoiding losses. Companies which have worse than expected loss records will have to pay more than the average premium.

In some states joint ventures are rated at 1.0 until sufficient history, usually three years, is available to develop an EMR for the venture. Other states average the EMRs of the firms involved to determine the rating for the combined venture. Since construction contractors are often involved in joint ventures, risk managers should be aware of the pitfalls and rewards available for the firm's EMR in these combined activities.

Loss Control

The best way of controlling costs associated with workers' compensation is by reducing injuries and illnesses at the worksite. The rationale of this for self-insured firms is obvious since the firm is the direct payer of all costs associated with the injury or illness. Insured firms also benefit by reducing injuries since future EMRs are based on past safety performance.

Most state laws set forth the method to be used in determining future EMRs and in many of these formulas the number of injuries and the incidence of injury is more important than the cost or extent of the injuries. For example, one injury costing $45,000 is less harmful to a firm's EMR than four injuries costing $10,000. The rationale for this policy is that once an incident occurs the extent of injury is often a matter of chance and therefore the $10,000 events could have easily been much more costly.

Another cost-saving technique for insured firms is to pay as many claims as the state law allows from the resources of the firm and not through the insurance company. Some states require all injuries for which a claim could be made to be reported, while others allow companies to pay modest claims without involving the insurance company or state. Since EMRs are determined based more on incidence than severity, avoiding the reporting of minor injuries will improve the firm's future EMR.

Return to work programs provide an important tool in controlling workers' compensation costs for both the self-insured and insured employer. Workers returned to work do not receive disability payments and, without medical expenses sufficient to require state notification, do not count in determining future EMRs. Employers with well-designed return to work programs have documented significant savings in worker's compensation costs and improvements in other costs of production. Improvements in injury and illness incidence rates have also been documented after the institution of a well-managed return-to-work program.

When an employee is injured, the responsibility for initiating the return to work program is the employer's. The employer must maintain timely and constant contact with the employee, the insurance company, and the employee's physician. The company must assure the physician that work is available which meets the limitations of the employee and that the company has in place a system to assure compliance with the limitations. The employer must also notify the employee of the availability of work within the limitations imposed by the physician and when to return to work.

The activities of workers who return to work must be monitored to ensure that they are working within the limitations set forth by their physician. Supervisors must understand the importance of the return to work program and be instructed not to allow employees to perform tasks beyond their restrictions, even on a voluntary basis. Supervisors must set a positive tone for the rest of the employees.

Employees also benefit from return to work programs. There is now significant medical documentation that testifies to the therapeutic value of employees returning to appropriate work as soon as possible after injury or illness.

SUMMARY

It is the employer's responsibility when a worker has an occupationally related injury or illness to

1.   Get medical treatment for the worker.

2.   Have the employee complete a notice of injury or illness.

3.   Make a report of the employee's injury, illness, or death.

4.   Receive the claims for the workers' compensation program.

5.   Assist the employees and their survivors in preparing claims.

Job safety and health and your workers' compensation program, if effectively managed, can be the tools to cost reduction and efficiency on your construction jobsites. To reduce your workers' compensation cost is to reduce the pain and suffering which your workforce is experiencing. Any responsible contractor, safety professional, or supervisor realizes that injured and ill workers are a problem which effects productivity, morale, and the bottom line. Thus, some effort must be expended to assure that your workers' compensation program is meeting the needs of your company. This chapter is only an introduction to workers' compensation. You will need to expand your knowledge and understanding beyond the scope of this chapter.

REFERENCES

Accident Prevention Manual for Business & Industry, Administration & Programs, National Safety Council, Chicago, IL, 1992, pp. 193-207.

Analysis of Workers' Compensation Laws., U.S. Chamber of Commerce, Washington, D.C., 1996.

Bear-Lehman, J., "Factors Affecting Return to Work After Hand Surgery," *American Journal of Occupational Therapy*, 37, 1983, pp. 189-194.

Caruso, LA., Chan, D.E., and Chan, A., "The Management of Work Related Back Pain," *American Journal of Occupational Therapy, 1987.*

Civitello, Donna, *Injured on the Job: A Handbook for Connecticut Workers,* ConnectiCOSH, Hartford, CT, 1992.

Dent, G.L., "Curing The Disabling Effects of Employee Injury," *Risk Management,* 31 (1), 1985, pp. 30-32.

Gice, J., "The Body is Willing, But is the Mind Able," *Risk Management*, 36 (5), 1989, pp. 30-35.

Gice, J.H., and Tompkins, K., "Cutting Costs with Return to Work Programs," *Risk Management*, 35 (4), 1988, pp. 62-65.

Green, J., "Employer Potential for Increased Control of Workers' Compensation Costs," *Professional Safety*, 34 (2), 1989, pp. 28-33.

Ritzel, D.O., and Allen, R.G., "Value of Work–A Way of Evaluating a Light Duty Program in a Work Setting," *Professional Safety*, 33 (11), 1988, pp. 23-25.

Chapter 16

RESOURCES AND INFORMATION ACCESS

Charles D. Reese

In recent years the sources of information for the construction industry, as well as other arenas, have mushroomed and undergone tremendous changes in their format and presentation. Sources of information are still found in published materials, such as books and trade journals, but with the growth of computers and the infamous internet, construction information has begun taking on a new look.

There is a wide variety of information sources for construction safety and health; these range from published books to trade journals, federal and state governments, professional associations and societies, educational institutions, and consultants.

Of course, books still provide a handy reference for accessing quick information regarding the construction industry or specific construction-related safety and health issues. Although these publications do exist, there are not a plethora of them.

Many of the construction trade associations conduct training, hold seminars, and develop construction-related materials. Most have frequently published journals which provide updates and sources of information and materials. These trade associations are often your contact when you need to locate individuals with unique areas of expertise.

Not only does federal OSHA develop, implement, and enforce the federal safety and health regulations, OSHA is the source for many publication, other relevant information, and a place where you can get answers from their staff, inspectors, and agency experts regarding compliance with safety and health regulations. OSHA can also help you solve the safety and health problems that you may face.

Often your state has an OSHA organization or some type of worker safety and health component and, under some state programs, free consultation is available to assist in achieving compliance with the federal regulations. These consultants may do a safety and health walk-through of your jobsite, take an industrial hygiene sampling, provide training, advise you on abatement actions that you can take, provide examples of written programs, and provide other helpful information. When you use this service, it will not result in your receiving any citations for the deficiencies that are found, unless you have a hazardous situation which could result in serious injury, illness, or death and you have refused to take corrective action to rid your worksite of these hazards. Then the consultant would be obligated to contact federal OSHA in order to protect the workers from serious harm.

Frequently there exist state and local organizations which act as vital sources of information and are attuned to the changes in the legislature and business practices within the state(s) where you are doing business. Through these organizations, you will have a voice at the grass root level and an impact on the decisions which will affect your construction business.

Professional associations and societies are organizations that have, as their primary purpose, the fostering of workplace safety and health. These organizations may be very broad in their focus, or have a specific area of safety and health, (i.e., ergonomics); regardless, they usually have access to individuals with a great deal of expertise. They may hold conventions, market publications, or develop and provide training programs. They tend to foster a collegial approach to addressing safety and health and cater to their membership with reduced cost and fees.

Another source of information you should never overlook is your state educational institutions. Universities and colleges may offer safety and health credit and/or non-credit courses. Often these courses can be specifically tailored for the construction industry and, thus, can address your unique safety and health needs. These training courses, whether credit or noncredit, can be geared to management, supervisors, or workers. Also, since many of the educational institutions employ staff who have specialties in construction safety and health, industrial hygiene, occupational medicine, occupational disease, chemical safety, safety engineering, and environmental safety and health, they often offer a wide variety of consultation services. You can use the internet to access the curriculum or training programs which are available through your colleges and universities.

Consulting services are an important function in providing assistance which is relevant to your safety and health issues. A consultant is always willing to provide safety and health services but, depending upon the urgency of your situation or the types of safety and health services needed, the cost may be quite high. Nevertheless, these costs are often justifiable because it is sometimes difficult to secure a consultant who has the expertise needed to assist with your specific problems. Once you locate a consultant, you will need to format a contractual agreement. The agreement should delineate the services to be provided, the outcomes or products to be attained, and the follow-up obligations you require.

THE COMPUTER

If you do not use a computer at the present time, it is time to start. You do not have to be a "computer whiz" or "computer geek" to put the computer to work for you. Most computers and their accompanying software or programs are reasonably friendly (easy to use) for new users. As you learn more about computer use, the sources of information available to you will also expand. The computer allows you to make use of electronic mail (e-mail) whereby you can maintain contact, leave messages, ask and answer questions, and receive and impart valuable information with your colleagues in the construction industry.

Your computer will allow you to store large amounts of information for future use and access new sources of information and help. You will be able to use many of the helpful information sources on the CD-ROM discs. They may contain training programs, templates for forms, and interpretation of safety and health issues, as well as give you access to OSHA information through the OSHA CD-ROM (price $39 for four quarterly updated discs).

By using the internet, you will be able to receive answers to questions from others all over the world who have the same interests in construction safety and health. There is also a multitude of government internet sites (locations) where all types of information about government programs and government information sources can be found. Furthermore, you can install special software (programs), or have someone else install them on your computer, which will be helpmates in solving problems or implementing safety and health programs.

PROFESSIONAL ORGANIZATIONS AND AGENCIES

These are national organizations that specialize in the many aspects of occupational safety and health. They have a wide range of resources as well as unique materials which have been developed by those with special expertise in occupational safety and health. Some key organizations and agencies are as follows:

Health and Environmental Assistance:

ABIH (American Board of
Industrial Hygiene)
4600 West Saginaw, Suite 101
Lansing, MI 48917
(517) 321-2638

ACGIH (American Conference of
Governmental Industrial Hygienists)
Building D-7
6500 Glenway Avenue
Cincinnati, OH 45211
(513) 661-7881

AIHA (American Industrial Hygiene
Association)
P.O. Box 8390
475 White Pond Drive
Akron, OH 44311
(216) 873-3300

**Safety and Engineering Consensus
Standards:**

ANSI (American National Standards
Institute)
11 West 42 Street
New York, NY 10038
(212) 354-3300

ASME (American Society of
Mechanical Engineers)
345 East 47th Street
New York, NY 10017
(212) 705-7722

ASTM (American Society for Testing
and Materials)
655 15th Street NW
Washington, D.C. 20005
639-4025

NSMS (National Safety Management
Society)
12 Pickens Lane
Weaverville, NC 28787
(800) 321-2910

Professional Safety Organizations:

ASSE (American Society for Safety
Engineers)
1800 East Oakton Street
Des Plaines, IL 60016
(708) 692-4121

BCSP (Board of Certified Safety
Professionals)
208 Burwash Ave.
Savoy, IL 61874
(217) 359-9263

HFS (Human Factors Society)
P.O. Box 1369
Santa Monica, CA 90406
(310) 394-1811

ISEA (Industrial Safety Equipment
Association)
1901 North Moore Street
Arlington, VA 22209
(703) 525-1695
Fax: (703) 528-2148

NSC (National Safety Council)
1121 Spring Lake Drive
Itasca, IL 60143-3201
(708) 285-1121

**Specialty Associations (With Specific
Expertise):**

AGA (American Gas Association)
1515 Wilson Blvd.
Arlington, VA 22209
(703) 841-8400

American Board of Toxicology
PO Box 30054
Raleigh, NC 27622
(919) 782-0036

API (American Petroleum Institute)
1220 L Street, NW
Washington, D.C. 20005
(202) 682-8000
FAX: (202) 682-8159

ASHRAE (American Society of Heating,
Refrigerating, and Air Conditioning
Engineers)
1791 Tullie Circle, NE
Atlanta, GA 30329
(404) 636-8400

ASTD (American Society for Training and
Development)
1640 King Street
P.O. Box 1443
Alexandria, VA 22313-2043
(703) 683-8129

AWS (American Welding Society)
P.O. Box 351040
550 LeJeune Road, NW
Miami, FL 33135
(305) 443-9353

Center For Atomic Radiation Studies
P.O. Box 381036
Cambridge, MA 02238
(508) 263-2065

CGA (Compressed Gas Association)
1235 Jefferson Davis Highway
Arlington, VA 22202
(703) 979-0900

**Human Factors and Ergonomics Society
   (HFES)**
P.O. Box 1369
Santa Monica, CA 90406-1369
(310) 394-1811

**Illuminating Engineering Society of
   North America**
345 East 47th Street
New York, NY 10017

Institute of Makers of Explosives
1120 19th Street, NW
Washington, D.C. 20036

Institute of Noise Control Engineering
P.O. Box 3206
Poughkeepsie, NY 12603
(914) 462-4006

**International Commission on Radiation
   Units and Measurements**
7910 Woodmont Ave.
Bethesda, MD 20814
(301) 657-2652

Laser Institute of America
12424 Research Parkway, Suite 130
Orlando, FL 32826
(407) 380-1553

National Propane Gas Association
1600 Eisenhower Lane
Lisle, IL 60532

**NFPA (National Fire Protection Associa-
   tion)**
1 Batterymarch Park
Quincy, MA 02269
(800) 344-3555

Noise Control Association
 680 Rainer Lane
 Port Ludlow, WA 98365-9775

**Scaffolding, Shoring, and Forming
   Institute**
1230 Keith Building
Cleveland, OH 44115

Society of Toxicology
1767 Business Center Dr.
Reston, VA 22090-5332
(703) 438-3115

The Chlorine Institute
2001 L Street
Washington, D.C. 20036
(202) 775-2790
FAX: (202) 223-7225

**Women's Occupational Health Resource
   Center**
117 St. Johns Place
Brooklyn, NY 11217
(718) 230-8822

FEDERAL GOVERNMENT SOURCES

The federal government is not the enemy, as many individuals surmise. It is a great resource for all types of information such as publications, training materials, compliance assistance, audio-visuals, access to experts, and other assorted occupational safety and health aids. In most cases, resources offered by the federal government are current and the response time is very reasonable. Asking for information does not act as trigger for your company to become a target for inspections or audits. The federal government would prefer to assist you in solving your safety and health issues prior to having them become problems. You will be pleasantly surprised by the help that you receive and all you need to do is ask. A listing of government agencies which have information regarding occupational safety and health is as follows:

BLS (Bureau of Labor Statistics)
U.S. Dept. of Labor
Occupational Safety and Health Statistics
441 G Street, NW
Washington, D.C. 20212
(202) 523-1382

CDC (Center for Disease Control)
U.S. Dept. of Health and Human Services
1600 Clifton Avenue, NE
Atlanta, GA 30333
(404) 329-3311

EPA (Environmental Protection Agency)
410 M Street, SW
Washington, D.C. 20460
(202) 382-4361

GPO (U.S. Government Printing Office)
Superintendent of Documents
732 N. Capitol Street, NW
Washington, D.C. 20402
(202) 512-1800

MSHA (Mine Safety and Health
Administration)
U.S. Department of Labor
4015 Wilson Blvd.
Arlington, VA 22203
(703) 235-1452

NAC (National Audio Visual Center)
National Archives and Records
Administration
Customer Services Section CL
8700 Edgewood Drive
Capitol Heights, MD 20743-3701
(301) 763-1896

**National Institute of Standards and
Technology**
U.S. Dept. of Commerce
National Engineering Laboratory
Route I-270 and Quince Orchard Road
Gaithersburg, MD 20899
(310) 921-3434

NIH (National Institutes of Health)
U.S. Department of Health and Human
Services
9000 Rockville Pike
Bethesda, MD 20205
(310) 496-5787

NIOSH (National Institute for Occupational
Safety and Health)
U.S. Dept. of Health and Human Services
Publications Dissemination
4676 Columbia Parkway
Cincinnati, OH 45226
(513) 533-8287 or (800) 35-NIOSH

NTIS (National Technical Information
Services)
U.S. Dept. of Commerce
5285 Port Royal Road
Springfield, VA 22161
(703) 487-4636

OSHA (Occupational Safety and Health
Administration – National Office)
U.S. Dept. of Labor
200 Constitution Avenue, NW
Washington, D.C. 20210
(202) 523-8151

OSHA (After Hours) National Hotline –
(800) 321-OSHA

OSHA's Training Institute
1555 Times Drive
Des Plaines, IL 60018
(708) 297-4913

OSHA Publications Office
Room N3101
Washington, D.C. 20210
(202) 219-9667

OSHRC (Occuaptional Safety and Health
Review Commission)
1825 K Street, NW
Washington, D.C. 20006
(203) 643-7943

CONSTRUCTION-RELATED ASSOCIATIONS

Many construction companies and contractors work within unique sectors (i.e., electrical). Only a small number of contractors try to be everything to everybody. Thus, with the specialization of contractors and subcontractors, a need has arisen for specific organizations that address the specialized needs related to newly developed techniques, new tools, changes in procedures, and safety and health. There are some associations that have expertise in all the specialties within the construction industry. These associations offer to its members a wide array of services which are related to their specialty. Also, there are state construction trade associations which address and dwell on state specific issues and work to enhance the construction industry and construction projects within the state. They have a wide interest which goes beyond dollars; their objective is to establish construction safety and health and accident/incident prevention on the projects and jobsites within your state. The following is a list of national organizations which have ties to the construction industry. (This list does not include any state construction trade associations.)

Air Conditioning Contractors of America
1712 New Hampshire Ave. NW
Washington, D.C. 20009
(202) 483-9370
FAX: (202) 232-8545

American Concrete Institute
P.O. Box 19150
Redford Station
Detroit, MI 48219
(313) 532-2600

American Concrete Pumping Association
279 E. H Street
Benecia, CA 94510
(707) 746-0730

**American Road and Transportation
    Builders Association**
The ARTBA Building
1010 Massachusetts Ave., NW
Washington, D.C. 20001
(202) 289-4434
FAX: (202) 289-4435

**American Society for Concrete
    Construction**
3330 Dundee Road
Northbrook, IL 60062
(708) 291-0270

American Subcontractors Association
1004 Duke Street
Alexandria, VA 22314
(703) 684-3450

Architectural Precast Association
825 E. 64th Street
Indianapolis, IN 46220
(317) 251-1214

Associated Builders and Contractors
1300 N 17th Street
Rosslyn, VA 22209
(703) 812-2000
FAX : (703) 812-8235

**Associated General Contractors of
    America, Inc.**
1957 E Street
Washington, D.C. 20006
(202) 373-2070

Brick Institute of America
11490 Commerce Park Drive
Reston, VA 22091
(703) 620-0010

Building Stone Institute
420 Lexington Ave.
New York, N Y 10170
(212) 490-2530

Concrete Foundation Association
P.O. Box 34745
North Kansas City, MS 64116
(816) 471-6686

Concrete Reinforcing Steel Institute
933 N. Plum Grove Road
Schaumburg, IL 60173-4758
(708) 517-1200

**Construction Industry Manufacturers
   Association**
111 E. Wisconsin Ave., Suite 1000
Milwaukee, WI 53202-4879
(414) 272-0943
FAX: (424) 272-1170

Home Builders Institute
1090 Vermont Ave., NW, Suite 600
Washington, D.C. 20005
(202) 371-0600
FAX: (202) 898-7777

Indiana Limestone Institute of America
Stone City National Bank Building
Bedford, IN 47421
(812) 275-4426

**Indiana Ready Mixed Concrete
   Association**
9860 North Michigan
Carmel, IN 46032
(317) 872-6302

**International Association of Concrete
   Repair Specialists**
Dulles International Airport
Washington, D.C. 20041
(703) 450-0116

**International Association of Drilling
   Contractors**
P.O. Box 4287
Houston, TX 77210

**International Council of Employers of
   Bricklayers and Allied Craftsman**
821 15th Street NW
Washington, D.C. 20005
(202) 783-3791

International Masonry Institute
823 15th Street NW
Washington, D.C. 20005
(202) 783-3908

Marble Institute of America
33505 State Street
Farmingham, MI 48024
(313) 476-5558

Mason Contractors Association of America
17W. 601 14th Street
Oak Brook Terrace, IL 60181
(708) 620-6767

Masonry Society
2619 Spruce Street
Boulder, CO 80302
(303) 939-9700

Metal Building Manufacturers Association
(216) 241-7333
FAX: (216) 241-0105

National Association of Home Builders
1201 15th Street, NW
Washington, D.C. 20005
(800) 368-5242

**National Association of Plumbing-
   Heating-Cooling Contractors**
P.O. Box 0800
1805 Washington Street
Fall Church, VA 22040
(703) 237-8100
FAX: (703) 237-7442

**National Association of Waterproofing
   Contractors**
25550 Chargrin Blvd., Suite 403
Cleveland, OH 44122
(216) 464-2484
FAX: (216) 595-8230

**National Association of Women in
   Construction**
327 S. Adams Street
Fort Worth, TX 76104-1081
(817) 877-5551
FAX: (817) 877-0324

National Concrete Masonry Association
2362 Horse Pen Road
Herndon, VA 22071
(703) 713-1900

National Constructors Association
1630 M Street, NW, Suite 900
Washington, D.C. 20036
(202) 466-8880
FAX: (202) 466-7512

National Electrical Contractors
   Association
3 Bethesda Metro Center, Suite 1100
Bethesda, MD 20814
(301) 657-3110
FAX: (301) 215-4500

National Precast Concrete Association
825 E. 64th Street
Indianapolis, IN 46220
(317) 253-0486

National Steel Erectors of America
P.O. Box 4891
Chapel Hill, NC 27515
(919) 933-5100
FAX: (919) 933-5101

Portland Cement Association
5420 Old Orchard Road
Skokie, IL 60077
(708) 858-9172

Precast/Prestressed Concrete Institute
175 W. Jackson Blvd.
Chicago, IL 60604
(312) 786-0300

Scaffolding Industry Association, Inc.
14029 Sherman Way, Suite 100
Van Nuys, CA 91405-2599
(818) 782-2012
FAX: (818) 786-3027

The National Roofing Contractors
   Association
10255 West Higgins Road, Suite 600
Rosemont, IL 60018-5667
(874) 299-9070
FAX: (874) 299-1183

Tilt-Up Concrete Association
P.O. Box 204
107 First Street West
Mount Vernon, IA 52314
(319) 895-6911

ELECTRONIC SOURCES (INTERNET)

Your computer must be connected to the internet; this means you must select an internet provider. This provider may be your local or long-distance phone company or it may be commercial services such as AOL, Prodigy, Erol, or Compuserve (just to name a few). Of course, it goes without saying; you will need a computer with a reasonably fast modem (the faster the better), a phone line, and some software such as Microsoft Explorer or Netscape Navigator (your internet provider usually provides this) which allows you to browse the internet.

Once you have access to the internet, there are several good "search engines" which are helpful in finding information (internet sites). These have names such as "Yahoo," "Lycos," "Alta Vista," or "InfoSeek." These "search engines" allow you to find the sites or locations of the information that you are interested in (i.e., fall protection, crane safety).

The internet sites have names which help you understand what they are. Some of the most common abbreviated names are

    http – this is a transfer protocol
    www – means world wide web and is a component of the internet
    com – means commercial
    edu – stands for education
    gov – means government
    org – stands for organization

Most internet sites start with http://www. – followed by an abbreviation for entity (company or institution) and other numbers, symbols, or abbreviations that seem to make no sense. The ending is usually com, gov, edu, or org. You can use set site addresses, such as the ones which follow, to access these specific locations:

Government

Address of Government Agencies – http://www.fedworld.gov

Agency for Toxic Substances and Disease Registry (ATSDR) – http://atsdr1.atsdr.cdc.gov:8080/

Bureau of Labor Statistics – http://www.bls.gov

Department of Transportation – http://www.dot.gov/

Environmental Protection Agency – http://www.epa.gov/

Mine Safety and Health Administration – http://www.msha.gov

National Institute for Occupational Safety and Health – http://www.cdc.gov/niosh/homepage.html

Occupational Safety and Health Administration – http://www.osha.gov

OSHA Ergonomics – http://www.osha.gov/ergo

Violence – http://www/osha.gov/workplace_violence/wrkplaveviolence.intro.html/

Other Sources

Accident Investigation – info@webcrawler. comAccident Investigation

Biomechanics – info@webcrawler.com"Biomechanics"

Canadian Centre for Occupational Health and Safety – http://ww.ccohs.ca

Construction Industry Institute – http://construction-institute.org/services/catalog/catalog.htm

Emergency Response – info@Yahoo. com/emergency Response

Ergonomics – info@webcrawler.com/ergonomics

Ergonomics – http://www.egroweb.com/ergonomics

Fire Protection – info@webcrawler.com/fire protection

Hazardous Waste – info@ webcrawler.com/air pollution/hazardous waste/

Injury Control Resources Information Network (ICRIN) – http://www.pitt.edu~hweiss/injury.html

Material Safety Data Sheets (MSDSs) – http://hazard.com/msds/msdsindex.html

Material Safety Data Sheets (MSDSs) – http://www.chem.uky.edu/resources/msds.html

Material Safety Data Sheets (MSDSs) – http://hazard.com

Material Safety Data Sheets (MSDSs) – http://www.phys.ksu.edu/~tipping/msds.html

Material Safety Data Sheets (MSDSs) – http://enviro-net.com/techical/msds

National Environmental Safety Compliance – http://www.albany.net/~nesc/

National Safety Council – http://nsc.org/nsc

Regulatory Compliance – info@webcrawler.com/regulatory compliance

Safety Inspections and Audits – info@webcrawler.com/safety inspections & audits

Safety Net Yellow Pages – http://www.tiac.net/users/dploss/html/olddir.html

Safety Online – http:www.safetyonline.net/home.htm

Technical Health and Safety Source – http://turva.me.tut.fi/~oshaweb

Consultants

California Safety Training Corp. – http://www.kern.com/cstc/

Castle Rock Safety Co. – http://www.safetyonline.net/safetcon/castle/home.htm

Hayes Consulting Group – http://www.midtown.net/~hcg/

Occupational Safety Consultants, Inc. – http://www.mindspring,com/~stephens/ oschome.html

Occupational Safety Services, Inc. – http://www.k2nesoft.com/~ossinc/

Operation Safe Sites (Construction) – http://ww.opsafesite.com/

Pathfinder Associates – http://ww.webcom.com/pathfndr/docs/internet.htm

Quest Consultants – http://www.questconsult.com

Reactive Management Corp. – http://safetyonline.net/safetcon/rmc/rmc.htm

Construction Organizations

Air Condition Contractors of America – http://www.acca.org/

American Road and Transportation Builders Association – http://www.artba.org/

American Subcontractors Association – http://www.naturalgas.org/association/asa/ index.htm

Associated Builders and Contractors – http://www.abc.org/

Construction Industry Manufacturers Association – http://www.cimanet.com/

Home Builders Institute – http://www.hbi.org/

Metal Building Manufacturers Association – http://www.taol/mbma/

National Association of Home Builders – http://www.nahb.com/pg2.html

National Association of Plumbing-Heating-Cooling Contractors – http:// www.naphcc.org/

National Association of Waterproofing Contractors – http://www.apk.net/nawc/

National Association of Women in Construction – http://www.nawic.org/

National Electrical Contractors Association — http://www.neca.net.org/

National Steel Erectors of America — http://users.southeast,net/~seaa/

Scaffolding Industry Association – email - sia@scaffold.org

The National Roofing Contractors Association – http://www.rooonline.org/ index.html

BOOKS

Books are always a quick and ready resource. Although there are many books related to occupational safety and health, there are fewer books which are directly related to construction safety and health. The following is not a comprehensive listing of the types of books available on safety and health, but a sampling of them:

Introduction to Occupational Epidemiology, Hernberg, Sven, Lewis Publishers/CRC Press, Boca Raton, FL (1992).

Introduction to Industrial Hygiene, Scott, Ronald M., Lewis Publishers/CRC Press, Boca Raton, FL (1995).

The Complete Guide to OSHA Compliance, Peterson, Ronald D. and Cohen, Joel M., Lewis Publishers/CRC Press, Boca Raton, FL (1996).

Occupational Health and Safety: Terms, Definitions and Abbreviations, Confer, Robert G. and Thomas R., Lewis Publishers/CRC Press, Boca Raton, FL (1994).

Accident Prevention and OSHA Compliance, Michaud, Patrick A., Lewis Publishers/CRC Press, Boca Raton, FL (1995).

Solutions: A Systems Approach, Parker, Kathryn G. and Imbus, Harold R., Lewis Publishers/ CRC Press, Boca Raton, FL (1992).

Illustrated Dictionary of Environmental Health and Occupational Safety, Koren, Herman, Lewis Publishers/CRC Press, Boca Raton, FL (1995).

Fundamentals of Occupational Safety and Health, Kohn, J.P., Friend, M.A., and Winterberger, C.A., Government Institutes, Inc., Rockville, MD (1996).

Supervisor's Safety Manual (9ᵗʰ Edition), National Safety Council, Itasca, IL (1997).

Motor Fleet Safety Manual (3ʳᵈ Edition), National Safety Council, Itasca, IL (1996).

Basic Guide to Industrial Hygiene, Vincoli, Jeffrey W., Van Nostrand Reinhold, New York (1995).

Construction Safety Planning, David V. MacCollum, New York: Van Nostrand Reinhold (1995).

Construction Safety Handbook., Mark McGuire Moran, Government Institutes, Inc., Rockville, MD (1996).

Engineering Physiology: Bases of Human Factors/Ergonomics, Kroemer, K.H.E. and Kroemer-Elbert, H.J., Van Nostrand Reinhold, New York (1990).

The Ergonomics of Workspaces and Machines: A Design Manual, Clark, T.S. and Corlett, E.N., Taylor & Francis, London (1984).

Best's Safety Directory, A.M. Best Co., Ambest Rd., Oldwick, NJ 08858, (908) 439-2200.

Ergonomics Sourcebook: A Guide To Human Factors Information, Kimberlie H. Pelsma, Editor, Lawrence, KS: Report Store (1987).

SELECTING CONSULTANTS

A comprehensive list of consultants is not provided in this book since you will be able to find a wide array of safety and health consultants in each state. The best that can be done is to try to provide some ideas and guidance regarding the evaluation of your need for a consultant and the selection of one who could meet your specific needs.

Consultants can draw on a wide and diverse experience base in helping to address your problems and issues. The consultant is not influenced by politics and allegiance and is, therefore, in a better position to make objective decisions. The professional consultant strives to give you cost effective solutions since his or her repeat business is based upon performance and professional reputation. Also, a consultant is a temporary employee and the usual personnel issues are not applicable to them. Consultants work at the times of the day, week, or month when you have the need and not at their convenience. Most consultants have become qualified to perform these services through either education or experience.

In determining the need for a consultant, you will want to consider whether using a consultant will be cost effective, faster, or more productive. You may also find the necessity for a consultant when you feel you need: outside advice; access to special instrumentation; an objective, unbiased opinion or solution; or, someone to back up your own assessment or convince others.

A consultant can address many of your safety and health issues. These issues can come from a wide variety of problems and be such things as the training of employees or the need for an engineering solution. The consultant can also act as an expert witness during legal actions. But, primarily, the consultant is hired to solve a problem. Thus, the consultant must be able to identify and define the existing problem and then attempt to solve that problem. The consultant has a wide array of resources and professional contacts which he or she can access in order to assist in solving the problem.

To get names and recommendations of potential consultants, you can contact professional organizations, colleagues, and insurance companies who often employ loss control and safety and health personnel. Furthermore, do not overlook your local colleges and universities; many times they also provide consultation services.

In selecting your consultant make sure that you ask for

- A complete resume which provides you with the consultant's formal education, as well as his or her years of experience.

- A listing of previous clients and permission to contact them.

- The length of time the individual has been a consultant and his or her current status regarding existing business obligations.

- Verification of professional training.

- Documentation of registrations or certifications, such as: registered professional engineer, certified safety professional, or certified industrial hygienist.

- A listing of the consultant's membership in professional associations.

- Any areas of specialization and ownership, or access to equipment and certified testing laboratories.

You should evaluate more than one consultant so that you will have some basis for comparison. This will help in your selection of the consultant who is best suited for your particular situation. Close attention should be paid to the consultant's compensation procedures. You will want to know: if he or she works hourly; what the estimated total cost will be; what will be used as a retainer structure; and how does the consultant bill for travel, shipping, report generation, computer time, and other services such as laboratory fees. Your decision should not be based purely upon a cost comparison. It may be like comparing cooking apples and eating apples. They are both apples, but only one has the qualities to meet your needs at that point in time and you will pay the price for the type of apple which meets your needs.

Although a verbal agreement may seem fine to you, it is recommended that you have a written agreement which spells out how much the services are going to cost, or the maximum number of hours which you will support. Also, you will need to include a scope of work which sets out the steps which will be followed in order to solve your problem. It is appropriate to include in this written document the output expectations from the process. A minimum output would be a written report, but you may not want to pay for this; if not, a verbal report would be the output in the agreement. All expected outputs should be part of this agreement. This may include drawings, step-by-step procedures, and the follow-up. You need to get all the information that is needed to solve your problem since, based upon the consultant's recommendations, you will have to construct, implement, or redesign. Unless you have some sort of a protective clause within your agreement, there is no guarantee that the consultant's recommendations will fix the problem. If you desire this guarantee, you should expect to pay more since the consultant will be responsible for the implementation of his or her recommendations.

If the consultant process is to work, it must be in full cooperation with the company; thus, the consultant can fail if the company is not an integral part of his team. For this reason it is important to provide all the information needed, clearly define the problem, set objectives, agree on realistic requirements, get a clear and complete proposal from the consultant, review the progress from time to time, and stay within the scope of the problem. The consultant can only be as effective as you allow him or her to be. You must do your part to assure success. Make sure you have done a cost benefit analysis before pursuing a consultant. Agree on cost and fees prior to starting.

The consultant is not your friend. You have hired him or her based upon your evaluation that this person is the best person to help you solve the problem which you cannot solve, do not have time to solve, or do not have the resources to solve. The steps the consultant should take are: define the problem, analyze the problem, and make recommendations for a solution. You can make the consultant a part of your team by providing full information and support. This will assure that you reap the greatest value from your investment.

Chapter 17

SUMMARY

Charles D. Reese

In the *Handbook of OSHA Construction Safety and Health*, a foundation has been set forth for a comprehensive approach to construction safety and health. As was said earlier, construction is a unique industry, but it goes without saying that the universal principles of occupational safety and health apply to construction, as well as they apply to all other industries. These five principles of occupational safety and health are

1. All accidents are preventable.

2. All levels of management are responsible for safety.

3. Every employee has the responsibility to himself/herself, their coworkers, and their families to work safely.

4. In order to eliminate accidents, management must ensure that all employees are properly trained on how to safely and efficiently perform every job task.

5. Every employee must be involved in every area of the safety and production process.

There is no justifiable reason in accepting death and carnage as an unavoidable consequence within the construction industry. Records show that companies and contractors who have accepted their responsibility for making safety and health a part of good business practices, have shown that there can be safe and healthy construction operations and jobsites.

For contractors it is more than applying safety and health principles; it is the development and fostering of an atmosphere which evokes all workers, including supervisors, to adhere to positive behavior toward safety and health. These attitudes and behaviors can be accomplished by following the ten commandments of safety and health, and they are to be applied, equally, to everyone on the construction worksite. These commandments are as follows:

1. LEARN the safe way to do your job before you start.

2. THINK safety, and ACT safety at all times,

3. OBEY safety rules and regulations – they are for your protection.

4. WEAR proper clothing and protective equipment.

5. CONDUCT oneself properly at all times – horseplay is prohibited.

6. OPERATE only the equipment you are authorized to use.

7. INSPECT tools and equipment for safe conditions before starting work.

8. ADVISE responsible parties promptly of any unsafe conditions or practices.

9. REPORT any injury immediately to the supervisor.

10. SUPPORT the safety and health program, and take an active part in safety meetings.

The attitudes and behaviors must be nurtured and they must be an integral part of the contractors approach to organized safety and health on their jobsites. But, little can be accomplished unless there exists the same technical expertise, regarding occupational safety and health, that exists for the technical aspects of the construction process.

The contractor must be knowledgeable in the technical aspects of occupational safety and health and must establish a plan for occupational safety and health. This unique plan should make use of the tools of safety and health, such as safety audits and job safety (hazard) analysis.

PLANNING THE SAFETY AND HEALTH INITIATIVE

When the planning process starts, commitment is the key to success. You must get the commitment of the supervisors and the workers. You need a foundation and that foundation is to be in the form of a written program which incorporates management's commitment, objectives, and investments (dollars committed to safety and health). There must be a process for a continuous evaluation of potential and existing hazards and there must be a process of control established and rules set in place for the correction of hazards. And, finally, the support of real and effective training for everyone is of utmost importance. If you do not set up the rules and requirements for jobsite or project safety and health, why should others worry about it. Someone must lead the way!

This does not mean that all of the planning process should fall on one person's shoulders. The more participation which you can elicit from others, the stronger the buy in, or ownership, that will exist. Do not expect the gaining of participation to be an easy process. Many people talk a good story, while fewer actually become part of the solution process.

This written plan should be dynamic so that you can make changes as the program matures.

THE PEOPLE IN CONSTRUCTION SAFETY AND HEALTH

There are four distinct and unique groups which heavily impact construction safety and health. They are the construction workers, the supervisors, the safety and health professionals, and the contractors. In the following paragraphs each group is separately addressed in an attempt to elicit their support and to delineate their vital role in achieving safer and healthier construction worksites.

Construction Workers

Each construction worker needs to give careful consideration to those working around them, especially to those workers who may not be familiar with their trade and its hazards, or to those workers who are new to the construction business.

The safety and health tools and principles presented in this book are the ones which help create a safer and healthier workplace for you and your fellow workers. You should learn about these safety and health techniques, keep an open mind, and help your supervisors, health and safety professionals, and contractors put them to use in order to provide for safety and health on your jobsites.

When you know the correct, safe, and healthy way to do a job, you must become part of the solution in preventing accidents, injuries, and illnesses on construction sites. With your knowledge of safety and health, you must now act responsibly and not look the other way when you see another worker or workers performing their jobs in an unsafe or unhealthy fashion. If you ever expect to raise the professional level of the construction industry and its workforce, you must become proactive for job safety and health; it must be as important to you as your desire is for attaining proficiency in the skills of your trade (see Figure 17-1).

Figure 17-1. Construction activities force construction workers to become team players

You will always run into a few individuals who will resist any change in the workplace, but one by one you can become a single unit whose goal is a safer, healthier, and more productive workplace; with this goal in mind, you will also experience less injuries, illnesses, and deaths.

Supervisors

Construction supervisors are one of the major components for a company's job safety and health program. The supervisors must do all that is in their power to assure that the workers are well protected on the job, even if at times this means "in spite of" the workers. The supervisors are accountable and responsible for identifying hazards, fixing hazards, and assuring that individuals are working in a safe and healthy manner. They must be honest, fair, and enforce safety and health on the jobsite in an equitable manner. How supervisors support workplace safety and health will determine the company's success or failure in regard to the company's efforts to reduce accidents, injuries, and illnesses on the jobsites. If supervisors do not present a positive attitude, nor act as a positive role model for good job safety and health, no one else will. This is an awesome burden to bear, but the supervisor/leader who truly cares about his/her workers will relish the opportunity to be an effective leader for construction jobsite safety and health (see Figure 17-2).

This may mean that you, as a supervisor, will need to learn and apply some of the tools and techniques put forth in this book. You may also need to make some revisions or adjustments to some of those safety and health tools and techniques in order to make them work for your unique operation.

Figure 17-2. Supervisor modeling safe work practices by wearing appropriate PPE

If those who work for you know that you are concerned about their safety and well-being, and that you are willing to take action when necessary, they will help you to accomplish the worksite safety and health goals.

As has been mentioned earlier, supervisors can make or break the safety and health effort. This is why you must support safety and health and be well trained. Almost every directive, policy, and rule is conveyed to the workforce by their supervisor; you are their most direct line of communication. To emphasize the importance of the supervisor's role in safety and health, the " Supervisor's Ten Commandments of Safety" are listed below. Read these ten commandments and verify if you are portraying a positive attitude toward achieving a safe and healthy workplace.

THE SUPERVISOR'S TEN COMMANDMENTS OF SAFETY

1. CARE for people at work as you would care for your family at home. Be sure each employee understands and accepts his/her personal responsibility for safety.

2. KNOW THE RULES for safety that apply to the work you supervise. Never let it be said that one of the workers you supervise was injured because you were not aware of the precautions required on their job.

3. ANTICIPATE the risks that may arise from changes in equipment or methods. Make use of the expert safety advice that is available to help you guard against such new hazards.

4. ENCOURAGE workers to discuss with you the hazards of their work. No job should proceed where a question concerning safety remains unanswered. When you are receptive to the ideas of workers, you tap a source of firsthand knowledge that will help you prevent needless loss and suffering.

5. INSTRUCT workers to work safely, as you would guide and council your own family – with persistence and patience.

6. FOLLOW UP your instructions consistently. See to it that workers make use of safeguards provided to them. If necessary, enforce safety rules by disciplinary action. Do not fail the company, which sanctioned these rules – or workers who need them.

7. SET A GOOD EXAMPLE. Demonstrate safety in your own work habits and personal conduct. Do not appear as a hypocrite in the eyes of the workers you supervise.

8. INVESTIGATE AND ANALYZE every accident – however slight – that befalls any of the workers you supervise. Where minor injuries go unheeded, crippling accidents may strike later.

9. COOPERATE FULLY with those in the organization who are actively concerned with worker safety. Their dedicated purpose is to keep workers fully able and on the job, and to cut down the heavy personal toll of accidents.

10. REMEMBER: Not only does accident prevention reduce human suffering and loss; from the practical viewpoint, it is no more than good business. Safety, therefore, is one of your prime obligations – to your company, to your fellow supervisors, and your fellow workers.

The supervisor must lead their workers into "thinking safety," as well as working safely day by day. The supervisor will win their loyal support and cooperation by showing genuine concern for their fellow employees. More than that, supervisors will gain in personal stature by modeling a good safety attitude and behavior. Good workers do good/safe work for a good leader (supervisor).

Safety and Health Professionals

As a safety and health professional/consultant you are often called upon to develop the safety and health initiative, develop specific programs, such as preventive maintenance programs, implement the company's safety and health program, recommend improvements to the safety and health effort, try to solve specific safety and health problems, or evaluate the effectiveness of the safety and health effort (see Figure 17-3).

Figure 17-3. Safety and health professional (on the right) discusses safety and health issues with a construction supervisor

Whether it be a safety and health issue or a construction related safety and health problem, this book can be your starting reference. No one source answers all the questions or provides all the information that you will need, but it helps to have at least one source which gets you going in the right direction. Your job description suggests that you know all there is to know about construction safety and health; this is seldom the case with any of us. It is certainly the intent of this book to provide as much help to you as possible within the confines of these pages.

You can also use the content of this text to train workers and supervisors. You can use different parts to convince or guide management regarding the development of safety and health on their construction jobsites and projects.

There are many aspects of this book which will provide you with the tools to assist you and your constituencies with many of the safety and health endeavors which are undertaken. You should encourage contractors that you are working for, or with, to read this book. As their interest in job safety and health grows, you can encourage them to become more familiar with the varied professional safety and health approaches, tools, and techniques available. The information found in these writings may also be instrumental in helping you gain more support for jobsite safety and health which is in your charge.

Contractors

If you are a contractor, you will want to read the following to decide if you are, in fact, addressing safety and health using many of the principles iterated within this book's chapters. Hopefully, this summary will motivate you into evaluating or assessing your company's commitment to occupational safety and health.

<u>People</u>

The foundation of your company consists of those individuals who work with you and for you. Their attitudes, morale, training, and expertise will determine your success. Actually, you, the contractor, are in less control of your company and your destiny than any individual within your company. You need to ask yourself if you are as effective a leader as you can be, and if you develop and provide a work environment where those who work with you and for you can successfully accomplish your contract within the parameters of the bid which you submitted.

This means, do I understand the individuals that I work with, do I effectively communicate my programs and goals, do I provide the motivational environment in which they can finish the construction project, and does my safety and health program convey my true concern for the welfare of my employees and for their safety and health on my jobsites? Do my employees, including my supervisors, support my commitment to on-the-job safety and health, as well as their welfare? Are they well trained regarding safety and hazard recognition so that they can be a help in assuring that I maintain a safe and healthy work environment? Do the employees know the safety and health policies and do they believe that I feel strongly enough about them to enforce them, or do they consider my policies, rules, and commitment a joke (see Figure 17-4)?

<u>Programs</u>

Without having an organized approach to occupational safety and health as a part of your business climate, you will be doomed for setbacks and, eventually, failure in your effort to control costly occupational safety and health incidents. You should have a written safety

Figure 17-4. The contractor's presence and communication demonstrate his/her commitment to a safe and productive construction site

and health program as the foundation of your safety and health initiative. As part of your commitment to jobsite safety and health, ask yourself the following questions regarding your approach to accident prevention.

Is my safety and health program well designed? Is it in parity with my other programs? Is it complete enough to assure a safe and healthy work environment in which my employees will be free from known hazards? Am I committed to my safety and health program and do I financially support it? Am I a role model for safety and health? Does my written safety and health program incorporate, at the least, the cardinal elements of management commitment and planning, hazard assessment, hazard correction and control, and safety and health training? Is my analysis of my safety and health initiative a dynamic and continuous process?

Do I make use of accident prevention tools, such as accident investigations, job safety analysis, worksite inspections, and preventive maintenance programs in my pursuit of accident prevention and loss control? Am I constantly on the lookout for other useable safety and health tools to improve my safety and health record? Have I asked others to be involved in making our accident prevention program more effective and better? Do those I ask include supervisors and workers?

Hazards

Anyone you ask will tell you that they think construction is a hazardous industry. As you walk over a construction site, you know that the dangers and potential hazards are there. But, just how much attention do you pay to them, and how aware of these dangers are you and your employees? You might want to ask yourself how aware am I, as well as my employees, of the construction hazards which can exist and impact our jobsites? Do I give thought to not only the safety hazards, but also the health hazards which may exist, or potentially exist? Are those hazards recognized, removed, or controlled on my jobsites? If personal protective equipment is one of my controls, do I assure that the equipment is provided and used as required on my jobsites, and do I also enforce my safety and health rules and policies by holding my supervisors truly accountable? Do I expect supervisors to follow my discipline policy even handily when such becomes necessary. Do I give my supervisors the power, authority, and right to correct, control, and enforce safety and health on jobsites? Do I convey a zero tolerance for hazardous action and hazardous conditions?

OSHA/ Regulations

More than likely you view OSHA regulations as a hindrance and burden to your overall operation. But, do you realize that OSHA is an integral part of your safety and health program? For without OSHA you would not have the foundation upon which to base your safety and health program, nor the mandates which provide you with someone (OSHA) to blame for the emphasis on occupational safety and health. You would not want to be considered the individual who set up all these safety and health rules to protect your workforce, would you? Do you realize that your workforce is probably more likely to adhere to your safety and health initiative in order to keep OSHA from coming in and finding numerous violations which could cost your company large sums of money and also jeopardize your cash flow and the workers' jobs?

Does your workforce work with you on safety and health issues in order to assure that OSHA won't get you? After all, it is like all things in life, without the structure for safety and health which is provided by OSHA and its regulations, you would not have the support or motivator which you need to rid your operation of those costly occupational injuries and illnesses which decrease your profit margin and drive up operating costs. Have you ever thought of OSHA as a helpmate and player in your safety and health goals?

Are you familiar with the OSHA regulations which impact your type of construction business? Do you make sure that your supervisors understand the regulations, and your workers know the essence of the regulations which are applicable to your jobsites? Do you expect each person who recognizes an OSHA violation to report it so that action can be taken to correct it as quickly as possible? If you are unsure a violation exists or you do not know how to correct it, do you call the nearest OSHA office or others with expertise to get answers for your questions? Do you know how to get help? There is no excuse for not seeking assistance from outside sources when you or your employees do not know the answer. OSHA would rather answer your questions than expend their limited resources to enforce safety and health on your projects.

THIS BOOK IS FOR YOU

The content of this book either provides answers and guidance, or directs you to sources of information. Do you realize it addresses most, if not all the safety and health issues which you will face when implementing safety and health on construction jobsites?

It goes without saying, if safety and health are not an integral part of a construction company's plans and policies, that company is either flirting with disaster, extremely lucky, or not making the profit which their efforts deserve. Usually when a company is unwilling to make safety and health an integral part of their business operation, it is just because they are not willing to serve two masters at the same time, production and safety. It seems foolish to jeopardize a company, the employees, and everyone's livelihood for what is deemed today, by most, as good business sense when a company makes workplace safety and health a partner within their construction business.

INCORPORATE SAFETY AND HEALTH

Accident prevention makes good business sense. For every dollar that you can see as direct cost, you are probably losing between three and ten dollars in indirect cost which cannot be as easily detected or accounted for. Safety awareness is not automatically attributed to the workforce; it must be carefully developed, if we truly care.

We must be SAFE

Search for hazards.
Assess risks.
Find solutions.
Enforce solutions.

There is always time and funding for job safety and health, because there always seems to be time to take a worker to the hospital, time and money to treat an eye injury, and money to pay for workers' compensation to cover back injuries or illnesses related to chemical exposure. If you don't have the $1,000,000 or time to deal with a worker's death, then you had better ensure that no worker dies on your jobsites. Because this is the kind of time and money that will be required of your company when a fatality does occur. It, therefore, seems logical that making the time and funds available for workplace safety and health makes good business sense.

APPENDIX A

NAICS VERSUS SIC

NAICS 1997			**SIC 1987**	
23	Construction			
233	Building, Developing and General Contracting			
2331	Land Subdivision and Land Development			
23311	Land Subdivision and Land Development	E	6552	Land Subdividers and Developers, Except Cemeteries
2332	Residential Building Construction			
23321	Single Family Housing Construction	R	1521	General Contractors–Single Family Houses
			*1531	Operative Builders (single-family housing construction)
23322	Multifamily Housing Construction	R	*1522	General Contractors–Residential Building, Other Than Single-Family (except hotel and motel construction)
			*1531	Operative Builders (multi-family housing construction)
2333	Non-residential Building Construction			
23331	Manufacturing and Industrial Building Construction	R	*1531	Operative Builders (manufacturing light industrial building construction)
			*1541	General Contractors–Industrial Buildings and Warehouses (except warehouse construction)
23332	Commercial and Institutional Building Construction		*1522	General Contractors–Residential Building Other than Single-Family (hotel and motel construction)
			*1521	Operative Builders (commercial and institutional building construction)
			*1541	General Contractors–Industrial Buildings and Warehouses (warehouse construction)

707

			1542	General Contractor–Nonresidential Buildings, Other than Industrial Buildings and Warehouses
234	Heavy Construction			
2341	Highway, Street, Bridge and Tunnel Construction			
23411	Highway and Street Construction	E	1611	Highway and Street Construction, Except Elevated Highways
23412	Bridge and Tunnel Construction	E	1622	Bridge, Tunnel, and Elevated Construction
2349	Other Heavy Construction			
23491	Water, Sewer, and Pipeline Construction	N	*1623	Water, Sewer, Pipeline, and Communications and Power Line Construction (water and sewer mains and pipelines)
23492	Power and Communication Transmission Line Construction	N	*1623	Water, Sewer, Pipelines, and Communications and Power Line Construction (communications and power line construction)
23493	Industrial Nonbuilding Structure Construction	N	*1629	Heavy Construction, NEC (industrial nonbuilding structures construction)
23499	All Other Heavy Construction	R	*1629	Heavy Construction, NEC (except industrial nonbuilding structures construction)
			*7353	Construction Equipment Rental and Leasing (with operator)
235	Special Trade Contractors			
2351	Plumbing, Heating and Air-Conditioning Contractors			
23511	Plumbing, Heating and Air-Conditioning Contractors	E	1711	Plumbing, Heating and Air-Conditioning
2352	Painting and Wall Covering Contractors			
23521	Painting and Wall Covering	R	1721	Painting and Paper Hanging
			*1799	Special Trade Contractors, NEC (paint and wallpaper, stripping and wallpaper removal contractors)
2353	Electrical Contractors			
23531	Electrical Contractors	R	*1731	Electrical Work (except burglar and fire alarm installation)
2354	Masonry, Drywall, Insulation, and Tile Contractors			
23541	Masonry and Stone Contractors	E	1741	Masonry, Stone Setting and Other Stone Work
23542	Drywall, Plastering, Acoustical and Insulation Contractors	R	1742	Plastering, Drywall, Acoustical, and Insulation Work
			*1743	Terrazzo, Tile, Marble and Mosaic work (fresco work)
			*1771	Concrete Work (stucco construction)

23543	Tile, Marble, Terrazzo and Mosaic Construction	R		*1743	Terrazzo, Tile, Marble, and Mosaic Work (except fresco work)
2355	Carpentry and Floor Contractors				
23551	Carpentry Contractors	E		1751	Carpentry Work
23552	Floor Laying and Other Floor Contractors	E		1752	Floor Laying and Other Floor Work, NEC
2356	Roofing, Siding and Sheet Metal Contractors				
23561	Roofing, Siding and Sheet Metal Contractors	E		1761	Roofing, Siding, and Sheet Metal Work
2357	Concrete Contractors				
23571	Concrete Contractors	R		*1771	Concrete Work (except stucco construction)
2358	Water Well Drilling Contractors				
23581	Water Well Drilling Contractors	E		1781	Water Well Drilling
2359	Other Special Trade Contractors				
23591	Structural Steel Erection Contractors	E		1791	Structural Steel Erection
23592	Glass and Glazing Contractors	R		1793	Glass and Glazing Work
				*1799	Specialty Trade Contractors, NEC (tinting glass work)
23593	Excavation Contractors	E		1794	Excavation Work
23594	Wrecking and Demolition Contractors	E		1795	Wrecking and Demolition Work
23595	Building Equipment and Other Machinery Installation Contractors	E		1796	Installation of Erection of Building Equipment, NEC
23599	All Other Special Trade Contractors	R		*1799	Special Trade Contractors, NEC (except paint and wallpaper stripping, wallpaper removal contractors, and tinting glass work)

* ABSTRACTED FROM INTERNET SITE: http://www.census.gov

APPENDIX B

Construction Training Requirements

The following training requirements have been excerpted from Title 29, Code of Federal Regulations Part 1926. Note that in addition to these requirements, some of Part 1910, relating to general industry, also contains applicable training standards.

SUBPART C –

General Safety for and Health Provisions 1926.20(b) (2) and (4)

(2) Such programs [as may be necessary to comply with this part] shall provide frequent and regular inspections of the jobsites, materials, and equipment to be made by competent persons [capable of identifying existing and predictable hazards in the surroundings or working conditions which are unsanitary, hazardous, or dangerous to employees, and who have authorization to take prompt corrective measures to eliminate them designated by the employers].

(4) The employer shall permit only those employees qualified [one who, by possession of a recognized degree, certificate, or professional standing, or who by extensive knowledge, training, and experience, has successfully demonstrated his ability to solve or resolve problems relating to the subject matter, the work, or the project] by training or experience to operate equipment and machinery.

Safety Training and Education 1926.21(a)

(a) General requirements. The Secretary shall, pursuant to section 107(f) of the Act, establish and supervise programs for the education and training of employers and employees in the recognition. Avoidance and prevention of unsafe conditions in employments covered by the Act.

192621(b)(1) through (6)

(1) The employer should avail himself of the safety and health training programs the Secretary provides.

(2) The employer shall instruct each employee in the recognition and avoidance of unsafe conditions and the regulations applicable to his work environment to control or eliminate any hazards or other exposure to illness or injury.

(3) Employees required to handle or use poisons, caustics, and other harmful substances shall be instructed regarding their safe handling and use, and be made aware of the potential hazards, personal hygiene, and personal protective measures required.

(4) In jobsite areas where harmful plants or animals are present, employees who may be exposed shall be instructed regarding the potential hazards and how to avoid injury, and the first aid procedures to be used in the event of injury.

(5) Employees required to handle or use flammable liquids, gases, or toxic materials shall be instructed in the safe handling and use of these materials and made aware of the specific requirements contained in Subparts D, F, and other applicable subparts of this part.

(6) (i) All employees required to enter into confined or enclosed spaces shall be instructed as to the nature of the hazards involved, the necessary precautions to be taken, and in the use of protective and emergency equipment required. The employer shall comply with any specific regulations that apply to work in dangerous or potentially dangerous areas.

(6) (ii) For purposes of subdivision (i) of this subparagraph, "confined or enclosed space" means any space having a limited means of egress, which is subject to the accumulation of toxic or flammable contaminants or has an oxygen deficient atmosphere. Confined or enclosed spaces include, but are not limited to, storage tanks, process vessels, bins, boilers, ventilation or exhaust ducts, sewers, underground utility vaults, tunnels, pipelines, and open top spaces more than 4 feet in depth such as pits, tubs, vaults, and vessels.

SUBPART D –

Medical Services and First Aid 1926.50(c)

(c) In the absence of an infirmary, clinic, hospital, or physician that is reasonably accessible in terms of time and distance to the worksite which is available for the treatment of injured employees, a person who has a valid certificate in first aid training from the U.S. Bureau of Mines, the American Red Cross, or equivalent training that can be verified by documentary evidence, shall be available at the worksite to render first aid.

Ionizing Radiation 1926.53(b)

(b) Any activity which involves the use of radioactive materials or X-rays, whether or not under license from the Atomic Energy Commission [Nuclear Regulatory Commission] shall be performed by competent persons specially trained in the proper and safe operation of such equipment. In the case of materials used under Commission license, only persons actually licensed, or competent persons under the direction and supervision of the licensee, shall perform such work.

Nonionizing Radiation
1926.54 (a) and (b)

(a) Only qualified and trained employees shall be assigned to install, adjust and operate laser equipment.

(b) Proof of qualification of the laser equipment operator shall be available and in possession of the operator at all times.

Gases, Vapors, Fumes,
Dusts, and Mists
1926.55(b)

(b) To achieve compliance with paragraph (a) of this section, administrative or engineering controls must first be implemented whenever feasible. When such controls are not feasible to achieve full compliance, protective equipment or other protective measures shall be used to keep the exposure of employees to air contaminants within the limits prescribed in this section. Any equipment and technical measures used for this purpose must first be approved for each particular use by a competent industrial hygienist or other technically qualified person. Whenever respirators are used, their use shall comply with 1926.103.

Asbestos, Tremolite,
Anthophyllite, and
Actinolite

(3) Employee information and training. (i) The employer shall institute a training program for all employees exposed to airborne concentrations of asbestos, tremolite, anthophyllite, actinolite, or a combination of these 1926.58(k)(3) minerals in excess of the action level and shall ensure their participation in the program.

(ii) Training shall be provided prior to or at the time of initial assignment, unless the employee has received equivalent training within the previous 12 months, and at least annually thereafter.

(iii) The training program shall be conducted in a manner that the employee is able to understand. The employer shall ensure that each such employee is informed of the following:

(A) Methods of recognizing asbestos, tremolite, anthophyllite, or actinolite exposure;

(B) The health effects associated with asbestos, tremolite, anthophyllite, or actinolite exposure;

(C) The relationship between smoking and asbestos, tremolite, anthophyllite, and actinolite in producing lung cancer;

(D) The nature of operations that could result in exposure to asbestos, tremolite, anthophyllite, or actinolite, the importance of necessary protective controls to minimize exposure including, as applicable, engineering controls, work practices, respirators, housekeeping procedures, hygiene facilities, protective clothing, decontamination procedures, emergency procedures, and waste disposal procedures, and any necessary instruction in the use of these controls and procedures;

(E) The purpose, proper use, fitting instructions, and limitations of respirators as required by 29 CFR 1910.134;

(F) The appropriate work practices for performing the asbestos, tremolite, anthophyllite, or actinolite job; and

(G) Medical surveillance program requirements.

(H) A review of this standard, including appendices.

(4) Access to training materials. (i) The employer shall make readily available to all affected employees without cost all written materials relating to the employee training program, including a copy of this regulation.

(ii) The employer shall provide to the Assistant Secretary and the Director, upon request, ail information and training materials relating to &e employee information and training program.

SUBPART E –

**Hearing Protection
1926.101(b)**

(b) Ear protective devices inserted in the ear shall be fitted or determined individually by competent persons.

**Respiratory Protection
1926.103(c)(1)**

(1) Employees required to use respiratory protective equipment approved for use in atmospheres immediately dangerous to life shall be thoroughly trained in its use. Employees required to use other types of respiratory protective equipment shall be instructed in the use and limitations of such equipment.

SUBPART F –

**Fire Protection
1926. 150(a)(5)**

(5) As warrantcd by the project, the employer shall provide a trained and equipped firefighting organization (Fire Brigade) to assure adequate protection to life. "Fire brigade" means an organized group of employees that are knowledgeable, trained, and skilled in the safe evacuation of employees during emergency situations and in assisting in firefighting operations.

1926.150(c)(l)(viii)

(viii) Portable fire extinguishers shall be inspected periodically and maintained in accordance with Maintenance and Use of Portable Fire Extinguishers, NFPA No. 10A-1970.

From ANSI Standard 10A-1970. The owner or occupant of a property in which fire extinguishers are located has an obligation for the care and use of these extinguishers at all times. By doing so, he is contributing to the protection of life and property. The nameplate(s) and instruction manual should be read and thoroughly understood by all persons who may be expected to use extinguishers.

"1120. To discharge this obligation he should give proper attention to the inspection, maintenance, and recharging of this fire protective equipment. He should also train his personnel in the correct use of fire extinguishers on the different types of fires which may occur on his property.

"3020. Persons responsible for performing maintenance operations come from three major groups:

Trained industrial safety or maintenance personnel.

Extinguisher service agencies.

Individual owners (e.g., self-employed...)."

SUBPART G –

Signaling
1926.201(a)(2)

(2) Signaling directions by flagmen shall conform to American National Standards Institute D6.1-1971, *Manual on Uniform Traffic Control Devices for Streets and Highways.*

SUBPART I –

Powder-Operated
Hand Tools
1926.302(e)(1) and (12)

(1) Only employees who have been trained in the operation ation of powder-actuated tools shall be allowed to use them.

(12) Powder-actuated tools used by employees shall meet all other applicable requirements of American National Standards Institute, A10.3-1970, Safety Requirements for Explosive Actuated Fastening Tools.

Woodworking Tools
1926.304(f)

(f) Other requirements. All woodworking tools and machinery shall meet other applicable requirements of American National Standards Institute, 01.1-1 961, Safety Code for Woodworking Machinery.

From ANSI Standard 01.1-1961, Selection and Training of Operators. Before a worker is permitted to operate any woodworking machine, he shall receive instructions in the hazards of the machine and the safe method of its operation. Refer to A9.7 of the Appendix.

"A9.7 Selection and Training of Operators. Operation of Machines, Tools, and Equipment. General."

"(1) Learn the machine's applications and limitations, as well as the specific potential hazards peculiar to this machine. Follow available operating instructions and safety rules carefully."

"(2) Keep working area clean and be sure adequate lighting is available."

"(3) Do not wear loose clothing, gloves, bracelets, necklaces, or ornaments. Wear face, eye, ear, respiratory, and body protection devices, as indicated for the operation or environment."

"(4) Do not use cutting tools larger or heavier than the machine is designed to accommodate. Never operate a cutting tool at greater speed than recommended."

"(5) Keep hands well away from saw blades and other cutting tools. Use a push stock or push block to hold or guide the work when working close to cutting tool."

"(6) Whenever possible, use properly locked clamps, jig, or vise to hold the work."

"(7) Combs (feather boards) shall be provided for use when an applicable guard cannot be used."

"(8) Never stand directly in line with a horizontally rotating cutting tool. This is particularly true when first starting a new tool, or a new tool is initially installed on the arbor."

"(9) Be sure the power is disconnected from the machine before tools are serviced."

"(10) Never leave the machine with the power on."

"(11) Be positive that hold-downs and antikickback devices are positioned properly, and that the workpiece is being fed through the cutting tool in the right direction."

"(12) Do not use a dull, gummy, bent, or cracked cutting tool."

"(13) Be sure that keys and adjusting wrenches have been removed before turning power on. "

"(14) Use only accessories designed for the machine."

"(15) Adjust the machine for minimum exposure of cutting tool necessary to perform the operation."

SUBPART J –

Gas Welding and Cutting 1926.350(d)(1) through (6)

(d) Use of fuel gas. The employer shall thoroughly instruct employees in the safe use of fuel gas as follows:

(1) Before a regulator to a cylinder valve is connected, the valve shall be opened slightly and closed immedi-

ately. (This action is generally termed "cracking" and is intended to clear the valve of dust or dirt that might otherwise enter the regulator.) The person cracking the valve shall stand to one side of the outlet, not in front of it. The valve of a fuel gas cylinder shall not be cracked where the gas would reach welding work, sparks, flame, or other possible sources of ignition.

(2) The cylinder valve shall always be opened slowly to prevent damage to the regulator. For quick closing, valves on fuel gas cylinders shall not be opened more than 1-1/2 turns. When a special wrench is required, it shall be left in position on the stem of the valve while the cylinder is in use so that the fuel gas flow can be shut off quickly in case of an emergency. In the case of manifolded or coupled cylinders, at least one each wrench shall always be available for immediate use. Nothing shall be placed on top of a fuel gas cylinder, when in use, which may damage the safety device or interfere with the quick closing of the valve.

(3) Fuel gas shall not be used from cylinders through torches or other devices which are equipped with shutoff valves without reducing the pressure through a suitable regulator attached to the cylinder valve or manifold.

(4) Before a regulator is removed from a cylinder valve, the cylinder valve shall always be closed and the gas released from the regulator.

(5) If, when the valve on a fuel gas cylinder is opened, there is found to be a leak around the valve stem, the valve shall be closed and the gland nut tightened. If this action does not stop the leak, the use of the cylinder shall be discontinued, and it shall be properly tagged and removed from the work area. In the event that fuel gas should leak from the cylinder valve, rather than from the valve stem, and the gas cannot be shut off, the cylinder shall be properly tagged and removed from the work area. If a regulator attached to a cylinder valve will effectively stop a leak through the valve seat, the cylinder need not be removed from the work area.

(6) If a leak should develop at a fuse plug or other safety device, the cylinder shall be removed from the work area.

1926.350(j)

(j) Additional rules. for Additional details not covered in this subpart, applicable technical portions of American National Standards Institute, Z49.1-1967, Safety in Welding and Cutting, shall apply.

From ANSI Standard Z49.1-1967, Fire Watch Duties. Fire watchers shall be trained in the use of fire extinguishing equipment. They shall be familiar with facilities for sounding an alarm in the event of a fire. They shall watch for fires in all exposed areas, try to extinguish them only when obviously within the capacity of the equipment available, or otherwise sound the alarm. A fire watch shall be maintained for at least a half hour after completion of welding or cutting operations to detect and extinguish possible smoldering fires.

Arc Welding and Cutting 1826.351(d)(1) through (5)

(d) Operating instructions. Employers shall instruct employees in the safe means of arc welding and cutting as follows:

(1) When electrode holders are to be left unattended, the electrodes shall be removed and the holders shall be so placed or protected that they cannot make electrical contact with employees or conducting objects.

(2) Hot electrode holders shall not be dipped in water; to do so may expose the arc welder or cutter to electric snack.

(3) When the arc welder or cutter has occasion to leave his work or to stop work for any appreciable length of time, or when the arc welding or cutting machine is to be moved, the power supply switch to the equipment shall be opened.

(4) Any faulty or defective equipment shall be reported to the supervisor.

(5) Other requirements, as outlined in Article 630, National Electrical Code. NFPA 70-1971; ANSI Cl-1971 (Rev. of 1968), Electric Welders, shall be used when applicable.

Fire Prevention 192S.352(e)

(e) When the welding, cutting, or heating operation is such that normal fire prevention precautions are not sufficient, additional personnel shall be assigned to guard against fire while the actual welding, cutting, or heating operation is being performed, and for a sufficient period of time after completion of the work to ensure that no possibility of fire exists. Such personnel shall be instructed as to the specific anticipated fire hazards and how the firefighting equipment provided is to be used.

Welding, Cutting and Heating in Way Preservative Coatings 1926.354(a)

(a) Before welding, cutting, or heating is commenced on any surface covered by a preservative coating of whose flammability is not known, a test shall be made by a competent person to determine its flammability.

(a) Preservative coatings shall be considered to be highly flammable when scraping burn with extreme rapidity.

Ground-Fault Protection 1926.404(b)(iii)(B)

(b) (iii)(B) The employer shall designate one or more competent persons (as CFR 1926.32) to implement the program.

SUBPART L –

Scaffolds
1926.454(a),
(b), and (c)

(a) The employer shall have each employee who performs work while on a scaffold trained by a person qualified in the subject matter to recognize the hazards associated with the type of scaffold being used and to understand the procedures to control or minimize those hazards. The training shall include the following areas, as applicable:

(1) The nature of any electrical hazards, fall hazards and falling object hazards in the work area;

(2) The correct procedures for dealing with electrical hazards and for erecting, maintaining, and disassembling the fall protection systems and falling object protection systems being used;

(3) The proper use of the scaffold, and the proper handling of materials on the scaffold;

(4) The maximum intended load and the load-carrying capacities of the scaffolds used; and

(5) Any other pertinent requirements of this subpart.

(b) The employer shall have each employee who is involved in erecting, disassembling, moving, operating, repairing, maintaining, or inspecting a scaffold trained by a competent person to recognize any hazards associated with the work in question. The training shall include the following topics, as applicable:

(1) The nature of scaffold hazards;

(2) The correct procedures for erecting, disassembling, moving, operating, repairing, inspecting, and maintaining the type of scaffold in question;

(3) The design criteria, maximum intended load-carrying capacity and intended use of the scaffold;

(4) Any other pertinent requirements of this subpart.

(c) When the employer has reason to believe that an employee lacks the skill or understanding needed for safe work involving the erection, use or dismantling of scaffolds, the employer shall retrain each such employee so that the requisite proficiency is regained. Retraining is required in at least the following situations:

(1) Where changes at the worksite present a hazard about which an employee has not been previously trained; or

(2) Where changes in the types of scaffolds, fall protection, falling object protection, or other equipment present a hazard about which an employee has not been previously trained; or

(3) Where inadequacies in an affected employee's work involving scaffolds indicate that the employee has not retained the requisite proficiency.

SUBPART M –

Fall Protection
1926.503 (a),
(b), and (c)

(a) "Training Program." (1) The employer shall provide a training program for each employee who might be exposed to fall hazards. The program shall enable each employee to recognize the hazards of falling and shall train each employee in the procedures to be followed in order to minimize these hazards.

(2) The employer shall assure that each employee has been trained, as necessary, by a competent person qualified in the following areas:

(i) The nature of fall hazards in the work area;

(ii) The correct procedures for erecting, maintaining, disassembling, and inspecting the fall protection systems to be used;

(iii) The use and operation of guardrail systems, personal fall arrest systems, safety net systems, warning line systems, safety monitoring systems, controlled access zones, and other protection to be used;

(iv) The role of each employee in the safety monitoring system when this system is used;

(v) The limitations on the use of mechanical equipment during the performance of roofing work on low-sloped roofs;

(vi) The correct procedures for the handling and storage of equipment and materials and the erection of overhead protection; and

(vii) The role of employees in fall protection plans;

(viii) The standards contained in this subpart.

(b) "Certification of training." (1) The employer shall verify compliance with paragraph (a) of this section by preparing a written certification record. The written certification record shall contain the name or other identity of the employee trained, the date(s) of the training, and the signature of the person who conducted the training or the signature of the employer. If the employer relies on training conducted by another employer or completed prior to the effective date of this section, the certification record shall indicate the date the employer determined the prior training was adequate rather than the date of actual training.

(2) The latest training certification shall be maintained.

(c) "Retraining." When the employer has reason to believe that any affected employee who has already been trained does not have the understanding and skill required by paragraph (a) of this section, the employer shall retrain each such employee. Circumstances where retraining is required include, but are not limited to, situations where:

(1) Changes in the workplace render previous training obsolete; or

(2) Changes in the types of fall protection systems or equipment to be used render previous training obsolete; or

(3) Inadequacies in an affected employee's knowledge or use of fall protection systems or equipment indicate that the employee has not retained the requisite understanding or skill.

SUBPART N –

Cranes and Derricks 1926.550(a) (1), (5), and (6)

(I) The employer shall comply with the manufacturer's specifications and limitations applicable to the operation of any and all cranes and derricks. Where manufacturer's specifications are not available, the limitations assigned to the equipment shall be based on the determinations of a qualified engineer competent in this field and such determinations will be appropriately documented and recorded. Attachments used with cranes shall not exceed the capacity, rating, or scope recommended by the manufacturer.

(5) The employer shall designate a competent person who shall inspect all machinery and equipment prior to each use, and during use, to make sure it is in safe operating condition. Any deficiencies shall be repaired, or defective parts replaced, before continued use.

(6) A thorough, annual inspection of the hoisting machinery shall be made by a competent person, or by a government or private agency recognized by the U.S. Department of Labor. The employer shall maintain a record of the dates and results of inspections for each hoisting machine and piece of equipment.

1926.550(g)(4)

(g) Personnel Platforms. The personnel platform and suspension system shall be designed by a qualified engineer or a qualified person competent in structural design.

1926.550(g) (5)(iv)

(g) A visual inspection of the crane or derrick, rigging, personnel platform and the crane or derrick base support or ground shall be conducted by a competent person immediately after the trial lift to determine whether the testing has exposed any defect or produced any adverse effect upon any component or structure.

Material, Hoist, Personnel Hoists, and Elevators 1926.552(a)(1)

(1) The employer shall comply with the manufacturer's specifications and limitations applicable to the operation of all hoists and elevators. Where manufacturer's specifi-

cations are not available, the limitations assigned to the equipment shall be based on the determinations of a professional engineer competent in the field.

1925.552(b)(7)

(7) All material hoist towers shall be designed by a licensed professional engineer.

1926.552(c)(15) and (17)(i)

(c) Personnel hoists.

(l5) Following assembly and erection of hoists, and before being put in service, an inspection and test of all functions and safety devices shall be made under the supervision of a competent person. A similar inspection and test is required following major alteration of an existing installation. All hoists shall be inspected and tested at not more than 3-month intervals. Records shall be maintained and kept on file for the duration of the job.

(17)(i) Personnel hoists used in bridge tower construction shall be approved by a registered professional engineer and erected under the supervision of a qualified engineer competent in this field.

SUBPART O –

Material Handling Equipment 1926.602(c)(1)(vi)

(c) Lifting and hauling equipment (other than equipment covered under Subpart N of this part).

(1)(vi) All industrial trucks in use shall meet the applicable requirements of design, construction, stability, inspection, testing, maintenance, and operation, as defined in American National Standards Institute B56.1-1969, Safety Standards for Powered Industrial Trucks.

From ANSI Standard B56.1-1969, "Operator Training. Only trained and authorized operators shall be permitted to operate a powered industrial truck. Methods shall be devised to train operators in the safe operation of powered industrial trucks. Badges or other visual indication of the operators' authorization should be displayed at all times during work period."

Site Clearing 1926.604(a)(1)

(1) Employees engaged in site clearing shall be protected from hazards of irritant and toxic plants and suitably instructed in the first aid treatment available.

SUBPART P –

Excavations General Requirements (Excavations, Trenching, and Shoring) 1926.650(i)

(i) Daily inspections of excavations shall be made by a competent person. If evidence of possible cave-ins or slides is apparent, all work in the excavation shall cease until the necessary precautions have been taken to safeguard the employees.

1926.651(1)(2)

(iii) A registered professional engineer has approved the determination that the structure is sufficiently removed from the excavation so as to be unaffected by the excavation activity; or

(iv) A registered professional engineer has approved the determination that such excavation work will not pose a hazard to employees.

(k) Support systems shall be planned and designed by a qualified person when excavation is in excess of 20 feet in depth, adjacent to structures or improvements, or subject to vibration or ground water.

(o) If the stability of adjoining buildings or walls is endangered by excavations, shoring bracing, or underpinning shall be provided as necessary to insure their safety. Such shoring, bracing, or underpinning shall be inspected daily or more often, as conditions warrant, by a competent person and the protection effectively maintained.

(x) Where ramps are used for employees or equipment, they shall be designed and constructed by qualified persons in accordance with accepted engineering requirements.

SUBPART Q –

Concrete and Masonry Construction 1926.701(a)

(a) No construction loads shall be placed on a concrete structure or portion of a concrete structure unless the employer determines, based on information received from from a person, who is qualified in structural design, that the structure or portion of the structure is capable of supporting the loads.

1926.703(b) (8)(i)

(i) The design of the shoring shall be prepared by a qualified designer and the erected shoring shall be inspected by an engineer qualified in structural design.

SUBPART R –

Bolting, Fitting-Up, and Plumbing-Up 1926.752(d)(4)

(4) Plumbing-up guys shall be removed only under the supervision of a riveting, competent person.

SUBPART S –

Underground Construction 1926.800(d)

(d) Safety instruction. All employees shall be instructed in the recognition and avoidance of hazards associated with underground construction activities including, where appropriate, the following subjects:

(1) Air monitoring;

(2) Ventilation;

(3) Illumination;

(4) Communications;

(5) Flood control;

(6) Mechanical equipment;

(7) Personal protective equipment;

(8) Fire prevention and protection;

(9) Explosives; and

(10) Emergency procedures, including evacuation plans and check-in/check-out systems.

Compressed Air 1926.803 (a)(1) and (2)

(1) There shall be present, at all times, at least one competent person designated by and representing the employer, who shall be familiar with this subpart in all respects, and responsible for full compliance with these and other applicable subparts.

(2) Every employee shall be instructed in the rules and regulations which concern his safety or the safety of others.

1926.803(b)(1) and (10)(xii)

(1) There shall be retained one or more licensed physicians familiar with and experienced in the physical requirements and the medical aspects of compressed air work and the treatment of decompression illness. He shall be available at all times while work is in progress in order to provide medical supervision of employees employed in compressed air work. He shall himself be physically qualified and be willing to enter a pressurized environment.

(10) The medical lock shall: (xii) Be in constant charge of an attendant under the direct control of the retained physician. The attendant shall be trained in the use of the lock and suitably instructed regarding steps to be taken in the treatment of employee exhibiting symptoms compatible with a diagnosis of decompression illness.

1926.803(e)(1)

(1) Every employee going under air pressure for the first time shall be instructed on how to avoid excessive discomfort.

1926.803 (h)(1)

(1) At all times there shall be a thoroughly experienced, competent, and reliable person on duty at the air control valves as a gauge tender who shall regulate the pressure in the working areas. During tunneling operations, one gauge tender may regulate the pressure in not more than two headings, provided that the gauge and controls are all in one location. In caisson work, there shall be a gauge tender for each caisson.

SUBPART T –

Demolition Preparatory Operations 1926.850(a)

(a) Prior to permitting employees to start demolition operations, an engineering survey shall be made, by a competent person, of the structure to determine the condition of the framing, floors, and walls, and possibility of unplanned collapse of any portion of the structure. Any adjacent structure where employees may be exposed shall also be similarly checked. The employer shall have in writing evidence that such a survey has been performed.

Chutes 1926.852(c)

(c) A substantial gate shall be installed in each chute at or near the discharge end. A competent employee shall be assigned to control the operation of the gate, and the backing and loading of trucks.

Mechanical Demolition 1926.859(g)

(g) During demolition, continuing inspections by a competent person shall be made as the work progresses to detect hazards resulting from weakened or deteriorated floors or walls, or loosened material. No employee shall be permitted to work where such hazards exist until they are corrected by shoring, bracing, or other effective means.

SUBPART U –

General Provisions (Blasting and Use of Explosives) 1926.900(a)

(a) The employer shall permit only authorized and qualified persons to handle and use explosives.

1926.900(k)(3)(i)

(i) The prominent display of adequate signs, warning against the use of mobile radio transmitters, on all roads within 1,000 feet of blasting operations. Whenever adherence to the 1,000-foot distance would create an operational handicap, a competent person shall be consulted to evaluate the particular situation, and alternative provisions may be made which are adequately designed to prevent any premature firing of electric blasting caps. A description of any such alternatives shall be reduced to writing and shall be certified as meeting the purposes of this subdivision by the competent person consulted. The description shall be maintained at the construction site during the duration of the work, and shall be available for inspection by representatives of the Secretary of Labor.

1926.900(q)

(q) All loading and firing shall be directed and supervised by competent persons thoroughly experienced in this field.

Blaster Qualifications 1926.901(c), (d), and (e)

(c) A blaster shall be qualified, by reason of training, knowledge, or experience, in the field of transporting, storing, handling, and use of explosives, and have a working knowledge of State and local laws and regulations which pertain to explosives.

(d) Blasters shall be required to furnish satisfactory evidence of competency in handling explosives and performing in a safe manner the type of blasting that will be required.

(e) The blaster shall be knowledgeable and competent in the use of each type of blasting method used.

Surface Transportation Explosives 1926.902 (b) and (i)

(b) Motor vehicles or conveyances transporting explosives shall only be driven by, and be in the charge of, a licensed driver who is physically fit. He shall be familiar with the local, State, and Federal regulation governing the transportation of explosives

(i) Each vehicle used for transportation of explosives shall be equipped with a fully charged fire extinguisher, in good condition. An Underwriters Laboratory-approved extinguisher of not less than 10-ABC rating will meet the minimum requirement. The driver shall be trained in the use of the extinguisher on his vehicle.

Firing the Blast 1926.909(a)

(a) A code of blasting signals equivalent to Table U-l, shall be posted on one or more conspicuous places at the operation, and all employees shall be required to familiarize themselves with the code and conform to it. Danger signs shall be placed at suitable locations.

Table U-1		
Warning	**Blast**	**All Clear**
A 1-minute series of long blasts 5 minutes prior to blast signal.	A series of short blasts 1 minute prior to the shot.	Prolonged blast following the inspection of blast area.

SUBPART V –

General (Power Requirements Transmission and Distribution 192S.950(d)(1) (ii), (vi), and (vii)

(1) When deenergizing lines and equipment operated in excess of 600 volts, and the means of disconnecting from electric end is not visibly open or visibly locked out, the provisions of subdivisions (i) through (vii) of this subparagraph shall be complied with.

(ii) Notification and assurance from the designated employee [a qualified person delegated to perform specific duties under the conditions existing shall be obtained that:

(a) All switches and disconnects through which electric energy may be supplied to the particular section of line or equipment to be worked have been deenergized;

(b) All switches and disconnectors are plainly tagged indicating that men are at work;

(c) And that where design of such switches and disconnectors permits, they have been rendered inoperable.

(vi) When more than one independent crew requires the same line or equipment to be deenergized, a prominent tag for each such independent crew shall be placed on the line or equipment by the designated employee in charge.

(vii) Upon completion of work on deenergized lines or equipment, each designated employee in charge shall determine that all employees in his crew are clear, that protective grounds installed by his crew have been removed, and he shall report to the designated authority that all tags protecting his crew may be removed.

**1926.950
(d)(2)(ii)**

(2) When a crew working on a line or equipment can clearly see that the means of disconnecting from electric energy are visibly open or visibly locked-out, the provisions of subdivisions (i) and (ii) of this paragraph shall apply:

(ii) Upon completion of work on deenergized lines or equipment, each designated employee in charge shall determine that all employees in his crew are clear, that protective grounds installed by his crew have been removed, and he shall report to the designated authority that all tags protecting his crew may be removed.

**1926.950(e),
(1), and (2)**

(1) The employer shall provide training or require that his employees are knowledgeable and proficient in:

(i) Procedures involving emergency situations, and

(ii) First aid fundamentals including resuscitation.

(2) In lieu of subparagraph (1) of this paragraph the employer may comply with the provisions of 5 1926.50(c) regarding first aid requirements.

**Overhead Cranes
1926.955(b)(3)(i)**

(3)(i) A designated employee shall be used in directing mobile equipment adjacent to footing excavations.

**1926.955(b)(8)
and (d)(l)**

(8) A designated employee shall be utilized to determine that required clearance is maintained in moving equipment under or near energized lines.

(1) Prior to stringing parallel to an existing energized transmission line, a competent determination shall be made to ascertain whether dangerous induced voltage buildups will occur, particularly during switching and ground fault conditions. When there is a possibility that such dangerous induced voltage may exist the employer shall comply with the provisions of subparagraphs (2) through (9) of this paragraph in addition to the provisions of paragraph (c) of this CFR 1926.955, unless the line is worked as energized.

1926.955(e)
(1) and (4)

(1) Employees shall be instructed and trained in the live line bare-hand technique and the safety requirements pertinent thereto before being permitted to use the technique on energized circuits.

(4) All work shall be personally supervised by a person trained and qualified to perform live-line bare-hand work.

Underground
Lines
1926.956(b)(1)

(1) While work is being performed in manholes, an employee shall be available in the immediate vicinity to render emergency assistance as may be required. This shall not prelude the employee in the immediate vicinity from occasionally entering a manhole to provide assistance, other than emergency. This requirement does not prelude a quailed employee [a person who by reason of experience or training is familiar with the operation to be performed and the hazards involved], working alone, from entering for brief periods time, a manhole where energized cables or equipment are in service, for the purpose of inspection, housekeeping, taking readings, or similar work if such work can be performed safely.

Construction
in Energized
Substations
1926.957(a)(1)

(1) When construction work is performed in an energized substation, authorization shall be obtained from the designated, authorized person [a qualified person delegated to perform specific duties under the conditions existing] before work is started.

1926.957(d)(1)

(1) Work on or adjacent to energized control panels shall be performed by designated employees.

1926.957(e)(l)

(1) Use of vehicles, gin poles, cranes, and other equipment in restricted or hazardous areas shall at all times be controlled by designated employees.

SUBPART X –

Ladders
1926.1053
(b)(15)

(15) Ladders shall be inspected by a competent person for visible defects on a periodic basis and after any occurrence that could affect their safe use.

1926.1060(a)
and (b)

(a) The employer shall provide a training program for each employee using ladders and stairways, as necessary. The program shall enable each employee to recognize hazards related to ladders and stairways, and shall train each employee in the procedures to be followed to minimize these hazards.

(1) The employer shall ensure that each employee has been trained by a competent person in the following areas, as applicable:

(i) The nature of fall hazards in the work area;

(ii) The correct procedures for erecting, maintaining, and disassembling the fall protection systems to be used;

(iii) The proper construction, use, placement, and care in handling of all stairways and ladders;

(iv) The maximum intended load-carrying capacities of ladders; and

(v) The standards contained in this subpart.

(b) Retaining shall be provided for each employee as necessary so that the employee maintains the understanding and knowledge acquired through compliance with this section.

Abstracted from the *Training Requirements in OSHA Standards and Training Guidelines,* 1992.

SAMPLE OF WRITTEN HAZARD COMMUNICATION PROGRAM

With Permission of the Laborers-AGC Education and Training Fund

This working guideline is offered to CAP Contributors to assist in the development of a compliance program set forth by the Federal Occupational Safety and Health Administration's Hazard Communication Standard (29 CFR 1926.59). It represents an application of the Standard which attempts to address the uniqueness of the construction industry, and the intent of the Law. (This proposed program should be supplemented by company personnel and professional guidance; tailor-fitting it to your particular company.) The Construction Advancement Program of Western Pennsylvania (CAP) provides no warranty or guarantee, either expressed or implied, for the program's performance for any particular purpose. No liability for inaccuracies or defects to the Hazard Communication Program can be assumed by CAP. It is rather a guideline, which, to the best of our knowledge, addresses the issues and intent of the Law.

COMPANY POLICY

People are our most important resource, and their health and safety is one of <u>Name of Company's</u> principal responsibilities and primary concerns. Safety must be made an integral part of all operations, whether it be in planning and development or contracting.

<u>Name of Company's</u> proven safety record reflects the management's commitment to employee welfare. However, I want to take this opportunity to reaffirm my belief in a strong safety and health program which will continue to create, and thus provide, a safe place for all employees to work. Accidents and chemical exposures cost time and money, but more importantly, they can cause personal loss and unnecessary injury, which has no price.

I hope you will share my concern for providing a safe working environment because a safety and health program's effectiveness depends upon the cooperation and concern of employees and management personnel who are knowledgeable of the potential hazards of chemical use in the workplace.

Name and Signature of Company President

COMPANY COMMITMENT

<u>Name of Company</u> is committed to providing each of its employees a safe and healthy work environment. Construction processes and other operations performed at the various <u>Name of Company</u> jobsites sometimes require the use of materials and chemicals that can be

731

hazardous, if not handled properly. When using these substances, it is important that workers are aware of the identity of the substances, as well as the toxic or other hazardous properties of the chemicals since an informed employee is more likely to be a careful employee. Therefore, in an effort to promote and maintain jobsites that are free from controllable safety and health hazards, the Name of Company has implemented this Hazard Communication Program to protect its employees.

EMPLOYEE RESPONSIBILITY

The success of this Hazard Communication Program depends to a great extent upon the cooperation of every employee. Employees should be alert to the potential hazards in their work areas; know and understand the hazards of those chemicals; consult the Material Safety Data Sheets (MSDS) for the specifics concerning the hazardous chemicals with which they work; follow the appropriate safe working procedures that have been established; wear required personal protective equipment; and actively participate in the training programs implemented to protect their health and safety.

Active employee participation in Name of Company Hazard Communication Program will result in the continued reduction of the incidence of chemical related illnesses and injuries at Name of Company jobsites and facilities. This written program will be available at the Location for review by any interested employee. A master copy of this program will also be maintained at the Name of Company main office.

PROGRAM MANAGER/COORDINATOR

The Program Manager/Coordinator of the OSHA Hazard Communication Program is Name/Title; who can be reached at the following telephone number _____ during the normal office hours of _____ to _____ Monday through Friday. In the case of emergencies outside the normal office hours, call _____.

The Program Manager/Coordinator is responsible for the effective implementation of this Program.

The Program Manager/Coordinator will periodically report to management regarding the Program's implementation, effectiveness and continuing update requirements. The Program Manager/Coordinator will establish the specifics of the "Hazard Communication Policy" and coordinate the efforts and activities of employees and management.

CHEMICAL INVENTORY

The Program Manager/Coordinator is responsible for developing and compiling a hazardous chemical inventory list which identifies all hazardous chemicals used by Name of Company employees. This list will be contained within the written program and updated periodically to ensure completeness.

The list will also be posted at location for employee information. Copies will be provided to any Name of Company or his/her authorized representative upon request to Person/position.

MATERIAL SAFETY DATA SHEET (MSDS) POLICY

The Program Manager/Coordinator has the overall responsibility for establishing and monitoring the MSDS Program. He/She will ensure that procedures are developed to obtain the necessary MSDSs for all substances and chemicals which are known to pose a health or physical hazard to employees who are exposed to them. He/She will review all incoming or updated MSDSs for new or significant health and safety information and pass it on to affected employees. A master copy of all complete and current MSDSs will be maintained and updated as necessary by the Program Manager/Coordinator. The Program Manager/Coordinator will assign responsibilities to appropriate on-site supervisory personnel for MSDS maintenance at each jobsite. He/She will ensure that current MSDSs for all chemicals and substances at the site are available to employees. If an MSDS is not available upon request, the Program Manager/Coordinator should be contacted. Periodically, site personnel shall check the MSDS collection for completeness.

CONTAINER LABELING POLICY

The Program Manager/Coordinator is responsible for adopting and enforcing <u>Name of Company</u> in-house labeling system. In general, <u>Name of Company</u> will rely on the manufacturer's applied labels whenever possible and will ensure that these labels are maintained in a legible condition. Containers which are not labeled or on which the manufacturer's label has been removed will be relabeled. <u>Name of Company</u> will ensure that each container is labeled with the identity of the hazardous chemical contained and any appropriate hazard warnings.

The <u>person/position</u> will verify that all containers received from manufacturers, suppliers, or importers used by employees are clearly and appropriately labeled. All chemicals on site will be stored in their original or approved containers with proper labels attached. Any container not labeled should be given to <u>person/position</u> for labeling or proper disposal. Workers may dispense chemicals from original containers only in small quantities intended for immediate use. Any chemical left after work is completed must be returned to the original container or <u>person/position</u> for proper handling. At no time should any unmarked containers of any size be left in the work area after work is completed. <u>Person/position</u> has the responsibility of monitoring and notifying the Program Manager/Coordinator of any on-site deficiencies which occur during day-to-day operations.

EMPLOYEE INFORMATION

The Hazard Communication Program Manager/Coordinator is responsible for the effective dissemination of <u>Name of Company</u> employee training and information programs.

Employees will be trained to work safely with hazardous chemicals. Employee training shall include

1. An overview of the requirements contained in the Hazard Communication Standard.

2. The locations of the MSDS file and written Hazard Communication Program.

3. The content and importance of Material Safety Data Sheets and labels.

4. The methods and observations that may be used to detect the presence or release of a hazardous chemical.

5. Physical and health hazards associated with chemicals.

6.   Methods used to protect employees.

7.   Safe work procedures, emergency responses and use of personal protective equipment.

Every Name of Company employee is required to actively participate in Name of Company hazard communication training programs.

HAZARDOUS NON-ROUTINE TASKS

Periodically, employees are required to perform hazardous non-routine work. It is Name of Company policy that such operations not be undertaken until the employees have been provided information on the possible undesirable effects that may arise during such operations. Prior to starting the work, each affected employee will be given this information by the Person/position. This information will include the chemical and physical hazards associated with the chemical, required personal protective equipment use, and the steps that the company is using to reduce the hazards. When necessary, areas will be posted to indicate the nature of the hazard involved.

EMERGENCY PROCEDURES

In the event of an overexposure to or spill of any hazardous chemical, the Name/position will be notified at once. The foreman or the immediate supervisors will be responsible for insuring that proper and appropriate emergency response actions are taken. The appropriate Material Safety Data Sheet pertaining to that chemical or substance will serve as reference for such actions.

MULTI-EMPLOYER JOBSITES

It is the policy of Name of Company to adequately apprise other contractors regarding the hazardous substance which their employees may be exposed to during the course of day-to-day construction activities. Contractors, whose employees may be exposed to hazardous substances used by Name of Company employees, will be given access to this Hazard Communication Program. This will provide all relevant chemical information necessary to protect their employees.

Contractors should be informed of conditions existing on-site which necessitate special precautionary measures through weekly safety or tool box meetings.

Other on-site employers working among Name of Company employees are also required to adhere to the provisions of the Hazard Communication Standard. They shall make available copies of MSDSs for all hazardous materials used by their employees which can be reviewed by Name of Company employees. MSDSs will be provided within a reasonable time period after such a request.

The Name of Company is firmly committed to its employees' health and safety, and as such requires all contractors whose employees work around and among Name of Company employees to also be knowledgeable on chemical safety and appropriate working procedures in an effort to reduce and eliminate chemical exposures to themselves, as well as the other craftsmen.

APPENDIX D

29 CFR 1926

CONSTRUCTION STANDARDS

JOBSITE AUDIT INSTRUMENT*

This audit instrument does not cover all facets of the construction regulatory requirements. It is meant to be a starting point for you to develop a construction company specific audit instrument by adding to and deleting from the items which are provided in this instrument. Note: This instrument or tool does not cover every section within each Subpart, but only the most common ones found on construction operations.

Subpart A – General – Not applicable to hazards on jobsite.

Subpart B – General Interpretations – Not applicable to hazards on jobsite.

Subpart C – General Safety and Health Provisions

1926.23   First aid and medical attention.
    _____ Are first aid provisions available to workers?

1926.24   Fire protection and prevention.
    _____ Does a fire protection and prevention program exist?

1926.25   Housekeeping.
    _____ Are stairs, work areas, and walkways free of hazards?
    _____ Are proper containers provided for trash, rags, etc.?
    _____ Is scrap and debris removed daily?
    _____ Are nails bent over or removed from scrap lumber?

Subpart D – Occupational Health and Environmental Controls

1926.50   Medical services and first aid.
    _____ Are first aid kits available and complete?

1926.51   Sanitation.
    _____ Are drinking water, cups, and receptacles provided?
    _____ Are there an adequate number of portable toilets provided?
    _____ Are washing facilities provided?

1926.52   Occupational noise exposure.
    _____ Have high noise areas been identified?
    _____ Is hearing protection provided and worn?
    _____ Are sound levels measured and are those produced by equipment known?
    _____ Is an audiometric testing program in effect?
    _____ Have engineering and administrative controls been instituted when possible?

1926.53   Ionizing radiation.
    _____ Are the Nuclear Regulatory Commission Standards for Protection Against Radiation (10CFR Part 20) followed?

1926.54 Nonionizing radiation.

_____ Are only trained and qualified workers assigned to operate laser equipment?

_____ Do laser operators carry proof of qualifications?

_____ Are workers who may be potentially exposed provided with antilaser eye protection?

_____ Are placards warning of lasers posted in the laser area?

_____ When the laser is not in use, are beam shutter or caps used or is the unit turned off?

_____ Is it a strict requirement that the laser beam shall never be directed at workers?

1926.55 Gases, vapors, fumes, dusts, and mists.

_____ Are exposures to vapors, fumes, dust, and mist controlled?

_____ Are agents identified which may cause harm by inhalation, ingestion, skin absorption, or contact?

_____ Is adequate ventilation provided?

_____ Is protective equipment used to protect against over-exposure?

_____ Are workers made knowledgeable of hazards when working with harmful agents?

1926.56 Illumination.

_____ Is work area lighting adequate according to Table D-1?

_____ Are 5 foot-candles of illumination provided throughout the general construction area?

_____ Are workshops and storerooms provided with 10 foot-candles of illumination?

_____ Are light guards provided where there is a possibility of breakage?

_____ Are light fixtures raised above worker's head?

_____ Are stairways, floor openings, and wall openings well illuminated?

1926.57 Ventilation.

_____ Is ventilation used as an engineering control?

1926.59 Hazard communication.

_____ Have hazard assessments been completed where necessary?

_____ Is the hazard communication program satisfactory?

_____ Are MSDSs available to employees?

_____ Are hazardous materials properly labeled?

_____ Are employees trained to work with hazardous materials?

1926.61 Retention of DOT markings, placards, and labels.

_____ Are DOT placards, markings, and labels in accordance with 49CFR Part 171 in use?

1926.65 Hazardous waste operations and emergency response.

_____ Is hazardous waste work being performed?

_____ Have the requirements of the regulations been followed?

Subpart E – Personal Protective and Life Saving Equipment

1926.95 Criteria for personal protective equipment.

_____ Have hazard assessments been completed where necessary?

_____ Is personal protective equipment provided and used?

_____ Have workers been trained in the use of PPE?

1926.96 Occupational foot protection.

_____ Are safety shoes worn to prevent foot injury when heavy pieces of work are handled?

_____ Are shoes designed and constructed to standard?

_____ Are ankles and legs protected where hazards exists?

_____ Is foot protection provided and used for wet conditions?

1926.100 Head protection.

_____ Are hardhats worn by all those in the construction zone?

_____ Are employees working in areas where there is a possible danger of head injury from impact, falling, or flying objects, or from electrical shock or burns protected by protective helmets?

_____ Does helmet design and construction meet standards?

1926.202 Eye and face protection.

_____ Are all workers wearing safety glasses with side shields?

_____ Is protective eye and face equipment provided and used where a reasonable probability of injury can occur and be prevented?

_____ Do protectors fit properly, are durable, capable of being disinfected/cleaned, and kept in good repair?

_____ Is eye protection distinctly marked, limitations or precautions transmitted to user, designed for construction site use, and used in accordance with ANSI Z87.1?

_____ Does protective equipment provide for the use of corrective lenses?

_____ Are filter lens and shade numbers standard for protection against radiant energy used?

_____ Is special eye protection provided for those exposed to laser beams, acids and chemical splashes, welding and cutting radiation, sparks, and flashes?

1926.103 Respiratory protection.

_____ Are approved respirators used when engineering and administrative controls are not feasible?

_____ Is approved respiratory equipment maintained and fit properly to assure effectiveness?

_____ Are medical examinations made available to workers and provided free of charge for workers who wear respirators or are exposed to asbestos or lead?

_____ Is there an established and maintained respiratory protective program?

1926.104 Safety belts, lifelines, and lanyards

_____ Are safety harnesses, belts, lifelines, and lanyards inspected and used?

1926.105 Safety nets

_____ Are safety nets in place where needed?

1926.106 Working over or near water.

_____ Are all workers wearing life jackets?

_____ Are ring buoys with 90 feet of line available at 200 feet intervals?

_____ Is at least one life saving skiff immediately available?

Subpart F – Fire Protection and Prevention

1926.150 Fire protection.

_____ Are combustible and flammable liquids stored in safety containers?

_____ Are fire extinguishers properly placed and inspected?

_____ Are fire warning signs posted where required?

_____ Is there access to fire hydrants and hoses?

_____ Are extinguishers inspected at least monthly for location, damage, and discharge?

1926.152 Flammable and combustible liquids.

_____ Are all flammable and combustible liquids used identified by Class, flash point, boiling point, vapor pressure, and major ingredients?

_____ Are locations where flammable vapors may be present protected from ignition sources by approved electrical installation and equipment, static electricity, bonding or grounding, no smoking warnings, no open flames or heating devices, no welding or cutting, no use of spark producing tools?

_____ Are flammable or combustible liquids stored in tanks or closed containers approved for the specific purpose by class of liquid, volume, and location? Does ventilation prevent vapor air mixtures in concentrations over one-fourth the lower flammable limit?

_____ Does the quantity of liquids kept in the operational vicinity not normally exceed supply for one day or one shift?

_____ Are safety containers used for storage and handling of 5 gallons or less?

_____ Do storage limitations outside of approved storage cabinets, inside cabinets, inside approved storage rooms, and in other types of occupancy or installations meet standard tables and requirements?

_____ Are suitable fire control devices which meet standard requirements provided?

_____ Is suitable spill control available?

1926.159 Employee alarm systems.

_____ Are alarms present to warn of hazards?

_____ Is annual inspection and testing of system done by a competent engineer, inspector, or person?

Subpart G – Signs, Signals, and Barricades

1926.200 Accident prevention signs and tags.

_____ Are signs and tags in place?

1926.201 Signaling.

_____ Are flagpersons provided with and made to wear orange or red warning garments while working in vehicular traffic?

_____ Are the warning garments reflectorized material?

_____ Are flagpersons knowledgeable of standard flag signals to control traffic effectively?

1926.202 Barricades.

_____ Do barricades conform to ANSI D6.1-1971?

Subpart H – Materials Handling, Storage, Use, and Disposal

1926.250 General requirements for storage.

_____ Are materials properly stored to prevent falling?

_____ Are maximum safe loads for floors posted and observed?

_____ Are aisles clear and of adequate size?

_____ Are bags, containers, bundles, pallets, boxes, lumber, bricks, etc. stored in tiers, blocked, interlocked, and limited in height or otherwise secured?

_____ Are storage areas free from accumulations of miscellaneous materials constituting hazards from tripping, fire, explosion, or pest harborage?

_____ Are all clearance signs posted?

_____ Are all commodities stored, handled, and piled with due regard to their fire characteristics?

1926.251 Rigging equipment for material handling.

_____ Are mechanical aids used when possible?

_____ Are safe working load limits not exceeded?

_____ Are wire rope clips used to form eyes of slings?

_____ Are special lifting devices labeled for maximum load?

_____ Is rigging equipment properly stored?

_____ Is rigging equipment in good condition?

_____ Is all handling gear such as ropes, slings, chains, hooks, and rings inspected before each use?

_____ Are safe working loads of ropes and slings determined for the size of the rope and angle of sling according to manufacturer's recommendations used?

_____ Are "U" bolt clips installed in the number recommended for the rope size?

_____ Is the allowable wear and stretch in chain links not exceeded?

_____ Are standard tables used to determine safe working loads?

_____ Are safe working loads of hooks and rings determined by the manufacturer's recommendations or tested to twice the intended load? Are bent or sprung hooks not used? Are safety type hooks used?

Subpart I – Tools – Hand and Power

1926.300 General requirements.

_____ Are tools maintained in good condition?

_____ Are damaged tools repaired, replaced, or removed from service?

_____ Are all safeguards in place?

1926.301 Hand tools.

_____ Are tool handles free of splits and cracks?

_____ Are handles wedged tightly in the heads of all tools?

_____ Are impact tools free of mushroomed heads?

_____ Are heads of chisels or punches ground periodically to prevent mushrooming?

_____ Are cutting edges kept sharp so the tool will move smoothly and efficiently without binding or skipping?

1926.302 Power-operated hand tools.

_____ Are power tools grounded or double insulated?

_____ Are pneumatic tools equipped with tool retainers (safety clips) and hose or connection restraints?

_____ Is compressed air in use equal to or less than 30 psi?

_____ Are powder-actuated tool operators properly trained?

_____ Are powder-actuated tools left unloaded until just prior to use?

_____ Are powder-actuated tools inspected daily?

_____ Are all power saws and similar equipment safe guarded?

_____ Are GFCIs used with electrically powered tools as a safe guard?

1926.303 Abrasive wheels and tools.

_____ Are grinders provided with safety guards?

_____ Are work rest and wheel guards on bench grinder adjusted properly?

_____ Are abrasive wheel grinders provided with safety guards which cover the spindle ends, nut, and flange projections?

1926.304 Woodworking tools.

_____ Are portable circular saws equipped with upper and lower blade guards and an automatic return guard on the lower blade?

_____ Do radial saws have an upper hood enclosure, lower portion of blade guarded to full diameter of the blade, non-kickback fingers or dogs, adjustable stops or other effective device to eliminate table over-run and automatic return provided?

_____ Are all gears, shafts, pulleys, belts, and moving parts covered?

_____ Are hand-fed circular rip and crosscut table saws above table level guarded by a hood and provided with non-kickback fingers or dogs and spreaders?

1926.306 Air Receivers.

_____ Is every air receiver equipped with a pressure gauge and with one or more automatic, spring-loaded safety valve?

_____ Is every air receiver provided with a drain pipe and valve at the lowest point for removal of accumulated oil and water?

_____ Is the air receiver frequently and periodically inspected, drained of oil and water, and determined to be operating effectively?

Subpart J – Welding and Cutting

General.

_____ Are areas established for welding and cutting equipment based on fire potentials?

_____ Are there designated individuals responsible for authorizing cutting or welding in non-welding areas?

_____ Are all cutters, welders, and supervisors trained in the safe operation and use of equipment and processes?

_____ Are combustible materials removed or protected from ignition?

_____ Is fire protection and extinguishing equipment properly located and available?

1926.350 Gas welding and cutting.

_____ Are gas cylinders secured in an upright position?

_____ Are valve caps in place when cylinders are not in use?

_____ Are special wrenches available when required by cylinders?

_____ Are cylinders being transported properly?

_____ Are oxygen and acetylene stored at a distance of 20 feet or a fire wall between them?

_____ Are cylinder valves closed and equipment purged when not in use?

_____ Are cylinders placed so sparks, hot slag, or flame cannot reach them, or fire resistant shields provided?

_____ Are cylinders not allowed to become part of electrical circuit?

_____ Do cylinders meet standard requirements for construction and maintenance?

_____ Are fuel gas and oxygen manifolds conspicuously and permanently marked according to contents?

_____ Are supply hose connections not interchangeable between fuel gas or oxygen supply headers?

_____ Are hose connections free of grease and oil?

_____ Are fuel gas hoses and oxygen hoses easily distinguishable from each other and not interchangeable?

_____ Are hoses inspected at the beginning of each shift?

1926.351 Arc welding and cutting.

_____ Are welding curtains used where needed?

_____ Are welding cables in good condition and properly insulated?

_____ Do equipment and apparatus comply with U.L. Standards or other applicable standards?

_____ Are manual electrode holders designed and insulated for arc welding and cutting?

_____ Are welding cables and connectors insulated and capable of handling maximum current?

_____ Are cables free from repair or splices for 10 feet from the electrode holders unless equal insulating quality and cable lug connectors are insulated?

_____ Are frames of arc welding and cutting machines grounded?

_____ Are workers assigned to arc welding or gas-shielded arc welding instructed and qualified?

_____ Are arc welding and cutting operations shielded by flameproof screens or located in bays or booths to protect from direct rays of the arc?

_____ Are operators specially protected from high intensities of ultraviolet radiation by screening or filter lenses?

_____ Is skin protected by clothing or other devices?

1926.352 Fire prevention.

_____ Are fire extinguishers immediately available during welding and cutting?

1926.353 Ventilation and protection in welding, cutting, and heating.

_____ Is there proper ventilation and/or respirators during welding and cutting operations?

Subpart K – Electrical

1926.402 Applicability.

_____ Are installations in accordance with the 1984 National Electric Code?

1926.403 General requirements.

_____ Are all exposed live parts guarded?

_____ Are current interrupters available for all equipment?

_____ Are all conductors and equipment "approved?"

_____ Are conductors joined and spliced according to requirements?

_____ Is all electrical equipment which presents a hazard to worker or fire hazard protected or enclosed?

_____ Are electrical tools and cords in good repair?

_____ Do proper working clearances exist around equipment in excess of 600 volts?

_____ Are workers protected from live parts/equipment in excess of 50 volts by enclosures, etc. so that only qualified workers (i.e., electricians) have access?

1926.404 Wiring design and protection.

_____ Are conductors used as grounds easily identified?

_____ Are ground-fault circuit interrupters, provided on all temporary wiring receptacles above 120 volts?

_____ Is an assured grounding conductor program established and implemented at the jobsite?

_____ Are portable and/or cord plug connections to equipment grounded?

_____ Do branch conductors have proper clearance in all directions?

_____ Do service disconnects de-energize all conductors?

_____ Are high voltage warning signs posted?

_____ Do circuits in excess of 600 volts comply with regulations?

_____ Are regulations for feeder and branch circuits in compliance?

_____ Does proper grounding exist for generators, transformers, metal cable trays, metal raceways, and metal enclosures?

_____ Are the metal parts of nonelectrical equipment grounded?

1926.405 Wiring methods, components, and equipment for general use.

_____ Are temporary lights provided with guards?

_____ Are temporary lights not suspended by their cords?

_____ Are cables and cords protected from damage?

_____ Are extension cords the three-wire type?

_____ Are flexible cords used in continuous lengths without splicing?

_____ Are portable hand lights provided with guards?

_____ Is twelve-volt portable lighting used in moist areas, drums, tanks, and vessels when no GFCI is present?

_____ Are special precautions for temporary wiring in excess of 600 volts taken to prevent contact?

_____ Are flexible cords used for fixed wiring or routed through walls, doors, windows, floor openings, or ceiling openings? (This is not acceptable.)

_____ Are motors, motor controls, or motor branch circuits properly wired and protected?

_____ Are all transformers properly installed and protected?

_____ Are capacitors capable of automatic drainage of stored charges?

1926.406 Specific purpose equipment and installations.

_____ Is electrical equipment used in connection with cranes, monorail hoist, hoist, and all runways properly installed?

_____ Are disconnects and limiting switches installed?

_____ Do elevators, moving sidewalks, and escalators have disconnects?

_____ Are X-ray or radiographic and fluoroscopic type equipment fitted with disconnects and high voltage conductors protected?

_____ Are all of these apparatuses effectively grounded?

1926.407 Hazardous (classified) locations.

_____ Are conductors or equipment located in damp or wet conditions, exposed to gases, fumes, liquids, other deteriorating factors, or exposed to excessive temperatures approved for the purpose and location?

_____ Are there Class I locations in which flammable gases or vapors are or may be present in quantities sufficient to produce explosive or ignitable mixtures?

_____ Are there Class II locations which are hazardous because of the presence of combustible dust?

_____ Are there Class III locations which are hazardous because of the presence of easily ignitable fibers or flyings?

1926.408 Special systems.

_____ Are the regulatory requirements met for systems operating over 660 volts?

_____ Do these systems have the appropriate interrupting and isolating devices?

_____ Are all systems effectively grounded?

SAFETY RELATED WORK PRACTICES

1926.416 General requirements.

_____ Are workers protected from electrical shock?

_____ Are the locations of underground electrical conductors known or employees protected?

_____ Are workers made aware of energized power lines prior to beginning work?

_____ Are barriers present to guard energized parts?

_____ Are electrical cords kept clear of walkways?

_____ Are leads maintained within the limits of the circuits wiring?

_____ Are fuse pullers used to extract fuses?

_____ Are defective extension cords or cables not used and not stapled or hung by nails or suspended by wire?

1926.417 Lockout and tagging of circuits.

_____ Are lockout/tagout procedures followed?

SAFETY-RELATED MAINTENANCE AND ENVIRONMENTAL CONSIDERATIONS

1926.431 Maintenance of equipment.

_____ Is all Class I, II, III equipment maintained in a dust tight, ignition proof, or explosive proof condition?

1926.432 Environmental deterioration of equipment.

_____ Is electrical equipment not exposed to adverse environmental conditions unless designed specifically for them?

SAFETY REQUIREMENTS FOR SPECIAL EQUIPMENT

1926.441 Battery locations and battery charging.

_____ Are battery recharging and storage areas well ventilated?

_____ Are racks and trays used during storage and recharging chemically resistant?

_____ Are workers provided with the proper PPE to protect against chemicals in batteries?

_____ Are water flushing facilities for neutralizing and fire prevention available?

_____ Is the battery charging area in a specific and protected area?

Subpart L – Scaffolding

1926.451 Scaffolding.

_____ Are scaffolds erected and dismantled under the direction of a competent person?

_____ Are there guardrails and toeboards on all open sides and ends of scaffold platforms 10 feet or more above the ground?

_____ Are tube and coupler scaffold posts accurately spaced, erected on suitable base, maintained plumb, and brace connections secured?

_____ Are planks secured not less than six inches nor more than twelve inches over the end support?

_____ Is a ladder or equivalent mean of access provided.

_____ Are scaffolds tied into structures when height and length requirements are met?

_____ Are lifelines and harnesses provided and used by each worker on swing and single-point adjustable suspension scaffolds?

_____ Are all footings or anchorages sound, rigid, and able to support the intended load?

_____ Are screens placed between toeboard and guardrail where persons pass under scaffolds?

_____ Is overhead protection provided for workers on scaffolds if overhead exposures exist?

_____ Can scaffold components support 4 times their intended load?

_____ Is planking overlapped 12 inches or secured and planks extend over end supports 6 to 12 inches?

_____ Are suspended or hanging scaffold components such as wire and fiber ropes protected from heat, chemicals, or corrosive substances?

_____ Can scaffolding ropes support 6 times the load?

_____ Are workers not permitted on scaffolds during storms, high winds, ice, or snow?

_____ Does the maximum work level height not exceed 4 times the least base dimension unless outrigger frames are used?

_____ Are scaffold platform widths not less than 18 inches?

1926.453 Aerial lifts

_____ Are safety harnesses with lanyards attached to the boom or basket and worn by occupants of aerial lifts?

_____ Are safety rails on all open sides of elevated work platforms?

Subpart M – Fall Protection

1926.501 Duty to have fall protection.

_____ Are floor, roof, and wall openings guarded by a standard guardrail and toeboard or covered securely?

_____ Are all open sided floors and platforms 6 feet or more aboveground guarded with or by a standard guardrail system?

_____ Are runways 6 feet or more aboveground properly guarded?

_____ Are standard guardrails approximately 42 inches in vertical height and can withstand 200 pounds force in any direction?

_____ Are workers protected from falls of six feet or greater?

1926.502 Fall protection systems criteria and practices.

_____ Are lifelines and safety harnesses provided for and used by workers exposed to the hazard of falling?

_____ Are lifelines secured above the point of operation to an anchorage or structural member capable of supporting a minimum dead weight of 5,000 pounds?

_____ Do lifelines have a minimum breaking strength of 5,000 pounds?

_____ Do safety harnesses/lanyards limit a fall to no greater than 6 feet?

_____ Are all safety harness and lanyard hardware drop forged or pressed steel meeting standard specifications and capable of withstanding tensile loading of 5,000 pounds without deformation?

_____ Do workers on two-point suspension scaffolds, boatswains chairs, float or ship scaffolds, needle beam scaffolds, roof or other work, protected by a safety harness and lanyard attached to a lifeline?

_____ Are workers on powered platforms provided with a safety harness attached to a lifeline?

_____ Are safety harnesses and lifelines worn by anyone entering closed vessels, tanks, chip bins and similar equipment and a person stationed outside in a position to handle the line and provide assistance in an emergency?

Subpart N – Cranes, Derricks, Hoists, Elevators, and Conveyors

1926.550 Cranes and derricks

_____ Are cranes certified annually?

_____ Are daily visual/physical inspections done and documented?

_____ Are initial and regular inspections established and maintained for cranes, derricks, and hoists?

_____ Are written dated and signed inspection reports and records readily available as well as modification or additions approved by manufacturer?

_____ Are standard hand signals or radio communications being used?

_____ Are load capacities posted on equipment?

_____ Is a fire extinguisher provided and maintained?

_____ Is the swing radius of the rear rotating superstructure of crane barricaded?

_____ Are power lines deenergized or is a safe distance maintained at all times?

_____ Is there a safety latch on the hook?

_____ Are nonconductive tag lines in use to control loads?

_____ Are workers prohibited from riding on hooks or loads?

_____ Is there a load rating chart in crane cab? Rated load marked on each side of crane? Boom angle radius indicator present? Audible warning signal?

_____ Are visible defects in wires, booms, etc. corrected immediately?

_____ Are belts, gears, sprockets, and other moving parts protected?

_____ Are requirements of ANSI Standards for cranes and derricks met and are manufacturer's directions followed?

1926.551 Helicopters.

_____ Does helicopter comply with FAA regulations?

_____ Are ground crews indoctrinated daily and provided protective eye wear and hardhats with chin straps?

_____ Are all loose materials and equipment 100 feet back from loading and un-loading areas?

_____ Do workers perform work under a hovering craft only to hook and unhook loads?

_____ Is constant communication maintained between pilot and designate ground crew member?

_____ Are there open fires in the area which could be spread by down wash?

1926.552 Material hoists, personnel hoists, and elevators.

_____ Are manufacturers' specifications and limitations followed?

_____ Are wire ropes inspected and removed from service if damaged or worn?

_____ Are operating rules written and posted in operators' compartment?

_____ Are workers allowed to ride on hoist?

_____ Is the operators' cab provided with overhead protection?

_____ Arc material hoists designed by a licensed professional engineer?

_____ Do material hoists conform to ANSI A10.5-1969.

_____ Are personnel hoists totally enclosed for the full height?

_____ Are all personnel cars full enclosed?

_____ Can the personnel hoist still move when doors or gates are not fully closed?

_____ Are there safety stopping devices in place as well as an emergency stop switch in the car?

_____ Are personnel hoists inspected and tested as required by the OSHA regulations?

1926.553 Base-mounted drum hoists.

_____ Are all moving parts guarded?

_____ Are all controls in easy access for the operator?

_____ Is there an emergency stopping mechanism?

1926.554 Overhead hoists.

_____ Is the safe working load exceeded?

_____ Is the operator clear of the load at all times?

_____ Is movement of the hoist restricted by other structures?

1926.555 Conveyors.

_____ Is the operator capable of stopping the conveyor?

_____ Is there an audible warning signal prior to conveyor starting?

_____ Is the emergency stop switch designed so that conveyor cannot be started again?

_____ Are conveyors guarded where workers could come in contact with moving parts?

_____ Do conveyors meet ANSI B20.1-1957?

_____ Are crossover, aisles, and passageways properly marked and overhead protection provided when needed?

Subpart O – Motor Vehicles, Mechanized Equipment, and Marine Operations

1926.600 Equipment.

_____ Do only authorized personnel operate equipment?

_____ Is equipment equipped with warning horn, whistle, gong, or other device which can be heard above normal noise level?

_____ Are safety tire racks or cage protection used for tire changing?

_____ Is heavy equipment blocked or cribbed during repair work?

_____ Are operators of vehicles properly trained and licensed?

1926.601 Motor vehicles.

_____ Is an overhead guard provided when operator of truck is exposed to danger of falling objects?

_____ Are motor vehicles checked at beginning of shift that all operating parts, equipment, and accessories are in safe operating condition and defects corrected?

_____ Do motor vehicles left unattended have lights, barricades, reflectors, night parking brake set, and wheels choked on inclines?

1926.602 Material handling equipment.

_____ Are scissors points guarded on front-end loaders, which constitute a hazard?

_____ Are braking systems, horn, seats belts, back-up alarms and overhead protection in good operating condition?

_____ Are bulldozer and scrapper blades, end loader buckets, dump bodies, and similar equipment fully lowered or blocked when not in use or being repaired?

_____ Is a reverse signal alarm or observer used when there is an obstructed view to rear?

_____ Is there a cab shield or canopy to protect operator from shifting or falling materials?

1926.603 Pile driving equipment.

_____ Do all boiler and pressure vessels meet appropriate regulations?

_____ Are stop blocks provided to prevent hammer from damaging the head block?

_____ Is blocking provide when workers are under the hammer?

_____ Is fall protection provide for workers working aloft on leads?

_____ Are steam hoses and the like provided with chains or cable to prevent whipping if they come loose?

1926.604 Site clearing.

_____ Are employees protected from toxic or irritant plants and instructed in first aid treatment?

_____ Does all equipment have rollover protection and overhead protection?

1926.605 Marine operations and equipment

Subpart P – Excavations

1926.651 Specific Excavation Requirements.

_____ Is there a person competent on excavations?

_____ Are workers protected from cave-ins?

_____ Are excavations five feet or greater in depth guarded by sloping, benching, shoring, shielding, or other equivalent means?

_____ Are slopes excavated to the angle of repose?

_____ Are spoil materials pulled back two feet from the edge of trench?

_____ Are ladders or ramps provided in trenches 4 feet or more in depth with no more than 25 feet of lateral travel required by workers?

_____ Are adjacent structures adequately shored or supported as well as roads and sidewalks?

_____ Are underground utilities properly supported when uncovered?

_____ Are daily inspections made to determine possibility of a cave-in?

Subpart Q – Concrete and Masonry Construction

1926.701 General requirements.

1926.702 Requirements for equipment and tools.

1926.703 Requirements for cast-in-place concrete.

_____ Are forms properly installed and braced?

_____ Do forms have adequate shoring, plumbed, and crossbraced?

_____ Is shoring maintained until strength is attained?

_____ Is all protruding re-bar guarded, capped, or covered?

_____ Do workers wear gloves and eye protection during form stripping operations?

_____ Are piplines and tremies used as a part of the concrete pumping systems secured by wire or equivalent means in addition to the regular couplings or connections?

_____ Is form lumber organized and stripped?

_____ Are employees working more than 6 feet above any adjacent working surfaces placing and tying reinforcing steel provided with a positioning safety belt and line or equivalent device?

_____ Are wire mesh rolls secured to prevent dangerous recoil action?

_____ Are there automatic shutoff switches on manually guided concrete troweling machines?

_____ Is there a positive fail-safe joint connector on pumpcrete air hoses?

_____ Do workers controlling nozzles wear head and face protection?

_____ Are there safety latches on concrete buckets?

_____ Are workers not allowed to ride on buckets or work under suspended buckets?

_____ Are all locking devices, braces, coupling pins, base plates, and other parts of metal tubular frames shoring in good condition, not overloaded and meet the shoring layout?

_____ Are drawings or plans showing jack layout, formwork, shoring, working decks, and scaffolding available at worksite?

_____ Is the loading designed and approved by engineer-architect?

1926.704 Requirements for precast concrete

_____ Are precast concrete wall units, structural framing, and tilt-up wall panels adequately supported to prevent overturning and to prevent collapse until permanent connections are completed?

_____ Are lifting inserts which are embedded or otherwise attached to tilt-up precast concrete members capable of supporting at least two times the maximum intended load applied or transmitted to them?

_____ Are lifting inserts which are embedded or otherwise attached to precast concrete members, other than the tilt-up members, capable of supporting at least four times the maximum intended load applied or transmitted to them?

_____ Is lifting hardware capable of supporting at least five times the maximum intended load applied transmitted to the lifting hardware?

_____ Are employees not allowed under precast concrete members being lifted or tilted into position except those employees required for the erection of those members?

1926.705 Requirements for lift-slab construction operations.

1926.706 Requirements of masonry construction.

_____ Is a limited access zone equal to the height of the wall plus 4 feet present?

_____ Is the limited access zone set up on the non-scaffold side?

_____ If the wall is over eight feet, is it braced?

Subpart R – Steel Erection

1926.750 Flooring requirements.

_____ Is safety railing around floor periphery, in wire rope flagged every six feet, or equal guarding?

_____ Are ladders or acceptable mean of access provided?

1926.751 Structural steel assembly.

_____ Are safety harnesses (belts for positioning) used?

_____ Is "Christmas Treeing" of steel according to policy?

1926.752 Bolting, riveting, fitting-up, and plumbing-up.

_____ Is there a means to keep bolts and drift pins from falling when being knock-out (during bolting)?

1926.75 Safety Nets.

_____ Are safety nets provided where required?

Subpart S – Tunnels and Shafts, Caissons, Cofferdams

Subpart T – Demolition

1926.850 Preparatory operations.

_____ Has an engineering survey been accomplished and discussed prior to demolition operation by a competent person?

_____ Is public and worker protection provided?

_____ Are utilities cut off or controlled?

_____ Is there clear operating space for trucks and equipment?

1926.851 Stairs, passageways, and ladders.

_____ Is there adequate access to ladders and stairs?

1926.852 Chutes.

_____ Are material chutes enclosed and guarded?

1926.853 Removal of materials through floor openings.

_____ Are floor openings covered?

Subpart U – Blasting and Use of Explosives

1926.900 General provisions.

_____ Are there written specific procedures for loading of explosives or blasting agents, initiation of blasting, use of fuses and detonating cord, inspection after blasting, misfires, and special underwater blasting, and blasting under compressed air?

1926.901 Blaster qualifications.

_____ Only trained and experienced authorized personnel handle or use explosives?

_____ Only authorized and qualified persons use, handle explosives, transport, prepare explosive charges, and conduct blasting operations?

_____ Are smoking, matches, open flames, liquor, narcotics, and drugs while handling or using explosives controlled and prohibited?

1926.902 Surface transportation of explosives.

_____ Are Department of Transportation requirements met for transporting explosives and are they stored according to Bureau of Alcohol, Tobacco, and Firearms regulations?

1926.904 Storage of explosives and blasting agents.

_____ Are all explosives accounted for, kept in locked magazine, inventoried, and use record kept?

Subpart V – Power Transmission and Distribution

1926.950 General requirements.

_____ Are existing conditions including energized lines and their voltage determined prior to starting work?

_____ Are safe working and clearance distances being maintained?

_____ Are proper procedures being followed to de-energize lines?

_____ Are workers trained in first aid and CPR?

1926.951 Tools and protective equipment.

_____ Is PPE inspected and tested prior to use?

_____ Are portable metal or conductive ladders not used near energized lines or equipment?

_____ Are live line tools certified and inspected prior to use?

1926.952 Mechanical equipment.

_____ Are aerial lift trucks near energized line or equipment grounded, barricaded, or insulated?

1926.954 Grounding for protection of employees.

_____ Are conductors and equipment treated as energized until tested, determined to be deenergized or grounded?

_____ Are grounding procedures followed?

1926.956 Underground lines.

_____ Are open manholes or vaults provided with warning signs and protected by barricades, temporary covers, or other guards?

_____ Are manholes and vaults provided with forced ventilation or found to be safe by testing for oxygen and flammable atmosphere prior to entry?

1926.957 Construction in energized substations.

_____ Are temporary fences provided when existing substation fence is removed or expanded for construction?

Subpart W – Rollover Protective Structures; Overhead Protection

1926.1000 Rollover protective structures (ROPS) for material handling equipment.

_____ Do all rubber-tired, self-propelled scrapers, rubber-tired front end loaders, rubber-tired dozers, wheel type agriculture and industrial tractors, crawler tractors, drawl-type loaders, and motor graders manufactured after July 1, 1969 have rollover protection which will support 2 times the weight of the equipment?

Subpart X – Stairways and Ladders

1926.1051 General Requirements.

_____ Are stairways or ladders provided at personnel points of access where there is a break in elevation of 19 inches or more and no other means of access is provided?

1926.1052 Stairways.

_____ Are risers' height and tread depth uniform?

_____ Are stairways free of slippery conditions or hazardous projections?

_____ Are stairs with four or more risers or rising more than 30 inches equipped with at least one handrail and one stairrail system along each unprotected side or edge?

_____ Are stairs not less than 36 inches from upper surface of the stairrail system to the surface of the tread?

_____ Are midrails, screens, mesh or equivalent installed between the top and bottom of stairrail system?

_____ Will handrails and top stairrails withstand at least 200 pounds of force?

_____ Are unprotected sides and edges of stairway landings provided with guardrails?

_____ Are stairs a minimum 22 inches width?

_____ Can stairs carry a load 5 times that anticipated (not less than 1000 pounds)?

_____ Is the angle of rise between 30 and 50 degrees to horizontal?

1926.1053 Ladders.

_____ Do extension ladders extend three feet above landing (or grab rail provided) and are secured against displacement?

_____ Are safety feet provided on extension ladders?

_____ Are ladders maintained at an angle of 4:1 height-to-base ratio?

_____ Do workers face the ladder while climbing and use three points of contact?

_____ Are metal spreaders locked in place when a step ladder is open during use?

_____ Are workers carrying objects up and down ladders which could fall or cause them to fall?

_____ Do portable wood ladders meet ANSI A14.1 specifications?

_____ Are ladders periodically inspected and maintained in good order?

_____ Are defective ladders tagged, repaired, or destroyed?

_____ Is special care taken to prevent short circuits or shock when metal ladders are used near electric circuits?

_____ Does the distance between rungs not exceed 12 inches and is it uniform throughout length of fixed ladder?

_____ Do fixed ladders meet ANSI A14.3 specifications?

_____ For fixed ladders, are the length of rungs or cleats 16 inches minimum and distance from rungs or steps to permanent objects in back 7 inches?

_____ For fixed ladders, do landing platforms exist every 30 feet with a cage, every 20 feet without cage?

_____ Do fixed ladder rails extend 42 inches above landing?

Subpart Y – Diving – Not a normal construction activity.

Subpart Z – Toxic and Hazardous Substances – Check Chapter 9 on Construction Health Hazards

APPENDIX E

ELECTRICAL HAZARDOUS LOCATION CLASSIFICATION

Class I - Highly Flammable Gases or Vapors

Division 1 – Locations where hazardous concentrations are probable, or where accidental occurrence should be simultaneous with failure of electrical equipment.

Division 2 – Locations where flammable concentrations are possible, but only in the event of process closures, rupture, ventilation failure, etc.

If the answer to any of the following questions is "yes," then the location would be a Class I:

- Are flammable liquids, vapors, or gases likely to be present?

- Are liquids having flash points at or above 100°F likely to be handled, processed, or stored at temperatures above their flash point?

Class I, Division 1

- Is a flammable mixture likely to be present under normal conditions?

- Is a flammable mixture likely to be present frequently because of repair, maintenance, or leaks?

- Would a failure of process, storage, handling, or other equipment be likely to cause and electrical failure coinciding with the release of flammable gas or liquid?

- Is the flammable liquid, vapor, or gas piping system in an inadequately ventilated location, and does the piping system contain valves, meters, or screwed or flanged fittings that are likely to leak?

- Is the zone below the surrounding elevation or grade such that flammable liquids or vapors may accumulate?

Some Examples of Class I, Division 1 Locations

- Locations where volatile flammable liquids or liquified flammable gases are transferred from one container to another.

- Interiors of paint spray booths and areas adjacent to paint spray booths and other spraying operations where volatile flammable solvents are used.

- Locations containing open tanks or vats of volatile flammable liquids.

- Drying rooms or compartments for the evaporation of flammable solvents.

751

- Cleaning and dyeing areas where hazardous liquids are used.

- Gas generator rooms and portions of gas manufacturing plants where flammable gas may escape.

- Inadequately ventilated pump and compressor rooms for flammable gas or volatile flammable liquids.

- All other locations where hazardous concentrations of flammable vapors or gases are likely to form in the course of normal operations.

Class I, Division 2

- Is the flammable liquid, vapor, or gas piping system in an inadequately ventilated location, but not likely to leak?

- Is the flammable liquid, vapor, or gas handled in an adequately ventilated location, and can the flammable substance escape only in the course of some abnormality such as failure of a gasket or packing?

- Is the location adjacent to Division 1 location, or can the flammable substance be conducted to the location through trenches, pipes, or ducts?

Equipment Required for Class I, Division 1 Hazardous Locations

A. Meters, relays, and instruments, such as voltage or current meters and pressure or temperature sensors, must be in enclosures approved for Class I, Division 1 locations. Such enclosures include explosion-proof and purged and pressurized enclosures. National Electrical Code (NEC), NFPA 70, 501-3(a).

B. Wiring methods acceptable for use in Class I, Division 1 locations include threaded rigid metal or steel intermediate metal conduit and type MI cable. Flexible fittings, such as motor terminations, must be approved for Class I locations. All boxes and enclosures must be explosion-proof and threaded for conduit or cable terminations. All joints must be wrenched tight with minimum of five threads engaged.

C. Sealing is required for conduit and cable systems to prevent the passage of gases, vapors, and flame from one part of the electrical installation to another through conduit. Type MI cable inherently prevents this from happening by its construction; however, it must be sealed to keep moisture and other fluids from entering the cable at terminations. NEC, 501-5.

    1. Seals are required where conduit passes from Division 1 to Division 2 or nonhazardous locations.

    2. Seals are required within 18 in. from enclosures containing arcing devices.

    3. Seals are required if conduit is 2 in. in diameter or larger entering an enclosure containing terminations, splices, or taps.

D. Drainage is required where liquid or condensed vapor may be trapped within an enclosure or raceway. An approved system of preventing accumulations or to permit periodic drainage are two methods to control condensation of vapors and liquid accumulation. NEC, 501-5(f).

E. Arcing devices, such as switched, circuit breakers, motor controllers, and fuses, must be approved for Class I locations. NEC, 501-6(a).

    F.   Motors shall be

        1.   approved for use in Class I, Division 2 locations,

        2.   totally enclosed with positive pressure ventilation,

        3.   totally enclosed inert-gas-filled with a positive pressure within the enclosure, or

        4.   submerged in a flammable liquid or gas. NEC, 501-8(a).

    G.   Lighting fixtures, both fixed and portable in rigid metal conduit, must be explosion-proof and guarded against physical damage. NEC, 501-9(a).

    H.   Flexible cords must be designed for extra hard usage, contain a grounding conductor, be supported so that there will be no tension on the terminal connections, and be provided with seals where they enter explosion-proof enclosures. NEC, 501-11.

Class II - Combustible Dusts

Division 1 – Locations where hazardous concentrations are probable, where their existence would be simultaneous with electrical equipment failure, or where electricity conducting dusts are involved.

Division 2 – Locations where hazardous concentrations are not likely, but where deposits of the dust might interfere with heat dissipation from electrical equipment, or be ignited by electrical equipment.

If the answer to any of the following questions is "yes," then the location would be a Class II:

- Are combustible dusts likely to be present?

- Are combustible dusts likely to ignite as a result of storage, handling, or other causes?

Class II, Division 1

- Is combustible dust likely to exist in suspension in air, under normal operating conditions, in sufficient quantities to produce explosive or ignitable mixture?

- Is combustible dust likely to exist in suspension in the air, because of maintenance or repair operations, in sufficient quantities to cause explosive or ignitable mixtures?

- Would failure of equipment be likely to cause an electrical system failure coinciding with the release of combustible dust in the air?

- Is combustible dust of an electrically conductive nature likely to be present?

Class II, Division 2

- Is the combustible dust likely to exist in suspension in air only under abnormal conditions, but can accumulations of dust be ignited by heat developed by electrical equipment, or by arcs, sparks, or burning materials expelled from electrical equipment?

- Are dangerous concentrations of ignitable dusts normally prevented by reliable dust-control equipment such as fans or filters?

- Is the location adjacent to a Division 1 location, and not separated by a fire wall?

- Are dust-producing materials stored or handled only in bags or containers and only stored-not used in the area?

Summary for Class II Hazardous Locations

A. Wiring methods for Class II, Division 1 locations: boxes and fittings containing arcing and sparking parts are required to be in dust-ignition-proof enclosures. Threaded metal conduit or type MI cable with approved terminations is required for Class II, Division 1 locations. National Electrical Code (NEC), NFPA 70, 502-4(a). In Class II, Division 2 locations, boxes and fittings are not required to be dust-ignition-proof but must be designed to minimize the entrance of dust and prevent the escape of sparks or burning material. In addition to the wiring systems suitable for Division 1 locations, the following systems are suitable for Division 2 locations: electrical metallic tubing, dust-tight wireways, and type MC and SNM cables. NEC, 502-4(b).

B. Suitable means of preventing the entrance of dust into a dust-ignition-proof enclosure must be provided where a raceway provides a path to the dust-ignition-proof enclosure from another enclosure that could allow the entrance of dust. NEC, 502-5.

C. In Class II, Division 1 locations, fixed and portable lighting must be dust-ignition proof. NEC, 502-11. Lighting fixtures in Class II, Division 2 locations must be designed to minimize accumulation of dust and must be enclosed to prevent the release of sparks or hot metal.

D. In Class II, Division 1 locations, motors, generators, and other rotating electrical machinery must be dust-ignition proof or totally enclosed pipe ventilated. NEC, 502-8.

E. Switches, circuit breakers, motor controllers, and fuses installed in Class II, Division 2 areas, enclosures for fuses, switches, circuit breakers, and motor controllers must be dust-tight.

F. Class II, Division 2 areas, rotating equipment must be one of the following types:

1. dust-ignition proof,

2. totally enclosed pipe ventilated,

3. totally enclosed non-ventilated, or

4. totally enclosed fan cooled.

G. Under certain conditions, standard open-type machines and self-cleaning squirrel-cage motors may be used. NEC 502-8(b).

H. In both divisions, each fixture must be clearly marked for the maximum wattage of the lamp, so that the maximum permissible surface temperature for the fixture is not exceeded. Additionally, fixtures must be protected from damage. NEC, 502-11.

I. Flexible cords in Division 1 and 2 are required to

1. be suitable for extra hard usage,

2. contain an equipment grounding conductor,

3. be connected to terminals in an approved manner,

4. be properly supported, and

5. be provided with suitable seals where necessary. NEC, 502-12.

J. Receptacles and attachment plugs used in Class II, Division 1 areas are required to be approved for Class II locations and provided with a connection for an equipment grounding conductor. NEC, 502-13.

K. In Division 2 areas, the receptacle must be designed so the connection to the supply circuit cannot be made or broken while the parts are exposed. This is commonly done with an interlocking arrangement between a circuit breaker and the receptacle. The plug cannot be removed until the circuit breaker is in the OFF position, and the breaker cannot be switched to the ON position unless the plug is inserted into the receptacle.

Class III - Combustible Fibers or Flyings

Division 1 – Locations in which easily ignitable fibers or materials producing combustible flyings are handled, manufactured, or used.

Division 2 – Locations in which such fibers of flyings are stored or handled, except in the process of manufacture.

If the answer to any of the following questions is "yes," then the location would be a Class III:

- Are easily ignitable fibers or flyings present, but not likely to be in suspension in the air in sufficient quantities to produce an ignitable mixture in the atmosphere?

Class III, Division 1

- Are easily ignitable fibers or materials producing combustible flyings handled, manufactured, or used?

Class III, Division 2

- Are easily ignitable fibers or flyings only handled and stored, and not processed?
- Is the location adjacent to Class III, Division 1 Location?

Summary of Class III Hazardous Locations

A. In Class III locations, wiring must be within a rigid metal or nonmetal conduit or be of type MI, MC, or SNM cable. Fittings and boxes shall be dust-tight. Flexible connectors are allowed. Equipment grounding shall be provided. National Electrical Code (NEC), NFPA 70, 503-3.

B. Switches, circuit breakers, motor controllers, and similar devices used in Class III hazardous locations must be within dust-tight enclosures. NEC, 503-4.

C. Motors, generators, and other rotating electrical machinery must be totally enclosed non-ventilated, totally enclosed pipe-ventilated, or totally enclosed fan-cooled. NEC, 503-6.

D. Lighting fixtures must have enclosures designed to minimize the entry of fibers, to prevent the escape of sparks or hot metal, and to have a maximum exposed surface temperature of less than 309°F (165°C). NEC, 503-9.

Note: Class III locations do not necessarily require equipment that is labeled as suitable for the location. Equipment in Class III locations is required only to meet certain performance requirements as outlined above.

Groups:

A – Atmospheres containing acetylene

B – Atmospheres containing hydrogen or gases or vapors of equivalent hazard

C – Atmospheres containing ethyl-ether vapors, ethylene, or cyclopropane

D – Atmospheres containing gasoline, hexane, naphtha, benzene, butane, propane, alcohol, acetone, benzol, or natural gas

E – Atmospheres containing metal dust, including aluminum, magnesium, and other metals of equally hazardous characteristics

F – Atmospheres containing carbon black, coke, or coal dust

G – Atmospheres containing flour, starch, or grain dusts

Adapted from the NFPA and NEC

APPENDIX F

RIGGING – THE CROSBY USER'S LIFTING GUIDE

**with
permission by the
Crosby Group, Inc.**

Crosby® USER'S GUIDE
LIFTING

VERSION (1/98)

1

RISK MANAGEMENT

DEFINITION

COMPREHENSIVE SET OF ACTIONS THAT REDUCES THE RISK OF A PROBLEM, A FAILURE, AN ACCIDENT

YOU NEED

- PRODUCT KNOWLEDGE
- APPLICATION KNOWLEDGE
- MANUFACTURER OF KNOWN CAPABILITY
- PRODUCTS THAT ARE CLEARLY IDENTIFIED WITH THE FOLLOWING;

1. MANUFACTURER'S NAME AND LOGO
2. LOAD RATING OR SIZE THAT REFERENCES RATINGS
3. TRACEABILITY CODE

A GOOD RISK MANAGEMENT PROGRAM RECOGNIZES

- PERFORMANCE REQUIREMENTS INCLUDE THE FOLLOWING:

1. LOAD RATED PRODUCTS
2. QUENCH AND TEMPERED
3. ABILITY TO DEFORM WHEN OVERLOADED.
4. ABILITY TO WITHSTAND REAL WORLD LOADING IN DAY TO DAY USE, TOUGHNESS.

TERMINOLOGY

WORKING LOAD LIMIT (WLL)

THE MAXIMUM MASS OR FORCE WHICH THE PRODUCT IS AUTHORIZED TO SUPPORT IN A PARTICULAR SERVICE.

PROOF TEST

A TEST APPLIED TO A PRODUCT SOLELY TO DETERMINE INJURIOUS MATERIAL OR MANU-FACTURING DEFECTS.

ULTIMATE STRENGTH

THE AVERAGE LOAD OR FORCE AT WHICH THE PRODUCT FAILS OR NO LONGER SUP-PORTS THE LOAD.

DESIGN FACTOR

AN INDUSTRIAL TERM DENOTING A PRODUCT'S THEORETICAL RESERVE CAPA-BILITY; USUALLY COMPUTED BY DIVIDING THE CATALOG ULTIMATE LOAD BY THE WORKING LOAD LIMIT. GENERALLY EXPRESSED AS A RATIO, e.g. 5 TO 1

INSPECTION OF FITTINGS

DEFORMATION

CROSBY RECOMMENDS THAT NO SIGNIFICANT DEFORMATION BE ALLOWED.

WEAR

ACCEPTABLE LIMITS:
5%   WEAR IN THE THROAT & EYE OF HOOKS AND OTHER CRITICAL SECTIONS OF ALL FITTINGS.
10%   WEAR IN OTHER AREAS.

CRACKS

REMOVE FITTINGS FROM SERVICE WITH CRACKS.

WELDING AND MODIFICATIONS

DO NOT WELD ON OR MODIFY FITTINGS OR BLOCKS.

FOR ADDITIONAL SUPPORT

the Crosby group, inc.

P.O. BOX 3128
TULSA OKLAHOMA 74101
(918)834-4611

WIRE ROPE SLING FACTS

2

INSPECTION AND REPLACEMENT PER ANSI B30.9

INSPECTION

ALL SLINGS SHALL BE VISUALLY INSPECTED BY THE PERSON HANDLING THE SLING EACH DAY THEY ARE USED: IN ADDITION, A PERIODIC INSPECTION SHALL BE PERFORMED BY A DESIGNATED PERSON, AT LEAST ANNUALLY, AND SHALL INCLUDE A RECORD OF THE INSPECTION.

- DISTORTION OF THE ROPE IN THE SLING SUCH AS KINKING, CRUSHING, UNSTRANDING, BIRDCAGING, MAIN STRAND DISPLACEMENT OR CORE PROTRUSION. LOSS OF ROPE DIAMETER IN SHORT ROPE LENGTHS OR UNEVENNESS OF OUTER STRANDS SHOULD PROVIDE EVIDENCE THE SLING SHOULD BE REPLACED.
- GENERAL CORROSION
- BROKEN OR CUT STRANDS
- NUMBER, DISTRIBUTION, AND TYPE OF VISIBLE BROKEN WIRES

REPLACEMENT

CONDITION SUCH AS THE FOLLOWING SHOULD BE SUFFICIENT REASON FOR CONSIDERATION OF SLING REPLACEMENT

- FOR STRAND LAID AND SINGLE PART SLINGS TEN RANDOMLY DISTRIBUTED BROKEN WIRES IN ONE ROPE LAY, OR FIVE BROKEN WIRES IN ONE ROPE STRAND IN ONE ROPE LAY.
- SEVERE LOCALIZED ABRASION OR SCRAPING
- KINKING, CRUSHING, BIRDCAGING, OR ANY DAMAGE RESULTING IN DISTORTION OF THE ROPE STRUCTURE.
- EVIDENCE OF HEAT DAMAGE
- END ATTACHMENTS THAT ARE CRACKED, DEFORMED, OR WORN TO THE EXTENT THAT THE STRENGTH OF THE SLING IS SUBSTANTIALLY AFFECTED.
- HOOKS SHOULD BE INSPECTED IN ACCORDANCE WITH ANSI B30.10
- SEVERE CORROSION OF THE ROPE OR END ATTACHMENTS

MULTI-PART REMOVAL CRITERIA FOR CABLE LAID AND BRAIDED SLINGS

SLING BODY	ALLOWABLE BROKEN WIRE PER LAY OR ONE BRAID	ALLOWABLE BROKEN STRANDS PER SLING LAY
LESS THAN 8 PER BRAID	20	1
CABLE LAID	20	1
8 PARTS AND MORE	40	1

REFER TO ANSI B30.9 FOR FULL DETAILS

WIRE ROPE SLING CAPACITIES (LBS.) – FLEMISH EYE – ANSI B30.9

6 X 19 AND 6 X 37
IMPROVED PLOW STEEL – IWRC 5/1 DESIGN FACTOR

WIRE ROPE SIZE	SHACKLE SIZE (Q & T CARBON SHACKLE — MINIMUM SHACKLE SIZE FOR A D/d > 1 AT LOAD CONNECTION)	VERTICAL (SINGLE LEG)	CHOKER (ANGLE 120)	TWO LEG OR BASKET HITCH (90°)	60 DEGREE SLING ANGLE	45 DEGREE SLING ANGLE	30 DEGREE SLING ANGLE
1/4	5/16	1120	820	2200	1940	1500	1100
5/16	3/8	1740	1280	3400	3000	2400	1700
3/8	7/16	2400	1840	4800	4200	3400	2400
7/16	1/2	3400	2400	6800	5800	4800	3400
1/2	5/8	4400	3200	8800	7600	6200	4400
9/16	5/8	5600	4000	11200	9600	7900	5600
5/8	3/4	6800	5000	13600	11800	9600	6800
3/4	7/8	9800	7200	19600	16900	13800	9800
7/8	1	13200	9600	26400	22800	18600	13200
1	1-1/8	17000	12600	34000	30000	24000	17000
1-1/8	1-1/4	20000	15800	40000	34600	28300	20000
1-1/4	1-3/8	26000	19400	52000	45000	36700	26000
1-3/8	1-1/2	30000	24000	60000	52000	42000	30000

- RATED CAPACITIES BASED ON PIN DIAMETER OR HOOK NO LARGER THAN THE NATURAL EYE WIDTH (1/2 X EYE LENGTH) OR LESS THAN THE NOMINAL SLING DIAMETER

REFER TO ANSI B30.9 FOR FULL DETAILS

HORIZONTAL SLING ANGLES OF LESS THAN 30 DEGREES ARE NOT RECOMMENDED

CHAIN SLING CAPACITIES (LBS.) – ANSI B30.9 DESIGN FACTOR 4/1

Crosby® QT ALLOY

CHAIN SIZE	VERTICAL (SINGLE LEG) 90°	TWO LEG OR BASKET HITCH	60 DEGREE SLING ANGLE 60°	45 DEGREE SLING ANGLE 45°	30 DEGREE SLING ANGLE 30°	SINGLE LEG MASTER LINK SIZE	DOUBLE LEG MASTER LINK SIZE
CHAIN GR – 8 DESIGN FACTOR 4/1							
1/4 - (9/32)	3500	7000	6050	4900	3500	1/2	1/2
3/8	7100	14200	12200	10000	7100	3/4	3/4
1/2	12000	24000	20750	16950	12000	1	1
5/8	18100	39200	31350	25500	18100	1-1/4	1-1/4
3/4	28300	56600	49000	40000	28300	1-1/4	1-1/2
7/8	34200	68400	59200	48350	34200	1-1/2	1-3/4
1	47700	95400	82600	67450	47700		
1-1/4	72300	144600	125200	102200	72300		

CHAIN – FACTS

INSPECTION AND REMOVAL FROM SERVICE PER ANSI N30.9

FREQUENT INSPECTION

- DAILY CHECK CHAIN AND ATTACHMENTS FOR WEAR, NICKS, CRACKS, BREAKS, GOUGES, STRETCH, BENDS, WELD SPLATTER, DISCOLORATION FROM EXCESSIVE TEMPERATURE, AND THROAT OPENINGS OF HOOKS.
  1. CHAIN LINKS AND ATTACHMENTS SHOULD HINGE FREELY TO ADJACENT LINKS.
  2. LATCHES ON HOOKS, IF PRESENT SHOULD HINGE FREELY AND SEAT PROPERLY WITHOUT EVIDENCE OF PERMANENT DISTORTION.

PERIODIC INSPECTION - INSPECTION RECORDS REQUIRED

- NORMAL SERVICE - YEARLY
- SEVERE SERVICE - MONTHLY

THIS INSPECTION SHALL INCLUDE EVERYTHING IN A FREQUENT INSPECTION PLUS EACH LINK AND END ATTACHMENT SHALL BE EXAMINED INDIVIDUALLY, TAKING CARE TO EXPOSE INNER LINK SURFACES OF THE CHAIN AND CHAIN ATTACHMENTS

1. WORN LINKS SHOULD NOT EXCEED VALUES GIVEN IN TABLE 1 OR RECOMMENDED BY THE MANUFACTURER
2. SHARP TRANSVERSE NICKS AND GOUGES SHOULD BE ROUNDED OUT BY GRINDING AND THE DEPTH OF THE GRINDING SHOULD NOT EXCEED VALUES IN TABLE 1
3. HOOKS SHOULD BE INSPECTED IN ACCORDANCE WITH ANSI B30.10
4. IF PRESENT, LATCHES ON HOOKS SHOULD SEAT PROPERLY, ROTATE FREELY, AND SHOW NO PERMANENT DISTORTION

TABLE 1

MAXIMUM ALLOWABLE WEAR AT ANY POINT OF LINK

NORMAL CHAIN OR COUPLING LINK CROSS SECTION	MAXIMUM ALLOWABLE WEAR DIAMETER INCHES
9/32	.037
3/8	.052
1/2	.069
5/8	.084
3/4	.105
7/8	.116
1	.137
1-1/4	.169

REFER TO ANSI B30.9 FOR FULL DETAILS

HORIZONTAL SLING ANGLES OF LESS THAN 30 DEGREES ARE NOT RECOMMENDED

WEB SLING CAPACITIES – ANSI B30.9 – DESIGN FACTOR 5/1

5

VERTICAL (SINGLE LEG)	CHOKER ANGLE 120	TWO LEG OR BASKET 90°	60 DEGREE SLING ANGLE 60°	45 DEGREE SLING ANGLE 45°	30 DEGREE SLING ANGLE 30°
100% OF SINGLE LEG	80% OF SINGLE LEG	200% OF SINGLE LEG	170% OF SINGLE LEG	140% OF SINGLE LEG	SAME AS SINGLE LEG

WEB SLING

INSPECTION AND REMOVAL FROM SERVICE PER ANSI B30.9

FREQUENT INSPECTION
THIS INSPECTION SHALL BE MADE BY THE PERSON HANDLING THE SLING EACH DAY THE SLING IS USED

PERIODIC INSPECTION WRITTEN INSPECTION RECORDS SHOULD BE KEPT FOR ALL SLINGS
THIS INSPECTION SHOULD BE CONDUCTED BY DESIGNATED PERSONNEL, FREQUENCY OF THE INSPECTION SHOULD BE BASED THE FOLLOWING:
1. FREQUENCY OF SLING USE
2. SEVERITY OF SERVICE CONDITIONS
3. EXPERIENCE GAINED ON THE SERVICE LIFE OF SLING USED IN SIMILAR APPLICATIONS
4. AT LEAST ANNUALLY

REMOVAL CRITERIA
1. ACID OR CAUSTIC BURNS
2. MELTING OR CHARRING OF ANY PART OF THE SLING
3. BROKEN, TEARS, CUTS, OR SNAGS
4. BROKEN OR WORN STITCHING IN LOAD BEARING SPLICES
5. EXCESSIVE ABRASIVE WEAR
6. KNOTS IN ANY PART OF THE SLING
7. EXCESSIVE PITTING OR CORROSION, OR CRACKED DISTORTED OR BROKEN FITTINGS
8. OTHER VISIBLE DAMAGE THAT CAUSES DOUBT AS TO THE STRENGTH OF THE SLING

- DO NOT "BUNCH", OR "PINCH" THE SLING IN FITTINGS
- DO NOT PLACE EYE OVER A PIN OR HOOK GREATER THAN 1/2 TIMES EYE LENGTH

REFER TO ANSI B30.9 FOR FULL DETAILS

HORIZONTAL SLING ANGLES OF LESS THAN 30 DEGREES ARE NOT RECOMMENDED

6

THE BASIC RIGGING PLAN

1. *WHO IS RESPONSIBLE (COMPETENT) FOR THE RIGGING? COMMUNICATIONS ESTABLISHED?*

2. *IS THE EQUIPMENT IN ACCEPTABLE CONDITION? APPROPRIATE TYPE, PROPER IDENTIFICATION?*

3. *ARE THE WORKING LOAD LIMITS ADEQUATE? WHAT IS WEIGHT OF LOAD? WHERE IS THE CENTER OF GRAVITY? WHAT IS THE SLING ANGLE? WILL THERE BE ANY SIDE LOADING? CAPACITY OF THE GEAR?*

4. *WILL THE LOAD BE UNDER CONTROL? TAG LINE AVAILABLE? IS THERE ANY POSSIBILITY OF FOULING? CLEAR OF PERSONNEL?*

5. *ARE THERE ANY UNUSUAL LOADING OR ENVIRONMENTAL CONDITIONS? WIND, TEMPERATURE, OTHER?*

6. *YOUR SPECIAL REQUIREMENTS?*

THE USERS RESPONSIBILITIES

UTILIZE APPROPRIATE RIGGING GEAR SUITABLE FOR OVERHEAD LIFTING

UTILIZE THE RIGGING GEAR WITHIN INDUSTRY STANDARDS AND THE MANUFACTURER'S RECOMMENDATIONS

CONDUCT REGULAR INSPECTION AND MAINTENANCE OF THE RIGGING GEAR

BASIC SLING OPERATING PRACTICES ANSI B30.9

WHENEVER ANY SLING IS USED, THE FOLLOWING PRACTICES SHALL BE OBSERVED.

1. SLINGS THAT ARE DAMAGED OR DEFECTIVE SHALL NOT BE USED.

2. SLINGS SHALL NOT BE SHORTENED WITH KNOTS OR BOLTS OR OTHER MAKESHIFT DEVICES.

3. SLING LEGS SHALL NOT BE KINKED.

4. SLINGS SHALL NOT BE LOADED IN EXCESS OF THEIR RATED CAPACITIES

5. SLINGS USED IN A BASKET HITCH SHALL HAVE THE LOADS BALANCED TO PREVENT SLIPPAGE.

6. SLINGS SHALL BE SECURELY ATTACHED TO THEIR LOAD.

7. SLINGS SHALL BE PADDED OR PROTECTED FROM THE SHARP EDGES OF THEIR LOADS.

8. SUSPENDED LOADS SHALL BE KEPT CLEAR OF ALL OBSTRUCTION.

9. ALL EMPLOYEES SHALL BE KEPT CLEAR OF LOADS ABOUT TO BE LIFTED AND OF SUSPENDED LOADS.

10. HANDS OR FINGERS SHALL NOT BE PLACED BETWEEN THE SLING AND ITS LOAD WHILE THE SLING IS BEING TIGHTENED AROUND THE LOAD.

11. SHOCK LOADING IS PROHIBITED!

12. A SLING SHALL NOT BE PULLED FROM UNDER A LOAD WHEN THE LOAD IS RESTING ON THE SLING.

INSPECTION: EACH DAY BEFORE BEING USED, THE SLING AND ALL FASTENINGS AND ATTACHMENTS SHALL BE INSPECTED FOR DAMAGE OR DEFECTS BY A COMPETENT PERSON DESIGNATED BY THE EMPLOYER. ADDITIONAL INSPECTIONS SHALL BE PERFORMED DURING SLING USE WHERE SERVICE CONDITIONS WARRANT. DAMAGED OR DEFECTIVE SLINGS SHALL BE IMMEDIATELY REMOVED FROM SERVICE.

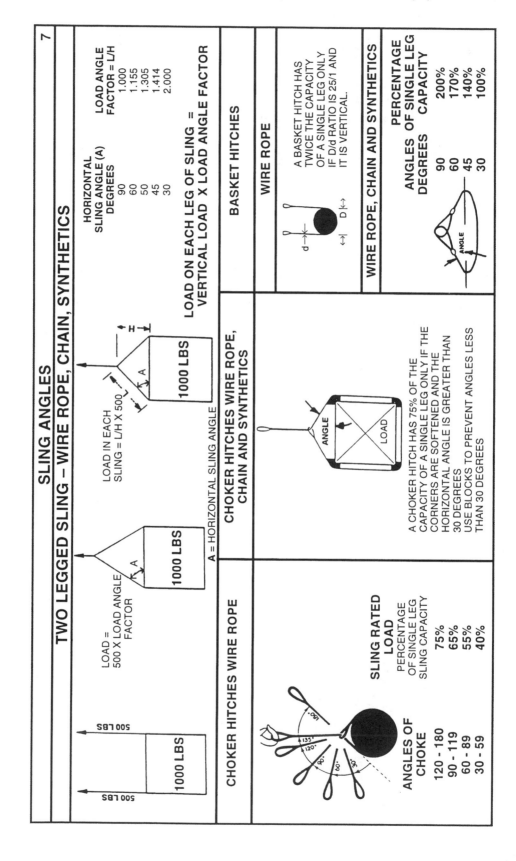

SLING ANGLES

TWO LEGGED SLING – WIRE ROPE, CHAIN, SYNTHETICS

7

LOAD = 500 X LOAD ANGLE FACTOR

1000 LBS

A = HORIZONTAL SLING ANGLE

LOAD IN EACH SLING = L/H X 500

1000 LBS

HORIZONTAL SLING ANGLE (A) DEGREES	LOAD ANGLE FACTOR = L/H
90	1.000
60	1.155
50	1.305
45	1.414
30	2.000

LOAD ON EACH LEG OF SLING = VERTICAL LOAD X LOAD ANGLE FACTOR

BASKET HITCHES

WIRE ROPE

A BASKET HITCH HAS TWICE THE CAPACITY OF A SINGLE LEG ONLY IF D/d RATIO IS 25/1 AND IT IS VERTICAL.

WIRE ROPE, CHAIN AND SYNTHETICS

ANGLES OF SINGLE LEG DEGREES	PERCENTAGE OF SINGLE LEG CAPACITY
90	200%
60	170%
45	140%
30	100%

CHOKER HITCHES WIRE ROPE, CHAIN AND SYNTHETICS

A CHOKER HITCH HAS 75% OF THE CAPACITY OF A SINGLE LEG ONLY IF THE CORNERS ARE SOFTENED AND THE HORIZONTAL ANGLE IS GREATER THAN 30 DEGREES

USE BLOCKS TO PREVENT ANGLES LESS THAN 30 DEGREES

CHOKER HITCHES WIRE ROPE

ANGLES OF CHOKE	SLING RATED LOAD PERCENTAGE OF SINGLE LEG SLING CAPACITY
120 - 180	75%
90 - 119	65%
60 - 89	55%
30 - 59	40%

LOAD DISTRIBUTION – RIGGING

8

LOADWALKING

LOAD ON SLING CALCULATED

TENSION 1= LOAD X D2 X S1/H(D1 + D2)
TENSION 2= LOAD X D1 X S2/H(D1 + D2)

UNEQUAL LEGS

LOAD ON SLING CALCULATED

TENSION 1= LOAD X D2 X S1/H(D1 + D2)
TENSION 2= LOAD X D1 X S2/H(D1 + D2)

TRIPLE AND QUAD LEG SLING

60°   60°   45°   45°
30°   30°   30°

TRIPLE LEG SLINGS HAVE 50% MORE CAPACITY THAN DOUBLE LEG ONLY IF THE CENTER OF GRAVITY IS IN CENTER OF CONNECTION POINT AND LEGS ARE ADJUSTED PROPERLY (EQUAL SHARE OF THE LOAD)

QUAD LEG SLINGS OFFER IMPROVED STABILITY BUT DO NOT PROVIDE INCREASED LIFTING CAPACITY.

TYPES OF HITCH CONSIDERATION

LOAD CONTROL
THE ABILITY OF THE SLING TO CONTROL THE MOVEMENT OF THE LOAD BEING LIFTED
CAPACITY
THE LOAD CAPACITY OF THE SLING AND TYPE OF HITCH
TYPE OF SLING
WIRE ROPE
CHAIN
WEBBING
CENTER OF GRAVITY
THE LOCATION OF THE CENTER OF THE LOAD'S WEIGHT

POSITIVE LOAD CONTROL

YES

REEVING INCREASES LOADS

REEVING THROUGH CONNECTIONS TO LOAD INCREASES LOAD ON CONNECTIONS FITTINGS BY AS MUCH AS TWICE— **DO NOT REEVE!**

Crosby® RIGGING HARDWARE

9

Crosby® SHACKLES — QUENCHED AND TEMPERED — QUIC-CHECK™

SCREW PIN AND BOLT TYPE — NOMINAL SIZE (IN) DIAMETER OF BOWS	CARBON SHACKLE DESIGN FACTOR 6/1 — CARBON MAXIMUM WORKING LOAD TONS	ALLOY SHACKLE DESIGN FACTOR 5/1 — ALLOY MAXIMUM WORKING LOAD TONS	INSIDE WIDTH AT PIN (INCHES)	DIAMETER OF PIN
3/16	1/3		.38	.25
1/4	1/2		.47	.31
5/16	3/4		.53	.38
3/8	1	2	.66	.44
7/16	1-1/2	2.6	.75	.50
1/2	2	3.3	.81	.63
5/8	3-1/4	5	1.06	.75
3/4	4-3/4	7	1.25	.88
7/8	6-1/2	9.5	1.44	1.00
1	8-1/2	12.5	1.69	1.13
1-1/8	9-1/2	15	1.81	1.25
1-1/4	12	18	2.03	1.38
1-3/8	13-1/2	21	2.25	1.50
1-1/2	17	30	2.38	1.63

- INSURE SCREW PIN TIGHT BEFORE EACH LIFT
- USE BOLT TYPE SHACKLE FOR PERMANENT INSTALLATION
- DO NOT SIDE LOAD ROUND PIN SHACKLE
- USE SCREW PIN OR BOLT TYPE TO COLLECT SLINGS

MAXIMUM INCLUDED ANGLE 120 DEGREES

REFER TO Crosby® CATALOG FOR ADDITIONAL INFORMATION

Crosby® HOOKS — QUENCHED AND TEMPERED — QUIC-CHECK™

SHANK HOOK SWIVEL HOOK EYE HOOK

DESIGN FACTOR
- EYEHOOKS – 5/1 (EXCEPT ALLOY 30 TON AND LARGER ARE 4-1/2 /1
- SHANK AND SWIVEL ARE 4-1/2 /1

CARBON MAXIMUM WORKING LOAD TONS	CODE	ALLOY MAXIMUM WORKING LOAD TONS	CODE	THROAT OPENING (INCHES)	DEFORMATION INDICATOR A - A
3/4	DC	1	DA	.88	1.50
1	FC	1-1/2	FA	.97	2.00
1-1/2	GC	2	GA	1.00	2.00
2	HC	3	HA	1.12	2.00
3	IC	*4-1/2 /5	IA	1.06	2.50
5	JC	7	JA	1.50	3.00
7-1/2	KC	11	KA	1.75	4.00
10	LC	15	LA	1.91	4.00
15	NC	22	NA	2.75	5.00
20	OC	30	OA	3.25	6.50
25	PC	37	PA	3.00	7.00
30	SC	45	SA	3.38	8.00
40	TC	60	TA	4.12	10.00

* 320N EYE HOOK IS NOW RATED AT 5 TONS

THROAT OPENING

MAXIMUM INCLUDED ANGLE 90 DEGREES

- DO NOT SIDELOAD
- DO NOT TIP LOAD
- DO NOT BACKLOAD

EYE HOOK

REFER TO Crosby® GROUP PRODUCT WARNING FOR ADDITIONAL INFORMATION

Crosby RIGGING HARDWARE

10

Crosby
WIRE ROPE CLIPS

SIZE	EFFICIENCY	NUMBER OF CLIPS	TURNBACK LENGTH (IN)	TORQUE FT – lbs
1/8	80%	2	3-1/4	4.5
3/16	80%	2	3-3/4	7.5
1/4	80%	2	4-3/4	15
5/16	80%	2	5-1/4	30
3/8	80%	2	6-1/2	45
7/16	80%	2	7	65
1/2	80%	3	11-1/2	65
9/16	80%	3	12	95
5/8	80%	3	12	95
3/4	80%	4	18	130
1	90%	5	26	225

APPLY U-BOLT OVER DEAD END OF THE WIRE ROPE LIVE END OF THE ROPE RESTS IN THE SADDLE A TERMINATION IS NOT COMPLETE UNTIL IT HAS BEEN FETORQUED A SECOND TIME NEVER SADDLE A DEAD HORSE!

FOR ADDITIONAL INFORMATION REFER TO THE Crosby PRODUCT WARNING

Crosby
TURNBUCKLE

SIZE	WORKING LOAD LIMIT JAW AND EYE 5/1 DESIGN FACTOR	WORKING LOAD LIMIT HOOK END FITTING 5/1 DESIGN FACTOR
1/4	500	400
5/16	800	700
3/8	1200	1000
1/2	2200	1500
5/8	3500	2250
3/4	5200	3000
7/8	7200	4000
1	10000	5000
1-1/4	15200	5000
1-1/2	21400	7500

THE USE OF LOCKNUTS OR MOUSING IS AN EFFECTIVE METHOD OF PREVENTING TURNBUCKLES FROM ROTATING

REFER TO THE Crosby CATALOG FOR ADDITIONAL INFORMATION

Crosby RIGGING HARDWARE

11

SHOULDER EYEBOLTS

QUENCHED AND TEMPERED

QUIC-CHECK

IN LINE

DESIGN FACTOR 5/1

SHANK DIAMETER	WORKING LOAD LIMITS IN LINE PULL (LBS.)	WORKING LOAD LIMITS 60 DEGREE SLING ANGLE (LBS.)	WORKING LOAD LIMITS 45 DEGREE SLING ANGLE (LBS.)	WORKING LOAD LIMITS ANGLE LESS THAN 45 DEGREES (LBS.)
1/4	650	420	195	160
5/16	1200	780	360	300
3/8	1550	1000	465	360
1/2	2600	1690	780	650
5/8	5200	3380	1560	1300
3/4	7200	4680	2160	1800
7/8	10600	6890	3180	2650
1	13300	8645	3990	3325
1-1/4	21000	13600	6300	5250
1-1/2	24000	15600	7200	6000

SHOULDER EYEBOLTS

- NEVER EXCEED WORKING LOAD LIMITS
- NEVER USE REGULAR NUT EYEBOLTS FOR ANGULAR LIFTS
- ALWAYS USE SHOULDER NUT EYEBOLTS FOR ANGULAR LIFTS
- FOR ANGULAR LIFTS, ADJUST WORKING LOAD AS SHOWN ABOVE
- ALWAYS TIGHTEN NUTS SECURELY AGAINST THE LOAD
- ALWAYS APPLY LOAD TO EYE BOLT IN THE PLANE OF THE EYE.

Sling Angle

SWIVEL HOIST RING

DESIGN FACTOR 5/1

WORKING LOAD LIMIT FULL 180 DEGREE PIVOT (LBS.)	THREAD SHANK SIZE U.N.C.	TORQUE FT – (LBS.)
800	5/16	7
1000	3/8	12
2500	1/2	28
4000	5/8	60
7000	3/4	100
8000	7/8	160
10000	1	230
15000	1-1/4	470
24000	1-1/2	800
30000	2	1100

SWIVEL HOIST RING

- WHEN USING LIFTING SLINGS OF TWO OR MORE LEGS MAKE SURE THE FORCES IN THE LEG ARE CALCULATED. SELECT THE PROPER SIZE SWIVEL HOIST RING TO ALLOW FOR LOAD IN SLING LEG.
- ALWAYS INSURE HOIST RING IS FREE TO ALIGN ITSELF WITH SLING
- ALWAYS INSURE HOIST RING IS PROPERLY TORQUED TO REQUIRED VALUE.

180° PIVOT

360° ROTATION

REFER TO THE Crosby GROUP WARNINGS FOR ADDITIONAL INFORMATION

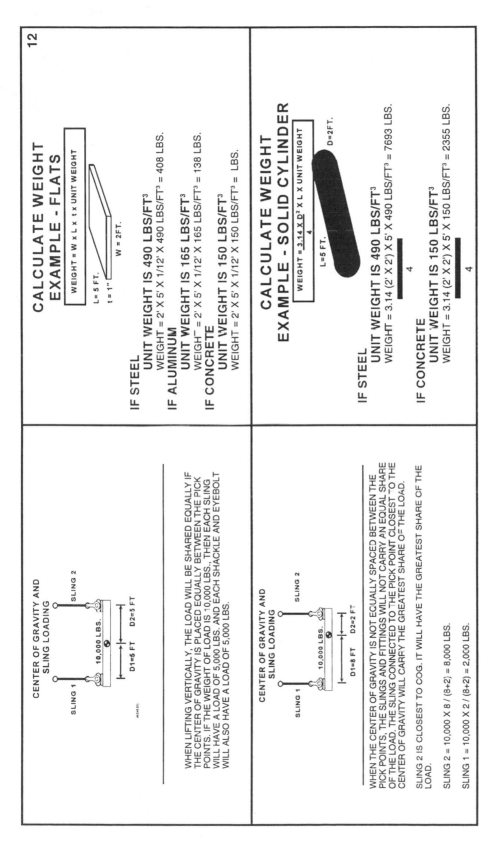

12

CALCULATE WEIGHT
EXAMPLE - FLATS

WEIGHT = W x L x t x UNIT WEIGHT

L = 5 FT.
t = 1"
W = 2 FT.

IF STEEL
UNIT WEIGHT IS 490 LBS/FT³
WEIGHT = 2' X 5' X 1/12' X 490 LBS/FT³ = 408 LBS.

IF ALUMINUM
UNIT WEIGHT IS 165 LBS/FT³
WEIGHT = 2' X 5' X 1/12' X 165 LBS/FT³ = 138 LBS.

IF CONCRETE
UNIT WEIGHT IS 150 LBS/FT³
WEIGHT = 2' X 5' X 1/12' X 150 LBS/FT³ = LBS.

CALCULATE WEIGHT
EXAMPLE - SOLID CYLINDER

WEIGHT = 3.14 X D² X L X UNIT WEIGHT / 4

L = 5 FT.
D = 2 FT.

IF STEEL
UNIT WEIGHT IS 490 LBS/FT³
WEIGHT = 3.14 (2' X 2') X 5' X 490 LBS/FT³ = 7693 LBS. / 4

IF CONCRETE
UNIT WEIGHT IS 150 LBS/FT³
WEIGHT = 3.14 (2' X 2') X 5' X 150 LBS/FT³ = 2355 LBS. / 4

CENTER OF GRAVITY AND SLING LOADING

SLING 1   SLING 2
10,000 LBS.
D1=5 FT   D2=5 FT

WHEN LIFTING VERTICALLY, THE LOAD WILL BE SHARED EQUALLY IF THE CENTER OF GRAVITY IS PLACED EQUALLY BETWEEN THE PICK POINTS. IF THE WEIGHT OF LOAD IS 10,000 LBS., THEN EACH SLING WILL HAVE A LOAD OF 5,000 LBS. AND EACH SHACKLE AND EYEBOLT WILL ALSO HAVE A LOAD OF 5,000 LBS.

CENTER OF GRAVITY AND SLING LOADING

SLING 1   SLING 2
10,000 LBS.
D1=8 FT   D2=2 FT

WHEN THE CENTER OF GRAVITY IS NOT EQUALLY SPACED BETWEEN THE PICK POINTS, THE SLINGS AND FITTINGS WILL NOT CARRY AN EQUAL SHARE OF THE LOAD. THE SLING CONNECTED TO THE PICK POINT CLOSEST TO THE CENTER OF GRAVITY WILL CARRY THE GREATEST SHARE OF THE LOAD.

SLING 2 IS CLOSEST TO COG. IT WILL HAVE THE GREATEST SHARE OF THE LOAD.

SLING 2 = 10,000 X 8 / (8+2) = 8,000 LBS.

SLING 1 = 10,000 X 2 / (8+2) = 2,000 LBS.

APPENDIX G

JOBSITE RULES

The following work rules are applicable to our operation and are enforceable on this job. This list is not considered all-inclusive since we cannot anticipate every possible situation in advance. Common sense is the overall basic rule. The pertinent requirements of OSHA regulation 29 CFR 1926, Safety and Health Regulations for Construction, along with applicable 29 CFR 1910, General Industry Regulations, also apply.

1. Report to work on time and in fit condition, prepared for work.

2. Wear suitable work clothes including long pants, shirt with sleeves, substantial leather shoes or boots. Do not wear sneakers, tennis shoes, soft soled athletic shoes, sandals, or similar shoes. Avoid loose clothing, ornaments, jewelry, and long hair, which can be caught in or on equipment.

3. Observe good personal hygiene. Use toilets provided on the job.

4. Sleeping, gambling, horseplay, fighting, theft, fireworks, weapons, and firearms are strictly prohibited on the job, and are grounds for immediate discharge and possible criminal prosecution.

5. Using or being under the influence of alcohol, narcotics, or other drugs or intoxicants are strictly prohibited. If a medical condition requires a prescription drug which may cause impairment, advise your supervisor.

6. Wear your personal protective equipment as prescribed for each task.

7. Hard hats must be worn at all times.

8. Wear safety glasses with side shields wherever dust, flying particles, or similar eye hazards exist. Most prescription glasses are NOT safety glasses. Add a full faceshield for grinding. Use proper tinted welding hoods or goggles for burning/cutting/welding. Sunglasses are NOT proper glasses for such purposes.

9. Wear safety full-body harnesses and shock-absorbing lanyards when instructed, especially on elevated work where there are no guardrails, on suspended scaffolds, and on aerial lifts. Lanyards length should allow for falls of six feet or less. Fasten lanyard at an anchorage point higher than the D-ring where it attaches to the harness if at all possible. Arrange your fall protection to not interfere with your work or be subject to damage. Inspect fall protection prior to each use for damage.

10. Housekeeping is everyone's responsibility. Discard scrap material, food containers, etc. in trash bins and barrels. Keep materials and equipment out of walkways and stairways. Store things neatly and securely, and away from open-

ings. Do not store material near the edges of roofs or open-sided floors, unless the material is tied or weighted down.

11. Use ladders properly and inspect them before use. Always stand facing the ladder when you are on it. Do not use a stepladder as a straight ladder. Do not sit on top a stepladder, straddle the two sides, or stand on the top two steps. Extend a straight extension ladder 3 feet above the top landing and secure it by tying or fastening the ladder against slipping or falling.

12. Handle all compressed gas cylinders with care. Keep cylinders upright at all times. Chair or tie them to prevent them from falling. Whenever possible, transport cylinders in cradles or carts.

13. Store oxygen at least 25 feet from flammable gases such as acetylene, propane, or hydrogen. Treat empty cylinders like full ones. Cylinders should be stored with valve protection caps in place. **Remember:** OSHA considers storage to mean any cylinders without regulators on them.

14. Powder-actuated tools. Hiliti guns shall be used only by workers who have completed certified training in their use. When not in use, guns should be unloaded and stored.

15. Use rigging properly and handle it carefully. Be sure slings are secured to the load and the load is balanced. Do not overload. Use a tag-line when appropriate. If slings or other rigging are frayed, worn, or otherwise unfit, take them out of service. Nylon slings have colored fiber inside of them. If these show through, the sling is unsafe for use. If a wire rope choker show more than 10 broken wires, do not use it.

16. Do not ride on lift trucks or on loads being lifted by cranes or hoists. Ride only in equipment or baskets built for carrying personnel. If you must ride in a pickup truck, sit in the truck bed, not on the side walls.

17. Follow all motor vehicle rules and regulations when driving company trucks, etc. Make sure load is secure before traveling. Wear seat belts. **Observe speed limits.** Report any vehicle defects or accidents to your supervisor.

18. Cover or barricade any floor opening more than 12 inches wide. You can use

    a. A cover of plywood on 2×4s must be fastened or cleated to prevent accidental removal.

    b. A barricade means a standard guardrail of steel cable or 2×4s. Loosely draped wire or waning tape around openings is **not** acceptable.

19. Do not enter a trench or other excavation unless it has been checked and made safe. Such excavations must be checked daily or whenever conditions change.

20. Confined spaces are tanks, sumps, boilers, and such. Do not enter unless the air quality has been properly checked and fresh air is provided. Be extra careful with electricity, gases, and chemicals while working inside such spaces.

21. Do not burn wood, scrap, etc. for disposal or heat.

22. Handle chemicals with care. This includes gases, fuel, solvents, cleaners, adhesives. Know and follow safety precautions. Use recommended personal protective equipment. Do not dispose of hazardous materials in drains, sewers, dumpsters, or by burning. Observe special disposal arrangements for such materials.

23. Observe "Danger," "Warning," "Caution," and "No Smoking" signs and notices. Stay out of restricted areas and equipment unless authorized to enter or work there.

24. Use and handle equipment, material, and safety devices with care. Use each for its intended purpose. Do not use anything that is defective. If appropriate, render it unusable. If not, tag and label it to warn others, and remove it from the work area.

25. Do not leave discharged fire extinguishers in the work area.

26. Do not expose yourself or others to dangerous conditions or actions. Report any unsafe conditions, correct it if possible, or at least warn others.

27. Report all injuries, no matter how serious, to your foreperson or superintendent immediately, even if you do not wish to get treated at the time. You must also inform them if you decide to get treatment at a later time.

28. Toolbox meetings are held regularly and attendance is mandatory for all workers. We encourage your participation and will gladly accept suggestions and questions.

29. When welding and cutting, have a fire extinguisher handy. Clear the work area of paper, plastic, etc. which may catch fire. Check the area below, and warn any workers there as well. Use a fire blanket or other protection to keep slag from falling into the area below.

30. When welding, use curtains or screen to reduce the hazard of welding flash to others. Check the welder leads for damage, keep them out of water, and arrange them safely so they are not subject to damage. This means out of the path of equipment or other traffic. Do not tamper with another's welding machine or controls.

31. Make sure each cylinder and regulator valve used in welding has a handle or wrench in place to close the valve if necessary. If you run hoses up to overhead work, support them properly. Do not wrap them around supports or tie them in knots. Fasten a rope (not wire) which you then tie to a support.

32. Before using an electrical power tool or extension cord, check it for damage. Insulation should not be cut or separated from the plug or socket. All plugs must have 3 prongs in good condition. You should use double-insulated tools when available.

33. Make sure all the extension cords are marked TYPE S, SJ, ST, or similar. Whenever possible, arrange cords overhead or out of walkways so they are not tripping hazards. Protect them from damage and wear. Do not lift or lower tools or materials by using extension cords tied to them.

34. OSHA requires either a Ground Fault Circuit Interrupter (GFCI) or an Assured Grounding Program (AGP) to be in place when electrically powered devices or tools are used on a construction site. AGP are required to be regularly tested every three months during normal continuous operation. Color markings are to be used to denote that up-to-date testing has occurred in the following manner:

   - January-March – White
   - July-September – Red
   - April-June – Green
   - October-December – Orange

   Do not use equipment if the current or next quarter's color is not in place.

35.   Be sure electrical circuits are properly deenergized before working on pumps, motors, circuits, equipment, etc. Use and observe the lockout/tagout procedures.

APPENDIX H

ERGONOMICS PROGRAM SURVEY INSTRUMENTS

Courtesy of NIOSH from Elements of Ergonomics Programs

Tray 4–A. Symptoms Survey Form

Symptoms Survey: *Ergonomics Program*

Date _____/_____/_____/

_____   _____   Job Name _____
Plant   Dept #

_____   _____   _____ years _____ months
Shift   Hours worked/week   Time on THIS Job

Other jobs you have done in the last year (for more than 2 weeks)

_____   _____   _____   _____ months _____ weeks
Plant   Dept #   Job Name   Time on THIS Job

_____   _____   _____   _____ months _____ weeks
Plant   Dept #   Job Name   Time on THIS Job

(If more than 2 jobs, include those you worked on the most)

Have you had any pain or discomfort during the last year?

☐ Yes   ☐ No (If NO, stop here)

If YES, carefully shade in area of the drawing which bothers you the MOST.

Front   Back

(Continued)

Tray 4–A (Continued).

(Complete a separate page for each area that bothers you)

Check Area: ☐ Neck ☐ Shoulder ☐ Elbow/Forearm ☐ Hand/Wrist ☐ Fingers
☐ Upper Back ☐ Low Back ☐ Thigh/Knee ☐ Low Leg ☐ Ankle/Foot

1. Please put a check by the words(s) that best describe your problem

☐ Aching ☐ Numbness (asleep) ☐ Tingling
☐ Burning ☐ Pain ☐ Weakness
☐ Cramping ☐ Swelling ☐ Other
☐ Loss of Color ☐ Stiffness

2. When did you first notice the problem?_____ (month)_____ (year)

3. How long does each episode last? (Mark an X along the line)

_____/_____/_____/_____/_____/
1 hour   1 day   1 week   1 month   6 months

4. How many separate episodes have you had in the last year?_____

5. What do you think caused the problem?_____

_____

6. Have you had this problem in the last 7 days? ☐ Yes ☐ No

7. How would you rate this problem? (mark an X on the line)
NOW

_____
None   Unbearable

When it is the WORST

_____
None   Unbearable

8. Have you had medical treatment for this problem? ☐ Yes ☐ No

8a. If NO, why not?_____

8a. If YES, where did you receive treatment?

☐ 1. Company Medical   Times in past year _____

☐ 2. Personal doctor   Times in past year _____

☐ 3. Other   Times in past year _____

Did treatment help? ☐ Yes ☐ No _____

9. How much time have you lost in the last year because of this problem?____ days

10. How many days in the last year were you on restricted or light duty because of this problem?
____ days

11. Please comment on what you think would improve your symptoms

_____

_____

Tray 5–A. General Ergonomic Risk Analysis Checklist

Check the box (❑) if your answer is "yes" to the question. A "yes" response indicates that an ergonomic risk factor may be present which requires further analysis.

Manual Material Handling

❑   Is there lifting of loads, tools, or parts?
❑   Is there lowering of tools, loads, or parts?
❑   Is there overhead reaching for tools, loads, or parts?
❑   Is there bending at the waist to handle tools, loads, or parts?
❑   Is there twisting at the waist to handle tools, loads, or parts?

For further analysis, refer to checklist 5–F.

Physical Energy Demands

❑   Do tools and parts weigh more than 10 lb?
❑   Is reaching greater than 20 in.?
❑   Is bending, stooping, or squatting a primary task activity?
❑   Is lifting or lowering loads a primary task activity?
❑   Is walking or carrying loads a primary task activity?
❑   Is stair or ladder climbing with loads a primary task activity?
❑   Is pushing or pulling loads a primary task activity?
❑   Is reaching overhead a primary task activity?
❑   Do any of the above tasks require five or more complete work cycles to be done within a minute?
❑   Do workers complain that rest breaks and fatigue allowances are insufficient?

For further analysis, refer to checklist 5–F.

Other Musculoskeletal Demands

❑   Do manual jobs require frequent, repetitive motions?
❑   Do work postures require frequent bending of the neck, shoulder, elbow, wrist, or finger joints?
❑   For seated work, do reaches for tools and materials exceed 15 in. from the worker's position?
❑   Is the worker unable to change his or her position often?
❑   Does the work involve forceful, quick, or sudden motions?
❑   Does the work involve shock or rapid buildup of forces?
❑   Is finger-pinch gripping used?
❑   Do job postures involve sustained muscle contraction of any limb?

For further analysis, refer to checklists 5–C, 5–D, and 5–E.

Computer Workstation

❑   Do operators use computer workstations for more than 4 hours a day?
❑   Are there complaints of discomfort from those working at these stations?
❑   Is the chair or desk nonadjustable?
❑   Is the display monitor, keyboard, or document holder nonadjustable?
❑   Does lighting cause glare or make the monitor screen hard to read?
❑   Is the room temperature too hot or too cold?
❑   Is there irritating vibration or noise?

For further analysis, refer to checklist 5–G.

*Adapted from The University of Utah Research Foundation "Checklist for General Ergonomic Risk Analysis," available from the ERGOWEB Internet site (http://ergoweb.com/).

Tray 5–A (Continued). General Ergonomic Risk Analysis Checklist

Environment

- ❑ Is the temperature too hot or too cold?
- ❑ Are the worker's hands exposed to temperatures less than 70 degrees Fahrenheit?
- ❑ Is the workplace poorly lit?
- ❑ Is there glare?
- ❑ Is there excessive noise that is annoying, distracting, or producing hearing loss?
- ❑ Is there upper extremity or whole body vibration?
- ❑ Is air circulation too high or too low?

General Workplace

- ❑ Are walkways uneven, slippery, or obstructed?
- ❑ Is housekeeping poor?
- ❑ Is there inadequate clearance or accessibility for performing tasks?
- ❑ Are stairs cluttered or lacking railings?
- ❑ Is proper footwear worn?

Tools

- ❑ Is the handle too small or too large?
- ❑ Does the handle shape cause the operator to bend the wrist in order to use the tool?
- ❑ Is the tool hard to access?
- ❑ Does the tool weigh more than 9 lb?
- ❑ Does the tool vibrate excessively?
- ❑ Does the tool cause excessive kickback to the operator?
- ❑ Does the tool become too hot or too cold?

For further analysis, refer to checklist 5–E.

Gloves

- ❑ Do the gloves require the worker to use more force when performing job tasks?
- ❑ Do the gloves provide inadequate protection?
- ❑ Do the gloves present a hazard of catch points on the tool or in the workplace?

Administration

- ❑ Is there little worker control over the work process?
- ❑ Is the task highly repetitive and monotonous?
- ❑ Does the job involve critical tasks with high accountability and little or no tolerance for error?
- ❑ Are work hours and breaks poorly organized?

Tray 5–B. Ergonomic Hazard Identification Checklist

Answer the following questions based on the primary job activities of workers in this facility.

Use the following responses to describe **how frequently** workers are exposed to the job conditions described below:

Never (worker is never exposed to the condition)
Sometimes (worker is exposed to the condition less than 3 times daily)
Usually (worker is exposed to the condition 3 times or more daily)

	Never	Sometimes	Usually	If *USUALLY*, list jobs to which answer applies here
1. Do workers perform tasks that are externally paced?				
2. Are workers required to exert force with their hands (e.g., gripping, pulling, pinching)?				
3. Do workers use handtools or handle parts or objects?				
4. Do workers stand continuously for periods of more than 30 min?				
5. Do workers sit for periods of more than 30 min without the opportunity to stand or move around freely?				
6. Do workers use electronic input devices (e.g., keyboards, mice, joysticks, track balls) for continuous periods of more than 30 min?				
7. Do workers kneel (one or both knees)?				
8. Do workers perform activities with hands raised above shoulder height?				

Tray 5–B (Continued).

	Never	Sometimes	Usually	If *USUALLY*, list jobs to which answer applies here
9. Do workers perform activities while bending or twisting at the waist?				
10. Are workers exposed to vibration?				
11. Do workers lift or lower objects between floor and waist height or above shoulder height?				
12. Do workers lift or lower objects more than once per min for continuous periods of more than 15 min?				
13. Do workers lift, lower, or carry large objects or objects that cannot be held close to the body?				
14. Do workers lift, lower, or carry objects weighing more than 50 lb?				

GLOSSARY OF TERMS

Facility: The location to which employees report each day for work. For situations in which employees do not report to any fixed location on a regular basis but are subject to common supervision, the facility may be defined as a central location where other OSHA records are maintained. (Note: Synonymous with establishment, as defined in OSHA recordkeeping requirements.)

Primary job activities: Job activities that make up a significant part of the work or are required for safety or contingency. Activities are not considered to be primary job activities if they make up a small percentage of the job (i.e., take up less than 10% of the worker's time), are not essential for safety or contingency, and can be readily accomplished in other ways (e.g., using equipment already available in the facility).

Externally paced activities: Work activities for which the worker does not have direct control of the rate of work. Externally paced work activities include activities for which (1) the worker must keep up with an assembly line or an independently-operating machine, (2) the worker must respond to a continuous queue (e.g., customers standing in line, phone calls at a switchboard), or (3) time standards are imposed on workers.

Tray 5–C. Workstation Checklist

"No" responses indicate potential problem areas which should receive further investigation.

1. Does the work space allow for full range of movement?   ❑ yes   ❑ no

2. Are mechanical aids and equipment available?   ❑ yes   ❑ no

3. Is the height of the work surface adjustable?   ❑ yes   ❑ no

4. Can the work surface be tilted or angled?   ❑ yes   ❑ no

5. Is the workstation designed to reduce or eliminate

      bending or twisting at the wrist?   ❑ yes   ❑ no
      reaching above the shoulder?   ❑ yes   ❑ no
      static muscle loading?   ❑ yes   ❑ no
      full extension of the arms?   ❑ yes .   ❑ no
      raised elbows?   ❑ yes   ❑ no

6. Are the workers able to vary posture?   ❑ yes   ❑ no

7. Are the hands and arms free from sharp edges on work surfaces?   ❑ yes   ❑ no

8. Is an armrest provided where needed?   ❑ yes   ❑ no

9. Is a footrest provided where needed?   ❑ yes   ❑ no

10. Is the floor surface free of obstacles and flat?   ❑ yes   ❑ no

11. Are cushioned floor mats provided for employees required to stand
    for long periods?   ❑ yes   ❑ no

12. Are chairs or stools easily adjustable and suited to the task?   ❑ yes   ❑ no

13. Are all task elements visible from comfortable positions?   ❑ yes   ❑ no

14. Is there a preventive maintenance program for mechanical aids, tools,
    and other equipment?   ❑ yes   ❑ no

Tray 5–D. Task Analysis Checklist

"No" responses indicate potential problem areas which should receive further investigation.

1. Does the design of the primary task reduce or eliminate

   bending or twisting of the back or trunk? ❑ yes ❑ no
   crouching? ❑ yes ❑ no
   bending or twisting the wrist? ❑ yes ❑ no
   extending the arms? ❑ yes ❑ no
   raised elbows? ❑ yes ❑ no
   static muscle loading? ❑ yes ❑ no
   clothes wringing motions? ❑ yes ❑ no
   finger pinch grip? ❑ yes ❑ no

2. Are mechanical devices used when necessary? ❑ yes ❑ no

3. Can the task be done with either hand? ❑ yes ❑ no

4. Can the task be done with two hands? ❑ yes ❑ no

5. Are pushing or pulling forces kept minimal? ❑ yes ❑ no

6. Are required forces judged acceptable by the workers? ❑ yes ❑ no

7. Are the materials

   able to be held without slipping? ❑ yes ❑ no
   easy to grasp? ❑ yes ❑ no
   free from sharp edges and corners? ❑ yes ❑ no

8. Do containers have good handholds? ❑ yes ❑ no

9. Are jigs, fixtures, and vises used where needed? ❑ yes ❑ no

10. As needed, do gloves fit properly and are they made of the proper fabric? ❑ yes ❑ no

11. Does the worker avoid contact with sharp edges when performing the task? ❑ yes ❑ no

12. When needed, are push buttons designed properly? ❑ yes ❑ no

13. Do the job tasks allow for ready use of personal equipment that
    may be required? ❑ yes ❑ no

14. Are high rates of repetitive motion avoided by

    job rotation? ❑ yes ❑ no
    self-pacing? ❑ yes ❑ no
    sufficient pauses? ❑ yes ❑ no
    adjusting the job skill level of the worker? ❑ yes ❑ no

15. Is the employee trained in

    proper work practices? ❑ yes ❑ no
    when and how to make adjustments? ❑ yes ❑ no
    recognizing signs and symptoms of potential problems? ❑ yes ❑ no

Tray 5–E. Handtool Analysis Checklist

"No" responses indicate potential problem areas which should receive further investigation.

1. Are tools selected to limit or minimize

    exposure to excessive vibration?   ❏ yes   ❏ no

    use of excessive force?   ❏ yes   ❏ no

    bending or twisting the wrist?   ❏ yes   ❏ no

    finger pinch grip?   ❏ yes   ❏ no

    problems associated with trigger finger?   ❏ yes   ❏ no

2. Are tools powered where necessary and feasible?   ❏ yes   ❏ no

3. Are tools evenly balanced?   ❏ yes   ❏ no

4. Are heavy tools suspended or counterbalanced in ways to facilitate use?   ❏ yes   ❏ no

5. Does the tool allow adequate visibility of the work?   ❏ yes   ❏ no

6. Does the tool grip/handle prevent slipping during use?   ❏ yes   ❏ no

7. Are tools equipped with handles of textured, non-conductive material?   ❏ yes   ❏ no

8. Are different handle sizes available to fit a wide range of hand sizes?   ❏ yes   ❏ no

9. Is the tool handle designed not to dig into the palm of the hand?   ❏ yes   ❏ no

10. Can the tool be used safely with gloves?   ❏ yes   ❏ no

11. Can the tool be used by either hand?   ❏ yes   ❏ no

12. Is there a preventive maintenance program to keep tools operating as designed?   ❏ yes   ❏ no

13. Have employees been trained

    in the proper use of tools?   ❏ yes   ❏ no

    when and how to report problems with tools?   ❏ yes   ❏ no

    in proper tool maintenance?   ❏ yes   ❏ no

Tray 5–F. Materials Handling Checklist

"No" responses indicate potential problem areas which should receive further investigation.

1. Are the weights of loads to be lifted judged acceptable by the workforce? ❑ yes ❑ no

2. Are materials moved over minimum distances? ❑ yes ❑ no

3. Is the distance between the object load and the body minimized? ❑ yes ❑ no

4. Are walking surfaces

    level? ❑ yes ❑ no
    wide enough? ❑ yes ❑ no
    clean and dry? ❑ yes ❑ no

5. Are objects

    easy to grasp? ❑ yes ❑ no
    stable? ❑ yes ❑ no
    able to be held without slipping? ❑ yes ❑ no

6. Are there handholds on these objects? ❑ yes ❑ no

7. When required, do gloves fit properly? ❑ yes ❑ no

8. Is the proper footwear worn? ❑ yes ❑ no

9. Is there enough room to maneuver? ❑ yes ❑ no

10. Are mechanical aids used whenever possible? ❑ yes ❑ no

11. Are working surfaces adjustable to the best handling heights? ❑ yes ❑ no

12. Does material handling avoid

    movements below knuckle height and above shoulder height? ❑ yes ❑ no
    static muscle loading? ❑ yes ❑ no
    sudden movements during handling? ❑ yes ❑ no
    twisting at the waist? ❑ yes ❑ no
    extended reaching? ❑ yes ❑ no

13. Is help available for heavy or awkward lifts? ❑ yes ❑ no

14. Are high rates of repetition avoided by

    job rotation? ❑ yes ❑ no
    self-pacing? ❑ yes ❑ no
    sufficient pauses? ❑ yes ❑ no

15. Are pushing or pulling forces reduced or eliminated? ❑ yes ❑ no

16. Does the employee have an unobstructed view of handling the task? ❑ yes ❑ no

17. Is there a preventive maintenance program for equipment? ❑ yes ❑ no

18. Are workers trained in correct handling and lifting procedures? ❑ yes ❑ no

Tray 5–G. Computer Workstation Checklist

"No" responses indicate potential problem areas which should receive further investigation.

1. Does the workstation ensure proper worker posture, such as

horizontal thighs?	☐ yes	☐ no
vertical lower legs?	☐ yes	☐ no
feet flat on floor or footrest?	☐ yes	☐ no
neutral wrists?	☐ yes	☐ no

2. Does the chair

adjust easily?	☐ yes	☐ no
have a padded seat with a rounded front?	☐ yes	☐ no
have an adjustable backrest?	☐ yes	☐ no
provide lumbar support?	☐ yes	☐ no
have casters?	☐ yes	☐ no

3. Are the height and tilt of the work surface on which the keyboard is located adjustable? ☐ yes ☐ no

4. Is the keyboard detachable? ☐ yes ☐ no

5. Do keying actions require minimal force? ☐ yes ☐ no

6. Is there an adjustable document holder? ☐ yes ☐ no

7. Are arm rests provided where needed? ☐ yes ☐ no

8. Are glare and reflections avoided? ☐ yes ☐ no

9. Does the monitor have brightness and contrast controls? ☐ yes ☐ no

10. Do the operators judge the distance between eyes and work to be satisfactory for their viewing needs? ☐ yes ☐ no

11. Is there sufficient space for knees and feet? ☐ yes ☐ no

12. Can the workstation be used for either right- or left-handed activity? ☐ yes ☐ no

13. Are adequate rest breaks provided for task demands? ☐ yes ☐ no

14. Are high stroke rates avoided by

job rotation?	☐ yes	☐ no
self-pacing?	☐ yes	☐ no
adjusting the job to the skill of the worker?	☐ yes	☐ no

15. Are employees trained in

proper postures?	☐ yes	☐ no
proper work methods?	☐ yes	☐ no
when and how to adjust their workstations?	☐ yes	☐ no
how to seek assistance for their concerns?	☐ yes	☐ no

Tray 5–H. Protocol for Videotaping Jobs for Risk Factors

The following is a guide to preparing a videotape and related task information for facilitating job analyses and assessments of risk factors for work-related musculoskeletal disorders.

Materials needed:

> Video camera and blank tapes
> Spare batteries (at least 2) and battery charger
> Clipboard, pens, paper, blank checklists
> Stopwatch, strain gauge (optional) for weighing objects

Videotaping Procedures:

1. To verify the accuracy of the video camera to record in real time, videotape a worker or job with a stopwatch running in the field of view for at least 1 min. The play-back of the tape should correspond to the lapsed time on the stopwatch.
2. Announce the name of the job on the voice channel of the video camera before the taping of any job. Restrict running time comments to the facts. Make no editorial comments.
3. Tape each job long enough to observe all aspects of the task. Tape 5 to 10 min for all jobs, including at least 10 complete cycles. Fewer cycles may be needed if all aspects of the job are recorded at least 3 to 4 times.
4. Hold the camera still, using a tripod if available. Don't walk unless absolutely necessary.
5. Begin taping each task with a whole-body shot of the worker. Include the seat/chair and the surface the worker is standing on. Hold this for 2 to 3 cycles, then zoom in on the hands/arms or other body parts which may be under stress due to the job task.
6. It is best to tape several workers to determine if workers of varying body size adopt different postures or are affected in other ways. If possible, try to tape the best and worst case situations in terms of worker "fit" to the job.
   The following suspected upper body problems suggest focusing on the parts indicated:

   > — wrist problems/complaints . . . . hands/wrists/forearms
   > — elbow problems/complaints . . . . arms/elbows
   > — shoulder problems/complaints . . . . . arms/shoulders

   For back and lower limb problems, the focus would be on movements of the trunk of the body and leg, knee, and foot areas under stress due to task loads or other requirements.
7. Video from whatever angles are needed to capture the body part(s) under stress.
8. Briefly tape the jobs performed before and after the one under actual study to see how the targeted job fits into the total department process.
9. For each taped task, obtain the following information to the maximum extent possible:

   > — if the task is continuous or sporadic
   > — if the worker performs the work for the entire shift, or if there is rotation with other workers
   > — measures of work surface heights and chair heights and whether adjustable
   > — weight, size and shape of handles and textures for tools in use; indications of vibration in power tool usage
   > — use of handwear
   > — weight of objects lifted, pushed, pulled, or carried
   > — nature of environment in which work is performed—(too cold or too hot?)

Tray 9–A. General Workstation Design Principles*

1. Make the workstation adjustable, enabling both large and small persons to fit comfortably and reach materials easily.

2. Locate all materials and tools in front of the worker to reduce twisting motions. Provide sufficient work space for the whole body to turn.

3. Avoid static loads, fixed work postures, and job requirements in which operators must frequently or for long periods

    — lean to the front or the side,
    — hold a limb in a bent or extended position,
    — tilt the head forward more than 15 degrees, or
    — support the body's weight with one leg.

4. Set the work surface above elbow height for tasks involving fine visual details and below elbow height for tasks requiring downward forces and heavy physical effort.

5. Provide adjustable, properly designed chairs with the following features

    — adjustable seat height,
    — adjustable up and down back rest, including a lumbar (lower-back) support,
    — padding that will not compress more than an inch under the weight of a seated individual, and a
    — chair that is stable to floor at all times (5-leg base).

6. Allow the workers, at their discretion, to alternate between sitting and standing. Provide floor mats or padded surfaces for prolonged standing.

7. Support the limbs: provide elbow, wrist, arm, foot, and back rests as needed and feasible.

8. Use gravity to move materials.

9. Design the workstation so that arm movements are continuous and curved. Avoid straight-line, jerking arm motions.

10. Design so arm movements pivot about the elbow rather than around the shoulder to avoid stress on shoulder, neck, and upper back.

11. Design the primary work area so that arm movements or extensions of more than 15 in. are minimized.

12. Provide dials and displays that are simple, logical, and easy to read, reach. and operate.

13. Eliminate or minimize the effects of undesirable environmental conditions such as excessive noise, heat, humidity, cold, and poor illumination.

*Adapted from design checklists developed by Dave Ridyard. CPE, CIH, CSP. Applied Ergonomics Technology. 270 Mather Road, Jenkintown. PA 19046–3129.

Tray 9–B. Design Principles for Repetitive Hand and Wrist Tasks·

1. Reduce the number of repetitions per shift. Where possible, substitute full or semi-automated systems.

2. Maintain neutral (handshake) wrist positions:

   - Design jobs and select tools to reduce extreme flexion or deviation of the wrist.
   - Avoid inward and outward rotation of the forearm when the wrist is bent to minimize elbow disorders (i.e., tennis elbow).

3. Reduce the force or pressure on the wrists and hands:

   - Wherever possible, reduce the weight and size of objects that must be handled repeatedly.
   - Avoid tools that create pressure on the base of the palm which can obstruct blood flow and nerve function.
   - Avoid repeated pounding with the base of the palm.
   - Avoid repetitive, forceful pressing with the finger tips.

4. Design tasks so that a power rather than a finger pinch grip can be used to grasp materials. Note that a pinch grip is five times more stressful than a power grip.

5. Avoid reaching more than 15 in. in front of the body for materials:

   - Avoid reaching above shoulder height, below waist level, or behind the body to minimize shoulder disorders.
   - Avoid repetitive work that requires full arm extension (i.e., the elbow held straight and the arm extended).

6. Provide support devices where awkward body postures (elevated hands or elbows and extended arms) must be maintained. Use fixtures to relieve stressful hand/arm positions.

7. Select power tools and equipment with features designed to control or limit vibration transmissions to the hands, or alternatively design work methods to reduce time or need to hold vibrating tools.

8. Provide for protection of the hands if working in a cold environment. Furnish a selection of glove sizes and sensitize users to problems of forceful overgripping when worn.

9. Select and use properly designed hand tools (e.g., grip size of tool handles should accommodate majority of workers).

Adapted from design checklists developed by Dave Ridyard. CPE, CIH. CSP. Applied Ergonomics Technology, 270 Mather Road, Jenkintown, PA 19046–3129.

Tray 9–C. Handtool Use and Selection Principles[*]

1. Maintain straight wrists. Avoid bending or rotating the wrists. Remember, bend the tool, not the wrist. A variety of bent-handle tools are commercially available.

2. Avoid static muscle loading. Reduce both the weight and size of the tool. Do not raise or extend elbows when working with heavy tools. Provide counter-balanced support devices for larger, heaver tools.

3. Avoid stress on soft tissues. Stress concentrations result from poorly designed tools that exert pressure on the palms or fingers. Examples include short-handled pliers and tools with finger grooves that do not fit the worker's hand.

4. Reduce grip force requirements. The greater the effort to maintain control of a handtool, the higher the potential for injury. A compressible gripping surface rather than hard plastic may alleviate this problem.

5. Whenever possible, select tools that use a full-hand power grip rather than a precision finger grip.

6. Maintain optimal grip span. Optimum grip spans for pliers, scissors, or tongs, measured from the fingers to the base of the thumb, range from 6 to 9 cm. The recommended handle diameters for circular-handle tools such as screwdrivers are 3 to 5 cm when a power grip is required, and 0.75 to 1.5 cm when a precision finger grip is needed.

7. Avoid sharp edges and pinch points. Select tools that will not cut or pinch the hands even when gloves are not worn.

8. Avoid repetitive trigger-finger actions. Select tools with large switches that can be operated with all four fingers. Proximity switches are the most desirable triggering mechanism.

9. Isolate hands from heat, cold, and vibration. Heat and cold can cause loss of manual dexterity and increased grip strength requirements. Excessive vibration can cause reduced blood circulation in the hands causing a painful condition known as white-finger syndrome.

10. Wear gloves that fit. Gloves reduce both strength and dexterity. Tight-fitting gloves can put pressure on the hands, while loose-fitting gloves reduce grip strength and pose other safety hazards (e.g., snagging).

[*]Adapted from design checklists developed by Dave Ridyard, CPE, CIH, CSP. Applied Ergonomics Technology, 270 Mather Road, Jenkintown, PA 19046–3129.

Tray 9–D. Design Principles for Lifting and Lowering Tasks*

1. Optimize material flow through the workplace by

— reducing manual lifting of materials to a minimum,
— establishing adequate receiving, storage, and shipping facilities, and
— maintaining adequate clearances in aisle and access areas.

2. Eliminate the need to lift or lower manually by

— increasing the weight to a point where it must be mechanically handled,
— palletizing handling of raw materials and products, and
— using unit load concept (bulk handling in large bins or containers).

3. Reduce the weight of the object by

— reducing the weight and capacity of the container,
— reducing the load in the container, and
— limiting the quantity per container to suppliers.

4. Reduce the hand distance from the body by

— changing the shape of the object or container so that it can be held closer to the body, and
— providing grips or handles for enabling the load to be held closer to the body.

5. Convert load lifting, carrying, and lowering movements to a push or pull by providing

— conveyors,
— ball caster tables,
— hand trucks, and
— four-wheel carts.

Adapted from design checklists developed by Dave Ridyard. CPE, CIH, CSP. Applied Ergonomics Technology, 270 Mather Road, Jenkintown. PA 19046–3129.

Tray 9–E. Design Principles for Pushing and Pulling Tasks

1. Eliminate the need to push or pull by using the following mechanical aids, when applicable:

    - Conveyors (powered and non-powered)
    - Powered trucks
    - Lift tables
    - Slides or chutes

2. Reduce the force required to push or pull by

    — reducing side and/or weight of load;
    — using four-wheel trucks or dollies;
    — using non-powered conveyors;
    — requiring that wheels and casters on hand-trucks or dollies have (1) periodic lubrication of bearings, (2) adequate maintenance, and (3) proper sizing (provide larger diameter wheels and casters);
    — maintaining the floors to eliminate holes and bumps; and
    — requiring surface treatment of floors to reduce friction.

3. Reduce the distance of the push or pull by

    — moving receiving, storage, production, or shipping areas closer to work production areas, and
    — improving the production process to eliminate unnecessary materials handling steps.

4. Optimize the technique of the push or pull by

    — providing variable-height handles so that both short and tall employees can maintain an elbow bend of 80 to 100 degrees,
    — replacing a pull with a push whenever possible, and
    — using ramps with a slope of less than 10%.

*Adapted from design checklists developed by Dave Ridyard, CPE, CIH, CSP. Applied Ergonomics Technology, 270 Mather Road, Jenkintown, PA 19046–3129.

Tray 9–F. Design Principles for Carrying Tasks[*]

1. Eliminate the need to carry by rearranging the workplace to eliminate unnecessary materials movement and using the following mechanical handling aids, when applicable:

    - Conveyors (all kinds)
    - Lift trucks and hand trucks
    - Tables or slides between workstations
    - Four-wheel carts or dollies
    - Air or gravity press ejection systems

2. Reduce the weight that is carried by

    — reducing the weight of the object,
    — reducing the weight of the container,
    — reducing the load in the container, and
    — reducing the quantity per container to suppliers.

3. Reduce the bulk of the materials that are carried by

    — reducing the size or shape of the object or container,
    — providing handles or hand-grips that allow materials to be held close to the body, and
    — assigning the job to two or more persons.

4. Reduce the carrying distance by

    — moving receiving, storage, or shipping areas closer to production areas, and
    — using powered and nonpowered conveyors.

5. Convert carry to push or pull by

    — using nonpowered conveyors, and
    — using hand trucks and push carts.

[*]Adapted from design checklists developed by Dave Ridyard, CPE, CIH, CSP. Applied Ergonomics Technology, 270 Mather Road, Jenkintown, PA 19046–3129.

OCCUPATIONAL SAFETY AND HEALTH ACT OF 1970

OSH Act of 1970, with revisions

INTRODUCTION

Contains the complete text of the Occupational Safety and Health Act of 1970. This Act was issued as Public Law 91-596, 91st Congress, S. 2193, approved on December 29, 1970, and amended by Public Law 101-552, section 3101, November 5, 1990. This file also contains changes to law based on a review by the Office of the Solicitor. This file is maintained by USDOL, OSHA, Directorate of Technical Support, Salt Lake Technical Center, Salt Lake City, Utah.

1 – **Introduction**
    Section Number 1

Section Title   Introduction
        Public Law 91 - 596
        91st Congress, S. 2193
        December 29, 1970
        As amended by Public Law 101-552,
        Section 3101, November 5, 1990

AN ACT

To assure safe and healthful working conditions for working men and women; by authorizing enforcement of the standards developed under the Act; by assisting and encouraging the State in their efforts to assure safe and healthful working conditions; by providing for research, information, education, and training in the field of occupational safety and health; and for other purposes.

Be it enacted by the Senate and House of Representatives of the United States of America in Congress assembled, that this Act may be cited as the "Occupational Safety and Health Act of 1970."

2 — **Congressional Findings and Purpose**
    Section Number 2

Section Title   Congressional Findings and Purpose

(a) The Congress finds that personal injuries and illnesses arising out of work situations impose a substantial burden upon, and are a hindrance to, interstate commerce in terms of lost production, wage loss, medical expenses, and disability compensation payments.

(b) The Congress declares it to be its purpose and policy, through the exercise of its powers to regulate commerce among the several States and with foreign nations and to provide for the general welfare, to assure so far as possible every working man and woman in the Nation safe and healthful working conditions and to preserve our human resources –

(1) by encouraging employers and employees in their efforts to reduce the number of occupational safety and health hazards at their places of employment, and to stimulate employers and employees to institute new and to perfect existing programs for providing safe and healthful working conditions;

(2) by providing that employers and employees have separate but dependent responsibilities and rights with respect to achieving safe and healthful working conditions;

(3) by authorizing the Secretary of Labor to set mandatory occupational safety and health standards applicable to businesses affecting interstate commerce, and by creating an Occupational Safety and Health Review Commission for carrying out adjudicatory functions under the Act;

(4) by building upon advances already made through employer and employee initiative for providing safe and healthful working conditions;

(5) by providing for research in the field of occupational safety and health, including the psychological factors involved, and by developing innovative methods, techniques, and approaches for dealing with occupational safety and health problems;

(6) by exploring ways to discover latent diseases, establishing causal connections between diseases and work in environmental conditions, and conducting other research relating to health problems, in recognition of the fact that occupational health standards present problems often different from those involved in occupational safety;

(7) by providing medical criteria which will assure insofar as practicable that no employee will suffer diminished health, functional capacity, or life expectancy as a result of his work experience;

(8) by providing for training programs to increase the number and competence of personnel engaged in the field of occupational safety and health;

(9) by providing for the development and promulgation of occupational safety and health standards;

(10) by providing an effective enforcement program which shall include a prohibition against giving advance notice of any inspection and sanctions for any individual violating this prohibition;

(11) by encouraging the States to assume the fullest responsibility for the administration and enforcement of their occupational safety and health laws by providing grants to the States to assist in identifying their needs and responsibilities in the area of occupational safety and health, to develop plans in accordance with the provisions of this Act, to improve the administration and enforcement of State occupational safety and health laws, and to conduct experimental and demonstration projects in connection therewith;

(12) by providing for appropriate reporting procedures with respect to occupational safety and health which procedures will help achieve the objectives of this Act and accurately describe the nature of the occupational safety and health problem;

(13) by encouraging joint labor management efforts to reduce injuries and disease arising out of employment.

3 — Definitions
Section Number 3

Section Title   Definitions

For the purposes of this Act –

(1) The term "Secretary" mean the Secretary of Labor.

(2) The term "Commission" means the Occupational Safety and Health Review Commission established under this Act.

(3) The term "commerce" means trade, traffic, commerce, transportation, or communication among the several States, or between a State and any place outside thereof, or within the District of Columbia, or a possession of the United States (other than the Trust Territory of the Pacific Islands), or between points in the same State but through a point outside thereof.

(4) The term "person" means one or more individuals, partnerships, associations, corporations, business trusts, legal representatives, or any organized group of persons.

(5) The term "employer" means a person engaged in a business affecting commerce who has employees, but does not include the United States or any State or political subdivision of a State.

(6) The term "employee" means an employee of an employer who is employed in a business of his employer which affects commerce.

(7) The term "State" includes a State of the United States, the District of Columbia, Puerto Rico, the Virgin Islands, American Samoa, Guam, and the Trust Territory of the Pacific Islands.

(8) The term "occupational safety and health standard" means a standard which requires conditions, or the adoption or use of one or more practices, means, methods, operations, or processes, reasonably necessary or appropriate to provide safe or healthful employment and places of employment.

(9) The term "national consensus standard" means any occupational safety and health standard or modification thereof which (1), has been adopted and promulgated by a nationally recognized standards-producing organization under procedures whereby it can be determined by the Secretary that persons interested and affected by the scope or provisions of the standard have reached substantial agreement on its adoption, (2) was formulated in a manner which afforded an opportunity for diverse views to be considered and (3) has been designated as such a standard by the Secretary, after consultation with other appropriate Federal agencies.

(10) The term "established Federal standard" means any operative occupational safety and health standard established by any agency of the United States and presently in effect, or contained in any Act of Congress in force on the date of enactment of this Act.

(11) The term "Committee" means the National Advisory Committee on Occupational Safety and Health established under this Act.

(12) The term "Director" means the Director of the National Institute for Occupational Safety and Health.

(13) The term "Institute" means the National Institute for Occupational Safety and Health established under this Act.

(14) The term "Workmen's Compensation Commission" means the National Commission on State Workmen's Compensation Laws established under this Act.

4 — Applicability of This Act
Section Number   4

Section Title   Applicability of This Act

(a) This Act shall apply with respect to employment performed in a workplace in a State, the District of Columbia, the Commonwealth of Puerto Rico, the Virgin Islands, American Samoa,

Guam, the Trust Territory of the Pacific Islands, Wake Island, Outer Continental Shelf Lands defined in the Outer Continental Shelf Lands Act, Johnston Island, and the Canal Zone. The Secretary of the Interior shall, by regulation, provide for judicial enforcement of this Act by the courts established for areas in which there are no United States district courts having jurisdiction.

(b)(1) Nothing in this Act shall apply to working conditions of employees with respect to which other Federal agencies, and State agencies acting under section 274 of the Atomic Energy Act of 1954, as amended (42 U.S.C. 2021), exercise statutory authority to prescribe or enforce standards or regulations affecting occupational safety or health.

(2) The safety and health standards promulgated under the Act of June 30, 1936, commonly known as the Walsh-Healey Act (41 U.S.C. 35 et seq.), the Service Contract Act of 1965 (41 U.S.C. 351 et seq.), Public Law 91-54, Act of August 9, 1969 (40 U.S.C. 333), Public Law 85-742, Act of August 23, 1958 (33 U.S.C. 941), and the National Foundation on Arts and Humanities Act (20 U.S.C. 951 et seq.) are superseded on the effective date of corresponding standards, promulgated under this Act, which are determined by the Secretary to be more effective. Standards issued under the laws listed in this paragraph and in effect on or after the effective date of this Act shall be deemed to be occupational safety and health standards issued under this Act, as well as under such other Acts.

(3) The Secretary shall, within three years after the effective date of this Act, report to the Congress his recommendations for legislation to avoid unnecessary duplication and to achieve coordination between this Act and other Federal laws.

(4) Nothing in this Act shall be construed to supersede or in any manner affect any workmen's compensation law or to enlarge or diminish or affect in any other manner the common law or statutory rights, duties, or liabilities of employers and employees under any law with respect to injuries, diseases, or death of employees arising out of, or in the course of, employment.

5 — Duties
Section Number   5

Section Title   **Duties**

(a) Each employer –

(1) shall furnish to each of his employees employment and a place of employment which are free from recognized hazards that are causing or are likely to cause death or serious physical harm to his employees;

(2) shall comply with occupational safety and health standards promulgated under this Act.

(b) Each employee shall comply with occupational safety and health standards and all rules, regulations, and orders issued pursuant to this Act which are applicable to his own actions and conduct.

6 — Occupational Safety and Health Standards
Section Number   6

Section Title   **Occupational Safety and Health Standards**

(a) Without regard to chapter 5 of title 5, United States Code, or to the other subsections of this section, the Secretary shall, as soon as practicable during the period beginning with the effective date of this Act and ending two years after such date, by rule promulgate as an occupational safety or health standard any national consensus standard, and any established Federal standard, unless he determines that the promulgation of such a standard would not result in improved safety or health for specifically designated employees. In the event of conflict among any such standards, the Secretary shall promulgate the standard which assures the greatest protection of the safety or health of the affected employees.

(b) The Secretary may by rule promulgate, modify, or revoke any occupational safety or health standard in the following manner:

(1) Whenever the Secretary, upon the basis of information submitted to him in writing by an interested person, a representative of any organization of employers or employees, a nationally recognized standards-producing organization, the Secretary of Health, Education, and Welfare, the National Institute for Occupational Safety and Health, or a State or political subdivision, or on the basis of information developed by the Secretary or otherwise available to him, determines that a rule should be promulgated in order to serve the objectives of this Act, the Secretary may request the recommendations of an advisory committee appointed under section 7 of this Act. The Secretary shall provide such an advisory committee with any proposals of his own or of the Secretary of Health, Education, and Welfare, together with all pertinent factual information developed by the Secretary or the Secretary of Health, Education, and Welfare, or otherwise available, including the results of research, demonstrations, and experiments. An advisory committee shall submit to the Secretary its recommendations regarding the rule to be promulgated within ninety days from the date of its appointment or within such longer or shorter period as may be prescribed by the Secretary, but in no event for a period which is longer than two hundred and seventy days.

(2) The Secretary shall publish a proposed rule promulgating, modifying, or revoking an occupational safety or health standard in the Federal Register and shall afford interested persons a period of thirty days after publication to submit written data or comments. Where an advisory committee is appointed and the Secretary determines that a rule should be issued, he shall publish the proposed rule within sixty days after the submission of the advisory committee's recommendations or the expiration of the period prescribed by the Secretary for such submission.

(3) On or before the last day of the period provided for the submission of written data or comments under paragraph (2), any interested person may file with the Secretary written objections to the proposed rule, stating the grounds therefor and requesting a public hearing on such objections. Within thirty days after the last day for filing such objections, the Secretary shall publish in the Federal Register a notice specifying the occupational safety or health standard to which objections have been filed and a hearing requested, and specifying a time and place for such hearing.

(4) Within sixty days after the expiration of the period provided for the submission of written data or comments under paragraph (2), or within sixty days after the completion of any hearing held under paragraph (3), the Secretary shall issue a rule promulgating, modifying, or revoking an occupational safety or health standard or make a determination that a rule should not be issued. Such a rule may contain a provision delaying its effective date for such period (not in excess of ninety days) as the Secretary determines may be necessary to insure that affected employers and employees will be informed of the existence of the standard and of its terms and that employers affected are given an opportunity to familiarize themselves and their employees with the existence of the requirements of the standard.

(5) The Secretary, in promulgating standards dealing with toxic materials or harmful physical agents under this subsection, shall set the standard which most adequately assures, to the extent feasible, on the basis of the best available evidence, that no employee will suffer material impairment of health or functional capacity even if such employee has regular exposure to the hazard dealt with by such standard for the period of his working life. Development of standards under this subsection shall be based upon research, demonstrations, experiments, and such other information as may be appropriate. In addition to the attainment of the highest degree of health and safety protection for the employee, other considerations shall be the latest available scientific data in the field, the feasibility of the standards, and experience gained under this and other health and safety laws. Whenever practicable, the standard promulgated shall be expressed in terms of objective criteria and of the performance desired.

(6)(A) Any employer may apply to the Secretary for a temporary order granting a variance from a standard or any provision thereof promulgated under this section. Such temporary order shall be granted only if the employer files an application which meets the requirements of clause (B) and establishes that (i) he is unable to comply with a standard by its effective date because of unavailability of professional or technical personnel or of materials and equipment needed to come into compliance with the standard or because necessary construction or alteration of facilities cannot be completed by the effective date, (ii) he is taking all available steps to safeguard his employees against the hazards covered by the standard, and (iii) he has an effective program for coming into compliance with the standard as quickly as practicable. Any temporary order issued under this paragraph shall prescribe the practices, means, methods, operations, and processes which the employer must adopt and use while the order is in effect and state in detail his program for coming into compliance with the standard. Such a temporary order may be granted only after notice to employees and an opportunity for a hearing: Provided, That the Secretary may issue one interim order to be effective until a decision is made on the basis of the hearing. No temporary order may be in effect for longer than the period needed by the employer to achieve compliance with the standard or one year, whichever is shorter, except that such an order may be renewed not more that twice (I) so long as the requirements of this paragraph are met and (II) if an application for renewal is filed at least 90 days prior to the expiration date of the order. No interim renewal of an order may remain in effect for longer than 180 days.

(B) An application for temporary order under this paragraph (6) shall contain:

    (i) a specification of the standard or portion thereof from which the employer seeks a variance, (ii) a representation by the employer, supported by representations from qualified persons having firsthand knowledge of the facts represented, that he is unable to comply with the standard or portion thereof and a detailed statement of the reasons therefor,

    (iii) a statement of the steps he has taken and will take (with specific dates) to protect employees against the hazard covered by the standard,

    (iv) a statement of when he expects to be able to comply with the standard and what steps he has taken and what steps he will take (with dates specified) to come into compliance with the standard, and

    (v) a certification that he has informed his employees of the application by giving a copy thereof to their authorized representative, posting a statement giving a summary of the application and specifying where a copy may be examined at the place or places where notices to employees are normally posted, and by other appropriate means.

A description of how employees have been informed shall be contained in the certification. The information to employees shall also inform them of their right to petition the Secretary for a hearing.

(C) The Secretary is authorized to grant a variance from any standard or portion thereof whenever he determines, or the Secretary of Health, Education, and Welfare certifies, that such variance is necessary to permit an employer to participate in an experiment approved by him or the Secretary of Health, Education, and Welfare designed to demonstrate or validate new and improved techniques to safeguard the health or safety of workers.

(7) Any standard promulgated under this subsection shall prescribe the use of labels or other appropriate forms of warning as are necessary to insure that employees are apprised of all hazards to which they are exposed, relevant symptoms and appropriate emergency treatment, and proper conditions and precautions of safe use or exposure. Where appropriate, such standard shall also prescribe suitable protective equipment and control or technological procedures to be used in connection with such hazards and shall provide for monitoring or measuring employee exposure at such locations and intervals, and in such manner as may be necessary for the protection of employees. In addition, where appropriate, any such standard shall prescribe the type and frequency of medical examinations or other tests which shall be made available, by the employer or at his cost, to employees exposed to such hazards in order to most effectively determine whether the health of such employees is adversely affected by such exposure. In the event such medical examinations are in the nature of research, as determined by the Secretary of Health, Education, and Welfare, such examinations may be furnished at the expense of the Secretary of Health, Education, and Welfare. The results of such examinations or tests shall be furnished only to the Secretary or the Secretary of Health, Education, and Welfare, and, at the request of the employee, to his physician. The Secretary, in consultation with the Secretary of Health, Education, and Welfare, may by rule promulgated pursuant to section 553 of title 5, United States Code, make appropriate modifications in the foregoing requirements relating to the use of labels or other forms of warning, monitoring or measuring, and medical examinations, as may be warranted by experience, information, or medical or technological developments acquired subsequent to the promulgation of the relevant standard.

(8) Whenever a rule promulgated by the Secretary differs substantially from an existing national consensus standard, the Secretary shall, at the same time, publish in the Federal Register a statement of the reasons why the rule as adopted will better effectuate the purposes of this Act than the national consensus standard.

(c)(1) The Secretary shall provide, without regard to the requirements of chapter 5, title 5, Unites States Code, for an emergency temporary standard to take immediate effect upon publication in the Federal Register if he determines (A) that employees are exposed to grave danger from exposure to substances or agents determined to be toxic or physically harmful or from new hazards, and (B) that such emergency standard is necessary to protect employees from such danger.

(2) Such standard shall be effective until superseded by a standard promulgated in accordance with the procedures prescribed in paragraph (3) of this subsection.

(3) Upon publication of such standard in the Federal Register the Secretary shall commence a proceeding in accordance with section 6(b) of this Act, and the standard as published shall also serve as a proposed rule for the proceeding. The Secretary shall promulgate a standard under this paragraph no later than six months after publication of the emergency standard as provided in paragraph (2) of this subsection.

(d) Any affected employer may apply to the Secretary for a rule or order for a variance from a standard promulgated under this section. Affected employees shall be given notice of each such application and an opportunity to participate in a hearing. The Secretary shall issue such rule or order if he determines on the record, after opportunity for an inspection where appropriate and a hearing, that the proponent of the variance has demonstrated by a preponderance of the evidence that the conditions, practices, means, methods, operations, or processes used or proposed to be used by an employer will provide employment and places of employment to his employees which are as safe and healthful as those which would prevail if he complied with the standard. The rule or order so issued shall prescribe the conditions the employer must maintain, and the practices, means, methods, operations, and processes which he must adopt and utilize to the extent they differ from the standard in question. Such a rule or order may be modified or revoked upon application by an employer, employees, or by the Secretary on his own motion, in the manner prescribed for its issuance under this subsection at any time after six months from its issuance.

(e) Whenever the Secretary promulgates any standard, makes any rule, order, or decision, grants any exemption or extension of time, or compromises, mitigates, or settles any penalty assessed under this Act, he shall include a statement of the reasons for such action, which shall be published in the Federal Register.

(f) Any person who may be adversely affected by a standard issued under this section may at any time prior to the sixtieth day after such standard is promulgated file a petition challenging the validity of such standard with the United States court of appeals for the circuit wherein such person resides or has his principal place of business, for a judicial review of such standard. A copy of the petition shall be forthwith transmitted by the clerk of the court to the Secretary. The filing of such petition shall not, unless otherwise ordered by the court, operate as a stay of the standard. The determinations of the Secretary shall be conclusive if supported by substantial evidence in the record considered as a whole.

(g) In determining the priority for establishing standards under this section, the Secretary shall give due regard to the urgency of the need for mandatory safety and health standards for particular industries, trades, crafts, occupations, businesses, workplaces or work environments. The Secretary shall also give due regard to the recommendations of the Secretary of Health, Education, and Welfare regarding the need for mandatory standards in determining the priority for establishing such standards.

7 – Advisory Committees; Administration
Section Number 7

Section Title   **Advisory Committees; Administration**

(a)(1) There is hereby established a National Advisory Committee on Occupational Safety and Health consisting of twelve members appointed by the Secretary, four of whom are to be designated by the Secretary of Health, Education, and Welfare, without regard to the provisions of title 5, United States Code, governing appointments in the competitive service, and composed of representatives of management, labor, occupational safety and occupational health professions, and of the public. The Secretary shall designate one of the public members as Chairman. The members shall be selected upon the basis of their experience and competence in the field of occupational safety and health.

(2) The Committee shall advise, consult with, and make recommendations to the Secretary and the Secretary of Health, Education, and Welfare on matters relating to the administration of the Act. The Committee shall hold no fewer than two meetings during each calendar year. All meetings of the Committee shall be open to the public and a transcript shall be kept and made available for public inspection.

(3) The members of the Committee shall be compensated in accordance with the provisions of section 3109 of title 5, United States Code.

(4) The Secretary shall furnish to the Committee an executive secretary and such secretarial, clerical, and other services as are deemed necessary to the conduct of its business.

(b) An advisory committee may be appointed by the Secretary to assist him in his standard setting functions under section 6 of this Act. Each such committee shall consist of not more than fifteen members and shall include as a member one of more designees of the Secretary of Health, Education, and Welfare, and shall include among its members an equal number of persons qualified by experience and affiliation to present the viewpoint of the employers involved, and of persons similarly qualified to present the viewpoint of the workers involved, as well as one or more representatives of health and safety agencies of the States. An advisory committee may also include such other persons as the Secretary may appoint who are qualified by knowledge and experience to make a useful contribution to the work of such committee, including one or more representatives of professional organizations of technicians or professionals specializing in occupational safety or health, and one or more representatives of nationally recognized standards-producing organizations, but the number of persons so appointed to any such advisory committee shall not exceed the number appointed to such committee as representatives of Federal and State agencies. Persons appointed to advisory committees from private life shall be compensated in the same manner as consultants or experts under section 3109 of title 5, United States Code. The Secretary shall pay to any State which is the employer of a member of such a committee who is a representative of the health or safety agency of that State, reimbursement sufficient to cover the actual cost to the State resulting from such representative's membership on such committee. Any meeting of such committee shall be open to the public and an accurate record shall be kept and made available to the public. No member of such committee (other than representatives of employers and employees) shall have an economic interest in any proposed rule.

(c) In carrying out his responsibilities under this Act, the Secretary is authorized to –

 (1) use, with the consent of any Federal agency, the services, facilities, and personnel of such agency, with or without reimbursement, and with the consent of any State or political subdivision thereof, accept and use the services, facilities, and personnel of any agency of such State of subdivision with reimbursement; and

 (2) employ experts and consultants or organizations thereof as authorized by section 3109 of title 5, United States Code, except that contracts for such employment may be renewed annually; compensate individuals so employed at rates not in excess of the rate specified at the time of service for grade GS-18 under section 5332 of title 5, United States Code, including traveltime, and allow them while away from their homes or regular places of business, travel expenses (including per diem in lieu of subsistence) as authorized by section 5703 of title 5, United States Code, for persons in the Government service employed intermittently, while so employed.

8 – Inspections, Investigations, and Recordkeeping
Section Number 8

Section Title   Inspections, Investigations, and Recordkeeping

(a) In order to carry out the purposes of this Act, the Secretary, upon presenting appropriate credentials to the owner, operator, or agent in charge, is authorized –

(1) to enter without delay and at reasonable times any factory, plant, establishment, construction site, or other area, workplace or environment where work is performed by an employee of an employer; and

(2) to inspect and investigate during regular working hours and at other reasonable times, and within reasonable limits and in a reasonable manner, any such place of employment and all

pertinent conditions, structures, machines, apparatus, devices, equipment, and materials therein, and to question privately any such employer, owner, operator, agent or employee.

(b) In making his inspections and investigations under this Act the Secretary may require the attendance and testimony of witnesses and the production of evidence under oath. Witnesses shall be paid the same fees and mileage that are paid witnesses in the courts of the United States. In case of a contumacy, failure, or refusal of any person to obey such an order, any district court of the United States or the United States courts of any territory orpossession, within the jurisdiction of which such person is found, orresides or transacts business, upon the application by the Secretary, shall have jurisdiction to issue to such person an order requiring such person to appear to produce evidence if, as, and when so ordered, and to give testimony relating to the matter under investigation or in question, and any failure to obey such order of the court may be punished by said court as a contempt thereof.

(c)(1) Each employer shall make, keep and preserve, and make available to the Secretary or the Secretary of Health, Education, and Welfare, such records regarding his activities relating to this Act as the Secretary, in cooperation with the Secretary of Health, Education, and Welfare,may prescribe by regulation as necessary or appropriate for the enforcement of this Act or for developing information regarding the causes and prevention of occupational accidents and illnesses. In order to carry out the provisions of this paragraph such regulations may include provisions requiring employers to conduct periodic inspections. The Secretary shall also issue regulations requiring that employers, through posting of notices or other appropriate means, keep their employees informed of their protections and obligations under this Act, including the provisions of applicable standards.

(2) The Secretary, in cooperation with the Secretary of Health, Education and Welfare, shall issue regulations requiring employers to maintain accurate records of, and to make periodic reports on, work-related deaths, injuries and illnesses other than minor injuries requiring only first aid treatment and which do not involve medical treatment, loss of consciousness, restriction of work or motion, or transfer to another job.

(3) The Secretary, in cooperation with the Secretary of Health, Education, and Welfare, shall issue regulations requiring employers to maintain accurate records of employee exposures to potentially toxic materials or harmful physical agents which are required to be monitored or measured under section 6. Such regulations shall provide employees or their representatives with an opportunity to observe such monitoring or measuring, and to have access to the records thereof. Such regulations shall also make appropriate provision for each employee or former employee to have access to such records as will indicate his own exposure to toxic materials or harmful physical agents. Each employer shall promptly notify any employee who has been or is being exposed to toxic materials or harmful physical agents in concentrations or at levels which exceed those prescribed by an applicable occupational safety and health standard promulgated under section 6, and shall inform any employee who is being thus exposed of the corrective action being taken.

(d) Any information obtained by the Secretary, the Secretary of Health, Education and Welfare, or a State agency under this Act shall be obtained with a minimum burden upon employers, especially those operating small businesses. Unnecessary duplication of efforts in obtaining information shall be reduced to the maximum extent feasible.

(e) Subject to regulations issued by the Secretary, a representative of the employer and a representative authorized by his employees shall be given an opportunity to accompany the Secretary or his authorized representative during the physical inspection of any workplace under subsection (a) for the purpose of aiding such inspection. Where there is no authorized employee representative, the Secretary or his authorized representative shall consult with a reasonable number of employees concerning matters of health and safety in the workplace.

(f)(1) Any employees or representative of employees who believe that a violation of a safety or health standard exists that threatens physical harm, or that an imminent danger exists, may request an inspection by giving notice to the Secretary or his authorized representative of such violation or danger. Any such notice shall be reduced to writing, shall set forth with reasonable particularity the grounds for the notice, and shall be signed by the employees or representative of employees, and a copy shall be provided the employer or his agent no later than at the time of inspection, except that, upon the request of the person giving such notice, his name and the names of individual employees referred to therein shall not appear in such copy or on any record published, released, or made available pursuant to subsection (g) of this section. If upon receipt of such notification the Secretary determines there are reasonable grounds to believe that such violation or danger exists, he shall make a special inspection in

accordance with the provisions of this section as soon as practicable, to determine if such violation or danger exists. If the Secretary determines there are no reasonable grounds to believe that a violation or danger exists he shall notify the employees or representative of the employees in writing of such determination.

(2) Prior to or during any inspection of a workplace, any employees or representative of employees employed in such workplace may notify the Secretary or any representative of the Secretary responsible for conducting the inspection, in writing, of any violation of this Act which they have reason to believe exists in such workplace. The Secretary shall, by regulation, establish procedures for informal review of any refusal by a representative of the Secretary to issue a citation with respect to any such alleged violation and shall furnish the employees or representative of employees requesting such review a written statement of the reasons for the Secretary's final disposition of the case.

(g)(1) The Secretary and Secretary of Health, Education, and Welfare are authorized to compile, analyze, and publish, either in summary or detailed form, all reports or information obtained under this section.

(2) The Secretary and the Secretary of Health, Education, and Welfare shall each prescribe such rules and regulations as he may deem necessary to carry out their responsibilities under this Act, including rules and regulations dealing with the inspection of an employer's establishment.

9 – Citations
Section Number   9

Section Title   Citations

(a) If, upon inspection or investigation, the Secretary or his authorized representative believes that an employer has violated a requirement of section 5 of this Act, of any standard, rule or order promulgated pursuant to section 6 of this Act, or of any regulations prescribed pursuant to this Act, he shall with reasonable promptness issue a citation to the employer. Each citation shall be in writing and shall describe with particularity the nature of the violation, including a reference to the provision of the Act, standard, rule, regulation, or order alleged to have been violated. In addition, the citation shall fix a reasonable time for the abatement of the violation. The Secretary may prescribe procedures for the issuance of a notice in lieu of a citation with respect to de minimis violations which have no direct or immediate relationship to safety or health.

(b) Each citation issued under this section, or a copy or copies thereof, shall be prominently posted, as prescribed in regulations issued by the Secretary, at or near each place a violation referred to in the citation occurred.

(c) No citation may be issued under this section after the expiration of six months following the occurrence of any violation.

10 – Procedure for Enforcement
Section Number 10

Section Title   **Procedure for Enforcement**

(a) If, after an inspection or investigation, the Secretary issues a citation under section 9(a), he shall, within a reasonable time after the termination of such inspection or investigation, notify the employer by certified mail of the penalty, if any, proposed to be assessed under section 17 and that the employer has fifteen working days within which to notify the Secretary that he wishes to contest the citation or proposed assessment of penalty. If, within fifteen working days from the receipt of the notice issued by the Secretary the employer fails to notify the Secretary that he intends to contest the citation or proposed assessment of penalty, and no notice is filed by any employees or representative of employees under subsection (c) within such time, the citation and the assessment, as proposed, shall be deemed a final order of the Commission and not subject to review by any court or agency.

(b) If the Secretary has reason to believe that an employer has failed to correct a violation for which a citation has been issued within the period permitted for its correction (which period shall not begin to run until the entry of a final order by the Commission in the case of any review proceedings under this section initiated by the employer in good faith and not solely for delay or avoidance of penalties), the Secretary shall notify the employer by certified mail of such failure and of the penalty proposed to be assessed under section 17 by reason of such failure, and that the employer has fifteen working days within which to notify the Secretary that he wishes to contest the Secretary's notification or the proposed assessment of penalty. If, within fifteen working days from the receipt of notification issued by the Secretary, the employer fails to notify the Secretary that he intends to contest the notification or proposed assessment of penalty, the notification and assessment, as proposed, shall be deemed a final order of the Commission and not subject to review by any court or agency.

(c) If an employer notifies the Secretary that he intends to contest a citation issued under section 9(a) or notification issued under subsection (a) or (b) of this section, or if, within fifteen working days of the issuance of a citation under section 9(a), any employee or representative of employees files a notice with the Secretary alleging that the period of time fixed in the citation for the abatement of the violation is unreasonable, the Secretary shall immediately advise the Commission of such notification, and the Commission shall afford an opportunity for a hearing (in accordance with section 554 of title 5, United States Code, but without regard to subsection (a)(3) of such section). The Commission shall thereafter issue an order, based on findings of fact, affirming, modifying, or vacating the Secretary's citation or proposed penalty, or directing other appropriate relief, and such order shall become final thirty days after its issuance. Upon a showing by an employer of a good faith effort to comply with the abatement requirements of a citation, and that abatement has not been completed because of factors beyond his reasonable control, the Secretary, after an opportunity for a hearing as provided in this subsection, shall issue an order affirming or modifying the abatement requirements in such citation. The rules of procedure prescribed by the Commission shall provide affected employees or representatives of affected employees an opportunity to participate as parties to hearings under this subsection.

11 – Judicial Review
    Section Number 11

Section Title   **Judicial Review**

(a) Any person adversely affected or aggrieved by an order of the Commission issued under subsection (c) of section 10 may obtain a review of such order in any United States court of appeals for the circuit in which the violation is alleged to have occurred or where the employer has its principal office, or in the Court of Appeals for the District of Columbia Circuit, by filing in such court within sixty days following the issuance of such order a written petition praying that the order be modified or set aside. A copy of such petition shall be forthwith transmitted by the clerk of the court to the Commission and to the other parties, and thereupon the Commission shall file in the court the record in the proceeding as provided in section 2112 of title 28, United States Code. Upon such filing, the court shall have jurisdiction of the proceeding and of the question determined therein, and shall have power to grant such temporary relief or restraining order as it deems just and proper, and to make and enter upon the pleadings, testimony, and proceedings set forth in such record a decree affirming, modifying, or setting aside in whole or in part, the order of the Commission and enforcing the same to the extent that such order is affirmed or modified. The commencement of proceedings under this subsection shall not, unless ordered by the court, operate as a stay of the order of the Commission. No objection that has not been urged before the Commission shall be considered by the court, unless the failure or neglect to urge such objection shall be excused because of extraordinary circumstances. The findings of the Commission with respect to questions of fact, if supported by substantial evidence on the record considered as a whole, shall be conclusive. If any party shall apply to the court for leave to adduce additional evidence and shall show to the satisfaction of the court that such additional evidence is material and that there were reasonable grounds for the failure to adduce such evidence in the hearing before the Commission, the court may order such additional evidence to be taken before the Commission and to be made a part of the record. The Commission may modify its findings as to the facts, or make new findings, by reason of additional evidence so taken and filed, and it shall file such modified or new findings, which findings with respect to questions of fact, if supported by substantial evidence on the record considered as a whole, shall be conclusive, and its recommendations, if any, for the modification or setting aside of its original order. Upon the filing of the record with it, the jurisdiction of the court shall be exclusive and its judgment and decree shall be final, except that the same shall be subject to review by the Supreme Court of the United States, as provided in section 1254 of title 28, United States Code.

(b) The Secretary may also obtain review or enforcement of any final order of the Commission by filing a petition for such relief in the United States court of appeals for the circuit in which the alleged violation occurred or in which the employer has its principal office, and the provisions of subsection (a) shall govern such proceedings to the extent applicable. If no petition for review, as provided in subsection (a), is filed within sixty days after service of the Commission's order, the Commission's findings of fact and order shall be conclusive in connection with any petition for enforcement which is filed by the Secretary after the expiration of such sixty-day period. In any such case, as well as in the case of a noncontested citation or notification by the Secretary which has become a final order of the Commission under subsection (a) or (b) of section 10, the clerk of the court, unless otherwise ordered by the court, shall forthwith enter a decree enforcing the order and shall transmit a copy of such decree to the Secretary and the employer named in the petition. In any contempt proceeding brought to enforce a decree of a court of appeals entered pursuant to this subsection or subsection (a), the court of appeals may assess the penalties provided in section 17, in addition to invoking any other available remedies.

(c)(1) No person shall discharge or in any manner discriminate against any employee because such employee has filed any complaint or instituted or caused to be instituted any proceeding under or related to this Act or has testified or is about to testify in any such proceeding or because of the exercise by such employee on behalf of himself or others of any right afforded by this Act.

(2) Any employee who believes that he has been discharged or otherwise discriminated against by any person in violation of this subsection may, within thirty days after such violation occurs, file a complaint with the Secretary alleging such discrimination. Upon receipt of such complaint, the Secretary shall cause such investigation to be made as he deems appropriate. If upon such investigation, the Secretary determines that the provisions of this subsection have been violated, he shall bring an action in any appropriate United States district court against such person. In any such action the United States district courts shall have jurisdiction, for cause shown to restrain violations of paragraph (1) of this subsection and order all appropriate relief including rehiring or reinstatement of the employee to his former position with back pay.

(3) Within 90 days of the receipt of a complaint filed under this subsection the Secretary shall notify the complainant of his determination under paragraph 2 of this subsection.

12 – The Occupational Safety and Health Review Commission
Section Number 12

Section Title   **The Occupational Safety and Health Review Commission**

(a) The Occupational Safety and Health Review Commission is hereby established. The Commission shall be composed of three members who shall be appointed by the President, by and with the advice and consent of the Senate, from among persons who by reason of training, education, or experience are qualified to carry out the functions of the Commission under this Act. The President shall designate one of the members of the Commission to serve as Chairman.

(b) The terms of members of the Commission shall be six years except that (1) the members of the Commission first taking office shall serve, as designated by the President at the time of appointment, one for a term of two years, one for a term of four years, and one for a term of six years, and (2) a vacancy caused by the death, resignation, or removal of a member prior to the expiration of the term for which he was appointed shall be filled only for the remainder of such unexpired term. A member of the Commission may be removed by the President for inefficiency, neglect of duty, or malfeasance in office.

(c)(1) Section 5314 of title 5, United States Code, is amended by adding at the end thereof the following new paragraph:

> *"(57) Chairman, Occupational Safety and Health Review Commission."*

(2) Section 5315 of title 5, United States Code, is amended by adding at the end thereof the following new paragraph:

> *"(94) Members, Occupational Safety and Health Review Commission."*

(d) The principal office of the Commission shall be in the District of Columbia. Whenever the Commission deems that the convenience of the public or of the parties may be promoted, or delay or expense may be minimized, it may hold hearings or conduct other proceedings at any other place.

(e) The Chairman shall be responsible on behalf of the Commission for the administrative operations of the Commission and shall appoint such administrative law judges and other employees as he deems necessary to assist in the performance of the Commission's functions and to fix their compensation in accordance with the provisions of chapter 51 and subchapter III of chapter 53 of title 5, United States Code, relating to classification and General Schedule pay rates: Provided, That assignment, removal and compensation of administrative law judges shall be in accordance with sections 3105, 3344, 5362, and 7521 of title 5, United States Code.

(f) For the purpose of carrying out its functions under this Act, two members of the Commission shall constitute a quorum and official action can be taken only on the affirmative vote of at least two members.

(g) Every official act of the Commission shall be entered of record, and its hearings and records shall be open to the public. The Commission is authorized to make such rules as are necessary for the orderly transaction of its proceedings. Unless the Commission has adopted a different rule, its proceedings shall be in accordance with the Federal Rules of Civil Procedure.

(h) The Commission may order testimony to be taken by deposition in any proceedings pending before it at any state of such proceeding. Any person may be compelled to appear and depose, and to produce books, papers, or documents, in the same manner as witnesses may be compelled to appear and testify and produce like documentary evidence before the Commission. Witnesses whose depositions are taken under this subsection, and the persons taking such depositions, shall be entitled to the same fees as are paid for like services in the courts of the United States.

(i) For the purpose of any proceeding before the Commission, the provisions of section 11 of the National Labor Relations Act (29 U.S.C. 161) are hereby made applicable to the jurisdiction and powers of the Commission.

(j) A administrative law judge appointed by the Commission shall hear, and make a determination upon, any proceeding instituted before the Commission and any motion in connection therewith, assigned to such administrative law judge by the Chairman of the Commission, and shall make a report of any such determination which constitutes his final disposition of the proceedings. The report of the administrative law judge shall become the final order of the Commission within thirty days after such report by the administrative law judge, unless within such period any Commission member has directed that such report shall be reviewed by the Commission.

(k) Except as otherwise provided in this Act, the administrative law judges shall be subject to the laws governing employees in the classified civil service, except that appointments shall be made without regard to section 5108 of title 5, United States Code. Each administrative law judge shall receive compensation at a rate not less than that prescribed for GS-16 under section 5332 of title 5, United States Code.

13 – Procedures to Counteract Imminent Dangers
Section Number 13

Section Title   **Procedures to Counteract Imminent Dangers**

(a) The United States district courts shall have jurisdiction, upon petition of the Secretary, to restrain any conditions or practices in any place of employment which are such that a danger exists which could reasonably be expected to cause death or serious physical harm immediately or before the imminence of such danger can be eliminated through the enforcement procedures otherwise provided by this Act. Any order issued under this section may require such steps to be taken as may be necessary to avoid, correct, or remove such imminent danger and prohibit the employment or presence of any individual in locations or under conditions where such imminent danger exists, except individuals whose presence is necessary to avoid, correct, or remove such imminent danger or to maintain the capacity of a continuous process operation to resume normal operations without a complete cessation of operations, or where a cessation of operations is necessary, to permit such to be accomplished in a safe and orderly manner.

(b) Upon the filing of any such petition the district court shall have jurisdiction to grant such injunctive relief or temporary restraining order pending the outcome of an enforcement proceeding pursuant to this Act. The proceeding shall be as provided by Rule 65 of the Federal Rules, Civil Procedure, except that no temporary restraining order issued without notice shall be effective for a period longer than five days.

(c) Whenever and as soon as an inspector concludes that conditions or practices described in subsection (a) exist in any place of employment, he shall inform the affected employees and employers of the danger and that he is recommending to the Secretary that relief be sought.

(d) If the Secretary arbitrarily or capriciously fails to seek relief under this section, any employee who may be injured by reason of such failure, or the representative of such employees, might bring an action against the Secretary in the United States district court for the district in which the imminent danger is alleged to exist or the employer has its principal office, or for the District of Columbia, for a writ of mandamus to compel the Secretary to seek such an order and for such further relief as may be appropriate.

14 – Representation in Civil Litigation
Section Number 14

Section Title   **Representation in Civil Litigation**

Except as provided in section 518(a) of title 28, United States Code, relating to litigation before the Supreme Court, the Solicitor of Labor may appear for and represent the Secretary in any civil litigation brought under this Act but all such litigation shall be subject to the direction and control of the Attorney General.

15 – Confidentiality of Trade Secrets
Section Number 15

Section Title   **Confidentiality of Trade Secrets**

All information reported to or otherwise obtained by the Secretary or his representative in connection with any inspection or proceeding under this Act which contains or which might reveal a trade secret referred to in section 1905 of title 18 of the United States Code shall be considered confidential for the purpose of that section, except that such information may be disclosed to other officers or employees concerned with carrying out this Act or when relevant in any proceeding under this Act. In any such proceeding the Secretary, the Commission, or the court shall issue such orders as may be appropriate to protect the confidentiality of trade secrets.

16 – Variations, Tolerances, and Exemptions
Section Number 16

Section Title   **Variations, Tolerances, and Exemptions**

The Secretary, on the record, after notice and opportunity for a hearing may provide such reasonable limitations and may make such rules and regulations allowing reasonable variations, tolerances, and exemptions to and from any or all provisions of this Act as he may find necessary and proper to avoid serious impairment of the national defense. Such action shall not be in effect for more than six months without notification to affected employees and an opportunity being afforded for a hearing.

17 – Penalties
Section Number 17

Section Title   Penalties

(a) Any employer who willfully or repeatedly violates the requirements of section 5 of this Act, any standard, rule, or order promulgated pursuant to section 6 of this Act, or regulations prescribed pursuant to this Act, may be assessed a civil penalty of not more than $70,000 for each violation, but not less than $5,000 for each willful violation.

(b) Any employer who has received a citation for a serious violation of the requirements of section 5 of this Act, of any standard, rule, or order promulgated pursuant to section 6 of this Act, or of any regulations prescribed pursuant to this Act, shall be assessed a civil penalty of up to $7,000 for each such violation.

(c) Any employer who has received a citation for a violation of the requirements of section 5 of this Act, of any standard, rule, or order promulgated pursuant to section 6 of this Act, or of regulations prescribed pursuant to this Act, and such violation is specifically determined not to be of a serious nature, may be assessed a civil penalty of up to $7,000 for each violation.

(d) Any employer who fails to correct a violation for which a citation has been issued under section 9(a) within the period permitted for its correction (which period shall not begin to run until the date of the final order of the Commission in the case of any review proceeding under section 10 initiated by the employer in good faith and not solely for delay or avoidance of penalties), may be assessed a civil penalty of not more than $7,000 for each day during which such failure or violation continues.

(e) Any employer who willfully violates any standard, rule, or order promulgated pursuant to section 6 of this Act, or of any regulations prescribed pursuant to this Act, and that violation caused death to any employee, shall, upon conviction, be punished by a fine of not more than $10,000 or by imprisonment for not more than six months, or by both; except that if the conviction is for a violation committed after a first conviction of such person, punishment shall be by a fine of not more than $20,000 or by imprisonment for not more than one year, or by both.

(f) Any person who gives advance notice of any inspection to be conducted under this Act, without authority from the Secretary or his designees, shall, upon conviction, be punished by a fine of not more than $1,000 or by imprisonment for not more than six months, or by both.

(g) Whoever knowingly makes any false statement, representation, or certification in any application, record, report, plan, or other document filed or required to be maintained pursuant to this Act shall, upon conviction, be punished by a fine of not more than $10,000, or by imprisonment for not more than six months, or by both.

(h)(1) Section 1114 of title 18, United States Code, is hereby amended by striking out "designated by the Secretary of Health, Education, and Welfare to conduct investigations, or inspections under the Federal Food, Drug, and Cosmetic Act" and inserting in lieu thereof "or of the Department of Labor assigned to perform investigative, inspection, or law enforcement functions".

(2) Notwithstanding the provisions of sections 1111 and 1114 of title 18, United States Code, whoever, in violation of the provisions of section 1114 of such title, kills a person while engaged in or on account of the performance of investigative, inspection, or law enforcement functions added to such section 1114 by paragraph (1) of this subsection, and who would otherwise be subject to the penalty provisions of such section 1111, shall be punished by imprisonment for any term of years or for life.

(i) Any employer who violates any of the posting requirements, as prescribed under the provisions of this Act, shall be assessed a civil penalty of up to $7,000 for each violation.

(j) The Commission shall have authority to assess all civil penalties provided in this section, giving due consideration to the appropriateness of the penalty with respect to the size of the business of the employer being charged, the gravity of the violation, the good faith of the employer, and the history of previous violations.

(k) For purposes of this section, a serious violation shall be deemed to exist in a place of employment if there is a substantial probability that death or serious physical harm could result from a condition which exists, or from one or more practices, means, methods, operations, or processes which have been adopted or are in use, in such place of employment unless the employer did not, and could not with the exercise of reasonable diligence, know of the presence of the violation.

(l) Civil penalties owed under this Act shall be paid to the Secretary for deposit into the Treasury of the United States and shall accrue to the United States and may be recovered in a civil action in the name of the United States brought in the United States district court for the district where the violation is alleged to have occurred or where the employer has its principal office.

18 – State Jurisdiction and State Plans
Section Number 18

Section Title   State Jurisdiction and State Plans

(a) Nothing in this Act shall prevent any State agency or court from asserting jurisdiction under State law over any occupational safety or health issue with respect to which no standard is in effect under section 6.

(b) Any State which, at any time, desires to assume responsibility for development and enforcement therein of occupational safety and health standards relating to any occupational safety or health issue with respect to which a Federal standard has been promulgated under section 6 shall submit a State plan for the development of such standards and their enforcement.

(c) The Secretary shall approve the plan submitted by a State under subsection (b), or any modification thereof, if such plan in his judgement –

    (1) designates a State agency or agencies as the agency or agencies responsible for administering the plan throughout the State,

    (2) provides for the development and enforcement of safety and health standards relating to one or more safety or health issues, which standards (and the enforcement of which standards) are or will be at least as effective in providing safe and healthful employment and places of employment as the standards promulgated under section 6 which relate to the same issues, and which standards, when applicable to products which are distributed or used in interstate commerce, are required by compelling local conditions and do not unduly burden interstate commerce,

(3) provides for a right of entry and inspection of all workplaces subject to the Act which is at least as effective as that provided in section 8, and includes a prohibition on advance notice of inspections,

(4) contains satisfactory assurances that such agency or agencies have or will have the legal authority and qualified personnel necessary for the enforcement of such standards,

(5) gives satisfactory assurances that such State will devote adequate funds to the administration and enforcement of such standards,

(6) contains satisfactory assurances that such State will, to the extent permitted by its law, establish and maintain an effective and comprehensive occupational safety and health program applicable to all employees of public agencies of the State and its political subdivisions, which program is as effective as the standards contained in an approved plan,

(7) requires employers in the State to make reports to the Secretary in the same manner and to the same extent as if the plan were not in effect, and

(8) provides that the State agency will make such reports to the Secretary in such form and containing such information, as the Secretary shall from time to time require.

(d) If the Secretary rejects a plan submitted under subsection (b), he shall afford the State submitting the plan due notice and opportunity for a hearing before so doing.

(e) After the Secretary approves a State plan submitted under subsection (b), he may, but shall not be required to, exercise his authority under sections 8, 9, 10, 13, and 17 with respect to comparable standards promulgated under section 6, for the period specified in the next sentence. The Secretary may exercise the authority referred to above until he determines, on the basis of actual operations under the State plan, that the criteria set forth in subsection (c) are being applied, but he shall not make such determination for at least three years after the plan's approval under subsection (c). Upon making the determination referred to in the preceding sentence, the provisions of sections 5(a)(2), 8 (except for the purpose of carrying out subsection (f) of this section), 9, 10, 13, and 17, and standards promulgated under section 6 of this Act, shall not apply with respect to any occupational safety or health issues covered under the plan, but the Secretary may retain jurisdiction under the above provisions in any proceeding commenced under section 9 or 10 before the date of determination.

(f) The Secretary shall, on the basis of reports submitted by the State agency and his own inspections make a continuing evaluation of the manner in which each State having a plan approved under this section is carrying out such plan. Whenever the Secretary finds, after affording due notice and opportunity for a hearing, that in the administration of the State plan there is a failure to comply substantially with any provision of the State plan (or any assurance contained therein), he shall notify the State agency of his withdrawal of approval of such plan and upon receipt of such notice such plan shall cease to be in effect, but the State may retain jurisdiction in any case commenced before the withdrawal of the plan in order to enforce standards under the plan whenever the issues involved do not relate to the reasons for the withdrawal of the plan.

(g) The State may obtain a review of a decision of the Secretary withdrawing approval of or rejecting its plan by the United States court of appeals for the circuit in which the State is located by filing in such court within thirty days following receipt of notice of such decision a petition to modify or set aside in whole or in part the action of the Secretary. A copy of such petition shall forthwith be served upon the Secretary, and thereupon the Secretary shall certify and file in the court the record upon which the decision complained of was issued as provided in section 2112 of title 28, United States Code. Unless the court finds that the Secretary's decision in rejecting a proposed State plan or withdrawing his approval of such a plan is not supported by substantial evidence the court shall affirm the Secretary's decision. The judgment of the court shall be subject to review by the Supreme Court of the United States upon certiorari or certification as provided in section 1254 of title 28, United States Code.

(h) The Secretary may enter into an agreement with a State under which the State will be permitted to continue to enforce one or more occupational health and safety standards in effect in such State until final action is taken by the Secretary with respect to a plan submitted by a State under subsection (b) of this section, or two years from the date of enactment of this Act, whichever is earlier.

19 – Federal Agency Safety Programs and Responsibilities
Section Number 19

Section Title   **Federal Agency Safety Programs and Responsibilities**

(a) It shall be the responsibility of the head of each Federal agency to establish and maintain an effective and comprehensive occupational safety and health program which is consistent with the standards promulgated under section 6. The head of each agency shall (after consultation with representatives of the employees thereof) –

 (1) provide safe and healthful places and conditions of employment, consistent with the standards set under section 6;

 (2) acquire, maintain, and require the use of safety equipment, personal protective equipment, and devices reasonably necessary to protect employees;

 (3) keep adequate records of all occupational accidents and illnesses for proper evaluation and necessary corrective action;

 (4) consult with the Secretary with regard to the adequacy as to form and content of records kept pursuant to subsection (a)(3) of this section; and

 (5) make an annual report to the Secretary with respect to occupational accidents and injuries and the agency's program under this section. Such report shall include any report submitted under section 7902(e)(2) of title 5, United States Code.

(b) The Secretary shall report to the President a summary or digest of reports submitted to him under subsection (a)(5) of this section, together with his evaluations of and recommendations derived from such reports.

(c) Section 7902(c)(1) of title 5, United States Code, is amended by inserting after "agencies" the following: "and of labor organizations representing employees".

(d) The Secretary shall have access to records and reports kept and filed by Federal agencies pursuant to subsections (a)(3) and (5) of this section unless those records and reports are specifically required by Executive order to be kept secret in the interest of the national defense or foreign policy, in which case the Secretary shall have access to such information as will not jeopardize national defense or foreign policy.

20 – Research and Related Activities
Section Number 20

Section Title   **Research and Related Activities**

(a)(1) The Secretary of Health, Education, and Welfare, after consultation with the Secretary and with other appropriate Federal departments or agencies, shall conduct (directly or by grants or contracts) research, experiments, and demonstrations relating to occupational safety and health, including studies of psychological factors involved, and relating to innovative methods, techniques, and approaches for dealing with occupational safety and health problems.

(2) The Secretary of Health, Education, and Welfare shall from time to time consult with the Secretary in order to develop specific plans for such research, demonstrations, and experiments as are necessary to produce criteria, including criteria identifying toxic substances, enabling the Secretary to meet his responsibility for the formulation of safety and health standards under this Act; and the Secretary of Health, Education, and Welfare, on the basis of such research, demonstrations, and experiments and any other information available to him, shall develop and publish at least annually such criteria as will effectuate the purposes of this Act.

(3) The Secretary of Health, Education, and Welfare, on the basis of such research, demonstrations, and experiments, and any other information available to him, shall develop criteria dealing with toxic materials and harmful physical agents and substances which will describe exposure levels that are safe for various periods of employment, including but not limited to the exposure levels at which no employee will suffer impaired health or functional capacities or diminished life expectancy as a result of his work experience.

(4) The Secretary of Health, Education, and Welfare shall also conduct special research, experiments, and demonstrations relating to occupational safety and health as are necessary to explore new problems, including those created by new technology in occupational safety and health, which may require ameliorative action beyond that which is otherwise provided for in the operating provisions of this Act. The Secretary of Health, Education, and Welfare shall also conduct research into the motivational and behavioral factors relating to the field of occupational safety and health.

(5) The Secretary of Health, Education, and Welfare, in order to comply with his responsibilities under paragraph (2), and in order to develop needed information regarding potentially toxic substances or harmful physical agents, may prescribe regulations requiring employers to measure, record, and make reports on the exposure of employees to substances or physical agents which the Secretary of Health, Education, and Welfare reasonably believes may endanger the health or safety of employees. The Secretary of Health, Education, and Welfare also is authorized to establish such programs of medical examinations and tests as may be necessary for determining the incidence of occupational illnesses and the susceptibility of employees to such illnesses. Nothing in this or any other provision of this Act shall be deemed to authorize or require medical examination, immunization, or treatment for those who object thereto on religious grounds, except where such is necessary for the protection of the health or safety of others. Upon the request of any employer who is required to measure and record exposure of employees to substances or physical agents as provided under this subsection, the Secretary of Health, Education, and Welfare shall furnish full financial or other assistance to such employer for the purpose of defraying any additional expense incurred by him in carrying out the measuring and recording as provided in this subsection.

(6) The Secretary of Health, Education, and Welfare shall publish within six months of enactment of this Act and thereafter as needed but at least annually a list of all known toxic substances by generic family or other useful grouping, and the concentrations at which such toxicity is known to occur. He shall determine following a written request by any employer or authorized representative of employees, specifying with reasonable particularity the grounds on which the request is made, whether any substance normally found in the place of employment has potentially toxic effects in such concentrations as used or found; and shall submit such determination both to employers and affected employees as soon as possible. If the Secretary of Health, Education, and Welfare determines that any substance is potentially toxic at the concentrations in which it is used or found in a place of employment, and such substance is not covered by an occupational safety or health standard promulgated under section 6, the

Secretary of Health, Education, and Welfare shall immediately submit such determination to the Secretary, together with all pertinent criteria.

(7) Within two years of enactment of the Act, and annually thereafter the Secretary of Health, Education, and Welfare shall conduct and publish industrywide studies of the effect of chronic or low-level exposure to industrial materials, processes, and stresses on the potential for illness, disease, or loss of functional capacity in aging adults.

(b) The Secretary of Health, Education, and Welfare is authorized to make inspections and question employers and employees as provided in section 8 of this Act in order to carry out his functions and responsibilities under this section.

(c) The Secretary is authorized to enter into contracts, agreements, or other arrangements with appropriate public agencies or private organizations for the purpose of conducting studies relating to his responsibilities under this Act. In carrying out his responsibilities under this subsection, the Secretary shall cooperate with the Secretary of Health, Education, and Welfare in order to avoid any duplication of efforts under this section.

(d) Information obtained by the Secretary and the Secretary of Health, Education, and Welfare under this section shall be disseminated by the Secretary to employers and employees and organizations thereof.

(e) The functions of the Secretary of Health, Education, and Welfare under this Act shall, to the extent feasible, be delegated to the Director of the National Institute for Occupational Safety and Health established by section 22 of this Act.

21 – Training and Employee Education
Section Number 21

Section Title   **Training and Employee Education**

(a) The Secretary of Health, Education, and Welfare, after consultation with the Secretary and with other appropriate Federal departments and agencies, shall conduct, directly or by grants or contracts (1) education programs to provide an adequate supply of qualified personnel to carry out the purposes of this Act, and (2) informational programs on the importance of and proper use of adequate safety and health equipment.

(b) The Secretary is also authorized to conduct, directly or by grants or contracts, short-term training of personnel engaged in work related to his responsibilities under this Act.

(c) The Secretary, in consultation with the Secretary of Health, Education, and Welfare, shall (1) provide for the establishment and supervision of programs for the education and training of employers and employees in the recognition, avoidance, and prevention of unsafe or unhealthful working conditions in employments covered by this Act, and (2) consult with and advise employers and employees, and organizations representing employers and employees as to effective means of preventing occupational injuries and illnesses.

22 – National Institute for Occupational Safety and Health
Section Number 22

Section Title   **National Institute for Occupational Safety and Health**

(a) It is the purpose of this section to establish a National Institute for Occupational Safety and Health in the Department of Health and Human Services in order to carry out the policy set forth in section 2 of this Act and to perform the functions of the Secretary of Health, Education, and Welfare under sections 20 and 21 of this Act.

(b) There is hereby established in the Department of Health and Human Services, a National Institute for Occupational Safety and Health. The Institute shall be headed by a Director who shall be appointed by the Secretary of Health, Education, and Welfare, and who shall serve for a term of six years unless previously removed by the Secretary of Health, Education, and Welfare.

(c) The Institute is authorized to –

    (1) develop and establish recommended occupational safety and health standards; and

    (2) perform all functions of the Secretary of Health, Education, and Welfare under sections 20 and 21 of this Act.

(d) Upon his own initiative, or upon the request of the Secretary of Health, Education, and Welfare, the Director is authorized (1) to conduct such research and experimental programs as he determines are necessary for the development of criteria for new and improved occupational safety and health standards, and (2) after consideration of the results of such research and experimental programs make recommendations concerning new or improved occupational safety and health standards. Any occupational safety and health standard recommended pursuant to this section shall immediately be forwarded to the Secretary of Labor, and to the Secretary of Health, Education, and Welfare.

(e) In addition to any authority vested in the Institute by other provisions of this section, the Director, in carrying out the functions of the Institute, is authorized to –

    (1) prescribe such regulations as he deems necessary governing the manner in which its functions shall be carried out;

    (2) receive money and other property donated, bequeathed, or devised, without condition or restriction other than that it be used for the purposes of the Institute and to use, sell, or otherwise dispose of such property for the purpose of carrying out its functions;

    (3) receive (and use, sell, or otherwise dispose of, in accordance with paragraph (2)), money and other property donated, bequeathed, or devised to the Institute with a condition or restriction, including a condition that the Institute use other funds of the Institute for the purposes of the gift;

    (4) in accordance with the civil service laws, appoint and fix the compensation of such personnel as may be necessary to carry out the provisions of this section;

    (5) obtain the services of experts and consultants in accordance with the provisions of section 3109 of title 5, United States Code;

    (6) accept and utilize the services of voluntary and noncompensated personnel and reimburse them for travel expenses including per diem, as authorized by section 5703 of title 5, United States Code;

    (7) enter into contracts, grants or other arrangements, or modifications thereof to carry out the provisions of this section, and such contracts or modifications thereof may be entered into without performance or other bonds, and without regard to section 3709 of the Revised Statutes, as amended (41 U.S.C. 5), or any other provision of law relating to competitive bidding;

    (8) make advance, progress, and other payments which the Director deems necessary under this title without regard to the provisions of section 3648 of the Revised Statutes, as amended (31 U.S.C. 529); and

    (9) make other necessary expenditures.

(f) The Director shall submit to the Secretary of Health, Education, and Welfare, to the President, and to the Congress an annual report of the operations of the Institute under this Act, which shall include a detailed statement of all private and public funds received and expended by it, and such recommendations as he deems appropriate.

23 – Grants to the States
Section Number 23

Section Title   Grants to the States

(a) The Secretary is authorized, during the fiscal year ending June 30, 1971, and the two succeeding fiscal years, to make grants to the States which have designated a State agency under section 18 to assist them –

(1) in identifying their needs and responsibilities in the area of occupational safety and health,

(2) in developing State plans under section 18, or

(3) in developing plans for –

(A) establishing systems for the collection of information concerning the nature and frequency of occupational injuries and diseases;

(B) increasing the expertise and enforcement capabilities of their personnel engaged in occupational safety and health programs; or

(C) otherwise improving the administration and enforcement of State occupational safety and health laws, including standards thereunder, consistent with the objectives of this Act.

(b) The Secretary is authorized, during the fiscal year ending June 30, 1971, and the two succeeding fiscal years, to make grants to the States for experimental and demonstration projects consistent with the objectives set forth in subsection (a) of this section.

(c) The Governor of the State shall designate the appropriate State agency for receipt of any grant made by the Secretary under this section.

(d) Any State agency designated by the Governor of the State desiring a grant under this section shall submit an application therefore to the Secretary.

(e) The Secretary shall review the application, and shall, after consultation with the Secretary of Health, Education, and Welfare, approve or reject such application.

(f) The Federal share for each State grant under subsection (a) or (b) of this section may not exceed 90 percentum of the total cost of the application. In the event the Federal share for all States under either such subsection is not the same, the differences among the States shall be established on the basis of objective criteria.

(g) The Secretary is authorized to make grants to the States to assist them in administering and enforcing programs for occupational safety and health contained in State plans approved by the Secretary pursuant to section 18 of this Act. The Federal share for each State grant under this subsection may not exceed 50 percentum of the total cost to the State of such a program. The last sentence of subsection (f) shall be applicable in determining the Federal share under this subsection.

(h) Prior to June 30, 1973, the Secretary shall, after consultation with the Secretary of Health, Education, and Welfare, transmit a report to the President and to the Congress, describing the experience under the grant programs authorized by this section and making any recommendations he may deem appropriate.

24 – Statistics
Section Number 24

Section Title   **Statistics**

(a) In order to further the purposes of this Act, the Secretary, in consultation with the Secretary of Health, Education, and Welfare, shall develop and maintain an effective program of collection, compilation, and analysis of occupational safety and health statistics. Such program may cover all employments whether or not subject to any other provisions of this Act but shall not cover employments excluded by section 4 of the Act. The Secretary shall compile accurate statistics on work injuries and illnesses which shall include all disabling, serious, or significant injuries and illnesses, whether or not involving loss of time from work, other than minor injuries requiring only first aid treatment and which do not involve medical treatment, loss of consciousness, restriction of work or motion, or transfer to another job.

(b) To carry out his duties under subsection (a) of this section, the Secretary may –

(1) promote, encourage, or directly engage in programs of studies, information and communication concerning occupational safety and health statistics;

(2) make grants to States or political subdivisions thereof in order to assist them in developing and administering programs dealing with occupational safety and health statistics; and

(3) arrange, through grants or contracts, for the conduct of such research and investigations as give promise of furthering the objectives of this section.

(c) The Federal share for each grant under subsection (b) of this section may be up to 50 percentum of the State's total cost.

(d) The Secretary may, with the consent of any State or political subdivision thereof, accept and use the services, facilities, and employees of the agencies of such State or political subdivision, with or without reimbursement, in order to assist him in carrying out his functions under this section.

(e) On the basis of the records made and kept pursuant to section 8(c) of this Act, employers shall file such reports with the Secretary as he shall prescribe by regulation, as necessary to carry out his functions under this Act.

(f) Agreements between the Department of Labor and States pertaining to the collection of occupational safety and health statistics already in effect on the effective date of this Act shall remain in effect until superseded by grants or contracts made under this Act.

25 – Audits
Section Number25

Section Title   **Audits**

(a) Each recipient of a grant under this Act shall keep such records as the Secretary or the Secretary of Health, Education, and Welfare shall prescribe, including records which fully disclose the amount and disposition by such recipient of the proceeds of such grant, the total cost of the project or undertaking in connection with which such grant is made or used, and the amount of that portion of the cost of the project or undertaking supplied by other sources, and such other records as will facilitate an effective audit.

(b) The Secretary or the Secretary of Health, Education, and Welfare, and the Comptroller General of the United States, or any of their duly authorized representatives, shall have access for the purpose of audit and examination to any books, documents, papers, and records of the recipients of any grant under this Act that are pertinent to any such grant.

26 – Annual Report
Section Number 26

Section Title   **Annual Report**

Within one hundred and twenty days following the convening of each regular session of each Congress, the Secretary and the Secretary of Health, Education, and Welfare shall each prepare and submit to the President for transmittal to the Congress a report upon the subject matter of this Act, the progress toward achievement of the purpose of this Act, the needs and require-ments in the field of occupational safety and health, and any other relevant information. Such reports shall include information regarding occupational safety and health standards, and crite-ria for such standards, developed during the preceding year; evaluation of standards and crite-ria previously developed under this Act, defining areas of emphasis for new criteria and stan-dards; an evaluation of the degree of observance of applicable occupational safety and health standards, and a summary of inspection and enforcement activity undertaken; analysis and evaluation of research activities for which results have been obtained under governmental and nongovernmental sponsorship; an analysis of major occupational diseases; evaluation of avail-able control and measurement technology for hazards for which standards or criteria have been developed during the preceding year; description of cooperative efforts undertaken between Government agencies and other interested parties in the implementation of this Act during the preceding year; a progress report on the development of an adequate supply of trained man-power in the field of occupational safety and health, including estimates of future needs and the efforts being made by Government and others to meet those needs; listing of all toxic substances in industrial usage for which labeling requirements, criteria, or standards have not yet been established; and such recommendations for additional legislation as are deemed nec-essary to protect the safety and health of the worker and improve the administration of this Act.

27 – National Commission on State Workmen's Compensation Laws
Section Number 27

Section Title   **National Commission on State Workmen's Compensation Laws**
(a)(1) The Congress hereby finds and declares that –

(A) the vast majority of American workers, and their families, are dependent on workmen's compensation for their basic economic security in the event such workers suffer dis-abling injury or death in the course of their employment; and that the full protection of American workers from job-related injury or death requires an adequate, prompt, and equitable system of workmen's compensation as well as an effective program of occupational health and safety regulation; and

(B) in recent years serious questions have been raised concerning the fairness and ad-equacy of present workmen's compensation laws in the light of the growth of the economy, the changing nature of the labor force, increases in medical knowledge, changes in the hazards associated with various types of employment, new technology creating new risks to health and safety, and increases in the general level of wages and the cost of living.

(2) The purpose of this section is to authorize an effective study and objective evaluation of State workmen's compensation laws in order to determine if such laws provide an adequate, prompt, and equitable system of compensation for injury or death arising out of or in the course of employment.

(b) There is hereby established a National Commission on State Workmen's Compensation Laws.

(c)(1) The Workmen's Compensation Commission shall be composed of fifteen members to be appointed by the President from among members of State workmen's compensation boards, representatives of insurance carriers, business, labor, members of the medical profession having experience in industrial medicine or in workmen's compensation cases, educators having special expertise in the field of workmen's compensation, and representatives of the general public. The Secretary, the Secretary of Commerce, and the Secretary of Health, Education, and Welfare shall be ex officio members of the Workmen's Compensation Commission:

(2) Any vacancy in the Workmen's Compensation Commission shall not affect its powers.

(3) The President shall designate one of the members to serve as Chairman and one to serve as Vice Chairman of the Workmen's Compensation Commission.

(4) Eight members of the Workmen's Compensation Commission shall constitute a quorum.

(d)(1) The Workmen's Compensation Commission shall undertake a comprehensive study and evaluation of State workmen's compensation laws in order to determine if such laws provide an adequate, prompt, and equitable system of compensation. Such study and evaluation shall include, without being limited to, the following subjects: (A) the amount and duration of permanent and temporary disability benefits and the criteria for determining the maximum limitations thereon, (B) the amount and duration of medical benefits and provisions insuring adequate medical care and free choice of physician, (C) the extent of coverage of workers, including exemptions based on numbers or type of employment, (D) standards for determining which injuries or diseases should be deemed compensable, (E) rehabilitation, (F) coverage under second or subsequent injury funds, (G) time limits on filing claims, (H) waiting periods, (I) compulsory or elective coverage, (J) administration, (K) legal expenses, (L) the feasibility and desirability of a uniform system of reporting information concerning job-related injuries and diseases and the operation of workmen's compensation laws, (M) the resolution of conflict of laws, extra territoriality and similar problems arising from claims with multistate aspects, (N) the extent to which private insurance carriers are excluded from supplying workmen's compensation coverage and the desirability of such exclusionary practices, to the extent they are found to exist, (O) the relationship between workmen's compensation on the one hand, and old-age, disability, and survivors insurance and other types of insurance, public or private, on the other hand, (P) methods of implementing the recommendations of the Commission.

(2) The Workmen's Compensation Commission shall transmit to the President and to the Congress not later than July 31, 1972, a final report containing a detailed statement of the findings and conclusions of the Commission, together with such recommendations as it deems advisable.

(e)(1) The Workmen's Compensation Commission or, on the authorization of the Workmen's Compensation Commission, any subcommittee or members thereof, may, for the purpose of carrying out the provisions of this title, hold such hearings, take such testimony, and sit and act at such times and places as the Workmen's Compensation Commission deems advisable. Any member authorized by the Workmen's Compensation Commission may administer oaths or affirmations to witnesses appearing before the Workmen's Compensation Commission or any subcommittee or members thereof.

(2) Each department, agency, and instrumentality of the executive branch of the Government, including independent agencies, is authorized and directed to furnish to the Workmen's Com-

pensation Commission, upon request made by the Chairman or Vice Chairman, such information as the Workmen's Compensation Commission deems necessary to carry out its functions under this section.

(f) Subject to such rules and regulations as may be adopted by the Workmen's Compensation Commission, the Chairman shall have the power to –

> (1) appoint and fix the compensation of an executive director, and such additional staff personnel as he deems necessary, without regard to the provisions of title 5, United States Code, governing appointments in the competitive service, and without regard to the provisions of chapter 51 and subchapter III of chapter 53 of such title relating to classification and General Schedule pay rates, but at rates not in excess of the maximum rate for GS-18 of the General Schedule under section 5332 of such title, and

> (2) procure temporary and intermittent services to the same extent as is authorized by section 3109 of title 5, United States Code.

(g) The Workmen's Compensation Commission is authorized to enter into contracts with Federal or State agencies, private firms, institutions, and individuals for the conduct of research or surveys the preparation of reports, and other activities necessary to the discharge of its duties.

(h) Members of the Workmen's Compensation Commission shall receive compensation for each day they are engaged in the performance of their duties as members of the Workmen's Compensation Commission at the daily rate prescribed for GS-18 under section 5332 of title 5, United States Code, and shall be entitled to reimbursement for travel, subsistence, and other necessary expenses incurred by them in the performance of their duties as members of the Workmen's Compensation Commission.

(i) There are hereby authorized to be appropriated such sums as may be necessary to carry out the provisions of this section.

(j) On the ninetieth day after the date of submission of its final report to the President, the Workmen's Compensation Commission shall cease to exist.

28 – Economic Assistance to Small Businesses
Section Number 28

Section Title   **Economic Assistance to Small Businesses**

(a) Section 7(b) of the Small Business Act, as amended, is amended –

> (1) by striking out the period at the end of "paragraph (5)" and inserting in lieu thereof "; and"; and;

> (2) by adding after paragraph (5) a new paragraph as follows:

>> "(6) to make such loans (either directly or in cooperation with banks or other lending institutions through agreements to participate on an immediate or deferred basis) as the Administration may determine to be necessary or appropriate to assist any small business concern in effecting additions to or alterations in the equipment, facilities,or methods of operation of such business in order to comply with the applicable standards promulgated pursuant to section 6 of the Occupational Safety and Health Act of 1970 or standards adopted by a State pursuant to a plan approved under section 18 of the Occupational Safety and Health Act of 1970, if the Administration determines that such concern is likely to suffer substantial economic injury without assistance under this paragraph."

(b) The third sentence of section 7(b) of the Small Business Act, as amended, is amended by striking out "or (5)" after "paragraph (3)" and inserting a comma followed by "(5) or (6)".

(c) Section 4(c)(1) of the Small Business Act, as amended, is amended by inserting "7(b)(6)," after "7(b)(5),".

(d) Loans may also be made or guaranteed for the purposes set forth in section 7(b)(6) of the Small Business Act, as amended, pursuant to the provisions of section 202 of the Public Works and Economic Development Act of 1965, as amended.

29 – Additional Assistant Secretary of Labor
Section Number 29

Section Title   Additional Assistant Secretary of Labor

(a) Section 2 of the Act of April 17, 1946 (60 Stat. 91) as amended (29 U.S.C. 553) is amended by –

(1) striking out "four" in the first sentence of such section and inserting in lieu thereof "five"; and

(2) adding at the end thereof the following new sentence, "One of such Assistant Secretaries shall be an Assistant Secretary of Labor for Occupational Safety and Health.".

(b) Paragraph (20) of section 5315 of title 5, United States Code, is amended by striking out "(4)" and inserting in lieu thereof "(5)".

30 – Additional Positions
Section Number 30

Section Title   Additional Positions

Section 5108(c) of title 5, United States Code, is amended by –

(1) striking out the word "and" at the end of paragraph (8);

(2) striking out the period at the end of paragraph (9) and inserting in lieu thereof a semicolon and the word "and"; and

(3) by adding immediately after paragraph (9) the following new paragraph:

"(10)(A) the Secretary of Labor, subject to the standards and procedures prescribed by this chapter, may place an additional twenty-five positions in the Department of Labor in GS-16, 17, and 18 for the purposes of carrying out his responsibilities under the Occupational Safety and Health Act of 1970;

"(B)   the Occupational Safety and Health Review Commission, subject to the standards and procedures prescribed by this chapter, may place ten positions in GS-16, 17, and 18 in carrying out its functions under the Occupational Safety and Health Act of 1970."

31 – Emergency Locator Beacons
Section Number 31

Section Title   Emergency Locator Beacons

Section 601 of the Federal Aviation Act of 1958 is amended by inserting at the end thereof a new subsection as follows:

"EMERGENCY LOCATOR BEACONS"

"(d)(1) Except with respect to aircraft described in paragraph (2) of this subsection, minimum standards pursuant to this section shall include a requirement that emergency locator beacons shall be installed –

"(A) on any fixed-wing, powered aircraft for use in air commerce the manufacture of which is completed, or which is imported into the United States, after one year following the date of enactment of this subsection; and

"(B) on any fixed-wing, powered aircraft used in air commerce after three years following such date.

"(2) The provisions of this subsection shall not apply to jet-powered aircraft; aircraft used in air transportation (other than air taxis and charter aircraft); military aircraft; aircraft used solely for training purposes not involving flights more than twenty miles from its base; and aircraft used for the aerial application of chemicals."

32 – Separability
Section Number 32

Section Title   Separability

If any provision of this Act, or the application of such provision to any person or circumstance, shall be held invalid, the remainder of this Act, or the application of such provision to persons or circumstances other than those as to which it is held invalid, shall not be affected thereby.

33 – Appropriations
Section Number 33

Section Title   Appropriations

There are authorized to be appropriated to carry out this Act for each fiscal year such sums as the Congress shall deem necessary.

34 – Effective Date
Section Number   34

Section Title   Effective Date

This Act shall take effect one hundred and twenty days after the date of its enactment.

>   Approved December 29, 1970.

>   Amended November 5, 1990.

LEGISLATIVE HISTORY:

HOUSE REPORTS: No. 91-1291 accompanying H.R. 16785
>   (Comm. on Education and Labor) and
>   No. 91-1765 (Comm. of Conference).

SENATE REPORT: **No. 91-1282**   (Comm. on Labor and Public Welfare).

CONGRESSIONAL RECORD, Vol. 116 (1970):

Oct. 13, Nov. 16, 17, considered and passed Senate.

Nov. 23, 24, considered and passed House, amended, in lieu of H.R. 16785.

Dec. 16, Senate agreed to conference report.

Dec. 17, House agreed to conference report.

APPENDIX J

OSHA REGIONAL OFFICES AND OFFICES OF STATE-APPROVED PLANS

REGION 1

Regional Office
JFK Federal Building, Room E340
Boston, Massachusetts 02203
(617) 565-9860
(617) 565-9827 FAX

Area Offices
Connecticut | Massachusetts | Maine | New Hampshire | Rhode Island | Vermont

In case of emergency call 1-800-321-OSHA

REGION 2

Regional Office
201 Varick Street, Room 670
New York, New York 10014
(212) 337-2378
(212) 337-2371 FAX

Area Offices
New Jersey | New York | Puerto Rico | Virgin Islands

In case of emergency call 1-800-321-OSHA

REGION 3

Regional Office
Gateway Building, Suite 2100
3535 Market Street
Philadelphia, Pennsylvania 19104
(215) 596-1201
(215) 596-4872 FAX

Area Offices
District of Columbia | Delaware | Maryland | Pennsylvania | Virginia | West Virginia

In case of emergency call 1-800-321-OSHA

827

REGION 4

Regional Office
61 Forsyth Street, SW
Atlanta, Georgia 30303
(404) 562-2300
(404) 562-2295 FAX

Area Offices
Alabama | Florida | Georgia | Kentucky | Mississippi | North Carolina | South Carolina | Tennessee

In case of emergency call 1-800-321-OSHA

REGION 5

Regional Office
230 South Dearborn Street, Room 3244
Chicago, Illinois 60604
(312) 353-2220
(312) 353-7774 FAX

Area Offices
Illinois | Indiana | Michigan | Minnesota | Ohio | Wisconsin

In case of emergency call 1-800-321-OSHA

REGION 6

Regional Office
525 Griffin Street, Room 602
Dallas, Texas
(214) 767-4731
(214) 767-4137 FAX

Area Offices
Arkansas | Louisiana | New Mexico | Oklahoma | Texas

In case of emergency call 1-800-321-OSHA

REGION 7

Regional Office
City Center Square
1100 Main Street, Suite 800
Kansas City, Missouri 64105
(816) 426-5861
(816) 426-2750 FAX

Area Offices
Iowa | Kansas | Missouri | Nebraska

In case of emergency call 1-800-321-OSHA

REGION 8

Regional Office
1999 Broadway, Suite 1690
Denver, Colorado 80202-5716
(303) 844-1600
(303) 844-1616 FAX

Area Offices
Colorado | Montana | North Dakota | South Dakota | Utah | Wyoming

In case of emergency call 1-800-321-OSHA

REGION 9

Regional Office
71 Stevenson Street, Room 420
San Francisco, California 94105
(415) 975-4310
(415) 744-4319 FAX

Area Offices
Arizona | California | Guam | Hawaii | Nevada

In case of emergency call 1-800-321-OSHA

REGION 10

Regional Office
1111 Third Avenue, Suite 715
Seattle, Washington 98101-3212
(206) 553-5930
(206) 553-6499 FAX

Area Offices
Alaska | Idaho | Oregon | Washington

In case of emergency call 1-800-321-OSHA

States with Approved OSHA Plans

Updated August 14, 1997

Alaska Department of Labor
1111 W. 8th Street, Room 306
Juneau, Alaska 99801
Tom Cashen, Commissioner (907) 465-2700 Fax: (907) 465-2784
Alan W. Dwyer, Program Director (907) 465-4855 Fax: (907) 465-3584

Industrial Commission of Arizona
800 W. Washington
Phoenix, Arizona 85007
Larry Etchechury, Director (602) 542-5795 Fax: (602) 542-1614
Derek Mullins, Program Director (602) 542-5795 Fax: Same as above

California Department of Industrial Relations
45 Fremont Street
San Francisco, California 94105
John Duncan, Acting Director (415) 972-8835 Fax: (415) 972-8848
Dr. John Howard, Chief (415) 972-8500 Fax: (415) 972-8513

Connecticut Department of Labor
200 Folly Brook Boulevard
Wethersfield, Connecticut 06109
James P. Butler, Commissioner (860) 566-5123 Fax: (860) 566-1520
Program Director's Office (860) 566-4550 Fax: (860) 566-6916

Hawaii Department of Labor and Industrial Relations
830 Punchbowl Street
Honolulu, Hawaii 96813
Lorraine H. Akiba, Director (808) 586-8844 Fax: (808) 586-9099
Jennifer Shishido, Administrator (808) 586-9116 Fax: (808) 586-9104

Indiana Department of Labor
State Office Building
402 West Washington Street, Room W195
Indianapolis, Indiana 46204
Timothy Joyce, Commissioner (317) 232-2378 Fax: (317) 233-3790
John Jones, Deputy Commissioner (317) 232-3325 Fax: Same as above

Iowa Division of Labor Services
1000 E. Grand Avenue
Des Moines, Iowa 50319
Byron K. Orton, Commissioner (515) 281-3447 Fax: (515) 242-5144
Mary L. Bryant, Administrator (515) 281-3469 Fax: (515) 281-7995

Kentucky Labor Cabinet
1047 U.S. Highway 127 South, Suite 2
Frankfort, Kentucky 40601
Joe Norsworthy, Secretary (502) 564-3070 Fax: (502) 564-5387
Steven A. Forbes, Federal\State Coordinator (502) 564-2300 Fax: (502) 564-1682

Maryland Division of Labor and Industry
Department of Licensing and Regulation
1100 North Eutaw Street, Room 613
Baltimore, Maryland 21201-2206
John P. O'Conner, Commissioner (410) 767-2215 Fax: (410) 767-2003
Ileana O'Brien, Deputy Commissioner (410) 767-2992 Fax: Same as above

Michigan Department of Consumer and Industry Services
3423 North Martin Luther King Boulevard
P.O. Box 30649
Lansing, Michigan 48909
Kathleen M. Wilbur, Director (517) 373-7230 Fax: (517) 373-2129
Douglas R. Earle, Program Director for Safety and Health (517) 322-1814 Fax: (517) 335-8010

Minnesota Department of Labor and Industry
443 Lafayette Road
St. Paul, Minnesota 55155
Gary Bastian, Commissioner (612) 296-2342 Fax: (612) 282-5405
Gail Blackstone, Assistant Commissioner (612) 296-6529 Fax: Same as above

Nevada Division of Industrial Relations
400 West King Street
Carson City, Nevada 97502
Ron Swirczek, Administrator (702) 687-3032 Fax: (702) 687-6305
Danny Evans, Assistant Administrator (702) 687-3250 Fax: (702) 687-6150

New Mexico Environment Department
1190 St. Francis Drive
P.O. Box 26110
Santa Fe, New Mexico 87502
Mark E. Weidler, Secretary (505) 827-2850 Fax: (505) 827-2836
Sam A. Rogers, Chief (505) 827-4230 Fax: Same as above

New York Department of Labor
W. Averell Harriman State Office Building - 12, Room 500
Albany, New York 12240
James T. Dillon, Acting Commissioner (518) 457-2741 Fax: (518) 457-6908
Richard Cuculo, Program Director (518) 457-3518 Fax: Same as above

North Carolina Department of Labor
319 Chapanoke Road
Raleigh, North Carolina 27603
Harry Payne, Commissioner (919) 662-4585 Fax: (919) 662-4582
Charles Jeffress, Deputy Commissioner (919) 662-4585 Fax: Same as above

Oregon Occupational Safety and Health Division
Department of Consumer & Business Services
350 Winter Street, NE, Room 430
Salem, Oregon 97310
Peter DeLuca, Administrator (503) 378-3272 Fax: (503) 378-4538
David Sparks, Deputy Administrator (503) 378-3272 Fax: Same as above

Puerto Rico Department of Labor and Human Resources
Prudencio Rivera Martinez Building
505 Munoz Rivera Avenue
Hato Rey, Puerto Rico 00918
Cesar J. Almodovar-Marchany, Secretary (787) 754-2119 Fax: (787) 753-9550
Assistant Secretary's Office (787) 754-2119/2171 Fax: (787) 767-6051

South Carolina Department of Labor, Licensing, and Regulation
Koger Office Park, Kingstree Building
110 Centerview Drive
P.O. Box 11329
Columbia, South Carolina 29210
Lewis Gossett, Director (803) 896-4300 Fax: (803) 896-4393
William Lybrand, Program Director (803) 734-9594 Fax: (803) 734-9772

Tennessee Department of Labor
710 James Robertson Parkway
Nashville, Tennessee 37243-0659
Alphonso R. Bodie, Commissioner (615) 741-2582 Fax: (615) 741-5078
Don Witt, Program Director (615) 741-2793 Fax: (615) 741-3325

Labor Commission of Utah
160 East 300 South, 3rd Floor
PO Box 146650
Salt Lake City, Utah 84114-6650
R. Lee Ellertson, Commissioner (801) 530-6898 Fax: (801) 530-6880
Jay W. Bagley, Administrator (801) 530-6898 Fax: (801) 530-7606

Vermont Department of Labor and Industry
National Life Building - Drawer 20
120 State Street
Montpelier, Vermont 05620
Stephen Jamsen, Commissioner (802) 828-2288 Fax: (802) 828-2748
Robert McLeod, Project Manager (802) 828-2765 Fax: Same as above

Virginia Department of Labor and Industry
Powers-Taylor Building
13 South 13th Street
Richmond, Virginia 23219
Theron Bell, Commissioner (804) 786-2377 Fax: (804) 371-6524
Charles Lahey, Deputy Commissioner (804) 786-2383 Fax: Same as above

Virgin Islands Department of Labor
2131 Hospital Street
Box 890, Christiansted
St. Croix, Virgin Islands 00820-4666
Carmelo Rivera, Commissioner (809) 773-1994 Fax: (809) 773-0094
Raymond Williams, Program Director (809) 772-1315 Fax: (809) 772-4323

Washington Department of Labor and Industries
General Administration Building
P.O. Box 44001
Olympia, Washington 98504-4001
Gary Moore, Director (360) 902-4200 Fax: (360) 902-4202
Michael Silverstein, Assistant Director (360) 902-5495 Fax: (360) 902-5529

Wyoming Department of Employment
Worker's Safety and Compensation Division
Herschler Building, 2nd Floor East
122 West 25th Street
Cheyenne, Wyoming 82002
Stephan R. Foster, Safety Administrator (307) 777-7786 Fax: (307) 777-5850

Other Relevant Addresses

1908 Consultation Training Coordinator
OSHA Training Institute
1555 Times Drive
Des Plaines, Illinois 60018
(847) 297-4810

29 CFR 1926 KEY WORDS
Courtesy of OSHA

LISTING OF KEY WORDS WITHIN THE CONSTRUCTION STANDARD AND THEIR LOCATION

SUBJECT	LOCATION
Accepted engineering practice	1926.650
Aluminum hydraulic shoring	1926.650
Anchorage	1926.500(b)
Approved	1926.32(c)
Barricade	1926.203
Barrier	1926.960
Bearer, scaffolding	1926.452(b)
Bell bottom pier hole	1926.650
Benching	1926.650
Block holing	1926.914
Boatswain's chair	1926.452(b)
Body belt	1926.500(b)
Body harness	1926.500(b)
Bond	1926.960
Bonding	1926.449
Bonding jumper	1926.449
Brace, scaffolding	1926.452(b)
Buckle	1926.500(b)
Buddy system	1926.65(a)
Caisson	1926.804
Carcinogen	1926.59, App A
Carpenter's bracket scaffold	1926.452(b)
Catastrophe	Chapt VIII, 2, FOM
Catastrophic release	1926.64(b)
Cave-in	1926.650
CAZ	1926.500(b)
Cemented soil	1926 Subpart P, Appendix A
Certified	1926.449
Chemical	1926.59(c)
Chemical manufacturer	1926.59(c)
Chemical name	1926.59(c)

SUBJECT	LOCATION
Chicken ladder	1926.452(b)
Chronic	1926.59, App A
Circuit	1926.960
Cleat	1926.1050
Cohesive soil	1926 Subpart P, Appendix A
Combustible liquid	1926.59(c)
Combustible liquids	1926.155
Compctent person	1926.32(f)
Competent person	1926.650
Competent person, lead	1926.62(b)
Compressed gas	1926.59(c)
Construction Work	1926.32(g)
Controlled Access Zone (CAZ)	1926.500(b)
Corrosive	1926.59, App A
Coupler, scaffold	1926.452(b)
Crawling board	1926.452(b)
Cross braces	1926.650
Current-carrying part	1926.960
Cutout	1926.449
Cutout box	1926.449
Dangerous equipment	1926.500(b)
Dead, deenergized	1926.960
Dead front	1926.449
Defect	1926.32(h)
Designated person	1926.960
Designated person	1926.32(i)
Designated representative	1926.59(c)
Distributor	1926.59(c)
Double cleat ladder	1926.1050
Double pole scaffold	1926.452(b)
Dry soil	1926 Subpart P, Appendix A

OTHER REFERENCES

Other definitions are contained in the OSHA Standards and may be located in the following Subparts/Topics.

APPENDIX L

DEFINITIONS FOR SUBPARTS OF 29 CFR 1926

SUBPART C – GENERAL SAFETY & HEALTH PROVISIONS

"Act" means section 107 of the Contract Work Hours and Safety Standards Act, commonly known as the Construction Safety Act (86 Stat. 96; 40 U.S.C. 333).

"Administration" means the Occupational Safety and Health Administration.

"ANSI" means American National Standards Institute.

"Approved" means sanctioned, endorsed, accredited, certified, or accepted as satisfactory by a duly constituted and nationally recognized authority or agency.

"Authorized person" means a person approved or assigned by the employer to perform a specific type of duty or duties or to be at a specific location or locations at the jobsite.

"Competent person" means one who is capable of identifying existing and predictable hazards in the surroundings or working conditions which are unsanitary, hazardous, or dangerous to employees, and who has authorization to take prompt corrective measures to eliminate them.

"Construction work." For purposes of this section, "Construction work" means work for construction, alteration, and/or repair, including painting and decorating.

"Defect" means any characteristic or condition which tends to weaken or reduce the strength of the tool, object, or structure of which it is a part.

"Designated person" means "authorized person" as defined in paragraph (d) of this section.

"Employee" means every laborer or mechanic under the Act regardless of the contractual relationship which may be alleged to exist between the laborer and mechanic and the contractor or subcontractor who engaged him. "Laborer and mechanic" are not defined in the Act, but the identical terms are used in the Davis-Bacon Act (40 U.S.C. 276a), which provides for minimum wage protection on Federal and federally assisted construction contracts. The use of the same term in a statute which often applies concurrently with section 107 of the Act has considerable presidential value in ascertaining the meaning of "laborer and mechanic" as used in the Act. "Laborer" generally means one who performs manual labor or who labors at an occupation requiring physical strength; "mechanic" generally means a worker skilled with tools. See 18 Comp. Gen. 341.

"Employer" means contractor or subcontractor within the meaning of the Act and of this part.

"Hazardous substance" means a substance which, by reason of being explosive, flammable, poisonous, corrosive, oxidizing, irritating, or otherwise harmful, is likely to cause death or injury.

"Qualified" means one who, by possession of a recognized degree, certificate, or professional standing, or who by extensive knowledge, training, and experience, has successfully demonstrated his ability to solve or resolve problems relating to the subject matter, the work, or the project.

"SAE" means Society of Automotive Engineers.

"Safety factor" means the ratio of the ultimate breaking strength of a member or piece of material or equipment to the actual working stress or safe load when in use.

"Secretary" means the Secretary of Labor.

"Shall" means mandatory.

"Should" means recommended.

"Suitable" means that which fits, and has the qualities or qualifications to meet a given purpose, occasion, condition, function, or circumstance.

SUBPART D – OCCUPATIONAL HEALTH AND ENVIRONMENTAL CONTROLS

1926.57 – Ventilation – Abrasive Blasting

"Abrasive." A solid substance used in an abrasive blasting operation.

"Abrasive blasting." The forcible application of an abrasive to a surface by pneumatic pressure, hydraulic pressure, or centrifugal force.

"Abrasive-blasting respirator." A continuous-flow air-line respirator constructed so that it will cover the wearer's head, neck, and shoulders to protect him from rebounding abrasive.

"Blast cleaning barrel." A complete enclosure which rotates on an axis, or which has an internal moving tread to tumble the parts, in order to expose various surfaces of the parts to the action of an automatic blast spray.

"Blast cleaning room." A complete enclosure in which blasting operations are performed and where the operator works inside of the room to operate the blasting nozzle and direct the flow of the abrasive material.

"Blasting cabinet." An enclosure where the operator stands outside and operates the blasting nozzle through an opening or openings in the enclosure.

"Clean air." Air of such purity that it will not cause harm or discomfort to an individual if it is inhaled for extended periods of time.

"Dust Collector." A device or combination of devices for separating dust from the air handled by an exhaust ventilation system.

"Exhaust ventilation system." A system for removing contaminated air from a space, comprising two or more of the following elements (A) enclosure or hood, (B) duct work, (C) dust collecting equipment, (D) exhauster, and (E) discharge stack.

"Particulate-filter respirator." An air purifying respirator, commonly referred to as a dust or a fume respirator, which removes most of the dust or fume from the air passing through the device.

"Respirable dust." Airborne dust in sizes capable of passing through the upper respiratory system to reach the lower lung passages.

"Rotary blast cleaning table." An enclosure where the pieces to be cleaned are positioned on a rotating table and are passed automatically through a series of blast sprays.

1926.57 – Ventilation – Grinding Wheels

"Abrasive cutting-off wheels." Organic-bonded wheels, the thickness of which is not more than one forty-eighth of their diameter for those up to, and including, 20 inches (50.8 cm) in diameter, and not more than one-sixtieth of their diameter for those larger than 20 inches (50.8 cm) in diameter, used for a multitude of operations variously known as cutting, cutting off, grooving, slotting, coping, and jointing, and the like. The wheels may be "solid" consisting of organic-bonded abrasive material throughout, "steel centered" consisting of a steel disc with a rim of organic-bonded material molded around the periphery, or of the "inserted tooth" type consisting of a steel disc with organic-bonded abrasive teeth or inserts mechanically secured around the periphery.

"Belts." All power-driven, flexible, coated bands used for grinding, polishing, or buffing purposes.

"Branch pipe." The part of an exhaust system piping that is connected directly to the hood or enclosure.

"Cradle." A movable fixture, upon which the part to be ground or polished is placed.

"Disc wheels." All power-driven rotatable discs faced with abrasive materials, artificial or natural, and used for grinding or polishing on the side of the assembled disc.

"Entry loss." The loss in static pressure caused by air flowing into a duct or hood. It is usually expressed in inches of water gauge.

"Exhaust system." A system consisting of branch pipes connected to hoods or enclosures, one or more header pipes, an exhaust fan, means for separating solid contaminants from the air flowing in the system, and a discharge stack to outside.

"Grinding wheels." All power-driven rotatable grinding or abrasive wheels, except disc wheels as defined in this standard, consisting of abrasive particles held together by artificial or natural bonds and used for peripheral grinding.

"Header pipe (main pipe)." A pipe into which one or more branch pipes enter and which connects such branch pipes to the remainder of the exhaust system.

"Hoods and enclosures." The partial or complete enclosure around the wheel or disc through which air enters an exhaust system during operation.

"Horizontal double-spindle disc grinder." A grinding machine carrying two power-driven, rotatable, coaxial, horizontal spindles upon the inside ends of which are mounted abrasive disc wheels used for grinding two surfaces simultaneously.

"Horizontal single-spindle disc grinder." A grinding machine carrying an abrasive disc wheel upon one or both ends of a power-driven, rotatable single horizontal spindle.

"Polishing and buffing wheels." All power-driven rotatable wheels composed all or in part of textile fabrics, wood, felt, leather, paper, and may be coated with abrasives on the periphery of the wheel for purposes of polishing, buffing, and light grinding.

"Portable grinder." Any power-driven rotatable grinding, polishing, or buffing wheel mounted in such manner that it may be manually manipulated.

"Scratch brush wheels." All power-driven rotatable wheels made from wire or bristles, and used for scratch cleaning and brushing purposes.

"Swing-frame grinder." Any power-driven rotatable grinding, polishing, or buffing wheel mounted in such a manner that the wheel with its supporting framework can be manipulated over stationary objects.

"Velocity pressure (vp)." The kinetic pressure in the direction of flow necessary to cause a fluid at rest to flow at a given velocity. It is usually expressed in inches of water gauge.

"Vertical spindle disc grinder." A grinding machine having a vertical, rotatable power-driven spindle carrying a horizontal abrasive disc wheel.

1926.57 – Ventilation – Spray Booths and Rooms

"Minimum maintained velocity" is the velocity of air movement which must be maintained in order to meet minimum specified requirements for health and safety.

"Spray booths" are defined and described in 1926.66(a). (See sections 103, 104, and 105 of the Standard for Spray Finishing Using Flammable and Combustible Materials, NFPA No. 33-1969.)

"Spray room" is a room in which spray-finishing operations not conducted in a spray booth are performed separately from other areas.

"Spray-finishing operations" are employment of methods wherein organic or inorganic materials are utilized in dispersed form for deposit on surfaces to be coated, treated, or cleaned. Such methods of deposit may involve either automatic, manual, or electrostatic deposition but do not include metal spraying or metallizing, dipping, flow coating, roller coating, tumbling, centrifuging, or spray washing and degreasing as conducted in self-contained washing and degreasing machines or systems.

1926.59 – Subpart D – Hazard Communication

"Article" means a manufactured item: (i) which is formed to a specific shape or design during manufacture; (ii) which has end use function(s) dependent in whole or in part upon its shape or design during end use; and (iii) which does not release, or otherwise result in exposure to, a hazardous chemical, under normal conditions of use.

"Assistant Secretary" means the Assistant Secretary of Labor for Occupational Safety and Health, U.S. Department of Labor, or designee.

"Chemical" means any element, chemical compound or mixture of elements and/or compounds.

"Chemical manufacturer" means an employer with a workplace where chemical(s) are produced for use or distribution.

"Chemical name" means the scientific designation of a chemical in accordance with the nomenclature system developed by the International Union of Pure and Applied Chemistry (IUPAC) or the Chemical Abstracts Service (CAS) rules of nomenclature, or a name which will clearly identify the chemical for the purpose of conducting a hazard evaluation.

"Combustible liquid" means any liquid having a flashpoint at or above 100 degree F (37.8 degree C), but below 200 degree F (93.3 degreeC), except any mixture having components with flashpoints of 200 degree F (93.3 degree C), or higher, the total volume of which make up 99 percent or more of the total volume of the mixture.

"Common name" means any designation or identification such as code name, code number, trade name, brand name or generic name used to identify a chemical other than by its chemical name.

"Compressed gas" means:

    (i)   A gas or mixture of gases having, in a container, an absolute pressure exceeding 40 psi at 70 degree F (21.1 degree C); or

(ii)   a gas or mixture of gases having, in a container, an absolute pressure exceeding 104 psi at 130 deg F (54.4 degree C) regardless of the pressure at 70 degree F (21.1 degree C); or

(iii)   A liquid having a vapor pressure exceeding 40 psi at 100 degree F (37.8 degree C) as determined by ASTM D-323-72.

"Container" means any bag, barrel, bottle, box, can, cylinder, drum, reaction vessel, storage tank, or the like that contains a hazardous chemical. For purposes of this section, pipes or piping systems, and engines, fuel tanks, or other operating systems in a vehicle, are not considered to be containers.

"Designated representative" means any individual or organization to whom an employee gives written authorization to exercise such employee's rights under this section. A recognized or certified collective bargaining agent shall be treated automatically as a designated representative without regard to written employee authorization.

"Director" means the Director, National Institute for Occupational Safety and Health, U.S. Department of Health and Human Services, or designee.

"Distributor" means a business, other than a chemical manufacturer or importer, which supplies hazardous chemicals to other distributors or to employers.

"Employee" means a worker who may be exposed to hazardous chemicals under normal operating conditions or in foreseeable emergencies. Workers such as office workers or bank tellers who encounter hazardous chemicals only in non-routine, isolated instances are not covered.

"Employer" means a person engaged in a business where chemicals are either used or distributed, or are produced for use or distribution, including a contractor or subcontractor.

"Explosive" means a chemical that causes a sudden, almost instantaneous release of pressure, gas, and heat when subjected to sudden shock, pressure, or high temperature.

"Exposure" or **"exposed"** means that an employee is subjected to a hazardous chemical in the course of employment through any route of entry (inhalation, ingestion, skin contact or absorption, etc.), and includes potential (e.g., accidental or possible) exposure.

"Flammable" means a chemical that falls into one of the following categories:

(i)   **"Aerosol, flammable"** means an aerosol that, when tested by the method described in 16 CFR 1500.45, yields a flame projection exceeding 18 inches at full valve opening, or a flashback (a flame extending back to the valve) at any degree of valve opening;

(ii)   **"Gas, flammable"** means:

(A)   A gas that, at ambient temperature and pressure, forms a flammable mixture with air at a concentration of thirteen (13) percent by volume or less; or

(B)   A gas that, at ambient temperature and pressure, forms a range of flammable mixtures with air wider than twelve (12) percent by volume, regardless of the lower limit;

(iii)   **"Liquid, flammable"** means any liquid having a flashpoint below 100 degree F (37.8 degree C), except any mixture having components with flashpoints of 100 degree F (37.8 degree C) or higher, the total of which make up 99 percent or more of the total volume of the mixture;

(iv)   **"Solid, flammable"** means a solid, other than a blasting agent or explosive as defined in 1910.109(a), that is liable to cause fire through friction, absorption of moisture, spontaneous chemical change, or retained heat from manufacturing or processing, or which can be ignited readily and when ignited burns so vigorously and persistently as to create a serious hazard. A chemical shall be considered to be a flammable solid if, when tested by the method described in 16 CFR 1500.44, it ignites and burns with a self-sustained flame at a rate greater than one-tenth of an inch per second along its major axis.

"Flashpoint" means the minimum temperature at which a liquid gives off a vapor in sufficient concentration to ignite when tested as follows:

   (i)   **Tagliabue Closed Tester** (See American National Standard Method of Test for Flash Point by Tag Closed Tester, Z11.24-1979 (ASTM D 56-79)) for liquids with a viscosity of less than 45 Saybolt University Seconds (SUS) at 100 degree F (37.8 degree C), that do not contain suspended solids and do not have a tendency to form a surface film under test; or

   (ii)   **Pensky-Martens Closed Tester** (See American National Standard Method of Test for Flash Point by Pensky-Martens Closed Tester, Z11.7-1979 (ASTM D 93-79)) for liquids with a viscosity equal to or greater than 45 SUS at 100 degree F (37.8 degree C), or that contain suspended solids, or that have a tendency to form a surface film under test; or

   (iii) **Setaflash Closed Tester** (see American National Standard Method of Test for Flash Point by Setaflash Closed Tester (ASTMD 3278-78)) Organic peroxides, which undergo autoaccelerating thermal decomposition, are excluded from any of the flashpoint determination methods specified above.

"Foreseeable emergency" means any potential occurrence such as, but not limited to, equipment failure, rupture of containers, or failure of control equipment which could result in an uncontrolled release of a hazardous chemical into the workplace.

"Hazard warning" means any words, pictures, symbols, or combination thereof appearing on a label or other appropriate form of warning which convey the hazard(s) of the chemical(s) in the container(s).

"Hazardous chemical" means any chemical which is a physical hazard or a health hazard.

"Health hazard" means a chemical for which there is statistically significant evidence based on at least one study conducted in accordance with established scientific principles that acute or chronic health effects may occur in exposed employees. The term "health hazard" includes chemicals which are carcinogens, toxic or highly toxic agents, reproductive toxins, irritants, corrosives, sensitizers, hepatotoxins, nephrotoxins, neurotoxins, agents which act on the hematopoietic system, and agents which damage the lungs, skin, eyes, or mucous membranes. Appendix A provides further definitions and explanations of the scope of health hazards covered by this section, and Appendix B describes the criteria to be used to determine whether or not a chemical is to be considered hazardous for purposes of this standard.

"Identity" means any chemical or common name which is indicated on the material safety data sheet (MSDS) for the chemical. The identity used shall permit cross-references to be made among the required list of hazardous chemicals, the label and the MSDS.

"Immediate use" means that the hazardous chemical will be under the control of and used only by the person who transfers it from a labeled container and only within the work shift in which it is transferred.

"Importer" means the first business with employees within the Customs Territory of the United States which receives hazardous chemicals produced in other countries for the purpose of supplying them to distributors or employers within the United States.

"Label" means any written, printed, or graphic material, displayed on or affixed to containers of hazardous chemicals.

"Material safety data sheet (MSDS)" means written or printed material concerning a hazardous chemical which is prepared in accordance with paragraph (g) of this section.

"Mixture" means any combination of two or more chemicals if the combination is not, in whole or in part, the result of a chemical reaction.

"Organic peroxide" means an organic compound that contains the bivalent -O-O-structure and which may be considered to be a structural derivative of hydrogen peroxide where one or both of the hydrogen atoms has been replaced by an organic radical.

"Oxidizer" means a chemical other than a blasting agent or explosive as defined in 1910.109(a), that initiates or promotes combustion in other materials, thereby causing fire either of itself or through the release of oxygen or other gases.

"Physical hazard" means a chemical for which there is scientifically valid evidence that it is a combustible liquid, a compressed gas, explosive, flammable, an organic peroxide, an oxidizer, pyrophoric, unstable (reactive) or water-reactive.

"Produce" means to manufacture, process, formulate, or repackage.

"Pyrophoric" means a chemical that will ignite spontaneously in air at a temperature of 130 degree F (54.4 degree C) or below.

"Responsible party" means someone who can provide additional information on the hazardous chemical and appropriate emergency procedures, if necessary.

"Specific chemical identity" means the chemical name, Chemical Abstracts Service (CAS) Registry Number, or any other information that reveals the precise chemical designation of the substance.

"Trade secret" means any confidential formula, pattern, process, device, information or compilation of information that is used in an employer's business, and that gives the employer an opportunity to obtain an advantage over competitors who do not know or use it. Appendix D sets out the criteria to be used in evaluating trade secrets.

"Unstable (reactive)" means a chemical which in the pure state, or as produced or transported, will vigorously polymerize, decompose, condense, or will become self-reactive under conditions of shocks, pressure or temperature.

"Use" means to package, handle, react, or transfer.

"Water-reactive" means a chemical that reacts with water to release a gas that is either flammable or presents a health hazard.

"Work area" means a room or defined space in a workplace where hazardous chemicals are produced or used, and where employees are present.

"Workplace" means an establishment, jobsite, or project, at one geographical location containing one or more work areas.

1926.62 – Subpart D – Lead

(b) "Definitions".

"Action level" means employee exposure, without regard to the use of respirators, to an airborne concentration of lead of 30 micrograms per cubic meter of air (30 ug/m(3)) calculated as an 8-hour time-weighted average (TWA).

"Assistant Secretary" means the Assistant Secretary of Labor for Occupational Safety and Health, U.S. Department of Labor, or designee.

"Competent person" means one who is capable of identifying existing and predictable lead hazards in the surroundings or working conditions and who has authorization to take prompt corrective measures to eliminate them.

"Director" means the Director, National Institute for Occupational Safety and Health (NIOSH), U.S. Department of Health and Human Services, or designee.

"Lead" means metallic lead, all inorganic lead compounds, and organic lead soaps. Excluded from this definition are all other organic lead compounds.

"This section" means this standard.

1926.64 – Subpart D – Process Safety Management

"Atmospheric tank" means a storage tank which has been designed to operate at pressures from atmospheric through 0.5 psig. (pounds per square inch gauge, 3.45 Kpa).

"Boiling point" means the boiling point of a liquid at a pressure of 14.7 pounds per square inch absolute (psia.) (760 mm.). For the purposes of this section, where an accurate boiling point is unavailable for the material in question, or for mixtures which do not have a constant boiling point, the 10 percent point of a distillation performed in accordance with the Standard Method of Test for Distillation of Petroleum Products, ASTM D-86-62, may be used as the boiling point of the liquid.

"Catastrophic release" means a major uncontrolled emission, fire, or explosion, involving one or more highly hazardous chemicals, that presents serious danger to employees in the workplace.

"Facility" means the buildings, containers or equipment which contain a process.

"Highly hazardous chemical" means a substance possessing toxic, reactive, flammable, or explosive properties and specified by paragraph (a)(1) of this section.

"Hot work" means work involving electric or gas welding, cutting, brazing, or similar flame or spark-producing operations.

"Normally unoccupied remote facility" means a facility which is operated, maintained or serviced by employees who visit the facility only periodically to check its operation and to perform necessary operating or maintenance tasks. No employees are permanently stationed at the facility. Facilities meeting this definition are not contiguous with, and must be geographically remote from all other buildings, processes or persons.

"Process" means any activity involving a highly hazardous chemical including any use, storage, manufacturing, handling, or the on-site movement of such chemicals, or combination of these activities. For purposes of this definition, any group of vessels which are interconnected and separate vessels which are located such that a highly hazardous chemical could be involved in a potential release shall be considered a single process.

"Replacement in kind" means a replacement which satisfies the design specification.

"Trade secret" means any confidential formula, pattern, process, device, information or compilation of information that is used in an employer's business, and that gives the employer an opportunity to obtain an advantage over competitors who do not know or use it. Appendix D contained in 1926.59 sets out the criteria to be used in evaluating trade secrets.

1926.65 – Subpart D – Hazardous Waste Operations

"Buddy system" means a system of organizing employees into work groups in such a manner that each employee of the work group is designated to be observed by at least one other employee in the work group. The purpose of the buddy system is to provide rapid assistance to employees in the event of an emergency.

"Clean-up operation" means an operation where hazardous substances are removed, contained, incinerated, neutralized, stabilized, cleared-up, or in any other manner processed or handled with the ultimate goal of making the site safer for people or the environment.

"Decontamination" means the removal of hazardous substances from employees and their equipment to the extent necessary to preclude the occurrence of foreseeable adverse health affects.

"Emergency response or responding to emergencies" means a response effort by employees from outside the immediate release area or by other designated responders (i.e., mutual-aid groups, local fire departments, etc.) to an occurrence which results, or is likely to result, in an uncontrolled release of a hazardous substance. Responses to incidental releases of hazardous substances where the substance can be absorbed, neutralized, or otherwise controlled at the time of release by employees in the immediate release area, or by maintenance personnel are not considered to be emergency responses within the scope of this standard. Responses to releases of hazardous substances where there is no potential safety or health hazard (i.e., fire, explosion, or chemical exposure) are not considered to be emergency responses.

"Facility" means (A) any building, structure, installation, equipment, pipe or pipeline (including any pipe into a sewer or publicly owned treatment works), well, pit, pond, lagoon, impoundment, ditch, storage container, motor vehicle, rolling stock, or aircraft, or (B) any site or area where a hazardous substance has been deposited, stored, disposed of, or placed, or otherwise come to be located; but does not include any consumer product in consumer use or any water-borne vessel.

"Hazardous materials response (HAZMAT) team" means an organized group of employees, designated by the employer, who are expected to perform work to handle and control actual or potential leaks or spills of hazardous substances requiring possible close approach to the substance. The team members perform responses to releases or potential releases of hazardous substances for the purpose of control or stabilization of the incident. A HAZMAT team is not a fire brigade nor is a typical fire brigade a HAZMAT team. A HAZMAT team, however, may be a separate component of a fire brigade or fire department.

"Hazardous substance" means any substance designated or listed under paragraphs (A) through (D) of this definition, exposure to which results or may result in adverse affects on the health or safety of employees:

(A)   Any substance defined under section 101(14) of CERCLA;

(B)   Any biological agent and other disease-causing agent which after release into the environment and upon exposure, ingestion, inhalation, or assimilation into any person, either directly from the environment or indirectly by ingestion through food chains, will or may reasonably be anticipated to cause death, disease, behavioral abnormalities, cancer, genetic mutation, physiological malfunctions (including malfunctions in reproduction) or physical deformations in such persons or their offspring;

(C)   Any substance listed by the U.S. Department of Transportation as hazardous materials under 49 CFR 172.101 and appendices; and

(D)   Hazardous waste as herein defined.

"Hazardous waste" means –

(A)   A waste or combination of wastes as defined in 40 CFR 261.3, or

(B)   Those substances defined as hazardous wastes in 49 CFR 171.8.

"Hazardous waste operation" means any operation conducted within the scope of this standard.

"Hazardous waste site" or **"Site"** means any facility or location within the scope of this standard at which hazardous waste operations take place.

"Health hazard" means a chemical, mixture of chemicals or a pathogen for which there is statistically significant evidence based on at least one study conducted in accordance with established scientific principles that acute or chronic health effects may occur in exposed employees. The term "health hazard" includes chemicals which are carcinogens, toxic or highly toxic agents, reproductive toxins, irritants, corrosives, sensitizers, heptaotoxins, nephrotoxins, neurotoxins, agents which act on the hematopoietic system, and agents which damage the lungs, skin, eyes, or mucous membranes. It also includes stress due to temperature extremes. Further definition of the terms used above can be found in appendix A to 29 CFR 1926.59.

"IDLH" or **"Immediately dangerous to life or health"** means an atmospheric concentration of any toxic, corrosive or asphyxiant substance that poses an immediate threat to life or would cause irreversible or delayed adverse health effects or would interfere with an individual's ability to escape from a dangerous atmosphere.

"Oxygen deficiency" means that concentration of oxygen by volume below which atmosphere supplying respiratory protection must be provided. It exists in atmospheres where the percentage of oxygen by volume is less than 19.5 percent oxygen.

"Permissible exposure limit" means the exposure, inhalation or dermal permissible exposure limit specified either in 1926.55, elsewhere in subpart D, or in other pertinent sections of this part.

"Post emergency response" means that portion of an emergency response performed after the immediate threat of a release has been stabilized or eliminated and clean-up of the site has begun. If post emergency response is performed by an employer's own employees who were part of the initial emergency response, it is considered to be part of the initial response and not post emergency response. However, if a group of an employer's own employees, separate from the group providing initial response, performs the clean-up operation, then the separate group of employees would be considered to be performing post-emergency response and subject to paragraph (q)(11) of this section.

"Published exposure level" means the exposure limits published in "NIOSH Recommendations for Occupational Health Standards" dated 1986 incorporated by reference, or if none is specified, the exposure limits published in the standards specified by the American Conference of Governmental Industrial Hygienists in their publication "Threshold Limit Values and Biological Exposure Indices for 1987-88" dated 1987 incorporated by reference.

"Qualified person" means a person with specific training, knowledge and experience in the area for which the person has the responsibility and the authority to control.

"Site safety and health supervisor (or official)" means the individual located on a hazardous waste site who is responsible to the employer and has the authority and knowledge necessary to implement the site safety and health plan and verify compliance with applicable safety and health requirements.

"Small quantity generator" means a generator of hazardous wastes who in any calendar month generates no more than 1,000 kilograms (2,205 pounds) of hazardous waste in that month.

"Uncontrolled hazardous waste site" means an area identified as an uncontrolled hazardous waste site by a governmental body, whether federal, state, local or other where an accumulation of hazardous substances creates a threat to the health and safety of individuals or the environment or both. Some sites are found on public lands such as those created by former municipal, county or state landfills where illegal or poorly managed waste disposal has taken place. Other sites are found on private property, often belonging to generators or former generators of hazardous substance wastes. Examples of such sites include, but are not limited to, surface impoundments, landfills, dumps, and tank or drum farms. Normal operations at TSD sites are not covered by this definition.

1926.66 – Subpart D – Criteria for Design and Construction of Spray Booths

"Aerated solid powders." Aerated powders shall mean any powdered material used as a coating material which shall be fluidized within a container by passing air uniformly from below. It is common practice to fluidize such materials to form a fluidized powder bed and then dip the part to be coated into the bed in a manner similar to that used in liquid dipping. Such beds are also used as sources for powder spray operations.

"Approved" shall mean approved and listed by a nationally recognized testing laboratory.

"Dry spray booth." A spray booth not equipped with a water washing system as described in paragraph (a)(4) of this section. A dry spray booth may be equipped with

(i)   Distribution or baffle plates to promote an even flow of air through the booth or cause the deposit of overspray before it enters the exhaust duct; or

(ii)   Overspray dry filters to minimize dusts; or

(iii)   Overspray dry filters to minimize dusts or residues entering exhaust ducts; or

(iv)   Overspray dry filter rolls designed to minimize dusts or residues entering exhaust ducts; or

where dry powders are being sprayed, with powder collection systems so arranged in the exhaust to capture oversprayed material.

"Electrostatic fluidized bed." A container holding powder coating material which is aerated from below so as to form an air-supported expanded cloud of such material which is electrically charged with a charge opposite to the charge of the object to be coated; such object is transported, through the container immediately above the charged and aerated materials in order to be coated.

"Fluidized bed." A container holding powder coating material which is aerated from below so as to form an air-supported expanded cloud of such material through which the preheated object to be coated is immersed and transported

"Listed." See "approved" in paragraph (a)(8) of this section.

"Spray booth." A power-ventilated structure provided to enclose or accommodate a spraying operation to confine and limit the escape of spray, vapor, and residue, and to safely conduct or direct them to an exhaust system.

"Spraying area." Any area in which dangerous quantities of flammable vapors or mists, or combustible residues, dusts, or deposits are present due to the operation of spraying processes.

"Waterwash spray booth." A spray booth equipped with a water washing system designed to minimize dusts or residues entering exhaust ducts and to permit the recovery of overspray finishing material.

Subpart E – Personal Protective and Life Saving Equipment

"Contaminant" means any material which by reason of its action upon, within, or to a person is likely to cause physical harm.

"Lanyard" means a rope, suitable for supporting one person. One end is fastened to a safety belt or harness and the other end is secured to a substantial object or a safety line.

"Lifeline" means a rope, suitable for supporting one person, to which a lanyard or safety belt (or harness) is attached.

"O.D." means optical density and refers to the light refractive characteristics of a lens.

"Radiant energy" means energy that travels outward in all directions from its sources.

"Safetybelt" means a device, usually worn around the waist which, by reason of its attachment to a lanyard and lifeline or a structure, will prevent a worker from falling.

1926.155 – Subpart F – Fire Protection & Prevention

"Approved," for the purpose of this subpart, means equipment that has been listed or approved by a nationally recognized testing laboratory such as Factory Mutual Engineering Corp., or Underwriters' Laboratories, Inc., or Federal agencies such as Bureau of Mines, or U.S. Coast Guard, which issue approvals for such equipment.

"Closed container" means a container so sealed by means of a lid or other device that neither liquid nor vapor will escape from it at ordinary temperatures.

"Combustible liquids" mean any liquid having a flash point at or above 140 degree F. (60 degree C.), and below 200 degree F. (93.4 degree C.).

"Combustion" means any chemical process that involves oxidation sufficient to produce light or heat.

"Fire brigade" means an organized group of employees that are knowledgeable, trained, and skilled in the safe evacuation of employees during emergency situations and in assisting in fire fighting operations.

"Fire resistance" means so resistant to fire that, for specified time and under conditions of a standard heat intensity, it will not fail structurally and will not permit the side away from the fire to become hotter than a specified temperature. For purposes of this part, fire resistance shall be determined by the Standard Methods of Fire Tests of Building Construction and Materials, NFPA 251-1969.

"Flammable" means capable of being easily ignited, burning intensely, or having a rapid rate of flame spread.

"Flammable liquids" means any liquid having a flash point below 140 degree F. and having a vapor pressure not exceeding 40 pounds per square inch (absolute) at 100 deg F.

"Flash point" of the liquid means the temperature at which it gives off vapor sufficient to form an ignitable mixture with the air near the surface of the liquid or within the vessel used as determined by appropriate test procedure and apparatus as specified below.

(1) The flash point of liquids having a viscosity less than 45 Saybolt Universal Second(s) at 100 degree F. (37.8 degree C.) and a flash point below 175 degree F. (79.4 degree C.) shall be determined in accordance with the Standard Method of Test for Flash Point by the Tag Closed Tester, ASTM D-56-69.

(2) The flash point of liquids having a viscosity of 45 Saybolt Universal Second(s) or more at 175 degree F. (79.4 deg C.) or higher shall be determined in accordance with the Standard Method of Test for Flash Point by the Pensky Martens Closed Tester, ASTM D-93-69.

"Liquefied petroleum gases," "LPG," and "LP Gas" mean and include any material which is composed predominantly of any of the following hydrocarbons, or mixtures of them, such as propane, propylene, butane (normal butane or iso-butane), and butylenes.

"Portable tank" means a closed container having a liquid capacity more than 60 U.S. gallons, and not intended for fixed installation.

"Safety can" means an approved closed container, of not more than 5 gallons capacity, having a flash-arresting screen, spring-closing lid and spout cover and so designed that it will safely relieve internal pressure when subjected to fire exposure.

"Vapor pressure" means the pressure, measured in pounds per square inch (absolute), exerted by a volatile liquid as determined by the "Standard Method of Test for Vapor Pressure of Petroleum Products (Reid Method)." (ASTM D-323-58.)

SUBPART G – SIGNS, SIGNALS, AND BARRICADES

"Barricade" means an obstruction to deter the passage of persons or vehicles.

"Signals" are moving signs, provided by workers, such as flagmen, or by devices, such as flashing lights, to warn of possible or existing hazards.

"Signs" are the warnings of hazard, temporarily or permanently affixed or placed, at locations where hazards exist.

"Tags" are temporary signs, usually attached to a piece of equipment or part of a structure, to warn of existing or immediate hazards.

SUBPART K – ELECTRICAL

"Accepted." An installation is "accepted" if it has been inspected and found to be safe by a qualified testing laboratory.

"Accessible." (As applied to wiring methods.) Capable of being removed or exposed without damaging the building structure or finish, or not permanently closed in by the structure or finish of the building. (See "concealed" and "exposed.")

"Accessible." (As applied to equipment.) Admitting close approach; not guarded by locked doors, elevation, or other effective means. (See "Readily accessible.")

"Ampacity." The current in amperes a conductor can carry continuously under the conditions of use without exceeding its temperature rating.

"Appliances." Utilization equipment, generally other than industrial, normally built in standardized sizes or types, which is installed or connected as a unit to perform one or more functions.

"Approved." Acceptable to the authority enforcing this Subpart. The authority enforcing this Subpart is the Assistant Secretary of Labor for Occupational Safety and Health. The definition of "acceptable" indicates what is acceptable to the Assistant Secretary of Labor, and therefore approved within the meaning of this Subpart.

"Askarel." A generic term for a group of nonflammable synthetic chlorinated hydrocarbons used as electrical insulating media. Askarels of various compositional types are used. Under arcing conditions the gases produced, while consisting predominantly of noncombustible hydrogen chloride, can include varying amounts of combustible gases depending upon the askarel type.

"Attachment plug (Plug cap)(Cap)." A device which, by insertion in a receptacle, establishes connection between the conductors of the attached flexible cord and the conductors connected permanently to the receptacle.

"Automatic." Self-acting, operating by its own mechanism when actuated by some impersonal influence, as for example, a change in current strength, pressure, temperature, or mechanical configuration.

"Bare conductor" See "Conductor."

"Bonding." The permanent joining of metallic parts to form an electrically conductive path which will assure electrical continuity and the capacity to conduct safely any current likely to be imposed.

"Bonding jumper." A reliable conductor to assure the required electrical conductivity between metal parts required to be electrically connected.

"Branch circuit." The circuit conductors between the final overcurrent device protecting the circuit and the outlet(s).

"Building." A structure which stands alone or which is cut off from adjoining structures by fire walls with all openings therein protected by approved fire doors.

"Cabinet." An enclosure designed either for surface or flush mounting, and provided with a frame, mat, or trim in which a swinging door or doors are or may be hung.

"Certified." Equipment is "certified" if it:

    (a)   Has been tested and found by a qualified testing laboratory to meet applicable test standards or to be safe for use in a specified manner, and

    (b)   Is of a kind whose production is periodically inspected by a qualified testing laboratory. Certified equipment must bear a label, tag, or other record of certification.

"Circuit breaker" –

    (a)   (600 volts nominal, or less.) A device designed to open and close a circuit by non-automatic means and to open the circuit automatically on a predetermined overcurrent without injury to itself when properly applied within its rating.

    (b)   (Over 600 volts, nominal.) A switching device capable of making, carrying, and breaking currents under normal circuit conditions, and also making, carrying for a specified time, and breaking currents under specified abnormal circuit conditions, such as those of short circuit.

"Class I locations." Class I locations are those in which flammable gases or vapors are or may be present in the air in quantities sufficient to produce explosive or ignitable mixtures. Class I locations include the following:

Class I, Division 1. A Class I, Division 1 location is a location

    (1)   In which ignitable concentrations of flammable gases or vapors may exist under normal operating conditions; or

    (2)   In which ignitable concentrations of such gases or vapors may exist frequently because of repair or maintenance operations or because of leakage; or

    (3)   In which breakdown or faulty operation of equipment or processes might release ignitable concentrations of flammable gases or vapors, and might also cause simultaneous failure of electric equipment.

NOTE: This classification usually includes locations where volatile flammable liquids or liquefied flammable gases are transferred from one container to another; interiors of spray booths and areas in the vicinity of spraying and painting operations where volatile flammable solvents are used; locations containing open tanks or vats of volatile flammable liquids; drying rooms or compartments for the evaporation of flammable solvents; inadequately ventilated pump rooms for flammable gas or for volatile flammable liquids; and all other locations where ignitable concentrations of flammable vapors or gases are likely to occur in the course of normal operations.

Class I, Division 2. A Class I, Division 2 location is a location

(1)   In which volatile flammable liquids or flammable gases are handled, processed, or used, but in which the hazardous liquids, vapors, or gases will normally be confined within closed containers or closed systems from which they can escape only in case of accidental rupture or breakdown of such containers or systems, or in case of abnormal operation of equipment; or

(2)   In which ignitable concentrations of gases or vapors are normally prevented by positive mechanical ventilation, and which might become hazardous through failure or abnormal operations of the ventilating equipment; or

(3)   That is adjacent to a Class I, Division 1 location, and to which ignitable concentrations of gases or vapors might occasionally be communicated unless such communication is prevented by adequate positive-pressure ventilation from a source of clean air, and effective safeguards against ventilation failure are provided.

NOTE: This classification usually includes locations where volatile flammable liquids or flammable gases or vapors are used, but which would become hazardous only in case of an accident or of some unusual operating condition. The quantity of flammable material that might escape in case of accident, the adequacy of ventilating equipment, the total area involved, and the record of the industry or business with respect to explosions or fires are all factors that merit consideration in determining the classification and extent of each location. Piping without valves, checks, meters, and similar devices would not ordinarily introduce a hazardous condition even though used for flammable liquids or gases. Locations used for the storage of flammable liquids or of liquefied or compressed gases in sealed containers would not normally be considered hazardous unless also subject to other hazardous conditions.

Electrical conduits and their associated enclosures separated from process fluids by a single seal or barrier are classed as a Division 2 location if the outside of the conduit and enclosures is a nonhazardous location.

"Class II locations." Class II locations are those that are hazardous because of the presence of combustible dust. Class II locations include the following:

Class II, Division 1. A Class II, Division 1 location is a location

(1)   In which combustible dust is or may be in suspension in the air under normal operating conditions, in quantities sufficient to produce explosive or ignitable mixtures; or

(2)   Where mechanical failure or abnormal operation of machinery or equipment might cause such explosive or ignitable mixtures to be produced, and might also provide a source of ignition through simultaneous failure of electric equipment, operation of protection devices, or from other causes; or

(3)   In which combustible dusts of an electrically conductive nature may be present.

NOTE: Combustible dusts which are electrically nonconductive include dusts produced in the handling and processing of grain and grain products, pulverized sugar and cocoa, dried egg and milk powders, pulverized spices, starch and pastes, potato and woodflour, oil meal from beans and seed, dried hay, and other organic materials which may produce combustible dusts when processed or handled. Dusts containing magnesium or aluminum are particularly hazardous and the use of extreme caution is necessary to avoid ignition and explosion.

Class II, Division 2. A Class II, Division 2 location is a location in which

(1)   Combustible dust will not normally be in suspension in the air in quantities sufficient to produce explosive or ignitable mixtures, and dust accumulations are normally insufficient to interfere with the normal operation of electrical equipment or other apparatus; or

(2)   Dust may be in suspension in the air as a result of infrequent malfunctioning of handling or processing equipment, and dust accumulations resulting therefrom may be ignitable by abnormal operation or failure of electrical equipment or other apparatus.

NOTE: This classification includes locations where dangerous concentrations of suspended dust would not be likely but where dust accumulations might form on or in the vicinity of electric equipment.

These areas may contain equipment from which appreciable quantities of dust would escape under abnormal operating conditions or be adjacent to a Class II Division 1 location, as described above, into which an explosive or ignitable concentration of dust may be put into suspension under abnormal operating conditions.

"Class III locations" Class III locations are those that are hazardous because of the presence of easily ignitable fibers or flyings but in which such fibers or flyings are not likely to be in suspension in the air in quantities sufficient to produce ignitable mixtures. Class 111 locations include the following:

" **Class III, Division 1. A Class III, Division 1"** location is a location in which easily ignitable fibers or materials producing combustible flyings are handled, manufactured, or used.

NOTE: Easily ignitable fibers and flyings include rayon, cotton (including cotton linters and cotton waste), sisal or henequen, istle, jute, hemp, tow, cocoa fiber, oakum, baled waste kapok, Spanish moss, excelsior, sawdust, woodchips, and other material of similar nature.

"Class III, Division 2. A Class III, Division 2" location is a location in which easily ignitable fibers are stored or handled, except in process of manufacture.

"Collector ring." A collector ring is an assembly of slip rings for transferring electrical energy from a stationary to a rotating member.

"Concealed." Rendered inaccessible by the structure or finish of the building. Wires in concealed raceways are considered concealed, even though they may become accessible by withdrawing them. [See "Accessible. (As applied to wiring methods.)"]

"Conductor" –

"Bare." A conductor having no covering or electrical insulation whatsoever.

"Covered." A conductor encased within material of composition or thickness that is not recognized as electrical insulation.

"Insulated." A conductor encased within material of composition and thickness that is recognized as electrical insulation.

"Controller." A device or group of devices that serves to govern, in some predetermined manner, the electric power delivered to the apparatus to which it is connected.

"Covered conductor." See "Conductor."

"Cutout." (Over 600 volts, nominal.) An assembly of a fuse support with either a fuseholder, fuse carrier, or disconnecting blade. The fuseholder or fuse carrier may include a conducting element (fuse link), or may act as the disconnecting blade by the inclusion of a nonfusible member.

"Cutout box." An enclosure designed for surface mounting and having swinging doors or covers secured directly to and telescoping with the walls of the box proper. (See "Cabinet.")

"Damp location." See "Location."

"Dead front." Without live parts exposed to a person on the operating side of the equipment.

"Device." A unit of an electrical system which is intended to carry but not utilize electric energy.

"Disconnecting means." A device, or group of devices, or other means by which the conductors of a circuit can be disconnected from their source of supply.

"Disconnecting (or Isolating) switch." (Over 600 volts, nominal.) A mechanical switching device used for isolating a circuit or equipment from a source of power.

"Dry location." See "Location."

"Enclosed." Surrounded by a case, housing, fence or walls which will prevent persons from accidentally contacting energized parts.

"Enclosure." The case or housing of apparatus, or the fence or walls surrounding an installation to prevent personnel from accidentally contacting energized parts, or to protect the equipment from physical damage.

"Equipment." A general term including material, fittings, devices, appliances, fixtures, apparatus, and the like, used as a part of, or in connection with, an electrical installation.

"Equipment grounding conductor." See "Grounding conductor, equipment."

"Explosion-proof apparatus." Apparatus enclosed in a case that is capable of withstanding an explosion of a specified gas or vapor which may occur within it and of preventing the ignition of a specified gas or vapor surrounding the enclosure by sparks, flashes, or explosion of the gas or vapor within, and which operates at such an external temperature that it will not ignite a surrounding flammable atmosphere.

"Exposed." (As applied to live parts.) Capable of being inadvertently touched or approached nearer than a safe distance by a person. It is applied to parts not suitably guarded, isolated, or insulated. (See "Accessible" and "Concealed.")

"Exposed." (As applied to wiring methods.) On or attached to the surface or behind panels designed to allow access. [See "Accessible. (As applied to wiring methods.)"]

"Exposed." (For the purposes of 1926.408(d), Communications systems.) Where the circuit is in such a position that in case of failure of supports or insulation, contact with another circuit may result.

"Externally operable." Capable of being operated without exposing the operator to contact with live parts.

"Feeder." All circuit conductors between the service equipment, or the generator switchboard of an isolated plant, and the final branch-circuit overcurrent device.

"Festoon lighting." A string of outdoor lights suspended between two points more than 15 feet (4.57 m) apart.

"Fitting." An accessory such as a locknut, bushing, or other part of a wiring system that is intended primarily to perform a mechanical rather than an electrical function.

"Fuse." (Over 600 volts, nominal.) An overcurrent protective device with a circuit opening fusible part that is heated and severed by the passage of overcurrent through it. A fuse comprises all the parts that form a unit capable of performing the prescribed functions. It may or may not be the complete device necessary to connect it into an electrical circuit.

"Ground." A conducting connection, whether intentional or accidental, between an electrical circuit or equipment and the earth, or to some conducting body that serves in place of the earth.

"Grounded." Connected to earth or to some conducting body that serves in place of the earth. Grounded, effectively (Over 600 volts, nominal.) Permanently connected to earth through a ground connection of sufficiently low impedance and having sufficient ampacity that ground fault current which may occur cannot build up to voltages dangerous to personnel.

"Grounded conductor." A system or circuit conductor that is intentionally grounded.

"Grounding conductor." A conductor used to connect equipment or the grounded circuit of a wiring system to a grounding electrode or electrodes.

"Grounding conductor, equipment." The conductor used to connect the noncurrent-carrying metal parts of equipment, raceways, and other enclosures to the system grounded conductor and/or the grounding electrode conductor at the service equipment or at the source of a separately derived system.

"Grounding electrode conductor." The conductor used to connect the grounding electrode to the equipment grounding conductor and/or to the grounded conductor of the circuit at the service equipment or at the source of a separately derived system.

"Ground-fault circuit interrupter." A device for the protection of personnel that functions to deenergize a circuit or portion thereof within an established period of time when a current to ground exceeds some predetermined value that is less than that required to operate the overcurrent protective device of the supply circuit.

"Guarded." Covered, shielded, fenced, enclosed, or otherwise protected by means of suitable covers, casings, barriers, rails, screens, mats, or platforms to remove the likelihood of approach to a point of danger or contact by persons or objects.

"Hoistway." Any shaftway, hatchway, well hole, or other vertical opening or space in which an elevator or dumbwaiter is designed to operate.

"Identified (conductors or terminals)." Identified, as used in reference to a conductor or its terminal, means that such conductor or terminal can be recognized as grounded.

"Identified (for the use)." Recognized as suitable for the specific purpose, function, use, environment, application, etc. where described as a requirement in this standard. Suitability of equipment for a specific purpose, environment, or application is determined by a qualified testing laboratory where such identification includes labeling or listing.

"Insulated conductor." See "Conductor."

"Interrupter switch." (Over 600 volts, nominal.) A switch capable of making, carrying, and interrupting specified currents.

"Intrinsically safe equipment and associated wiring." Equipment and associated wiring in which any spark or thermal effect, produced either normally or in specified fault conditions, is incapable, under certain prescribed test conditions, of causing ignition of a mixture of flammable or combustible material in air in its most easily ignitable concentration.

"Isolated." Not readily accessible to persons unless special means for access are used.

"Isolated power system." A system comprising an isolating transformer or its equivalent, a line isolation monitor, and its ungrounded circuit conductors.

"Labeled." Equipment or materials to which has been attached a label, symbol or other identifying mark of a qualified testing laboratory which indicates compliance with appropriate standards or performance in a specified manner.

"Lighting outlet." An outlet intended for the direct connection of a lampholder, a lighting fixture, or a pendant cord terminating in a lampholder.

"Listed." Equipment or materials included in a list published by a qualified testing laboratory whose listing states either that the equipment or material meets appropriate standards or has been tested and found suitable for use in a specified manner.

"Location" –

(a)   Damp location. Partially protected locations under canopies, marquees, roofed open porches, and like locations, and interior locations subject to moderate degrees of moisture, such as some basements.

(b)   Dry location. A location not normally subject to dampness or wetness. A location classified as dry may be temporarily subject to dampness or wetness, as in the case of a building under construction.

(c)   Wet location. Installations underground or in concrete slabs or masonry in direct contact with the earth, and locations subject to saturation with water or other liquids, such as locations exposed to weather and unprotected.

"Mobile X-ray." X-ray equipment mounted on a permanent base with wheels and/or casters for moving while completely assembled.

"Motor control center." An assembly of one or more enclosed sections having a common power bus and principally containing motor control units.

"Outlet." A point on the wiring system at which current is taken to supply utilization equipment.

"Overcurrent." Any current in excess of the rated current of equipment or the ampacity of a conductor. It may result from overload (see definition), short circuit, or ground fault. A current in excess of rating may be accommodated by certain equipment and conductors for a given set of conditions. Hence the rules for overcurrent protection are specific for particular situations.

"Overload." Operation of equipment in excess of normal, full load rating, or of a conductor in excess of rated ampacity which, when it persists for a sufficient length of time, would cause damage or dangerous overheating. A fault, such as a short circuit or ground fault, is not an overload. (See "Overcurrent.")

"Panelboard." A single panel or group of panel units designed for assembly in the form of a single panel; including buses, automatic overcurrent devices, and with or without switches for the control of light, heat, or power circuits; designed to be placed in a cabinet or cutout box placed in or against a wall or partition and accessible only from the front. (See "Switchboard.")

"Portable X-ray." X-ray equipment designed to be hand-carried.

"Power fuse." (Over 600 volts, nominal.) See "Fuse."

"Power outlet." An enclosed assembly which may include receptacles, circuit breakers, fuseholders, fused switches, buses and watt-hour meter mounting means; intended to serve as a means for distributing power required to operate mobile or temporarily installed equipment.

"Premises wiring system." That interior and exterior wiring, including power, lighting, control, and signal circuit wiring together with all of its associated hardware, fittings, and wiring devices, both permanently and temporarily installed, which extends from the load end of the service drop, or load end of the service lateral conductors to the outlet(s). Such wiring does not include wiring internal to appliances, fixtures, motors, controllers, motor control centers, and similar equipment.

"Qualified person." One familiar with the construction and operation of the equipment and the hazards involved.

"Qualified testing laboratory." A properly equipped and staffed testing laboratory which has capabilities for and which provides the following services:

(a) Experimental testing for safety of specified items of equipment and materials referred to in this standard to determine compliance with appropriate test standards or performance in a specified manner;

(b) Inspecting the run of such items of equipment and materials at factories for product evaluation to assure compliance with the test standards;

(c) Service-value determinations through field inspections to monitor the proper use of labels on products and with authority for recall of the label in the event a hazardous product is installed;

(d) Employing a controlled procedure for identifying the listed and/or labeled equipment or materials tested; and

(e) Rendering creditable reports or findings that are objective and without bias of the tests and test methods employed.

"Raceway." A channel designed expressly for holding wires, cables, or busbars, with additional functions as permitted in this subpart. Raceways may be of metal or insulating material, and the term includes rigid metal conduit, rigid nonmetallic conduit, intermediate metal conduit, liquidtight flexible metal conduit, flexible metallic tubing, flexible metal conduit, electrical metallic tubing, underfloor raceways, cellular concrete floor raceways, cellular metal floor raceways, surface raceways, wireways, and busways.

"Readily accessible." Capable of being reached quickly for operation, renewal, or inspections, without requiring those to whom ready access is requisite to climb over or remove obstacles or to resort to portable ladders, chairs, etc. (See "Accessible.")

"Receptacle." A receptacle is a contact device installed at the outlet for the connection of a single attachment plug. A single receptacle is a single contact device with no other contact device on the same yoke. A multiple receptacle is a single device containing two or more receptacles.

"Receptacle outlet." An outlet where one or more receptacles are installed.

"Remote-control circuit." Any electric circuit that controls any other circuit through a relay or an equivalent device.

"Sealable equipment." Equipment enclosed in a case or cabinet that is provided with a means of sealing or locking so that live parts cannot be made accessible without opening the enclosure. The equipment may or may not be operable without opening the enclosure.

"Separately derived system." A premises wiring system whose power is derived from generator, transformer, or converter windings and has no direct electrical connection, including a solidly connected grounded circuit conductor, to supply conductors originating in another system.

"Service." The conductors and equipment for delivering energy from the electricity supply system to the wiring system of the premises served.

"Service conductors." The supply conductors that extend from the street main or from transformers to the service equipment of the premises supplied.

"Service drop." The overhead service conductors from the last pole or other aerial support to and including the splices, if any, connecting to the service-entrance conductors at the building or other structure.

"Service-entrance conductors, overhead system." The service conductors between the terminals of the service equipment and a point usually outside the building, clear of building walls, where joined by tap or splice to the service drop.

"Service-entrance conductors, underground system." The service conductors between the terminals of the service equipment and the point of connection to the service lateral. Where service equipment is located outside the building walls, there may be no service-entrance conductors, or they may be entirely outside the building.

"Service equipment." The necessary equipment, usually consisting of a circuit breaker or switch and fuses, and their accessories, located near the point of entrance of supply conductors to a building or other structure, or an otherwise defined area, and intended to constitute the main control and means of cutoff of the supply.

"Service raceway." The raceway that encloses the service-entrance conductors.

"Signaling circuit." Any electric circuit that energizes signaling equipment.

"Switchboard." A large single panel, frame, or assembly of panels which have switches, buses, instruments, overcurrent and other protective devices mounted on the face or back or both. Switchboards are generally accessible from the rear as well as from the front and are not intended to be installed in cabinets. (See "Panelboard.")

"Switches" –

    (a)   General-use switch. A switch intended for use in general distribution and branch circuits. It is rated in amperes, and it is capable of interrupting its rated current at its rated voltage.

    (b)   General-use snap switch. A form of general-use switch so constructed that it can be installed in flush device boxes or on outlet box covers, or otherwise used in conjunction with wiring systems recognized by this subpart.

    (c)   Isolating switch. A switch intended for isolating an electric circuit from the source of power. It has no interrupting rating, and it is intended to be operated only after the circuit has been opened by some other means.

    (d)   Motor-circuit switch. A switch, rated in horsepower, capable of interrupting the maximum operating overload current of a motor of the same horsepower rating as the switch at the rated voltage.

"Switching devices." (Over 600 volts, nominal.) Devices designed to close and/or open one or more electric circuits. Included in this category are circuit breakers, cutouts, disconnecting (or isolating) switches, disconnecting means, and interrupter switches.

"Transportable X-ray." X-ray equipment installed in a vehicle or that may readily be disassembled for transport in a vehicle.

"Utilization equipment." Utilization equipment means equipment which utilizes electric energy for mechanical, chemical, heating, lighting, or similar useful purpose.

"Utilization system." A utilization system is a system which provides electric power and light for employee workplaces, and includes the premises wiring system and utilization equipment.

"Ventilated." Provided with a means to permit circulation of air sufficient to remove an excess of heat, fumes, or vapors.

"Volatile flammable liquid." A flammable liquid having a flash point below 38 degrees C (100 degrees F) or whose temperature is above its flash point, or a Class II combustible liquid having a vapor pressure not exceeding 40 psia (276 kPa) at 38 degree C (100 degree F) whose temperature is above its flash point.

"Voltage." (Of a circuit.) The greatest root-mean-square (effective) difference of potential between any two conductors of the circuit concerned.

"Voltage, nominal." A nominal value assigned to a circuit or system for the purpose of conveniently designating its voltage class (as 120/240, 480Y/277, 600, etc.). The actual voltage at which a circuit operates can vary from the nominal within a range that permits satisfactory operation of equipment.

"Voltage to ground." For grounded circuits, the voltage between the given conductor and that point or conductor of the circuit that is grounded; for ungrounded circuits, the greatest voltage between the given conductor and any other conductor of the circuit.

"Watertight." So constructed that moisture will not enter the enclosure.

"Weatherproof." So constructed or protected that exposure to the weather will not interfere with successful operation. Rainproof, raintight, or watertight equipment can fulfill the requirements for weatherproof where varying weather conditions other than wetness, such as snow, ice, dust, or temperature extremes, are not a factor.

"Wet location." See "Location."

SUBPART L – SCAFFOLDS

"Adjustable suspension scaffold" means a suspension scaffold equipped with a hoist(s) that can be operated by an employee(s) on the scaffold.

"Bearer (putlog)" means a horizontal transverse scaffold member (which may be supported by ledgers or runners) upon which the scaffold platform rests and which joins scaffold uprights, posts, poles, and similar members.

"Boatswains' chair" means a single-point adjustable suspension scaffold consisting of a seat or sling designed to support one employee in a sitting position.

"Body belt (safety belt)" means a strap with means both for securing it about the waist and for attaching it to a lanyard, lifeline, or deceleration device.

"Body harness" means a design of straps which may be secured about the employee in a manner to distribute the fall arrest forces over at least the thighs, pelvis, waist, chest and shoulders, with means for attaching it to other components of a personal fall arrest system.

"Brace" means a rigid connection that holds one scaffold member in a fixed position with respect to another member, or to a building or structure.

"Bricklayers' square scaffold" means a supported scaffold composed of framed squares which support a platform.

"Carpenters' bracket scaffold" means a supported scaffold consisting of a platform supported by brackets attached to building or structural walls.

"Catenary scaffold" means a suspension scaffold consisting of a platform supported by two essentially horizontal and parallel ropes attached to structural members of a building or other structure. Additional support may be provided by vertical pickups.

"Chimney hoist" means a multipoint adjustable suspension scaffold used to provide access to work inside chimneys. (See "Multipoint adjustable suspension scaffold.")

"Cleat" means a structural block used at the end of a platform to prevent the platform from slipping off its supports. Cleats are also used to provide footing on sloped surfaces such as crawling boards.

"Competent person" means one who is capable of identifying existing and predictable hazards in the surroundings or working conditions which are unsanitary, hazardous, or dangerous to employees, and who has authorization to take prompt corrective measures to eliminate them.

"Continuous run scaffold (Run scaffold)" means a two-point or multipoint adjustable suspension scaffold constructed using a series of interconnected braced scaffold members or supporting structures erected to form a continuous scaffold.

"Coupler" means a device for locking together the tubes of a tube and coupler scaffold.

"Crawling board (chicken ladder)" means a supported scaffold consisting of a plank with cleats spaced and secured to provide footing, for use on sloped surfaces such as roofs.

"Deceleration device" means any mechanism, such as a rope grab, rip-stitch lanyard, specially-woven lanyard, tearing or deforming lanyard, or automatic self-retracting lifeline lanyard, which dissipates a substantial amount of energy during a fall arrest or limits the energy imposed on an employee during fall arrest.

"Double pole (independent pole) scaffold" means a supported scaffold consisting of a platform(s) resting on cross beams (bearers) supported by ledgers and a double row of uprights independent of support (except ties, guys, braces) from any structure.

"Equivalent" means alternative designs, materials, or methods to protect against a hazard which the employer can demonstrate will provide an equal or greater degree of safety for employees than the methods, materials, or designs specified in the standard.

"Exposed power lines" means electrical power lines which are accessible to employees and which are not shielded from contact. Such lines do not include extension cords or power tool cords.

"Eye or Eye splice" means a loop with or without a thimble at the end of a wire rope.

"Fabricated decking and planking" means manufactured platforms made of wood (including laminated wood, and solid sawn wood, planks), metal, or other materials.

"Fabricated frame scaffold (tubular welded frame scaffold)" means a scaffold consisting of a platform(s) supported on fabricated end frames with integral posts, horizontal bearers, and intermediate members.

"Failure" means load refusal, breakage, or separation of component parts. Load refusal is the point where the ultimate strength is exceeded.

"Float (ship) scaffold" means a suspension scaffold consisting of a braced platform resting on two parallel bearers and hung from overhead supports by ropes of fixed length.

"Form scaffold" means a supported scaffold consisting of a platform supported by brackets attached to formwork.

"Guardrail system" means a vertical barrier, consisting of, but not limited to, toprails, midrails, and posts, erected to prevent employees from falling off a scaffold platform or walkway to lower levels.

"Hoist" means a manual or power-operated mechanical device to raise or lower a suspended scaffold.

"Horse scaffold" means a supported scaffold consisting of a platform supported by construction horses (saw horses). Horse scaffolds constructed of metal are sometimes known as trestle scaffolds.

"Independent pole scaffold" (see "Double pole scaffold").

"Interior hung scaffold" means a suspension scaffold consisting of a platform suspended from the ceiling or roof structure by fixed length supports.

"Ladder jack scaffold" means a supported scaffold consisting of a platform resting on brackets attached to ladders.

"Ladder stand" means a mobile, fixed-size, self-supporting ladder consisting of a wide flat tread ladder in the form of stairs.

"Landing" means a platform at the end of a flight of stairs.

"Large area scaffold" means a pole scaffold, tube and coupler scaffold, systems scaffold, or fabricated frame scaffold erected over substantially the entire work area. For example, a scaffold erected over the entire floor area of a room.

"Lean-to scaffold" means a supported scaffold which is kept erect by tilting it toward and resting it against a building or structure.

"Lifeline" means a component consisting of a flexible line that connects to an anchorage at one end to hang vertically (vertical lifeline), or that connects to anchorages at both ends to stretch horizontally (horizontal lifeline), and which serves as a means for connecting other components of a personal fall arrest system to the anchorage.

"Lower levels" means areas below the level where the employee is located and to which an employee can fall. Such areas include, but are not limited to, ground levels, floors, roofs, ramps, runways, excavations, pits, tanks, materials, water, and equipment.

"Masons' adjustable supported scaffold" (see "Self-contained adjustable scaffold").

"Masons' multipoint adjustable suspension scaffold" means a continuous run suspension scaffold designed and used for masonry operations.

"Maximum intended load" means the total load of all persons, equipment, tools, materials, transmitted loads, and other loads reasonably anticipated to be applied to a scaffold or scaffold component at any one time.

"Mobile scaffold" means a powered or unpowered, portable, caster or wheel-mounted supported scaffold.

"Multi-level suspended scaffold" means a two-point or multipoint adjustable suspension scaffold with a series of platforms at various levels resting on common stirrups.

"Multipoint adjustable suspension scaffold" means a suspension scaffold consisting of a platform(s) which is suspended by more than two ropes from overhead supports and equipped with means to raise and lower the platform to desired work levels. Such scaffolds include chimney hoists.

"Needle beam scaffold" means a platform suspended from needle beams.

"Open sides and ends" means the edges of a platform that are more than 14 inches (36 cm) away horizontally from a sturdy, continuous, vertical surface (such as a building wall) or a sturdy, continuous horizontal surface (such as a floor), or a point of access. Exception: For plastering and lathing operations the horizontal threshold distance is 18 inches (46 cm).

"Outrigger" means the structural member of a supported scaffold used to increase the base width of a scaffold in order to provide support for and increased stability of the scaffold.

"Outrigger beam (Thrustout)" means the structural member of a suspension scaffold or outrigger scaffold which provides support for the scaffold by extending the scaffold point of attachment to a point out and away from the structure or building.

"Outrigger scaffold" means a supported scaffold consisting of a platform resting on outrigger beams (thrustouts) projecting beyond the wall or face of the building or structure, the inboard ends of which are secured inside the building or structure.

"Overhand bricklaying" means the process of laying bricks and masonry units such that the surface of the wall to be jointed is on the opposite side of the wall from the mason, requiring the mason to lean over the wall to complete the work. It includes mason tending and electrical installation incorporated into the brick wall during the overhand bricklaying process.

"Personal fall arrest system" means a system used to arrest an employee's fall. It consists of an anchorage, connectors, a body belt or body harness and may include a lanyard, deceleration device, lifeline, or combinations of these.

"Platform" means a work surface elevated above lower levels. Platforms can be constructed using individual wood planks, fabricated planks, fabricated decks, and fabricated platforms.

"Pole scaffold" (see definitions for "Single-pole scaffold" and "Double (independent) pole scaffold").

"Power operated hoist" means a hoist which is powered by other than human energy.

"Pump jack scaffold" means a supported scaffold consisting of a platform supported by vertical poles and movable support brackets.

"Qualified" means one who, by possession of a recognized degree, certificate, or professional standing, or who by extensive knowledge, training, and experience, has successfully demonstrated his/her ability to solve or resolve problems related to the subject matter, the work, or the project.

"Rated load" means the manufacturer's specified maximum load to be lifted by a hoist or to be applied to a scaffold or scaffold component.

"Repair bracket scaffold" means a supported scaffold consisting of a platform supported by brackets which are secured in place around the circumference or perimeter of a chimney, stack, tank, or other supporting structure by one or more wire ropes placed around the supporting structure.

"Roof bracket scaffold" means a rooftop supported scaffold consisting of a platform resting on angular-shaped supports.

"Runner (ledger or ribbon)" means the lengthwise horizontal spacing or bracing member which may support the bearers.

"Scaffold" means any temporary elevated platform (supported or suspended) and its supporting structure (including points of anchorage), used for supporting employees or materials or both.

"Self-contained adjustable scaffold" means a combination supported and suspension scaffold consisting of an adjustable platform(s) mounted on an independent supporting frame(s) not a part of the object being worked on, and which is equipped with a means to permit the raising and lowering of the platform(s). Such systems include rolling roof rigs, rolling outrigger systems, and some masons' adjustable supported scaffolds.

"Shore scaffold" means a supported scaffold which is placed against a building or structure and held in place with props.

"Single-point adjustable suspension scaffold" means a suspension scaffold consisting of a platform suspended by one rope from an overhead support and equipped with means to permit the movement of the platform to desired work levels.

"Single-pole scaffold" means a supported scaffold consisting of a platform(s) resting on bearers, the outside ends of which are supported on runners secured to a single row of posts or uprights, and the inner ends of which are supported on or in a structure or building wall.

"Stair tower (Scaffold stairway/tower)" means a tower comprised of scaffold components and which contains internal stairway units and rest platforms. These towers are used to provide access to scaffold platforms and other elevated points such as floors and roofs.

"Stall load" means the load at which the prime-mover of a power-operated hoist stalls or the power to the prime-mover is automatically disconnected.

"Step, platform, and trestle ladder scaffold" means a platform resting directly on the rungs of step ladders or trestle ladders.

"Stilts" means a pair of poles or similar supports with raised footrests, used to permit walking above the ground or working surface.

"Stonesetters' multipoint adjustable suspension scaffold" means a continuous run suspension scaffold designed and used for stonesetters' operations.

"Supported scaffold" means one or more platforms supported by outrigger beams, brackets, poles, legs, uprights, posts, frames, or similar rigid support.

"Suspension scaffold" means one or more platforms suspended by ropes or other non-rigid means from an overhead structure(s).

"System scaffold" means a scaffold consisting of posts with fixed connection points that accept runners, bearers, and diagonals that can be interconnected at predetermined levels.

"Tank builders' scaffold" means a supported scaffold consisting of a platform resting on brackets that are either directly attached to a cylindrical tank or attached to devices that are attached to such a tank.

"Top plate bracket scaffold" means a scaffold supported by brackets that hook over or are attached to the top of a wall. This type of scaffold is similar to carpenters' bracket scaffolds and form scaffolds and is used in residential construction for setting trusses.

"Tube and coupler scaffold" means a supported or suspended scaffold consisting of a platform(s) supported by tubing, erected with coupling devices connecting uprights, braces, bearers, and runners.

"Tubular welded frame scaffold" (see "Fabricated frame scaffold").

"Two-point suspension scaffold (swing stage)" means a suspension scaffold consisting of a platform supported by hangers (stirrups) suspended by two ropes from overhead supports and equipped with means to permit the raising and lowering of the platform to desired work levels.

"Unstable objects" means items whose strength, configuration, or lack of stability may allow them to become dislocated and shift and therefore may not properly support the loads imposed on them. Unstable objects do not constitute a safe base support for scaffolds, platforms, or employees. Examples include, but are not limited to, barrels, boxes, loose brick, and concrete blocks.

"Vertical pickup" means a rope used to support the horizontal rope in catenary scaffolds.

"Walkway" means a portion of a scaffold platform used only for access and not as a work level.

"Window jack scaffold" means a platform resting on a bracket or jack which projects through a window opening.

"Working load" means a load imposed by men, materials, and equipment.

SUBPART M – FALL PROTECTION

"Anchorage" means a secure point of attachment for lifelines, lanyards or deceleration devices.

"Body belt (safety belt)" means a strap with means both for securing it about the waist and for attaching it to a lanyard, lifeline, or deceleration device.

"Body harness" means straps which may be secured about the employee in a manner that will distribute the fall arrest forces over at least the thighs, pelvis, waist, chest, and shoulders with means for attaching it to other components of a personal fall arrest system.

"Buckle" means any device for holding the body belt or body harness closed around the employee's body.

"Connector" means a device which is used to couple (connect) parts of the personal fall arrest system and positioning device systems together. It may be an independent component of the system, such as a carabineer, or it may be an integral component of part of the system (such as a buckle or deering sewn into a body belt or body harness, or a snap-hook spliced or sewn to a lanyard or self-retracting lanyard).

"Controlled access zone (CAZ)" means an area in which certain work (e.g., overhand bricklaying) may take place without the use of guardrail systems, personal fall arrest systems, or safety net systems and access to the zone is controlled.

"Dangerous equipment" means equipment (such as pickling or galvanizing tanks, degreasing units, machinery, electrical equipment, and other units) which, as a result of form or function, may be hazardous to employees who fall onto or into such equipment.

"Deceleration device" means any mechanism, such as a rope grab, lanyards, automatic self-retracting lifelines/lanyards, etc., which serves to dissipate a substantial amount of energy during a fall arrest, or otherwise limit the energy imposed on an employee during fall arrest.

"Deceleration distance" means the additional vertical distance a falling employee travels, excluding lifeline elongation and free fall distance, before stopping, from the point at which the deceleration device begins to operate. It is measured as the distance between the location of an employee's body belt or body harness attachment point at the moment of activation (at the onset of fall arrest forces) of the deceleration device during a fall, and the location of that attachment point after the employee comes to a full stop.

"Equivalent" means alternative designs, materials, or methods to protect against a hazard which the employer can demonstrate will provide an equal or greater degree of safety for employees than the methods, materials, or designs specified in the standard.

"Failure" means load refusal, breakage, or separation of component parts. Load refusal is the point where the ultimate strength is exceeded.

"Free fall" means the act of falling before a personal fall arrest system begins to apply force to arrest the fall.

"Free fall distance" means the vertical displacement of the fall arrest attachment point on the employee's body belt or body harness between onset of the fall and just before the system begins to apply force to arrest the fall. This distance excludes deceleration distance, and lifeline/lanyard elongation, but includes any deceleration device slide distance or self-retracting lifeline/lanyard extension before they operate and fall arrest forces occur.

"Guardrail system" means a barrier erected to prevent employees from falling to lower levels.

"Hole" means a gap or void 2 inches (5.1 cm) or more in its least dimension, in a floor, roof, or other walking/working surface.

"Infeasible" means that it is impossible to perform the construction work using a conventional fall protection system (i.e., guardrail system, safety net system, or personal fall arrest system) or that it is technologically impossible to use any one of these systems to provide fall protection.

"Lanyard" means a flexible line of rope, wire rope, or strap which generally has a connector at each end for connecting the body belt or body harness to a deceleration device, lifeline, or anchorage.

"Leading edge" means the edge of a floor, roof, or formwork for a floor or other walking/working surface (such as the deck) which changes location as additional floor, roof, decking, or formwork sections are placed, formed, or constructed. A leading edge is considered to be an "unprotected side and edge" during periods when it is not actively and continuously under construction.

"Lifeline" means a component consisting of a flexible line for connection to an anchorage at one end to hang vertically (vertical lifeline), or for connection to anchorages at both ends to stretch horizontally (horizontal lifeline), and which serves as a means for connecting other components of a personal fall arrest system to the anchorage.

"Low-slope roof" means a roof having a slope less than or equal to 4 in 12 (vertical to horizontal).

"Lower levels" means those areas or surfaces to which an employee can fall. Such areas or surfaces include, but are not limited to, ground levels, floors, platforms, ramps, runways, excavations, pits, tanks, material, water, equipment, structures, or portions thereof.

"Mechanical equipment" means all motor or human propelled wheeled equipment used for roofing work, except wheelbarrows and mopcarts.

"Opening" means a gap or void 30 inches (76 cm) or more high and 18 inches (48 cm) or more wide, in a wall or partition, through which employees can fall to a lower level.

"Overhand bricklaying and related work" means the process of laying bricks and masonry units such that the surface of the wall to be jointed is on the opposite side of the wall from the mason, requiring the mason to lean over the wall to complete the work. Related work includes mason tending and electrical installation incorporated into the brick wall during the overhand bricklaying process.

"Personal fall arrest system" means a system used to arrest an employee in a fall from a working level. It consists of an anchorage, connectors, a body belt or body harness and may include a lanyard, deceleration device, lifeline, or suitable combinations of these. As of January 1, 1998, the use of a body belt for fall arrest is prohibited.

"Positioning device system" means a body belt or body harness system rigged to allow an employee to be supported on an elevated vertical surface, such as a wall, and work with both hands free while leaning.

"Rope grab" means a deceleration device which travels on a lifeline and automatically, by friction, engages the lifeline and locks so as to arrest the fall of an employee. A rope grab usually employs the principle of inertial locking, cam/level locking, or both.

"Roof" means the exterior surface on the top of a building. This does not include floors or formwork which, because a building has not been completed, temporarily become the top surface of a building.

"Roofing work" means the hoisting, storage, application, and removal of roofing materials and equipment, including related insulation, sheet metal, and vapor barrier work, but not including the construction of the roof deck.

"Safety-monitoring system" means a safety system in which a competent person is responsible for recognizing and warning employees of fall hazards.

"Self-retracting lifeline/lanyard" means a deceleration device containing a drum-wound line which can be slowly extracted from, or retracted onto, the drum under slight tension during normal employee movement, and which, after onset of a fall, automatically locks the drum and arrests the fall.

"Snaphook" means a connector comprised of a hook-shaped member with a normally closed keeper, or similar arrangement, which may be opened to permit the hook to receive an object and, when released, automatically closes to retain the object. Snaphooks are generally one of two types:

(1)   The locking type with a self-closing, self-locking keeper which remains closed and locked until unlocked and pressed open for connection or disconnection; or

(2)   The non-locking type with a self-closing keeper which remains closed until pressed open for connection or disconnection. As of January 1, 1998, the use of a non-locking snaphook as part of personal fall arrest systems and positioning device systems is prohibited.

"Steep roof" means a roof having a slope greater than 4 in 12 (vertical to horizontal).

"Toeboard" means a low protective barrier that will prevent the fall of materials and equipment to lower levels and provide protection from falls for personnel.

"Unprotected sides and edges" means any side or edge (except at entrances to points of access) of a walking/working surface, e.g., floor, roof, ramp, or runway where there is no wall or guardrail system at least 39 inches (1.0 m) high.

"Walking/working surface" means any surface, whether horizontal or vertical on which an employee walks or works, including, but not limited to, floors, roofs, ramps, bridges, runways, formwork and concrete reinforcing steel but not including ladders, vehicles, or trailers, on which employees must be located in order to perform their job duties.

"Warning line system" means a barrier erected on a roof to warn employees that they are approaching an unprotected roof side or edge, and which designates an area in which roofing work may take place without the use of guardrail, body belt, or safety net systems to protect employees in the area.

"Work area" means that portion of a walking/working surface where job duties are being performed.

SUBPART N – CRANES AND DERRICKS

"Failure" means load refusal, breakage, or separation of components.

"Hoist" (or hoisting) means all crane or derrick functions such as lowering, lifting, swinging, booming in and out or up and down, or suspending a personnel platform.

"Load refusal" means the point where the ultimate strength is exceeded.

"Maximum intended load" means the total load of all employees, tools, materials, and other loads reasonably anticipated to be applied to a personnel platform or personnel platform component at any one time.

"Runway" means a firm, level surface designed, prepared, and designated as a path of travel for the weight and configuration of the crane being used to lift and travel with the crane suspended platform. An existing surface may be used as long as it meets these criteria.

SUBPART O – MOTOR VEHICLES, MECHANIZED EQUIPMENT, AND MARINE OPERATIONS

"Apron." The area along the waterfront edge of the pier or wharf.

"Bulwark." The side of a ship above the upper deck.

"Coaming." The raised frame, as around a hatchway in the deck, to keep out water.

"Jacob's ladder." A marine ladder of rope or chain with wooden or metal rungs.

"Rail" for the purpose of 1926.605, means a light structure serving as a guard at the outer edge of a ship's deck.

SUBPART P – EXCAVATIONS

"Accepted engineering practices" means those requirements which are compatible with standards of practice required by a registered professional engineer.

"Aluminum Hydraulic Shoring" means a pre-engineered shoring system comprised of aluminum hydraulic cylinders (crossbraces) used in conjunction with vertical rails (uprights) or horizontal rails (wales). Such system is designed specifically to support the sidewalls of an excavation and prevent cave-ins.

"Bell-bottom pier hole" means a type of shaft or footing excavation, the bottom of which is made larger than the cross section above to form a belled shape.

"Benching (Benching system)" means a method of protecting employees from cave-ins by excavating the sides of an excavation to form one or a series of horizontal levels or steps, usually with vertical or near-vertical surfaces between levels.

"Cave-in" means the separation of a mass of soil or rock material from the side of an excavation, or the loss of soil from under a trench shield or support system, and its sudden movement into the excavation, either by falling or sliding, in sufficient quantity so that it could entrap, bury, or other wise injure and immobilize a person.

"Competent person" means one who is capable of identifying existing and predictable hazards in the surroundings, or working conditions which are unsanitary, hazardous, or dangerous to employees, and who has authorization to take prompt corrective measures to eliminate them.

"Cross braces" mean the horizontal members of a shoring system installed perpendicular to the sides of the excavation, the ends of which bear against either uprights or wales.

"Excavation" means any man-made cut, cavity, trench, or depression in an earth surface, formed by earth removal.

"Faces or sides" means the vertical or inclined earth surfaces formed as a result of excavation work.

"Failure" means the breakage, displacement, or permanent deformation of a structural member or connection so as to reduce its structural integrity and its supportive capabilities.

"Hazardous atmosphere" means an atmosphere which by reason of being explosive, flammable, poisonous, corrosive, oxidizing, irritating, oxygen deficient, toxic, or otherwise harmful, may cause death, illness, or injury.

"Kickout" means the accidental release or failure of a cross brace.

"Protective system" means a method of protecting employees from cave-ins, from material that could fall or roll from an excavation face or into an excavation, or from the collapse of adjacent structures. Protective systems include support systems, sloping and benching systems, shield systems, and other systems that provide the necessary protection.

"Ramp" means an inclined walking or working surface that is used to gain access to one point from another, and is constructed from earth or from structural materials such as steel or wood.

"Registered Professional Engineer" means a person who is registered as a professional engineer in the state where the work is to be performed. However, a professional engineer, registered in any state is deemed to be a "registered professional engineer" within the meaning of this standard when approving designs for "manufactured protective systems" or "tabulated data" to be used in interstate commerce.

"Sheeting" means the members of a shoring system that retain the earth in position and in turn are supported by other members of the shoring system.

"Shield (Shield system)" means a structure that is able to withstand the forces imposed on it by a cave-in and thereby protect employees within the structure. Shields can be permanent structures or can be designed to be portable and moved along as work progresses. Additionally, shields can be either premanufactured or job-built in accordance with 1926.652(c)(3) or (c)(4). Shields used in trenches are usually referred to as "trench boxes" or "trench shields."

"Shoring (Shoring system)" means a structure such as a metal hydraulic, mechanical or timber shoring system that supports the sides of an excavation and which is designed to prevent cave-ins.

"Sides" See "Faces."

"Sloping (Sloping system)" means a method of protecting employees from cave-ins by excavating to form sides of an excavation that are inclined away from the excavation so as to prevent cave-ins. The angle of incline required to prevent a cave-in varies with differences in such factors as the soil type, environmental conditions of exposure, and application of surcharge loads.

"Stable rock" means natural solid mineral material that can be excavated with vertical sides and will remain intact while exposed. Unstable rock is considered to be stable when the rock material on the side or sides of the excavation is secured against caving-in or movement by rock bolts or by another protective system that has been designed by a registered professional engineer.

"Structural ramp" means a ramp built of steel or wood, usually used for vehicle access. Ramps made of soil or rock are not considered structural ramps.

"Support system" means a structure such as underpinning, bracing, or shoring, which provides support to an adjacent structure, underground installation, or the sides of an excavation.

"Tabulated data" means tables and charts approved by a registered professional engineer and used to design and construct a protective system.

"Trench (Trench excavation)" means a narrow excavation (in relation to its length) made below the surface of the ground. In general, the depth is greater than the width, but the width of a trench (measured at the bottom) is not greater than 15 feet (4.6 m) or less (measured at the bottom of the excavation), the excavation is also considered to be a trench.

"Trench box" See "Shield."

"Trench shield" See "Shield."

"Uprights" means the vertical members of a trench shoring system placed in contact with the earth and usually positioned so that individual members do not contact each other. Uprights placed so that individual members are closely spaced, in contact with or interconnected to each other, are often called "sheeting."

"Wales" means horizontal members of a shoring system placed parallel to the excavation face whose sides bear against the vertical members of the shoring system or earth.

1926, Subpt. P, App. A – Soil Classification

"Cemented soil" means a soil in which the particles are held together by a chemical agent, such as calcium carbonate, such that a hand-size sample cannot be crushed into powder or individual soil particles by finger pressure.

"Cohesive soil" means clay (fine grained soil), or soil with a high clay content, which has cohesive strength. Cohesive soil does not crumble, can be excavated with vertical sidesloped, and is plastic when moist. Cohesive soil is hard to break up when dry, and exhibits significant cohesion when submerged. Cohesive soils include clayey silt, sandy clay, silty clay, clay, and organic clay.

"Dry soil" means soil that does not exhibit visible signs of moisture content.

"Fissured" means a soil material that has a tendency to break along definite planes of fracture with little resistance, or a material that exhibits open cracks, such as tension cracks, in an exposed surface.

"Granular soil" means gravel, sand, or silt (coarse grained soil) with little or no clay content. Granular soil has no cohesive strength. Some moist granular soils exhibit apparent cohesion. Granular soil cannot be molded when moist and crumbles easily when dry.

"Layered system" means two or more distinctly different soil or rock types arranged in layers. Micaceous seams or weakened planes in rock or shale are considered layered.

"Moist soil" means a condition in which a soil looks and feels damp. Moist cohesive soil can easily be shaped into a ball and rolled into small diameter threads before crumbling. Moist granular soil that contains some cohesive material will exhibit signs of cohesion between particles.

"Plastic" means a property of a soil which allows the soil to be deformed or molded without cracking, or appreciable volume change.

"Saturated soil" means a soil in which the voids are filled with water. Saturation does not require flow. Saturation, or near saturation, is necessary for the proper use of instruments such as a pocket penetrometer or sheer vane.

"Soil classification system" means, for the purpose of this subpart, a method of categorizing soil and rock deposits in a hierarchy of Stable Rock, Type A, Type B, and Type C, in decreasing order of stability. The categories are determined based on an analysis of the properties and performance characteristics of the deposits and the characteristics of the deposits and the environmental conditions of exposure.

"Stable rock" means natural solid mineral matter that can be excavated with vertical sides and remain intact while exposed.

"Submerged soil" means soil which is underwater or is free seeping.

"Type A" means cohesive soils with an unconfined, compressive strength of 1.5 ton per square foot (tsf) (144 kPa) or greater. Examples of cohesive soils are: clay, silty clay, sandy clay, clay loam and, in some cases, silty clay loam and sandy clay loam. Cemented soils such as caliche and hardpan are also considered Type A. However, no soil is Type A if:

(i)　　The soil is fissured; or

(ii)　　The soil is subject to vibration from heavy traffic, pile driving, or similar effects; or

(iii)　　The soil has been previously disturbed; or

(iv)　　The soil is part of a sloped, layered system where the layers dip into the excavation on a slope of four horizontal to one vertical (4H:1V) or greater; or

(v)　　The material is subject to other factors that would require it to be classified as a less stable material.

"Type B" means:

(i)   Cohesive soil with an unconfined compressive strength greater than 0.5 tsf (48 kPa) but less than 1.5 tsf (144 kPa); or

(ii) Granular cohesionless soils including: angular gravel (similar to crushed rock), silt, silt loam, sandy loam and, in some cases, silty clay loam and sandy clay loam.

(iii) Previously disturbed soils except those which would otherwise be classed as Type C soil.

(iv) Soil that meets the unconfined compressive strength or cementation requirements for Type A, but is fissured or subject to vibration; or

(v)   Dry rock that is not stable; or

(vi) Material that is part of a sloped, layered system where the layers dip into the excavation on a slope less steep than four horizontal to one vertical (4H:1V), but only if the material would otherwise be classified as Type B.

"Type C" means:

(i)   Cohesive soil with an unconfined compressive strength of 0.5 tsf (48 kPa) or less; or

(ii) Granular soils including gravel, sand, and loamy sand; or

(iii) Submerged soil or soil from which water is freely seeping; or

(iv) Submerged rock that is not stable; or

(v)   Material in a sloped, layered system where the layers dip into the excavation or a slope of four horizontal to one vertical (4H:1V) or steeper.

"Unconfined compressive strength" means the load per unit area at which a soil will fail in compression. It can be determined by laboratory testing, or estimated in the field using a pocket penetrometer, by thumb penetration tests, and other methods.

"Wet soil" means soil that contains significantly more moisture than moist soil, but in such a range of values that cohesive material will slump or begin to flow when vibrated. Granular material that would exhibit cohesive properties when moist will lose those cohesive properties when wet.

1926, Subpt. P, App. B – Sloping and Benching

"Actual slope" means the slope to which an excavation face is excavated.   Distress means that the soil is in a condition where a cave-in is imminent or is likely to occur. Distress is evidenced by such phenomena as the development of fissures in the face of or adjacent to an open excavation; the subsidence of the edge of an excavation; the slumping of material from the face or the bulging or heaving of material from the bottom of an excavation; the spalling of material from the face of an excavation; and ravelling, i.e., small amounts of material such as pebbles or little clumps of material suddenly separating from the face of an excavation and trickling or rolling down into the excavation.

"Maximum allowable slope" means the steepest incline of an excavation face that is acceptable for the most favorable site conditions as protection against cave-ins, and is expressed as the ratio of horizontal distance to vertical rise (H:V).

"Short-term exposure" means a period of time less than or equal to 24 hours that an excavation is open.

SUBPART Q – CONCRETE AND MASONRY CONSTRUCTION

"Bull float" means a tool used to spread out and smooth concrete.

"Formwork" means the total system of support for freshly placed or partially cured concrete, including the mold or sheeting (form) that is in contact with the concrete as well as all supporting members including shores, reshores, hardware, braces, and related hardware.

"Jacking operation" means the task of lifting a slab (or group of slabs vertically from one location to another (e.g., from the casting location to a temporary (parked) location, or to its final location in the structure), during the construction of a building/structure where the lift-slab process is being used.

"Lift slab" means a method of concrete construction in which floor and roof slabs are cast on or at ground level and, using jacks, lifted into position.

"Limited access zone" means an area alongside a masonry wall, which is under construction, and which is clearly demarcated to limit access by employees.

"Precast concrete" means concrete members (such as walls, panels, slabs, columns, and beams) which have been formed, cast, and cured prior to final placement in a structure.

"Reshoring" means the construction operation in which shoring equipment (also called reshores or reshoring equipment) is placed, as the original forms and shores are removed, in order to support partially cured concrete and construction loads.

"Shore" means a supporting member that resists a compressive force imposed by a load.

"Vertical slip forms" means forms which are jacked vertically during the placement of concrete.

SUBPART S – UNDERGROUND CONSTRUCTION, CAISSONS, COFFERDAMS, AND COMPRESSED AIR

"Absolute pressure" (psia.). The sum of the atmospheric pressure and gauge pressure (psig.).

"Atmospheric pressure." The pressure of air at sea level, usually 14.7 psia. (1 atmosphere), or 0 psig.

"Bulkhead." An airtight structure separating the working chamber from free air or from another chamber under a lesser pressure than the working pressure.

"Caisson." A wood, steel, concrete, or reinforced concrete, air- and water-tight chamber in which it is possible for men to work under air pressure greater than atmospheric pressure to excavate material below water level.

"Decanting." A method used for decompressing under emergency circumstances. In this procedure, the employees are brought to atmospheric pressure with a very high gas tension in the tissues and then immediately recompressed in a second and separate chamber or lock.

"Emergency locks." A lock designed to hold and permit the quick passage of an entire shift of employees.

"Gauge pressure" (psig.). Pressure measured by a gauge and indicating the pressure exceeding atmospheric.

"High air." Air pressure used to supply power to pneumatic tools and devices.

"Low air." Air supplied to pressurize working chambers and locks.

"Man lock." A chamber through which men pass from one air pressure environment into another.

"Materials lock." A chamber through which materials and equipment pass from one air pressure environment into another.

"Medical lock." A special chamber in which employees are treated for decompression illness. It may also be used in preemployment physical examinations to determine the adaptability of the prospective employee to changes in pressure.

"Normal condition." One during which exposure to compressed air is limited to a single continuous working period followed by a single decompression in any given 24-hour period; the total time of exposure to compressed air during the single continuous working period is not interrupted by exposure to normal atmospheric pressure, and a second exposure to compressed air does not occur until at least 12 consecutive hours of exposure to normal atmospheric pressure has elapsed since the employee has been under pressure.

"Pressure." A force acting on a unit area. Usually shown as pounds per square inch. (psi)

"Safety screen." An air- and water-tight diaphragm placed across the upper part of a compressed air tunnel between the face and bulkhead, in order to prevent flooding the crown of the tunnel between the safety screen and the bulkhead, thus providing a safe means of refuge and exit from a flooding or flooded tunnel.

"Special decompression chamber." A chamber to provide greater comfort of employees when the total decompression time exceeds 75 minutes.

"Working chamber." The space or compartment under air pressure in which the work is being done.

SUBPART U – BLASTING AND THE USE OF EXPLOSIVES

"American Table of Distances" (also known as Quantity Distance Tables) means American Table of Distances for Storage of Explosives as revised and approved by the Institute of the Makers of Explosives, June 5, 1964.

"Approved storage facility." A facility for the storage of explosive materials conforming to the requirements of this part and covered by a license or permit issued under authority of the Bureau of Alcohol, Tobacco and Firearms. (See 27 CFR part 55.)

"Blast area." The area in which explosives loading and blasting operations are being conducted.

"Blaster." The person or persons authorized to use explosives for blasting purposes and meeting the qualifications contained in 1926.901.

"Blasting agent." A blasting agent is any material or mixture consisting of a fuel and oxidizer used for blasting, but not classified an explosive and in which none of the ingredients is classified as an explosive provided the furnished (mixed) product cannot be detonated with a No. 8 test blasting cap when confined. A common blasting agent presently in use is a mixture of ammonium nitrate ($NH(4)NO(3)$) and carbonaceous combustibles, such as fuel oil or coal, and may either be procured, premixed and packaged from explosives companies or mixed in the field.

"Blasting cap." A metallic tube closed at one end, containing a charge of one or more detonating compounds, and designed for and capable of detonation from the sparks or flame from a safety fuse inserted and crimped into the open end.

"Block holing." The breaking of boulders by firing a charge of explosives that has been loaded in a drill hole.

"Conveyance." Any unit for transporting explosives or blasting agents, including but not limited to trucks, trailers, rail cars, barges, and vessels.

"Detonating cord." A flexible cord containing a center core of high explosives which when detonated, will have sufficient strength to detonate other cap-sensitive explosives with which it is in contact.

"Detonator." Blasting caps, electric blasting caps, delay electric blasting caps, and nonelectric delay blasting caps.

"Electric blasting cap." A blasting cap designed for and capable of detonation by means of an electric current.

"Electric blasting circuitry." (1) Bus wire. An expendable wire, used in parallel or series, in parallel circuits, to which are connected the leg wires of electric blasting caps. (2) Connecting wire. An insulated expendable wire used between electric blasting caps and the leading wires or between the bus wire and the leading wires. (3) Leading wire. An insulated wire used between the electric power source and the electric blasting cap circuit. (4) Permanent blasting wire. A permanently mounted insulated wire used between the electric power source and the electric blasting cap circuit.

"Electric delay blasting caps." Caps designed to detonate at a predetermined period of time after energy is applied to the ignition system.

"Explosives" –

(1) Any chemical compound, mixture, or device, the primary or common purpose of which is to function by explosion; that is, with substantially instantaneous release of gas and heat, unless such compound, mixture, or device is otherwise specifically classified by the U.S. Department of Transportation.

(2) All material which is classified as Class A, Class B, and Class C Explosives by the U.S. Department of Transportation.

(3) Classification of explosives by the U.S. Department of Transportation is as follows:

Class A Explosives. Possessing detonating hazard, such as dynamite, nitroglycerin, picric acid, lead azide, fulminate of mercury, black powder, blasting caps, and detonating primers.

Class B Explosives. Possessing flammable hazard, such as propellant explosives, including some smokeless propellants.

Class C Explosives. Include certain types of manufactured articles which contain Class A or Class B explosives, or both, as components, but in restricted quantities.

"Fuse lighters." Special devices for the purpose of igniting safety fuse.

"Magazine." Any building or structure, other than an explosives manufacturing building, used for the storage of explosives.

"Misfire." An explosive charge which failed to detonate.

"Mud-capping." (sometimes known as bulldozing, adobe blasting, or dobying). The blasting of boulders by placing a quantity of explosives against a rock, boulder, or other object without confining the explosives in a drill hole.

"Nonelectric delay blasting cap." A blasting cap with an integral delay element in conjunction with and capable of being detonated by a detonation impulse or signal from miniaturized detonating cord.

"Primary blasting." The blasting operation by which the original rock formation is dislodged from its natural location.

"Primer." A cartridge or container of explosives into which a detonator or detonating cord is inserted or attached.

"Safety fuse." A flexible cord containing an internal burning medium by which fire is conveyed at a continuous and uniform rate for the purpose of firing blasting caps.

"Secondary blasting." The reduction of oversize material by the use of explosives to the dimension required for handling, including mudcapping and blockholing.

"Stemming." A suitable inert incombustible material or device used to confine or separate explosives in a drill hole, or to cover explosives in mud-capping.

"Springing." The creation of a pocket in the bottom of a drill hole by the use of a moderate quantity of explosives in order that larger quantities or explosives may be inserted therein.

"Water gels, or slurry explosives." A wide variety of materials used for blasting. They all contain substantial proportions of water and high proportions of ammonium nitrate, some of which is in solution in the water. Two broad classes of water gels are: (1) Those which are sensitized by a material classed as an explosive, such as TNT or smokeless powder, and (2) those which contain no ingredient classified as an explosive; these are sensitized with metals such as aluminum or with other fuels. Water gels may be premixed at an explosives plant or mixed at the site immediately before delivery into the bore hole.

"Semiconductive hose." A hose with an electrical resistance high enough to limit flow of stray electric currents to safe levels, yet not so high as to prevent drainage of static electric charges to ground; hose of not more than 2 megohms resistance over its entire length and of not less than 5,000 ohms per foot meets the requirement.

SUBPART V – POWER TRANSMISSION AND DISTRIBUTION

"Alive or live (energized)" means electrically connected to a source of potential difference, or electrically charged so as to have a potential significantly different from that of the earth in the vicinity. The term "live" is sometimes used in place of the term "current-carrying," where the intent is clear, to avoid repetition of the longer term.

"Automatic circuit recloser" means a self-controlled device for automatically interrupting and reclosing an alternating current circuit with a predetermined sequence of opening and reclosing followed by resetting, hold closed, or lockout operation.

"Barrier" means a physical obstruction which is intended to prevent contact with energized lines or equipment.

"Barricade" means a physical obstruction such as tapes, screens, or cones intended to warn and limit access to a hazardous area.

"Bond" means an electrical connection from one conductive element to another for the purpose of minimizing potential differences or providing suitable conductivity for fault current or for mitigation of leakage current and electrolytic action.

"Bushing" means an insulating structure including a through conductor, or providing a passageway for such a conductor, with provision for mounting on a barrier, conducting or otherwise, for the purpose of insulating the conductor from the barrier and conducting current from one side of the barrier to the other.

"Cable" means a conductor with insulation, or a stranded conductor with or without insulation and other coverings (single-conductor cable) or a combination of conductors insulated from one another (multiple-conductor cable).

"Cable sheath" means a protective covering applied to cables. NOTE: A cable sheath may consist of multiple layers of which one or more is conductive.

"Circuit" means a conductor or system of conductors through which an electric current is intended to flow.

"Communication lines" means the conductors and their supporting or containing structures which are used for public or private signal or communication service, and which operate at potentials not exceeding 400 volts to ground or 750 volts between any two points of the circuit, and the transmitted power of which does not exceed 150 watts. When operating at less than 150 volts no limit is placed on the capacity of the system. NOTE: Telephone, telegraph, railroad signal, data, clock, fire, police-alarm, community television antenna, and other systems conforming with the above are included. Lines used for signaling purposes, but not included under the above definition, are considered as supply lines of the same voltage and are to be so run.

"Conductor" means a material, usually in the form of a wire, cable, or bus bar suitable for carrying an electric current.

"Conductor shielding" means an envelope which encloses the conductor of a cable and provides an equipotential surface in contact with the cable insulation.

"Current-carrying part" means a conducting part intended to be connected in an electric circuit to a source of voltage. Non-current-carrying parts are those not intended to be so connected.

"Dead (deenergized)" means free from any electrical connection to a source of potential difference and from electrical charges: Not having a potential difference from that of earth. NOTE: The term is used only with reference to current-carrying parts which are sometimes alive (energized).

"Designated employee" means a qualified person delegated to perform specific duties under the conditions existing.

"Effectively grounded" means intentionally connected to earth through a ground connection or connections of sufficiently low impedance and having sufficient current-carrying capacity to prevent the buildup of voltages which may result in undue hazard to connected equipment or to persons.

"Electric line trucks" means a truck used to transport men, tools, and material, and to serve as a traveling workshop for electric power line construction and maintenance work. It is sometimes equipped with a boom and auxiliary equipment for setting poles, digging holes, and elevating material or men.

"Electric supply lines" means those conductors used to transmit electric energy and their necessary supporting or containing structures. Signal lines of more than 400 volts to ground are always supply lines within the meaning of the rules, and those of less than 400 volts to ground may be considered as supply lines, if so run and operated throughout.

"Enclosed" means surrounded by a case, cage, or fence, which will protect the contained equipment and prevent accidental contact of a person with live parts.

"Equipment." This is a general term which includes fittings, devices, appliances, fixtures, apparatus, and the like, used as part of, or in connection with, an electrical power transmission and distribution system, or communication systems.

"Exposed" means not isolated or guarded.

"Guarded" means protected by personnel, covered, fenced, or enclosed by means of suitable casings, barrier rails, screens, mats, platforms, or other suitable devices in accordance with standard barricading techniques designed to prevent dangerous approach or contact by persons or objects.

NOTE: Wires, which are insulated but not otherwise protected, are not considered as guarded.

"Ground. (Reference)" means that conductive body, usually earth, to which an electric potential is referenced.

"Ground (as a noun)" means a conductive connection whether intentional or accidental, by which an electric circuit or equipment is connected to reference ground.

"Ground (as a verb)" means the connecting or establishment of a connection, whether by intention or accident of an electric circuit or equipment to reference ground.

"Grounded conductor" means a system or circuit conductor which is intentionally grounded.

"Grounded system" means a system of conductors in which at least one conductor or point (usually the middle wire, or neutral point of transformer or generator windings) is intentionally grounded, either solidly or through a current-limiting device (not a current-interrupting device).

"Grounding electrode (ground electrode)" grounding electrode means a conductor embedded in the earth, used for maintaining ground potential on conductors connected to it, and for dissipating into the earth current conducted to it.

"Grounding electrode resistance" means the resistance of the grounding electrode to earth.

"Grounding electrode conductor (grounding conductor)" means a conductor used to connect equipment or the grounded circuit of a wiring system to a grounding electrode.

"Hotline tools and ropes" means those tools and ropes which are especially designed for work on energized high voltage lines and equipment. Insulated aerial equipment especially designed for work on energized high voltage lines and equipment shall be considered hot line.

"Insulated" means separated from other conducting surfaces by a dielectric substance (including air space) offering a high resistance to the passage of current. NOTE: When any object is said to be insulated, it is understood to be insulated in suitable manner for the conditions to which it is subjected. Otherwise, it is within the purpose of this subpart, uninsulated. Insulating covering of conductors is one means of making the conductor insulated.

"Insulation (as applied to cable)" means that which is relied upon to insulate the conductor from other conductors or conducting parts or from ground.

"Insulation shielding" means an envelope which encloses the insulation of a cable and provides an equipotential surface in contact with cable insulation.

"Isolated" means an object that is not readily accessible to persons unless special means of access are used.

"Manhole" means a subsurface enclosure which personnel may enter and which is used for the purpose of installing, operating, and maintaining equipment and/or cable.

"(kk) Pulling tension" means the longitudinal force exerted on a cable during installation.

"Qualified person" means a person who by reason of experience or training is familiar with the operation to be performed and the hazards involved.

"Switch" means a device for opening and closing or changing the connection of a circuit. In these rules, a switch is understood to be manually operable, unless otherwise stated.

"Tag" means a system or method of identifying circuits, systems, or equipment for the purpose of alerting persons that the circuit, system or equipment is being worked on.

"Unstable material" means earth material, other than running, that because of its nature or the influence of related conditions, cannot be depended upon to remain in place without extra support, such as would be furnished by a system of shoring.

"Vault" means an enclosure above or below ground which personnel may enter and is used for the purpose of installing, operating, and/or maintaining equipment and/or cable.

"Voltage" means the effective (rms) potential difference between any two conductors or between a conductor and ground. Voltages are expressed in nominal values. The nominal voltage of a system or circuit is the value assigned to a system or circuit of a given voltage class for the purpose of convenient designation. The operating voltage of the system may vary above or below this value.

"Voltage of an effectively grounded circuit" means the voltage between any conductor and ground unless otherwise indicated.

"Voltage of a circuit not effectively grounded" means the voltage between any two conductors. If one circuit is directly connected to and supplied from another circuit of higher voltage (as in the case of an autotransformer), both are considered as of the higher voltage, unless the circuit of lower voltage is effectively grounded, in which case its voltage is not determined by the circuit of higher voltage. Direct connection implies electric connection as distinguished from connection merely through electromagnetic or electrostatic induction.

1926.1050 – SUBPART X – STAIRWAYS AND LADDERS

"Cleat" means a ladder crosspiece of rectangular cross section placed on edge upon which a person may step while ascending or descending a ladder.

"Double-cleat ladder" means a ladder similar in construction to a single-cleat ladder, but with a center rail to allow simultaneous two-way traffic for employees ascending or descending.

"Equivalent" means alternative designs, materials, or methods that the employer can demonstrate will provide an equal or greater degree of safety for employees than the method or item specified in the standard.

"Extension trestle ladder" means a self-supporting portable ladder, adjustable in length consisting of a trestle ladder base and a vertically adjustable extension section, with a suitable means for locking the ladders together.

"Failure" means load refusal, breakage or separation of component parts. Load refusal is the point where the structural members lose their ability to carry the loads.

"Fixed-ladder" means a ladder that cannot be readily moved or carried because it is an integral part of a building or structure. A side-step fixed ladder is a fixed ladder that requires a person getting off at the top to step to the side of the ladder side rails to reach the landing. A through fixed ladder is a fixed ladder that requires a person getting off at the top to step between the side rails of the ladder to reach the landing.

"Handrail" means a rail used to provide employees with a handhold for support.

"Individual-rung/step ladders" means ladders without a side rail or center rail support. Such ladders are made by mounting individual steps or rungs directly to the side or wall of the structure.

"Job-made ladder" means a ladder that is fabricated by employees, typically at the construction site, and is not commercially manufactured. This definition does not apply to any individual-rung/step ladders.

"Ladder stand." A mobile fixed size self-supporting ladder consisting of a wide flat tread ladder in the form of stairs. The assembly may include handrails.

"Lower levels" means those areas to which an employee can fall from a stairway or ladder. Such areas include ground levels, floors, roofs, ramps, runways, excavations, pits, tanks, material, water, equipment, and similar surfaces. It does not include the surface from which the employee falls.

"Maximum intended load" means the total load of all employees, equipment, tools, materials, transmitted loads, and other loads anticipated to be applied to a ladder component at any one time.

"Nosing" means that portion of a tread projecting beyond the face of the riser immediately below.

"Point of access" means all areas used by employees for work related passage from one area or level to another. Such open areas include doorways, passageways, stairway openings, studded walls, and various other permanent or temporary openings used for such travel.

"Portable ladder" means a ladder that can be readily moved or carried.

"Riser height" means the vertical distance from the top of a tread to the top of the next higher tread or platform/landing or the distance from the top of a platform/landing to the top of the next higher tread or platform/landing.

"Side-step fixed ladder" See "Fixed ladder."

"Single-cleat ladder" means a ladder consisting of a pair of side rails, connected together by cleats, rungs, or steps.

"Single-rail ladder" means a portable ladder with rungs, cleats, or steps mounted on a single rail instead of the normal two rails used on most other ladders.

"Spiral stairway" means a series of steps attached to a vertical pole and progressing upward in a winding fashion within a cylindrical space.

"Stairrail system" means a vertical barrier erected along the unprotected sides and edges of a stairway to prevent employees from falling to lower levels. The top surface of a stairrail system may also be a "handrail."

"Step stool (ladder type)" means a self-supporting, foldable, portable ladder, nonadjustable in length, 32 inches or less in overall size, with flat steps and without a pail shelf, designed to be climbed on the ladder top cap as well as all steps. The side rails may continue above the top cap.

"Through fixed ladder" See "Fixed ladder."

"Tread depth" means the horizontal distance from front to back of a tread (excluding nosing, if any).

"Unprotected sides and edges" means any side or edge (except at entrances to points of access) of a stairway where there is no stairrail system or wall 36 inches (.9 m) or more in height, and any side or edge (except at entrances to points of access) of a stairway landing, or ladder platform where there is no wall or guardrail system 39 inches (1 m) or more in height.

SUBPART Y – DIVING

"Acfm." Actual cubic feet per minute.

"ASME Code or equivalent." ASME (American Society of Mechanical Engineers) Boiler and Pressure Vessel Code, Section VIII, or an equivalent code which the employer can demonstrate to be equally effective.

"ATA." Atmosphere absolute.

"Bell." An enclosed compartment, pressurized (closed bell) or unpressurized (open bell), which allows the diver to be transported to and from the underwater work area and which may be used as a temporary refuge during diving operations.

"Bottom time." The total elapsed time measured in minutes from the time when the diver leaves the surface in descent to the time that the diver begins ascent.

"Bursting pressure." The pressure at which a pressure containment device would fail structurally.

"Cylinder." A pressure vessel for the storage of gases.

"Decompression chamber." A pressure vessel for human occupancy such as a surface decompression chamber, closed bell, or deep diving system used to decompress divers and to treat decompression sickness.

"Decompression sickness." A condition with a variety of symptoms which may result from gas or bubbles in the tissues of divers after pressure reduction.

"Decompression table." A profile or set of profiles of depth-time relationships for ascent rates and breathing mixtures to be followed after a specific depth-time exposure or exposures.

"Dive location." A surface or vessel from which a diving operation is conducted.

"Dive-location reserve breathing gas." A supply system of air or mixed-gas (as appropriate) at the dive location which is independent of the primary supply system and sufficient to support divers during the planned decompression.

"Dive team." Divers and support employees involved in a diving operation, including the designated person-in-charge.

"Diver." An employee working in water using underwater apparatus which supplies compressed breathing gas at the ambient pressure.

"Diver-carried reserve breathing gas." A diver-carried supply of air or mixed gas (as appropriate) sufficient under standard operating conditions to allow the diver to reach the surface, or another source of breathing gas, or to be reached by a standby diver.

"Diving mode." A type of diving requiring specific equipment, procedures, and techniques (SCUBA, surface-supplied air, or mixed gas).

"Fsw." Feet of seawater (or equivalent static pressure head).

"Heavy gear." Diver-worn deep-sea dress including helmet, breastplate, dry suit, and weighted shoes.

"Hyperbaric conditions." Pressure conditions in excess of surface pressure.

"Inwater stage." A suspended underwater platform which supports a diver in the water.

"Liveboating." The practice of supporting a surfaced-supplied air or mixed gas diver from a vessel which is underway.

"Mixed-gas diving." A diving mode in which the diver is supplied in the water with a breathing gas other than air.

"No-decompression limits." The depth-time limits of the "no-decompression limits and repetitive dive group designation table for no-decompression air dives," U.S. Navy Diving Manual or equivalent limits which the employer can demonstrate to be equally effective.

"Psi(g)." Pounds per square inch (gauge).

"Scientific diving." means diving performed solely as a necessary part of a scientific, research, or educational activity by employees whose sole purpose for diving is to perform scientific research tasks. Scientific diving does not include performing any tasks usually associated with commercial diving such as: Placing or removing heavy objects underwater; inspection of pipelines and similar objects; construction; demolition; cutting or welding; or the use of explosives.

"SCUBA diving." A diving mode independent of surface supply in which the diver uses open circuit self-contained underwater breathing apparatus.

"Standby diver." A diver at the dive location available to assist a diver in the water.

"Surface-supplied air diving." A diving mode in which the diver in the water is supplied from the dive location with compressed air for breathing.

"Treatment table." A depth-time and breathing gas profile designed to treat decompression sickness.

"Umbilical." The composite hose bundle between a dive location and a diver or bell, or between a diver and a bell, which supplies the diver or bell with breathing gas, communications, power, or heat as appropriate to the diving mode or conditions, and includes a safety line between the diver and the dive location.

"Volume tank." A pressure vessel connected to the outlet of a compressor and used as an air reservoir.

"Working pressure." The maximum pressure to which a pressure containment device may be exposed under standard operating conditions.

• *REFERENCE: SEE CODE OF FEDERAL REGULATION TITLE 29 PART 1926.*

APPENDIX M

CRANE ILLUSTRATIONS

from
OSHA Technical Manual – Chapter 13

Courtesy of OSHA

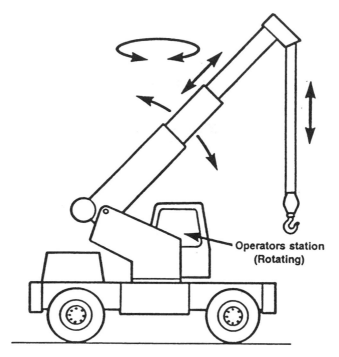

Figure 13-A1. Wheel-mounted crane: telescoping boom (single control station)

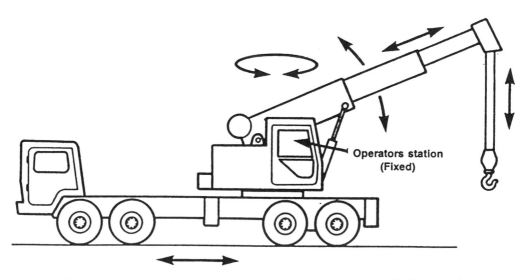

Figure 13-A2. Wheel-mounted crane telescoping hydraulic boom (multiple control station)

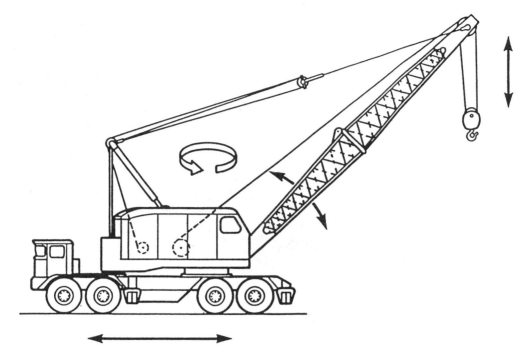

Figure 13-A3. Wheel-mounted crane latticework boom (multiple control station)

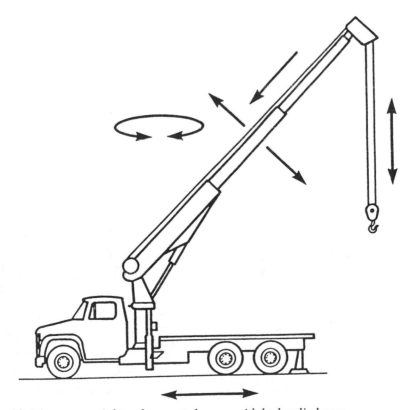

Figure 13-A4. commercial truck-mounted crane with hydraulic boom

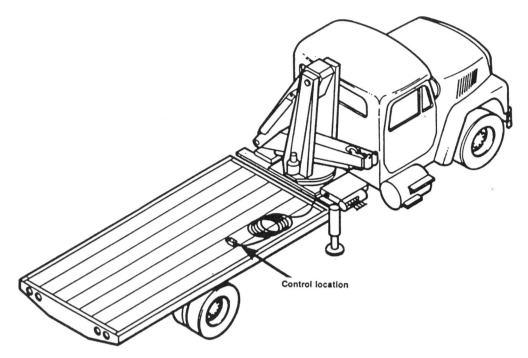

Figure 13-A5. Commercial truck-mounted crane with articulated boom

Figure 13-A6. Commercial truck-mounted crane with trolley boom

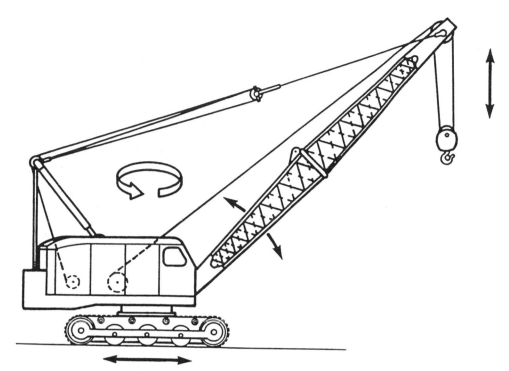

Figure 13-A7. Crawler-mounted latticework boom crane

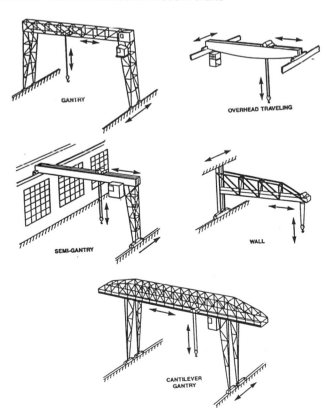

Figure 13-A8. Overhead track-mounted cranes

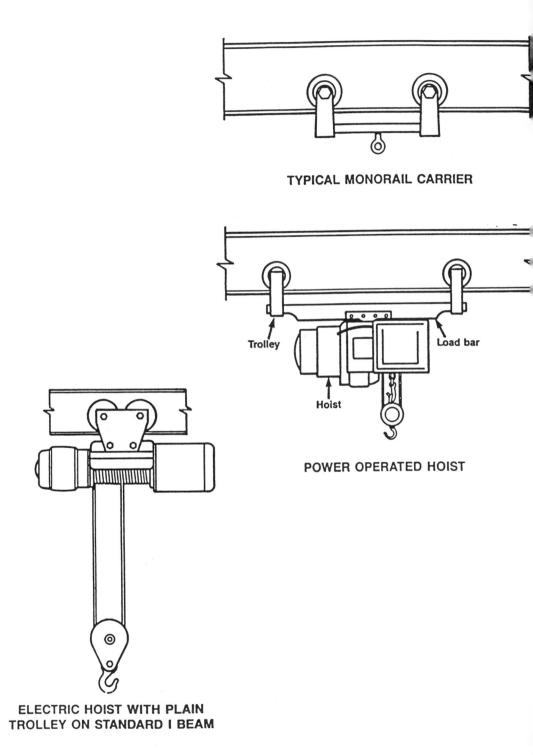

TYPICAL MONORAIL CARRIER

Trolley

Load bar

Hoist

POWER OPERATED HOIST

**ELECTRIC HOIST WITH PLAIN
TROLLEY ON STANDARD I BEAM**

Figure 13-A9. monorails and underhung cranes

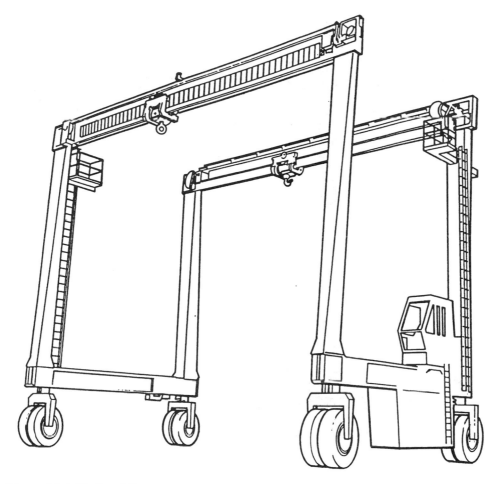

Figure 13-A10. Straddle cranes

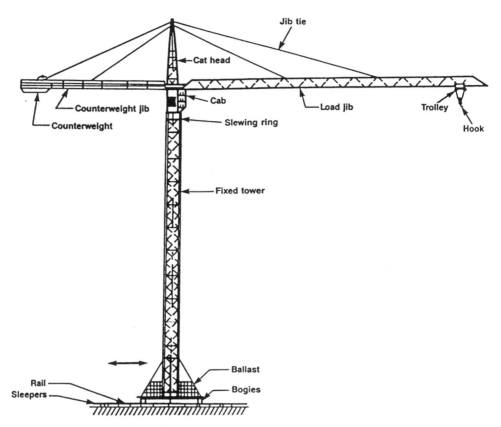

Figure 13-A11. Hammerhead tower cranes

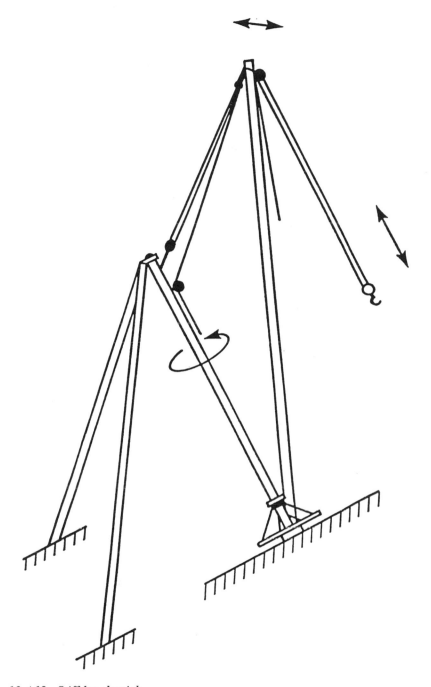

Figure 13-A12. Stiff-leg derrick

APPENDIX N

RESPIRATORY PROTECTION
(29 CFR 1910.134)

This section applies to General Industry (part 1910), Shipyards (part 1915), Marine Terminals (part 1917), Longshoring (part 1918), and Construction (part 1926).

(a) Permissible Practice

(1) In the control of those occupational diseases caused by breathing air contaminated with harmful dusts, fogs, fumes, mists, gases, smokes, sprays, or vapors, the primary objective shall be to prevent atmospheric contamination. This shall be accomplished as far as feasible by accepted engineering control measures (for example, enclosure or confinement of the operation, general and local ventilation, and substitution of less toxic materials). When effective engineering controls are not feasible, or while they are being instituted, appropriate respirators shall be used pursuant to this section.

(2) Respirators shall be provided by the employer when such equipment is necessary to protect the health of the employee. The employer shall provide the respirators which are applicable and suitable for the purpose intended. The employer shall be responsible for the establishment and maintenance of a respiratory protection program which shall include the requirements outlined in paragraph (c) of this section.

(b) Definitions

The following definitions are important terms used in the respiratory protection standard in this section.

"Air-purifying respirator" means a respirator with an air-purifying filter, cartridge, or canister that removes specific air contaminants by passing ambient air through the air-purifying element.

"Assigned protection factor (APF)" [Reserved]

"Atmosphere-supplying respirator" means a respirator that supplies the respirator user with breathing air from a source independent of the ambient atmosphere, and includes supplied-air respirators (SARs) and self-contained breathing apparatus (SCBA) units.

"Canister or cartridge" means a container with a filter, sorbent, or catalyst, or combination of these items, which removes specific contaminants from the air passed through the container.

"Demand respirator" means an atmosphere-supplying respirator that admits breathing air to the facepiece only when a negative pressure is created inside the facepiece by inhalation.

"Emergency situation" means any occurrence such as, but not limited to, equipment failure, rupture of containers, or failure of control equipment that may or does result in an uncontrolled significant release of an airborne contaminant.

"Employee exposure" means exposure to a concentration of an airborne contaminant that would occur if the employee were not   using respiratory protection.

"End-of-service-life indicator (ESLI)" means a system that warns the respirator user of the approach of the end of adequate respiratory protection, for example, that the sorbent is approaching saturation or is no longer effective.

"Escape-only respirator" means a respirator intended to be used only for emergency exit.

"Filter or air purifying element" means a component used in respirators to remove solid or liquid aerosols from the inspired air.

"Filtering facepiece (dust mask)" means a negative pressure particulate respirator with a filter as an integral part of the facepiece or with the entire facepiece composed of the filtering medium.

"Fit factor" means a quantitative estimate of the fit of a particular respirator to a specific individual, and typically estimates the ratio of the concentration of a substance in ambient air to its concentration inside the respirator when worn.

"Fit test" means the use of a protocol to qualitatively or quantitatively evaluate the fit of a respirator on an individual. (See also Qualitative fit test QLFT and Quantitative fit test QNFT.)

"Helmet" means a rigid respiratory inlet covering that also provides head protection against impact and penetration.

"High efficiency particulate air (HEPA) filter" means a filter that is at least 99.97% efficient in removing monodisperse particles of 0.3 micrometers in diameter. The equivalent NIOSH 42 CFR 84 particulate filters are the N100, R100, and P100 filters.

"Hood" means a respiratory inlet covering that completely covers the head and neck and may also cover portions of the shoulders and torso.

"Immediately dangerous to life or health (IDLH)" means an atmosphere that poses an immediate threat to life, would cause irreversible adverse health effects, or would impair an individual's ability to escape from a dangerous atmosphere.

"Interior structural fire fighting" means the physical activity of fire suppression, rescue, or both, inside of buildings or enclosed structures which are involved in a fire situation beyond the incipient stage. (See 29 CFR 1910.155.)

"Loose-fitting facepiece" means a respiratory inlet covering that is designed to form a partial seal with the face. Maximum use concentration (MUC) [Reserved].

"Negative pressure respirator (tight fitting)" means a respirator in which the air pressure inside the facepiece is negative during inhalation with respect to the ambient air pressure outside the respirator.

"Oxygen deficient atmosphere" means an atmosphere with an oxygen content below 19.5% by volume.

"Physician or other licensed health care professional (PLHCP)" means an individual whose legally permitted scope of practice (i.e., license, registration, or certification) allows him or her to independently provide, or be delegated the responsibility to provide, some or all of the health care services required by paragraph (e) of this section.

"Positive pressure respirator" means a respirator in which the pressure inside the respiratory inlet covering exceeds the ambient air pressure outside the respirator.

"Powered air-purifying respirator (PAPR)" means an air-purifying respirator that uses a blower to force the ambient air through air-purifying elements to the inlet covering.

"Pressure demand respirator" means a positive pressure atmosphere-supplying respirator that admits breathing air to the facepiece when the positive pressure is reduced inside the facepiece by inhalation.

"Qualitative fit test (QLFT)" means a pass/fail fit test to assess the adequacy of respirator fit that relies on the individual's response to the test agent.

"Quantitative fit test (QNFT)" means an assessment of the adequacy of respirator fit by numerically measuring the amount of leakage into the respirator.

"Respiratory inlet covering" means that portion of a respirator that forms the protective barrier between the user's respiratory tract and an air-purifying device or breathing air source, or both. It may be a facepiece, helmet, hood, suit, or a mouthpiece respirator with nose clamp.

"Self-contained breathing apparatus (SCBA)" means an atmosphere-supplying respirator for which the breathing air source is designed to be carried by the user.

"Service life" means the period of time that a respirator, filter or sorbent, or other respiratory equipment provides adequate protection to the wearer.

"Supplied-air respirator (SAR) or airline respirator" means an atmosphere supplying respirator for which the source of breathing air is not designed to be carried by the user.

This section means this respiratory protection standard. Tight-fitting facepiece means a respiratory inlet covering that forms a complete seal with the face. User seal check means an action conducted by the respirator user to determine if the respirator is properly seated to the face.

(c) Respiratory Protection Program

This paragraph requires the employer to develop and implement a written respiratory protection program with required worksite-specific procedures and elements for required respirator use. The program must be administered by a suitably trained program administrator. In addition, certain program elements may be required for voluntary use to prevent potential hazards associated with the use of the respirator. The *Small Entity Compliance Guide* contains criteria for the selection of a program administrator and a sample program that meets the requirements of this paragraph. Copies of the *Small Entity Compliance Guide* will be available on or about April 8,1998 from the Occupational Safety and Health Administration's Office of Publications, Room N 3101, 200 Constitution Avenue, NW, Washington, D.C., 20210 (202-210 4667).

(1) In any workplace where respirators are necessary to protect the health of the employee or whenever respirators are required by the employer, the employer shall establish and implement a written respiratory protection program with worksite specific procedures. The program shall be updated as necessary to reflect those changes in workplace conditions that affect respirator use. The employer shall include in the program the following provisions of this section, as applicable:

(i) Procedures for selecting respirators for use in the workplace;

(ii) Medical evaluations of employees required to use respirators;

(iii) Fit testing procedures for tightfitting respirators;

(iv) Procedures for proper use of respirators in routine and reasonably foreseeable emergency situations;

(v) Procedures and schedules for cleaning, disinfecting, storing, inspecting, repairing, discarding, and otherwise maintaining respirators;

(vi) Procedures to ensure adequate air quality, quantity, and flow of breathing air for atmosphere-supplying respirators;

(vii)   Training of employees in the respiratory hazards to which they are potentially exposed during routine and emergency situations;

(viii) Training of employees in the proper use of respirators, including putting on and removing them, any limitations on their use, and their maintenance; and

(ix)   Procedures for regularly evaluating the effectiveness of the program.

(2)   Where respirator use is not required.

(i)   An employer may provide respirators at the request of employees or permit employees to use their own respirators, if the employer determines that such respirator use will not in itself create a hazard. If the employer determines that any voluntary respirator use is permissible, the employer shall provide the respirator users with the information contained in Appendix D to this section ("Information for Employees Using Respirators When Not Required Under the Standard"); and

(ii)   In addition, the employer must establish and implement those elements of a written respiratory protection program necessary to ensure that any employee using a respirator voluntarily is medically able to use that respirator, and that the respirator is cleaned, stored, and maintained so that its use does not present a health hazard to the user. Exception: Employers are not required to include in a written respiratory protection program those employees whose only use of respirators involves the voluntary use of filtering facepieces (dust masks).

(3)   The employer shall designate a program administrator who is qualified by appropriate training or experience that is commensurate with the complexity of the program to administer or oversee the respiratory protection program and conduct the required evaluations of program effectiveness.

(4)   The employer shall provide respirators, training, and medical evaluations at no cost to the employee.

(d) Selection of Respirators

This paragraph requires the employer to evaluate respiratory hazard(s) in the workplace, identify relevant workplace and user factors, and base respirator selection on these factors. The paragraph also specifies appropriately protective respirators for use in IDLH atmospheres, and limits the selection and use of air-purifying respirators.

(1)   General requirements.

(i)   The employer shall select and provide an appropriate respirator based on the respiratory hazard(s) to which the worker is exposed and workplace and user factors that affect respirator performance and reliability.

(ii)   The employer shall select a NIOSH-certified respirator. The respirator shall be used in compliance with the conditions of its certification.

(iii) The employer shall identify and evaluate the respiratory hazard(s) in the workplace; this evaluation shall include a reasonable estimate of employee exposures to respiratory hazard(s) and an identification of the contaminant's chemical state and physical form. Where the employer cannot identify or reasonably estimate the employee exposure, the employer shall consider the atmosphere to be IDLH.

(iv)   The employer shall select respirators from a sufficient number of respirator models and sizes so that the respirator is acceptable to, and correctly fits, the user.

(2)   Respirators for IDLH atmospheres.

(i)   The employer shall provide the following respirators for employee use in IDLH atmospheres:

(A)   A full facepiece pressure demand SCBA certified by NIOSH for a minimum service life of thirty minutes, or

(B)   A combination full facepiece pressure demand supplied-air respirator (SAR) with auxiliary self-contained air supply.

[ii)   Respirators provided only for escape from IDLH atmospheres shall be NIOSH-certified for escape from the atmosphere in which they will be used.

(iii)   All oxygen-deficient atmospheres shall be considered IDLH. Exception: If the employer demonstrates that, under all foreseeable conditions, the oxygen concentration can be maintained within the ranges specified in Table II of this section (i.e., for the altitudes set out in the table), then any atmosphere supplying respirator may be used.

(3)   Respirators for atmospheres that are not IDLH.

(i)   The employer shall provide a respirator that is adequate to protect the health of the employee and ensure compliance with all other OSHA statutory and regulatory requirements, under routine and reasonably foreseeable emergency situations.

(A)   Assigned Protection Factors (APFs) [Reserved]

(B)   Maximum Use Concentration (FMUC)) [Reserved]

(ii)   The respirator selected shall be appropriate for the chemical state and physical form of the contaminant.

(iii)   For protection against gases and vapors, the employer shall provide:

(A)   An atmosphere-supplying respirator, or

(B)   An air-purifying respirator, provided that:

(1)   The respirator is equipped with an end-of-service-life indicator (ESLI) certified by NIOSH for the contaminant; or

(2)   If there is no ESLI appropriate for conditions in the employer's workplace, the employer implements a change schedule for canisters and cartridges that is based on objective information or data that will ensure that canisters and cartridges are changed before the end of their service life. The employer shall describe in the respirator program the information and data relied upon and the basis for the canister and cartridge change schedule and the basis for reliance on the data.

(iv)   For protection against particulates, the employer shall provide:

(A)   An atmosphere-supplying respirator: or

(B)   An air-purifying respirator equipped with a filter certified by NIOSH under 30 CFR part 11 as a high efficiency particulate air (HEPA) filter, or an air-purifying respirator equipped with filter certified for particulates by NIOSH under 42 CFR part 84; or

(C)   For contaminants consisting primarily of particles with mass median aerodynamic diameters (MMAD) of at least 2 micrometers, an air-purifying respirator equipped with any filter certified for particulates by NIOSH.

TABLE I	
Altitude (ft.)	Oxygen deficient Atmospheres (% O2) for which the employer may rely on atmosphere supplying respirators
Less than 3,001	16.0-19.5
3,001-4,000	16.4-19.5
4,001-5,000	17.1-19.5
5,001-6,000	17.8-19.5
6,001-7,000	18.5-19.5
7,001-8,001	19.3-19.5

Note: Above 8,000 feet the exception does not apply. Oxygen-enriched breathing air must be supplied above 14,000 fcct.

(e) Medical Evaluation

Using a respirator may place a physiological burden on employees that varies with the type of respirator worn, the job and workplace conditions in which the respirator is used, and the medical status of the employee. Accordingly, this paragraph specifies the minimum requirements for medical evaluation that employers must implement to determine the employee's ability to use a respirator.

(1)   General. The employer shall provide a medical evaluation to determine the employee's ability to use a respirator, before the employee is fit tested or required to use the respirator in the workplace. The employer may discontinue an employee's medical evaluations when the employee is no longer required to use a respirator.

(2)   Medical evaluation procedures.

(i)   The employer shall identify a physician or other licensed health care professional (PLHCP) to perform medical evaluations using a medical questionnaire or an initial medical examination that obtains the same information as the medical questionnaire.

(ii)   The medical evaluation shall obtain the information requested by the questionnaire in Sections 1 and 2, Part A of Appendix C of this section.

(3)   Follow-up medical examination.

    (i)   The employer shall ensure that a follow-up medical examination is provided for an employee who gives a positive response to any question among questions 1 through 8 in Section 2, Part A of Appendix C or whose initial medical examination demonstrates the need for a follow-up medical examination.

    (ii)   The follow-up medical examination shall include any medical tests, consultations, or diagnostic procedures that the PLHCP deems necessary to make a final determination.

(4)   Administration of the medical questionnaire and examinations.

    (i)   The medical questionnaire and examinations shall be administered confidentially during the employee's normal working hours or at a time and place convenient to the employee. The medical questionnaire shall be administered in a manner that ensures that the employee understands its content.

    (ii)   The employer shall provide the employee with an opportunity to discuss the questionnaire and examination results with the PLHCP.

(5)   Supplemental information for the PLHCP.

    (i)   The following information must be provided to the PLHCP before the PLHCP makes a recommendation concerning an employee's ability to use a respirator:

        (A)   The type and weight of the respirator to be used by the employee;

        (B)   The duration and frequency of respirator use (including use for rescue and escape);

        (C)   The expected physical work effort;

        (D)   Additional protective clothing and equipment to be worn; and

        (E)   Temperature and humidity extremes that may be encountered.

    (ii)   Any supplemental information provided previously to the PLHCP regarding an employee need not be provided for a subsequent medical evaluation if the information and the PLHCP remain the same.

    (iii)   The employer shall provide the PLHCP with a copy of the written respiratory protection program and a copy of this section.

Note to Paragraph: (e)(5)(iii) When the employer replaces a PLHCP, the employer must ensure that the new PLHCP obtains this information, either by providing the documents directly to the PLHCP or having the documents transferred from the former PLHCP to the new PLHCP. However, OSHA does not expect employers to have employees medically reevaluated solely because a new PLHCP has been selected.

(6)   Medical determination.

In determining the employee's ability to use a respirator, the employer shall:

    (i)   Obtain a written recommendation regarding the employee's ability to use the respirator from the PLHCP. The recommendation shall provide only the following information:

        (A)   Any limitations on respirator use related to the medical condition of the employee, or relating to the workplace conditions in which the respirator will be used, including whether or not the employee is medically able to use the respirator,

(B)   The need, if any, for follow-up medical evaluations, and

(C)   A statement that the PLHCP has provided the employee with a copy of the PLHCP's written recommendation.

(ii)   If the respirator is a negative pressure respirator and the PLHCP finds a medical condition that may place the employee's health at increased risk if the respirator is used, the employer shall provide a PAPR if the PLHCP's medical evaluation finds that the employee can use such a respirator; if a subsequent medical evaluation finds that the employee is medically able to use a negative pressure respirator, then the employer is no longer required to provide a PAPR.

(7)   Additional medical evaluations. At a minimum, the employer shall provide additional medical evaluations that comply with the requirements of this section if:

(i)   An employee reports medical signs or symptoms that are related to ability to use a respirator;

(ii)   A PLHCP, supervisor, or the respirator program administrator informs the employer that an employee needs to be reevaluated;

(iii)   Information from the respiratory protection program, including observations made during fit testing and program evaluation, indicates a need for employee reevaluation; or

(iv)   A change occurs in workplace conditions (e.g., physical work effort, protective clothing, temperature) that may result in a substantial increase in the physiological burden placed on an employee.

(f) Fit Testing

This paragraph requires that, before an employee may be required to use any respirator with a negative or positive pressure tight-fitting facepiece, the employee must be fit tested with the same make, model, style, and size of respirator that will be used. This paragraph specifies the kinds of fit tests allowed, the procedures for conducting them, and how the results of the fit tests must be used.

(1)   The employer shall ensure that employees using a tight-fitting facepiece respirator pass an appropriate qualitative fit test (QLFT) or quantitative fit test (QNFT) as stated in this paragraph.

(2)   The employer shall ensure that an employee using a tight-fitting facepiece respirator is fit tested prior to initial use of the respirator, whenever a different respirator facepiece (size, style, model, or make) is used, and at least annually thereafter.

(3)   The employer shall conduct an additional fit test whenever the employee reports, or the employer, PLHCP, supervisor, or program administrator makes visual observations of, changes in the employee's physical condition that could affect respirator fit. Such conditions include, but are not limited to, facial scarring, dental changes, cosmetic surgery, or an obvious change in body weight.

(4)   If after passing a QLFT or QNFT, the employee subsequently notifies the employer, program administrator, supervisor, or PLHCP that the fit of the respirator is unacceptable, the employee shall be given a reasonable opportunity to select a different respirator facepiece and to be retested.

(5)   The fit test shall be administered using an OSHA-accepted QLFT or QNFT

protocol. The OSHA-accepted QLFT and QNFT protocols and procedures are contained in Appendix A of this section.

(6)   QLFT may only be used to fit test negative pressure air-purifying respirators that must achieve a fit factor of 100 or less.

(7)   If the fit factor, as determined through an OSHA-accepted QNFT protocol, is equal to or greater than 100 for tight-fitting half facepieces, or equal to or greater than 500 for tight-fitting full facepieces, the QNFT has been passed with that respirator.

(8)   Fit testing of tight-fitting atmosphere-supplying respirators and tight-fitting powered air-purifying respirators shall be accomplished by performing quantitative or qualitative fit testing in the negative pressure mode, regardless of the mode of operation (negative or positive pressure) that is used for respiratory protection.

    (i)   Qualitative fit testing of these respirators shall be accomplished by temporarily converting the respirator user's actual facepiece into a negative pressure respirator with appropriate filters, or by using an identical negative pressure air-purifying respirator facepiece with the same sealing surfaces as a surrogate for the atmosphere supplying or powered air-purifying respirator facepiece.

    (ii)   Quantitative fit testing of these respirators shall be accomplished by modifying the facepiece to allow sampling inside the facepiece in the breathing zone of the user, midway between the nose and mouth. This requirement shall be   accomplished by installing a permanent sampling probe onto a surrogate facepiece, or by using a sampling adapter designed to temporarily provide a means of sampling air from inside the facepiece.

    (iii)   Any modifications to the respirator facepiece for fit testing shall be completely removed, and the facepiece restored to NIOSH-approved configuration, before that facepiece can be used in the workplace.

(g) Use of Respirators

This paragraph requires employers to establish and implement procedures for the proper use of respirators. These requirements include prohibiting conditions that may result in facepiece seal leakage, preventing employees from removing respirators in hazardous environments, taking actions to ensure continued effective respirator operation throughout the work shift, and establishing procedures for the use of respirators in IDLH atmospheres or in interior structural fire fighting situations.

(1)   Facepiece seal protection.

    (i)   The employer shall not permit respirators with tight-fitting facepieces to be worn by employees who have:

        (A)   Facial hair that comes between the sealing surface of the facepiece and the face or that interferes with valve function; or

        (B)   Any condition that interferes with the face-to-facepiece seal or valve function.

    (ii)   If an employee wears corrective glasses or goggles or other personal protective equipment, the employer shall ensure that such equipment is worn in a manner that does not interfere with the seal of the facepiece to the face of the user.

(iii)   For all tight-fitting respirators, the employer shall ensure that employees perform a user seal check each time they put on the respirator using the procedures in Appendix B-1 or procedures recommended by the respirator manufacturer that the employer demonstrates are as effective as those in Appendix B-1 of this section.

(2)   Continuing respirator effectiveness.

(i)   Appropriate surveillance shall be maintained of work area conditions and degree of employee exposure or stress. When there is a change in work area conditions or degree of employee exposure or stress that may affect respirator effectiveness, the employer shall reevaluate the continued effectiveness of the respirator.

(ii)   The employer shall ensure that employees leave the respirator use area:

(A) To wash their faces and respirator facepieces as necessary to prevent eye or skin irritation associated with respirator use; or

(B) If they detect vapor or gas breakthrough, changes in breathing resistance, or leakage of the facepiece; or

(C) To replace the respirator or the filter, cartridge, or canister elements.

(iii)   If the employee detects vapor or gas breakthrough, changes in breathing resistance, or leakage of the facepiece, the employer must replace or repair the respirator before allowing the employee to return to the work area.

(3)   Procedures for IDLH atmospheres. For all IDLH atmospheres, the employer shall ensure that:

(i)   One employee or, when needed, more than one employee is located outside the IDLH atmosphere;

(ii)   Visual, voice, or signal line communication is maintained between the employee(s) in the IDLH atmosphere and the employee(s) located outside the IDLH atmosphere;

(iii)   The employee(s) located outside the IDLH atmosphere are trained and equipped to provide effective emergency rescue;

(iv)   The employer or designee is notified before the employee(s) located outside the IDLH atmosphere enter the IDLH atmosphere to provide emergency rescue;

(v)   The employer or designee authorized to do so by the employer, once notified, provides necessary assistance appropriate to the situation;

(vi)   Employee(s) located outside the IDLH atmospheres are equipped with:

(A)   Pressure demand or other positive pressure SCBAs, or a pressure demand or other positive pressure supplied-air respirator with auxiliary SCBA; and either

(B)   Appropriate retrieval equipment for removing the employee(s) who enter(s) these hazardous atmospheres where retrieval equipment would contribute to the rescue of the employee(s) and would not increase the overall risk resulting from entry; or

(C)   Equivalent means for rescue where retrieval equipment is not required under paragraph (g)(3)(vi)(B).

(4)   Procedures for interior structural fire fighting. In addition to the requirements

set forth under paragraph (g)(3), in interior structural fires, the employer shall ensure that:

(i)   At least two employees enter the IDLH atmosphere and remain in visual or voice contact with one another at all times;

(ii)   At least two employees are located outside the IDLH atmosphere, and

(iii)   All employees engaged in interior structural fire fighting use SCBAs.

Note 1 to paragraph (g): One of the two individuals located outside the IDLH atmosphere may be assigned to an additional role, such as incident commander in charge of the emergency or safety officer, so long as this individual is able to perform assistance or rescue activities without jeopardizing the safety or health of any firefighter working at the incident.

Note 2 to paragraph (g): Nothing in this section is meant to preclude firefighters from performing emergency rescue activities before an entire team has assembled.

(h) Maintenance and Care of Respirators

This paragraph requires the employer to provide for the cleaning and disinfecting, storage, inspection, and repair of respirators used by employees.

(1)   Cleaning and disinfecting. The employer shall provide each respirator user with a respirator that is clean, sanitary, and in good working order. The employer shall ensure that respirators are cleaned and disinfected using the procedures in Appendix B-2 of this section, or procedures recommended by the respirator manufacturer, provided that such procedures are of equivalent effectiveness. The respirators shall be cleaned and disinfected at the following intervals:

(i)   Respirators issued for the exclusive use of an employee shall be cleaned and disinfected as often as necessary to be maintained in a sanitary condition;

(ii)   Respirators issued to more than one employee shall be cleaned and disinfected before being worn by different individuals;

(iii)   Respirators maintained for emergency use shall be cleaned and disinfected after each use; and

(iv)   Respirators used in fit testing and training shall be cleaned and disinfected after each use.

(2)   Storage. The employer shall ensure that respirators are stored as follows:

(i)   All respirators shall be stored to protect them from damage, contamination, dust, sunlight, extreme temperatures, excessive moisture, and damaging chemicals, and they shall be packed or stored to prevent deformation of the facepiece and exhalation valve.

(ii)   In addition to the requirements of paragraph (h)(2)(i) of this section, emergency respirators shall be:

(A)   Kept accessible to the work area;

(B)   Stored in compartments or in covers that are clearly marked as containing emergency respirators; and

(C)   Stored in accordance with any applicable manufacturer instructions.

(3) Inspection.

    (i) The employer shall ensure that respirators are inspected as follows:

        (A) All respirators used in routine situations shall be inspected before each use and during cleaning;

        (B) All respirators maintained for use in emergency situations shall be inspected at least monthly and in accordance with the manufacturer's recommendations, and shall be checked for proper function before and after each use; and

        (C) Emergency escape-only respirators shall be inspected before being carried into the workplace for use.

    (ii) The employer shall ensure that respirator inspections include the following:

        (A) A check of respirator function, tightness of connections, and the condition of the various parts including, but not limited to, the facepiece, head straps, valves, connecting tube, and cartridges, canisters or filters; and

        (B) A check of elastomeric parts for pliability and signs of deterioration.

    (iii) In addition to the requirements of paragraphs (h)(3)(i) and (ii) of this section, self-contained breathing apparatus shall be inspected monthly.

Air and oxygen cylinders shall be maintained in a fully charged state and shall be recharged when the pressure falls to 90% of the manufacturer's recommended pressure level. The employer shall determine that the regulator and warning devices function properly.

    (iv) For respirators maintained for emergency use, the employer shall:

        (A) Certify the respirator by documenting the date the inspection was performed, the name (or signature) of the person who made the inspection, the findings, required remedial action, and a serial number or other means of identifying the inspected respirator; and

        (B) Provide this information on a tag or label that is attached to the storage compartment for the respirator, is kept with the respirator, or is included in inspection reports stored as paper or electronic files. This information shall be maintained until replaced following a subsequent certification.

(4) Repairs. The employer shall ensure that respirators that fail an inspection or are otherwise found to be defective are removed from service, and are discarded or repaired or adjusted in accordance with the following procedures:

    (i) Repairs or adjustments to respirators are to be made only by persons appropriately trained to perform such operations and shall use only the respirator manufacturer's NIOSH-approved parts designed for the respirator;

    (ii) Repairs shall be made according to the manufacturer's recommendations and specifications for the type and extent of repairs to be performed; and

    (iii) Reducing and admission valves, regulators, and alarms shall be adjusted or repaired only by the manufacturer or a technician trained by the manufacturer.

(iv)   Breathing Air Quality and Use

This paragraph requires the employer to provide employees using atmosphere supplying respirators (supplied-air and SCBA) with breathing gases of high purity.

(1)   The employer shall ensure that compressed air, compressed oxygen, liquid air, and liquid oxygen used for respiration accords with the following specifications:

    (i)   Compressed and liquid oxygen shall meet the United States Pharmacopoeia requirements for medical or breathing oxygen; and

    (ii)   Compressed breathing air shall meet at least the requirements for Type 1-Grade D breathing air described in ANSI/Compressed Gas Association Commodity Specification for Air, 7.1-1989, to include:

        (A)   Oxygen content (v/v) of 19.5 - 23.5%;

        (B)   Hydrocarbon (condensed) content of 5 milligrams per cubic meter of air or less;

        (C)   Carbon monoxide (CO) content of 10 ppm or less;

        (D)   Carbon dioxide content of 1,000 ppm or less; and

        (E)   Lack of noticeable odor.

(2)   The employer shall ensure that compressed oxygen is not used in atmosphere-supplying respirators that have previously used compressed air.

(3)   The employer shall ensure that oxygen concentrations greater than 23.5% are used only in equipment designed for oxygen service or distribution.

(4)   The employer shall ensure that cylinders used to supply breathing air to respirators meet the following requirements:

    (i)   Cylinders are tested and maintained as prescribed in the Shipping Container Specification Regulations of the Department of Transportation (49 CFR part 173 and part 178);

    (ii)   Cylinders of purchased breathing air have a certificate of analysis from the supplier that the breathing air meets the requirements for Type 1 grade D breathing air; and

    (iii)   The moisture content in the cylinder does not exceed a dew point of – 50° F (– 45.6° C) at 1 atmosphere pressure.

(5)   The employer shall ensure that compressors used to supply breathing air to respirators are constructed and situated so as to:

    (i)   Prevent entry of contaminated air into the air-supply system.

    (ii)   Minimize moisture content so that the dew point at 1 atmosphere pressure is 10 degrees F (5.56° C) below the ambient temperature.

    (iii)   Have suitable in-line air-purifying sorbent beds and filters to further ensure breathing air quality. Sorbent beds and filters shall be maintained and replaced or refurbished periodically following the manufacturer's instructions.

    (iv)   Have a tag containing the most recent change date and the signature of the person authorized by the employer to perform the change. The tag shall be maintained at the compressor.

(6)   For compressors that are not oil lubricated, the employer shall ensure that carbon monoxide levels in the breathing air do not exceed 10 ppm.

(7)   For oil-lubricated compressors, the employer shall use a high-temperature or carbon monoxide alarm, or both, to monitor carbon monoxide levels. If only high-temperature alarms are used, the air supply shall be monitored at intervals sufficient to prevent carbon monoxide in the breathing air from exceeding 10 ppm.

(8)   The employer shall ensure that breathing air couplings are incompatible with outlets for nonrespirable worksite air or other   gas systems. No asphyxiating substance shall be introduced into breathing air lines.

(9)   The employer shall use breathing gas containers marked in accordance with the NIOSH respirator certification standard, 42CFR part 84.

(j) Identification of Filters, Cartridges, and Canisters

The employer shall ensure that all filters, cartridges, and canisters used in the workplace are labeled and color coded with the NIOSH approval label and that the label is not removed and remains legible.

(k) Training and Information

This paragraph requires the employer to provide effective training to employees who are required to use respirators. The training must be comprehensive, understandable, and recur annually, and more often if necessary. This paragraph also requires the employer to provide the basic information on respirators in Appendix D of this section to employees who wear respirators when not required by this section or by the employer to do so.

(1)   The employer shall ensure that each employee can demonstrate knowledge of at least the following:

   (i)   Why the respirator is necessary and how improper fit, usage, or maintenance can compromise the protective effect of the respirator;

   (ii)   What the limitations and capabilities of the respirator are;

   (iii)   How to use the respirator effectively in emergency situations, including situations in which the respirator malfunctions;

   (iv)   How to inspect, put on and remove, use, and check the seals of the respirator;

   (v)   What the procedures are for maintenance and storage of the respirator;

   (vi)   How to recognize medical signs and symptoms that may limit or prevent the effective use of respirators; and

   (vii)   The general requirements of this section.

(2)   The training shall be conducted in a manner that is understandable to the employee.

(3)   The employer shall provide the training prior to requiring the employee to use a respirator in the workplace.

(4)   An employer who is able to demonstrate that a new employee has received training within the last 12 months that addresses the elements specified in paragraph (k)(1)(i) through (vii) is not required to repeat such training provided that, as required by paragraph (k)(1), the employee can demonstrate knowledge of those element(s). Previous training not repeated initially by the em-

ployer must be provided no later than 12 months from the date of the previous training.

(5) Retraining shall be administered annually, and when the following situations occur:

(i) Changes in the workplace or the type of respirator render previous training obsolete;

(ii) Inadequacies in the employee's knowledge or use of the respirator indicate that the employee has not retained the requisite understanding or skill; or

(iii) Any other situation arises in which retraining appears necessary to ensure safe respirator use.

(6) The basic advisory information on respirators, as presented in Appendix D of this section, shall be provided by the employer in any written or oral format, to employees who wear respirators when such use is not required by this section or by the employer.

(l) Program Evaluation

This section requires the employer to conduct evaluations of the workplace to ensure that the written respiratory protection program is being properly implemented, and to consult employees to ensure that they are using the respirators properly.

(1) The employer shall conduct evaluations of the workplace as necessary to ensure that the provisions of the current written program are being effectively implemented and that it continues to be effective.

(2) The employer shall regularly consult employees required to use respirators to assess the employees' views on program effectiveness and to identify any problems. Any problems that are identified during this assessment shall be corrected. Factors to be assessed include, but are not limited to:

(i) Respirator fit (including the ability to use the respirator without interfering with effective workplace performance);

(ii) Appropriate respirator selection for the hazards to which the employee is exposed;

(iii) Proper respirator use under the workplace conditions the employee encounters; and

(iv) Proper respirator maintenance.

(m) Record Keeping

This section requires the employer to establish and retain written information regarding medical evaluations, fit testing, and the respirator program. This information will facilitate employee involvement in the respirator program, assist the employer in auditing the adequacy of the program, and provide a record for compliance determinations by OSHA.

(1) Medical evaluation. Records of medical evaluations required by this section must be retained and made available in accordance with 29 CFR 1910.1020.

(2) Fit testing.

(i) The employer shall establish a record of the qualitative and quantitative fit tests administered to an employee including:

(A)   The name or identification of the employee tested;

(B)   Type of fit test performed;

(C)   Specific make, model, style, and size of respirator tested;

(D)   Date of test; and

(E)   The pass/fail results for QLFTs or the fit factor and strip chart recording or other recording of the test results for QNFTs.

(ii)   Fit test records shall be retained for respirator users until the next fit test is administered.

(3)   A written copy of the current respirator program shall be retained by the employer.

(4)   Written materials required to be retained under this paragraph shall be made available upon request to affected employees and to the Assistant Secretary or designee for examination and copying.

(n) Dates

(1)   Effective date. This section is effective April 8,1998. The obligations imposed by this section commence on the effective date unless otherwise noted in this paragraph. Compliance with obligations that do not commence on the effective date shall occur no later than the applicable startup date.

(2)   Compliance dates. All obligations of this section commence on the effective date except as follows:

(i)   The determination that respirator use is required (paragraph (a)) shall be completed no later than September 8, 1998.

(ii)   Compliance with provisions of this section for all other provisions shall be completed no later than October 5, 1998.

(3)   The provisions of 29 CFR 1910.134 and 29 CFR 1926.103, contained in the 29 CFR parts 1900 to 1910.99 and the 29 CFR part 1926 editions, revised as of July 1, 1997, are in effect and enforceable until April 8, 1998, or during any administrative or judicial stay of the provisions of this section.

(4)   Existing Respiratory Protection Programs. If, in the 12 month period preceding April 8, 1998, the employer has conducted annual respirator training, fit testing, respirator program evaluation, or medical evaluations, the employer may use the results of those activities to comply with the corresponding provisions of this section, providing that these activities were conducted in a manner that meets the requirements of this section.

(o) Appendices

(1)   Compliance with Appendix A, Appendix B-1, Appendix B-2, and Appendix C of this section is mandatory.

(2)   Appendix D of this section is nonmandatory and is not intended to create any additional obligations not otherwise imposed or to detract from any existing obligations.

Appendix A to §1910.134: Fit Testing Procedures   (Mandatory)

Part 1. OSHA-Accepted Fit Test Protocols

A. Fit Testing Procedures – General Requirements

The employer shall conduct fit testing using the following procedures. The requirements in this appendix apply to all OSHA-accepted fit test methods, both QLFT and QNFT.

1. The test subject shall be allowed to pick the most acceptable respirator from a sufficient number of respirator models and sizes so that the respirator is acceptable to, and correctly fits, the user.

2. Prior to the selection process, the test subject shall be shown how to put on a respirator, how it should be positioned on the face, how to set strap tension, and how to determine an acceptable fit. A mirror shall be available to assist the subject in evaluating the fit and positioning of the respirator. This instruction may not constitute the subject's formal training on respirator use, because it is only a review.

3. The test subject shall be informed that he/she is being asked to select the respirator that provides the most acceptable fit. Each respirator represents a different size and shape, and if fitted and used properly, will provide adequate protection.

4. The test subject shall be instructed to hold each chosen facepiece up to the face and eliminate those that obviously do not give an acceptable fit.

5. The more acceptable facepieces are noted in case the one selected proves unacceptable, the most comfortable mask is donned and worn at least five minutes to assess comfort. Assistance in assessing comfort can be given by discussing the points in the following item A.6. If the test subject is not familiar with using a particular respirator, the test subject shall be directed to don the mask several times and to adjust the straps each time to become adept at setting proper tension on the straps.

6. Assessment of comfort shall include a review of the following points with the test subject and allowing the test subject adequate time to determine the comfort of the respirator:

    (a) Position of the mask on the nose

    (b) Room for eye protection

    (c) Room to talk

    (d) Position of mask on face and cheeks

7. The following criteria shall be used to help determine the adequacy of the respirator fit:

    (a) Chin properly placed;

    (b) Adequate strap tension, not overly tightened;

    (c) Fit across nose bridge;

    (d) Respirator of proper size to span distance from nose to chin;

    (e) Tendency of respirator to slip;

    (f) Self-observation in mirror to evaluate fit and respirator position.

8. The test subject shall conduct a user seal check, either the negative and positive pressure seal checks described in Appendix B-1 of this section or those

recommended by the respirator manufacturer which provide equivalent protection to the procedures in Appendix B-1. Before conducting the negative and positive pressure checks, the subject shall be told to seat the mask on the face by moving the head from side to side and up and down slowly while taking in a few slow deep breaths. Another facepiece shall be selected and retested if the test subject fails the user seal check tests.

9. The test shall not be conducted if there is any hair growth between the skin and the facepiece sealing surface, such as stubble beard growth, beard, mustache, or sideburns which cross the respirator sealing surface. Any type of apparel which interferes with a satisfactory fit shall be altered or removed.

10. If a test subject exhibits difficulty in breathing during the tests, she or he shall be referred to a physician or other licensed health care professional, as appropriate, to determine whether the test subject can wear a respirator while performing her or his duties.

11. If the employee finds the fit of the respirator unacceptable, the test subject shall be given the opportunity to select a different respirator and to be retested.

12. Exercise regimen. Prior to the commencement of the fit test, the test subject shall be given a description of the fit test and the test subject's responsibilities during the test procedure. The description of the process shall include a description of the test exercises that the subject will be performing. The respirator to be tested shall be worn for at least 5 minutes before the start of the fit test.

13. The fit test shall be performed while the test subject is wearing any applicable safety equipment that may be worn during actual respirator use which could interfere with respirator fit.

14. Test Exercises.

    (a) The following test exercises are to be performed for all fit testing methods prescribed in this appendix, except for the CNP method. A separate fit testing exercise regimen is contained in the CNP protocol.

The test subject shall perform exercises, in the test environment, in the following manner:

    (1) Normal breathing. In a normal standing position, without talking, the subject shall breathe normally.

    (2) Deep breathing. In a normal standing position, the subject shall breathe slowly and deeply, taking caution so as not to hyperventilate.

    (3) Turning head side to side. Standing in place, the subject shall slowly turn his/her head from side to side between the extreme positions on each side. The head shall be held at each extreme momentarily so the subject can inhale at each side.

    (4) Moving head up and down. Standing in place, the subject shall slowly move his/her head up and down. The subject shall be instructed to inhale in the up position (i.e., when looking toward the ceiling).

    (5) Talking. The subject shall talk out loud slowly and loud enough so as to be heard clearly by the test conductor. The subject can read from a prepared text such as the Rainbow Passage, count backward from 100, or recite a memorized poem or song.

Rainbow Passage

When the sunlight strikes raindrops in the air, they act like a prism and form a rainbow. The rainbow is a division of white light into many beautiful colors. These take the shape of a long round arch, with its path high above, and its two ends apparently beyond the horizon. There is, according to legend, a boiling pot of gold at one end. People look, but no one ever finds it. When a man looks for something beyond reach, his friends say he is looking for the pot of gold at the end of the rainbow.

(6) Grimace. The test subject shall grimace by smiling or frowning. (This applies only to QNFT testing; it is not performed for QLFT)

(7) Bending over. The test subject shall bend at the waist as if he/she were to touch his/her toes. Jogging in place shall be substituted for this exercise in those test environments such as shroud type QNFT or QLFT units that do not permit bending over at the waist.

(8) Normal breathing. Same as exercise (1).

(b) Each test exercise shall be performed for one minute except for the grimace exercise which shall be performed for 15 seconds. The test subject shall be questioned by the test conductor regarding the comfort of the respirator upon completion of the protocol. If it has become unacceptable, another model of respirator shall be tried. The respirator shall not be adjusted once the fit test exercises begin. Any adjustment voids the test, and the fit test must be repeated.

B. Qualitative Fit Test (QlFT) Protocols

1. <u>General</u>

(a) The employer shall ensure that persons administering QLFT are able to prepare test solutions, calibrate equipment and perform tests properly, recognize invalid tests, and ensure that test equipment is in proper working order.

(b) The employer shall ensure that QLFT equipment is kept clean and well maintained so as to operate within the parameters for which it was designed.

2. <u>Isoamyl Acetate Protocol</u>

Note: This protocol is not appropriate to use for the fit testing of particulate respirators. If used to fit test particulate respirators, the respirator must be equipped with an organic vapor filter.

(a) Odor Threshold Screening

Odor threshold screening, performed without wearing a respirator, is intended to determine if the individual tested can detect the odor of isoamyl acetate at low levels.

(1) Three 1 liter glass jars with metal lids are required.

(2) Odor-free water (e.g., distilled or spring water) at approximately 25°C (77°F) shall be used for the solutions.

(3) The isoamyl acetate (IAA) (also known at isopentyl acetate) stock solution is prepared by adding 1 ml of pure IAA to 800 ml of odor-free water in a 1 liter jar, closing the lid and shaking for 30 seconds. A new solution shall be prepared at least weekly.

(4)   The screening test shall be conducted in a room separate from the room used for actual fit testing. The two rooms shall be well ventilated to prevent the odor of IAA from becoming evident in the general room air where testing takes place.

(5)   The odor test solution is prepared in a second jar by placing 0.4 ml of the stock solution into 500 ml of odor-free water using a clean dropper or pipette. The solution shall be shaken for 30 seconds and allowed to stand for two to three minutes so that the IAA concentration above the liquid may reach equilibrium. This solution shall be used for only one day.

(6)   A test blank shall be prepared in a third jar by adding B 500 cc of odor-free water.

(7)   The odor test and test blank jar lids shall be labeled (e.g., 1 and 2) for jar identification. Labels shall be placed on the lids so that they can be peeled off periodically and switched to maintain the integrity of the test.

(8)   The following instruction shall be typed on a card and placed on the table in front of the two test jars (i.e., 1 and 2): "The purpose of this test is to determine if you can smell banana oil at a low concentration. The two bottles in front of you contain water. One of these bottles also contains a small amount of banana oil. Be sure the covers are tight, then shake each bottle for two seconds. Unscrew the lid of each bottle, one at a time, and sniff at the mouth of the bottle. Indicate to the test conductor which bottle contains banana oil."

(9)   The mixtures used in the IAA odor detection test shall be prepared in an area separate from where the test is performed, in order to prevent olfactory fatigue in the subject.

(10)   If the test subject is unable to correctly identify the jar containing the odor test solution, the IAA qualitative fit test shall not be performed.

(11)   If the test subject correctly identifies the jar containing the odor test solution, the test subject may proceed to respirator selection and fit testing.

(b) Isoamyl Acetate Fit Test

(1)   The fit test chamber shall be a clear 55 gallon drum liner suspended inverted over a 2-foot diameter frame so that the top of the chamber is about 6 inches above the test subject's head. If no drum liner is available, a similar chamber shall be constructed using plastic sheeting. The inside top center of the chamber shall have a small hook attached.

(2)   Each respirator used for the fitting and fit testing shall be equipped with organic vapor cartridges or offer protection against organic vapors.

(3)   After selecting, donning, and properly adjusting a respirator, the test subject shall wear it to the fit testing room. This room shall be separate from the room used for odor threshold screening and respirator selection, and shall be well ventilated, as by an exhaust fan or lab hood, to prevent general room contamination.

(4)   A copy of the test exercises and any prepared text from which the subject is to read shall be taped to the inside of the test chamber.

(5)   Upon entering the test chamber, the test subject shall be given a 6-inch by 5-inch piece of paper towel, or other porous, absorbent, single-ply material, folded in half and wetted with 0.75 ml of pure IAA. The test subject shall hang the wet towel on the hook at the top of the chamber. An IAA test swab or ampule may be substituted for the IAA wetted paper towel provided it has been demonstrated that the alternative IAA source will generate an IAA test atmosphere with a concentration equivalent to that generated b the paper towel method.

(6)   Allow two minutes for the IAA test concentration to stabilize before starting the fit test exercises. This would be an appropriate time to talk with the test subject; to explain the fit test, the importance of his/ her cooperation, and the purpose for the test exercises; or to demonstrate some of the exercises.

(7)   If at any time during the test, the subject detects the banana-like odor of IAA, the test is failed. The subject shall quickly exit from the test chamber and leave the test area to avoid olfactory fatigue.

(8)   If the test is failed, the subject shall return to the selection room and remove the respirator. The test subject shall repeat the odor sensitivity test, select and put on another respirator, return to the test area and again begin the fit test procedure described in (b) (1) through (7) above. The process continues until a respirator that fits well has been found. Should the odor sensitivity test be failed, the subject shall wait at least 5 minutes before retesting. Odor sensitivity will usually have returned by this time.

(9)   If the subject passes the test, the efficiency of the test procedure shall be demonstrated by having the subject break the respirator face seal and take a breath before exiting the chamber.

(10)   When the test subject leaves the chamber, the subject shall remove the saturated towel and return it to the person conducting the test, so that there is no significant IAA concentration buildup in the chamber during subsequent tests. The used towels shall be kept in a self-sealing plastic bag to keep the test area from being contaminated.

3.   Saccharin Solution Aerosol Protocol

The entire screening and testing procedure shall be explained to the test subject prior to the conduct of the screening test.

(a)   Taste threshold screening. The saccharin taste threshold screening, performed without wearing a respirator, is intended to determine whether the individual being tested can detect the taste of saccharin.

(1) During threshold screening as well as during fit testing, subjects shall wear an enclosure about the head and shoulders that is approximately 12 inches in diameter by 14 inches tall with at least the front portion clear and that allows free movements of the head when a respirator is worn. An enclosure substantially similar to the 3M hood assembly, parts # FT 14 and # FT 15 combined, is adequate.

(2) The test enclosure shall have a 3/4 inch (1.9 cm) hole in front of the test subject's nose and mouth area to accommodate the nebulizer nozzle.

(3) The test subject shall don the test enclosure. Throughout the threshold screening test, the test subject shall breathe through his/her slightly open mouth with tongue extended. The subject is instructed to report when he/she detects a sweet taste.

(4) Using a DeVilbiss Model 40 Inhalation Medication Nebulizer or equivalent, the test conductor shall spray the threshold check solution into the enclosure. The nozzle is directed away from the nose and mouth of the person. This nebulizer shall be clearly marked to distinguish it from the fit test solution nebulizer.

(5) The threshold check solution is prepared by dissolving 0.83 gram of sodium saccharin USP in 100 ml of warm water. It can be prepared by putting 1 ml of the fit test solution (see (b)(5) below) in 100 ml of distilled water.

(6) To produce the aerosol, the nebulizer bulb is firmly squeezed so that it collapses completely, then released and allowed to fully expand.

(7) Ten squeezes are repeated rapidly and then the test subject is asked whether the saccharin can be tasted. If the test subject reports tasting the sweet taste during the ten squeezes, the screening test is completed. The taste threshold is noted as ten regardless of the number of squeezes actually completed.

(8) If the first response is negative, ten more squeezes are repeated rapidly and the test subject is again asked whether the saccharin is tasted. If the test subject reports tasting the sweet taste during the second ten squeezes, the screening test is completed. The taste threshold is noted as twenty regardless of the number of squeezes actually completed.

(9) If the second response is negative, ten more squeezes are repeated rapidly and the test subject is again asked whether the saccharin is tasted. If the test subject reports tasting the sweet taste during the third set of ten squeezes, the screening test is completed. The taste threshold is noted as thirty regardless of the number of squeezes actually completed.

(10) The test conductor will take note of the number of squeezes required to solicit a taste response.

(11) If the saccharin is not tasted after 30 squeezes (step 10), the test subject is unable to taste saccharin and may not perform the saccharin fit test.

Note to paragraph 3.(a): If the test subject eats or drinks something sweet before the screening test, he/she may be unable to taste the weak saccharin solution.

(12)   If a taste response is elicited, the test subject shall be asked to take note of the taste for reference in the fit test.

(13)   Correct use of the nebulizer means that approximately 1 ml of liquid is used at a time in the nebulizer body.

(14)   The nebulizer shall be thoroughly rinsed in water, shaken dry, and refilled at least each morning and afternoon or at least every four hours.

(b) Saccharin solution aerosol fit test procedure.

(1)   The test subject may not eat, drink (except plain water), smoke, or chew gum for 15 minutes before the test.

(2)   The fit test uses the same enclosure described in 3.(a) above.

(3)   The test subject shall don the enclosure while wearing the respirator selected in Section 1.A. of this Appendix. The respirator shall be properly adjusted and equipped with a particulate filter(s).

(4)   A second DeVilbiss Model 40 Inhalation Medication Nebulizer or equivalent is used to spray the fit test solution into the enclosure. This nebulizer shall be clearly marked to distinguish it from the screening test solution nebulizer.

(5)   The fit test solution is prepared by adding 83 grams of sodium saccharin to 100 ml of warm water.

(6)   As before, the test subject shall breathe through the slightly open mouth with tongue extended, and report if he/she tastes the sweet taste of saccharin.

(7)   The nebulizer is inserted into the hole in the front of the enclosure and an initial concentration of saccharin fit test solution is sprayed into the enclosure using the same number of squeezes (either 10, 20, or 30 squeezes) based on the number of squeezes required to elicit a taste response as noted during the screening test. A minimum of 10 squeezes is required.

(8)   After generating the aerosol, the test subject shall be instructed to perform the exercises in Section 1.A.14. of this Appendix.

(9)   Every 30 seconds the aerosol concentration shall be replenished using one half the original number of squeezes used initially (e.g., 5,10, or 15).

(10)   The test subject shall indicate to the test conductor if at any time during the fit test the taste of saccharin is detected. If the test subject does not report tasting the saccharin, the test is passed.

(11)   If the taste of saccharin is detected, the fit is deemed unsatisfactory and the test is failed. A different respirator shall be tried and the entire test procedure is repeated (taste threshold screening and fit testing).

(12)   Since the nebulizer has a tendency to clog during use, the test operator must make periodic checks of the nebulizer to ensure

that it is not clogged. If clogging is found at the end of the test session, the test is invalid.

Bitrex™ (Denatonium Benzoate) Solution Aerosol Qualitative Fit Test Protocol

The Bitrex™ (Denatonium benzoate) solution aerosol QLFT protocol uses the published saccharin test protocol because that protocol is widely accepted. Bitrex is routinely used as a taste aversion agent in household liquids which children should not be drinking and is endorsed by the American Medical Association, the National Safety Committee, and the American Association of Poison Control Centers. The entire screening and testing procedure shall be explained to the test subject prior to the conduct of the screening test.

(a) Taste Threshold Screening.

The Bitrex taste threshold screening, performed without wearing a respirator, is intended to determine whether the individual being tested can detect the taste of Bitrex.

(1)   During threshold screening as well as during fit testing, subjects shall wear an enclosure about the head and shoulders that is approximately 12 inches (30.5 cm) in diameter by 14 inches (35.6 cm) tall. The front portion of the enclosure shall be clear from the respirator and allow free movement of the head when a respirator is worn. An enclosure substantially similar to the 3M hood assembly, parts #14 and #15 combined, is adequate.

(2)   The test enclosure shall have a 3/4 inch (1.9 cm) hole in front of the test subject's nose and mouth area to accommodate the nebulizer nozzle.

(3)   The test subject shall don the test enclosure. Throughout the threshold screening test, the test subject shall breathe through his or her slightly open mouth with tongue extended. The subject is instructed to report when he/she detects a bitter taste.

(4)   Using a DeVilbiss Model 40 Inhalation Medication Nebulizer or equivalent, the test conductor shall spray the Threshold Check Solution into the enclosure. This Nebulizer shall be clearly marked to distinguish it from the fit test solution nebulizer.

(5)   The Threshold Check Solution is prepared by adding 13.5 milligrams of Bitrex to 100 ml of 5% salt (NaCl) solution in distilled water.

(6)   To produce the aerosol, the nebulizer bulb is firmly squeezed so that the bulb collapses completely, and is then released and allowed to fully expand.

(7)   An initial ten squeezes are repeated rapidly and then the test subject is asked whether the Bitrex can be tasted. If the test subject reports tasting the bitter taste during the ten squeezes, the screening test is completed. The taste threshold is noted as ten regardless of the number of squeezes actually completed.

(8)   If the first response is negative, ten more squeezes are repeated rapidly and the test subject is again asked whether the Bitrex is tasted. If the test subject reports tasting the bitter taste during the second ten squeezes, the screening test is completed. The taste threshold is noted as twenty regardless of the number of squeezes actually completed.

(9) If the second response is negative, ten more squeezes are repeated rapidly and the test subject is again asked whether the Bitrex is tasted. If the test subject reports tasting the bitter taste during the third set of ten squeezes, the screening test is completed. The taste threshold is noted as thirty regardless of the number of squeezes actually completed.

(10) The test conductor will take note of the number of squeezes required to solicit a taste response.

(11) If the Bitrex is not tasted after 30 squeezes (step 10), the test subject is unable to taste Bitrex and may not perform the Bitrex fit test.

(12) If a taste response is elicited, the test subject shall be asked to take note of the taste for reference in the fit test.

(13) Correct use of the nebulizer means that approximately 1 ml of liquid is used at a time in the nebulizer body.

(14) The nebulizer shall be thoroughly rinsed in water, shaken to dry, and refilled at least each morning and afternoon or at least every four hours.

(b) Bitrex Solution Aerosol Fit Test Procedure.

(1) The test subject may not eat, drink (except plain water), smoke, or chew gum for 15 minutes before the test.

(2) The fit test uses the same enclosure as that described in 4. (a) above.

(3) The test subject shall don the enclosure while wearing the respirator selected according to section 1. A. of this appendix. The respirator shall be properly adjusted and equipped with any type particulate filter(s).

(4) A second DeVilbiss Model 40 Inhalation Medication Nebulizer or equivalent is used to spray the fit test solution into the enclosure. This nebulizer shall be clearly marked to distinguish it from the screening test solution nebulizer.

(5) The fit test solution is prepared by adding 337.5 mg of Bitrex to 200 ml of a 5% salt (NaCI) solution in warm water.

(6) As before, the test subject shall breathe through his or her slightly open mouth with tongue extended, and be instructed to report if he/she tastes the bitter taste of Bitrex.

(7) The nebulizer is inserted into the hole in the front of the enclosure and an initial concentration of the fit test solution is sprayed into the enclosure using the same number of squeezes (either 10, 20 or 30 squeezes) based on the number of squeezes required to elicit a taste response as noted during the screening test.

(8) After generating the aerosol, the test subject shall be instructed to perform the exercises in section 1. A. 14. of this appendix.

(9) Every 30 seconds the aerosol concentration shall be replenished using one half the number of squeezes used initially (e.g., 5, 10, or 15).

(10) The test subject shall indicate to the test conductor if at any time during the fit test the taste of Bitrex is detected. If the test subject does not report tasting the Bitrex, the test is passed.

(11)   If the taste of Bitrex is detected, the fit is deemed unsatisfactory and the test is failed. A different respirator shall be tried and the entire test procedure is repeated (taste threshold screening and fit testing).

4.   Irritant Smoke (Stannic Chloride) Protocol

This qualitative fit test uses a person's response to the irritating chemicals released in the "smoke" produced by a stannic chloride ventilation smoke tube to detect leakage into the respirator.

(a)   General Requirements and Precautions

(1)   The respirator to be tested shall be equipped with high efficiency particulate air (HEPA) or P100 series filter(s).

(2)   Only stannic chloride smoke tubes shall be used for this protocol.

(3)   No form of test enclosure or hood for the test subject shall be used.

(4)   The smoke can be irritating to the eyes, lungs, and nasal passages. The test conductor shall take precautions to minimize the test subject's exposure to irritant smoke. Sensitivity varies, and certain individuals may respond to a greater degree to irritant smoke. Care shall be taken when performing the sensitivity screening checks that determine whether the test subject can detect irritant smoke to use only the minimum amount of smoke necessary to elicit a response from the test subject.

(5)   The fit test shall be performed in an area with adequate ventilation to prevent exposure of the person conducting the fit test or the build-up of irritant smoke in the general atmosphere.

(b)   Sensitivity Screening Check

The person to be tested must demonstrate his or her ability to detect a weak concentration of the irritant smoke.

(1)   The test operator shall break both ends of a ventilation smoke tube containing stannic chloride and attach one end of the smoke tube to a low flow air pump set to deliver 200 milliliters per minute, or an aspirator squeeze bulb. The test operator shall cover the other end of the smoke tube with a short piece of tubing to prevent potential injury from the jagged end of the smoke tube.

(2)   The test operator shall advise the test subject that the smoke can be irritating to the eyes, lungs, and nasal passages and instruct the subject to keep his/her eyes closed while the test is performed.

(3)   The test subject shall be allowed to smell a weak concentration of the irritant smoke before the respirator is donned to become familiar with its irritating properties and to determine if he/she can detect the irritating properties of the smoke. The test operator shall carefully direct a small amount of the irritant smoke in the test subject's direction to determine that he/she can detect it.

(c)   Irritant Smoke Fit Test Procedure

(1)   The person being fit tested shall don the respirator without assistance, and perform the required user seal check(s).

(2)   The test subject shall be instructed to keep his/her eyes closed.

(3)   The test operator shall direct the stream of irritant smoke from the smoke tube toward the faceseal area of the test subject, using the low flow pump or the squeeze bulb. The test operator shall begin at least 12 inches from the facepiece and move the smoke stream around the whole perimeter of the mask. The operator shall gradually make two more passes around the perimeter of the mask, moving to within six inches of the respirator.

(4)   If the person being tested has not had an involuntary response and/or detected the irritant smoke, proceed with the test exercises.

(5)   The exercises identified in Section l.A.14. of this Appendix shall be performed by the test subject while the respirator seal is being continually challenged by the smoke, directed around the perimeter of the respirator at a distance of six inches.

(6)   If the person being fit tested reports detecting the irritant smoke at any time, the test is failed. The person being retested must repeat the entire sensitivity check and fit test procedure.

(7)   Each test subject passing the irritant smoke test without evidence of a response (involuntary cough, irritation) shall be given a second sensitivity screening check, with the smoke from the same smoke tube used during the fit test, once the respirator has been removed, to determine whether he/she still reacts to the smoke. Failure to evoke a response shall void the fit test.

(8)   If a response is produced during this second sensitivity check, then the fit test is passed.

C. Quantitative Fit Test (QNFT) Protocols

The following quantitative fit testing procedures have been demonstrated to be acceptable: Quantitative fit testing using a non-hazardous test aerosol (such as corn oil polyethylene glycol 400 IPEG 4001, all-2-ethyl hexyl sebacate (DEHS], or sodium chloride) generated in a test chamber, and employing instrumentation to quantify the fit of the respirator; Quantitative fit testing using ambient aerosol as the test agent and appropriate instrumentation (condensation nuclei counter) to quantify the respirator fit; Quantitative fit testing using controlled negative pressure and appropriate instrumentation to measure the volumetric leak rate of a facepiece to quantify the respirator fit.

1.   General

(a)   The employer shall ensure that persons administering QNFT are able to calibrate equipment and perform tests properly, recognize invalid tests, calculate fit factors properly, and ensure that test equipment is in proper working order.

(b)   The employer shall ensure that QNFT equipment is kept clean, and is maintained and calibrated according to the manufacturer's instructions so as to operate at the parameters for which it was designed.

2.   Generated Aerosol Quantitative Fit Testing Protocol

(a)   Apparatus.

(1)   Instrumentation. Aerosol generation, dilution, and measurement systems using particulates (corn oil, polyethylene glycol 400 IPEG 4001, all-2-ethyl hexyl sebacate IDEHSI or sodium chloride) as test aerosols shall be used for quantitative fit testing.

(2)   Test chamber. The test chamber shall be large enough to permit all test subjects to perform freely all required exercises without disturbing the test agent concentration or the measurement apparatus. The test chamber shall be equipped and constructed so that the test agent is effectively isolated from the ambient air, yet uniform in concentration throughout the chamber.

(3)   When testing air-purifying respirators the normal filter or cartridge element shall be replaced with a high efficiency particulate air (HEPA) or P100 series filter supplied by the same manufacturer.

(4)   The sampling instrument shall be selected so that a computer record or strip chart record may be made of the test showing the rise and fall of the test agent concentration with each inspiration and expiration at fit factors of at least 2,000.

Integrators or computers that integrate the amount of test agent penetration leakage into the respirator for each exercise may be used provided a record of the readings is made.

(5)   The combination of substitute air purifying elements, test agent and test agent concentration shall be such that the test subject is not exposed in excess of an established exposure limit for the test agent at any time during the testing process, based upon the length of the exposure and the exposure limit duration.

(6)   The sampling port on the test specimen respirator shall be placed and constructed so that no leakage occurs around the port (e.g., where the respirator is probed), a free air flow is allowed into the sampling line at all times, and there is no interference with the fit or performance of the respirator. The in-mask sampling device (probe) shall be designed and used so that the air sample is drawn from the breathing zone of the test subject, midway between the nose and mouth and with the probe extending into the facepiece cavity at least 1/4 inch.

(7)   The test setup shall permit the person administering the test to observe the test subject inside the chamber during the test.

(8)   The equipment generating the test atmosphere shall maintain the concentration of test agent constant to within a 10 percent variation for the duration of the test.

(9)   The time lag (interval between an event and the recording of the event on the strip chart or computer or integrator) shall be kept to a minimum. There shall be a clear association between the occurrence of an event and its being recorded.

(10)   The sampling line tubing for the test chamber atmosphere and for the respirator sampling port shall be of equal diameter and of the same material. The length of the two lines shall be equal.

(11)   The exhaust flow from the test chamber shall pass through an appropriate filter (i.e., high efficiency particulate or P100 series filter) before release.

(12) When sodium chloride aerosol is used, the relative humidity inside the test chamber shall not exceed 50 percent.

(13) The limitations of instrument detection shall be taken into account when determining the fit factor.

(14) Test respirators shall be maintained in proper working order and be inspected regularly for deficiencies such as cracks or missing valves and gaskets.

(b) Procedural Requirements.

(1) When performing the initial user seal check using a positive or negative pressure check, the sampling line shall be crimped closed in order to avoid air pressure leakage during either of these pressure checks.

(2) The use of an abbreviated screening QLFT test is optional. Such a test may be utilized in order to quickly identify poor fitting respirators that passed the positive and/or negative pressure test and reduce the amount of QNFT time. The use of the CNC QNFT instrument in the count mode is another optional method to obtain a quick estimate of fit and eliminate poor fitting respirators before going on to perform a full QNFT.

(3) A reasonably stable test agent concentration shall be measured in the test chamber prior to testing. For canopy or shower curtain types of test units, the determination of the test agent's stability may be established after the test subject has entered the test environment.

(4) Immediately after the subject enters the test chamber, the test agent concentration inside the respirator shall be measured to ensure that the peak penetration does not exceed 5 percent for a half mask or 1 percent for a full facepiece respirator.

(5) A stable test agent concentration shall be obtained prior to the actual start of testing.

(6) Respirator restraining straps shall not be over-tightened for testing. The straps shall be adjusted by the wearer without assistance from other persons to give a reasonably comfortable fit typical of normal use. The respirator shall not be adjusted once the fit test exercises begin.

(7) The test shall be terminated whenever any single peak penetration exceeds 5 percent for half masks and 1 percent for full facepiece respirators. The test subject shall be refitted and retested.

(8) Calculation of fit factors.

(i) The fit factor shall be determined for the quantitative fit test by taking the ratio of the average chamber concentration to the concentration measured inside the respirator for each test exercise except the grimace exercise.

(ii) The average test chamber concentration shall be calculated as the arithmetic average of the concentration measured before and after each test (i.e., 7 exercises) or the arithmetic average of the concentration measured before and after each exercise or the true average measured continuously during the respirator sample.

(iii) The concentration of the challenge agent inside the respirator shall be determined by one of the following methods:

(A) Average peak penetration method means the method of determining test agent penetration into the respirator utilizing a strip chart recorder, integrator, or computer. The agent penetration is determined by an average of the peak heights on the graph or by computer integration, for each exercise except the grimace exercise. Integrators or computers that calculate the actual test agent penetration into the respirator for each exercise will also be considered to meet the requirements of the average peak penetration method.

(B) Maximum peak penetration method means the method of determining test agent penetration in the respirator as determined by strip chart recordings of the test. The highest peak penetration for a given exercise is taken to be representative of average penetration into the respirator for that exercise.

(C) Integration by calculation of the area under the individual peak for each exercise except the grimace exercise. This includes computerized integration.

(D) The calculation of the overall fit factor using individual exercise fit factors involves first converting the exercise fit factors to penetration values, determining the average, and then converting that result back to a fit factor. This procedure is described in the following equation:

$$\text{Overall Fit Factor} = \frac{\text{Number of Exercises}}{1/ff1 + 1/ff2 + 1/ff3 + 1/ff4 + 1/ff5 + 1/ff7 + 1/ff8}$$

Where ff1, ff2, ff3, etc. are the fit factors for exercises 1, 2, 3, etc.

(9) The test subject shall not be permitted to wear a half mask or quarter facepiece respirator unless a minimum fit factor of 100 is obtained, or a full facepiece respirator unless a minimum fit factor of 500 is obtained.

(10) Filters used for quantitative fit testing shall be replaced whenever increased breathing resistance is encountered, or when the test agent has altered the integrity of the filter media.

3. Ambient aerosol condensation nuclei counter (CNC) quantitative fit testing protocol.

The ambient aerosol condensation nuclei counter (CNC) quantitative fit testing

(Portacount) protocol quantitatively fit tests respirators with the use of a probe. The probed respirator is only used for quantitative fit tests. A probed respirator has a special sampling device, installed on the respirator, that allows the probe to sample the air from inside the mask. A probed respirator is required for each make, style, model, and size that the employer uses and can be obtained from the respirator manufacturer or distributor. The CNC instrument manufacturer, TSI Inc., also provides probe attachments (TSI sampling adapters) that permit fit testing in an employee's own respirator. A minimum fit factor pass level of at least 100 is necessary for a half-mask respirator and a minimum fit factor pass level of at least 500 is required for a full facepiece negative pressure respirator. The entire screening and testing procedure shall be explained to the test subject prior to the conduct of the screening test.

(a)   Portacount Fit Test Requirements.

(1)   Check the respirator to make sure the respirator is fitted with a high-efficiency filter and that the sampling probe and line are properly attached to the facepiece.

(2)   Instruct the person to be tested to don the respirator for five minutes before the fit test starts. This purges the ambient particles trapped inside the respirator and permits the wearer to make certain the respirator is comfortable. This individual shall already have been trained on how to wear the respirator properly.

(3)   Check the following conditions for the adequacy of the respirator fit: chin properly placed; adequate strap tension, not overly tightened; fit across nose bridge; respirator of proper size to span distance from nose to chin; tendency of the respirator to slip; self-observation in a mirror to evaluate fit and respirator position.

(4)   Have the person wearing the respirator do a user seal check. If leakage is detected, determine the cause. If leakage is from a poorly fitting facepiece, try another size of the same model respirator, or another model of respirator.

(5)   Follow the manufacturer's instructions for operating the Portacount and proceed with the test.

(6)   The test subject shall be instructed to perform the exercises in Section 1.A.14. of this Appendix.

(7)   After the test exercises, the test subject shall be questioned by the test conductor regarding the comfort of the respirator upon completion of the protocol. If it has become unacceptable, another model of respirator shall be tried.

(b)   Portacount Test Instrument.

(1)   The Portacount will automatically stop and calculate the overall fit factor for the entire set of exercises. The overall fit factor is what counts. The Pass or Fail message will indicate whether or not the test was successful. If the test was a Pass, the fit test is over.

(2)   Since the pass or fail criterion of the Portacount is user programmable, the test operator shall ensure that the pass or fail criterion meet the requirements for minimum respirator performance in this Appendix.

(3)   A record of the test needs to be kept on file, assuming the fit test was successful. The record must contain the test subject's name overall fit factor; make, model, style, and size of respirator used; and date tested.

(4)   Controlled negative pressure (CNP) quantitative fit testing protocol.

The CNP protocol provides an alternative to aerosol fit test methods. The CNP fit test method technology is based on exhausting air from a temporarily sealed respirator facepiece to generate and then maintain a constant negative pressure inside the facepiece. The rate of air exhaust is controlled so that a constant negative pressure is maintained in the respirator during the fit test. The level of pressure is selected to replicate the mean inspiratory pressure that causes leakage into the respirator under normal use conditions. With pressure held constant, air flow out of the respirator is equal to air flow into the respirator. Therefore, measurement of the exhaust stream that is required to hold the pressure in the temporarily sealed respirator constant yields a direct measure of leakage air flow into the respirator. The CNP fit test method measures leak rates through the facepieces as a method for determining the facepiece fit for negative pressure respirators. The CNP instrument manufacturer Dynatech Nevada also provides attachments (sampling manifolds) that replace the filter cartridges to permit fit testing in an employee's own respirator. To perform the test, the test subject closes his or her mouth and holds his/her breath, after which an air pump removes air from the respirator facepiece at a pre-selected constant pressure. The facepiece fit is expressed as the leak rate through the facepiece, expressed as milliliters per minute. The quality and validity of the CNP fit tests are determined by the degree to which the in-mask pressure tracks the test pressure during the system measurement time of approximately five seconds. Instantaneous feedback in the form of a real-time pressure trace of the in-mask pressure is provided and used to determine test validity and quality. A minimum fit factor pass level of 100 is necessary for a halfmask respirator and a minimum fit factor of at least 500 is required for a full facepiece respirator. The entire screening and testing procedure shall be explained to the test subject prior to the conduct of the screening test.

(a)   CNP Fit Test Requirements.

(1)   The instrument shall have a nonadjustable test pressure of 15.0 mm water pressure.

(2)   The CNP system defaults selected for test pressure shall be set at –1. 5 mm of water (–0.58 inches of water) and the modeled inspiratory flow rate shall be 53.8 liters per minute for performing fit tests.

(*Note:* CNP systems have built-in capability to conduct fit testing that is specific to unique work rate, mask, and gender situations that might apply in a specific workplace. Use of system default values which were selected to represent respirator wear with medium cartridge resistance at a low-moderate work rate, will allow inter-test comparison of the respirator fit.)

(3)   The individual who conducts the CNP fit testing shall be thoroughly trained to perform the test.

(4)   The respirator filter or cartridge needs to be replaced with the CNP test manifold. The inhalation valve downstream from the manifold either needs to be temporarily removed or propped open.

(5)   The test subject shall be trained to hold his or her breath for at least 20 seconds.

(6)   The test subject shall don the test respirator without any assistance from the individual who conducts the CNP fit test.

(7) The QNFT protocol shall be followed according to Section 1.C.1. of this Appendix with an exception for the CNP test exercises.

(b) CNP Test Exercises.

(1) Normal breathing. In a normal standing position, without talking, the subject shall breathe normally for 1 minute. After the normal breathing exercise, the subject needs to hold head straight ahead and hold his or her breath for 10 seconds during the test measurement.

(2) Deep breathing. In a normal standing position, the subject shall breathe slowly and deeply for 1 minute, being careful not to hyperventilate. After the deep breathing exercise, the subject shall hold his or her head straight ahead and hold his or her breath for 10 seconds during test measurement.

(3) Turning head side-to-side. Standing in place, the subject shall slowly turn his or her head from side to side between the extreme positions on each side for 1 minute. The head shall be held at each extreme momentarily so the subject can inhale at each side. After the turning head side-to-side exercise, the subject needs to hold head full left and hold his or her breath for 10 seconds during test measurement. Next, the subject needs to hold head full right and hold his or her breath for 10 seconds during test measurement.

(4) Moving head up and down. Standing in place, the subject shall slowly move his or her head up and down for 1 minute. The subject shall be instructed to inhale in the up position (i.e., when looking toward the ceiling). After the moving head up-and-down exercise, the subject shall hold his or her head full up and hold his or her breath for 10 seconds during test measurement. Next the subject shall hold his or her head full down and hold his or her breath for 10 seconds during test measurement.

(5) Talking. The subject shall talk out loud slowly and loud enough so as to be heard clearly by the test conductor. The subject can read from a prepared text such as the Rainbow Passage, count backward from 100, or recite a memorized poem or song for 1 minute, After the talking exercise, the subject shall hold his or her head straight ahead and hold his or her breath for 10 seconds during the test measurement.

(6) Grimace. The test subject shall grimace by smiling or frowning for 15 seconds.

(7) Bending Over. The test subject shall bend at the waist as if he or she were to touch his or her toes for 1 minute. Jogging in place shall be substituted for this exercise in those test environments such as shroud-type QNFT units that prohibit bending at the waist. After the bending-over exercise, the subject shall hold his or her head straight ahead and hold his or her breath for 10 seconds during the test measurement.

(8) Normal Breathing. The test subject shall remove and re-don the respirator within a one-minute period. Then, in a normal standing position, without talking, the subject shall breathe normally for 1 minute. After the normal breathing exercise, the subject shall hold his or her head straight ahead and hold his or her breath for 10 seconds during the test measurement. After the test exercises, the test subject shall be questioned by the test conductor regarding the comfort of the respirator upon completion of the protocol. If it has become unacceptable, another model of a respirator shall be tried.

(c) CNP Test Instrument.

(1)   The test instrument shall have an effective audio warning device when the test subject fails to hold his or her breath during the test. The test shall be terminated whenever the test subject failed to hold his or her breath. The test subject may be refitted and retested.

(2)   A record of the test shall be kept on file, assuming the fit test was successful. The record must contain the test subject's name; overall fit factor; make, model, style, and size of respirator used; and date tested.

Part II. New Fit Test Protocols

A.   Any person may submit to OSHA an application for approval of a new fit test protocol. If the application meets the following criteria, OSHA will initiate a rulemaking proceeding under section 6(b)(7) of the OSH Act to determine whether to list the new protocol as an approved protocol in this Appendix A.

B.   The application must include a detailed description of the proposed new fit test protocol. This application must be supported by either:

1.   A test report prepared by an independent government research laboratory (e.g., Lawrence Livermore National Laboratory, Los Alamos National Laboratory, the National Institute for Standards and Technology) stating that the laboratory has tested the protocol and had found it to be accurate and reliable; or

2.   An article that has been published in a peer-reviewed industrial hygiene journal describing the protocol and explaining how test data support the protocol's accuracy and reliability.

C.   If OSHA determines that additional information is required before the Agency commences a rulemaking proceeding under this section, OSHA will so notify the applicant and afford the applicant the opportunity to submit the supplemental Information. Initiation of a rulemaking proceeding will be deferred until OSHA has received and evaluated the supplemental information.

Appendix B-1 to §1910.134: User Seal Check Procedures (Mandatory)

The individual who uses a tight-fitting respirator is to perform a user seal check to ensure that an adequate seal is achieved each time the respirator is put on. Either the positive and negative pressure checks listed in this appendix, or the respirator manufacturer's recommended user seal check method shall be used. User seal checks are not substitutes for qualitative or quantitative fit tests.

I.   Facepiece Positive and/or Negative Pressure Checks

A.   Positive pressure check. Close off the exhalation valve and exhale gently into the facepiece. The face fit is considered satisfactory if a slight positive pressure can be built up inside the facepiece without any evidence of outward leakage of air at the seal. For most respirators this method of leak testing requires the wearer to first remove the exhalation valve cover before closing off the exhalation valve and then carefully replacing it after the test.

B.   Negative pressure check. Close off the inlet opening of the canister or cartridge(s) by covering with the palm of the hand(s) or by replacing the filter seal(s), inhale gently so that the facepiece collapses slightly, and hold the breath for ten seconds. The design of the inlet opening of some cartridges cannot be effectively covered with the palm of the hand. The test can be performed by covering the inlet opening of the cartridge with a thin latex or nitrile glove. If the facepiece remains in its slightly collapsed condition and no inward leakage of air is detected, the tightness of the respirator is considered satisfactory.

II. Manufacturer's Recommended User Seal Check Procedures

The respirator manufacturer's recommended procedures for performing a user seal check may be used instead of the positive and/or negative pressure check procedures provided that the employer demonstrates that the manufacturer's procedures are equally effective.

Appendix B-2 to §1910.134: Respirator Cleaning Procedures (Mandatory)

These procedures are provided for employer use when cleaning respirators. They are general in nature, and the employer as an alternative may use the cleaning recommendations provided by the manufacturer of the respirators used by their employees, provided such procedures are as effective as those listed here in Appendix B-2. Equivalent effectiveness simply means that the procedures used must accomplish the objectives set forth in Appendix B-2, i.e., must ensure that the respirator is properly cleaned and disinfected in a manner that prevents damage to the respirator and does not cause harm to the user.

I.   Procedures for Cleaning Respirators

A.   Remove filters, cartridges, or canisters. Disassemble facepieces by removing speaking diaphragms, demand and pressure-demand valve assemblies, hoses, or any components recommended by the manufacturer. Discard or repair any defective parts.

B.   Wash components in warm (43° C,110° Fl maximum) water with a mild detergent or with a cleaner recommended by the manufacturer. A stiff bristle (not wire) brush may be used to facilitate the removal of dirt.

C.   Rinse components thoroughly in clean, warm (43°C 1110° F] maximum), preferably running water. Drain.

D.   When the cleaner used does not contain a disinfecting agent, respirator components should be immersed for two minutes in one of the following:

1.   Hypochlorite solution (50 ppm of chlorine) made by adding approximately one milliliter of laundry bleach to one liter of water at 43° C (110° F); or,

2.   Aqueous solution of iodine (50 ppm iodine) made by adding approximately 0.8 milliliters of tincture of iodine (6-8 grams ammonium and/or potassium iodide/100 cc of 45% alcohol) to one liter of water at 43° C (110° F); or,

3.   Other commercially available cleansers of equivalent disinfectant quality when used as directed, if their use is recommended or approved by the respirator manufacturer.

E.   Rinse components thoroughly in clean, warm (43° C [110° F] maximum), preferably running water. Drain. The importance of thorough rinsing cannot be overemphasized. Detergents or disinfectants that dry on facepieces may result in dermatitis. In addition, some disinfectants may cause deterioration of rubber or corrosion of metal parts if not completely removed.

F.   Components should be hand-dried with a clean lint-free cloth or air-dried.

G.   Reassemble facepiece, replacing filters, cartridges, and canisters where necessary.

H.   Test the respirator to ensure that all components work properly.

Appendix C to §1910.134: OSHA Respirator

Medical Evaluation Questionnaire (Mandatory)

To the employer: Answers to questions in Section 1, and to question 9 in Section 2 of Part A, do not require a medical examination.

To the employee:

Can you read (circle one): Yes / No

Your employer must allow you to answer this questionnaire during normal working hours, or at a time and place that is convenient to you. To maintain your confidentiality, your employer or supervisor must not look at or review your answers, and your employer must tell you how to deliver or send this questionnaire to the health care professional who will review it.

Part A. Section 1. (Mandatory) The following information must be provided by every employee who has been selected to use any type of respirator (please print).

1.   Today's date:

2.   Your name:

3.   Your age (to nearest year):

4.   Sex (circle one): Male/Female

5.   Your height: ft.   in.

6.   Your weight: lbs.

7.   Your job title:

8.   A phone number where you can be reached by the health care professional who reviews this questionnaire (include the Area Code):

9.   The best time to phone you at this number:

10. Has your employer told you how to contact the health care professional who will review this questionnaire (circle one): Yes / No

11. Check the type of respirator you will use (you can check more than one category):

    a.   ____ N, R, or P disposable respirator (filter-mask, non-cartridge type only).

    b.   ____ Other type (for example, half- or full-facepiece type, powered-air purifying, supplied-air, self-contained breathing apparatus).

12. Have you worn a respirator (circle one): Yes / No

    If "yes," what type(s):

Part A. Section 2. (Mandatory) Questions 1 through 9 below must be answered by every employee who has been selected to use any type of respirator (please circle "yes" or "no").

1. Do you currently smoke tobacco, or have you smoked tobacco in the last month:
Yes / No

2. Have you ever had any of the following conditions?
   a. Seizures (fits): Yes / No
   b. Diabetes (sugar disease): Yes / No
   c. Allergic reactions that interfere with your breathing: Yes / No
   d. Claustrophobia (fear of closed-in places): Yes / No
   e. Trouble smelling odors: Yes / No

3. Have you ever had any of the following pulmonary or lung problems?
   a. Asbestosis: Yes / No
   b. Asthma: Yes / No
   c. Chronic bronchitis: Yes / No
   d. Emphysema: Yes / No
   e. Pneumonia: Yes / No
   f. Tuberculosis: Yes / No
   g. Silicosis: Yes / No
   h. Pneumothorax (collapsed lung): Yes / No
   i. Lung cancer: Yes / No
   j. Broken ribs: Yes / No
   k. Any chest injuries or surgeries: Yes / No
   l. Any other lung problem that you've been told about: Yes / No

4. Do you currently have any of the following symptoms of pulmonary or lung illness?
   a. Shortness of breath: Yes / No
   b. Shortness of breath when walking fast on level ground or walking up a slight hill or incline: Yes / No
   c. Shortness of breath when walking with other people at an ordinary pace on level ground: Yes / No
   d. Have to stop for breath when walking at your own pace on level ground: Yes / No
   e. Shortness of breath when washing or dressing yourself: Yes / No
   f. Shortness of breath that interferes with your job: Yes / No
   g. Coughing that produces phlegm (thick sputum): Yes / No
   h. Coughing that wakes you early in the morning: Yes / No
   i. Coughing that occurs mostly when you are lying down: Yes / No
   j. Coughing up blood in the last month: Yes / No
   k. Wheezing: Yes / No
   l. Wheezing that interferes with your job: Yes / No
   m. Chest pain when you breathe deeply: Yes / No
   n. Any other symptoms that you think may be related to lung problems: Yes / No

5. Have you ever had any of the following cardiovascular or heart problems?
    a. Heart attack: Yes / No
    b. Stroke: Yes / No
    c. Angina: Yes / No
    d. Heart failure: Yes / No
    e. Swelling in your legs or feet (not caused by walking): Yes / No
    f. Heart arrhythmia (heart beating irregularly): Yes / No
    g. High blood pressure: Yes / No
    h. Any other heart problem that you've been told about: Yes / No

6. Have you ever had any of the following cardiovascular or heart symptoms?
    a. Frequent pain or tightness in your chest: Yes / No
    b. Pain or tightness in your chest during physical activity: Yes / No
    c. Pain or tightness in your chest that interferes with your job: Yes / No
    d. In the past two years, have you noticed your heart skipping or missing a beat: Yes / No
    e. Heartburn or indigestion that is not related to eating: Yes / No
    f. Any other symptoms that you think may be related to heart or circulation problems: Yes / No

7. Do you currently take medication for any of the following problems?
    a. Breathing or lung problems: Yes / No
    b. Heart trouble: Yes / No
    c. Blood pressure: Yes / No
    d. Seizures (fits): Yes / No

8. If you've used a respirator, have you ever had any of the following problems? (If you've never used a respirator, check the following space and go to question 9:)
    a. Eye irritation: Yes / No
    b. Skin allergies or rashes: Yes / No
    c. Anxiety: Yes / No
    d. General weakness or fatigue: Yes / No
    e. Any other problem that interferes with your use of a respirator: Yes / No

9. Would you like to talk to the health care professional who will review this questionnaire about your answers to this questionnaire: Yes / No

Questions 10 to 15 below must be answered by every employee who has been asked or chosen to use either a full-facepiece respirator or a self contained breathing apparatus (SCBA). For employees who have been selected to use other types of respirators, answering these questions is voluntary.

10. Have you ever lost vision in either eye (temporarily or permanently): Yes / No

11. Do you currently have any of the following vision problems?
    a. Wear contact lenses: Yes / No
    b. Wear glasses: Yes / No
    c. Color blind: Yes / No
    d. Any other eye or vision problem: Yes / No

12. Have you ever had an injury to your ears including a broken ear drum: Yes / No

13. Do you currently have any of the following hearing problems?

    a. Difficulty hearing: Yes / No

    b. Wear a hearing aid: Yes / No

    c. Any other hearing or ear problem: Yes / No

14. Have you ever had a back injury: Yes / No

15. Do you currently have any of the following musculoskeletal problems?

    a. Weakness in any of your arms, hands, legs, or feet: Yes / No

    b. Back pain: Yes / No

    c. Difficulty fully moving your arms and legs: Yes / No

    d. Pain or stiffness when you lean forward or backward at the waist: Yes / No

    e. Difficulty fully moving your head up or down: Yes / No

    f. Difficulty fully moving your head side to side: Yes / No

    g. Difficulty bending at your knees: Yes / No

    h. Difficulty squatting to the ground: Yes / No

    i. Climbing a flight of stairs or a ladder carrying more than 25 lbs: Yes / No

    j. Any other muscle or skeletal problem that interferes with using a respirator: Yes / No

Part B. Any of the following questions, and other questions not listed, may be added to the questionnaire at the discretion of the health care professional who will review the questionnaire.

1. In your present job, are you working at high altitudes (over 5,000 feet) or in a place that has lower than normal amounts of oxygen: Yes / No

   If "yes," do you have feelings of dizziness, shortness of breath, pounding in your chest, or other symptoms when you're working under these conditions: Yes / No

2. At work or at home, have you ever been exposed to hazardous solvents, hazardous airborne chemicals (e.g., gases, fumes, or dust), or have you come into skin contact with hazardous chemicals: Yes / No

   If "yes," name the chemicals if you know them:

3. Have you ever worked with any of the materials, or under any of the conditions listed below:

   a. Asbestos: Yes / No

   b. Silica (e.g., in sandblasting): Yes / No

   c. Tungsten/cobalt (e.g., grinding or welding this material): Yes / No

   d. Beryllium: Yes / No

   e. Aluminum:. Yes / No

   f. Coal (for example, mining): Yes / No

   g. Iron: Yes / No

   h. Tin: Yes / No

   i. Dusty environments: Yes / No

   j. Any other hazardous exposures: Yes / No

      If "yes," describe these exposures:

4. List any second jobs or side businesses you have:

5. List your previous occupations:

6. List your current and previous hobbies:

7. Have you been in the military services? Yes / No. If "yes," were you exposed to biological or chemical agents (either in training or combat):   Yes / No

8. Have you ever worked on a HAZMAT team? Yes / No

9. Other than medications for breathing and lung problems, heart trouble, blood pressure, and seizures mentioned earlier in this questionnaire, are you taking any other medications for any reason (including over-the-counter medications): Yes / No

    If "yes," name the medications if you know them:

10. Will you be using any of the following items with your respirator(s)? a) HEPA Filters: Yes / No. b) Canisters (for example, gas masks):   Yes / No. c) Cartridges: Yes / No

11. How often are you expected to use the respirator(s)? (circle "yes" or "no" for all answers that apply to you):

    a. Escape only (no rescue): Yes / No

    b. Emergency rescue only: Yes / No

    c. Less than 5 hours per week: Yes / No

    d. Less than 2 hours per day: Yes / No

    e. 2 to 4 hours per day: Yes / No

    f. Over 4 hours per day: Yes / No

12. During the period you are using the respirator(s), is your work effort:

    a. Light (less than 200 kcal per hour): Yes / No

    If "yes," how long does this period last during the average shift: hrs.   mins.

    Examples of a light work effort are sitting while writing, typing, drafting, or performing light assembly work; or standing while operating a drill press (1-3 lbs.) or controlling machines.

    b. Moderate (200 to 350 kcal per hour): Yes / No

    If "yes," how long does this period last during the average shift: hrs.   mins.

    Examples of moderate work effort are sitting while nailing or filing; driving a truck or bus in urban traffic; standing while drilling, nailing, performing assembly work, or transferring a moderate load (about 35 lbs.) at trunk level; walking on a level surface about 2 mph or down a 5-degree grade about 3 mph: or pushing a wheelbarrow with a heavy load (about 100 lbs.) on a level surface.

    c. Heavy (above 350 kcal per hour): Yes / No

    If "yes," how long does this period last during the average shift: hrs.   mins.

    Examples of heavy work are lifting a heavy load (about 50 lbs.) from the floor to your waist or shoulder; working on a loading dock; shoveling; standing while bricklaying or chipping castings; walking up an 8-degree grade about 2 mph; climbing stairs with a heavy load (about 50 lbs.).

13. Will you be wearing protective clothing and/or equipment (other than the respirator) when you're using your respirator: Yes / No

    If "yes," describe this protective clothing and/or equipment:

14. Will you be working under hot conditions (temperature exceeding 77 degrees F): Yes / No

15. Will you be working under humid conditions: Yes / No

16. Describe the work you'll be doing while you're using your respirator(s):

17. Describe any special or hazardous conditions you might encounter when you're using your respirator(s) (for example, confined spaces, life threatening gases):

18. Provide the following information, if you know it, for each toxic substance that you'll be exposed to when you're using your respirator(s):

Name of the first toxic substance:

Estimated maximum exposure level per shift:

Duration of exposure per shift:

Name of the second toxic substance:

Estimated maximum exposure level per shift:

Duration of exposure per shift:

Name of the third toxic substance:

Estimated maximum exposure level per shift:

Duration of exposure per shift:

The name of any other toxic substances that you'll be exposed to while using your respirator.

19. Describe any special responsibilities you'll have while using your respirator(s) that may affect the safety and well-being of others (for example, rescue, security):

Appendix D to §1910.134)

INFORMATION for Employees Using Respirators When Not Required Under the Standard (Non-Mandatory)

Respirators are an effective method of protection against designated hazards when properly selected and worn. Respirator use is encouraged, even when exposures are below the exposure limit, to provide an additional level of comfort and protection for workers. However, if a respirator is used improperly or not kept clean, the respirator itself can become a hazard to the worker. Sometimes, workers may wear respirators to avoid exposures to hazards, even if the amount of hazardous substance does not exceed the limits set by OSHA standards. If your employer provides respirators for your voluntary use, of if you provide your own respirator, you need to take certain precautions to be sure that the respirator itself does not present a hazard.

You should do the following:

1. Read and heed all instructions provided by the manufacturer on use, maintenance, cleaning and care, and warnings regarding the respirators limitations.

2. Choose respirators certified for use to protect against the contaminant of concern. NIOSH, the National Institute for Occupational Safety and Health of the U.S. Department of Health and Human Services, certifies respirators. A label or statement of certification should appear on the respirator or respirator packaging. It will tell you what the respirator is designed for and how much it will protect you.

3. Do not wear your respirator into atmospheres containing contaminants for which your respirator is not designed to protect against. For example, a respirator designed to filter dust particles will not protect you against gases, vapors, or very small solid particles of fumes or smoke.

4.   Keep track of your respirator so that you do not mistakenly use someone else's respirator.

SUBPART L—[Amended]

8.   The authority citation for Subpart L of Part 1910 is revised to read as follows:

Authority: Secs. 4,6, and 8 of the Occupational Safety and Health Act of 1970 (29 U.S.C. 653, 655,657); Secretary of Labor's Orders 12-71 (36 FR 8754), 8-76 (41 FR 25059), 9-83 (48 FR 35736), 1-90 (55 FR 9033), or 6-96 (62 FR 111), as applicable.

9.   Section 1910.156 is amended by revising paragraphs (fl(1)(i) and (fl(1)(v) as follows:

§1910.156 Fire brigades.

(f)   Respiratory protection. (1) General.

(i)   The employer must ensure that respirators are provided to, and used by, fire brigade members, and that the respirators meet the requirements of 29 CFR 1910.134 and this paragraph.

(v)   Self-contained breathing apparatuses must have a minimum service-life rating of 30 minutes in accordance with the methods and requirements specified by NIOSH under 42 CFR part 84, except for escape self-contained breathing apparatus (ESCBAs) used only for emergency escape purposes.

SUBPART Q [Amended]

10. The authority citation for Subpart Q of Part 1910 is revised to read as follows:

Authority: Secs. 4, 6, and 8 of the Occupational Safety and Health Act of 1970 (29 U.S.C. 653, 655, 657): Secretary of Labor's Orders 12-71 (36 FR 8754), 8-76 (41 FR 25059), 9-83 (48 FR 35736),1-90 (55 FR 9033), or 6-96 (62 FR 111), as applicable; and 29 CFR part 19.

11. Section 1910.252 is amended by revising paragraphs (c)(4)(ii), (c)(4)(iii), (c)(7)(iii), (c)(9)(i), and (c)(10) as follows:

§1910.252 General requirements.

(c) (4)

(ii) Airline respirators. In circumstances for which it is impossible to provide such ventilation, airline respirators or hose masks approved for this purpose by the National Institute for Occupational Safety and Health (NIOSH) under 42 CFR part 84 must be used.

(iii) Self-contained units. In areas immediately hazardous to life, a full facepiece, pressure-demand, self-contained breathing apparatus or a combination full-facepiece, pressure demand supplied-air respirator with an auxiliary, self-contained air supply approved by NIOSH under 42 CFR part 84 must be used.

(7) (iii) Local ventilation. In confined spaces or indoors, welding or cutting operations involving metals containing lead, other than as an impurity, or metals coated with lead-bearing materials, including paint, must be done using local exhaust ventilation or airline respirators. Such operations, when done outdoors, must be done using respirators approved for this purpose by NIOSH under 42 CFR part 84. In all cases, workers in the immediate vicinity of the cutting operation must be protected by local exhaust ventilation or airline respirators.

(9)   (i)   General. In confined spaces or indoors, welding or cutting operations involving cadmium-bearing or cadmium-coated base metals must be done using local exhaust ventilation or airline respirators unless atmospheric tests under the most adverse conditions show that employee exposure is within the acceptable concentrations specified by 29 CFR 1910,1000, Such operations, when done outdoors, must be done using respirators, such as fume respirators, approved for this purpose by NIOSH under 42 CFR part 84.

(10)   Mercury. In confined spaces or indoors, welding or cutting operations involving metals coated with mercury bearing materials, including paint, must be done using local exhaust ventilation or airline respirators unless atmospheric tests under the most adverse conditions show that employee exposure is within the acceptable concentrations specified by 29 CFR 1910.1000. Such operations, when done outdoors, must be done using respirators approved for this purpose by NIOSH under 42 CFR part 84.

SUBPART R—[Amended]

12.   The authority citation for Subpart R of Part 1910 is revised as follows:

Authority: Sections 4, 6, and 8 of the Occupational Safety and Health Act of 1970 (29 U.S.C. 653, 6s5, 6s7); Secretary of Labor's Orders 12-71 (36 FR 87s4), 8-76 (41 FR 25059), 9-83 (48 FR 35736), 1-90 (55 FR 9033), or 6-96 (62 FR 111), as applicable; and 29 CFR part 11. Sections 1910.261, 1910.262, 1910.265 through 1910.269, 1910.274, and 1910.275 also issued under 29 CFR part 1911.

13.   Section 1910.261 is amended by revising paragraphs (b)(2), (g)(10), (h)(2)(iii), and (h)(2)(iv) as follows:

§1910.261 Pulp, paper, and paperboard mills.

(b)   (2) Personal protective clothing and equipment. Foot protection, shin guards, hard hats, noise-attenuation devices, and other personal protective clothing and equipment must be worn when the extent of the hazard warrants their use. Such equipment must be worn when specifically required by other paragraphs of this section, and must be maintained in accordance with applicable American National Standards Institute standards. Respirators, goggles, protective masks, rubber gloves, rubber boots, and other such equipment must be cleaned and disinfected before being used by another employee. Required eye, head, and ear protection must conform to American National Standards Institute standards Z24.221957, Z87.1-1968, and Z89.1-1969. Respiratory protection must conform to the requirements of 29 CFR 1910.134.

(g)   (10) Gas masks (digester building). Gas masks must be available, and they must furnish adequate protection against sulfurous acid and chlorine gases and be inspected and repaired in accordance with 29 CFR 1910.134.

(h) (2)(iii)   Gas masks must be provided for emergency use in accordance with 29 CFR 1910.134.

(iv)   For emergency and rescue operations, the employer must provide employees with self-contained breathing apparatuses or supplied-air respirators, and ensure that employees use these respirators, in accordance with the requirements of 29 CFR 1910.134.

SUBPART Z —[Amended]

14.   The general authority citation for Subpart Z of 29 CFR Part 1910 is revised to read as follows:

Authority: Secs. 4, 6, and 8 of the Occupational Safety and Health Act (29 U.S.C. 653, 65s, and 657); Secretary of Labor's Orders 12-71 (36 FR 8754), 8-76 (41 FR 2s059), 9-83 (48 FR 35736),1-90 (56 FR 9033), or 6-96 (62 FR 111), as applicable; and 29 CFR Part 1911.

TABLE 1 – RESPIRATORY PROTECTION FOR ASBESTOS FIBERS

15.   Section 1910.1001 is amended by removing Appendix C and revising paragraph (g), to read as follows:

§ 1910.1001 Asbestos.

(g)   Respiratory protection.

(1)   General. For employees who use respirators required by this section, the employer must provide respirators that comply   with the requirements of this paragraph. Respirators must be used during:

(i)   Periods necessary to install or implement feasible engineering and work-practice controls.

(ii)   Work operations, such as maintenance and repair activities, for which engineering and work-practice controls are not feasible.

(iii)   Work operations for which feasible engineering and work-practice controls are not yet sufficient to reduce employee exposure to or below the TWA and/or excursion limit.

(iv)   Emergencies.

(2)   Respirator program.

(i)   The employer must implement a respiratory protection program in accordance with 29 CFR 1910.134 (b) through (d) (except (d)(1)(iii)), and (fl through (m).

(ii)   The employer must provide a tight-fitting, powered, air-purifying respirator instead of any negative pressure respirator specified in Table 1 of this section when an employee chooses to use this type of respirator and the respirator provides adequate protection to the employee.

(iii)   No employee must be assigned to tasks requiring the use of respirators if, based on their most recent medical examination, the examining physician determines that the employee will be unable to function normally using a respirator, or that the safety or health of the employee or other employees will be impaired by the use of a respirator. Such employees must be assigned to another job or given the opportunity to transfer to a different position, the duties of which they can perform. If such a transfer position is available, the position must be with the same employer, in the same geographical area, and with the same seniority, status, and rate of pay the employee had just prior to such transfer.

(3)   Respirator selection. The employer must select and provide the appropriate respirator from Table 1 of this section

TABLE 1	
RESPIRATORY PROTECTION FOR ASBESTOS FIBERS	
Airborne Concentration of Asbestos or Condition of Use	Required Respirator
Not in excess of 1 f/cc (10 X PEL)	Half-mask air purifying respirator other than a disposable respirator, equipped with high efficiency filters.
Not in excess of 5 f/cc (50 X PEL)	Full facepiece air-purifying respirator equipped with high efficiency filters.
Not in excess of 10 f/cc (100 X PEL)	Any powered air-purifying respirator equipped with high efficiency filters or any supplied air respirator operated in continuous flow mode.
Not in excess of 100 f/cc (1,000 X PEL)	Full facepiece supplied air respirator operated in pressure demand mode.
Greater than 100 f/cc (1,000 X PEL) or unknown concentration.	Full facepiece supplied air respirator operated in pressure demand mode, equipped with an auxiliary positive pressure self-contained breathing apparatus.

NOTE:

a. Respirators assigned for high environmental concentrations may be used at lower concentrations, or when required respirator use is independent of concentration.

b. A high efficiency filter means a filter that is at least 99.97 percent efficient against mono-dispersed particles of 0.3 micrometers in diameter or larger.

16.　　§ 1910.1025 Lead.

　　(f)　　Respiratory protection.

　　　　(1)　General. For employees who use respirators required by this section, the employer must provide respirators that comply with the requirements of this paragraph. Respirators must be used during:

　　　　　　(i)　Periods necessary to install or implement engineering or work-practice controls. Except that no employer can require an employee to use a respirator longer than 4.4 hours per day.

　　　　　　(ii)　Work operations for which engineering and work-practice controls are not sufficient to reduce employee exposures to or below the permissible exposure limit.

　　　　　　(iii)　Periods when an employee requests a respirator.

　　　　(2)　Respirator program.

　　　　　　(i)　The employer must implement a respiratory protection program in accordance with 29 CFR 1910.134 (b) through (d) (except (d)(l)(iii)), and (f) through (m).

　　　　　　(ii)　If an employee has breathing difficulty during fit testing or respirator use, the employer must provide the employee with a medical examination in accordance with paragraph (j)(3~(i)(C) of this section to determine whether or not the employee can use a respirator while performing the required duty.

TABLE II	
RESPIRATORY PROTECTION FOR LEAD AEROSOLS	
Airborne Concentration of Lead or Condition of Use	Required Respirator
Not in excess of 0.5 mg/m3 (10X PEL)	Half-mask air purifying respirator other than a disposable respirator, equipped with high efficiency filters.
Not in excess of 2.5 mg/m3 (50X PEL)	Full facepiece air-purifying respirator equipped with high efficiency filters.
Not in excess of 50 mg/m3 (1000X PEL)	1. Any powered air-purifying respirator equipped with high efficiency filters, or 2. Half-mask supplied air respirator operated in positive-pressure mode.
Not in excess of 100 mg/m3 (2000 X PEL)	Supplied-air respirators with full facepiece, hood, helmet, or suit, operated in positive pressure mode.
Greater than 100 mg/m3, unknown concentration or fire fighting	Full facepiece, self-contained breathing apparatus operated in positive pressure mode.

Note:
Respirators specified for high concentrations can be used at lower concentrations of lead.
Full facepiece is required if the lead aerosols cause eye or skin irritation at the use concentrations.
A high efficiency particulate filter means 99.97 percent efficient against 0.3 micron size particles.

    (3)   Respirator selection.

        (i)   The employer must select the appropriate respirator or combination of respirators from Table II of this section.

        (ii)   The employer must provide a powered air-purifying respirator instead of the respirator specified in Table II of this section when an employee chooses to use this type of respirator and such a respirator provides adequate protection to the employee.

Respirator Protection Factors

    Respirators offer varying degrees of protection against lead, asbestos, and other toxic chemicals. Understanding the differences between types of respirators (air-purifying, powered air-purifying, air-supplied, etc.) is necessary to determine the amount of protection given to the wearer. To compare these, one must understand the concept of a protection factor (PF).

    A protection factor is a number ratio obtained when the concentration of a contaminant outside the mask is divided by the concentration found inside the mask. In the case of lead, it is a comparison of the milligrams per cubic meter outside of the mask versus the milligrams of lead inside the mask.

$$PROTECTION\ FACTOR\ (PF) = \frac{CONCENTRATION\ OUTSIDE\ MASK}{CONCENTRATION\ INSIDE\ MASK}$$

The protection factor depends greatly on the fit of the mask to the wearer's face. The protection constantly changes depending upon the worker's activities and even shaving habits. A worker who forgot to shave one morning will not receive as much protection that day since the mask will not fit as well to the face. The mask must fit properly!

Since it is impossible to measure the concentration inside the mask (where the worker is breathing) for each worker, all the time, during all the various activities he or she may be conducting, protection factors, based on extensive research, have been developed for different categories of respirators. You can use the protection factors to determine what type of respirator is needed to maintain the concentration of lead inside the mask below a certain level.

Using the OSHA General Industry standard, the contractor may select the appropriate respirator to maintain the concentration inside the respirator below personal exposure limits. It should be noted that the protection factors for powered-air purifying respirators are estimated on the most recent data available.

Requirements For An Acceptable Respiratory Program

Following are program requirements listed in OSHA 1910.134 which is referenced in the construction standards.

(1) Written standard operating procedures governing the selection and use of respirators shall be established.

(2) Respirators shall be selected on the basis of hazards to which the worker is exposed.

(3) The user shall be instructed and trained in the proper use of respirators and their limitations.

(4) [Reserved]

(5) Respirators shall be regularly cleaned and disinfected. Those used by more than one worker shall be thoroughly cleaned and disinfected after each use.

(6) Respirators shall be stored in a convenient, clean, and sanitary location.

(7) Respirators used routinely shall be inspected during cleaning. Worn or deteriorated parts shall be replaced. Respirators for emergency use such as self-contained devices shall be thoroughly inspected at least once a month and after each use.

(8) Appropriate surveillance of work area conditions and degree of employee exposure or stress shall be maintained.

(9) There shall be regular inspection and evaluation to determine the continued effectiveness of the program.

(10) Persons should not be assigned to tasks requiring use of respirators unless it has been determined that they are physically able to perform the work and use the equipment. The local physician shall determine what health and physical conditions are pertinent. The respirator user's medical status should be reviewed periodically (for instance, annually).

(11)   Approved or accepted respirators shall be used when they are available. The respirator furnished shall provide adequate respiratory protection against the particular hazard for which it is designed in accordance with standards established by competent authorities. The U.S. Department of Interior, Bureau of Mines and the U.S. Department of Agriculture are recognized as such authorities. Although respirators listed by the U.S. Department of Agriculture continue to be acceptable for protection against specified pesticides, the U.S. Department of the Interior, Bureau of Mines, is the agency now responsible for testing and approving pesticide respirators.

Appendix O

SCAFFOLDS AND AERIAL LIFT ILLUSTRATIONS

Permission by Scaffolding Industry Association, Inc.

SCAFFOLDING WORK SURFACES

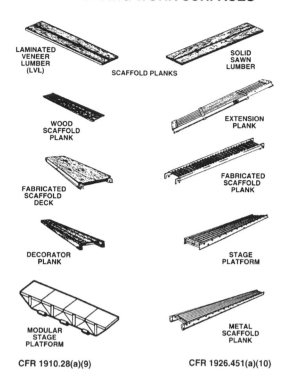

LAMINATED VENEER LUMBER (LVL)

SOLID SAWN LUMBER

SCAFFOLD PLANKS

WOOD SCAFFOLD PLANK

EXTENSION PLANK

FABRICATED SCAFFOLD DECK

FABRICATED SCAFFOLD PLANK

DECORATOR PLANK

STAGE PLATFORM

MODULAR STAGE PLATFORM

METAL SCAFFOLD PLANK

CFR 1910.28(a)(9)   CFR 1926.451(a)(10)

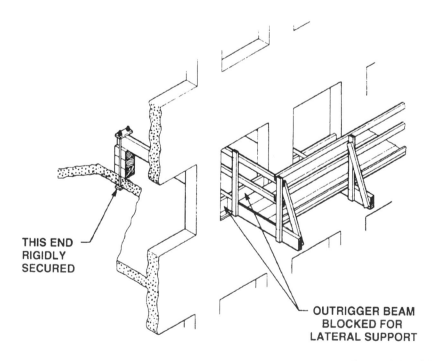

THIS END RIGIDLY SECURED

OUTRIGGER BEAM BLOCKED FOR LATERAL SUPPORT

CFR 1910.28(e)   CFR 1926.451(g)

FABRICATED TUBULAR FRAME
MANUALLY PROPELLED
MOBILE SCAFFOLD

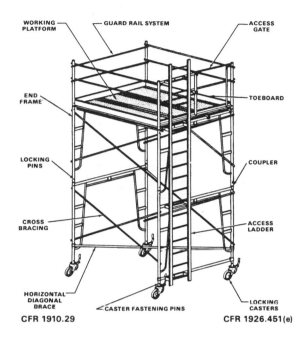

WORKING PLATFORM

GUARD RAIL SYSTEM

ACCESS GATE

END FRAME

TOEBOARD

LOCKING PINS

COUPLER

CROSS BRACING

ACCESS LADDER

HORIZONTAL DIAGONAL BRACE

CASTER FASTENING PINS

LOCKING CASTERS

CFR 1910.29

CFR 1926.451(e)

PREFABRICATED MOBILE TOWER UNIT

THIS SCAFFOLD IS NORMALLY MANUFACTURED AS
COMPLETE UNITS/TOWERS FOR USE AS MANUALLY
PROPELLED MOBILE SCAFFOLD WITH SUPPLIERS
IDENTIFICATION SYMBOL

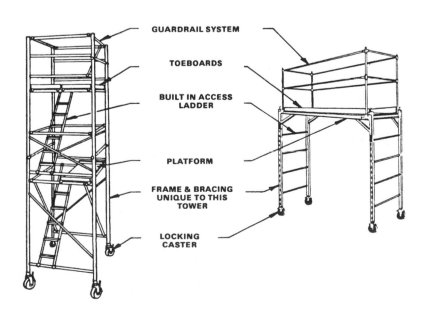

GUARDRAIL SYSTEM

TOEBOARDS

BUILT IN ACCESS LADDER

PLATFORM

FRAME & BRACING UNIQUE TO THIS TOWER

LOCKING CASTER

WOOD POLE SCAFFOLD

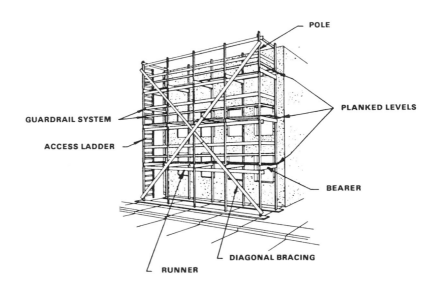

SEE TABLES IN OSHA STANDARDS FOR SIZE & SPACING OF MEMBERS

CFR 1910.28(b)   1926.451(b)

FRAME SCAFFOLD ACCESS

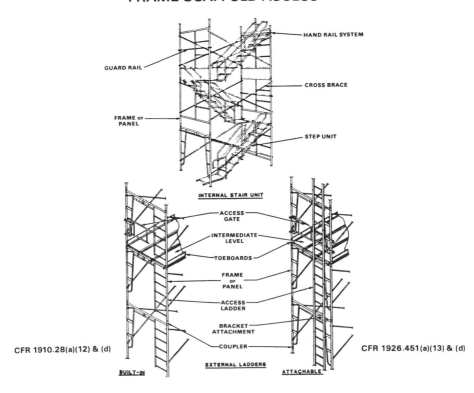

CFR 1910.28(a)(12) & (d)   CFR 1926.451(a)(13) & (d)

SYSTEM SCAFFOLD

VARIOUS INDUSTRY
JOINT CONNECTIONS

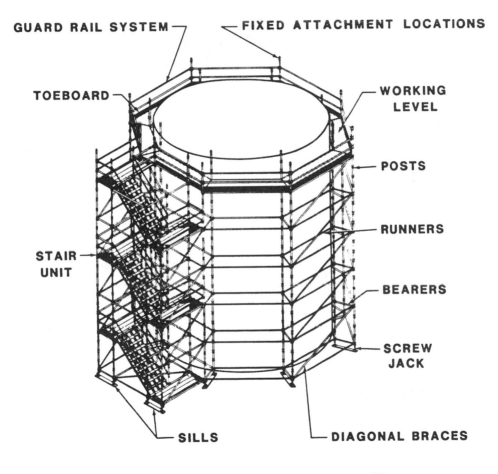

GUARD RAIL SYSTEM

FIXED ATTACHMENT LOCATIONS

TOEBOARD

WORKING LEVEL

POSTS

RUNNERS

STAIR UNIT

BEARERS

SCREW JACK

SILLS

DIAGONAL BRACES

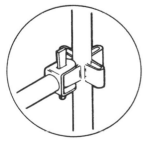

VARIOUS INDUSTRY
JOINT CONNECTIONS

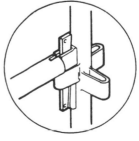

TUBE AND COUPLER SCAFFOLD

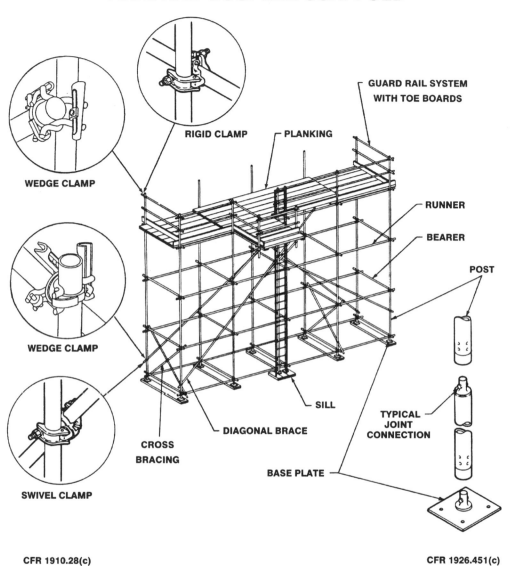

RIGID CLAMP

PLANKING

GUARD RAIL SYSTEM
WITH TOE BOARDS

WEDGE CLAMP

RUNNER

BEARER

POST

WEDGE CLAMP

SILL

TYPICAL
JOINT
CONNECTION

DIAGONAL BRACE

CROSS
BRACING

SWIVEL CLAMP

BASE PLATE

CFR 1910.28(c)

CFR 1926.451(c)

TWO-POINT SUSPENDED SCAFFOLD GROUND RIGGED SWAY CONTROL ILLUSTRATION

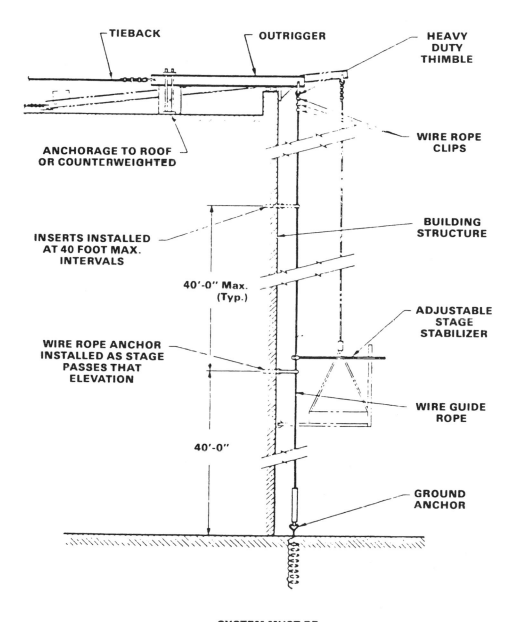

**SYSTEM MUST BE
DESIGNED BY
QUALIFIED PERSONNEL**

TWO POINT SUSPENDED SCAFFOLD

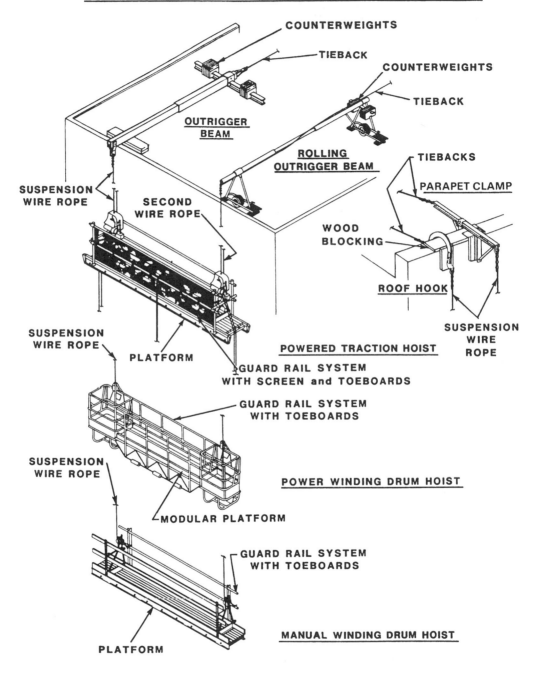

CFR 1910.28(g)   CFR 1926.451(i)

MULTI-LEVEL SUSPENDED SCAFFOLD
WITH POWERED HOISTS

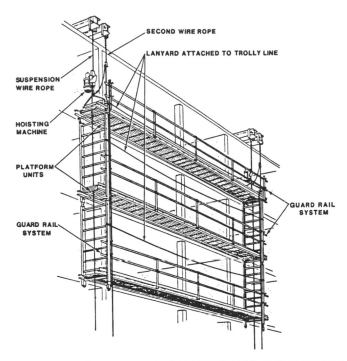

SECOND WIRE ROPE

LANYARD ATTACHED TO TROLLY LINE

SUSPENSION
WIRE ROPE

HOISTING
MACHINE

PLATFORM
UNITS

GUARD RAIL
SYSTEM

GUARD RAIL
SYSTEM

MULTI-POINT SUSPENDED SCAFFOLD

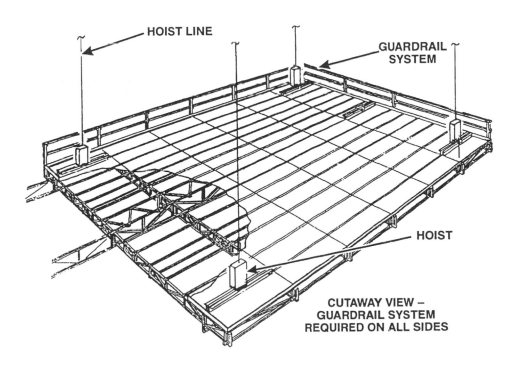

HOIST LINE

GUARDRAIL
SYSTEM

HOIST

CUTAWAY VIEW –
GUARDRAIL SYSTEM
REQUIRED ON ALL SIDES

SUSPENDED PLATFORM
WELDING PRECAUTIONS

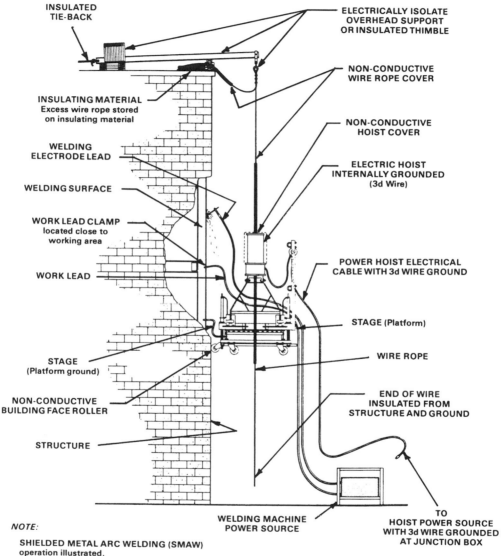

INSULATED
TIE-BACK

ELECTRICALLY ISOLATE
OVERHEAD SUPPORT
OR INSULATED THIMBLE

INSULATING MATERIAL
Excess wire rope stored
on insulating material

NON-CONDUCTIVE
WIRE ROPE COVER

WELDING
ELECTRODE LEAD

NON-CONDUCTIVE
HOIST COVER

WELDING SURFACE

ELECTRIC HOIST
INTERNALLY GROUNDED
(3d Wire)

WORK LEAD CLAMP
located close to
working area

WORK LEAD

POWER HOIST ELECTRICAL
CABLE WITH 3d WIRE GROUND

STAGE (Platform)

STAGE
(Platform ground)

WIRE ROPE

NON-CONDUCTIVE
BUILDING FACE ROLLER

END OF WIRE
INSULATED FROM
STRUCTURE AND GROUND

STRUCTURE

NOTE:

WELDING MACHINE
POWER SOURCE

TO
HOIST POWER SOURCE
WITH 3d WIRE GROUNDED
AT JUNCTION BOX

SHIELDED METAL ARC WELDING (SMAW)
operation illustrated.

Same procedure to be used for:
— GAS METAL ARC WELDING (GMAW)
— FLUX CORED ARC WELDING (FCAW)

STONE SETTERS' ADJUSTABLE
MULTI-POINT SUSPENDED SCAFFOLD
WITH MANUAL WINDING
DRUM HOISTS

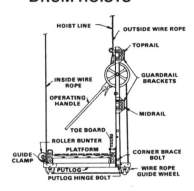

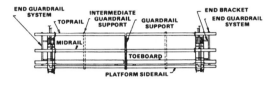

CFR 1910.28(h)

CFR 1926.451(j)

WORK CAGES

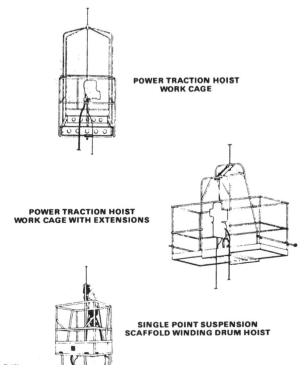

POWER TRACTION HOIST
WORK CAGE

POWER TRACTION HOIST
WORK CAGE WITH EXTENSIONS

SINGLE POINT SUSPENSION
SCAFFOLD WINDING DRUM HOIST

CFR 1910.28(i)

CFR 1926.451(k)

MASONS' ADJUSTABLE MULTI-POINT SUSPENSION SCAFFOLD WITH WINDING DRUM HOISTS

ALTERNATE BOLT & SPECIAL ANCHOR
IMBEDDED IN CONCRETE AT TIME OF POUR

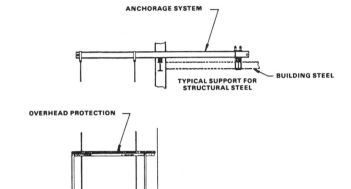

ANCHORAGE SYSTEM

BUILDING STEEL

TYPICAL SUPPORT FOR
STRUCTURAL STEEL

OVERHEAD PROTECTION

GUARDRAIL SYSTEM
WITH SCREEN

CFR 1910.28(f)   CFR 1926.451(h)

BOATSWAIN'S CHAIR

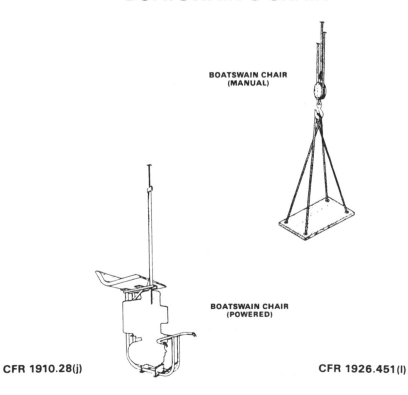

BOATSWAIN CHAIR
(MANUAL)

BOATSWAIN CHAIR
(POWERED)

CFR 1910.28(j)   CFR 1926.451(l)

INTERIOR HUNG SCAFFOLD

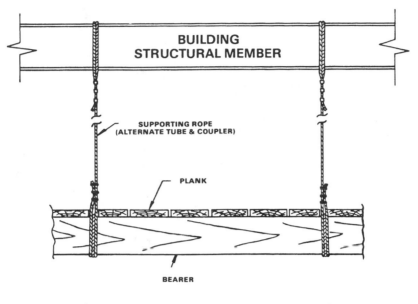

BUILDING STRUCTURAL MEMBER

SUPPORTING ROPE
(ALTERNATE TUBE & COUPLER)

PLANK

BEARER

CFR 1910.28(p)   CFR 1926.451(r)

CATENARY SCAFFOLD

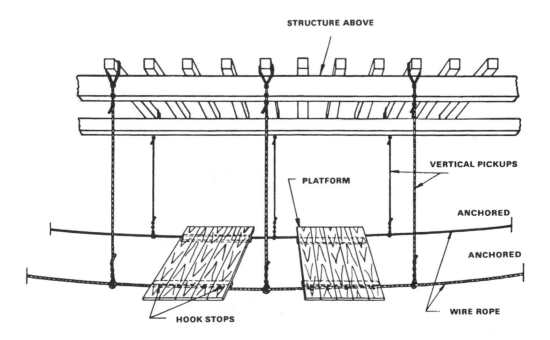

STRUCTURE ABOVE

VERTICAL PICKUPS

PLATFORM

ANCHORED

ANCHORED

HOOK STOPS

WIRE ROPE

INTERIOR HUNG

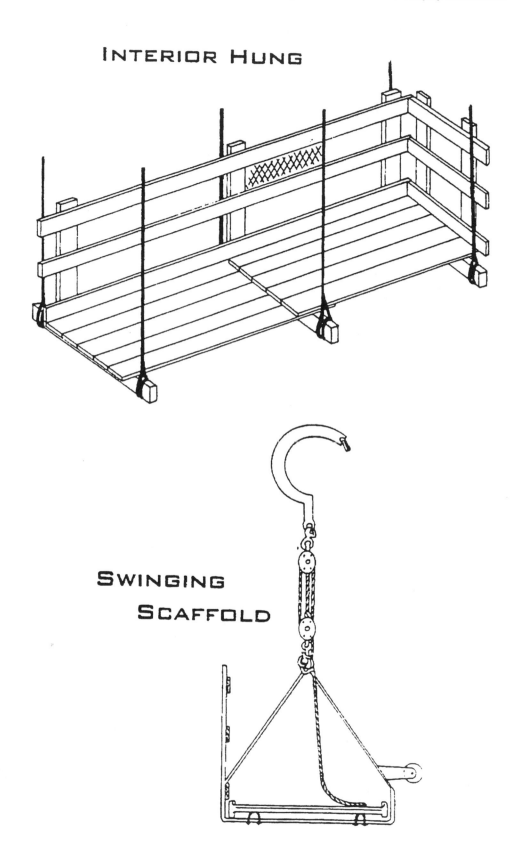

SWINGING

SCAFFOLD

NEEDLE BEAM SCAFFOLD

STRUCTURAL MEMBER ABOVE

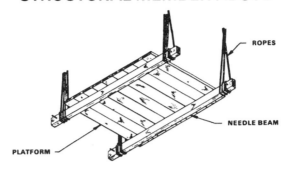

ROPES

NEEDLE BEAM

PLATFORM

CFR 1910.28(n) CFR 1926.451(p)

FLOAT SCAFFOLD

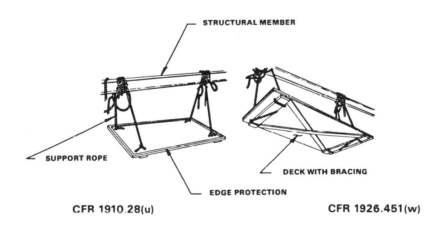

STRUCTURAL MEMBER

SUPPORT ROPE

DECK WITH BRACING

EDGE PROTECTION

CFR 1910.28(u) CFR 1926.451(w)

WINDOW JACK SCAFFOLD

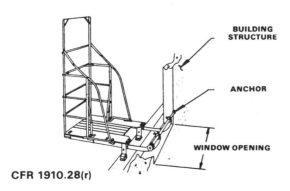

BUILDING STRUCTURE

ANCHOR

WINDOW OPENING

CFR 1910.28(r) CFR 1926.451(t)

CARPENTER'S BRACKET SCAFFOLD

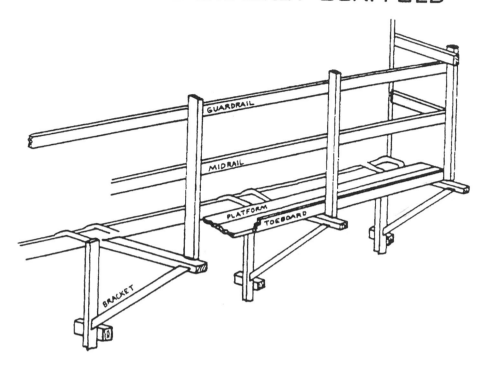

WINDOW JACK SCAFFOLD

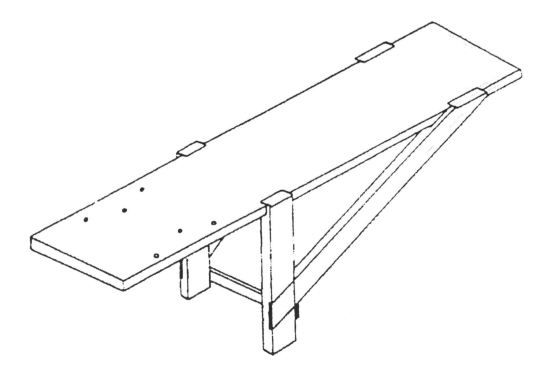

Roofing Brackets

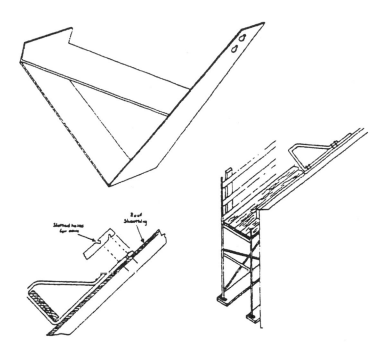

Chicken Scaffold

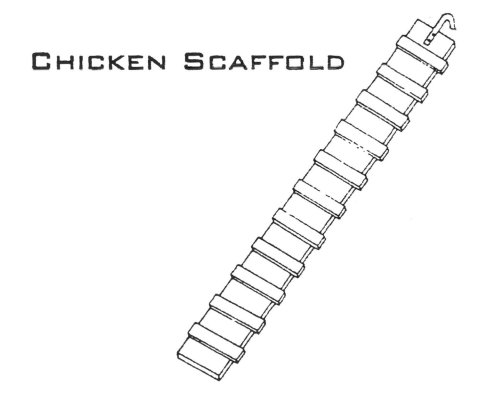

PLASTERERS DECORATORS SCAFFOLD

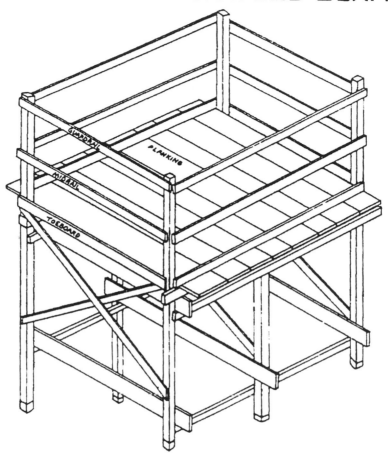

METAL
CARPENTER BRACKET

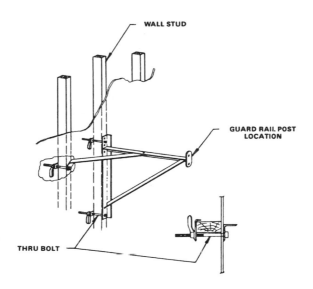

CFR 1910.28(k)   CFR 1926.451(m)

PUMP JACK SCAFFOLD

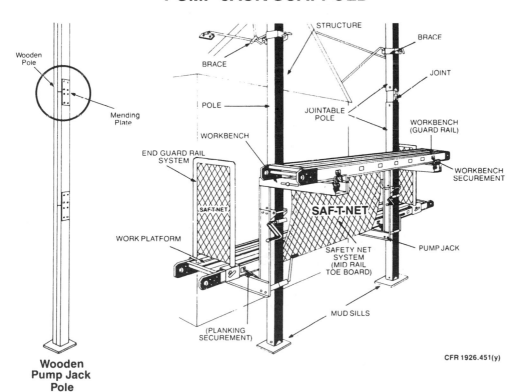

**Wooden
Pump Jack
Pole**

CFR 1926.451(y)

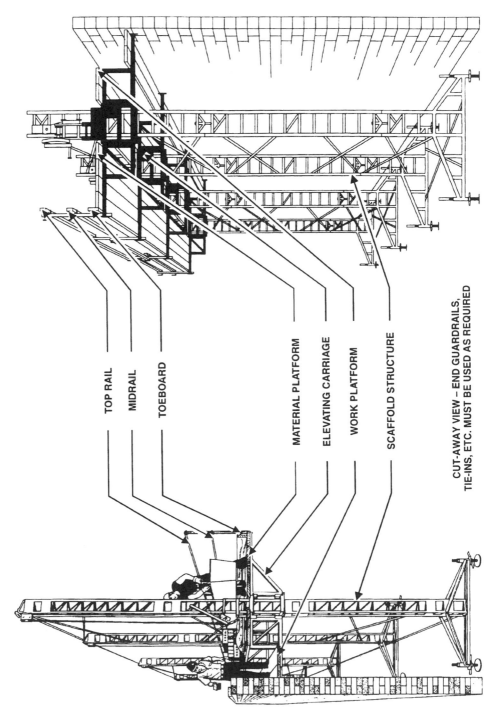

ADJUSTABLE SCAFFOLDS

TOP RAIL

MIDRAIL

TOEBOARD

MATERIAL PLATFORM

ELEVATING CARRIAGE

WORK PLATFORM

SCAFFOLD STRUCTURE

CUT-AWAY VIEW – END GUARDRAILS,
TIE-INS, ETC. MUST BE USED AS REQUIRED

HORSE SCAFFOLD

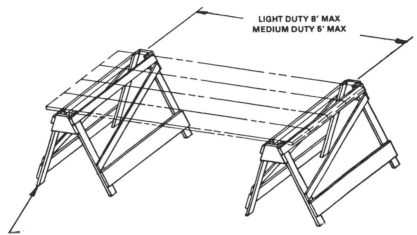

LIGHT DUTY 8' MAX
MEDIUM DUTY 5' MAX

**SEE APPROPRIATE OSHA STANDARDS TABLES FOR
MEMBER SIZES & PLANK SIZES**

CFR 1910.28(m)   CFR 1926.451(o)

BRICKLAYERS' SQUARE SCAFFOLD

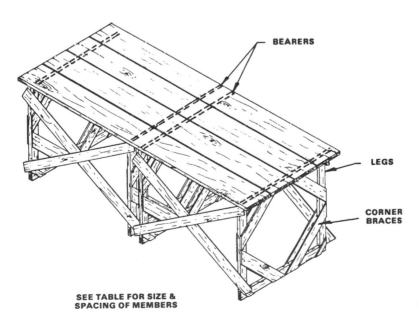

BEARERS

LEGS

**CORNER
BRACES**

**SEE TABLE FOR SIZE &
SPACING OF MEMBERS**

CFR 1926.451(n)

LADDER JACK SCAFFOLD

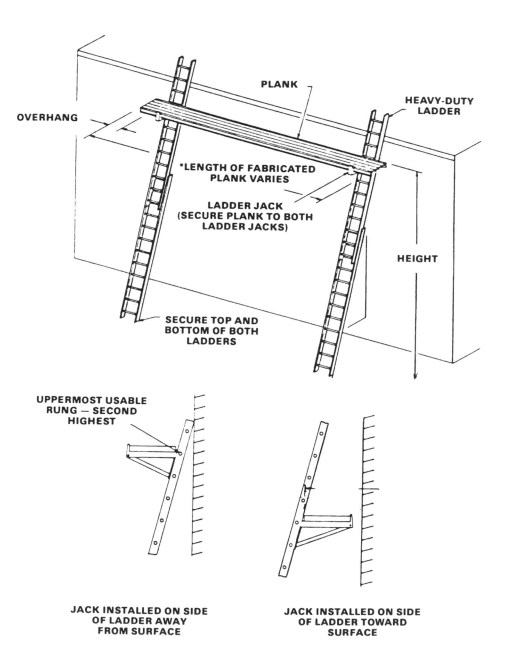

PLANK

HEAVY-DUTY LADDER

OVERHANG

*LENGTH OF FABRICATED PLANK VARIES

LADDER JACK (SECURE PLANK TO BOTH LADDER JACKS)

HEIGHT

SECURE TOP AND BOTTOM OF BOTH LADDERS

UPPERMOST USABLE RUNG — SECOND HIGHEST

JACK INSTALLED ON SIDE OF LADDER AWAY FROM SURFACE

JACK INSTALLED ON SIDE OF LADDER TOWARD SURFACE

*See OSHA requirements regarding width, height, spans and types of ladders.

CFR 1910.28(q)

CFR 1926.451(s)

EXTENSION TRESTLE LADDER SUPPORTED SCAFFOLD

PLATFORM NO HIGHER THAN THIRD RUNG FROM TOP

USE OF EXTENSION TRESTLE LADDERS TO SUPPORT SCAFFOLD PLANK

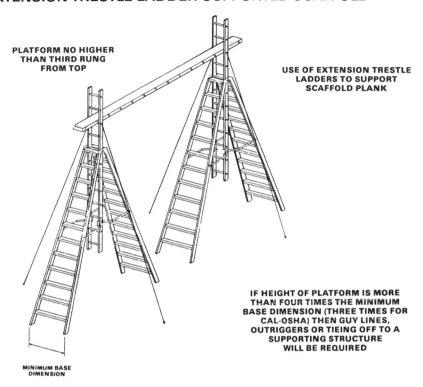

IF HEIGHT OF PLATFORM IS MORE THAN FOUR TIMES THE MINIMUM BASE DIMENSION (THREE TIMES FOR CAL-OSHA) THEN GUY LINES, OUTRIGGERS OR TIEING OFF TO A SUPPORTING STRUCTURE WILL BE REQUIRED

MINIMUM BASE DIMENSION

FREE STANDING LADDER SUPPORTED SCAFFOLD

PLATFORM NO HIGHER THAN SECOND STEP FROM TOP

USE OF STEPLADDERS OR TRESTLE LADDERS TO SUPPORT SCAFFOLD

TYPICAL EXAMPLES OF VEHICLE-MOUNTED
ELEVATING AND ROTATING AERIAL DEVICES
COVERED IN ANSI/SIA A92.2 STANDARD

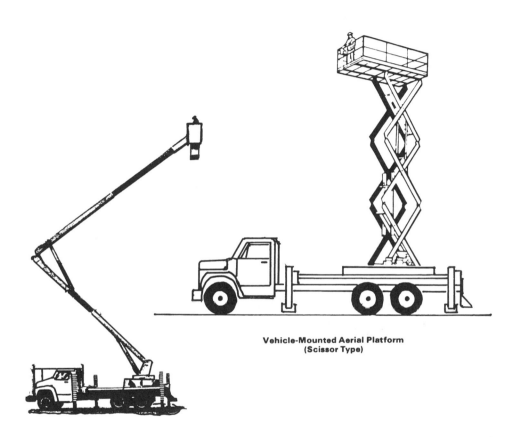

**Vehicle-Mounted Aerial Platform
(Scissor Type)**

**Vehicle-Mounted Aerial Platform
with Telescoping and Rotating Boom**

TYPICAL EXAMPLES OF MANUALLY PROPELLED ELEVATING AERIAL PLATFORMS COVERED IN ANSI/SIA A92.3 STANDARD

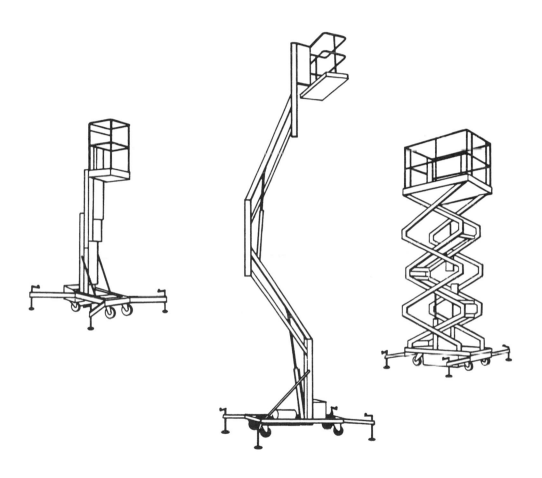

TYPICAL EXAMPLES OF BOOM-SUPPORTED ELEVATING
WORK PLATFORMS COVERED IN ANSI/SIA A92.5 STANDARD

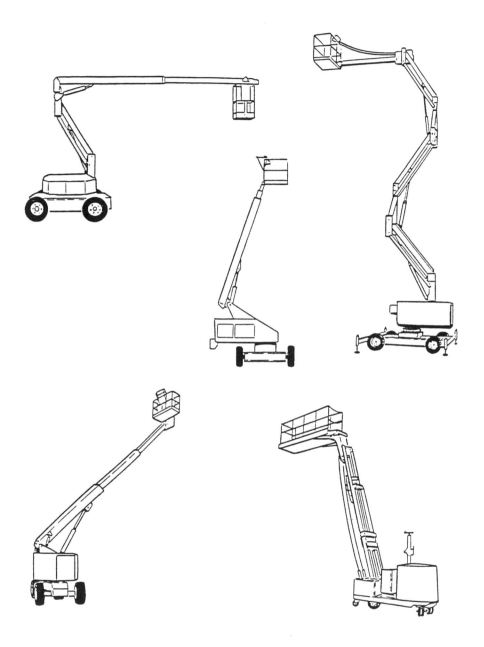

TYPICAL EXAMPLES OF SELF-PROPELLED ELEVATING
WORK PLATFORMS COVERED IN ANSI/SIA A92.6 STANDARD

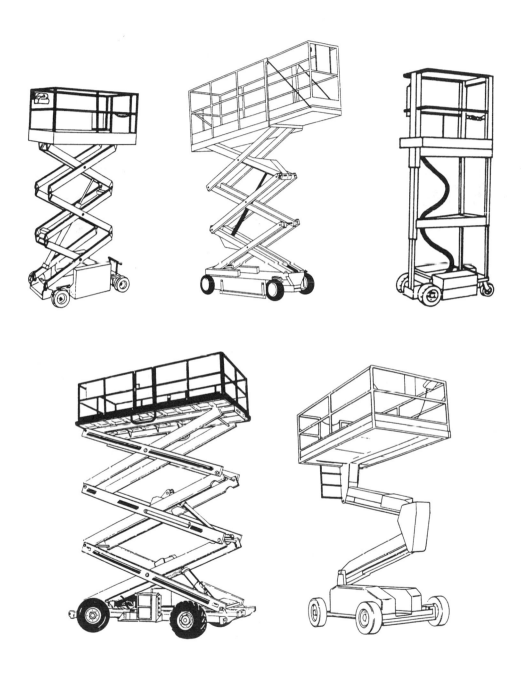

TYPICAL EXAMPLES OF AIRLINE GROUND SUPPORT
VEHICLE-MOUNTED VERTICAL LIFT DEVICES
COVERED IN ANSI/SIA A92.7 STANDARD

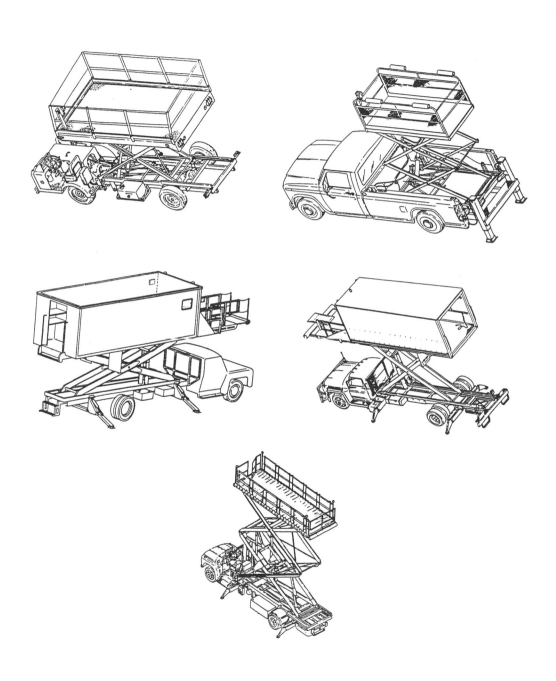

**TYPICAL EXAMPLE OF ARTICULATED BOOM FOR
VEHICLE-MOUNTED BRIDGE INSPECTION
AND MAINTENANCE DEVICES
COVERED IN ANSI/SIA A92.8 STANDARD**

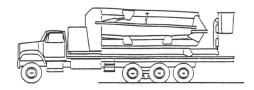

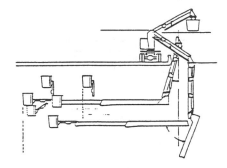

**TYPICAL EXAMPLES OF TOWER-TYPE CONSTRUCTION
FOR VEHICLE-MOUNTED BRIDGE INSPECTION
AND MAINTENANCE DEVICES
COVERED IN ANSI/SIA A92.8 STANDARD**

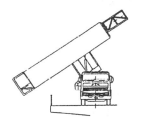

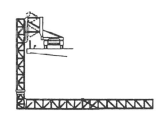

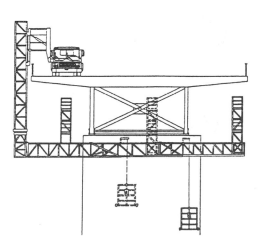

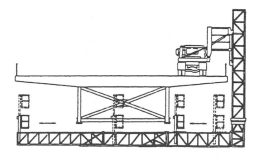

TYPICAL EXAMPLES OF MAST-CLIMBING WORK PLATFORMS
COVERED IN ANSI/SIA A92.9 STANDARD

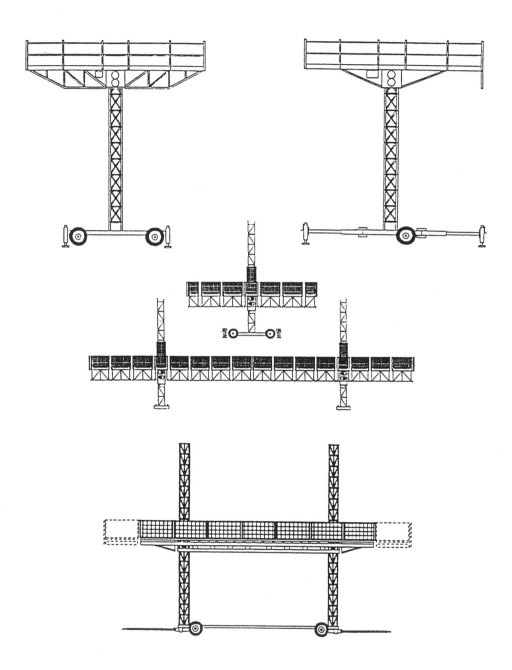

INDEX

E

H

I

K

L

M

S

T